VECTOR IDENTITIES

If $\mathbf{u} = u_1\mathbf{i} + u_2\mathbf{j} + u_3\mathbf{k}$ then (dot product) $\mathbf{u} \bullet \mathbf{v} = u_1v_1 + u_2v_2 + u_3v_3$

$\quad \mathbf{v} = v_1\mathbf{i} + v_2\mathbf{j} + v_3\mathbf{k}$

$\quad \mathbf{w} = w_1\mathbf{i} + w_2\mathbf{j} + w_3\mathbf{k}$ (cross product) $\mathbf{u} \times \mathbf{v} = \begin{vmatrix} \mathbf{i} & \mathbf{j} & \mathbf{k} \\ u_1 & u_2 & u_3 \\ v_1 & v_2 & v_3 \end{vmatrix} = (u_2v_3 - u_3v_2)\mathbf{i} + (u_3v_1 - u_1v_3)\mathbf{j} + (u_1v_2 - u_2v_1)\mathbf{k}$

length of $\mathbf{u} = |\mathbf{u}| = \sqrt{\mathbf{u} \bullet \mathbf{u}} = \sqrt{u_1^2 + u_2^2 + u_3^2}$ angle between $\mathbf{u}$ and $\mathbf{v} = \cos^{-1}\left(\dfrac{\mathbf{u} \bullet \mathbf{v}}{|\mathbf{u}||\mathbf{v}|}\right)$

triple product identities $\mathbf{u} \bullet (\mathbf{v} \times \mathbf{w}) = \mathbf{v} \bullet (\mathbf{w} \times \mathbf{u}) = \mathbf{w} \bullet (\mathbf{u} \times \mathbf{v})$ $\mathbf{u} \times (\mathbf{v} \times \mathbf{w}) = (\mathbf{u} \bullet \mathbf{w})\mathbf{v} - (\mathbf{u} \bullet \mathbf{v})\mathbf{w}$

IDENTITIES INVOLVING GRADIENT, DIVERGENCE, CURL, AND LAPLACIAN

$\nabla = \mathbf{i}\dfrac{\partial}{\partial x} + \mathbf{j}\dfrac{\partial}{\partial y} + \mathbf{k}\dfrac{\partial}{\partial z}$ ("del" or "nabla" operator)

$\nabla \phi(x, y, z) = \mathbf{grad}\,\phi(x, y, z) = \dfrac{\partial \phi}{\partial x}\mathbf{i} + \dfrac{\partial \phi}{\partial y}\mathbf{j} + \dfrac{\partial \phi}{\partial z}\mathbf{k}$

$\mathbf{F}(x, y, z) = F_1(x, y, z)\mathbf{i} + F_2(x, y, z)\mathbf{j} + F_3(x, y, z)\mathbf{k}$

$\nabla \bullet \mathbf{F}(x, y, z) = \mathbf{div}\,\mathbf{F}(x, y, z) = \dfrac{\partial F_1}{\partial x} + \dfrac{\partial F_2}{\partial y} + \dfrac{\partial F_3}{\partial z}$

$\nabla \times \mathbf{F}(x, y, z) = \mathbf{curl}\,\mathbf{F}(x, y, z) = \begin{vmatrix} \mathbf{i} & \mathbf{j} & \mathbf{k} \\ \dfrac{\partial}{\partial x} & \dfrac{\partial}{\partial y} & \dfrac{\partial}{\partial z} \\ F_1 & F_2 & F_3 \end{vmatrix}$

$\quad = \left(\dfrac{\partial F_3}{\partial y} - \dfrac{\partial F_2}{\partial z}\right)\mathbf{i} + \left(\dfrac{\partial F_1}{\partial z} - \dfrac{\partial F_3}{\partial x}\right)\mathbf{j} + \left(\dfrac{\partial F_2}{\partial x} - \dfrac{\partial F_1}{\partial y}\right)\mathbf{k}$

$\nabla(\phi\psi) = \phi\nabla\psi + \psi\nabla\phi$

$\nabla \bullet (\phi\mathbf{F}) = (\nabla\phi) \bullet \mathbf{F} + \phi(\nabla \bullet \mathbf{F})$

$\nabla \times (\phi\mathbf{F}) = (\nabla\phi) \times \mathbf{F} + \phi(\nabla \times \mathbf{F})$

$\nabla \times (\nabla\phi) = 0$ (**curl grad** $= 0$)

$\nabla^2\phi(x, y, z) = \nabla \bullet \nabla\phi(x, y, z) = \mathbf{div}\,\mathbf{grad}\,\phi = \dfrac{\partial^2\phi}{\partial x^2} + \dfrac{\partial^2\phi}{\partial y^2} + \dfrac{\partial^2\phi}{\partial z^2}$

$\nabla \bullet (\mathbf{F} \times \mathbf{G}) = (\nabla \times \mathbf{F}) \bullet \mathbf{G} - \mathbf{F} \bullet (\nabla \times \mathbf{G})$

$\nabla \times (\mathbf{F} \times \mathbf{G}) = \mathbf{F}(\nabla \bullet \mathbf{G}) - \mathbf{G}(\nabla \bullet \mathbf{F}) - (\mathbf{F} \bullet \nabla)\mathbf{G} + (\mathbf{G} \bullet \nabla)\mathbf{F}$

$\nabla(\mathbf{F} \bullet \mathbf{G}) = \mathbf{F} \times (\nabla \times \mathbf{G}) + \mathbf{G} \times (\nabla \times \mathbf{F}) + (\mathbf{F} \bullet \nabla)\mathbf{G} + (\mathbf{G} \bullet \nabla)\mathbf{F}$

$\nabla \bullet (\nabla \times \mathbf{F}) = 0$ (**div curl** $= 0$)

$\nabla \times (\nabla \times \mathbf{F}) = \nabla(\nabla \bullet \mathbf{F}) - \nabla^2\mathbf{F}$ (**curl curl** $=$ **grad div** $-$ laplacian)

VERSIONS OF THE FUNDAMENTAL THEOREM OF CALCULUS

$\displaystyle\int_a^b f'(t)\,dt = f(b) - f(a)$ (the one-dimensional **Fundamental Theorem**)

$\displaystyle\int_C \mathbf{grad}\,\phi \bullet d\mathbf{r} = \phi\big(\mathbf{r}(b)\big) - \phi\big(\mathbf{r}(a)\big)$ if C is the curve $\mathbf{r} = \mathbf{r}(t),\ (a \le t \le b)$.

$\displaystyle\iint_R \left(\dfrac{\partial F_2}{\partial x} - \dfrac{\partial F_1}{\partial y}\right) dA = \oint_C \mathbf{F} \bullet d\mathbf{r} = \oint_C F_1(x, y)\,dx + F_2(x, y)\,dy$ where C is the positively oriented boundary of R (**Green's Theorem**)

$\displaystyle\iint_S \mathbf{curl}\,\mathbf{F} \bullet \hat{\mathbf{N}}\,dS = \oint_C \mathbf{F} \bullet d\mathbf{r} = \oint_C F_1(x, y, z)\,dx + F_2(x, y, z)\,dy + F_3(x, y, z)\,dz$ where C is the oriented boundary of S.   (**Stokes's Theorem**)

Three-dimensional versions: S is the closed boundary of D, with outward normal $\hat{\mathbf{N}}$

$\displaystyle\iiint_D \mathbf{div}\,\mathbf{F}\,dV = \oiint_S \mathbf{F} \bullet \hat{\mathbf{N}}\,dS$ **Divergence Theorem**

$\displaystyle\iiint_D \mathbf{curl}\,\mathbf{F}\,dV = -\oiint_S \mathbf{F} \times \hat{\mathbf{N}}\,dS$

$\displaystyle\iiint_D \mathbf{grad}\,\phi\,dV = \oiint_S \phi\hat{\mathbf{N}}\,dS$

FORMULAS RELATING TO CURVES IN 3-SPACE

Curve: $\mathbf{r} = \mathbf{r}(t) = x(t)\mathbf{i} + y(t)\mathbf{j} + z(t)\mathbf{k}$

Velocity: $\mathbf{v} = \dfrac{d\mathbf{r}}{dt} = v\hat{\mathbf{T}}$

Speed: $v = |\mathbf{v}| = \dfrac{ds}{dt}$

Arc length: $s = \displaystyle\int_{t_0}^t v\,dt$

Acceleration: $\mathbf{a} = \dfrac{d\mathbf{v}}{dt} = \dfrac{d^2\mathbf{r}}{dt^2}$

Tangential and normal components: $\mathbf{a} = \dfrac{dv}{dt}\hat{\mathbf{T}} + v^2\kappa\hat{\mathbf{N}}$

Unit tangent: $\hat{\mathbf{T}} = \dfrac{\mathbf{v}}{v}$

Binormal: $\hat{\mathbf{B}} = \dfrac{\mathbf{v} \times \mathbf{a}}{|\mathbf{v} \times \mathbf{a}|}$

Normal: $\hat{\mathbf{N}} = \hat{\mathbf{B}} \times \hat{\mathbf{T}} = \dfrac{d\hat{\mathbf{T}}/dt}{|d\hat{\mathbf{T}}/dt|}$

Curvature: $\kappa = \dfrac{|\mathbf{v} \times \mathbf{a}|}{v^3}$

Radius of curvature: $\rho = \dfrac{1}{\kappa}$

Torsion: $\tau = \dfrac{(\mathbf{v} \times \mathbf{a}) \bullet (d\mathbf{a}/dt)}{|\mathbf{v} \times \mathbf{a}|^2}$

The Frenet-Serret formulas: $\dfrac{d\hat{\mathbf{T}}}{ds} = \kappa\hat{\mathbf{N}},\qquad \dfrac{d\hat{\mathbf{N}}}{ds} = -\kappa\hat{\mathbf{T}} + \tau\hat{\mathbf{B}},\qquad \dfrac{d\hat{\mathbf{B}}}{ds} = -\tau\hat{\mathbf{N}}$

ORTHOGONAL CURVILINEAR COORDINATES

transformation: $x = x(u, v, w), \quad y = y(u, v, w), \quad z = z(u, v, w)$

position vector: $\mathbf{r} = x(u, v, w)\mathbf{i} + y(u, v, w)\mathbf{j} + z(u, v, w)\mathbf{k}$

scale factors: $h_u = \left|\dfrac{\partial \mathbf{r}}{\partial u}\right|, \quad h_v = \left|\dfrac{\partial \mathbf{r}}{\partial v}\right|, \quad h_w = \left|\dfrac{\partial \mathbf{r}}{\partial w}\right|$

local basis: $\hat{\mathbf{u}} = \dfrac{1}{h_u}\dfrac{\partial \mathbf{r}}{\partial u}, \quad \hat{\mathbf{v}} = \dfrac{1}{h_v}\dfrac{\partial \mathbf{r}}{\partial v}, \quad \hat{\mathbf{w}} = \dfrac{1}{h_w}\dfrac{\partial \mathbf{r}}{\partial w}$

volume element: $dV = h_u h_v h_w \, du \, dv \, dw$

scalar field: $f(u, v, w)$

vector field: $\mathbf{F}(u, v, w) = F_u(u, v, w)\hat{\mathbf{u}} + F_v(u, v, w)\hat{\mathbf{v}} + F_w(u, v, w)\hat{\mathbf{w}}$

gradient: $\nabla f = \dfrac{1}{h_u}\dfrac{\partial f}{\partial u}\hat{\mathbf{u}} + \dfrac{1}{h_v}\dfrac{\partial f}{\partial v}\hat{\mathbf{v}} + \dfrac{1}{h_w}\dfrac{\partial f}{\partial w}\hat{\mathbf{w}}$

divergence: $\nabla \bullet \mathbf{F} = \dfrac{1}{h_u h_v h_w}\left[\dfrac{\partial}{\partial u}\left(h_v h_w F_u\right) + \dfrac{\partial}{\partial v}\left(h_u h_w F_v\right) + \dfrac{\partial}{\partial w}\left(h_u h_v F_w\right)\right]$

$$\nabla^2 f = \dfrac{1}{h_u h_v h_w}\left[\dfrac{\partial}{\partial u}\left(\dfrac{h_v h_w}{h_u}\dfrac{\partial f}{\partial u}\right) + \dfrac{\partial}{\partial v}\left(\dfrac{h_u h_w}{h_v}\dfrac{\partial f}{\partial v}\right) + \dfrac{\partial}{\partial w}\left(\dfrac{h_u h_v}{h_w}\dfrac{\partial f}{\partial w}\right)\right]$$

curl: $\nabla \times \mathbf{F} = \dfrac{1}{h_u h_v h_w}\begin{vmatrix} h_u\hat{\mathbf{u}} & h_v\hat{\mathbf{v}} & h_w\hat{\mathbf{w}} \\ \dfrac{\partial}{\partial u} & \dfrac{\partial}{\partial v} & \dfrac{\partial}{\partial w} \\ F_u h_u & F_v h_v & F_w h_w \end{vmatrix}$

PLANE POLAR COORDINATES

transformation: $x = r\cos\theta, \quad y = r\sin\theta$

position vector: $\mathbf{r} = r\cos\theta\,\mathbf{i} + r\sin\theta\,\mathbf{j}$

scale factors: $h_r = \left|\dfrac{\partial \mathbf{r}}{\partial r}\right| = 1, \quad h_\theta = \left|\dfrac{\partial \mathbf{r}}{\partial \theta}\right| = r$

local basis: $\hat{\mathbf{r}} = \cos\theta\,\mathbf{i} + \sin\theta\,\mathbf{j}, \quad \hat{\boldsymbol{\theta}} = -\sin\theta\,\mathbf{i} + \cos\theta\,\mathbf{j}$

area element: $dA = r\,dr\,d\theta$

scalar field: $f(r, \theta)$

vector field: $\mathbf{F}(r, \theta) = F_r(r, \theta)\hat{\mathbf{r}} + F_\theta(r, \theta)\hat{\boldsymbol{\theta}}$

gradient: $\nabla f = \dfrac{\partial f}{\partial r}\hat{\mathbf{r}} + \dfrac{1}{r}\dfrac{\partial f}{\partial \theta}\hat{\boldsymbol{\theta}}$

divergence: $\nabla \bullet \mathbf{F} = \dfrac{\partial F_r}{\partial r} + \dfrac{1}{r}F_r + \dfrac{1}{r}\dfrac{\partial F_\theta}{\partial \theta}$

laplacian: $\nabla^2 f = \dfrac{\partial^2 f}{\partial r^2} + \dfrac{1}{r}\dfrac{\partial f}{\partial r} + \dfrac{1}{r^2}\dfrac{\partial^2 f}{\partial \theta^2}$

curl: $\nabla \times \mathbf{F} = \left[\dfrac{\partial F_\theta}{\partial r} + \dfrac{F_\theta}{r} - \dfrac{1}{r}\dfrac{\partial F_r}{\partial \theta}\right]\mathbf{k}$

CYLINDRICAL COORDINATES

transformation: $x = r\cos\theta, \quad y = r\sin\theta, \quad z = z$

position vector: $\mathbf{r} = r\cos\theta\,\mathbf{i} + r\sin\theta\,\mathbf{j} + z\mathbf{k}$

scale factors: $h_r = \left|\dfrac{\partial \mathbf{r}}{\partial r}\right| = 1, \quad h_\theta = \left|\dfrac{\partial \mathbf{r}}{\partial \theta}\right| = r, \quad h_z = \left|\dfrac{\partial \mathbf{r}}{\partial z}\right| = 1$

local basis: $\hat{\mathbf{r}} = \cos\theta\,\mathbf{i} + \sin\theta\,\mathbf{j}, \quad \hat{\boldsymbol{\theta}} = -\sin\theta\,\mathbf{i} + \cos\theta\,\mathbf{j}, \quad \hat{\mathbf{z}} = \mathbf{k}$

volume element: $dV = r\,dr\,d\theta\,dz$

surface area element (on $r = a$): $dS = a\,d\theta\,dz$

scalar field: $f(r, \theta, z)$

vector field: $\mathbf{F}(r, \theta, z) = F_r(r, \theta, z)\hat{\mathbf{r}} + F_\theta(r, \theta, z)\hat{\boldsymbol{\theta}} + F_z(r, \theta, z)\mathbf{k}$

gradient: $\nabla f = \dfrac{\partial f}{\partial r}\hat{\mathbf{r}} + \dfrac{1}{r}\dfrac{\partial f}{\partial \theta}\hat{\boldsymbol{\theta}} + \dfrac{\partial f}{\partial z}\mathbf{k}$

divergence: $\nabla \bullet \mathbf{F} = \dfrac{\partial F_r}{\partial r} + \dfrac{1}{r}F_r + \dfrac{1}{r}\dfrac{\partial F_\theta}{\partial \theta} + \dfrac{\partial F_z}{\partial z}$

laplacian: $\nabla^2 f = \dfrac{\partial^2 f}{\partial r^2} + \dfrac{1}{r}\dfrac{\partial f}{\partial r} + \dfrac{1}{r^2}\dfrac{\partial^2 f}{\partial \theta^2} + \dfrac{\partial^2 f}{\partial z^2}$

curl: $\nabla \times \mathbf{F} = \dfrac{1}{r}\begin{vmatrix} \hat{\mathbf{r}} & r\hat{\boldsymbol{\theta}} & \mathbf{k} \\ \dfrac{\partial}{\partial r} & \dfrac{\partial}{\partial \theta} & \dfrac{\partial}{\partial z} \\ F_r & rF_\theta & F_z \end{vmatrix}$

SPHERICAL COORDINATES

transformation: $x = \rho\sin\phi\cos\theta, \quad y = \rho\sin\phi\sin\theta, \quad z = \rho\cos\phi$

position vector: $\mathbf{r} = \rho\sin\phi\cos\theta\,\mathbf{i} + \rho\sin\phi\sin\theta\,\mathbf{j} + \rho\cos\phi\,\mathbf{k}$

scale factors: $h_\rho = \left|\dfrac{\partial \mathbf{r}}{\partial \rho}\right| = 1, \quad h_\phi = \left|\dfrac{\partial \mathbf{r}}{\partial \phi}\right| = \rho, \quad h_\theta = \left|\dfrac{\partial \mathbf{r}}{\partial \theta}\right| = \rho\sin\phi$

local basis: $\hat{\boldsymbol{\rho}} = \sin\phi\cos\theta\,\mathbf{i} + \sin\phi\sin\theta\,\mathbf{j} + \cos\phi\,\mathbf{k}, \quad \hat{\boldsymbol{\phi}} = \cos\phi\cos\theta\,\mathbf{i} + \cos\phi\sin\theta\,\mathbf{j} - \sin\phi\,\mathbf{k}, \quad \hat{\boldsymbol{\theta}} = -\sin\theta\,\mathbf{i} + \cos\theta\,\mathbf{j}$

volume element: $dV = \rho^2\sin\phi\,d\rho\,d\phi\,d\theta$

surface area element (on $\rho = a$): $dS = a^2\sin\phi\,d\theta\,d\phi$

scalar field: $f(\rho, \phi, \theta)$

vector field: $\mathbf{F}(\rho, \phi, \theta) = F_\rho(\rho, \phi, \theta)\hat{\boldsymbol{\rho}} + F_\phi(\rho, \phi, \theta)\hat{\boldsymbol{\phi}} + F_\theta(\rho, \phi, \theta)\hat{\boldsymbol{\theta}}$

gradient: $\nabla f = \dfrac{\partial f}{\partial \rho}\hat{\boldsymbol{\rho}} + \dfrac{1}{\rho}\dfrac{\partial f}{\partial \phi}\hat{\boldsymbol{\phi}} + \dfrac{1}{\rho\sin\phi}\dfrac{\partial f}{\partial \theta}\hat{\boldsymbol{\theta}}$

divergence: $\nabla \bullet \mathbf{F} = \dfrac{\partial F_\rho}{\partial \rho} + \dfrac{2}{\rho}F_\rho + \dfrac{1}{\rho}\dfrac{\partial F_\phi}{\partial \phi} + \dfrac{\cot\phi}{\rho}F_\phi + \dfrac{1}{\rho\sin\phi}\dfrac{\partial F_\theta}{\partial \theta}$

laplacian: $\nabla^2 f = \dfrac{\partial^2 f}{\partial \rho^2} + \dfrac{2}{\rho}\dfrac{\partial f}{\partial \rho} + \dfrac{1}{\rho^2}\dfrac{\partial^2 f}{\partial \phi^2} + \dfrac{\cot\phi}{\rho^2}\dfrac{\partial f}{\partial \phi} + \dfrac{1}{\rho^2\sin^2\phi}\dfrac{\partial^2 f}{\partial \theta^2}$

curl: $\nabla \times \mathbf{F} = \dfrac{1}{\rho^2\sin\phi}\begin{vmatrix} \hat{\boldsymbol{\rho}} & \rho\hat{\boldsymbol{\phi}} & \rho\sin\phi\hat{\boldsymbol{\theta}} \\ \dfrac{\partial}{\partial \rho} & \dfrac{\partial}{\partial \phi} & \dfrac{\partial}{\partial \theta} \\ F_\rho & \rho F_\phi & \rho\sin\phi F_\theta \end{vmatrix}$

Calculus
Several Variables
Sixth Edition

Robert A. Adams
University of British Columbia

PEARSON

Addison
Wesley

Toronto

Library and Archives Canada Cataloguing in Publication

Adams, Robert A. (Robert Alexander), 1940-
    Calculus : several variables / Robert A. Adams. — 6th ed.

Includes index.
First-4th eds published under title: Calculus of several variables.
ISBN 0-321-30716-X

1. Functions of several real variables—Textbooks. 2. Vector
analysis—Textbooks. I. Adams, Robert A. (Robert Alexander), 1940-
II. Title.

QA303.A3 2006   515'.84   C2005-906071-9

ISBN 0-321-30716-X

Vice President, Editorial Director: Michael J. Young
Acquisitions Editor: Kelly Torrance
Marketing Manager: Janet Piper
Developmental Editor: Remo Celio
Production Editor: Mary Ann McCutcheon
Copy Editor: Betty Robinson
Proofreader: Bonnie Di Malta
Production Coordinator: Andrea Falkenberg
Permissions Research: Christina Beamish
Art Director: Mary Opper
Cover Design: Jennifer Stimson
Cover Image: Doris Mitsch

1 2 3 4 5   10 09 08 07 06

Printed and bound in the United States.

PEARSON
Addison
Wesley

Credits

The Chapter 17 epigraph used by permission of Oxford University Press; the Appendix 2 epigraph used by permission of Springer, *The Mathematical Intelligencer*, v. 13, 1991; the Appendix 4 epigraph used by permission of the American Mathematical Society, J.J. Sylvester, *The Collected Mathematical Papers*, Vol. II, p. 214 (footnote), Chelsea New York (1973).

To Anne

Contents

Preface

The main part of this book is extracted from the latter half (Chapters 9 through 17 and all Appendices) of the author's text *Calculus: a Complete Course, 6th Edition*.

By the time a textbook reaches its sixth edition, readers have come to have certain expectations regarding content and style that an author cannot afford to ignore. The purpose of this book remains to present the calculus in clear, coherent, and readable form, and above all in a form that readers will find *interesting*. Completeness has also been a goal; it is better to cover too many topics with the foreknowledge that instructors and students will have to skip some, than not to have enough to meet the needs of users with a variety of interests and backgrounds. Two peripheral topics that were present in the fourth edition and omitted from the fifth, namely, very brief introductions to Fourier Series (Section 9.9) and to functions of a complex variable (Appendix II), have been restored at the request of some users who were sorry to find them gone.

Little change in the ordering of topics presented has been made in this edition. Some material has been added (see the listing below), while nothing significant has been removed. Nevertheless, the usual rewriting and fine-tuning of the presentation and a small increase in page size have resulted in a significantly lower page count. Major changes in the text are few and are as follows:

- A preliminary chapter (Chapter R) has been included to review the main topics covered in *Single-Variable Calculus, 6th Edition*, the first half of *Calculus: a Complete Course*. This was also present in the fourth edition but absent from the fifth.

- A full chapter (Chapter 17) on Differential Equations has been reinstated.

- The use of Maple for dealing with vectors, matrices, and vector-valued functions now makes use of the newer "LinearAlgebra" and "VectorCalculus" packages. This simplifies things greatly. It affects chapters from 10 on and has resulted in a complete rewriting of Section 10.7.

- Section 9.6 (Taylor and Maclaurin Series) has had many changes and has absorbed what was a separate section revisiting Taylor's Formula. Thus the whole approach to Taylor's Theorem is more unified, although it is still split between Chapter 4 (not present in this volume) and Chapter 9.

In addition to the above, numerous small local changes have been made here and there to improve the text. Some awkward examples have been removed or replaced with more appropriate ones. Some exercises have been removed and others added. A few new figures have been added and others improved. Some notations have been changed. Open intervals are now written (a, b) again; $]a, b[$, although logical, just doesn't look right. The process of making a book more reader-friendly is ongoing and relies heavily on comments from readers.

To the Student

When I took my first course in calculus, there was no prescribed textbook, but a book called *Calculus Made Easy* was recommended to those of us who felt we needed something beyond the notes we could take during lectures. I bought the book, hoping that it would live up to its title and I would have an easy time learning calculus. It didn't and I didn't.

Is multivariable calculus a very difficult subject? No, not really, but it sometimes seems that way to students, especially at the beginning, because of the new ideas and techniques involved, and because success in mastering it depends on having a very solid basis in single-variable calculus to build upon. You may want to review the background material in Chapter R even if your instructor does not refer to it in class. Learning multivariable and vector calculus will provide you with many useful tools for analyzing problems in numerous fields of interest, especially those regarded as "scientific." It is worth your while to acquire those tools, but, like any other worthwhile task, this one requires much effort on your part. No book or instructor can remove that requirement.

In writing this book I have tried to organize material in such a way as to make it as easy as possible, but not at the expense of "sweeping the real difficulties under the rug." You may find some of the concepts difficult to understand when they are first introduced. If so, *reread* the material slowly, if necessary several times; *think about it*; formulate questions to ask fellow students, your TA, or your instructor. Don't delay. It is important to resolve your problems as soon as possible. If you don't understand today's topic, you may not understand how it applies to tomorrow's either. Mathematics is a "linear discipline"; it builds from one idea to the next.

Doing exercises is the best way to deepen your understanding of calculus and to convince yourself that you understand it. There are numerous exercises in this text — too many for you to try them all. Some are "drill" exercises to help you develop your skills in calculation. More important, however, are the problems that develop reasoning skills and your ability to apply the techniques you have learned to concrete situations. In some cases you will have to plan your way through a problem that requires several diffferent "steps" before you can get to the answer. Other exercises are designed to extend the theory developed in the text and therefore enhance your understanding of the concepts of calculus.

The exercises vary greatly in difficulty. Usually, the more difficult ones occur toward the end of exercise sets, but these sets are not strictly graded in this way because exercises on a specific topic tend to be grouped together. Some exercises in the regular sets are marked with an asterisk *. This symbol indicates that the exercise is *either* somewhat more theoretical *or* somewhat more difficult than most. The theoretical ones need not be difficult; sometimes they are quite easy. Most of the problems in the *Challenging Problems* section forming part of the *Chapter Review* at the end of most chapters are also on the difficult side although these are not typically marked with a *.

Do not be discouraged if you can't do *all* the exercises. Some are very difficult indeed, and only a few very gifted students will be able to do them. However, you should be able to do the vast majority of the exercises. Some will take much more effort than others. When you encounter difficulties with problems, proceed as follows:

1. Read and reread the problem until you understand exactly what information it gives you and what you are expected to find or do.

2. If appropriate, draw a diagram illustrating the relationships between the quantities involved in the problem.

3. If necessary, introduce symbols to represent the quantities in the problem. Use appropriate letters (e.g., V for volume, t for time). Don't feel you have to use x or y for everything.

4. Develop a plan of attack. This is usually the hardest part. Look for known relationships; try to recognize patterns; look back at the worked examples in the current or relevant previous sections; try to think of possible connections between the problem at hand and others you have seen before. Can the problem be simplified by making some extra assumptions? If you can solve a simplified version, it may help you decide what to do with the given problem. Can the problem be broken down into several cases, each of which is a simpler problem? When reading the examples in the text, be alert for methods that may turn out to be useful in other contexts later.

5. Try to carry out the steps in your plan. If you have trouble with one or more of them, you may have to alter the plan.

6. When you arrive at an answer for a problem, always ask yourself whether it is *reasonable*. If it isn't, look back to determine places where you may have made an error.

Answers for most of the odd-numbered exercises are provided at the back of the book. Exceptions are exercises that don't have short answers, for example "Prove that ..." or "Show that ..." problems where the answer is the whole solution. A *Student Solutions Manual* that contains detailed solutions to even-numbered exercises is available.

Besides asterisks used to mark more difficult and theoretical problems, the following symbols are used to mark exercises of special types:

◈ exercises pertaining to differential equations and initial-value problems. (It is not used in sections that are wholly concerned with DEs.)

▦ problems requiring the use of a calculator. Often a scientific calculator is needed. Some such problems may require a programmable calculator.

▨ problems requiring the use of either a graphing calculator or mathematical graphing software on a personal computer.

▥ problems requiring the use of a computer. Typically, these will require either computer algebra software (e.g., Maple, Mathematica) or a spreadsheet program (e.g., Lotus 123, Microsoft Excel, Quattro Pro).

To the Instructor

As its title suggests, this book is intended to cover all the material usually encountered in the latter part of a three- or four-semester real-variable calculus program, after the differential and integral calculus for real-valued functions of a single real variable has been completed. (This material is treated in the companion volume *Single-Variable Calculus*; the material of both volumes comprise the complete book *Calculus: a Complete Course.*) In this text we study vector-valued functions of a single real variable (Chapter 11), real-valued functions of several real variables (Chapters 12–14), and vector-valued functions of several real variables (Chapters 15–16). Chapter 9 concerns sequences and series, and is included in both this volume and *Single-Variable Calculus* because the position of this material in an overall calculus program is somewhat arbitrary. Chapter 10 contains necessary background on vectors and geometry in 3-space as well as a bit of linear algebra that is useful, though not absolutely essential, for the understanding of subsequent multivariable material. There is a wealth of material here — too much to include in any course. You must select what material to include and what to omit, taking into account the background and needs of your students.

At the University of British Columbia, where the author taught for 34 years, calculus is divided into four trimesters, the first two covering single-variable calculus (Chapters 1–9), the third covering functions of several variables (part of Chapter 10, and Chapters 12–14) taken by somewhat fewer students, and the fourth covering vector calculus (part of Chapter 10, and Chapters 11, 15 and 16) taken by a relatively small number of students, mainly in mathematics, engineering, and the physical sciences. In none of these courses is there enough time to cover all the material in the appropriate chapters; some sections are always omitted. However, the wide selection of topics and applications provides for a rich and varied learning experience for students.

The text is designed for general calculus courses, especially those for science and engineering students. Most of the material requires only a a good background in the differential and integral calculus of real-valued functions of a single real variable. Some optional material is more subtle and/or theoretical, and is intended mainly for stronger students.

Several supplements are available for use with *Calculus: A Complete Course, 6th Edition.* Those available to students include:

- a **Companion Website** (www.pearsoned.ca/adams): This website includes valuable student resources, such as Multiple Choice Quizzes, Web-Based Tutorials, Maple Worksheets, links to related websites, and more.

- a **Student Solutions Manual** (ISBN: 0-321-30713-5): This manual contains detailed solutions to all the even-numbered exercises, prepared by the author.

- **Calculus: The Maple Way, Second Edition** by Robert Israel (University of British Columbia) (ISBN: 0-201-61383-2): This lab manual contains instructions, examples, and problems dealing with the effective use of the computer algebra system Maple as a tool for calculus.

Supplements available to instructors include an **Instructor's Resource CD-ROM** (ISBN: 0-321-31529-4): This resource CD includes the following instructor supplements:

- **Instructor's Solutions Manual**: This manual includes detailed solutions to all the exercises, prepared by the author.

- **Text Solutions in online-publishable form**: Detailed solutions to all exercises individually in .pdf format.

- **Pearson TestGen**: Over 1500 test questions are provided in TestGen format. TestGen is testing software that enables instructors to view and edit the existing questions, add questions, generate tests, and distribute the tests in a variety of formats. Powerful search and sort functions make it easy to locate questions and arrange them in any order desired. TestGen also enables instructors to administer tests on a local area network, have the tests graded electronically, and have the results prepared in electronic or printed reports. TestGen is compatible with Windows and Macintosh operating systems and can be downloaded from the TestGen website located at www.pearsoned.com/testgen. Contact your local sales representative for details and access.

- **Image Library**: All of the figures in the text provided as individual enlarged .pdf files suitable for printing to transparencies.

Most of these instructor supplements are also available for download from a password-protected section of Pearson Education Canada's online catalogue (vig.pearsoned.ca). Navigate to this book's catalogue page to view a list of those supplements that are available. See your local sales representative for details and access.

Acknowledgments

The first five editions of this material have been used for classes of general science, engineering, and mathematics majors and honours students at the University of British Columbia. I am grateful to many colleagues and students, at UBC and at many other institutions where these books have been used, for their encouragement and useful comments and suggestions.

In preparing this edition I have had guidance from several dedicated reviewers who provided new insight and direction to my writing. I am especially grateful to the following people:

Pekka Alestalo	Helsinki University of Technology	Finland
Helge Dahle	University of Bergen	Norway
Marius Irgens	Norwegian University of Science & Technology	Norway
Michael Lamoureux	University of Calgary	Canada
Tom Lindstrøm	University of Oslo	Norway
Marlene McCallum	Carleton University	Canada
Graeme Pettet	Queensland University of Technology	Australia
Viena Stastna	University of Calgary	Canada
Martin Stynes	National University of Ireland, Cork	Ireland
Brian Wetton	University of British Columbia	Canada
Eric Woolgar	University of Alberta	Canada
Frank Zorzitto	University of Waterloo	Canada

Finally, I wish to thank the sales and marketing staff of all Addison-Wesley (now Pearson Canada) divisions around the world for making the previous editions so successful, the editorial and production staff in Toronto, in particular, Meaghan Eley, Michael Kelley, Kelly Torrance, Dave Ward, and Mary Ann McCutcheon for their assistance and encouragement, and Betty R. Robinson for her careful copy editing of the typeset manuscript.

I typeset this volume using TEX and PostScript on a PC running Linux-Mandrake, and also generated all of the figures in PostScript using the mathematical graphics software package **MG** developed by my colleague Robert Israel and myself.

The expunging of errors and obscurities in a text is an ongoing and asymptotic process; hopefully each edition is better than the previous one. Nevertheless, some such imperfections always remain, and I will be grateful to any readers who call them to my attention, or give me any other suggestions for future improvements.

R.A.A.
Vancouver, Canada
April 2005
adms@math.ubc.ca

CHAPTER R

A Brief Review of Single-Variable Calculus

"Reeling and Writing, of course, to begin with,"
the Mock Turtle replied, "and the different branches
of Arithmetic — Ambition, Distraction, Uglification,
and Derision."

Lewis Carroll (Charles Lutwidge Dodgson) 1832–1898
from *Alice's Adventures in Wonderland*

Introduction Single-variable calculus is concerned with studying the behaviour of functions of a single real variable. Such a function f is a rule or *machine* that generates a real number output $f(x)$ from each real number x in its *domain* of possible input values. Calculus builds upon algebra and geometry, which also study functions. What distinguishes calculus is the introduction of a new concept called a *limit* that enables calculus to study the way function values change in a dynamic way. Calculus is divided into two basic branches, differential calculus and integral calculus, both of which have limits at their cores.

Limits and Continuity

We say that the function $f(x)$ has **limit** L as x approaches x_0, and we write

$$\lim_{x \to x_0} f(x) = L,$$

provided that we can ensure that the number $f(x)$ is *as close as we want* to the number L by taking x *close enough* (but not equal to) x_0. Expressed in terms of symbols, this condition says that for every positive number ϵ (no matter how small), there exists a positive number δ (depending on ϵ) such that

$$0 < |x - x_0| < \delta \quad \Rightarrow \quad |f(x) - L| < \epsilon.$$

We can also define the **left limit** and **right limit** of f at x_0 by requiring that $x < x_0$ or $x > x_0$ in the above definition. The limit exists if and only if both the left and right limits exist and the two are equal.

Limits of combinations of functions can be calculated by using the following **limit rules**: if $\lim_{x \to a} f(x) = L$ and $\lim_{x \to a} g(x) = M$, then

$$\lim_{x \to a} [f(x) + g(x)] = L + M \qquad \lim_{x \to a} kf(x) = kL$$

$$\lim_{x \to a} [f(x) - g(x)] = L - M \qquad \lim_{x \to a} \frac{f(x)}{g(x)} = \frac{L}{M}, \quad \text{if } M \neq 0$$

$$\lim_{x \to a} f(x)g(x) = LM \qquad \lim_{x \to a} [f(x)]^r = L^r, \quad \text{if } L > 0.$$

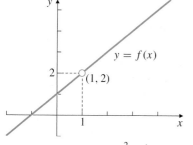

Figure R.1 If $f(x) = \dfrac{x^2 - 1}{x - 1}$, then $\lim_{x \to 1} f(x) = 2$ even though $f(1)$ is not defined

If $b < a < c$, $f(x) \le g(x) \le h(x)$ on $[b, c]$, and $\lim_{x \to a} f(x) = \lim_{x \to a} h(x) = L$, then $\lim_{x \to a} g(x) = L$ also.

The limit concept can be extended to allow for **limits at infinity** and **infinite limits**. For example,

$$\lim_{x \to 1} \frac{1}{(x-1)^2} = \infty, \qquad \lim_{x \to \infty} \frac{3x+2}{x+5} = \lim_{x \to \infty} \frac{3+(2/x)}{1+(5/x)} = \frac{3}{1} = 3.$$

A function f is **continuous** at a point c in its domain if

$$\lim_{x \to c} f(x) = f(c).$$

(If $f(x)$ is only defined on one side of c, the limit need only be one-sided.) Functions that are continuous wherever they are defined are called continuous functions.

Figure R.2

(a) f is continuous at c

(b) f is not continuous at c because $\lim_{x \to c} f(x) \ne f(c)$

(c) f is not continuous at c because $\lim_{x \to c} f(x)$ does not exist

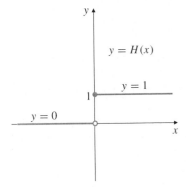

Figure R.3 $H(x)$ is discontinuous at $x = 0$ because $\lim_{x \to 0-} H(x) = 0$ and $\lim_{x \to 0+} H(x) = 1$

Most functions encountered previously are continuous. For example, the following functions are continuous wherever they are defined:

(a) all polynomials

(b) all rational functions

(c) all rational powers $x^{m/n} = \sqrt[n]{x^m}$

(d) the trigonometric functions sine, cosine, tangent, secant, cosecant, and cotangent

(e) the absolute value function $|x|$.

The Heaviside function $H(x) = \begin{cases} 1 & \text{if } x \ge 0 \\ 0 & \text{if } x < 0 \end{cases}$ is defined for all real x, but is discontinuous at $x = 0$.

Sums, products, constant multiples, and compositions of continuous functions are continuous. So are quotients, except, of course, at points where the denominator is zero. The following theorems are very important ones in calculus.

THEOREM **1**

The Max-Min Theorem

If $f(x)$ is continuous on the closed, finite interval $[a, b]$, then there exist numbers x_1 and x_2 in $[a, b]$ such that for all x in $[a, b]$,

$$f(x_1) \le f(x) \le f(x_2).$$

Thus f has a unique absolute minimum value $m = f(x_1)$ and a unique absolute maximum value $M = f(x_2)$. The points x_1 and x_2 are not necessarily unique.

THEOREM **2**

The Intermediate-Value Theorem

If $f(x)$ is continuous on the interval $[a, b]$ and if s is a number between $f(a)$ and $f(b)$, then there exists a number c in $[a, b]$ such that $f(c) = s$.

Tangents and Derivatives

The line joining the two points $P = (x_0, f(x_0))$ and $Q = (x_0 + h, f(x_0 + h))$ on the graph $y = f(x)$ of the function f has slope

$$\frac{f(x_0 + h) - f(x_0)}{h}.$$

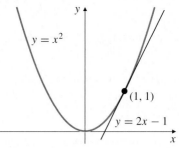

Figure R.4 The tangent to $y = x^2$ at $(1, 1)$

If this slope has a limit m as h approaches 0 (i.e., as Q approaches P along the graph), then the line through P with slope m is **tangent** to the curve $y = f(x)$ at P. For example, $y = 2x - 1$ is tangent to $y = x^2$ at $(1, 1)$ because it passes through that point and has slope

$$\lim_{h \to 0} \frac{(1 + h)^2 - 1^2}{h} = \lim_{h \to 0} \frac{2h + h^2}{h} = \lim_{h \to 0} (2 + h) = 2.$$

If the (two-sided) limit of the slope of PQ does not exist but is either ∞ or $-\infty$, then the vertical line $x = x_0$ through P is tangent to the graph $y = f(x)$ at P. Curves have tangents only at points where they are "smooth."

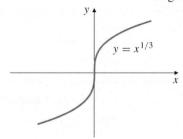

Figure R.5 The y-axis is tangent to $y = x^{1/3}$ at the origin

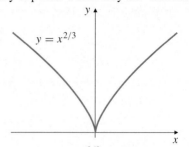

Figure R.6 $y = x^{2/3}$ has no tangent at the origin

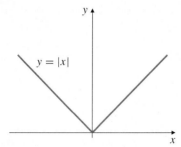

Figure R.7 $y = |x|$ has no tangent at the origin

If a curve has a nonvertical tangent at a point P, then the slope of the curve at P is the slope of that tangent. In general, the slope of the curve $y = f(x)$ varies from point to point on the curve, so is itself a function of x. We call this function the **derivative of** f and denote it by f'.

The **derivative** of a function f is the function f' defined by

$$f'(x) = \lim_{h \to 0} \frac{f(x + h) - f(x)}{h}$$

at all points x for which the limit exists (i.e., is a finite real number). If $f'(x)$ exists, we say that f is **differentiable** at x.

A function is continuous at any point where it is differentiable. Its graph is smooth at such a point and has a nonvertical tangent there.

Table 1. Some elementary functions and their derivatives

$f(x)$	$f'(x)$
C (constant)	0
x	1
x^2	$2x$
$\dfrac{1}{x}$	$-\dfrac{1}{x^2} \quad (x \neq 0)$
$\sqrt{x}$	$\dfrac{1}{2\sqrt{x}} \quad (x > 0)$
x^r	$r\,x^{r-1} \quad (x^{r-1} \text{ real})$
$|x|$	$\dfrac{x}{|x|} = \operatorname{sgn} x$

If $y = f(x)$ we can use the dependent variable y to represent the function, and we can denote the derivative of the function in any of the following ways:

$$D_x y = y' = \frac{dy}{dx} = \frac{d}{dx} f(x) = f'(x) = Df(x).$$

The value of the derivative of a function at a particular number x_0 in its domain can also be expressed in several ways:

$$D_x y \Big|_{x=x_0} = y' \Big|_{x=x_0} = \frac{dy}{dx} \Big|_{x=x_0} = \frac{d}{dx} f(x) \Big|_{x=x_0} = f'(x_0) = Df(x_0).$$

The calculation of derivatives of combinations of functions is carried out with the aid of the following differentiation rules:

THEOREM **3**

Differentiation Rules

If f and g are differentiable at x, then so are $f + g$, $f - g$, Cf (where C is a constant), fg, and $1/g$ and f/g (provided $g(x) \neq 0$). Moreover,

$(f + g)'(x) = f'(x) + g'(x)$	**The Sum Rule**
$(f - g)'(x) = f'(x) - g'(x)$	**The Difference Rule**
$(Cf)'(x) = Cf'(x)$	**The Constant Multiple Rule**
$(fg)'(x) = f'(x)g(x) + f(x)g'(x)$	**The Product Rule**
$\left(\dfrac{1}{g}\right)'(x) = \dfrac{-g'(x)}{(g(x))^2}$	**The Reciprocal Rule**
$\left(\dfrac{f}{g}\right)'(x) = \dfrac{g(x)f'(x) - f(x)g'(x)}{(g(x))^2}$	**The Quotient Rule**

If $f(u)$ is differentiable at $u = g(x)$, and $g(x)$ is differentiable at x, then the composite function $f \circ g(x) = f(g(x))$ is differentiable at x, and

$$(f \circ g)'(x) = f'(g(x))g'(x) \qquad \textbf{The Chain Rule}$$

In terms of Leibniz notation, if $y = f(u)$, where $u = g(x)$, then

$$\frac{dy}{dx} = \frac{dy}{du}\frac{du}{dx}, \qquad \text{where} \quad \frac{dy}{du} \quad \text{is evaluated at } u = g(x).$$

Implicit Differentiation

An equation involving two variables x and y can be regarded as defining one of the variables, say y, implicitly as a function of the other, and hence differentiated by the Chain Rule. For instance,

$$y^2 + x^3 = 5xy \qquad \Rightarrow \qquad 2y\frac{dy}{dx} + 3x^2 = 5x\frac{dy}{dx} + 5y.$$

The Trigonometric Functions

The functions $\cos t$ and $\sin t$ are defined to be, respectively, the x- and y-coordinates of the point P_t on the circle $x^2 + y^2 = 1$ obtained by starting at the point $A = (1, 0)$ on the circle and proceeding a distance $|t|$ measured along the circle in a counterclockwise direction if $t > 0$, or in a clockwise direction if $t < 0$. The number t is the radian measure of the angle AOP_t. Since the circumference of the circle is 2π, we have

$$\pi \text{ radians} = 180°.$$

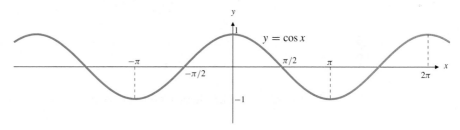

Figure R.8 Angle $AOP_t = t$ radians, where t is the length of arc AP_t

The graphs of cosine and sine indicate that these functions are defined and continuous at all real numbers x, and are periodic with period 2π.

Figure R.9 The graph of $\cos x$

Figure R.10 The graph of $\sin x$

Cosine and sine satisfy numerous identities, among them $\cos^2 t + \sin^2 t = 1$, and

---

Addition formulas for cosine and sine

$$\cos(s + t) = \cos s \cos t - \sin s \sin t$$
$$\sin(s + t) = \sin s \cos t + \cos s \sin t$$
$$\cos(s - t) = \cos s \cos t + \sin s \sin t$$
$$\sin(s - t) = \sin s \cos t - \cos s \sin t.$$

Double- and half-angle formulas

$$\sin 2t = 2 \sin t \cos t \qquad \text{and}$$
$$\cos 2t = \cos^2 t - \sin^2 t$$
$$= 2 \cos^2 t - 1 \qquad (\text{using } \sin^2 t + \cos^2 t = 1)$$
$$= 1 - 2 \sin^2 t$$
$$\cos^2 t = \frac{1 + \cos 2t}{2}, \qquad \sin^2 t = \frac{1 - \cos 2t}{2}.$$

Four other trigonometric functions are defined in terms of cosine and sine:

$$\tan t = \frac{\sin t}{\cos t} \qquad\qquad \sec t = \frac{1}{\cos t}$$

$$\cot t = \frac{\cos t}{\sin t} = \frac{1}{\tan t} \qquad\qquad \csc t = \frac{1}{\sin t}.$$

The derivatives of the trigonometric functions follow from the addition formulas for sine and cosine, and the important limit

$$\lim_{\theta \to 0} \frac{\sin \theta}{\theta} = 1 \qquad \text{(where } \theta \text{ is in radians)}.$$

They are as follows:

$$\frac{d}{dx} \sin x = \cos x \qquad\qquad \frac{d}{dx} \cos x = -\sin x$$

$$\frac{d}{dx} \tan x = \sec^2 x \qquad\qquad \frac{d}{dx} \sec x = \sec x \tan x$$

$$\frac{d}{dx} \cot x = -\csc^2 x \qquad\qquad \frac{d}{dx} \csc x = -\csc x \cot x.$$

The Mean-Value Theorem

THEOREM 4

The Mean-Value Theorem

Suppose that the function f is continuous on the closed, finite interval $[a, b]$ and that it is differentiable on the open interval (a, b). Then there exists a point c in the open interval (a, b) such that

$$\frac{f(b) - f(a)}{b - a} = f'(c).$$

This says that the slope of the chord line joining the points $(a, f(a))$ and $(b, f(b))$ is equal to the slope of the tangent line to the curve $y = f(x)$ at the point $(c, f(c))$, so the two lines are parallel.

Figure R.11   There is a point C on the curve where the tangent is parallel to the chord AB

Higher-Order Derivatives

If the derivative $y' = f'(x)$ of a function $y = f(x)$ is itself differentiable at x, we can calculate *its* derivative, which we call the **second derivative** of f and denote by $y'' = f''(x)$. Some of the more common notations for second derivatives are

$$y'' = f''(x) = \frac{d^2 y}{dx^2} = \frac{d}{dx} \frac{d}{dx} f(x) = \frac{d^2}{dx^2} f(x) = D_x^2 y = D_x^2 f(x).$$

Similarly, third-, fourth-, and in general nth-order derivatives can be calculated. The prime notation is inconvenient for derivatives of high order, so we denote the order by a superscript in parentheses (to distinguish it from an exponent): the nth derivative of $y = f(x)$ is

$$y^{(n)} = f^{(n)}(x) = \frac{d^n y}{dx^n} = \frac{d^n}{dx^n} f(x) = D_x^n y = D_x^n f(x),$$

and it is defined to be the derivative of the $(n - 1)$st derivative. For $n = 1, 2$, and 3, primes are still normally used: $f^{(2)}(x) = f''(x)$, $f^{(3)}(x) = f'''(x)$. It is sometimes convenient to denote $f^{(0)}(x) = f(x)$, that is, to regard a function as its own zeroth-order derivative.

Antiderivatives and Indefinite Integrals

An **antiderivative** of a function f on an interval I is another function F satisfying

$$F'(x) = f(x) \quad \text{for } x \text{ in } I.$$

Antiderivatives are not unique, but any two antiderivatives of the same function on the same interval differ by a constant; if $f'(x) = 0$ on an interval I, then $f(x) = C$ (a constant function) on I. The **indefinite integral** of $f(x)$ on interval I is the *general* antiderivative of f on I, that is,

$$\int f(x)\, dx = F(x) + C \qquad \text{on } I,$$

where $F'(x) = f(x)$ for all x in I. Here is a list of some elementary indefinite integrals:

(a) $\displaystyle \int dx = \int 1\, dx = x + C$ (b) $\displaystyle \int x\, dx = \frac{x^2}{2} + C$

(c) $\displaystyle \int x^2\, dx = \frac{x^3}{3} + C$ (d) $\displaystyle \int \frac{1}{x^2}\, dx = -\frac{1}{x} + C$

(e) $\displaystyle \int \frac{1}{\sqrt{x}}\, dx = 2\sqrt{x} + C$ (f) $\displaystyle \int x^r\, dx = \frac{x^{r+1}}{r+1} + C \ (r \neq -1)$

(g) $\displaystyle \int \sin x\, dx = -\cos x + C$ (h) $\displaystyle \int \cos x\, dx = \sin x + C$

(i) $\displaystyle \int \sec^2 x\, dx = \tan x + C$ (j) $\displaystyle \int \csc^2 x\, dx = -\cot x + C$

(k) $\displaystyle \int \sec x \tan x\, dx = \sec x + c$ (l) $\displaystyle \int \csc x \cot x\, dx = -\csc x + C.$

Differential Equations and Initial-Value Problems

A **differential equation** (DE) is an equation involving one or more derivatives of an unknown function. Any function whose derivatives satisfy the differential equation *identically on an interval* is called a **solution** of the equation on that interval. For instance, the function $y = x^3 - x$ is a solution of the differential equation

$$\frac{dy}{dx} = 3x^2 - 1$$

on the whole real line. This differential equation has more than one solution; in fact, $y = x^3 - x + C$ is a solution for any value of the constant C; this is called a **general solution** of the equation.

The **order** of a differential equation is the order of the highest-order derivative appearing in the equation. The general solution of a DE of order n typically involves n arbitrary constants.

An **initial-value problem** (IVP) is a problem that consists of

(i) a differential equation (to be solved for an unknown function) and

(ii) prescribed values for the solution and enough of its derivatives at a particular point (the initial point) to determine values for all the arbitrary constants in the general solution of the DE and so yield a unique particular solution.

Inverse Functions

A function f is **one-to-one** if $f(x_1) \neq f(x_2)$ whenever x_1 and x_2 belong to the domain of f and $x_1 \neq x_2$. Such a function f has an **inverse function**, denoted f^{-1}. The value of $f^{-1}(x)$ is the unique number y in the domain of f for which $f(y) = x$. Thus

$$y = f^{-1}(x) \quad \Longleftrightarrow \quad x = f(y).$$

Inverse functions have the following properties:

1. $y = f^{-1}(x) \quad \Longleftrightarrow \quad x = f(y).$
2. The domain of f^{-1} is the range of f.
3. The range of f^{-1} is the domain of f.
4. $f^{-1}(f(x)) = x$ for all x in the domain of f.
5. $f(f^{-1}(x)) = x$ for all x in the domain of f^{-1}.
6. $(f^{-1})^{-1}(x) = f(x)$ for all x in the domain of f.
7. The graph of f^{-1} is the reflection of the graph of f in the line $x = y$.
8. $\dfrac{d}{dx} f^{-1}(x) = \dfrac{1}{f'\left(f^{-1}(x)\right)}.$

The Natural Logarithm and Exponential

One important pair of inverse functions consists of the **natural logarithm** function $\ln x$, defined for $x > 0$, and the **exponential** function, $\exp x = e^x$, where $e = \exp(1) = 2.7\,1828\,1828\,45\,90\,45 \ldots$

The natural logarithm satisfies

$$\ln 1 = 0 \qquad \text{and} \qquad \frac{d}{dx}\ln x = \frac{1}{x},$$

from which the following properties of logarithms can be deduced:

(i) $\ln(xy) = \ln x + \ln y$ (ii) $\ln\left(\dfrac{1}{x}\right) = -\ln x$

(iii) $\ln\left(\dfrac{x}{y}\right) = \ln x - \ln y$ (iv) $\ln\left(x^r\right) = r \ln x$

(v) $\displaystyle\lim_{x \to \infty} \ln x = \infty$ (vi) $\displaystyle\lim_{x \to 0+} \ln x = -\infty.$

On any interval not containing 0, we have $\displaystyle\int \frac{1}{x}\,dx = \ln|x| + C.$

Being the inverse of the natural logarithm, the exponential function satisfies

$$y = \exp x = e^x \quad \Longleftrightarrow \quad x = \ln y \quad (y > 0).$$

Therefore, we have $e^0 = 1$ and

(i) $e^{x+y} = e^x e^y$ (ii) $(e^x)^r = e^{rx}$

(iii) $e^{-x} = \dfrac{1}{e^x}$ (iv) $e^{x-y} = \dfrac{e^x}{e^y}$

(v) $\displaystyle\lim_{x \to -\infty} e^x = 0$ (vi) $\displaystyle\lim_{x \to \infty} e^x = \infty.$

Also

$$\frac{d}{dx} e^x = e^x, \qquad \text{and} \qquad \int e^x\,dx = e^x + C \quad \text{on any interval.}$$

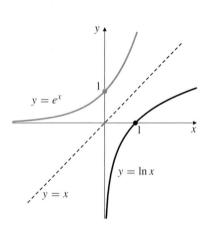

Figure R.12   The graphs of $y = e^x$ and $y = \ln x$

For every real number x, $e^x = \lim_{n \to \infty} \left(1 + \dfrac{x}{n}\right)^n$.

The initial-value problem $\begin{cases} \dfrac{dy}{dt} = ky \\ y(0) = y_0 \end{cases}$ has solution $y = y_0 e^{kt}$.

For $a > 0$, the **general exponential function** a^x is defined by $a^x = e^{x \ln a}$. Its inverse is the **base a logarithm**, $\log_a x = \dfrac{\ln x}{\ln a}$. We have

$$\frac{d}{dx} a^x = a^x \ln a \quad \text{and} \quad \frac{d}{dx} \log_a x = \frac{1}{x \ln a}.$$

The Inverse Trigonometric Functions

The **inverse sine** function, $\sin^{-1} x$ or $\arcsin x$, and the **inverse tangent** function, $\tan^{-1} x$ or $\arctan x$, are defined by

$$y = \sin^{-1} x \iff x = \sin y \quad \text{and} \quad -\frac{\pi}{2} \le y \le \frac{\pi}{2}$$
$$y = \tan^{-1} x \iff x = \tan y \quad \text{and} \quad -\frac{\pi}{2} < y < \frac{\pi}{2}.$$

We have

$$\frac{d}{dx} \sin^{-1} x = \frac{1}{\sqrt{1 - x^2}} \qquad \int \frac{1}{\sqrt{a^2 - x^2}} \, dx = \sin^{-1} \frac{x}{a} + C \quad (a > 0)$$
$$\frac{d}{dx} \tan^{-1} x = \frac{1}{1 + x^2} \qquad \int \frac{dx}{a^2 + x^2} = \frac{1}{a} \tan^{-1}\left(\frac{x}{a}\right) + C.$$

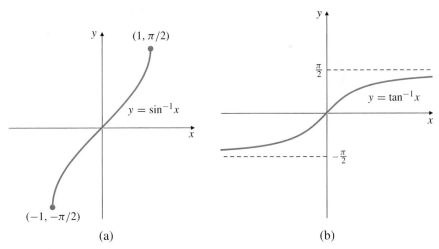

Figure R.13

(a) The graph of $\sin^{-1} x$

(b) The graph of $\tan^{-1} x$

The other inverse trigonometric functions can be expressed in terms of $\sin^{-1}$ and $\tan^{-1}$. For example,

$$\cos^{-1} x = \frac{\pi}{2} - \sin^{-1} x \qquad \text{for} \quad -1 \le x \le 1$$
$$\sec^{-1} x = \cos^{-1}\left(\frac{1}{x}\right) = \frac{\pi}{2} - \sin^{-1}\left(\frac{1}{x}\right) \qquad \text{for} \quad |x| \ge 1.$$

The derivatives of these functions are

$$\frac{d}{dx} \cos^{-1} x = -\frac{1}{\sqrt{1 - x^2}}, \qquad \frac{d}{dx} \sec^{-1} x = \frac{1}{|x|\sqrt{x^2 - 1}}.$$

Hyperbolic Functions

The hyperbolic cosine, sine, and tangent functions are defined in terms of the exponential functions:

$$\cosh x = \frac{e^x + e^{-x}}{2}, \qquad \sinh x = \frac{e^x - e^{-x}}{2}, \qquad \tanh x = \frac{e^x - e^{-x}}{e^x + e^{-x}}.$$

It follows that $\cosh^2 x - \sinh^2 x = 1$, $(d/dx)\cosh x = \sinh x$, and $(d/dx)\sinh x = \cosh x$. The three other hyperbolic functions—sech, csch, and coth—are the reciprocals of cosh, sinh, and tanh, respectively.

The inverses of the hyperbolic functions can be expressed in terms of natural logarithms:

$$\cosh^{-1} x = \ln\left(x + \sqrt{x^2 - 1}\right) \qquad \text{for } x \geq 1,$$

$$\sinh^{-1} x = \ln\left(x + \sqrt{x^2 + 1}\right) \qquad \text{for all } x,$$

$$\tanh^{-1} x = \frac{1}{2}\ln\left(\frac{1+x}{1-x}\right) \qquad \text{for } -1 < x < 1.$$

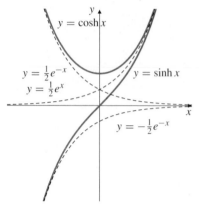

Figure R.14 The graphs $\cosh x$ and $\sinh x$ and some exponential graphs to which they are asymptotic

Maximum and Minimum Values

The function f has an **absolute maximum value** $f(x_0)$ at the point x_0 in its domain if $f(x) \leq f(x_0)$ holds for every x in the domain of f. Similarly, f has an **absolute minimum value** $f(x_1)$ at the point x_1 in its domain if $f(x) \geq f(x_1)$ holds for every x in the domain of f. Theorem 1 implies that a continuous function defined on a finite, closed interval has absolute maximum and minimum values.

The function f has a local maximum (or minimum) value at x_0 if it has an absolute maximum (or minimum) value at x_0 when its domain is restricted to points sufficiently near x_0. Geometrically, the graph of f is at least as high (or low) at x_0 as it is at nearby points.

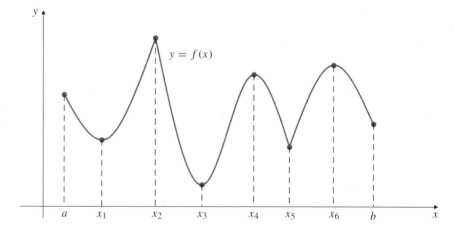

Figure R.15 f has local maximum values at a, x_2, x_4, and x_6, and local minimum values at x_1, x_3, x_5, and b. Its absolute maximum value is $f(x_2)$, and its absolute minimum value is $f(x_3)$

If f has a local maximum or minimum value at x in its domain, then x must be either

(i) a **critical point** of f (a point where $f'(x) = 0$),

(ii) a **singular point** of f (a point where $f'(x)$ is not defined), or

(iii) an **endpoint** of the domain of f (a point in $D(f)$ that does not belong to any open interval contained in $D(f)$).

Linear Approximations and Newton's Method

The **linearization**, or **best linear approximation**, of the function f near $x = a$ is the function $L(x)$ defined by

$$L(x) = f(a) + f'(a)(x - a).$$

If $|f''(t)| \le K$ for all t in an interval containing a and x, then the error $E(x) = f(x) - L(x)$ in the linear approximation $f(x) \approx L(x)$ satisfies

$$|E(x)| \le \frac{K}{2}(x - a)^2.$$

Newton's method is a procedure for approximating a **root** of the equation $f(x) = 0$, that is, a number r such that $f(r) = 0$, where f is differentiable near the root. Start by guessing an initial approximation x_0 for the root, and then calculating successive approximations $x_1, x_2, x_3, \ldots$ according to the formula

$$x_{n+1} = x_n - \frac{f(x_n)}{f'(x_n)}.$$

Note that x_{n+1} is the x-intercept of the line tangent to $y = f(x)$ at $x = x_n$.

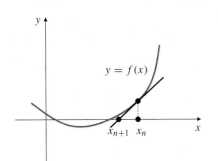

Figure R.16 The geometric construction of the Newton's method formula

The Shape of a Graph

Examining the sign of the derivative of a function gives information about the behaviour of that function. For example, if f is continuous on an interval I and differentiable on the open interval J consisting of the points of I that are not endpoints of I, then the following assertions are valid:

(a) If $f'(x) = 0$ for all x in J, then f is constant on I.

(b) If $f'(x) > 0$ for all x in J, then f is increasing on I.

(c) If $f'(x) < 0$ for all x in J, then f is decreasing on I.

(d) If $f'(x) \ge 0$ for all x in J, then f is nondecreasing on I.

(e) If $f'(x) \le 0$ for all x in J, then f is nonincreasing on I.

All of these follow from the Mean-Value Theorem.

Knowledge of where a function is increasing and decreasing can tell us whether the function has a local maximum or local minimum value (or neither) at a critical point, a singular point, or an endpoint of its domain. For example, if f is continuous at x_0, where $a < x_0 < b$, and $f'(x) > 0$ on (a, x_0) and $f'(x) < 0$ on (x_0, b), then f must have a local maximum value at x_0. If $f'(x) < 0$ on (a, x_0) and $f'(x) > 0$ on (x_0, b), then f must have a local minimum value at x_0. This way of classifying critical or singular points of f is called the **first derivative test**.

We say that the function f is **concave up** on an open interval I if it is differentiable there and the derivative f' is an increasing function on I. Similarly, f is **concave down** on I if f' exists and is decreasing on I.

We say that the point $(x_0, f(x_0))$ is an **inflection point** of the curve $y = f(x)$ (or that the function $f(x)$ has an inflection point at $x = x_0$) if the following two conditions are satisfied:

(a) the graph $y = f(x)$ has a tangent line at x_0, and

(b) the concavity of f is opposite on opposite sides of x_0.

Concavity is determined by the sign of the second derivative of a function:

(a) If $f''(x) > 0$ on interval I, then f is concave up on I.

(b) If $f''(x) < 0$ on interval I, then f is concave down on I.

If f has an inflection point at x_0, and if $f''(x_0)$ exists, then $f''(x_0) = 0$.

The **second derivative test** can be used to classify a critical point of a function:

(a) If $f'(x_0) = 0$ and $f''(x_0) < 0$, then f has a local maximum value at x_0.

(b) If $f'(x_0) = 0$ and $f''(x_0) > 0$, then f has a local minimum value at x_0.

(c) If $f'(x_0) = 0$ and $f''(x_0) = 0$, no conclusion can be drawn; f may have a local maximum at x_0 or a local minimum, or it may have an inflection point instead.

Higher-Order Approximations

The **Taylor polynomial of degree** n for $f(x)$ at $x = a$ is the polynomial $P_n(x)$ having degree $\leq n$ that best approximates the behaviour of $f(x)$ near $x = a$. It is given by

$$P_n(x) = f(a) + f'(a)(x-a) + \frac{f''(a)}{2!}(x-a)^2 + \cdots + \frac{f^{(n)}(a)}{n!}(x-a)^n,$$

since, at $x = a$, the values of this polynomial and all its derivatives of order up to n are equal to the corresponding values of f and its derivatives. The following theorem provides a means for estimating the error $E_n(x) = f(x) - P_n(x)$ in the approximation $f(x) \approx P_n(x)$.

THEOREM 5

Taylor's Theorem with Lagrange Remainder

If the $(n+1)$st-order derivative, $f^{(n+1)}(t)$, exists for all t in an interval containing a and x, then

$$f(x) = f(a) + f'(a)(x-a) + \frac{f''(a)}{2!}(x-a)^2 + \cdots$$
$$+ \frac{f^{(n)}(a)}{n!}(x-a)^n + E_n(x), \quad \textbf{Taylor's Formula},$$

where the error term $E_n(x)$ is given by

$$E_n(x) = \frac{f^{(n+1)}(X)}{(n+1)!}(x-a)^{n+1}, \quad \textbf{Lagrange Remainder},$$

for some number X between a and x.

Indeterminate Forms and l'Hôpital's Rule

The laws for calculating limits are indecisive when they lead to values such as 0/0 or $\infty - \infty$, which are undefined in the real number system. Table 2 gives examples of various types of **indeterminate forms**:

Table 2. Types of indeterminate forms

Type	Example
$[0/0]$	$\displaystyle\lim_{x \to 1} \frac{\ln x}{x - 1}$
$[\infty/\infty]$	$\displaystyle\lim_{x \to 0} \frac{\ln(1/x^2)}{\cot(x^2)}$
$[0 \cdot \infty]$	$\displaystyle\lim_{x \to 0+} x \ln \frac{1}{x}$
$[\infty - \infty]$	$\displaystyle\lim_{x \to (\pi/2)-} \left(\tan x - \frac{1}{\pi - 2x} \right)$
$[0^0]$	$\displaystyle\lim_{x \to 0+} x^x$
$[\infty^0]$	$\displaystyle\lim_{x \to (\pi/2)-} (\tan x)^{\cos x}$
$[1^\infty]$	$\displaystyle\lim_{x \to \infty} \left(1 + \frac{1}{x} \right)^x$

Limits of type $[0/0]$ or $[\infty/\infty]$ can sometimes be evaluated using the following theorem:

THEOREM 6

l'Hôpital's Rule

Suppose the functions f and g are differentiable on the interval (a, b), and $g'(x) \neq 0$ there. Suppose also that the following two conditions hold:

(i) either $\lim_{x \to a+} f(x) = \lim_{x \to a+} g(x) = 0$, or $\lim_{x \to a+} f(x) = \pm\infty$ and $\lim_{x \to a+} g(x) = \pm\infty$

(ii) $\lim_{x \to a+} \dfrac{f'(x)}{g'(x)} = L$ (where L is finite or ∞ or $-\infty$).

Then $\lim_{x \to a+} \dfrac{f(x)}{g(x)} = L$.

Similar results hold for $\lim_{x \to b-}$ and for $\lim_{x \to c}$, and the cases $a = -\infty$ and $b = \infty$ are allowed.

The other types of indeterminate forms in the table can usually be put in the form $[0/0]$ or $[\infty/\infty]$ by algebraic manipulation and the taking of logarithms.

Riemann Sums and the Definite Integral

If m and n are integers with $m \leq n$, and if f is a function defined at the integers m, $m+1, m+2, \ldots, n$, the symbol $\sum_{i=m}^{n} f(i)$ represents the sum of the values of f at those integers:

$$\sum_{i=m}^{n} f(i) = f(m) + f(m+1) + f(m+2) + \cdots + f(n).$$

The explicit sum appearing on the right side of this equation is the **expansion** of the sum represented in sigma notation on the left side. The values of some particular sums can be expressed as simple formulas:

(a) $\displaystyle\sum_{i=1}^{n} 1 = \underbrace{1 + 1 + 1 + \cdots + 1}_{n \text{ terms}} = n$

(b) $\displaystyle\sum_{i=1}^{n} i = 1 + 2 + 3 + \cdots + n = \dfrac{n(n+1)}{2}$

(c) $\displaystyle\sum_{i=1}^{n} i^2 = 1^2 + 2^2 + 3^2 + \cdots + n^2 = \dfrac{n(n+1)(2n+1)}{6}$

(d) $\displaystyle\sum_{i=1}^{n} r^{i-1} = 1 + r + r^2 + r^3 + \cdots + r^{n-1} = \dfrac{r^n - 1}{r - 1}$ if $r \neq 1$.

A **partition** of the interval $[a, b]$ is a finite set of points arranged in order between a and b on the real line, say,

$$P = \{x_0, x_1, x_2, x_3, \ldots, x_{n-1}, x_n\},$$

where $a = x_0 < x_1 < x_2 < x_3 < \cdots < x_{n-1} < x_n = b$. It divides $[a, b]$ into n subintervals of which the ith, $[x_{i-1}, x_i]$, has length $\Delta x_i = x_i - x_{i-1}$. Let f be continuous on $[a, b]$, and let c_i be a point in the ith subinterval of P. The sum

$$R(f, P) = \sum_{i=1}^{n} f(c_i) \, \Delta x_i$$

$$= f(c_1) \, \Delta x_1 + f(c_2) \, \Delta x_2 + f(c_3) \, \Delta x_3 + \cdots + f(c_n) \, \Delta x_n$$

is called a **Riemann sum** of f on $[a, b]$.

Figure R.17 The Riemann sum $R(f, P)$ is the sum of areas of the rectangles shaded in colour minus the sum of the areas of the rectangles shaded in grey

Because f is continuous on $[a, b]$, there is a single number I such that we can ensure that $R(f, P)$ is as close as we want to I, independently of how the numbers c_i are chosen, by choosing P so that all its subintervals are sufficiently short. We call I the definite integral of f on $[a, b]$, and write it

$$I = \int_a^b f(x)\,dx.$$

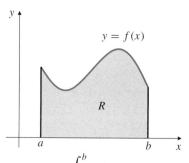

Figure R.18 $\int_a^b f(x)\,dx$ = area of R

The definite integral has the following properties:

(a) If $f(x) \geq 0$ on $[a, b]$, then $\int_a^b f(x)\,dx$ represents the area bounded by $y = f(x)$, $y = 0$, $x = a$, and $x = b$.

(b) An integral over an interval of zero length is zero. $\int_a^a f(x)\,dx = 0$.

(c) Reversing the limits of integration changes the sign of the integral.

$$\int_b^a f(x)\,dx = -\int_a^b f(x)\,dx.$$

(d) An integral depends linearly on the integrand. If A and B are constants, then

$$\int_a^b \big(Af(x) + Bg(x)\big)\,dx = A\int_a^b f(x)\,dx + B\int_a^b g(x)\,dx.$$

(e) An integral depends additively on the interval of integration.

$$\int_a^b f(x)\,dx + \int_b^c f(x)\,dx = \int_a^c f(x)\,dx.$$

(f) If $f(x) \leq g(x)$ and $a \leq b$, then $\int_a^b f(x)\,dx \leq \int_a^b g(x)\,dx$.

(g) The **triangle inequality** for definite integrals: if $a \leq b$, then

$$\left| \int_a^b f(x)\,dx \right| \leq \int_a^b |f(x)|\,dx.$$

(h) If f is an odd function ($f(-x) = -f(x)$), then $\int_{-a}^a f(x)\,dx = 0$.

(i) If f is an even function ($f(-x) = f(x)$), then $\int_{-a}^a f(x)\,dx = 2\int_0^a f(x)\,dx$.

If f is continuous on $[a, b]$, then there exists a point c in $[a, b]$ such that

$$\int_a^b f(x)\,dx = (b - a)f(c).$$

We call $f(c)$ the **average value** or **mean value** of f on $[a, b]$, and denote it by $\overline{f}$.

$$\overline{f} = \frac{1}{b - a} \int_a^b f(x)\,dx.$$

A function f is **piecewise continuous** on $[a, b]$ if there exists a partition $a = c_0 < c_1 < c_2 < \cdots < c_n = b$ of $[a, b]$ and functions $F_i(x)$ continuous on $[c_{i-1}, c_i]$ such that $f(x) = F_i(x)$ for $c_{i-1} < x < c_i$. The definite integral can be extended to such functions f by defining

$$\int_a^b f(x)\,dx = \sum_{i=1}^n \int_{c_{i-1}}^{c_i} F_i(x)\,dx.$$

THEOREM 7

The Fundamental Theorem of Calculus

Suppose that the function f is continuous on an interval I containing the point a.

PART I. Let the function F be defined on I by

$$F(x) = \int_a^x f(t)\,dt.$$

Then F is differentiable on I and $F'(x) = f(x)$ there; that is, F is an antiderivative of f on I. Expressed in terms of symbols,

$$\frac{d}{dx} \int_a^x f(t)\,dt = f(x).$$

PART II. If $G(x)$ is *any* antiderivative of $f(x)$ on I, so that $G'(x) = f(x)$ on I, then for any b in I we have

$$\int_a^b f(x)\,dx = G(b) - G(a).$$

Areas Between Curves

If $a < b$ and $f(x) \le g(x)$ on $[a, b]$, then the area A of the region R bounded by the curves $y = f(x)$ and $y = g(x)$ between the vertical lines $x = a$ and $x = b$ is given by

$$A = \int_a^b \big(g(x) - f(x)\big)\,dx.$$

It is useful to regard this formula as expressing A as the "sum" (i.e., the integral) of *infinitely many* **area elements**

$$dA = (g(x) - f(x))\,dx,$$

corresponding to values of x between a and b.

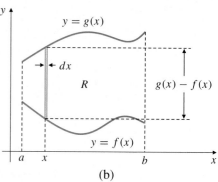

Figure R.19

(a) The region R lying between two graphs

(b) An area element of the region R

(a)   (b)

Integration Formulas and Techniques

Some elementary integrals

1. $\displaystyle\int 1\,dx = x + C$

2. $\displaystyle\int x\,dx = \frac{1}{2}x^2 + C$

3. $\displaystyle\int x^2\,dx = \frac{1}{3}x^3 + C$

4. $\displaystyle\int \frac{1}{x^2}\,dx = -\frac{1}{x} + C$

5. $\displaystyle\int \sqrt{x}\,dx = \frac{2}{3}x^{3/2} + C$

6. $\displaystyle\int \frac{1}{\sqrt{x}}\,dx = 2\sqrt{x} + C$

7. $\displaystyle\int x^r\,dx = \frac{1}{r+1}x^{r+1} + C \quad (r \neq -1)$

8. $\displaystyle\int \frac{1}{x}\,dx = \ln|x| + C$

9. $\displaystyle\int \sin ax\,dx = -\frac{1}{a}\cos ax + C$

10. $\displaystyle\int \cos ax\,dx = \frac{1}{a}\sin ax + C$

11. $\displaystyle\int \sec^2 ax\,dx = \frac{1}{a}\tan ax + C$

12. $\displaystyle\int \csc^2 ax\,dx = -\frac{1}{a}\cot ax + C$

13. $\displaystyle\int \sec ax\,\tan ax\,dx = \frac{1}{a}\sec ax + C$

14. $\displaystyle\int \csc ax\,\cot ax\,dx = -\frac{1}{a}\csc ax + C$

15. $\displaystyle\int \frac{1}{\sqrt{a^2 - x^2}}\,dx = \sin^{-1}\frac{x}{a} + C \quad (a > 0)$

16. $\displaystyle\int \frac{1}{a^2 + x^2}\,dx = \frac{1}{a}\tan^{-1}\frac{x}{a} + C$

17. $\displaystyle\int e^{ax}\,dx = \frac{1}{a}e^{ax} + C$

18. $\displaystyle\int b^{ax}\,dx = \frac{1}{a\ln b}b^{ax} + C$

19. $\displaystyle\int \cosh ax\,dx = \frac{1}{a}\sinh ax + C$

20. $\displaystyle\int \sinh ax\,dx = \frac{1}{a}\cosh ax + C$

The following techniques can be used to evaluate more complicated integrals.

The Method of Substitution. Let $u = g(x)$. Then $du = g'(x)\,dx$, and

$$\int f\big(g(x)\big)\,g'(x)\,dx = \int f(u)\,du.$$

Suppose that g is a differentiable function on $[a, b]$ and satisfies $g(a) = A$ and $g(b) = B$. Also suppose that f is continuous on the range of g. Then

$$\int_a^b f\big(g(x)\big)\,g'(x)\,dx = \int_A^B f(u)\,du.$$

Substitution and trigonometric identities provide values for the following integrals:

$$\int \tan x \, dx = \ln|\sec x| + C,$$

$$\int \cot x \, dx = \ln|\sin x| + C = -\ln|\csc x| + C,$$

$$\int \sec x \, dx = \ln|\sec x + \tan x| + C,$$

$$\int \csc x \, dx = -\ln|\csc x + \cot x| + C = \ln|\csc x - \cot x| + C,$$

$$\int \cos^2 x \, dx = \frac{1}{2}(x + \sin x \cos x) + C,$$

$$\int \sin^2 x \, dx = \frac{1}{2}(x - \sin x \cos x) + C.$$

Integration by Parts. If $U = U(x), dV = V'(x)\,dx, dU = U'(x)\,dx$, and $V = V(x)$, then $d(UV) = [U'(x)V(x) + U(x)V'(x)]\,dx$, and so

$$\int U(x)V'(x)\,dx = \int U\,dV = UV - \int V\,dU = U(x)V(x) - \int V(x)U'(x)\,dx.$$

Inverse Trigonometric Substitutions.

1. Integrals involving $\sqrt{a^2 - x^2}$ (where $a > 0$) can frequently be reduced to a simpler form by means of the substitution

$$x = a\sin\theta \quad \text{or, equivalently,} \quad \theta = \sin^{-1}\frac{x}{a}.$$

   In this case, $\sqrt{a^2 - x^2} = a\cos\theta$ and $dx = a\cos\theta\,d\theta$.

2. Integrals involving $\sqrt{a^2 + x^2}$ or $\dfrac{1}{x^2 + a^2}$ (where $a > 0$) are often simplified by the substitution

$$x = a\tan\theta \quad \text{or, equivalently,} \quad \theta = \tan^{-1}\frac{x}{a}.$$

   In this case, $\sqrt{a^2 + x^2} = a\sec\theta$ and $dx = a\sec^2\theta\,d\theta$.

3. Integrals involving $\sqrt{x^2 - a^2}$ (where $a > 0$) can frequently be simplified by using the substitution

$$x = a\sec\theta \quad \text{or, equivalently,} \quad \theta = \sec^{-1}\frac{x}{a}.$$

   If $x \geq a$, then $\sqrt{x^2 - a^2} = a\tan\theta$; if $x \leq -a$, then $\sqrt{x^2 - a^2} = -a\tan\theta$. In either case $dx = a\sec\theta\tan\theta\,d\theta$.

Integrals of Rational Functions. If $P(x)$ and $Q(x)$ are polynomials such that the degree of P is less than the degree of Q and Q can be factored into a product of linear and quadratic factors, then the fraction P/Q can be written as a sum of other such fractions whose denominators are the factors of Q. For example,

$$\frac{1}{x^2 - a^2} = \frac{1}{2a}\frac{1}{x - a} - \frac{1}{2a}\frac{1}{x + a}.$$

The integral of $P(x)/Q(x)$ can thereby be expressed as a sum of simpler integrals. The following integrals are useful in this context:

$$\int \frac{1}{ax+b}\, dx = \frac{1}{a}\, \ln |ax+b| + C,$$

$$\int \frac{x\, dx}{x^2 + a^2} = \frac{1}{2}\, \ln(x^2 + a^2) + C,$$

$$\int \frac{x\, dx}{x^2 - a^2} = \frac{1}{2}\, \ln |x^2 - a^2| + C,$$

$$\int \frac{dx}{x^2 + a^2} = \frac{1}{a}\, \tan^{-1} \frac{x}{a} + C,$$

$$\int \frac{dx}{x^2 - a^2} = \frac{1}{2a}\, \ln \left| \frac{x-a}{x+a} \right| + C.$$

Improper Integrals

Improper integrals are ones that correspond to areas of unbounded regions in the xy-plane. Such integrals occur if either (or both) of the limits of integration is infinite or if f is unbounded near a point of $[a, b]$ (and hence is not continuous on $[a, b]$). In this context, the ordinary definite integral of a continuous function over a closed, finite interval is called a **proper integral**.

Improper Integrals of Type I. If f is continuous on $[a, \infty)$, we define the improper integral of f over $[a, \infty)$ as a limit of proper integrals:

$$\int_a^\infty f(x)\, dx = \lim_{R \to \infty} \int_a^R f(x)\, dx.$$

Similarly, if f is continuous on $(-\infty, b]$, we define

$$\int_{-\infty}^b f(x)\, dx = \lim_{R \to -\infty} \int_R^b f(x)\, dx.$$

In either case, if the limit exists (is a finite number), we say that the improper integral **converges**; if the limit does not exist, we say that the improper integral **diverges**. If the limit is ∞ (or $-\infty$), we say the improper integral **diverges to infinity** (or **diverges to negative infinity**).

Improper Integrals of Type II. If f is continuous on the interval $(a, b]$ and is possibly unbounded near a, we define the improper integral

$$\int_a^b f(x)\, dx = \lim_{c \to a+} \int_c^b f(x)\, dx.$$

Similarly, if f is continuous on $[a, b)$ and is possibly unbounded near b, we define

$$\int_a^b f(x)\, dx = \lim_{c \to b-} \int_a^c f(x)\, dx.$$

These improper integrals may converge, diverge, diverge to infinity, or diverge to negative infinity.

For example, if $0 < a < \infty$, then

$$\int_a^\infty x^{-p}\, dx \quad \begin{cases} \text{converges to } \dfrac{a^{1-p}}{p-1} & \text{if } p > 1 \\ \text{diverges to } \infty & \text{if } p \le 1 \end{cases}$$

$$\int_0^a x^{-p}\,dx \quad \begin{cases} \text{converges to } \dfrac{a^{1-p}}{1-p} & \text{if } p < 1 \\ \text{diverges to } \infty & \text{if } p \geq 1. \end{cases}$$

To determine convergence or divergence and estimate the size of improper integrals for which no antiderivative can be found, you can compare the integral with a simpler one that can be evaluated:

Let $-\infty \leq a < b \leq \infty$ and suppose that functions f and g are continuous on the interval (a, b) and satisfy $0 \leq f(x) \leq g(x)$. If $\int_a^b g(x)\,dx$ converges, then so does $\int_a^b f(x)\,dx$, and

$$\int_a^b f(x)\,dx \leq \int_a^b g(x)\,dx.$$

Equivalently, if $\int_a^b f(x)\,dx$ diverges to ∞, then so does $\int_a^b g(x)\,dx$.

Numerical Integration

We assume that $f(x)$ is continuous on $[a, b]$ and subdivide $[a, b]$ into n subintervals of equal length $h = (b - a)/n$ using the $n + 1$ points

$$x_0 = a, \quad x_1 = a + h, \quad x_2 = a + 2h, \quad \ldots, \quad x_n = a + nh = b.$$

We also assume that the value of $f(x)$ at each of these points is known:

$$y_0 = f(x_0), \quad y_1 = f(x_1), \quad y_2 = f(x_2), \quad \ldots, \quad y_n = f(x_n).$$

The n-subinterval **Trapezoid Rule** approximation to $\displaystyle\int_a^b f(x)\,dx$, denoted T_n, is given by

$$T_n = h\left(\frac{1}{2}y_0 + y_1 + y_2 + y_3 + \cdots + y_{n-1} + \frac{1}{2}y_n\right)$$

$$= \frac{h}{2}\left(y_0 + 2\sum_{j=1}^{n-1} y_j + y_n\right).$$

Let $m_j = a + \left(j - \frac{1}{2}\right)h$ for $1 \leq j \leq n$. The **Midpoint Rule** approximation to $\int_a^b f(x)\,dx$, denoted M_n, is given by

$$M_n = h\big(f(m_1) + f(m_2) + \cdots + f(m_n)\big) = h\sum_{j=1}^{n} f(m_j).$$

Observe that $T_{2n} = \frac{1}{2}(T_n + M_n)$.

The **Simpson's Rule** approximation to $\int_a^b f(x)\,dx$ based on a subdivision of $[a, b]$ into an even number n of subintervals of equal length $h = (b - a)/n$ is denoted S_n, and is given by

$$\int_a^b f(x)\,dx \approx S_n$$

$$= \frac{h}{3}\left(y_0 + 4y_1 + 2y_2 + 4y_3 + 2y_4 + \cdots + 2y_{n-2} + 4y_{n-1} + y_n\right)$$

$$= \frac{h}{3}\left(y_{\text{``ends''}} + 4y_{\text{``odds''}} + 2y_{\text{``evens''}}\right)$$

The numbers y_i are as specified for the Trapezoid Rule. Observe that $S_{2n} = \dfrac{T_n + 2M_n}{3}$.

The following error estimates hold for these approximations: If f has a continuous second derivative on $[a, b]$ satisfying $|f''(x)| \leq K$ there, then

$$\left| \int_a^b f(x)\, dx - T_n \right| \leq \frac{K(b-a)}{12} h^2 = \frac{K(b-a)^3}{12n^2},$$

$$\left| \int_a^b f(x)\, dx - M_n \right| \leq \frac{K(b-a)}{24} h^2 = \frac{K(b-a)^3}{24n^2}.$$

If f has a continuous fourth derivative on the interval $[a, b]$, satisfying $|f^{(4)}(x)| \leq K$ there, then

$$\left| \int_a^b f(x)\, dx - S_n \right| \leq \frac{K(b-a)}{180} h^4 = \frac{K(b-a)^5}{180n^4}.$$

Volumes

The volume V of a solid between $x = a$ and $x = b$ having cross-sectional area $A(x)$ at position x is

$$V = \int_a^b A(x)\, dx.$$

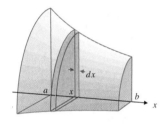

Figure R.20   The volume element

Here the integral "sums" volume elements $dV = A(x)\, dx$, which are the volumes of "infinitely thin slices" perpendicular to the x-axis.

For **solids of revolution** the cross-sections are circular disks. If a vertical strip $0 < y < f(x)$ between x and $x + dx$ is rotated about the x-axis, it generates a circular disk of radius $f(x)$ and thickness dx. If the strip is rotated about the y-axis, it generates a cylindrical shell of height $f(x)$, radius x, and thickness dx.

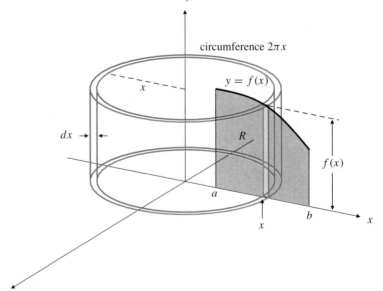

Figure R.21   When rotated around the y-axis, the area element of width dx under $y = f(x)$ at x generates a cylindrical shell of height $f(x)$, circumference $2\pi x$, and hence volume $dV = 2\pi x\, f(x)\, dx$

The volume of the solid obtained by rotating the plane region $0 \leq y \leq f(x)$, $a < x < b$ about the x-axis is

$$V = \pi \int_a^b \left(f(x) \right)^2 dx.$$

The volume of the solid obtained by rotating the same region about the y-axis is

$$V = 2\pi \int_a^b x\, f(x)\, dx.$$

Arc Length and Surface Area

The arc length s of the curve $y = f(x)$ from $x = a$ to $x = b$ is given by

$$s = \int_a^b \sqrt{1 + \left(f'(x)\right)^2}\, dx = \int_a^b \sqrt{1 + \left(\frac{dy}{dx}\right)^2}\, dx.$$

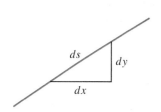

Figure R.22   A differential triangle

You can regard this integral formula as giving the arc length s of the curve as a "sum" of **arc length elements**

$$s = \int_{x=a}^{x=b} ds, \qquad \text{where} \qquad ds = \sqrt{(dx)^2 + (dy)^2} = \sqrt{1 + \left(\frac{dy}{dx}\right)^2}\, dx.$$

When a plane curve is rotated (in three dimensions) about a line in the plane of the curve, it sweeps out a **surface of revolution**. If the radius of rotation of an arc length element ds on the curve is r, then it generates, on rotation, a circular band of width ds and length (circumference) $2\pi r$. The area of this band is, therefore,

$$dS = 2\pi r\, ds.$$

For example, if $f'(x)$ is continuous on $[a, b]$ and the curve $y = f(x)$ is rotated about the x-axis, then the area of the surface of revolution so generated is

$$S = 2\pi \int_{x=a}^{x=b} |y|\, ds = 2\pi \int_a^b |f(x)| \sqrt{1 + (f'(x))^2}\, dx.$$

If the rotation is about the y-axis, the surface area is

$$S = 2\pi \int_{x=a}^{x=b} |x|\, ds = 2\pi \int_a^b |x| \sqrt{1 + (f'(x))^2}\, dx.$$

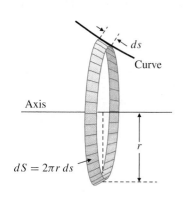

Figure R.23   The circular band generated by rotating arc length element ds about the axis

Centres of Mass and Centroids

Suppose that matter is distributed on the interval $[a, b]$ with *line density* (mass per unit length) $\delta(x)$, which depends on position x in the interval. The mass, m, and moment about $x = 0$, $M_{x=0}$, of the matter in the interval, are given by

$$m = \int_a^b \delta(x)\, dx, \qquad M_{x=0} = \int_a^b x\delta(x)\, dx.$$

The centre of mass of the distribution is the point $\bar{x} = M_{x=0}/m$. If the whole mass m were placed at $\bar{x}$, it would have the same moment about $x = 0$. If the density is constant, the centre of mass is the midpoint of the interval. We call it the **centroid** of the interval.

The centroid of any region is the centre of mass of an object of constant density $\delta = 1$ occupying that region. The centroid of the plane region $a \le x \le b, 0 \le y \le f(x)$, is $(\bar{x}, \bar{y})$, where $\bar{x} = \dfrac{M_{x=0}}{A}, \bar{y} = \dfrac{M_{y=0}}{A}$, and

$$A = \int_a^b f(x)\, dx, \qquad M_{x=0} = \int_a^b x f(x)\, dx, \qquad M_{y=0} = \frac{1}{2} \int_a^b \left(f(x)\right)^2 dx.$$

For example, the coordinates of the centroid of a triangle are the averages of the corresponding coordinates of the three vertices of the triangle. The triangle with vertices (x_1, y_1), (x_2, y_2), and (x_3, y_3) has centroid

$$(\bar{x}, \bar{y}) = \left(\frac{x_1 + x_2 + x_3}{3}, \frac{y_1 + y_2 + y_3}{3} \right).$$

THEOREM   **8**   **Pappus's Theorem**

(a) If a plane region R lies on one side of a line L in that plane, and if R is rotated about L to generate a solid of revolution, the volume V of that solid is the product of the area of R and the distance travelled by the centroid of R under the rotation; that is,

$$V = 2\pi \bar{r} A,$$

where A is the area of R, and $\bar{r}$ is the perpendicular distance from the centroid of R to L.

(b) If a plane curve C, lying on one side of a line L in the plane, is rotated about that line to generate a surface of revolution, then the area S of that surface is the length of C times the distance travelled by the centroid of C:

$$S = 2\pi \bar{r} s,$$

where s is the length of the curve C, and $\bar{r}$ is the perpendicular distance from the centroid of C to the line L.

Conics

A **parabola** consists of points in the plane that are equidistant from a given point (the **focus** of the parabola) and a given straight line (the **directrix** of the parabola). The line through the focus perpendicular to the directrix is called the **principal axis** (or simply the **axis**) of the parabola. The **vertex** of the parabola is the point where the parabola crosses its principal axis. It is on the axis halfway between the focus and the directrix.

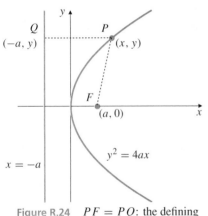

Figure R.24 $PF = PQ$: the defining property of a parabola

Table 3.   Some parabolas with vertex at the origin

Focus	Directrix	Equation
$(a, 0)$	$x = -a$	$y^2 = 4ax$
$(-a, 0)$	$x = a$	$y^2 = -4ax$
$(0, a)$	$y = -a$	$x^2 = 4ay$
$(0, -a)$	$y = a$	$x^2 = -4ay$

An **ellipse** consists of all points in the plane, the sum of whose distances from two fixed points (the **foci**) is constant.

If $a > b > 0$, then the equation $\dfrac{x^2}{a^2} + \dfrac{y^2}{b^2} = 1$ represents an ellipse with centre at the origin and foci at $(\pm c, 0)$, where $c^2 = a^2 - b^2$. The sum of the distances from $P = (x, y)$ on the ellipse to the two foci is $2a$.

a is the **semi-major axis**

b is the **semi-minor axis**

$c = \sqrt{a^2 - b^2}$ is the **semi-focal separation**

$\epsilon = \dfrac{c}{a} < 1$ is the **eccentricity**.

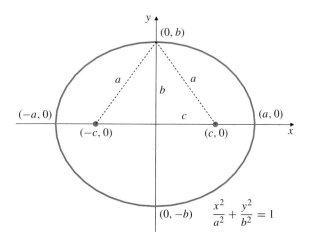

Figure R.25   An ellipse

A **hyperbola** consists of all points in the plane, the difference between whose distances from two fixed points (the **foci**) is constant.

The equation $\dfrac{x^2}{a^2} - \dfrac{y^2}{b^2} = 1$ represents a hyperbola with centre at the origin and foci at $(\pm c, 0)$, where $c^2 = a^2 + b^2$. The difference between the distances from $P = (x, y)$ on the hyperbola to the two foci is $2a$. The points $(a, 0)$ and $(-a, 0)$ are called the **vertices** of the hyperbola. The line through the centre, the vertices, and the foci is the **transverse axis**. The line through the centre perpendicular to the transverse axis is the **conjugate axis**. The hyperbola has two asymptotes with equations $(x/a) \pm (y/b) = 0$. A **rectangular** hyperbola is one whose asymptotes are perpendicular to each other.

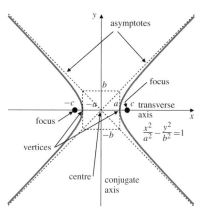

Figure R.26   Terms associated with a hyperbola

a is the **semi-transverse axis**

b is the **semi-conjugate axis**

$c = \sqrt{a^2 + b^2}$ is the **semi-focal separation**

$\epsilon = \dfrac{c}{a} > 1$ is the **eccentricity**.

Parametric Curves

A **parametric curve** $\mathcal{C}$ in the plane consists of an ordered pair (f, g) of continuous functions each defined on the same interval I. The equations

$$x = f(t), \qquad y = g(t), \qquad \text{for } t \text{ in } I,$$

are called the **parametric equations** of the curve $\mathcal{C}$. The independent variable t is called the **parameter**. We often regard the parametric curve as being *the set of points* $\big(f(t), g(t)\big)$ corresponding to values of t in I rather than as the pair of functions itself.

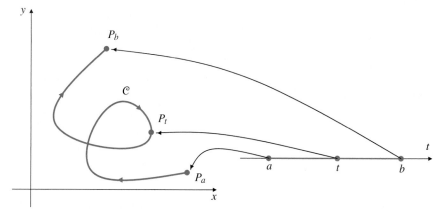

Figure R.27   As t increases from a to b, the point $(x, y) = \big(f(t), g(t)\big)$ traces the curve $\mathcal{C}$ in the xy-plane in the direction indicated

For example, the path followed by a particle moving in the xy-plane can be regarded as a parametric curve with parametric equations giving the position of the particle at time t.

The arc length element for a parametric curve is given by

$$ds = \sqrt{\left(\frac{dx}{dt}\right)^2 + \left(\frac{dy}{dt}\right)^2}\, dt,$$

so that the length of the curve $\mathcal{C}$: $x = f(t)$, $y = g(t)$, $(a \le t \le b)$, is

$$s = \int_{t=a}^{t=b} ds = \int_a^b \sqrt{\left(f'(t)\right)^2 + \left(g'(t)\right)^2}\, dt.$$

Figure R.28   Relating Cartesian and polar coordinates of a point

Polar Coordinates

Instead of using Cartesian coordinates (x, y), we can locate a point in the plane by using **polar coordinates** $[r, \theta]$, where

(i) r is the distance from O to P, and

(ii) θ is the angle that the ray OP makes with the positive x-axis (counterclockwise angles being considered positive).

The polar and Cartesian coordinates of a point are linked by the following equations:

$$\begin{aligned} x &= r\cos\theta & r^2 &= x^2 + y^2 \\ y &= r\sin\theta & \tan\theta &= \frac{y}{x}. \end{aligned}$$

The equations for r^2 and $\tan\theta$ cannot be solved uniquely for r and θ; the Cartesian coordinates of a point are unique, but its polar coordinates are not.

Polar graphs typically have equations of the form $r = f(\theta)$. Such a graph can approach the origin only from those directions θ for which $f(\theta) = 0$.

At any point P other than the origin on the polar curve $r = f(\theta)$, the angle ψ between the radial line from the origin to P and the tangent to the curve satisfies

$$\tan\psi = \frac{f(\theta)}{f'(\theta)}.$$

In particular, $\psi = \pi/2$ if $f'(\theta) = 0$.

If $f(\theta_0) = 0$ and the curve has a tangent line at θ_0, then that tangent line has polar equation $\theta = \theta_0$.

The region bounded by $r = f(\theta)$ and the rays $\theta = \alpha$ and $\theta = \beta$, $(\alpha < \beta)$, has area

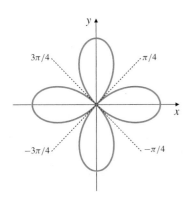

Figure R.29   The polar curve $r = \cos(2\theta)$

$$A = \frac{1}{2}\int_\alpha^\beta \left(f(\theta)\right)^2 d\theta.$$

The arc length element for the polar curve $r = f(\theta)$ is

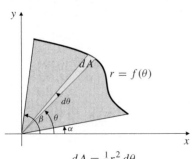

$dA = \frac{1}{2}r^2\, d\theta$

Figure R.30

$= \frac{1}{2}\left(f(\theta)\right)^2 d\theta$

$$ds = \sqrt{\left(\frac{dr}{d\theta}\right)^2 + r^2}\, d\theta = \sqrt{\left(f'(\theta)\right)^2 + \left(f(\theta)\right)^2}\, d\theta.$$

CHAPTER 9

Sequences, Series, and Power Series

"Then you should say what you mean," the March Hare went on.
"I do," Alice hastily replied; "at least — at least I mean what I say
— that's the same thing, you know."
"Not the same thing a bit!" said the Hatter. "Why, you might just
as well say that 'I see what I eat' is the same thing as 'I eat what I
see!' "

Lewis Carroll (Charles Lutwidge Dodgson) 1832–1898
from *Alice's Adventures in Wonderland*

Introduction   An infinite series is a sum that involves infinitely many terms. Since addition is carried out on two numbers at a time, the evaluation of the sum of an infinite series necessarily involves finding a limit. Complicated functions $f(x)$ can frequently be expressed as series of simpler functions. For example, many of the transcendental functions we have encountered can be expressed as series of powers of x so that they resemble polynomials of infinite degree. Such series can be differentiated and integrated term by term, and they play a very important role in the study of calculus.

9.1 Sequences and Convergence

By a **sequence** (or an **infinite sequence**) we mean an ordered list having a first element but no last element. For our purposes, the elements (called **terms**) of a sequence will always be real numbers, although much of our discussion could be applied to complex numbers as well. Examples of sequences are:

$\{1,\ 2,\ 3,\ 4,\ 5,\ \ldots\}$ the sequence of positive integers,

$\left\{-\dfrac{1}{2},\ \dfrac{1}{4},\ -\dfrac{1}{8},\ \dfrac{1}{16},\ \ldots\right\}$ the sequence of positive integer powers of $-\dfrac{1}{2}$.

The terms of a sequence are usually listed in braces as shown. The ellipsis points ($\ldots$) should be read "and so on."

An infinite sequence is a special kind of function, one whose domain is a set of integers extending from some starting integer to infinity. The starting integer is usually 1, so the domain is the set of positive integers. The sequence $\{a_1,\ a_2,\ a_3,\ a_4,\ \ldots\}$ is the function f that takes the value $f(n) = a_n$ at each positive integer n. A sequence can be specified in three ways:

(i) We can list the first few terms followed by $\ldots$ *if the pattern is obvious.*

(ii) We can provide a formula for the **general term** a_n as a function of n.

(iii) We can provide a formula for calculating the term a_n as a function of earlier terms $a_1,\ a_2,\ \ldots,\ a_{n-1}$ and specify enough of the beginning terms so the process of computing higher terms can begin.

In each case it must be possible to determine any term of the sequence, although it may be necessary to calculate all the preceding terms first.

Example 1   (Some examples of sequences)

(a) $\{n\} = \{1,\ 2,\ 3,\ 4,\ 5,\ \ldots\}$

(b) $\left\{\left(-\dfrac{1}{2}\right)^n\right\} = \left\{-\dfrac{1}{2},\ \dfrac{1}{4},\ -\dfrac{1}{8},\ \dfrac{1}{16},\ \ldots\right\}$

(c) $\left\{\dfrac{n-1}{n}\right\} = \left\{0,\ \dfrac{1}{2},\ \dfrac{2}{3},\ \dfrac{3}{4},\ \dfrac{4}{5},\ \ldots\right\}$

(d) $\{(-1)^{n-1}\} = \{\cos((n-1)\pi)\} = \{1,\ -1,\ 1,\ -1,\ 1,\ \ldots\}$

(e) $\left\{\dfrac{n^2}{2^n}\right\} = \left\{\dfrac{1}{2},\ 1,\ \dfrac{9}{8},\ 1,\ \dfrac{25}{32},\ \dfrac{36}{64},\ \dfrac{49}{128},\ \ldots\right\}$

(f) $\left\{\left(1+\dfrac{1}{n}\right)^n\right\} = \left\{2,\ \left(\dfrac{3}{2}\right)^2,\ \left(\dfrac{4}{3}\right)^3,\ \left(\dfrac{5}{4}\right)^4,\ \ldots\right\}$

(g) $\left\{\dfrac{\cos(n\pi/2)}{n}\right\} = \left\{0,\ -\dfrac{1}{2},\ 0,\ \dfrac{1}{4},\ 0,\ -\dfrac{1}{6},\ 0,\ \dfrac{1}{8},\ 0,\ \ldots\right\}$

(h) $a_1 = 1,\quad a_{n+1} = \sqrt{6+a_n},\quad (n = 1,\ 2,\ 3,\ \ldots)$

In this case $\{a_n\} = \{1,\ \sqrt{7},\ \sqrt{6+\sqrt{7}},\ \ldots\}$. Note that there is no *obvious* formula for a_n as an explicit function of n here, but we can still calculate a_n for any desired value of n provided we first calculate all the earlier values $a_2,\ a_3,\ \ldots,\ a_{n-1}$.

(i) $a_1 = 1,\ a_2 = 1,\ a_{n+2} = a_n + a_{n+1},\quad (n = 1,\ 2,\ 3,\ \ldots)$

Here $\{a_n\} = \{1,\ 1,\ 2,\ 3,\ 5,\ 8,\ 13,\ 21,\ \ldots\}$. This is called the **Fibonacci sequence**. Each term after the second is the sum of the previous two terms.

In parts (a)–(g) of Example 1, the formulas on the left sides define the general term of each sequence $\{a_n\}$ as an explicit function of n. In parts (h) and (i) we say the sequence $\{a_n\}$ is defined **recursively** or **inductively**; each term must be calculated from previous ones rather than directly as a function of n.

We introduce terminology used to describe various properties of sequences.

DEFINITION 1

Terms for describing sequences

(a) The sequence $\{a_n\}$ is **bounded below** by L, and L is a **lower bound** for $\{a_n\}$, if $a_n \geq L$ for every $n = 1,\ 2,\ 3,\ \ldots$. The sequence is **bounded above** by M, and M is an **upper bound**, if $a_n \leq M$ for every such n.

The sequence $\{a_n\}$ is **bounded** if it is both bounded above and bounded below. In this case there is a constant K such that $|a_n| \leq K$ for every $n = 1, 2, 3, \ldots$. (We can take K to be the larger of $-L$ and M.)

(b) The sequence $\{a_n\}$ is **positive** if it is bounded below by zero, that is, if $a_n \geq 0$ for every $n = 1,\ 2,\ 3,\ \ldots$; it is **negative** if $a_n \leq 0$ for every n.

(c) The sequence $\{a_n\}$ is **increasing** if $a_{n+1} \geq a_n$ for every $n = 1, 2, 3, \ldots$; it is **decreasing** if $a_{n+1} \leq a_n$ for every such n. The sequence is said to be **monotonic** if it is either increasing or decreasing. (The terminology here is looser than that used for functions, where we would have used *nondecreasing* and *nonincreasing* to describe this behaviour. The distinction between $a_{n+1} > a_n$ and $a_{n+1} \geq a_n$ is not as important for sequences as it is for functions defined on intervals.)

(d) The sequence $\{a_n\}$ is **alternating** if $a_n a_{n+1} < 0$ for every $n = 1, 2, \ldots$, that is, if any two consecutive terms have opposite sign. Note that this definition requires $a_n \neq 0$ for each n.

Example 2   (**Describing some sequences**)

(a) The sequence $\{n\} = \{1,\ 2,\ 3,\ \ldots\}$ is positive, increasing, and bounded below. A lower bound for the sequence is 1 or any smaller number. The sequence is not bounded above.

(b) $\left\{\dfrac{n-1}{n}\right\} = \left\{0,\ \dfrac{1}{2},\ \dfrac{2}{3},\ \dfrac{3}{4},\ \ldots\right\}$ is positive, bounded, and increasing. Here, 0 is a lower bound and 1 is an upper bound.

(c) $\left\{\left(-\dfrac{1}{2}\right)^n\right\} = \left\{-\dfrac{1}{2},\ \dfrac{1}{4},\ -\dfrac{1}{8},\ \dfrac{1}{16},\ \ldots\right\}$ is bounded and alternating. Here, $-1/2$ is a lower bound and $1/4$ is an upper bound.

(d) $\{(-1)^n n\} = \{-1,\ 2,\ -3,\ 4,\ -5,\ \ldots\}$ is alternating but not bounded either above or below.

When you want to show that a sequence is increasing, you can try to show that the inequality $a_{n+1} - a_n \geq 0$ holds for $n \geq 1$. Alternatively, if $a_n = f(n)$ for a differentiable function $f(x)$, you can show that f is a nondecreasing function on $[1, \infty)$ by showing that $f'(x) \geq 0$ there. Similar approaches are useful for showing that a sequence is decreasing.

Example 3   If $a_n = \dfrac{n}{n^2 + 1}$, show that the sequence $\{a_n\}$ is decreasing.

Solution   Since $a_n = f(n)$, where $f(x) = \dfrac{x}{x^2 + 1}$ and

$$f'(x) = \frac{(x^2 + 1)(1) - x(2x)}{(x^2 + 1)^2} = \frac{1 - x^2}{(x^2 + 1)^2} \leq 0 \quad \text{for } x \geq 1,$$

the function $f(x)$ is decreasing on $[1, \infty)$; therefore, $\{a_n\}$ is a decreasing sequence.

The sequence $\left\{\dfrac{n^2}{2^n}\right\} = \left\{\dfrac{1}{2},\ 1,\ \dfrac{9}{8},\ 1,\ \dfrac{25}{32},\ \dfrac{36}{64},\ \dfrac{49}{128},\ \ldots\right\}$ is positive and therefore bounded below. It seems clear that from the fourth term on, all the terms are getting smaller. However, $a_2 > a_1$ and $a_3 > a_2$. Since $a_{n+1} \leq a_n$ only if $n \geq 3$, we say that this sequence is **ultimately decreasing**. The adverb *ultimately* is used to describe any termwise property of a sequence that the terms have from some point on, but not necessarily at the beginning of the sequence. Thus, the sequence

$$\{n - 100\} = \{-99,\ -98,\ \ldots,\ -2,\ -1,\ 0,\ 1,\ 2,\ 3,\ \ldots\}$$

is *ultimately positive* even though the first 99 terms are negative, and the sequence

$$\left\{(-1)^n + \frac{4}{n}\right\} = \left\{3,\ 3,\ \frac{1}{3},\ 2,\ -\frac{1}{5},\ \frac{5}{3},\ -\frac{3}{7},\ \frac{3}{2},\ \ldots\right\}$$

is *ultimately alternating* even though the first few terms do not alternate.

Convergence of Sequences

Central to the study of sequences is the notion of convergence. The concept of the limit of a sequence is a special case of the concept of the limit of a function $f(x)$ as $x \to \infty$. We say that the sequence $\{a_n\}$ **converges to the limit** L, and we write $\lim_{n \to \infty} a_n = L$, provided the distance from a_n to L on the real line approaches 0 as n increases toward ∞. We state this definition more formally as follows:

DEFINITION **2**

Limit of a sequence

We say that sequence $\{a_n\}$ converges to the limit L, and we write $\lim_{n\to\infty} a_n = L$, if for every positive real number ϵ there exists an integer N (which may depend on ϵ) such that if $n \geq N$, then $|a_n - L| < \epsilon$.

This definition is illustrated in Figure 9.1.

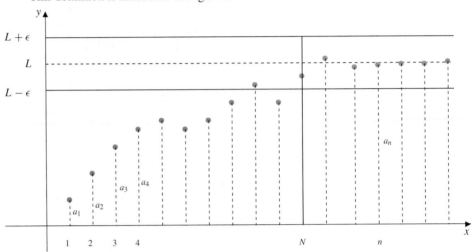

Figure 9.1   A convergent sequence

Example 4   Show that $\lim_{n\to\infty} \dfrac{c}{n^p} = 0$ for any real number c and any $p > 0$.

Solution   Let $\epsilon > 0$ be given. Then

$$\left| \frac{c}{n^p} \right| < \epsilon \quad \text{if} \quad n^p > \frac{|c|}{\epsilon},$$

that is, if $n \geq N$, the least integer greater than $(|c|/\epsilon)^{1/p}$. Therefore, by Definition 2, $\lim_{n\to\infty} \dfrac{c}{n^p} = 0$.

Every sequence $\{a_n\}$ must either **converge** to a finite limit L or **diverge**. That is, either $\lim_{n\to\infty} a_n = L$ exists (is a real number) or $\lim_{n\to\infty} a_n$ does not exist. If $\lim_{n\to\infty} a_n = \infty$, we can say that the sequence diverges to ∞; if $\lim_{n\to\infty} a_n = -\infty$, we can say that it diverges to $-\infty$. If $\lim_{n\to\infty} a_n$ simply does not exist (but is not ∞ or $-\infty$), we can only say that the sequence diverges.

Example 5   **(Examples of convergent and divergent sequences)**

(a) $\{(n-1)/n\}$ converges to 1; $\lim_{n\to\infty}(n-1)/n = \lim_{n\to\infty}\big(1 - (1/n)\big) = 1$.

(b) $\{n\} = \{1, 2, 3, 4, \ldots\}$ diverges to ∞.

(c) $\{-n\} = \{-1, -2, -3, -4, \ldots\}$ diverges to $-\infty$.

(d) $\{(-1)^n\} = \{-1, 1, -1, 1, -1, \ldots\}$ simply diverges.

(e) $\{(-1)^n n\} = \{-1, 2, -3, 4, -5, \ldots\}$ diverges (but not to ∞ or $-\infty$ even though $\lim_{n\to\infty} |a_n| = \infty$).

The limit of a sequence is equivalent to the limit of a function as its argument approaches infinity:

If $\lim_{x\to\infty} f(x) = L$ and $a_n = f(n)$, then $\lim_{n\to\infty} a_n = L$.

Because of this, the standard rules for limits of functions (Theorems 2 and 4 of Section 1.2) also hold for limits of sequences, with the appropriate changes of notation. Thus, if $\{a_n\}$ and $\{b_n\}$ converge, then

$$\lim_{n\to\infty} (a_n \pm b_n) = \lim_{n\to\infty} a_n \pm \lim_{n\to\infty} b_n,$$

$$\lim_{n\to\infty} ca_n = c \lim_{n\to\infty} a_n,$$

$$\lim_{n\to\infty} a_n b_n = \left(\lim_{n\to\infty} a_n\right)\left(\lim_{n\to\infty} b_n\right),$$

$$\lim_{n\to\infty} \frac{a_n}{b_n} = \frac{\lim\limits_{n\to\infty} a_n}{\lim\limits_{n\to\infty} b_n} \qquad \text{assuming } \lim_{n\to\infty} b_n \neq 0.$$

If $a_n \le b_n$ ultimately, then $\lim\limits_{n\to\infty} a_n \le \lim\limits_{n\to\infty} b_n$.

If $a_n \le b_n \le c_n$ ultimately, and $\lim\limits_{n\to\infty} a_n = L = \lim\limits_{n\to\infty} c_n$, then $\lim\limits_{n\to\infty} b_n = L$.

The limits of many explicitly defined sequences can be evaluated using these properties in a manner similar to the methods used for limits of the form $\lim_{x\to\infty} f(x)$ in Section 1.3.

Example 6 Calculate the limits of the sequences

(a) $\left\{\dfrac{2n^2 - n - 1}{5n^2 + n - 3}\right\}$, (b) $\left\{\dfrac{\cos n}{n}\right\}$, and (c) $\{\sqrt{n^2 + 2n} - n\}$.

Solution

(a) We divide the numerator and denominator of the expression for a_n by the highest power of n in the denominator, that is, by n^2:

$$\lim_{n\to\infty} \frac{2n^2 - n - 1}{5n^2 + n - 3} = \lim_{n\to\infty} \frac{2 - (1/n) - (1/n^2)}{5 + (1/n) - (3/n^2)} = \frac{2 - 0 - 0}{5 + 0 - 0} = \frac{2}{5},$$

since $\lim_{n\to\infty} 1/n = 0$ and $\lim_{n\to\infty} 1/n^2 = 0$. The sequence converges and its limit is 2/5.

(b) Since $|\cos n| \le 1$ for every n, we have

$$-\frac{1}{n} \le \frac{\cos n}{n} \le \frac{1}{n} \quad \text{for} \quad n \ge 1.$$

Now, $\lim_{n\to\infty} -1/n = 0$ and $\lim_{n\to\infty} 1/n = 0$. Therefore, by the sequence version of the Squeeze Theorem, $\lim_{n\to\infty}(\cos n)/n = 0$. The given sequence converges to 0.

(c) For this sequence we multiply the numerator and the denominator (which is 1) by the conjugate of the expression in the numerator:

$$\lim_{n\to\infty} (\sqrt{n^2 + 2n} - n) = \lim_{n\to\infty} \frac{(\sqrt{n^2 + 2n} - n)(\sqrt{n^2 + 2n} + n)}{\sqrt{n^2 + 2n} + n}$$

$$= \lim_{n\to\infty} \frac{2n}{\sqrt{n^2 + 2n} + n} = \lim_{n\to\infty} \frac{2}{\sqrt{1 + (2/n)} + 1} = 1.$$

The sequence converges to 1.

Example 7   Evaluate $\lim\limits_{n\to\infty} n \tan^{-1}\left(\dfrac{1}{n}\right)$.

Solution   For this example it is best to replace the nth term of the sequence by the corresponding function of a real variable x and take the limit as $x \to \infty$. We use l'Hôpital's Rule:

$$
\begin{aligned}
\lim_{n\to\infty} n \tan^{-1}\left(\frac{1}{n}\right) &= \lim_{x\to\infty} x \tan^{-1}\left(\frac{1}{x}\right) \\
&= \lim_{x\to\infty} \frac{\tan^{-1}\left(\dfrac{1}{x}\right)}{\dfrac{1}{x}} \qquad \left[\frac{0}{0}\right] \\
&= \lim_{x\to\infty} \frac{\dfrac{1}{1+(1/x^2)}\left(-\dfrac{1}{x^2}\right)}{-\left(\dfrac{1}{x^2}\right)} = \lim_{x\to\infty} \frac{1}{1+\dfrac{1}{x^2}} = 1.
\end{aligned}
$$

THEOREM 1

If $\{a_n\}$ converges, then $\{a_n\}$ is bounded.

PROOF   Suppose $\lim_{n\to\infty} a_n = L$. According to Definition 2, for $\epsilon = 1$ there exists a number N such that if $n > N$, then $|a_n - L| < 1$; therefore $|a_n| < 1 + |L|$ for such n. (Why is this true?) If K denotes the largest of the numbers $|a_1|, |a_2|, \ldots, |a_N|$, and $1 + |L|$, then $|a_n| \le K$ for every $n = 1, 2, 3, \ldots$. Hence $\{a_n\}$ is bounded.

The converse of Theorem 1 is false; the sequence $\{(-1)^n\}$ is bounded but does not converge.

The *completeness property* of the real number system (see Section P.1) can be reformulated in terms of sequences to read as follows:

> **Bounded monotonic sequences converge**
>
> If the sequence $\{a_n\}$ is bounded above and is (ultimately) increasing, then it converges. The same conclusion holds if $\{a_n\}$ is bounded below and is (ultimately) decreasing.

Thus, a bounded, ultimately monotonic sequence is convergent. (See Figure 9.2.)

Figure 9.2   An ultimately increasing sequence that is bounded above

There is a subtle point to note in this solution. Showing that $\{a_n\}$ is increasing is pretty obvious, but how did we know to try and show that 3 (rather than some other number) was an upper bound? The answer is that we actually did the last part first and showed that *if* $\lim a_n = a$ exists, *then* $a = 3$. It then makes sense to try and show that $a_n < 3$ for all n.

Example 8 Let a_n be defined recursively by

$$a_1 = 1, \qquad a_{n+1} = \sqrt{6 + a_n} \qquad (n = 1, 2, 3, \ldots).$$

Show that $\lim_{n \to \infty} a_n$ exists and find its value.

Solution Observe that $a_2 = \sqrt{6 + 1} = \sqrt{7} > a_1$. If $a_{k+1} > a_k$, then $a_{k+2} = \sqrt{6 + a_{k+1}} > \sqrt{6 + a_k} = a_{k+1}$, so $\{a_n\}$ is increasing, by induction. Now observe that $a_1 = 1 < 3$. If $a_k < 3$, then $a_{k+1} = \sqrt{6 + a_k} < \sqrt{6 + 3} = 3$, so $a_n < 3$ for every n by induction. Since $\{a_n\}$ is increasing and bounded above, $\lim_{n \to \infty} a_n = a$ exists, by completeness. Since $\sqrt{6 + x}$ is a continuous function of x, we have

$$a = \lim_{n \to \infty} a_{n+1} = \lim_{n \to \infty} \sqrt{6 + a_n} = \sqrt{6 + \lim_{n \to \infty} a_n} = \sqrt{6 + a}.$$

Thus $a^2 = 6 + a$, or $a^2 - a - 6 = 0$, or $(a - 3)(a + 2) = 0$. This quadratic has roots $a = 3$ and $a = -2$. Since $a_n \geq 1$ for every n, we must have $a \geq 1$. Therefore, $a = 3$ and $\lim_{n \to \infty} a_n = 3$.

Example 9 Does $\left\{\left(1 + \dfrac{1}{n}\right)^n\right\}$ converge or diverge?

Solution We could make an effort to show that the given sequence is, in fact, increasing and bounded above. (See Exercise 32 at the end of this section.) However, we already know the answer. The sequence converges by Theorem 6 of Section 3.4:

$$\lim_{n \to \infty} \left(1 + \frac{1}{n}\right)^n = e^1 = e.$$

THEOREM 2

If $\{a_n\}$ is (ultimately) increasing, then either it is bounded above, and therefore convergent, or it is not bounded above and diverges to infinity.

The proof of this theorem is left as an exercise. A corresponding result holds for (ultimately) decreasing sequences.

The following theorem evaluates two important limits that find frequent application in the study of series.

THEOREM 3

(a)   If $|x| < 1$, then $\lim\limits_{n \to \infty} x^n = 0$.

(b)   If x is any real number, then $\lim\limits_{n \to \infty} \dfrac{x^n}{n!} = 0$.

PROOF For part (a) observe that

$$\lim_{n \to \infty} \ln |x|^n = \lim_{n \to \infty} n \ln |x| = -\infty,$$

since $\ln |x| < 0$ when $|x| < 1$. Accordingly, since e^x is continuous,

$$\lim_{n \to \infty} |x|^n = \lim_{n \to \infty} e^{\ln |x|^n} = e^{\lim_{n \to \infty} \ln |x|^n} = 0.$$

Since $-|x|^n \leq x^n \leq |x|^n$, we have $\lim_{n \to \infty} x^n = 0$ by the Squeeze Theorem.

For part (b), pick any x and let N be an integer such that $N > |x|$. If $n > N$ we have

$$\left|\frac{x^n}{n!}\right| = \frac{|x|}{1} \frac{|x|}{2} \frac{|x|}{3} \cdots \frac{|x|}{N-1} \frac{|x|}{N} \frac{|x|}{N+1} \cdots \frac{|x|}{n}$$

$$< \frac{|x|^{N-1}}{(N-1)!} \frac{|x|}{N} \frac{|x|}{N} \frac{|x|}{N} \cdots \frac{|x|}{N}$$

$$= \frac{|x|^{N-1}}{(N-1)!} \left(\frac{|x|}{N}\right)^{n-N+1} = K \left(\frac{|x|}{N}\right)^n,$$

where $K = \dfrac{|x|^{N-1}}{(N-1)!}\left(\dfrac{|x|}{N}\right)^{1-N}$ is a constant that is independent of n. Since $|x|/N < 1$, we have $\lim_{n\to\infty}(|x|/N)^n = 0$ by part (a). Thus $\lim_{n\to\infty}|x^n/n!| = 0$, so $\lim_{n\to\infty} x^n/n! = 0$.

Example 10 Find $\lim_{n\to\infty}\dfrac{3^n + 4^n + 5^n}{5^n}$.

Solution $\displaystyle\lim_{n\to\infty}\dfrac{3^n + 4^n + 5^n}{5^n} = \lim_{n\to\infty}\left[\left(\dfrac{3}{5}\right)^n + \left(\dfrac{4}{5}\right)^n + 1\right] = 0 + 0 + 1 = 1$, by Theorem 3(a).

Exercises 9.1

In Exercises 1–13, determine whether the given sequence is (a) bounded (above or below), (b) positive or negative (ultimately), (c) increasing, decreasing, or alternating, and (d) convergent, divergent, divergent to ∞ or $-\infty$.

1. $\left\{\dfrac{2n^2}{n^2+1}\right\}$

2. $\left\{\dfrac{2n}{n^2+1}\right\}$

3. $\left\{4 - \dfrac{(-1)^n}{n}\right\}$

4. $\left\{\sin\dfrac{1}{n}\right\}$

5. $\left\{\dfrac{n^2-1}{n}\right\}$

6. $\left\{\dfrac{e^n}{\pi^n}\right\}$

7. $\left\{\dfrac{e^n}{\pi^{n/2}}\right\}$

8. $\left\{\dfrac{(-1)^n n}{e^n}\right\}$

9. $\left\{\dfrac{2^n}{n^n}\right\}$

10. $\left\{\dfrac{(n!)^2}{(2n)!}\right\}$

11. $\left\{n\cos\left(\dfrac{n\pi}{2}\right)\right\}$

12. $\left\{\dfrac{\sin n}{n}\right\}$

13. $\{1, 1, -2, 3, 3, -4, 5, 5, -6, \ldots\}$

In Exercises 14–29, evaluate, wherever possible, the limit of the sequence $\{a_n\}$.

14. $a_n = \dfrac{5 - 2n}{3n - 7}$

15. $a_n = \dfrac{n^2 - 4}{n + 5}$

16. $a_n = \dfrac{n^2}{n^3 + 1}$

17. $a_n = (-1)^n \dfrac{n}{n^3 + 1}$

18. $a_n = \dfrac{n^2 - 2\sqrt{n} + 1}{1 - n - 3n^2}$

19. $a_n = \dfrac{e^n - e^{-n}}{e^n + e^{-n}}$

20. $a_n = n\sin\dfrac{1}{n}$

21. $a_n = \left(\dfrac{n-3}{n}\right)^n$

22. $a_n = \dfrac{n}{\ln(n+1)}$

23. $a_n = \sqrt{n+1} - \sqrt{n}$

24. $a_n = n - \sqrt{n^2 - 4n}$

25. $a_n = \sqrt{n^2 + n} - \sqrt{n^2 - 1}$

26. $a_n = \left(\dfrac{n-1}{n+1}\right)^n$

27. $a_n = \dfrac{(n!)^2}{(2n)!}$

28. $a_n = \dfrac{n^2 2^n}{n!}$

29. $a_n = \dfrac{\pi^n}{1 + 2^{2n}}$

30. Let $a_1 = 1$ and $a_{n+1} = \sqrt{1 + 2a_n}$ ($n = 1, 2, 3, \ldots$). Show that $\{a_n\}$ is increasing and bounded above. (*Hint:* Show that 3 is an upper bound.) Hence, conclude that the sequence converges, and find its limit.

*31. Repeat Exercise 30 for the sequence defined by $a_1 = 3$, $a_{n+1} = \sqrt{15 + 2a_n}$, $n = 1, 2, 3, \ldots$. This time you will have to guess an upper bound.

*32. Let $a_n = \left(1 + \dfrac{1}{n}\right)^n$ so that $\ln a_n = n\ln\left(1 + \dfrac{1}{n}\right)$. Use properties of the logarithm function to show that (a) $\{a_n\}$ is increasing and (b) e is an upper bound for $\{a_n\}$.

*33. Prove Theorem 2. Also, state an analogous theorem pertaining to ultimately decreasing sequences.

34. If $\{|a_n|\}$ is bounded, prove that $\{a_n\}$ is bounded.

35. If $\lim_{n\to\infty}|a_n| = 0$, prove that $\lim_{n\to\infty} a_n = 0$.

*36. Which of the following statements are TRUE and which are FALSE? Justify your answers.

(a) If $\lim_{n\to\infty} a_n = \infty$ and $\lim_{n\to\infty} b_n = L > 0$, then $\lim_{n\to\infty} a_n b_n = \infty$.

(b) If $\lim_{n\to\infty} a_n = \infty$ and $\lim_{n\to\infty} b_n = -\infty$, then $\lim_{n\to\infty}(a_n + b_n) = 0$.

(c) If $\lim_{n\to\infty} a_n = \infty$ and $\lim_{n\to\infty} b_n = -\infty$, then $\lim_{n\to\infty} a_n b_n = -\infty$.

(d) If neither $\{a_n\}$ nor $\{b_n\}$ converges, then $\{a_n b_n\}$ does not converge.

(e) If $\{|a_n|\}$ converges, then $\{a_n\}$ converges.

9.2 Infinite Series

An **infinite series**, usually just called a **series**, is a formal sum of infinitely many terms; for instance, $a_1 + a_2 + a_3 + a_4 + \cdots$ is a series formed by adding the terms of the sequence $\{a_n\}$. This series is also denoted $\sum_{n=1}^{\infty} a_n$:

$$\sum_{n=1}^{\infty} a_n = a_1 + a_2 + a_3 + a_4 + \cdots$$

For example,

$$\sum_{n=1}^{\infty} \frac{1}{n} = 1 + \frac{1}{2} + \frac{1}{3} + \frac{1}{4} + \cdots$$

$$\sum_{n=1}^{\infty} \frac{(-1)^{n-1}}{2^{n-1}} = 1 - \frac{1}{2} + \frac{1}{4} - \frac{1}{8} + \frac{1}{16} - \cdots.$$

It is sometimes necessary or useful to start the sum from some index other than 1:

$$\sum_{n=0}^{\infty} a^n = 1 + a + a^2 + a^3 + \cdots$$

$$\sum_{n=2}^{\infty} \frac{1}{\ln n} = \frac{1}{\ln 2} + \frac{1}{\ln 3} + \frac{1}{\ln 4} + \cdots.$$

Note that the latter series would make no sense if we had started the sum from $n = 1$; the first term would have been undefined.

When necessary, we can change the index of summation to start at a different value. This is accomplished by a substitution as illustrated in Example 3 of Section 5.1. For instance, using the substitution $n = m - 2$, we can rewrite $\sum_{n=1}^{\infty} a_n$ in the form $\sum_{m=3}^{\infty} a_{m-2}$. Both sums give rise to the same expansion

$$\sum_{n=1}^{\infty} a_n = a_1 + a_2 + a_3 + \cdots = \sum_{m=3}^{\infty} a_{m-2}.$$

Addition is an operation that is carried out on two numbers at a time. If we want to calculate the finite sum $a_1 + a_2 + a_3$, we could proceed by adding $a_1 + a_2$ and then adding a_3 to this sum, or else we might first add $a_2 + a_3$ and then add a_1 to the sum. Of course, the associative law for addition assures us we will get the same answer both ways. This is the reason the symbol $a_1 + a_2 + a_3$ makes sense; we would otherwise have to write $(a_1 + a_2) + a_3$ or $a_1 + (a_2 + a_3)$. This reasoning extends to any sum $a_1 + a_2 + \cdots + a_n$ of finitely many terms, but it is not obvious what should be meant by a sum with infinitely many terms:

$$a_1 + a_2 + a_3 + a_4 + \cdots.$$

We no longer have any assurance that the terms can be added up in any order to yield the same sum. In fact, we will see in Section 9.4 that in certain circumstances, changing the order of terms in a series can actually change the sum of the series. The interpretation we place on the infinite sum is that of adding from left to right, as suggested by the grouping

$$\cdots ((((a_1 + a_2) + a_3) + a_4) + a_5) + \cdots.$$

We accomplish this by defining a new sequence $\{s_n\}$, called the **sequence of partial sums** of the series $\sum_{n=1}^{\infty} a_n$, so that s_n is the sum of the first n terms of the series:

$$s_1 = a_1$$
$$s_2 = s_1 + a_2 = a_1 + a_2$$
$$s_3 = s_2 + a_3 = a_1 + a_2 + a_3$$
$$\vdots$$
$$s_n = s_{n-1} + a_n = a_1 + a_2 + a_3 + \cdots + a_n = \sum_{j=1}^{n} a_j.$$
$$\vdots$$

We then define the sum of the infinite series to be the limit of this sequence of partial sums.

DEFINITION 3

> **Convergence of a series**
>
> We say that the series $\sum_{n=1}^{\infty} a_n$ **converges to the sum** s, and we write
>
> $$\sum_{n=1}^{\infty} a_n = s,$$
>
> if $\lim_{n \to \infty} s_n = s$, where s_n is the nth partial sum of $\sum_{n=1}^{\infty} a_n$:
>
> $$s_n = a_1 + a_2 + a_3 + \cdots + a_n = \sum_{j=1}^{n} a_j.$$

Thus, a *series* converges if and only if the *sequence* of its partial sums converges.

Similarly, a series is said to diverge to infinity, diverge to negative infinity, or simply diverge if its sequence of partial sums does so. It must be stressed that the convergence of the series $\sum_{n=1}^{\infty} a_n$ depends on the convergence of the sequence $\{s_n\} = \{\sum_{j=1}^{n} a_j\}$, *not* the sequence $\{a_n\}$.

Geometric Series

DEFINITION 4

> **Geometric series**
>
> A series of the form $\sum_{n=1}^{\infty} a\, r^{n-1} = a + ar + ar^2 + ar^3 + \cdots$, whose nth term is $a_n = a\, r^{n-1}$, is called a **geometric series**. The number a is the first term. The number r is called the **common ratio** of the series, since it is the value of the ratio of the $(n+1)$st term to the nth term for any $n \geq 1$:
>
> $$\frac{a_{n+1}}{a_n} = \frac{ar^n}{ar^{n-1}} = r, \qquad n = 1, 2, 3, \ldots.$$

The nth partial sum s_n of a geometric series is calculated as follows:

$$s_n = a + ar + ar^2 + ar^3 + \cdots + ar^{n-1}$$
$$r s_n = \quad\;\; ar + ar^2 + ar^3 + \cdots + ar^{n-1} + ar^n.$$

The second equation is obtained by multiplying the first by r. Subtracting these two equations (note the cancellations), we get $(1-r)s_n = a - ar^n$. If $r \neq 1$, we can divide by $1 - r$ and get a formula for s_n.

Partial sums of geometric series

If $r = 1$, then the nth partial sum of a geometric series $\sum_{n=1}^{\infty} ar^{n-1}$ is $s_n = a + a + \cdots + a = na$. If $r \neq 1$, then

$$s_n = a + ar + ar^2 + \cdots + ar^{n-1} = \frac{a(1-r^n)}{1-r}.$$

If $a = 0$, then $s_n = 0$ for every n, and $\lim_{n\to\infty} s_n = 0$. Now suppose $a \neq 0$. If $|r| < 1$, then $\lim_{n\to\infty} r^n = 0$, so $\lim_{n\to\infty} s_n = a/(1-r)$. If $r > 1$, then $\lim_{n\to\infty} r^n = \infty$, and $\lim_{n\to\infty} s_n = \infty$ if $a > 0$, or $\lim_{n\to\infty} s_n = -\infty$ if $a < 0$. The same conclusion holds if $r = 1$, since $s_n = na$ in this case. If $r \leq -1$, $\lim_{n\to\infty} r^n$ does not exist and neither does $\lim_{n\to\infty} s_n$. Hence we conclude that

$$\sum_{n=1}^{\infty} ar^{n-1} \begin{cases} \text{converges to } 0 & \text{if } a = 0 \\ \text{converges to } \dfrac{a}{1-r} & \text{if } |r| < 1 \\ \text{diverges to } \infty & \text{if } r \geq 1 \text{ and } a > 0 \\ \text{diverges to } -\infty & \text{if } r \geq 1 \text{ and } a < 0 \\ \text{diverges} & \text{if } r \leq -1 \text{ and } a \neq 0. \end{cases}$$

The representation of the function $1/(1-x)$ as the sum of a geometric series,

$$\frac{1}{1-x} = \sum_{n=0}^{\infty} x^n = 1 + x + x^2 + x^3 + \cdots \quad \text{for } -1 < x < 1,$$

will be important in our discussion of power series later in this chapter.

Example 1 **(Examples of geometric series and their sums)**

(a) $1 + \dfrac{1}{2} + \dfrac{1}{4} + \dfrac{1}{8} + \cdots = \displaystyle\sum_{n=1}^{\infty} \left(\dfrac{1}{2}\right)^{n-1} = \dfrac{1}{1 - \dfrac{1}{2}} = 2.$ Here $a = 1$ and $r = \dfrac{1}{2}$.

Since $|r| < 1$, the series converges.

(b) $\pi - e + \dfrac{e^2}{\pi} - \dfrac{e^3}{\pi^2} + \cdots = \displaystyle\sum_{n=1}^{\infty} \pi \left(-\dfrac{e}{\pi}\right)^{n-1}$ Here $a = \pi$ and $r = -\dfrac{e}{\pi}$.

$$= \frac{\pi}{1 - \left(-\dfrac{e}{\pi}\right)} = \frac{\pi^2}{\pi + e}.$$

The series converges since $\left|-\dfrac{e}{\pi}\right| < 1$.

(c) $1 + 2^{1/2} + 2 + 2^{3/2} + \cdots = \displaystyle\sum_{n=1}^{\infty} (\sqrt{2})^{n-1}$. This series diverges to ∞ since $a = 1 > 0$ and $r = \sqrt{2} > 1$.

(d) $1 - 1 + 1 - 1 + 1 - \cdots = \displaystyle\sum_{n=1}^{\infty} (-1)^{n-1}$. This series diverges since $r = -1$.

(e) Let $x = 0.32\,32\,32 \cdots = 0.\overline{32}$; then

$$x = \frac{32}{100} + \frac{32}{100^2} + \frac{32}{100^3} + \cdots = \sum_{n=1}^{\infty} \frac{32}{100} \left(\frac{1}{100}\right)^{n-1} = \frac{32}{100} \frac{1}{1 - \dfrac{1}{100}} = \frac{32}{99}.$$

This is an alternative to the method of Example 1 of Section P.1 for representing repeating decimals as quotients of integers.

Example 2 If money earns interest at a constant effective rate of 5% per year, how much should you pay today for an annuity that will pay you (a) $1,000 at the end of each of the next 10 years and (b) $1,000 at the end of every year forever?

Solution A payment of $1,000 that is due to be received n years from now has present value $1,000 \times \left(\dfrac{1}{1.05}\right)^n$ (since A would grow to $A(1.05)^n$ in n years). Thus $1,000 payments at the end of each of the next n years are worth s_n at the present time, where

$$
s_n = 1,000\left[\frac{1}{1.05} + \left(\frac{1}{1.05}\right)^2 + \cdots + \left(\frac{1}{1.05}\right)^n\right]
$$

$$
= \frac{1,000}{1.05}\left[1 + \frac{1}{1.05} + \left(\frac{1}{1.05}\right)^2 + \cdots + \left(\frac{1}{1.05}\right)^{n-1}\right].
$$

$$
= \frac{1,000}{1.05}\,\frac{1 - \left(\dfrac{1}{1.05}\right)^n}{1 - \dfrac{1}{1.05}} = \frac{1,000}{0.05}\left[1 - \left(\frac{1}{1.05}\right)^n\right]
$$

(a) The present value of 10 future payments is $s_{10} = \$7,721.73$.

(b) The present value of future payments continuing forever is

$$
\$ \lim_{n\to\infty} s_n = \frac{\$1,000}{0.05} = \$20,000.
$$

Telescoping Series and Harmonic Series

Example 3 Show that the series

$$
\sum_{n=1}^{\infty} \frac{1}{n(n+1)} = \frac{1}{1 \times 2} + \frac{1}{2 \times 3} + \frac{1}{3 \times 4} + \frac{1}{4 \times 5} + \cdots
$$

converges and find its sum.

Solution Since $\dfrac{1}{n(n+1)} = \dfrac{1}{n} - \dfrac{1}{n+1}$, we can write the partial sum s_n in the form

$$
s_n = \frac{1}{1 \times 2} + \frac{1}{2 \times 3} + \frac{1}{3 \times 4} + \cdots + \frac{1}{(n-1)n} + \frac{1}{n(n+1)}
$$

$$
= \left(1 - \frac{1}{2}\right) + \left(\frac{1}{2} - \frac{1}{3}\right) + \left(\frac{1}{3} - \frac{1}{4}\right) + \cdots
$$

$$
\cdots + \left(\frac{1}{n-1} - \frac{1}{n}\right) + \left(\frac{1}{n} - \frac{1}{n+1}\right)
$$

$$
= 1 - \frac{1}{2} + \frac{1}{2} - \frac{1}{3} + \frac{1}{3} - \cdots - \frac{1}{n} + \frac{1}{n} - \frac{1}{n+1}
$$

$$
= 1 - \frac{1}{n+1}.
$$

Therefore, $\lim_{n\to\infty} s_n = 1$ and the series converges to 1:

$$
\sum_{n=1}^{\infty} \frac{1}{n(n+1)} = 1.
$$

This is an example of a **telescoping series**, so called because the partial sums *fold up* into a simple form when the terms are expanded in partial fractions. Other examples can be found in the exercises at the end of this section. As these examples show, the method of partial fractions can be a useful tool for series as well as for integrals.

Example 4   Show that the **harmonic series**

$$\sum_{n=1}^{\infty} \frac{1}{n} = 1 + \frac{1}{2} + \frac{1}{3} + \frac{1}{4} + \cdots$$

diverges to infinity.

Solution   If s_n is the nth partial sum of the harmonic series, then

$$
\begin{aligned}
s_n &= 1 + \frac{1}{2} + \frac{1}{3} + \cdots + \frac{1}{n} \\
&= \text{sum of areas of rectangles shaded in Figure 9.3} \\
&> \text{area under } y = \frac{1}{x} \text{ from } x = 1 \text{ to } x = n + 1 \\
&= \int_{1}^{n+1} \frac{dx}{x} = \ln(n + 1).
\end{aligned}
$$

Now $\lim_{n\to\infty} \ln(n + 1) = \infty$. Therefore, $\lim_{n\to\infty} s_n = \infty$ and

$$\sum_{n=1}^{\infty} \frac{1}{n} = 1 + \frac{1}{2} + \frac{1}{3} + \cdots \qquad \text{diverges to infinity.}$$

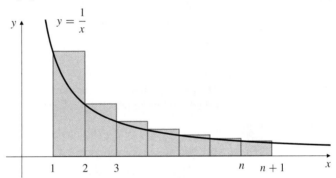

Figure 9.3   A partial sum of the harmonic series

Like geometric series, the harmonic series will often be encountered in subsequent sections.

Some Theorems About Series

THEOREM 4

If $\sum_{n=1}^{\infty} a_n$ converges, then $\lim_{n\to\infty} a_n = 0$.

PROOF   If $s_n = a_1 + a_2 + \cdots + a_n$, then $s_n - s_{n-1} = a_n$. If $\sum_{n=1}^{\infty} a_n$ converges, then $\lim_{n\to\infty} s_n = s$ exists, and $\lim_{n\to\infty} s_{n-1} = s$. Hence $\lim_{n\to\infty} a_n = s - s = 0$.

Remark   Theorem 4 is *very important* for the understanding of infinite series. Students often err either in forgetting that *a series cannot converge if its terms do not approach zero* or in confusing this result with its *converse*, which is false. The converse would say that if $\lim_{n\to\infty} a_n = 0$, then $\sum_{n=1}^{\infty} a_n$ must converge. The harmonic series is a counterexample showing the falsehood of this assertion:

$$\lim_{n\to\infty} \frac{1}{n} = 0 \qquad \text{but} \qquad \sum_{n=1}^{\infty} \frac{1}{n} \text{ diverges to infinity.}$$

When considering whether a given series converges, the first question you should ask yourself is: "Does the nth term approach 0 as n approaches ∞?" If the answer is *no*, then the series does *not* converge. If the answer is *yes*, then the series *may or may not* converge. If the sequence of terms $\{a_n\}$ tends to a nonzero limit L, then $\sum_{n=1}^{\infty} a_n$ diverges to infinity if $L > 0$ and diverges to negative infinity if $L < 0$.

Example 5

(a) $\displaystyle\sum_{n=1}^{\infty} \frac{n}{2n-1}$ diverges to infinity since $\lim_{n\to\infty} \dfrac{n}{2n-1} = 1/2 > 0.$

(b) $\sum_{n=1}^{\infty}(-1)^n n \sin(1/n)$ diverges since

$$\lim_{n\to\infty}\left|(-1)^n n \sin \frac{1}{n}\right| = \lim_{n\to\infty} \frac{\sin(1/n)}{1/n} = \lim_{x\to 0+} \frac{\sin x}{x} = 1 \neq 0.$$

The following theorem asserts that it is only the *ultimate* behaviour of $\{a_n\}$ that determines whether $\sum_{n=1}^{\infty} a_n$ converges. Any finite number of terms can be dropped from the beginning of a series without affecting the convergence; the convergence depends only on the *tail* of the series. Of course, the actual sum of the series depends on *all* the terms.

THEOREM 5

$\sum_{n=1}^{\infty} a_n$ converges if and only if $\sum_{n=N}^{\infty} a_n$ converges for any integer $N \geq 1$.

THEOREM 6

If $\{a_n\}$ is ultimately positive, then the series $\sum_{n=1}^{\infty} a_n$ must either converge (if its partial sums are bounded above) or diverge to infinity (if its partial sums are not bounded above).

The proofs of these two theorems are posed as exercises at the end of this section. The following theorem is just a reformulation of standard laws of limits.

THEOREM 7

If $\sum_{n=1}^{\infty} a_n$ and $\sum_{n=1}^{\infty} b_n$ converge to A and B, respectively, then

(a) $\sum_{n=1}^{\infty} c a_n$ converges to cA (where c is any constant);

(b) $\sum_{n=1}^{\infty} (a_n \pm b_n)$ converges to $A \pm B$;

(c) if $a_n \leq b_n$ for all $n = 1, 2, 3, \ldots$, then $A \leq B$.

Example 6   Find the sum of the series $\displaystyle\sum_{n=1}^{\infty} \frac{1 + 2^{n+1}}{3^n}.$

Solution   The given series is the sum of two geometric series,

$$\sum_{n=1}^{\infty} \frac{1}{3^n} = \sum_{n=1}^{\infty} \frac{1}{3}\left(\frac{1}{3}\right)^{n-1} = \frac{1/3}{1-(1/3)} = \frac{1}{2} \quad \text{and}$$

$$\sum_{n=1}^{\infty} \frac{2^{n+1}}{3^n} = \sum_{n=1}^{\infty} \frac{4}{3}\left(\frac{2}{3}\right)^{n-1} = \frac{4/3}{1-(2/3)} = 4.$$

Thus its sum is $\dfrac{1}{2} + 4 = \dfrac{9}{2}$ by Theorem 7(b).

Exercises 9.2

In Exercises 1–18, find the sum of the given series, or show that the series diverges (possibly to infinity or negative infinity). Exercises 11–14 are telescoping series and should be done by partial fractions as suggested in Example 3 in this section.

1. $\dfrac{1}{3} + \dfrac{1}{9} + \dfrac{1}{27} + \cdots = \displaystyle\sum_{n=1}^{\infty} \frac{1}{3^n}$

2. $3 - \dfrac{3}{4} + \dfrac{3}{16} - \dfrac{3}{64} + \cdots = \displaystyle\sum_{n=1}^{\infty} 3\left(-\frac{1}{4}\right)^{n-1}$

3. $\displaystyle\sum_{n=5}^{\infty} \frac{1}{(2+\pi)^{2n}}$

4. $\displaystyle\sum_{n=0}^{\infty} \frac{5}{10^{3n}}$

5. $\displaystyle\sum_{n=2}^{\infty} \frac{(-5)^n}{8^{2n}}$

6. $\displaystyle\sum_{n=0}^{\infty} \frac{1}{e^n}$

7. $\displaystyle\sum_{k=0}^{\infty} \frac{2^{k+3}}{e^{k-3}}$

8. $\displaystyle\sum_{j=1}^{\infty} \pi^{j/2} \cos(j\pi)$

9. $\displaystyle\sum_{n=1}^{\infty} \frac{3+2^n}{2^{n+2}}$

10. $\displaystyle\sum_{n=0}^{\infty} \frac{3+2^n}{3^{n+2}}$

11. $\displaystyle\sum_{n=1}^{\infty} \frac{1}{n(n+2)} = \frac{1}{1 \times 3} + \frac{1}{2 \times 4} + \frac{1}{3 \times 5} + \cdots$

12. $\displaystyle\sum_{n=1}^{\infty} \frac{1}{(2n-1)(2n+1)} = \frac{1}{1 \times 3} + \frac{1}{3 \times 5} + \frac{1}{5 \times 7} + \cdots$

13. $\displaystyle\sum_{n=1}^{\infty} \frac{1}{(3n-2)(3n+1)} = \frac{1}{1 \times 4} + \frac{1}{4 \times 7} + \frac{1}{7 \times 10} + \cdots$

*14. $\displaystyle\sum_{n=1}^{\infty} \frac{1}{n(n+1)(n+2)}$

$= \dfrac{1}{1 \times 2 \times 3} + \dfrac{1}{2 \times 3 \times 4} + \dfrac{1}{3 \times 4 \times 5} + \cdots$

15. $\displaystyle\sum_{n=1}^{\infty} \frac{1}{2n-1}$

16. $\displaystyle\sum_{n=1}^{\infty} \frac{n}{n+2}$

17. $\displaystyle\sum_{n=1}^{\infty} n^{-1/2}$

18. $\displaystyle\sum_{n=1}^{\infty} \frac{2}{n+1}$

19. Obtain a simple expression for the partial sum s_n of the series $\sum_{n=1}^{\infty}(-1)^n$, and use it to show that the series diverges.

20. Find the sum of the series

$$\frac{1}{1} + \frac{1}{1+2} + \frac{1}{1+2+3} + \frac{1}{1+2+3+4} + \cdots.$$

21. When dropped, an elastic ball bounces back up to a height three-quarters of that from which it fell. If the ball is dropped from a height of 2 m and allowed to bounce up and down indefinitely, what is the total distance it travels before coming to rest?

22. If a bank account pays 10% simple interest into an account once a year, what is the balance in the account at the end of 8 years if $1,000 is deposited into the account at the beginning of each of the 8 years? (Assume there was no balance in the account initially.)

*23. Prove Theorem 5.   ***24.** Prove Theorem 6.

*25. State a theorem analogous to Theorem 6 but for a negative sequence.

In Exercises 26–31, decide whether the given statement is TRUE or FALSE. If it is true, prove it. If it is false, give a counterexample showing the falsehood.

*26. If $a_n = 0$ for every n, then $\sum a_n$ converges.

*27. If $\sum a_n$ converges, then $\sum(1/a_n)$ diverges to infinity.

*28. If $\sum a_n$ and $\sum b_n$ both diverge, then so does $\sum(a_n + b_n)$.

*29. If $a_n \geq c > 0$ for every n, then $\sum a_n$ diverges to infinity.

*30. If $\sum a_n$ diverges and $\{b_n\}$ is bounded, then $\sum a_n b_n$ diverges.

*31. If $a_n > 0$ and $\sum a_n$ converges, then $\sum(a_n)^2$ converges.

9.3 Convergence Tests for Positive Series

In the previous section we saw a few examples of convergent series (geometric and telescoping series) whose sums could be determined exactly because the partial sums s_n could be expressed in closed form as explicit functions of n whose limits as $n \to \infty$ could be evaluated. It is not usually possible to do this with a given series, and therefore it is not usually possible to determine the sum of the series exactly. However, there are many techniques for determining whether a given series converges and, if it does, for approximating the sum to any desired degree of accuracy.

In this section we deal exclusively with *positive series*, that is, series of the form

$$\sum_{n=1}^{\infty} a_n = a_1 + a_2 + a_3 + \cdots,$$

where $a_n \geq 0$ for all $n \geq 1$. As noted in Theorem 6, such a series will converge if its partial sums are bounded above and will diverge to infinity otherwise. All our results apply equally well to *ultimately* positive series since convergence or divergence depends only on the *tail* of a series.

The Integral Test

The integral test provides a means for determining whether an ultimately positive series converges or diverges by comparing it with an improper integral that behaves similarly.

Example 4 in Section 9.2 is an example of the use of this technique. We formalize the method in the following theorem.

THEOREM **8**

The integral test

Suppose that $a_n = f(n)$, where f is positive, continuous, and nonincreasing on an interval $[N, \infty)$ for some positive integer N. Then

$$\sum_{n=1}^{\infty} a_n \qquad \text{and} \qquad \int_{N}^{\infty} f(t)\, dt$$

either both converge or both diverge to infinity.

PROOF   Let $s_n = a_1 + a_2 + \cdots + a_n$. If $n > N$, we have

$$
\begin{aligned}
s_n &= s_N + a_{N+1} + a_{N+2} + \cdots + a_n \\
&= s_N + f(N+1) + f(N+2) + \cdots + f(n) \\
&= s_N + \text{ sum of areas of rectangles shaded in Figure 9.4(a)} \\
&\le s_N + \int_{N}^{\infty} f(t)\, dt.
\end{aligned}
$$

If the improper integral $\int_{N}^{\infty} f(t)\, dt$ converges, then the sequence $\{s_n\}$ is bounded above and $\sum_{n=1}^{\infty} a_n$ converges.

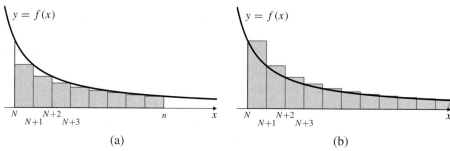

Figure 9.4

(a)   (b)

Conversely, suppose that $\sum_{n=1}^{\infty} a_n$ converges to the sum s. Then

$$
\begin{aligned}
\int_{N}^{\infty} f(t)\, dt &= \text{ area under } y = f(t) \text{ above } y = 0 \text{ from } t = N \text{ to } t = \infty \\
&\le \text{ sum of areas of shaded rectangles in Figure 9.4(b)} \\
&= a_N + a_{N+1} + a_{N+2} + \cdots \\
&= s - s_{N-1} < \infty,
\end{aligned}
$$

so the improper integral represents a finite area and is thus convergent. (We omit the remaining details showing that $\lim_{R \to \infty} \int_{N}^{R} f(t)\, dt$ exists; like the series case, the argument depends on the completeness of the real numbers.)

Remark   If $a_n = f(n)$, where f is positive, continuous, and nonincreasing on $[1, \infty)$, then Theorem 8 assures us that $\sum_{n=1}^{\infty} a_n$ and $\int_{1}^{\infty} f(x)\, dx$ both converge or both diverge to infinity. It does *not* tell us that the sum of the series is equal to the value of the integral. The two are not likely to be equal in the case of convergence. However, as we see below, integrals can help us approximate the sum of a series.

The principal use of the integral test is to establish the result of the following example concerning the series $\sum_{n=1}^{\infty} n^{-p}$, which is called a **p-series**. This result should be memorized; we will frequently compare the behaviour of other series with p-series later in this and subsequent sections.

Example 1 (*p*-series) Show that

$$\sum_{n=1}^{\infty} n^{-p} = \sum_{n=1}^{\infty} \frac{1}{n^p} \begin{cases} \text{converges if } p > 1 \\ \text{diverges to infinity if } p \le 1. \end{cases}$$

Solution   Observe that if $p > 0$, then $f(x) = x^{-p}$ is positive, continuous, and decreasing on $[1, \infty)$. By the integral test, the *p*-series converges for $p > 1$ and diverges for $0 < p \le 1$ by comparison with $\int_1^{\infty} x^{-p} \, dx$. (See Theorem 2(a) of Section 6.5.) If $p \le 0$, then $\lim_{n \to \infty} (1/n^p) \ne 0$, so the series cannot converge in this case. Being a positive series, it must diverge to infinity.

Remark   The harmonic series $\sum_{n=1}^{\infty} n^{-1}$ (the case $p = 1$ of the *p*-series) is on the borderline between convergence and divergence, although it diverges. While its terms decrease toward 0 as n increases, they do not decrease *fast enough* to allow the sum of the series to be finite. If $p > 1$, the terms of $\sum_{n=1}^{\infty} n^{-p}$ decrease toward zero fast enough that their sum is finite. We can refine the distinction between convergence and divergence at $p = 1$ by using terms that decrease faster than $1/n$, but not as fast as $1/n^q$ for any $q > 1$. If $p > 0$, the terms $1/(n(\ln n)^p)$ have this property since $\ln n$ grows more slowly than any positive power of n as n increases. The question now arises whether $\sum_{n=2}^{\infty} 1/(n(\ln n)^p)$ converges. It does, provided again that $p > 1$; you can use the substitution $u = \ln x$ to check that

$$\int_2^{\infty} \frac{dx}{x(\ln x)^p} = \int_{\ln 2}^{\infty} \frac{du}{u^p},$$

which converges if $p > 1$ and diverges if $0 < p \le 1$. This process of fine-tuning Example 1 can be extended even further. (See Exercise 36 below.)

Using Integral Bounds to Estimate the Sum of a Series

Suppose that $a_k = f(k)$ for $k = n + 1, n + 2, n + 3, \ldots$, where f is a positive, continuous function, decreasing at least on the interval $[n, \infty)$. We have:

$$s - s_n = \sum_{k=n+1}^{\infty} f(k)$$

$$= \text{sum of areas of rectangles shaded in Figure 9.5(a)}$$

$$\le \int_n^{\infty} f(x) \, dx.$$

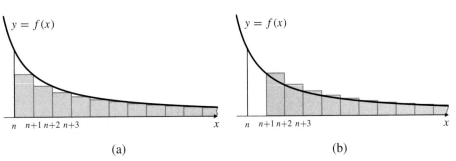

Figure 9.5   (a)   (b)

Similarly,

$$s - s_n = \text{sum of areas of rectangles in Figure 9.5(b)}$$

$$\ge \int_{n+1}^{\infty} f(x) \, dx.$$

If we define

$$A_n = \int_n^\infty f(x)\,dx,$$

then we can combine the above inequalities to obtain

$$A_{n+1} \leq s - s_n \leq A_n,$$

or, equivalently:

$$s_n + A_{n+1} \leq s \leq s_n + A_n.$$

The error in the approximation $s \approx s_n$ satisfies $0 \leq s - s_n \leq A_n$. However, since s must lie in the interval $[s_n + A_{n+1}, s_n + A_n]$, we can do better by using the midpoint s_n^* of this interval as an approximation for s. The error is then less than half the length $A_n - A_{n+1}$ of the interval:

A better integral approximation

The error $|s - s_n^*|$ in the approximation

$$s \approx s_n^* = s_n + \frac{A_{n+1} + A_n}{2}, \qquad \text{where} \quad A_n = \int_n^\infty f(x)\,dx,$$

satisfies $\quad |s - s_n^*| \leq \dfrac{A_n - A_{n+1}}{2}.$

(Whenever a quantity is known to lie in a certain interval, the midpoint of that interval can be used to approximate the quantity, and the absolute value of the error in that approximation does not exceed half the length of the interval.)

Example 2   Find the best approximation s_n^* to the sum s of the series $\sum_{n=1}^\infty 1/n^2$, making use of the partial sum s_n of the first n terms. How large would n have to be to ensure that the approximation $s \approx s_n^*$ has error less than 0.001 in absolute value? How large would n have to be to ensure that the approximation $s \approx s_n$ has error less than 0.001 in absolute value?

Solution   Since $f(x) = 1/x^2$ is positive, continuous, and decreasing on $[1, \infty)$ for any $n = 1, 2, 3, \ldots$, we have

$$s_n + A_{n+1} \leq s \leq s_n + A_n,$$

where

$$A_n = \int_n^\infty \frac{dx}{x^2} = \lim_{R \to \infty} \left(-\frac{1}{x} \right) \Big|_n^R = \frac{1}{n}.$$

The best approximation to s using s_n is

$$s_n^* = s_n + \frac{1}{2} \left(\frac{1}{n+1} + \frac{1}{n} \right) = s_n + \frac{2n+1}{2n(n+1)}$$

$$= 1 + \frac{1}{4} + \frac{1}{9} + \cdots + \frac{1}{n^2} + \frac{2n+1}{2n(n+1)}.$$

The error in this approximation satisfies

$$|s - s_n^*| \leq \frac{1}{2} \left(\frac{1}{n} - \frac{1}{n+1} \right) = \frac{1}{2n(n+1)} \leq 0.001,$$

provided $2n(n + 1) \geq 1/0.001 = 1,000$. It is easily checked that this condition is satisfied if $n \geq 22$; the approximation

$$s \approx s_{22}^* = 1 + \frac{1}{4} + \frac{1}{9} + \cdots + \frac{1}{22^2} + \frac{45}{44 \times 23}$$

will have error with absolute value not exceeding 0.001. Had we used the approximation $s \approx s_n$ we could only have concluded that

$$0 \leq s - s_n \leq A_n = \frac{1}{n} < 0.001,$$

provided $n > 1,000$; we would need $1,000$ terms of the series to get the desired accuracy.

■

Comparison Tests

The next test we consider for positive series is analogous to the comparison theorem for improper integrals. (See Theorem 3 of Section 6.5.) It enables us to determine the convergence or divergence of one series by comparing it with another series that is known to converge or diverge.

THEOREM 9

A comparison test

Let $\{a_n\}$ and $\{b_n\}$ be sequences for which there exists a positive constant K such that, ultimately, $0 \leq a_n \leq Kb_n$.

(a) If the series $\sum_{n=1}^{\infty} b_n$ converges, then so does the series $\sum_{n=1}^{\infty} a_n$.

(b) If the series $\sum_{n=1}^{\infty} a_n$ diverges to infinity, then so does the series $\sum_{n=1}^{\infty} b_n$.

PROOF Since a series converges if and only if its tail converges (Theorem 5), we can assume, without loss of generality, that the condition $0 \leq a_n \leq Kb_n$ holds for all $n \geq 1$. Let $s_n = a_1 + a_2 + \cdots + a_n$ and $S_n = b_1 + b_2 + \cdots + b_n$. Then $s_n \leq KS_n$. If $\sum b_n$ converges, then $\{S_n\}$ is convergent and hence is bounded by Theorem 1. Hence $\{s_n\}$ is bounded above. By Theorem 6, $\sum a_n$ converges. Since the convergence of $\sum b_n$ guarantees that of $\sum a_n$, if the latter series diverges to infinity, then the former cannot converge either, so it must diverge to infinity too.

⬛ BEWARE

Theorem 9 does *not* say that if $\sum a_n$ converges, then $\sum b_n$ converges. It is possible that the *smaller* sum may be finite while the *larger* one is infinite. (Do not confuse a theorem with its converse.)

Example 3 Which of the following series converge? Give reasons for your answers.

(a) $\displaystyle\sum_{n=1}^{\infty} \frac{1}{2^n + 1}$, (b) $\displaystyle\sum_{n=1}^{\infty} \frac{3n + 1}{n^3 + 1}$, (c) $\displaystyle\sum_{n=2}^{\infty} \frac{1}{\ln n}$.

Solution In each case we must find a suitable comparison series that we already know converges or diverges.

(a) Since $0 < \dfrac{1}{2^n + 1} < \dfrac{1}{2^n}$ for $n = 1, 2, 3, \ldots$, and since $\sum_{n=1}^{\infty} \dfrac{1}{2^n}$ is a convergent geometric series, the series $\sum_{n=1}^{\infty} \dfrac{1}{2^n + 1}$ also converges by comparison.

(b) Observe that $\dfrac{3n + 1}{n^3 + 1}$ behaves like $\dfrac{3}{n^2}$ for large n, so we would expect to compare the series with the convergent p-series $\sum_{n=1}^{\infty} n^{-2}$. We have, for $n \geq 1$,

$$\frac{3n + 1}{n^3 + 1} = \frac{3n}{n^3 + 1} + \frac{1}{n^3 + 1} < \frac{3n}{n^3} + \frac{1}{n^3} < \frac{3}{n^2} + \frac{1}{n^2} = \frac{4}{n^2}.$$

Thus, the given series converges by Theorem 9.

(c) For $n = 2, 3, 4, \ldots$, we have $0 < \ln n < n$. Thus $\dfrac{1}{\ln n} > \dfrac{1}{n}$. Since $\sum_{n=2}^{\infty} \dfrac{1}{n}$ diverges to infinity (it is a harmonic series), so does $\sum_{n=2}^{\infty} \dfrac{1}{\ln n}$ by comparison.   ∎

The following theorem provides a version of the comparison test that is not quite as general as Theorem 9 but is often easier to apply in specific cases.

THEOREM 10

A limit comparison test

Suppose that $\{a_n\}$ and $\{b_n\}$ are positive sequences and that

$$\lim_{n \to \infty} \frac{a_n}{b_n} = L,$$

where L is either a nonnegative finite number or $+\infty$.

(a) If $L < \infty$ and $\sum_{n=1}^{\infty} b_n$ converges, then $\sum_{n=1}^{\infty} a_n$ also converges.

(b) If $L > 0$ and $\sum_{n=1}^{\infty} b_n$ diverges to infinity, then so does $\sum_{n=1}^{\infty} a_n$.

PROOF   If $L < \infty$, then for n sufficiently large, we have $b_n > 0$ and

$$0 \le \frac{a_n}{b_n} \le L + 1,$$

so $0 \le a_n \le (L + 1)b_n$. Hence $\sum_{n=1}^{\infty} a_n$ converges if $\sum_{n=1}^{\infty} b_n$ converges, by Theorem 9(a).

If $L > 0$, then for n sufficiently large

$$\frac{a_n}{b_n} \ge \frac{L}{2}.$$

Therefore, $0 < b_n \le (2/L)a_n$, and $\sum_{n=1}^{\infty} a_n$ diverges to infinity if $\sum_{n=1}^{\infty} b_n$ does, by Theorem 9(b).

Example 4   Which of the following series converge?   Give reasons for your answers.

(a) $\displaystyle\sum_{n=1}^{\infty} \frac{1}{1 + \sqrt{n}}$,   (b) $\displaystyle\sum_{n=1}^{\infty} \frac{n + 5}{n^3 - 2n + 3}$.

Solution   Again we must make appropriate choices for comparison series.

(a) The terms of this series decrease like $1/\sqrt{n}$. Observe that

$$L = \lim_{n \to \infty} \frac{\dfrac{1}{1 + \sqrt{n}}}{\dfrac{1}{\sqrt{n}}} = \lim_{n \to \infty} \frac{\sqrt{n}}{1 + \sqrt{n}} = \lim_{n \to \infty} \frac{1}{(1/\sqrt{n}) + 1} = 1.$$

Since the p-series $\sum_{n=1}^{\infty} \dfrac{1}{\sqrt{n}}$ diverges to infinity ($p = 1/2$), so does the series $\sum_{n=1}^{\infty} \dfrac{1}{1 + \sqrt{n}}$, by the limit comparison test.

(b) For large n, the terms behave like n/n^3, so let us compare the series with the p-series $\sum_{n=1}^{\infty} 1/n^2$, which we know converges.

$$L = \lim_{n \to \infty} \frac{\dfrac{n + 5}{n^3 - 2n + 3}}{\dfrac{1}{n^2}} = \lim_{n \to \infty} \frac{n^3 + 5n^2}{n^3 - 2n + 3} = 1.$$

Since $L < \infty$, the series $\sum\limits_{n=1}^{\infty} \dfrac{n+5}{n^3 - 2n + 3}$ also converges by the limit comparison test.

■

In order to apply the original version of the comparison test (Theorem 9) successfully, it is important to have an intuitive feeling for whether the given series converges or diverges. The form of the comparison will depend on whether you are trying to prove convergence or divergence. For instance, if you did not know intuitively that

$$\sum_{n=1}^{\infty} \frac{1}{100n + 20,000}$$

would have to diverge to infinity, you might try to argue that

$$\frac{1}{100n + 20,000} < \frac{1}{n} \qquad \text{for } n = 1, 2, 3, \ldots.$$

While true, this doesn't help at all. $\sum_{n=1}^{\infty} 1/n$ diverges to infinity; therefore Theorem 9 yields no information from this comparison. We could, of course, argue instead that

$$\frac{1}{100n + 20,000} \geq \frac{1}{20,100n} \qquad \text{if } n \geq 1,$$

and conclude by Theorem 9 that $\sum_{n=1}^{\infty}(1/(100n + 20,000))$ diverges to infinity by comparison with the divergent series $\sum_{n=1}^{\infty} 1/n$. An easier way is to use Theorem 10 and the fact that

$$L = \lim_{n\to\infty} \frac{\dfrac{1}{100n + 20,000}}{\dfrac{1}{n}} = \lim_{n\to\infty} \frac{n}{100n + 20,000} = \frac{1}{100} > 0.$$

However, the limit comparison test Theorem 10 has a disadvantage when compared to the ordinary comparison test Theorem 9. It can fail in certain cases because the limit L does not exist. In such cases it is possible that the ordinary comparison test may still work.

Example 5   Test the series $\sum\limits_{n=1}^{\infty} \dfrac{1 + \sin n}{n^2}$ for convergence.

Solution   Since

$$\lim_{n\to\infty} \frac{\dfrac{1 + \sin n}{n^2}}{\dfrac{1}{n^2}} = \lim_{n\to\infty} (1 + \sin n)$$

does not exist, the limit comparison test gives us no information. However, since $\sin n \leq 1$, we have

$$0 \leq \frac{1 + \sin n}{n^2} \leq \frac{2}{n^2} \qquad \text{for } n = 1, 2, 3, \ldots.$$

The given series does, in fact, converge by comparison with $\sum_{n=1}^{\infty} 1/n^2$, using the ordinary comparison test.

■

The Ratio and Root Tests

THEOREM 11

The ratio test

Suppose that $a_n > 0$ (ultimately) and that $\rho = \lim\limits_{n \to \infty} \dfrac{a_{n+1}}{a_n}$ exists or is $+\infty$.

(a) If $0 \le \rho < 1$, then $\sum_{n=1}^{\infty} a_n$ converges.

(b) If $1 < \rho \le \infty$, then $\lim_{n \to \infty} a_n = \infty$ and $\sum_{n=1}^{\infty} a_n$ diverges to infinity.

(c) If $\rho = 1$, this test gives no information; the series may either converge or diverge to infinity.

PROOF   Here ρ is the lowercase Greek letter "rho" (pronounced "roh").

(a) Suppose $\rho < 1$. Pick a number r such that $\rho < r < 1$. Since we are given that $\lim_{n \to \infty} a_{n+1}/a_n = \rho$, we have $a_{n+1}/a_n \le r$ for n sufficiently large; that is, $a_{n+1} \le r a_n$ for $n \ge N$, say. In particular,

$$a_{N+1} \le r a_N$$

$$a_{N+2} \le r a_{N+1} \le r^2 a_N$$

$$a_{N+3} \le r a_{N+2} \le r^3 a_N$$

$$\vdots$$

$$a_{N+k} \le r^k a_N \qquad (k = 0, 1, 2, 3, \ldots)$$

Hence, $\sum_{n=N}^{\infty} a_n$ converges by comparison with the convergent geometric series $\sum_{k=0}^{\infty} r^k$. It follows that $\sum_{n=1}^{\infty} a_n = \sum_{n=1}^{N-1} a_n + \sum_{n=N}^{\infty} a_n$ must also converge.

(b) Now suppose that $\rho > 1$. Pick a number r such that $1 < r < \rho$. Since $\lim_{n \to \infty} a_{n+1}/a_n = \rho$, we have $a_{n+1}/a_n \ge r$ for n sufficiently large, say for $n \ge N$. We assume N is chosen large enough that $a_N > 0$. It follows by an argument similar to that used in part (a) that $a_{N+k} \ge r^k a_N$ for $k = 0, 1, 2, \ldots$, and since $r > 1$, $\lim_{n \to \infty} a_n = \infty$. Therefore $\sum_{n=1}^{\infty} a_n$ diverges to infinity.

(c) If ρ is computed for the series $\sum_{n=1}^{\infty} 1/n$ and $\sum_{n=1}^{\infty} 1/n^2$, we get $\rho = 1$ in each case. Since the first series diverges to infinity and the second converges, the ratio test cannot distinguish between convergence and divergence if $\rho = 1$.

All p-series fall into the indecisive category where $\rho = 1$, as does $\sum_{n=1}^{\infty} a_n$, where a_n is any rational function of n. The ratio test is most useful for series whose terms decrease at least exponentially fast. The presence of factorials in a term also suggests that the ratio test might be useful.

Example 6   Test the following series for convergence:

(a) $\displaystyle\sum_{n=1}^{\infty} \frac{99^n}{n!}$,   (b) $\displaystyle\sum_{n=1}^{\infty} \frac{n^5}{2^n}$,   (c) $\displaystyle\sum_{n=1}^{\infty} \frac{n!}{n^n}$,   (d) $\displaystyle\sum_{n=1}^{\infty} \frac{(2n)!}{(n!)^2}$.

Solution   We use the ratio test for each of these series.

(a) $\rho = \lim\limits_{n \to \infty} \dfrac{99^{n+1}}{(n+1)!} \bigg/ \dfrac{99^n}{n!} = \lim\limits_{n \to \infty} \dfrac{99}{n+1} = 0 < 1.$

Thus $\sum_{n=1}^{\infty} (99^n / n!)$ converges.

(b) $\rho = \lim\limits_{n \to \infty} \dfrac{(n+1)^5}{2^{n+1}} \bigg/ \dfrac{n^5}{2^n} = \lim\limits_{n \to \infty} \dfrac{1}{2} \left(\dfrac{n+1}{n} \right)^5 = \dfrac{1}{2} < 1.$

Hence $\sum_{n=1}^{\infty} (n^5 / 2^n)$ converges.

(c) $\rho = \lim\limits_{n\to\infty} \dfrac{(n+1)!}{(n+1)^{n+1}} \bigg/ \dfrac{n!}{n^n} = \lim\limits_{n\to\infty} \dfrac{(n+1)!n^n}{(n+1)^{n+1}n!} = \lim\limits_{n\to\infty} \left(\dfrac{n}{n+1}\right)^n$

$$= \lim_{n\to\infty} \frac{1}{\left(1+\dfrac{1}{n}\right)^n} = \frac{1}{e} < 1.$$

Thus $\sum_{n=1}^{\infty}(n!/n^n)$ converges.

(d) $\rho = \lim\limits_{n\to\infty} \dfrac{(2(n+1))!}{((n+1)!)^2} \bigg/ \dfrac{(2n)!}{(n!)^2} = \lim\limits_{n\to\infty} \dfrac{(2n+2)(2n+1)}{(n+1)^2} = 4 > 1.$

Thus $\sum_{n=1}^{\infty}(2n)!/(n!)^2$ diverges to infinity. ∎

The following theorem is very similar to the ratio test but is less frequently used. Its proof is left as an exercise. (See Exercise 37.) For examples of series to which it can be applied, see Exercises 38 and 39.

THEOREM 12

The root test

Suppose that $a_n > 0$ (ultimately) and that $\sigma = \lim_{n\to\infty}(a_n)^{1/n}$ exists or is $+\infty$.

(a) If $0 \le \sigma < 1$, then $\sum_{n=1}^{\infty} a_n$ converges.

(b) If $1 < \sigma \le \infty$, then $\lim_{n\to\infty} a_n = \infty$ and $\sum_{n=1}^{\infty} a_n$ diverges to infinity.

(c) If $\sigma = 1$, this test gives no information; the series may either converge or diverge to infinity.

Using Geometric Bounds to Estimate the Sum of a Series

Suppose that an inequality of the form

$$0 \le a_k \le Kr^k$$

holds for $k = n+1, n+2, n+3, \ldots$, where K and r are constants and $r < 1$. We can then use a geometric series to bound the tail of $\sum_{n=1}^{\infty} a_n$.

$$0 \le s - s_n = \sum_{k=n+1}^{\infty} a_k \le \sum_{k=n+1}^{\infty} Kr^k$$
$$= Kr^{n+1}(1 + r + r^2 + \cdots)$$
$$= \frac{Kr^{n+1}}{1-r}.$$

Since $r < 1$, the series converges and the error approaches 0 at an exponential rate as n increases.

Example 7   In Section 9.6 we will show that

$$e = \frac{1}{0!} + \frac{1}{1!} + \frac{1}{2!} + \frac{1}{3!} + \cdots = \sum_{n=0}^{\infty} \frac{1}{n!}.$$

(Recall that $0! = 1$.) Estimate the error if the sum s_n of the first n terms of the series is used to approximate e. Find e to 3-decimal-place accuracy using the series.

Solution   We have

$$s_n = \frac{1}{0!} + \frac{1}{1!} + \frac{1}{2!} + \frac{1}{3!} + \cdots + \frac{1}{(n-1)!}$$
$$= 1 + 1 + \frac{1}{2} + \frac{1}{6} + \frac{1}{24} + \cdots + \frac{1}{(n-1)!}.$$

(Since the series starts with the term for $n = 0$, the nth term is $1/(n-1)!$.) We can estimate the error in the approximation $s \approx s_n$ as follows:

$$0 < s - s_n = \frac{1}{n!} + \frac{1}{(n+1)!} + \frac{1}{(n+2)!} + \frac{1}{(n+3)!} + \cdots$$

$$= \frac{1}{n!} \left(1 + \frac{1}{n+1} + \frac{1}{(n+1)(n+2)} + \frac{1}{(n+1)(n+2)(n+3)} + \cdots \right)$$

$$< \frac{1}{n!} \left(1 + \frac{1}{n+1} + \frac{1}{(n+1)^2} + \frac{1}{(n+1)^3} + \cdots \right)$$

since $n + 2 > n + 1$, $n + 3 > n + 1$, and so on. The latter series is geometric, so

$$0 < s - s_n < \frac{1}{n!} \frac{1}{1 - \dfrac{1}{n+1}} = \frac{n+1}{n!n}.$$

If we want to evaluate e accurately to 3 decimal places, then we must ensure that the error is less than 5 in the fourth decimal place, that is, that the error is less than 0.0005. Hence we want

$$\frac{n+1}{n} \frac{1}{n!} < 0.0005 = \frac{1}{2,000}.$$

Since $7! = 5,040$ but $6! = 720$, we can use $n = 7$ but no smaller. We have

$$e \approx s_7 = 1 + 1 + \frac{1}{2!} + \frac{1}{3!} + \frac{1}{4!} + \frac{1}{5!} + \frac{1}{6!}$$

$$= 2 + \frac{1}{2} + \frac{1}{6} + \frac{1}{24} + \frac{1}{120} + \frac{1}{720} \approx 2.718 \quad \text{to 3 decimal places.}$$

It is appropriate to use geometric series to bound the tails of positive series whose convergence would be demonstrated by the ratio test. Such series converge ultimately faster than any p-series $\sum_{n=1}^{\infty} n^{-p}$, for which the limit ratio is $\rho = 1$.

Exercises 9.3

In Exercises 1–26, determine whether the given series converges or diverges by using any appropriate test. The p-series can be used for comparison, as can geometric series. Be alert for series whose terms do not approach 0.

1. $\displaystyle\sum_{n=1}^{\infty} \frac{1}{n^2 + 1}$

2. $\displaystyle\sum_{n=1}^{\infty} \frac{n}{n^4 - 2}$

3. $\displaystyle\sum_{n=1}^{\infty} \frac{n^2 + 1}{n^3 + 1}$

4. $\displaystyle\sum_{n=1}^{\infty} \frac{\sqrt{n}}{n^2 + n + 1}$

5. $\displaystyle\sum_{n=1}^{\infty} \left| \sin \frac{1}{n^2} \right|$

6. $\displaystyle\sum_{n=8}^{\infty} \frac{1}{\pi^n + 5}$

7. $\displaystyle\sum_{n=2}^{\infty} \frac{1}{(\ln n)^3}$

8. $\displaystyle\sum_{n=1}^{\infty} \frac{1}{\ln(3n)}$

9. $\displaystyle\sum_{n=1}^{\infty} \frac{1}{\pi^n - n^\pi}$

10. $\displaystyle\sum_{n=0}^{\infty} \frac{1+n}{2+n}$

11. $\displaystyle\sum_{n=1}^{\infty} \frac{1 + n^{4/3}}{2 + n^{5/3}}$

12. $\displaystyle\sum_{n=1}^{\infty} \frac{n^2}{1 + n\sqrt{n}}$

13. $\displaystyle\sum_{n=3}^{\infty} \frac{1}{n \ln n \sqrt{\ln \ln n}}$

14. $\displaystyle\sum_{n=2}^{\infty} \frac{1}{n \ln n (\ln \ln n)^2}$

15. $\displaystyle\sum_{n=1}^{\infty} \frac{1 - (-1)^n}{n^4}$

16. $\displaystyle\sum_{n=1}^{\infty} \frac{1 + (-1)^n}{\sqrt{n}}$

17. $\displaystyle\sum_{n=1}^{\infty} \frac{1}{2^n (n+1)}$

18. $\displaystyle\sum_{n=1}^{\infty} \frac{n^4}{n!}$

19. $\displaystyle\sum_{n=1}^{\infty} \frac{n!}{n^2 e^n}$

20. $\displaystyle\sum_{n=1}^{\infty} \frac{(2n)! 6^n}{(3n)!}$

21. $\displaystyle\sum_{n=2}^{\infty} \frac{\sqrt{n}}{3^n \ln n}$

22. $\displaystyle\sum_{n=0}^{\infty} \frac{n^{100} 2^n}{\sqrt{n!}}$

23. $\displaystyle\sum_{n=1}^{\infty} \frac{(2n)!}{(n!)^3}$

24. $\displaystyle\sum_{n=1}^{\infty} \frac{1 + n!}{(1+n)!}$

25. $\displaystyle\sum_{n=4}^{\infty} \frac{2^n}{3^n - n^3}$ **26.** $\displaystyle\sum_{n=1}^{\infty} \frac{n^n}{\pi^n n!}$

In Exercises 27–30, use s_n and integral bounds to find the smallest interval that you can be sure contains the sum s of the series. If the midpoint s_n^* of this interval is used to approximate s, how large should n be chosen to ensure that the error is less than 0.001?

27. $\displaystyle\sum_{k=1}^{\infty} \frac{1}{k^4}$ **28.** $\displaystyle\sum_{k=1}^{\infty} \frac{1}{k^3}$

29. $\displaystyle\sum_{k=1}^{\infty} \frac{1}{k^{3/2}}$ **30.** $\displaystyle\sum_{k=1}^{\infty} \frac{1}{k^2 + 4}$

For each positive series in Exercises 31–34, find the best upper bound you can for the error $s - s_n$ encountered if the partial sum s_n is used to approximate the sum s of the series. How many terms of each series do you need to be sure that the approximation has error less than 0.001?

31. $\displaystyle\sum_{k=1}^{\infty} \frac{1}{2^k k!}$ **32.** $\displaystyle\sum_{n=1}^{\infty} \frac{1}{(2n-1)!}$

33. $\displaystyle\sum_{n=0}^{\infty} \frac{2^n}{(2n)!}$ **34.** $\displaystyle\sum_{n=1}^{\infty} \frac{1}{n^n}$

35. Use the integral test to show that $\displaystyle\sum_{n=1}^{\infty} \frac{1}{1 + n^2}$ converges. Show that the sum s of the series is less than $\pi/2$.

*36. Show that $\sum_{n=3}^{\infty}(1/(n \ln n(\ln\ln n)^p)$ converges if and only if $p > 1$. Generalize this result to series of the form

$$\sum_{n=N}^{\infty} \frac{1}{n(\ln n)(\ln\ln n)\cdots(\ln_j n)(\ln_{j+1} n)^p},$$

where $\ln_j n = \underbrace{\ln\ln\ln\ln\cdots\ln}_{j \ \ln's} n$.

*37. Prove the root test. *Hint:* Mimic the proof of the ratio test.

38. Use the root test to show that $\displaystyle\sum_{n=1}^{\infty} \frac{2^{n+1}}{n^n}$ converges.

*39. Use the root test to test the following series for convergence:

$$\sum_{n=1}^{\infty} \left(\frac{n}{n+1}\right)^{n^2}.$$

40. Repeat Exercise 38, but use the ratio test instead of the root test.

*41. Try to use the ratio test to determine whether $\displaystyle\sum_{n=1}^{\infty} \frac{2^{2n}(n!)^2}{(2n)!}$ converges. What happens? Now observe that

$$\frac{2^{2n}(n!)^2}{(2n)!} = \frac{[2n(2n-2)(2n-4)\cdots 6 \times 4 \times 2]^2}{2n(2n-1)(2n-2)\cdots 4 \times 3 \times 2 \times 1}$$

$$= \frac{2n}{2n-1} \times \frac{2n-2}{2n-3} \times \cdots \times \frac{4}{3} \times \frac{2}{1}.$$

Does the given series converge? Why or why not?

*42. Determine whether the series $\displaystyle\sum_{n=1}^{\infty} \frac{(2n)!}{2^{2n}(n!)^2}$ converges. *Hint:* Proceed as in Exercise 41. Show that $a_n \geq 1/(2n)$.

*43. (a) Show that if $k > 0$ and n is a positive integer, then

$$n < \frac{1}{k}(1 + k)^n.$$

(b) Use the estimate in (a) with $0 < k < 1$ to obtain an upper bound for the sum of the series $\sum_{n=0}^{\infty} n/2^n$. For what value of k is this bound least?

(c) If we use the sum s_n of the first n terms to approximate the sum s of the series in (b), obtain an upper bound for the error $s - s_n$ using the inequality from (a). For given n, find k to minimize this upper bound.

*44. (Improving the convergence of a series) We know that $\sum_{n=1}^{\infty} 1/(n(n+1)) = 1$. (See Example 3 of Section 9.2.) Since

$$\frac{1}{n^2} = \frac{1}{n(n+1)} + c_n, \quad \text{where} \quad c_n = \frac{1}{n^2(n+1)},$$

we have $\displaystyle\sum_{n=1}^{\infty} \frac{1}{n^2} = 1 + \sum_{n=1}^{\infty} c_n.$

The series $\sum_{n=1}^{\infty} c_n$ converges more rapidly than does $\sum_{n=1}^{\infty} 1/n^2$ because its terms decrease like $1/n^3$. Hence, fewer terms of that series will be needed to compute $\sum_{n=1}^{\infty} 1/n^2$ to any desired degree of accuracy than would be needed if we calculated with $\sum_{n=1}^{\infty} 1/n^2$ directly. Using integral upper and lower bounds, determine a value of n for which the modified partial sum s_n^* for the series $\sum_{n=1}^{\infty} c_n$ approximates the sum of that series with error less than 0.001 in absolute value. Hence, determine $\sum_{n=1}^{\infty} 1/n^2$ to within 0.001 of its true value.

(The technique exibited in this exercise is known as *improving the convergence* of a series. It can be applied to estimating the sum $\sum a_n$ if we know the sum $\sum b_n$ and if $a_n - b_n = c_n$, where $|c_n|$ decreases faster than $|a_n|$ as n tends to infinity.)

45. Consider the series $s = \sum_{n=1}^{\infty} 1/(2^n + 1)$, and the partial sum s_n of its first n terms.

(a) How large need n be taken to ensure that the error in the approximation $s \approx s_n$ is less than 0.001 in absolute value?

(b) The geometric series $\sum_{n=1}^{\infty} 1/2^n$ converges to 1. If

$$b_n = \frac{1}{2^n} - \frac{1}{2^n + 1}$$

for $n = 1, 2, 3, \ldots$, how many terms of the series $\sum_{n=1}^{\infty} b_n$ are needed to calculate its sum to within 0.001?

(c) Use the result of part (b) to calculate the $\sum_{n=1}^{\infty} 1/(2^n + 1)$ to within 0.001.

9.4 Absolute and Conditional Convergence

All of the series $\sum_{n=1}^{\infty} a_n$ considered in the previous section were ultimately positive; that is, $a_n \geq 0$ for n sufficiently large. We now drop this restriction and allow arbitrary real terms a_n. We can, however, always obtain a positive series from any given series by replacing all the terms with their absolute values.

DEFINITION 5

> **Absolute convergence**
>
> The series $\sum_{n=1}^{\infty} a_n$ is said to be **absolutely convergent** if $\sum_{n=1}^{\infty} |a_n|$ converges.

The series

$$s = \sum_{n=1}^{\infty} \frac{(-1)^n}{n^2} = -1 + \frac{1}{4} - \frac{1}{9} + \frac{1}{16} - \cdots$$

converges absolutely since

$$S = \sum_{n=1}^{\infty} \left| \frac{(-1)^n}{n^2} \right| = \sum_{n=1}^{\infty} \frac{1}{n^2} = 1 + \frac{1}{4} + \frac{1}{9} + \frac{1}{16} + \cdots$$

converges. It seems reasonable that the first series must converge, and its sum s should satisfy $-S \leq s \leq S$. In general, the cancellation that occurs because some terms are negative and others positive makes it *easier* for a series to converge than if all the terms are of one sign. We verify this insight in the following theorem.

THEOREM 13

If a series converges absolutely, then it converges.

PROOF Let $\sum_{n=1}^{\infty} a_n$ be absolutely convergent, and let $b_n = a_n + |a_n|$ for each n. Since $-|a_n| \leq a_n \leq |a_n|$, we have $0 \leq b_n \leq 2|a_n|$ for each n. Thus $\sum_{n=1}^{\infty} b_n$ converges by the comparison test. Therefore, $\sum_{n=1}^{\infty} a_n = \sum_{n=1}^{\infty} b_n - \sum_{n=1}^{\infty} |a_n|$ also converges.

Again you are cautioned not to confuse the statement of Theorem 13 with the converse statement, which is false. We will show later in this section that the **alternating harmonic series**

$$\sum_{n=1}^{\infty} \frac{(-1)^{n-1}}{n} = 1 - \frac{1}{2} + \frac{1}{3} - \frac{1}{4} + \frac{1}{5} - \cdots$$

BEWARE

Although absolute convergence implies convergence, convergence does *not* imply absolute convergence.

converges, although it does not converge absolutely. If we replace all the terms by their absolute values, we get the divergent harmonic series

$$\sum_{n=1}^{\infty} \frac{1}{n} = 1 + \frac{1}{2} + \frac{1}{3} + \frac{1}{4} + \cdots = \infty.$$

DEFINITION 6

> **Conditional convergence**
>
> If $\sum_{n=1}^{\infty} a_n$ is convergent, but not absolutely convergent, then we say that it is **conditionally convergent** or that it **converges conditionally**.

The alternating harmonic series is an example of a conditionally convergent series.

The comparison tests, the integral test, and the ratio test can each be used to test for absolute convergence. They should be applied to the series $\sum_{n=1}^{\infty} |a_n|$. For the ratio test we calculate $\rho = \lim_{n\to\infty} |a_{n+1}/a_n|$. If $\rho < 1$, then $\sum_{n=1}^{\infty} a_n$ converges absolutely. If $\rho > 1$, then $\lim_{n\to\infty} |a_n| = \infty$, so both $\sum_{n=1}^{\infty} |a_n|$ and $\sum_{n=1}^{\infty} a_n$ must diverge. If $\rho = 1$, we get no information; the series $\sum_{n=1}^{\infty} a_n$ may converge absolutely, it may converge conditionally, or it may diverge.

Example 1 Test the following series for absolute convergence:

(a) $\displaystyle\sum_{n=1}^{\infty} \frac{(-1)^{n-1}}{2n-1}$, (b) $\displaystyle\sum_{n=1}^{\infty} \frac{n\cos(n\pi)}{2^n}$.

Solution

(a) $\displaystyle\lim_{n\to\infty} \left| \frac{(-1)^{n-1}}{2n-1} \right| \bigg/ \frac{1}{n} = \lim_{n\to\infty} \frac{n}{2n-1} = \frac{1}{2} > 0$. Since the harmonic series $\sum_{n=1}^{\infty}(1/n)$ diverges to infinity, the comparison test assures us that $\sum_{n=1}^{\infty}((-1)^{n-1}/(2n-1))$ does not converge absolutely.

(b) $\displaystyle\rho = \lim_{n\to\infty} \left| \frac{(n+1)\cos((n+1)\pi)}{2^{n+1}} \bigg/ \frac{n\cos(n\pi)}{2^n} \right| = \lim_{n\to\infty} \frac{n+1}{2n} = \frac{1}{2} < 1$.

(Note that $\cos(n\pi)$ is just a fancy way of writing $(-1)^n$.) Therefore (ratio test) $\sum_{n=1}^{\infty}((n\cos(n\pi))/2^n)$ converges absolutely. ∎

The Alternating Series Test

We cannot use any of the previously developed tests to show that the alternating harmonic series converges; all of those tests apply only to (ultimately) positive series, so they can test only for absolute convergence. Demonstrating convergence that is not absolute is generally more difficult to do. We present only one test that can establish such convergence; this test can only be used on a very special kind of series.

THEOREM 14

The alternating series test

Suppose $\{a_n\}$ is a sequence whose terms satisfy, for some positive integer N,

  (i) $a_n a_{n+1} < 0$ for $n \geq N$,

  (ii) $|a_{n+1}| \leq |a_n|$ for $n \geq N$, and

  (iii) $\lim_{n\to\infty} a_n = 0$,

that is, the terms are ultimately alternating in sign and decreasing in size, and the sequence has limit zero. Then the series $\sum_{n=1}^{\infty} a_n$ converges.

PROOF Without loss of generality we can assume $N = 1$ because convergence only depends on the tail of a series. We also assume $a_1 > 0$; the proof if $a_1 < 0$ is similar. If $s_n = a_1 + a_2 + \cdots + a_n$ is the nth partial sum of the series, it follows from the alternation of $\{a_n\}$ that $a_{2n+1} > 0$ and $a_{2n} < 0$ for each n. Since the terms decrease in size, $a_{2n+1} \geq -a_{2n+2}$. Therefore $s_{2n+2} = s_{2n} + a_{2n+1} + a_{2n+2} \geq s_{2n}$ for $n = 1, 2, 3, \ldots$; the even partial sums $\{s_{2n}\}$ form an increasing sequence. Similarly $s_{2n+1} = s_{2n-1} + a_{2n} + a_{2n+1} \leq s_{2n-1}$, so the odd partial sums $\{s_{2n-1}\}$ form a decreasing sequence. Since $s_{2n} = s_{2n-1} + a_{2n} \leq s_{2n-1}$, we can say, for any n, that

$$s_2 \leq s_4 \leq s_6 \leq \cdots \leq s_{2n} \leq s_{2n-1} \leq s_{2n-3} \leq \cdots \leq s_5 \leq s_3 \leq s_1.$$

Hence, s_2 is a lower bound for the decreasing sequence $\{s_{2n-1}\}$, and s_1 is an upper bound for the increasing sequence $\{s_{2n}\}$. Both of these sequences therefore converge by the completeness of the real numbers:

$$\lim_{n\to\infty} s_{2n-1} = s_{\text{odd}}, \qquad \lim_{n\to\infty} s_{2n} = s_{\text{even}}.$$

Now $a_{2n} = s_{2n-1} - s_{2n}$, so $0 = \lim_{n\to\infty} a_{2n} = \lim_{n\to\infty}(s_{2n-1} - s_{2n}) = s_{\text{odd}} - s_{\text{even}}$. Therefore $s_{\text{odd}} = s_{\text{even}} = s$, say. Every partial sum s_n is either of the form s_{2n-1} or of the form s_{2n}. Thus, $\lim_{n\to\infty} s_n = s$ exists and the series $\sum(-1)^{n-1}a_n$ converges to this sum s.

⚠ BEWARE

Read this proof slowly and think about why each statement is true.

Remark The proof of Theorem 14 shows that the sum s of the series always lies between any two consecutive partial sums of the series:

  either $s_n < s < s_{n+1}$ or $s_{n+1} < s < s_n$.

This proves the following theorem.

THEOREM **15**

Error estimate for alternating series

If the sequence $\{a_n\}$ satisfies the conditions of the alternating series test (Theorem 14), so that the series $\sum_{n=1}^{\infty} a_n$ converges to the sum s, then the error in the approximation $s \approx s_n$ (where $n \geq N$) has the same sign as the first omitted term $a_{n+1} = s_{n+1} - s_n$, and it size is no greater than the size of that term:

$$|s - s_n| \leq |s_{n+1} - s_n| = |a_{n+1}|.$$

Example 2   How many terms of the series $\sum_{n=1}^{\infty} \frac{(-1)^n}{1 + 2^n}$ are needed to compute the sum of the series with error less than 0.001?

Solution   This series satisfies the hypotheses for Theorem 15. If we use the partial sum of the first n terms of the series to approximate the sum of the series, the error will satisfy

$$|\text{error}| \leq |\text{first omitted term}| = \frac{1}{1 + 2^{n+1}}.$$

This error is less than 0.001 if $1 + 2^{n+1} > 1,000$. Since $2^{10} = 1,024$, $n + 1 = 10$ will do; we need 9 terms of the series to compute the sum to within 0.001 of its actual value.

When determining the convergence of a given series, it is best to consider first whether the series converges absolutely. If it does not, then there remains the possibility of conditional convergence.

Example 3   Test the following series for absolute and conditional convergence:

(a) $\sum_{n=1}^{\infty} \frac{(-1)^{n-1}}{n}$,   (b) $\sum_{n=2}^{\infty} \frac{\cos(n\pi)}{\ln n}$,   (c) $\sum_{n=1}^{\infty} \frac{(-1)^{n-1}}{n^4}$.

Solution   The absolute values of the terms in series (a) and (b) are $1/n$ and $1/(\ln n)$, respectively. Since $1/(\ln n) > 1/n$, and $\sum_{n=1}^{\infty} 1/n$ diverges to infinity, neither series (a) nor (b) converges absolutely. However, both series satisfy the requirements of Theorem 14 and so both converge. Each of these series is conditionally convergent.

Series (c) is absolutely convergent because $|(-1)^{n-1}/n^4| = 1/n^4$, and $\sum_{n=1}^{\infty} 1/n^4$ is a convergent p-series ($p = 4 > 1$). We could establish its convergence using Theorem 14, but there is no need to do that since every absolutely convergent series is convergent (Theorem 13).

Example 4   For what values of x does the series $\sum_{n=1}^{\infty} \frac{(x - 5)^n}{n \, 2^n}$ converge absolutely? converge conditionally? diverge?

Solution   For such series whose terms involve functions of a variable x, it is usually wisest to begin testing for absolute convergence with the ratio test. We have

$$\rho = \lim_{n \to \infty} \left| \frac{(x - 5)^{n+1}}{(n + 1)2^{n+1}} \middle/ \frac{(x - 5)^n}{n \, 2^n} \right| = \lim_{n \to \infty} \frac{n}{n + 1} \left| \frac{x - 5}{2} \right| = \left| \frac{x - 5}{2} \right|.$$

The series converges absolutely if $|(x - 5)/2| < 1$. This inequality is equivalent to $|x - 5| < 2$ (the distance from x to 5 is less than 2), that is, $3 < x < 7$. If $x < 3$ or $x > 7$, then $|(x - 5)/2| > 1$. The series diverges; its terms do not approach zero.

If $x = 3$, the series is $\sum_{n=1}^{\infty} \left((-1)^n / n \right)$, which converges conditionally (it is an alternating harmonic series); if $x = 7$, the series is the harmonic series $\sum_{n=1}^{\infty} 1/n$, which diverges to infinity. Hence, the given series converges absolutely on the open interval $(3, 7)$, converges conditionally at $x = 3$, and diverges everywhere else.

Example 5   For what values of x does the series $\displaystyle\sum_{n=0}^{\infty} (n+1)^2 \left(\frac{x}{x+2} \right)^n$ converge absolutely? converge conditionally? diverge?

Solution   Again we begin with the ratio test.

$$\rho = \lim_{n \to \infty} \left| (n+2)^2 \left(\frac{x}{x+2} \right)^{n+1} \bigg/ (n+1)^2 \left(\frac{x}{x+2} \right)^n \right|$$

$$= \lim_{n \to \infty} \left(\frac{n+2}{n+1} \right)^2 \left| \frac{x}{x+2} \right| = \left| \frac{x}{x+2} \right| = \frac{|x|}{|x+2|}$$

The series converges absolutely if $|x|/|x+2| < 1$. This condition says that the distance from x to 0 is less than the distance from x to -2. Hence $x > -1$. The series diverges if $|x|/|x+2| > 1$, that is, if $x < -1$. If $x = -1$, the series is $\sum_{n=0}^{\infty} (-1)^n (n+1)^2$, which diverges. We conclude that the series converges absolutely for $x > -1$, converges conditionally nowhere, and diverges for $x \leq -1$.

When using the alternating series test, it is important to verify (at least mentally) that *all three conditions* (i)–(iii) are satisfied.

Example 6   Test the following series for convergence:

(a) $\displaystyle\sum_{n=1}^{\infty} (-1)^{n-1} \frac{n+1}{n}$,

(b) $\displaystyle 1 - \frac{1}{4} + \frac{1}{3} - \frac{1}{16} + \frac{1}{5} - \cdots = \sum_{n=1}^{\infty} a_n$,   where

$$a_n = \begin{cases} 1/n & \text{if } n \text{ is odd,} \\ -1/n^2 & \text{if } n \text{ is even.} \end{cases}$$

Solution

(a) Here, the terms a_n alternate and decrease in size as n increases. However, $\lim_{n \to \infty} a_n = 1 \neq 0$. The alternating series test does not apply. In fact, the given series diverges because its terms do not approach 0.

(b) This series alternates and its terms have limit zero, However, the terms are not decreasing in size (even ultimately). Once again, the alternating series test cannot be applied. In fact, since

$$-\frac{1}{4} - \frac{1}{16} - \cdots - \frac{1}{(2n)^2} - \cdots \qquad \text{converges, and}$$

$$1 + \frac{1}{3} + \frac{1}{5} + \cdots + \frac{1}{2n-1} + \cdots \qquad \text{diverges to infinity,}$$

it is readily seen that the given series diverges to infinity.

Rearranging the Terms in a Series

The basic difference between absolute and conditional convergence is that when a series $\sum_{n=1}^{\infty} a_n$ converges absolutely, it does so because its terms $\{a_n\}$ decrease in size fast enough that their sum can be finite even if no cancellation occurs due to terms of opposite sign. If cancellation is required to make the series converge (because the terms decrease slowly), then the series can only converge conditionally.

Consider the alternating harmonic series

$$1 - \frac{1}{2} + \frac{1}{3} - \frac{1}{4} + \frac{1}{5} - \frac{1}{6} + \cdots.$$

This series converges, but only conditionally. If we take the subseries containing only the positive terms, we get the series

$$1 + \frac{1}{3} + \frac{1}{5} + \frac{1}{7} + \cdots,$$

which diverges to infinity. Similarly, the subseries of negative terms

$$-\frac{1}{2} - \frac{1}{4} - \frac{1}{6} - \frac{1}{8} - \cdots$$

diverges to negative infinity.

If a series converges absolutely, the subseries consisting of positive terms and the subseries consisting of negative terms must each converge to a finite sum. If a series converges conditionally, the positive and negative subseries will both diverge, to ∞ and $-\infty$, respectively.

Using these facts we can answer a question raised at the beginning of Section 9.2. If we rearrange the terms of a convergent series so that they are added in a different order, must the rearranged series converge, and if it does will it converge to the same sum as the original series? The answer depends on whether the original series was absolutely convergent or merely conditionally convergent.

THEOREM 16

Convergence of rearrangements of a series

(a) If the terms of an absolutely convergent series are rearranged so that addition occurs in a different order, the rearranged series still converges to the same sum as the original series.

(b) If a series is conditionally convergent, and L is any real number, then the terms of the series can be rearranged so as to make the series converge (conditionally) to the sum L. It can also be rearranged so as to diverge to ∞ or to $-\infty$, or just to diverge.

Part (b) shows that conditional convergence is a rather suspect kind of convergence, being dependent on the order in which the terms are added. We will not present a formal proof of the theorem but will give an example suggesting what is involved. (See also Exercise 30 below.)

Example 7 In Section 9.5 we will show that the alternating harmonic series

$$\sum_{n=1}^{\infty} \frac{(-1)^{n-1}}{n} = 1 - \frac{1}{2} + \frac{1}{3} - \frac{1}{4} + \frac{1}{5} - \frac{1}{6} + \frac{1}{7} - \cdots$$

converges (conditionally) to the sum $\ln 2$. Describe how to rearrange its terms so that it converges to 8 instead.

Solution   Start adding terms of the positive subseries

$$1 + \frac{1}{3} + \frac{1}{5} + \cdots,$$

and keep going until the partial sum exceeds 8. (It will, eventually, because the positive subseries diverges to infinity.) Then add the first term $-1/2$ of the negative subseries

$$-\frac{1}{2} - \frac{1}{4} - \frac{1}{6} - \cdots.$$

This will reduce the partial sum below 8 again. Now resume adding terms of the positive subseries until the partial sum climbs above 8 once more. Then add the second term of the negative subseries and the partial sum will drop below 8. Keep repeating this procedure, alternately adding terms of the positive subseries to force the sum above 8 and then terms of the negative subseries to force it below 8. Since both subseries have infinitely many terms and diverge to ∞ and $-\infty$, respectively, eventually every term of the original series will be included, and the partial sums of the new series will oscillate back and forth around 8, converging to that number. Of course, any number other than 8 could also be used in place of 8.

Exercises 9.4

Determine whether the series in Exercises 1–12 converge absolutely, converge conditionally, or diverge.

1. $\displaystyle\sum_{n=1}^{\infty} \frac{(-1)^{n-1}}{\sqrt{n}}$

2. $\displaystyle\sum_{n=1}^{\infty} \frac{(-1)^{n}}{n^2 + \ln n}$

3. $\displaystyle\sum_{n=1}^{\infty} \frac{\cos(n\pi)}{(n+1)\ln(n+1)}$

4. $\displaystyle\sum_{n=1}^{\infty} \frac{(-1)^{2n}}{2^n}$

5. $\displaystyle\sum_{n=0}^{\infty} \frac{(-1)^{n}(n^2 - 1)}{n^2 + 1}$

6. $\displaystyle\sum_{n=1}^{\infty} \frac{(-2)^{n}}{n!}$

7. $\displaystyle\sum_{n=1}^{\infty} \frac{(-1)^{n}}{n\pi^n}$

8. $\displaystyle\sum_{n=0}^{\infty} \frac{-n}{n^2 + 1}$

9. $\displaystyle\sum_{n=1}^{\infty}(-1)^{n} \frac{20n^2 - n - 1}{n^3 + n^2 + 33}$

10. $\displaystyle\sum_{n=1}^{\infty} \frac{100 \cos(n\pi)}{2n + 3}$

11. $\displaystyle\sum_{n=1}^{\infty} \frac{n!}{(-100)^n}$

12. $\displaystyle\sum_{n=10}^{\infty} \frac{\sin(n + 1/2)\pi}{\ln \ln n}$

For the series in Exercises 13–16, find the smallest integer n that ensures that the partial sum s_n approximates the sum s of the series with error less than 0.001 in absolute value.

13. $\displaystyle\sum_{n=1}^{\infty}(-1)^{n-1} \frac{n}{n^2 + 1}$

14. $\displaystyle\sum_{n=0}^{\infty} \frac{(-1)^{n}}{(2n)!}$

15. $\displaystyle\sum_{n=1}^{\infty}(-1)^{n-1} \frac{n}{2^n}$

16. $\displaystyle\sum_{n=0}^{\infty}(-1)^{n} \frac{3^n}{n!}$

Determine the values of x for which the series in Exercises 17–24 converge absolutely, converge conditionally, or diverge.

17. $\displaystyle\sum_{n=0}^{\infty} \frac{x^n}{\sqrt{n+1}}$

18. $\displaystyle\sum_{n=1}^{\infty} \frac{(x-2)^n}{n^2 2^{2n}}$

19. $\displaystyle\sum_{n=0}^{\infty}(-1)^{n} \frac{(x-1)^n}{2n+3}$

20. $\displaystyle\sum_{n=1}^{\infty} \frac{1}{2n-1}\left(\frac{3x+2}{-5}\right)^n$

21. $\displaystyle\sum_{n=2}^{\infty} \frac{x^n}{2^n \ln n}$

22. $\displaystyle\sum_{n=1}^{\infty} \frac{(4x+1)^n}{n^3}$

23. $\displaystyle\sum_{n=1}^{\infty} \frac{(2x+3)^n}{n^{1/3} 4^n}$

24. $\displaystyle\sum_{n=1}^{\infty} \frac{1}{n}\left(1 + \frac{1}{x}\right)^n$

∗25. Does the alternating series test apply directly to the series $\sum_{n=1}^{\infty}(1/n)\sin(n\pi/2)$? Determine whether the series converges.

∗26. Show that the series $\sum_{n=1}^{\infty} a_n$ converges absolutely if $a_n = 10/n^2$ for even n and $a_n = -1/10n^3$ for odd n.

∗27. Which of the following statements are TRUE and which are FALSE? Justify your assertion of truth, or give a counterexample to show falsehood.

(a) If $\sum_{n=1}^{\infty} a_n$ converges, then $\sum_{n=1}^{\infty}(-1)^n a_n$ converges.

(b) If $\sum_{n=1}^{\infty} a_n$ converges and $\sum_{n=1}^{\infty}(-1)^n a_n$ converges, then $\sum_{n=1}^{\infty} a_n$ converges absolutely.

(c) If $\sum_{n=1}^{\infty} a_n$ converges absolutely, then

$$\sum_{n=1}^{\infty}(-1)^n a_n \text{ converges absolutely.}$$

∗28. (a) Use a Riemann sum argument to show that

$$\ln n! \geq \int_1^n \ln t \, dt = n \ln n - n + 1.$$

(b) For what values of x does the series $\sum_{n=1}^{\infty} \frac{n! x^n}{n^n}$ converge absolutely? converge conditionally? diverge? (*Hint:* First use the ratio test. To test the cases where $\rho = 1$, you may find the inequality in part (a) useful.)

*29. For what values of x does the series $\sum_{n=1}^{\infty} \frac{(2n)!x^n}{2^{2n}(n!)^2}$ converge absolutely? converge conditionally? diverge? *Hint:* See Exercise 42 of Section 9.3.

*30. Devise procedures for rearranging the terms of the alternating harmonic series so that the rearranged series (a) diverges to ∞,   (b) converges to -2.

9.5 Power Series

This section is concerned with a special kind of infinite series called a *power series*, which may be thought of as a polynomial of infinite degree.

DEFINITION 7

> **Power series**
>
> A series of the form
>
> $$\sum_{n=0}^{\infty} a_n(x - c)^n = a_0 + a_1(x - c) + a_2(x - c)^2 + a_3(x - c)^3 + \cdots$$
>
> is called a **power series in powers of $x = c$** or a **power series about c**. The constants $a_0, a_1, a_2, \ldots$ are called the **coefficients** of the power series.

Since the terms of a power series are functions of a variable x, the series may or may not converge for each value of x. For those values of x for which the series does converge, the sum defines a function of x. For example, if $-1 < x < 1$, then

$$1 + x + x^2 + x^3 + \cdots = \frac{1}{1 - x}.$$

The geometric series on the left side is a power series *representation* of the function $1/(1 - x)$ in powers of x (or about 0). Note that the representation is valid only in the open interval $(-1, 1)$ even though $1/(1 - x)$ is defined for all real x except $x = 1$. For $x = -1$ and for $|x| > 1$ the series does not converge, so it cannot represent $1/(1 - x)$ at these points.

The point c is the **centre of convergence** of the power series $\sum_{n=0}^{\infty} a_n(x - c)^n$. The series certainly converges (to a_0) at $x = c$. (All the terms except possibly the first are 0.) Theorem 17 below shows that if the series converges anywhere else, then it converges on an interval (possibly infinite) centred at $x = c$, and it converges absolutely everywhere on that interval except possibly at one or both of the endpoints if the interval is finite. The geometric series

$$1 + x + x^2 + x^3 + \cdots$$

is an example of this behaviour. It has centre of convergence $c = 0$, and converges only on the interval $(-1, 1)$, centred at 0. The convergence is absolute at every point of the interval. Another example is the series

$$\sum_{n=1}^{\infty} \frac{1}{n\,2^n}(x - 5)^n = \frac{x - 5}{2} + \frac{(x - 5)^2}{2 \times 2^2} + \frac{(x - 5)^3}{3 \times 2^3} + \cdots,$$

which we discussed in Example 4 of Section 9.4. We showed that this series converges on the interval $[3, 7)$, an interval with centre $x = 5$, and that the convergence is absolute on the open interval $(3, 7)$ but is only conditional at the endpoint $x = 3$.

THEOREM 17

For any power series $\sum_{n=0}^{\infty} a_n(x - c)^n$ one of the following alternatives must hold:

(i) the series may converge only at $x = c$,

(ii) the series may converge at every real number x, or

(iii) there may exist a positive real number R such that the series converges at every x satisfying $|x - c| < R$ and diverges at every x satisfying $|x - c| > R$. In this case the series may or may not converge at either of the two *endpoints* $x = c - R$ and $x = c + R$.

In each of these cases the convergence is absolute except possibly at the endpoints $x = c - R$ and $x = c + R$ in case (iii).

PROOF   We observed above that every power series converges at its centre of convergence; only the first term can be nonzero so the convergence is absolute. To prove the rest of this theorem, it suffices to show that if the series converges at any number $x_0 \neq c$, then it converges absolutely at every number x closer to c than x_0 is, that is, at every x satisfying $|x - c| < |x_0 - c|$. This means that convergence at any $x_0 \neq c$ implies absolute convergence on $(c - x_0, c + x_0)$, so the set of points x where the series converges must be an interval centred at c.

Suppose, therefore, that $\sum_{n=0}^{\infty} a_n (x_0 - c)^n$ converges. Then $\lim a_n (x_0 - c)^n = 0$, so $|a_n (x_0 - c)^n| \leq K$ for all n, where K is some constant (Theorem 1 of Section 9.1). If $r = |x - c| / |x_0 - c| < 1$, then

$$\sum_{n=0}^{\infty} |a_n (x - c)^n| = \sum_{n=0}^{\infty} |a_n (x_0 - c)^n| \left| \frac{x - c}{x_0 - c} \right|^n \leq K \sum_{n=0}^{\infty} r^n = \frac{K}{1 - r} < \infty.$$

Thus $\sum_{n=0}^{\infty} a_n (x - c)^n$ converges absolutely.

By Theorem 17, the set of values x for which the power series $\sum_{n=0}^{\infty} a_n (x - c)^n$ converges is an interval centred at $x = c$. We call this interval the **interval of convergence** of the power series. It must be of one of the following forms:

(i) the isolated point $x = c$ (a degenerate closed interval $[c, c]$),

(ii) the entire line $(-\infty, \infty)$,

(iii) a finite interval centred at c:
$[c - R, c + R]$, or $[c - R, c + R)$, or $(c - R, c + R]$, or $(c - R, c + R)$.

The number R in (iii) is called the **radius of convergence** of the power series. In case (i) we say the radius of convergence is $R = 0$; in case (ii) it is $R = \infty$.

The radius of convergence, R, can often be found by using the ratio test on the power series: if

$$\rho = \lim_{n \to \infty} \left| \frac{a_{n+1}(x - c)^{n+1}}{a_n (x - c)^n} \right| = \left(\lim_{n \to \infty} \left| \frac{a_{n+1}}{a_n} \right| \right) |x - c|$$

exists, then the series $\sum_{n=0}^{\infty} a_n (x - c)^n$ converges absolutely where $\rho < 1$, that is, where

$$|x - c| < R = 1 \left/ \lim_{n \to \infty} \left| \frac{a_{n+1}}{a_n} \right| \right. .$$

The series diverges if $|x - c| > R$.

Radius of convergence

Suppose that $L = \lim_{n \to \infty} \left| \frac{a_{n+1}}{a_n} \right|$ exists or is ∞. Then the power series $\sum_{n=0}^{\infty} a_n (x - c)^n$ has radius of convergence $R = 1/L$. (If $L = 0$, then $R = \infty$; if $L = \infty$, then $R = 0$.)

Example 1   Determine the centre, radius, and interval of convergence of

$$\sum_{n=0}^{\infty} \frac{(2x + 5)^n}{(n^2 + 1)3^n}.$$

Solution   The series can be rewritten

$$\sum_{n=0}^{\infty} \left(\frac{2}{3}\right)^n \frac{1}{n^2+1} \left(x+\frac{5}{2}\right)^n.$$

The centre of convergence is $x = -5/2$. The radius of convergence, R, is given by

$$\frac{1}{R} = L = \lim \left| \frac{\left(\dfrac{2}{3}\right)^{n+1} \dfrac{1}{(n+1)^2+1}}{\left(\dfrac{2}{3}\right)^n \dfrac{1}{n^2+1}} \right| = \lim \frac{2}{3} \frac{n^2+1}{(n+1)^2+1} = \frac{2}{3}.$$

Thus $R = 3/2$. The series converges absolutely on $(-5/2 - 3/2, -5/2 + 3/2) = (-4, -1)$, and it diverges on $(-\infty, -4)$ and on $(-1, \infty)$. At $x = -1$ the series is $\sum_{n=0}^{\infty} 1/(n^2+1)$; at $x = -4$ it is $\sum_{n=0}^{\infty} (-1)^n/(n^2+1)$. Both series converge (absolutely). The interval of convergence of the given power series is therefore $[-4, -1]$. ∎

Example 2   Determine the radii of convergence of the series

(a) $\displaystyle\sum_{n=0}^{\infty} \frac{x^n}{n!}$ and   (b) $\displaystyle\sum_{n=0}^{\infty} n!x^n.$

Solution

(a) $L = \left| \lim \dfrac{1}{(n+1)!} \middle/ \dfrac{1}{n!} \right| = \lim \dfrac{n!}{(n+1)!} = \lim \dfrac{1}{n+1} = 0.$ Thus $R = \infty$.

This series converges (absolutely) for all x. The sum is e^x, as will be shown in Example 1 in the next section.

(b) $L = \left| \lim \dfrac{(n+1)!}{n!} \right| = \lim(n+1) = \infty.$ Thus $R = 0$.

This series converges only at its centre of convergence, $x = 0$. ∎

Algebraic Operations on Power Series

To simplify the following discussion, we will consider only power series with centre of convergence 0, that is, series of the form

$$\sum_{n=0}^{\infty} a_n x^n = a_0 + a_1 x + a_2 x^2 + a_3 x^3 + \cdots.$$

Any properties we demonstrate for such series extend automatically to power series of the form $\sum_{n=0}^{\infty} a_n(y-c)^n$ via the change of variable $x = y - c$.

First, we observe that series having the same centre of convergence can be added or subtracted on whatever interval is common to their intervals of convergence. The following theorem is a simple consequence of Theorem 7 of Section 9.2 and does not require a proof.

THEOREM 18

Let $\sum_{n=0}^{\infty} a_n x^n$ and $\sum_{n=0}^{\infty} b_n x^n$ be two power series with radii of convergence R_a and R_b, respectively, and let c be a constant. Then

(i) $\sum_{n=0}^{\infty} (ca_n) x^n$ has radius of convergence R_a, and

$$\sum_{n=0}^{\infty} (ca_n) x^n = c \sum_{n=0}^{\infty} a_n x^n$$

wherever the series on the right converges.

(ii) $\sum_{n=0}^{\infty}(a_n + b_n)\,x^n$ has radius of convergence R at least as large as the smaller of R_a and R_b ($R \ge \min\{R_a, R_b\}$), and

$$\sum_{n=0}^{\infty}(a_n + b_n)\,x^n = \sum_{n=0}^{\infty} a_n\,x^n + \sum_{n=0}^{\infty} b_n\,x^n$$

wherever both series on the right converge.

The situation regarding multiplication and division of power series is more complicated. We will mention only the results and will not attempt any proofs of our assertions. A textbook in mathematical analysis will provide more details.

Long multiplication of the form

$$(a_0 + a_1 x + a_2 x^2 + \cdots)(b_0 + b_1 x + b_2 x^2 + \cdots)$$
$$= a_0 b_0 + (a_0 b_1 + a_1 b_0)x + (a_0 b_2 + a_1 b_1 + a_2 b_0)x^2 + \cdots$$

leads us to conjecture the formula

$$\left(\sum_{n=0}^{\infty} a_n x^n\right)\left(\sum_{n=0}^{\infty} b_n x^n\right) = \sum_{n=0}^{\infty} c_n x^n,$$

where

$$c_n = a_0 b_n + a_1 b_{n-1} + \cdots + a_n b_0 = \sum_{j=0}^{n} a_j b_{n-j}.$$

The series $\sum_{n=0}^{\infty} c_n x^n$ is called the **Cauchy product** of the series $\sum_{n=0}^{\infty} a_n x^n$ and $\sum_{n=0}^{\infty} b_n x^n$. Like the sum, the Cauchy product also has radius of convergence at least equal to the lesser of those of the factor series.

Example 3 Since

$$\frac{1}{1-x} = 1 + x + x^2 + x^3 + \cdots = \sum_{n=0}^{\infty} x^n$$

holds for $-1 < x < 1$, we can determine a power series representation for $1/(1-x)^2$ by taking the Cauchy product of this series with itself. Since $a_n = b_n = 1$ for $n = 0, 1, 2, \ldots$, we have

$$c_n = \sum_{j=0}^{n} 1 = n + 1 \qquad \text{and}$$

$$\frac{1}{(1-x)^2} = 1 + 2x + 3x^2 + 4x^3 + \cdots = \sum_{n=0}^{\infty}(n+1)x^n,$$

which must also hold for $-1 < x < 1$. The same series can be obtained by direct long multiplication of the series:

$$
\begin{array}{rcccccccl}
& 1 & + & x & + & x^2 & + & x^3 & + & \cdots \\
\times & 1 & + & x & + & x^2 & + & x^3 & + & \cdots \\
\hline
& 1 & + & x & + & x^2 & + & x^3 & + & \cdots \\
& & & x & + & x^2 & + & x^3 & + & \cdots \\
& & & & & x^2 & + & x^3 & + & \cdots \\
& & & & & & & x^3 & + & \cdots \\
& & & & & & & & & \cdots \\
\hline
& 1 & + & 2x & + & 3x^2 & + & 4x^3 & + & \cdots
\end{array}
$$

Long division can also be performed on power series, but there is no simple rule for determining the coefficients of the quotient series. The radius of convergence of the quotient series is not less than the least of the three numbers R_1, R_2, and R_3, where R_1 and R_2 are the radii of convergence of the numerator and denominator series and R_3 is the distance from the centre of convergence to the nearest *complex number* where the denominator series has sum equal to 0. To illustrate this point, observe that 1 and $1 - x$ are both power series with infinite radii of convergence:

$$1 = 1 + 0x + 0x^2 + 0x^3 + \cdots \qquad \text{for all } x,$$

$$1 - x = 1 - x + 0x^2 + 0x^3 + \cdots \qquad \text{for all } x.$$

Their quotient, $1/(1-x)$, however, only has radius of convergence 1, the distance from the centre of convergence $x = 0$ to the point $x = 1$ where the denominator vanishes:

$$\frac{1}{1-x} = 1 + x + x^2 + x^3 + \cdots \qquad \text{for} \quad |x| < 1.$$

Differentiation and Integration of Power Series

If a power series has a positive radius of convergence, it can be differentiated or integrated term by term. The resulting series will converge to the appropriate derivative or integral of the sum of the original series everywhere except possibly at the endpoints of the interval of convergence of the original series. This very important fact ensures that, for purposes of calculation, power series behave just like polynomials, the easiest functions to differentiate and integrate. We formalize the differentiation and integration properties of power series in the following theorem.

THEOREM 19

Term-by-term differentiation and integration of power series

If the series $\sum_{n=0}^{\infty} a_n x^n$ converges to the sum $f(x)$ on an interval $(-R, R)$, where $R > 0$, that is,

$$f(x) = \sum_{n=0}^{\infty} a_n x^n = a_0 + a_1 x + a_2 x^2 + a_3 x^3 + \cdots, \qquad (-R < x < R),$$

then f is differentiable on $(-R, R)$ and

$$f'(x) = \sum_{n=1}^{\infty} n a_n x^{n-1} = a_1 + 2a_2 x + 3a_3 x^2 + \cdots, \qquad (-R < x < R).$$

Also, f is integrable over any closed subinterval of $(-R, R)$, and if $|x| < R$, then

$$\int_0^x f(t)\, dt = \sum_{n=0}^{\infty} \frac{a_n}{n+1} x^{n+1} = a_0 x + \frac{a_1}{2} x^2 + \frac{a_2}{3} x^3 + \cdots.$$

PROOF Let x satisfy $-R < x < R$ and choose $H > 0$ such that $|x| + H < R$. By Theorem 17 we then have[1]

$$\sum_{n=1}^{\infty} |a_n|(|x| + H)^n = K < \infty.$$

The Binomial Theorem (see Section 9.9) shows that if $n \geq 1$, then

$$(x + h)^n = x^n + nx^{n-1} h + \sum_{k=2}^{n} \binom{n}{k} x^{n-k} h^k.$$

While understanding the statement of this theorem is very important for what follows, understanding the proof is not. Feel free to skip the proof and go on to the applications.

---

[1] This proof is due to R. Výborný, *American Mathematical Monthly*, April 1987.

Therefore, if $|h| \le H$ we have

$$
\begin{aligned}
|(x+h)^n - x^n - nx^{n-1}h| &= \left| \sum_{k=2}^{n} \binom{n}{k} x^{n-k} h^k \right| \\
&\le \sum_{k=2}^{n} \binom{n}{k} |x|^{n-k} \frac{|h|^k}{H^k} H^k \\
&\le \frac{|h|^2}{H^2} \sum_{k=0}^{n} \binom{n}{k} |x|^{n-k} H^k \\
&= \frac{|h|^2}{H^2} (|x| + H)^n .
\end{aligned}
$$

Also,

$$
|nx^{n-1}| = \frac{n|x|^{n-1} H}{H} \le \frac{1}{H} (|x| + H)^n .
$$

Thus,

$$
\sum_{n=1}^{\infty} |na_n x^{n-1}| \le \frac{1}{H} \sum_{n=1}^{\infty} |a_n| (|x| + H)^n = \frac{K}{H} < \infty ,
$$

so the series $\sum_{n=1}^{\infty} na_n x^{n-1}$ converges (absolutely) to $g(x)$, say. Now

$$
\begin{aligned}
\left| \frac{f(x+h) - f(x)}{h} - g(x) \right| &= \left| \sum_{n=1}^{\infty} \frac{a_n(x+h)^n - a_n x^n - na_n x^{n-1}h}{h} \right| \\
&\le \frac{1}{|h|} \sum_{n=1}^{\infty} |a_n| |(x+h)^n - x^n - nx^{n-1}h| \\
&\le \frac{|h|}{H^2} \sum_{n=1}^{\infty} |a_n| (|x| + H)^n \le \frac{K|h|}{H^2} .
\end{aligned}
$$

Letting h approach zero, we obtain $|f'(x) - g(x)| \le 0$, so $f'(x) = g(x)$, as required.

Now observe that since $|a_n/(n+1)| \le |a_n|$, the series

$$
h(x) = \sum_{n=0}^{\infty} \frac{a_n}{n+1} x^{n+1}
$$

converges (absolutely) at least on the interval $(-R, R)$. Using the differentiation result proved above, we obtain

$$
h'(x) = \sum_{n=0}^{\infty} a_n x^n = f(x).
$$

Since $h(0) = 0$, we have

$$
\int_0^x f(t)\, dt = \int_0^x h'(t)\, dt = h(t) \Big|_0^x = h(x),
$$

as required.

Together, these results imply that the termwise differentiated or integrated series have the same radius of convergence as the given series. In fact, as the following examples illustrate, the interval of convergence of the differentiated series is the same as that of the original series except for the *possible* loss of one or both endpoints if the original series converges at endpoints of its interval of convergence. Similarly, the integrated series will converge everywhere on the interval of convergence of the original series and possibly at one or both endpoints of that interval, even if the original series does not converge at the endpoints.

Being differentiable on $(-R, R)$, where R is the radius of convergence, the sum $f(x)$ of a power series is necessarily continuous on that open interval. If the series happens to converge at either or both of the endpoints $-R$ and R, then f is also continuous (on one side) up to these endpoints. This result is stated formally in the following theorem. We will not prove it here; the interested reader is referred to textbooks on mathematical analysis for a proof.

THEOREM 20

Abel's Theorem

The sum of a power series is a continuous function everywhere on the interval of convergence of the series. In particular, if $\sum_{n=0}^{\infty} a_n R^n$ converges for some $R > 0$, then

$$\lim_{x \to R-} \sum_{n=0}^{\infty} a_n x^n = \sum_{n=0}^{\infty} a_n R^n,$$

and if $\sum_{n=0}^{\infty} a_n(-R)^n$ converges, then

$$\lim_{x \to -R+} \sum_{n=0}^{\infty} a_n x^n = \sum_{n=0}^{\infty} a_n(-R)^n.$$

The following examples show how the above theorems are applied to obtain power series representations for functions.

Example 4 Find power series representations for the functions

(a) $\dfrac{1}{(1-x)^2}$, (b) $\dfrac{1}{(1-x)^3}$, and (c) $\ln(1+x)$

by starting with the geometric series

$$\frac{1}{1-x} = \sum_{n=0}^{\infty} x^n = 1 + x + x^2 + x^3 + \cdots \qquad (-1 < x < 1)$$

and using differentiation, integration, and substitution. Where is each series valid?

Solution

(a) Differentiate the geometric series term by term to obtain

$$\frac{1}{(1-x)^2} = \sum_{n=1}^{\infty} n x^{n-1} = 1 + 2x + 3x^2 + 4x^3 + \cdots \qquad (-1 < x < 1).$$

This is the same result obtained by multiplication of series in Example 3 above.

(b) Differentiate again to get, for $-1 < x < 1$,

$$\frac{2}{(1-x)^3} = \sum_{n=2}^{\infty} n(n-1) x^{n-2} = (1 \times 2) + (2 \times 3)x + (3 \times 4)x^2 + \cdots.$$

Now divide by 2:

$$\frac{1}{(1-x)^3} = \sum_{n=2}^{\infty} \frac{n(n-1)}{2} x^{n-2} = 1 + 3x + 6x^2 + 10x^3 + \cdots \quad (-1 < x < 1).$$

(c) Substitute $-t$ in place of x in the original geometric series:

$$\frac{1}{1+t} = \sum_{n=0}^{\infty} (-1)^n t^n = 1 - t + t^2 - t^3 + t^4 - \cdots \quad (-1 < t < 1).$$

Integrate from 0 to x, where $|x| < 1$, to get

$$\ln(1+x) = \int_0^x \frac{dt}{1+t} = \sum_{n=0}^{\infty} (-1)^n \int_0^x t^n \, dt$$

$$= \sum_{n=0}^{\infty} (-1)^n \frac{x^{n+1}}{n+1} = x - \frac{x^2}{2} + \frac{x^3}{3} - \frac{x^4}{4} + \cdots \quad (-1 < x \le 1).$$

Note that the latter series converges (conditionally) at the endpoint $x = 1$ as well as on the interval $-1 < x < 1$. Since $\ln(1 + x)$ is continuous at $x = 1$, Theorem 20 assures us that the series must converge to that function at $x = 1$ also. In particular, therefore, the alternating harmonic series converges to $\ln 2$:

$$\ln 2 = 1 - \frac{1}{2} + \frac{1}{3} - \frac{1}{4} + \frac{1}{5} - \cdots = \sum_{n=0}^{\infty} \frac{(-1)^n}{n+1}.$$

This would not, however, be a very useful formula for calculating the value of $\ln 2$. (Why not?)

Example 5 Use the geometric series of the previous example to find a power series representation for $\tan^{-1} x$.

Solution Substitute $-t^2$ for x in the geometric series. Since $0 \le t^2 < 1$ whenever $-1 < t < 1$, we obtain

$$\frac{1}{1+t^2} = 1 - t^2 + t^4 - t^6 + t^8 - \cdots \quad (-1 < t < 1).$$

Now integrate from 0 to x, where $|x| < 1$:

$$\tan^{-1} x = \int_0^x \frac{dt}{1+t^2} = \int_0^x (1 - t^2 + t^4 - t^6 + t^8 - \cdots) \, dt$$

$$= x - \frac{x^3}{3} + \frac{x^5}{5} - \frac{x^7}{7} + \frac{x^9}{9} - \cdots$$

$$= \sum_{n=0}^{\infty} (-1)^n \frac{x^{2n+1}}{2n+1} \quad (-1 < x < 1).$$

However, note that the series also converges (conditionally) at $x = -1$ and 1. Since $\tan^{-1}$ is continuous at ± 1, the above series representation for $\tan^{-1} x$ also holds for these values, by Theorem 20. Letting $x = 1$ we get another interesting result:

$$\frac{\pi}{4} = 1 - \frac{1}{3} + \frac{1}{5} - \frac{1}{7} + \frac{1}{9} - \cdots.$$

Again, however, this would not be a good formula with which to calculate a numerical value of π. (Why not?)

Example 6 Find the sum of the series $\displaystyle\sum_{n=1}^{\infty} \frac{n^2}{2^n}$ by first finding the sum of the power series

$$\sum_{n=1}^{\infty} n^2 x^n = x + 4x^2 + 9x^3 + 16x^4 + \cdots.$$

Solution Observe in Example 4(a) how the process of differentiating the geometric series produces a series with coefficients 1, 2, 3, ... Start with the series obtained for $1/(1-x)^2$ and multiply it by x to obtain

$$\sum_{n=1}^{\infty} nx^n = x + 2x^2 + 3x^3 + 4x^4 + \cdots = \frac{x}{(1-x)^2}.$$

Now differentiate again to get a series with coefficients 1^2, 2^2, 3^2, ...:

$$\sum_{n=1}^{\infty} n^2 x^{n-1} = 1 + 4x + 9x^2 + 16x^3 + \cdots = \frac{d}{dx} \frac{x}{(x-1)^2} = \frac{1+x}{(1-x)^3}.$$

Multiplication by x again gives the desired power series:

$$\sum_{n=1}^{\infty} n^2 x^n = x + 4x^2 + 9x^3 + 16x^4 + \cdots = \frac{x(1+x)}{(1-x)^3}.$$

Differentiation and multiplication by x do not change the radius of convergence, so this series converges to the indicated function for $-1 < x < 1$. Putting $x = 1/2$, we get

$$\sum_{n=1}^{\infty} \frac{n^2}{2^n} = \frac{\frac{1}{2} \times \frac{3}{2}}{\frac{1}{8}} = 6.$$

The following example illustrates how substitution can be used to obtain power series representations of functions with centres of convergence different from 0.

Example 7 Find a series representation of $f(x) = 1/(2+x)$ in powers of $x - 1$. What is the interval of convergence of this series?

Solution Let $t = x - 1$ so that $x = t + 1$. We have

$$\frac{1}{2+x} = \frac{1}{3+t} = \frac{1}{3} \frac{1}{1 + \frac{t}{3}}$$

$$= \frac{1}{3} \left(1 - \frac{t}{3} + \frac{t^2}{3^2} - \frac{t^3}{3^3} + \cdots \right) \qquad (-1 < t/3 < 1)$$

$$= \sum_{n=0}^{\infty} (-1)^n \frac{t^n}{3^{n+1}} \qquad (-3 < t < 3)$$

$$= \sum_{n=0}^{\infty} (-1)^n \frac{(x-1)^n}{3^{n+1}} \qquad (-2 < x < 4).$$

Note that the radius of convergence of this series is 3, the distance from the centre of convergence, 1, to the point -2 where the denominator is 0. We could have predicted this in advance.

Maple Calculations

Maple can find the sums of many kinds of series, including absolutely and conditionally convergent numerical series and many power series. Even when Maple can't find the formal sum of a (convergent) series, it can provide a decimal approximation to the precision indicated by the current value of its variable `Digits`, which defaults to 10. Here are some examples.

```
>   sum(n^4/2^n, n=1..infinity);
```

$$150$$

```
>   sum(1/n^2, n=1..infinity);
```

$$\frac{1}{6}\pi^2$$

```
>   sum(exp(-n^2), n=0..infinity);
```

$$\sum_{n=0}^{\infty} e^{(-n^2)}$$

```
>   evalf(%);
```

$$1.386\,318\,602$$

```
>   f := x -> sum(x^(n-1)/n, n=1..infinity);
```

$$f := x \rightarrow \sum_{n=1}^{\infty}\left(\frac{x^{(n-1)}}{n}\right)$$

```
>   f(1); f(-1); f(1/2);
```

$$\infty$$
$$\ln(2)$$
$$2\ln(2)$$

Exercises 9.5

Determine the centre, radius, and interval of convergence of each of the power series in Exercises 1–8.

1. $\displaystyle\sum_{n=0}^{\infty} \frac{x^{2n}}{\sqrt{n+1}}$

2. $\displaystyle\sum_{n=0}^{\infty} 3n\,(x+1)^n$

3. $\displaystyle\sum_{n=1}^{\infty} \frac{1}{n}\left(\frac{x+2}{2}\right)^n$

4. $\displaystyle\sum_{n=1}^{\infty} \frac{(-1)^n}{n^4 2^{2n}}\,x^n$

5. $\displaystyle\sum_{n=0}^{\infty} n^3 (2x-3)^n$

6. $\displaystyle\sum_{n=1}^{\infty} \frac{e^n}{n^3}\,(4-x)^n$

7. $\displaystyle\sum_{n=0}^{\infty} \frac{(1+5^n)}{n!}\,x^n$

8. $\displaystyle\sum_{n=1}^{\infty} \frac{(4x-1)^n}{n^n}$

9. Use multiplication of series to find a power series representation of $1/(1-x)^3$ valid in the interval $(-1, 1)$.

10. Determine the Cauchy product of the series $1 + x + x^2 + x^3 + \cdots$ and $1 - x + x^2 - x^3 + \cdots$. On what interval and to what function does the product series converge?

11. Determine the power series expansion of $1/(1-x)^2$ by formally dividing $1 - 2x + x^2$ into 1.

Starting with the power series representation

$$\frac{1}{1-x} = 1 + x + x^2 + x^3 + \cdots, \qquad (-1 < x < 1),$$

determine power series representations for the functions indicated in Exercises 12–20. On what interval is each representation valid?

12. $\dfrac{1}{2-x}$ in powers of x

13. $\dfrac{1}{(2-x)^2}$ in powers of x

14. $\dfrac{1}{1+2x}$ in powers of x

15. $\ln(2-x)$ in powers of x

16. $\dfrac{1}{x}$ in powers of $x - 1$

17. $\dfrac{1}{x^2}$ in powers of $x + 2$

18. $\dfrac{1-x}{1+x}$ in powers of x

19. $\dfrac{x^3}{1-2x^2}$ in powers of x

20. $\ln x$ in powers of $x - 4$

Determine the interval of convergence and the sum of each of the series in Exercises 21–26.

21. $1 - 4x + 16x^2 - 64x^3 + \cdots = \displaystyle\sum_{n=0}^{\infty}(-1)^n(4x)^n$

*22. $3 + 4x + 5x^2 + 6x^3 + \cdots = \sum_{n=0}^{\infty} (n+3)x^n$

*23. $\dfrac{1}{3} + \dfrac{x}{4} + \dfrac{x^2}{5} + \dfrac{x^3}{6} + \cdots = \sum_{n=0}^{\infty} \dfrac{x^n}{n+3}$

*24. $1 \times 3 - 2 \times 4x + 3 \times 5x^2 - 4 \times 6x^3 + \cdots$
$$= \sum_{n=0}^{\infty} (-1)^n (n+1)(n+3)\, x^n$$

*25. $2 + 4x^2 + 6x^4 + 8x^6 + 10x^8 + \cdots = \sum_{n=0}^{\infty} 2(n+1)\, x^{2n}$

*26. $1 - \dfrac{x^2}{2} + \dfrac{x^4}{3} - \dfrac{x^6}{4} + \dfrac{x^8}{5} - \cdots = \sum_{n=0}^{\infty} \dfrac{(-1)^n x^{2n}}{n+1}$

Use the technique (or the result) of Example 6 to find the sums of the numerical series in Exercises 27–32.

27. $\displaystyle\sum_{n=1}^{\infty} \dfrac{n}{3^n}$

28. $\displaystyle\sum_{n=0}^{\infty} \dfrac{n+1}{2^n}$

*29. $\displaystyle\sum_{n=0}^{\infty} \dfrac{(n+1)^2}{\pi^n}$

*30. $\displaystyle\sum_{n=1}^{\infty} \dfrac{(-1)^n n(n+1)}{2^n}$

31. $\displaystyle\sum_{n=1}^{\infty} \dfrac{(-1)^{n-1}}{n2^n}$

32. $\displaystyle\sum_{n=3}^{\infty} \dfrac{1}{n2^n}$

9.6 Taylor and Maclaurin Series

If a power series $\sum_{n=0}^{\infty} a_n (x-c)^n$ has a positive radius of convergence R, then the sum of the series defines a function $f(x)$ on the interval $(c-R, c+R)$. We say that the power series is a **representation** of $f(x)$ on that interval. What relationship exists between the function $f(x)$ and the coefficients $a_0,\ a_1,\ a_2,\ \ldots$ of the power series? The following theorem answers this question.

THEOREM 21

Suppose the series

$$f(x) = \sum_{n=0}^{\infty} a_n (x-c)^n = a_0 + a_1(x-c) + a_2(x-c)^2 + a_3(x-c)^3 + \cdots$$

converges to $f(x)$ for $c - R < x < c + R$, where $R > 0$. Then

$$a_k = \dfrac{f^{(k)}(c)}{k!} \qquad \text{for } k = 0, 1, 2, 3, \ldots.$$

PROOF   This proof requires that we differentiate the series for $f(x)$ term by term several times, a process justified by Theorem 19 (suitably reformulated for powers of $x - c$):

$$f'(x) = \sum_{n=1}^{\infty} n a_n (x-c)^{n-1} = a_1 + 2a_2(x-c) + 3a_3(x-c)^2 + \cdots$$

$$f''(x) = \sum_{n=2}^{\infty} n(n-1) a_n (x-c)^{n-2} = 2a_2 + 6a_3(x-c) + 12a_4(x-c)^2 + \cdots$$

$$\vdots$$

$$f^{(k)}(x) = \sum_{n=k}^{\infty} n(n-1)(n-2)\cdots(n-k+1) a_n (x-c)^{n-k}$$

$$= k! a_k + \dfrac{(k+1)!}{1!} a_{k+1}(x-c) + \dfrac{(k+2)!}{2!} a_{k+2}(x-c)^2 + \cdots.$$

Each series converges for $c - R < x < c + R$. Setting $x = c$, we obtain $f^{(k)}(c) = k! a_k$, which proves the theorem.

Theorem 21 shows that a function $f(x)$ that has a power series representation with centre at c and positive radius of convergence must have derivatives of all orders in an interval around $x = c$, and it can have only one representation as a power series in powers of $x - c$, namely

$$f(x) = \sum_{n=0}^{\infty} \frac{f^{(n)}(c)}{n!}(x - c)^n = f(c) + f'(c)(x - c) + \frac{f''(c)}{2!}(x - c)^2 + \cdots.$$

Such a series is called a Taylor series or, if $c = 0$, a Maclaurin series.

DEFINITION 8

Taylor and Maclaurin series

If $f(x)$ has derivatives of all orders at $x = c$ (i.e., if $f^{(k)}(c)$ exists for $k = 0, 1, 2, 3, \ldots$), then the series

$$\sum_{k=0}^{\infty} \frac{f^{(k)}(c)}{k!}(x - c)^k$$

$$= f(c) + f'(c)(x - c) + \frac{f''(c)}{2!}(x - c)^2 + \frac{f^{(3)}(c)}{3!}(x - c)^3 + \cdots$$

is called the **Taylor series of f about c** (or the **Taylor series of f in powers of $x - c$**). If $c = 0$, the term **Maclaurin series** is usually used in place of Taylor series.

Note that the partial sums of such Taylor (or Maclaurin) series are just the Taylor (or Maclaurin) polynomials studied in Section 4.8.

The Taylor series is a power series as defined in the previous section. Theorem 17 implies that c must be the centre of any interval on which such a series converges, but the definition of Taylor series makes no requirement that the series should converge anywhere except at the point $x = c$ where the series is just $f(c) + 0 + 0 + \cdots$. The series exists provided all the derivatives of f exist at $x = c$; in practice this means that each derivative must exist in an open interval containing $x = c$. (Why?) However, the series may converge nowhere except at $x = c$, and if it does converge elsewhere, it may converge to something other than $f(x)$. (See Exercise 40 at the end of this section for an example where this happens.) If the Taylor series does converge to $f(x)$ in an open interval containing c, then we will say that f is analytic at c.

DEFINITION 9

Analytic functions

A function f is **analytic at c** if f has a Taylor series at c and that series converges to $f(x)$ in an open interval containing c. If f is analytic at each point of an open interval, then we say it is analytic on that interval.

Most, but not all, of the elementary functions encountered in calculus are analytic wherever they have derivatives of all orders. On the other hand, whenever a power series in powers of $x - c$ converges for all x in an open interval containing c, then its sum $f(x)$ is analytic at c, and the given series is the Taylor series of f about c.

Maclaurin Series for Some Elementary Functions

Calculating Taylor and Maclaurin series for a function f directly from Definition 8 is practical only when we can find a formula for the nth derivative of f. Examples of such functions include $(ax + b)^r$, e^{ax+b}, $\ln(ax + b)$, $\sin(ax + b)$, $\cos(ax + b)$, and sums of such functions.

Example 1   Find the Taylor series for e^x about $x = c$. Where does the series converge to e^x? Where is e^x analytic? What is the Maclaurin series for e^x?

Solution   Since all the derivatives of $f(x) = e^x$ are e^x, we have $f^{(n)}(c) = e^c$ for every integer $n \geq 0$. Thus the Taylor series for e^x about $x = c$ is

$$\sum_{n=0}^{\infty} \frac{e^c}{n!}(x - c)^n = e^c + e^c(x - c) + \frac{e^c}{2!}(x - c)^2 + \frac{e^c}{3!}(x - c)^3 + \cdots.$$

The radius of convergence R of this series is given by

$$\frac{1}{R} = \lim_{n \to \infty} \left| \frac{e^c/(n + 1)!}{e^c/n!} \right| = \lim_{n \to \infty} \frac{n!}{(n + 1)!} = \lim_{n \to \infty} \frac{1}{n + 1} = 0.$$

Thus the radius of convergence is $R = \infty$ and the series converges for all x.
Suppose the sum is $g(x)$:

$$g(x) = e^c + e^c(x - c) + \frac{e^c}{2!}(x - c)^2 + \frac{e^c}{3!}(x - c)^3 + \cdots.$$

By Theorem 19, we have

$$g'(x) = 0 + e^c + \frac{e^c}{2!}2(x - c) + \frac{e^c}{3!}3(x - c)^2 + \cdots$$

$$= e^c + e^c(x - c) + \frac{e^c}{2!}(x - c)^2 + \cdots = g(x).$$

Also, $g(c) = e^c + 0 + 0 + \cdots = e^c$. Since $g(x)$ satisfies the differential equation $g'(x) = g(x)$ of exponential growth, we have $g(x) = Ce^x$. Substituting $x = c$ gives $e^c = g(c) = Ce^c$, so $C = 1$. Thus the Taylor series for e^x in powers of $x - c$ converges to e^x for every real number x:

$$e^x = \sum_{n=0}^{\infty} \frac{e^c}{n!}(x - c)^n$$

$$= e^c + e^c(x - c) + \frac{e^c}{2!}(x - c)^2 + \frac{e^c}{3!}(x - c)^3 + \cdots \qquad \text{(for all } x\text{)}.$$

In particular, e^x is analytic on the whole real line $\mathbb{R}$. Setting $c = 0$ we obtain the Maclaurin series for e^x:

$$e^x = \sum_{n=0}^{\infty} \frac{x^n}{n!} = 1 + x + \frac{x^2}{2!} + \frac{x^3}{3!} + \cdots \qquad \text{(for all } x\text{)}.$$

Example 2   Find the Maclaurin series for (a) $\sin x$ and (b) $\cos x$. Where does each series converge?

Solution   Let $f(x) = \sin x$. Then we have $f(0) = 0$ and

$$f'(x) = \cos x \qquad\qquad f'(0) = 1$$
$$f''(x) = -\sin x \qquad\qquad f''(0) = 0$$
$$f^{(3)}(x) = -\cos x \qquad\qquad f^{(3)}(0) = -1$$
$$f^{(4)}(x) = \sin x \qquad\qquad f^{(4)}(0) = 0$$
$$f^{(5)}(x) = \cos x \qquad\qquad f^{(5)}(0) = 1$$

$$\vdots \qquad\qquad\qquad\qquad \vdots$$

Thus, the Maclaurin series for $\sin x$ is

$$g(x) = 0 + x + 0 - \frac{x^3}{3!} + 0 + \frac{x^5}{5!} + 0 - \cdots$$

$$= x - \frac{x^3}{3!} + \frac{x^5}{5!} - \frac{x^7}{7!} + \cdots = \sum_{n=0}^{\infty} \frac{(-1)^n}{(2n+1)!} x^{2n+1}.$$

We have denoted the sum by $g(x)$ since we don't yet know whether the series converges to $\sin x$. The series does converge for all x by the ratio test:

$$\lim_{n \to \infty} \left| \frac{\dfrac{(-1)^{n+1}}{(2(n+1)+1)!} x^{2(n+1)+1}}{\dfrac{(-1)^n}{(2n+1)!} x^{2n+1}} \right| = \lim_{n \to \infty} \frac{(2n+1)!}{(2n+3)!} |x|^2$$

$$= \lim_{n \to \infty} \frac{|x|^2}{(2n+3)(2n+2)} = 0.$$

Now we can differentiate the function $g(x)$ twice to get

$$g'(x) = 1 - \frac{x^2}{2!} + \frac{x^4}{4!} - \frac{x^6}{6!} + \cdots$$

$$g''(x) = -x + \frac{x^3}{3!} - \frac{x^5}{5!} + \frac{x^7}{7!} - \cdots = -g(x).$$

Thus, $g(x)$ satisfies the differential equation $g''(x) + g(x) = 0$ of simple harmonic motion. The general solution of this equation, as observed in Section 3.7, is

$$g(x) = A \cos x + B \sin x.$$

Observe, from the series, that $g(0) = 0$ and $g'(0) = 1$. These values determine that $A = 0$ and $B = 1$. Thus, $g(x) = \sin x$ and $g'(x) = \cos x$ for all x.

We have therefore demonstrated that

$$\sin x = \sum_{n=0}^{\infty} \frac{(-1)^n}{(2n+1)!} x^{2n+1} = x - \frac{x^3}{3!} + \frac{x^5}{5!} - \frac{x^7}{7!} + \cdots \quad \text{(for all } x\text{)},$$

$$\cos x = \sum_{n=0}^{\infty} \frac{(-1)^n}{(2n)!} x^{2n} = 1 - \frac{x^2}{2!} + \frac{x^4}{4!} - \frac{x^6}{6!} + \cdots \quad \text{(for all } x\text{)}.$$

Theorem 21 shows that we can use any available means to find a power series converging to a given function on an interval, and the series obtained will turn out to be the Taylor series. In Section 9.5 several series were constructed by manipulating a geometric series. These include:

Some Maclaurin series

$$\frac{1}{1-x} = \sum_{n=0}^{\infty} x^n = 1 + x + x^2 + x^3 + \cdots \qquad (-1 < x < 1)$$

$$\frac{1}{(1-x)^2} = \sum_{n=1}^{\infty} n x^{n-1} = 1 + 2x + 3x^2 + 4x^3 + \cdots \qquad (-1 < x < 1)$$

$$\ln(1+x) = \sum_{n=1}^{\infty} \frac{(-1)^{n-1}}{n} x^n = x - \frac{x^2}{2} + \frac{x^3}{3} - \frac{x^4}{4} + \cdots \qquad (-1 < x \le 1)$$

$$\tan^{-1} x = \sum_{n=0}^{\infty} \frac{(-1)^n}{2n+1} x^{2n+1} = x - \frac{x^3}{3} + \frac{x^5}{5} - \frac{x^7}{7} + \cdots \qquad (-1 \le x \le 1)$$

These series, together with the intervals on which they converge, are frequently used hereafter and should be memorized.

Other Maclaurin and Taylor Series

Series can be combined in various ways to generate new series. For example, we can find the Maclaurin series for e^{-x} by replacing x with $-x$ in the series for e^x:

$$e^{-x} = \sum_{n=0}^{\infty} \frac{(-1)^n}{n!} x^n = 1 - x + \frac{x^2}{2!} - \frac{x^3}{3!} + \cdots \quad \text{(for all x)}.$$

The series for e^x and e^{-x} can then be subtracted or added and the results divided by 2 to obtain Maclaurin series for the hyperbolic functions $\sinh x$ and $\cosh x$:

$$\sinh x = \frac{e^x - e^{-x}}{2} = \sum_{n=0}^{\infty} \frac{x^{2n+1}}{(2n+1)!} = x + \frac{x^3}{3!} + \frac{x^5}{5!} + \cdots \quad \text{(for all x)}$$

$$\cosh x = \frac{e^x + e^{-x}}{2} = \sum_{n=0}^{\infty} \frac{x^{2n}}{(2n)!} = 1 + \frac{x^2}{2!} + \frac{x^4}{4!} + \cdots \quad \text{(for all x)}.$$

Remark Observe the similarity between the series for $\sin x$ and $\sinh x$ and between those for $\cos x$ and $\cosh x$. If we were to allow complex numbers (numbers of the form $z = x + iy$, where $i^2 = -1$ and x and y are real; see Appendix I) as arguments for our functions, and if we were to demonstrate that our operations on series could be extended to series of complex numbers, we would see that $\cos x = \cosh(ix)$ and $\sin x = -i \sinh(ix)$. In fact, $e^{ix} = \cos x + i \sin x$ and $e^{-ix} = \cos x - i \sin x$, so

$$\cos x = \frac{e^{ix} + e^{-ix}}{2}, \quad \text{and} \quad \sin x = \frac{e^{ix} - e^{-ix}}{2i}.$$

Such formulas are encountered in the study of functions of a complex variable (see Appendix II); from the complex point of view the trigonometric and exponential functions are just different manifestations of the same basic function, a complex exponential $e^z = e^{x+iy}$. We content ourselves here with having mentioned the interesting relationships above and invite the reader to verify them formally by calculating with series. (Such formal calculations do not, of course, constitute a proof, since we have not established the various rules covering series of complex numbers.)

Example 3 Obtain Maclaurin series for the following functions:

(a) $e^{-x^2/3}$, (b) $\dfrac{\sin(x^2)}{x}$, (c) $\sin^2 x$.

Solution

(a) We substitute $-x^2/3$ for x in the Maclaurin series for e^x:

$$e^{-x^2/3} = 1 - \frac{x^2}{3} + \frac{1}{2!}\left(\frac{x^2}{3}\right)^2 - \frac{1}{3!}\left(\frac{x^2}{3}\right)^3 + \cdots$$

$$= \sum_{n=0}^{\infty} (-1)^n \frac{1}{3^n n!} x^{2n} \quad \text{(for all real x)}.$$

(b) For all $x \neq 0$ we have

$$\frac{\sin x^2}{x} = \frac{1}{x}\left(x^2 - \frac{(x^2)^3}{3!} + \frac{(x^2)^5}{5!} - \cdots\right)$$

$$= x - \frac{x^5}{3!} + \frac{x^9}{5!} - \cdots = \sum_{n=0}^{\infty} (-1)^n \frac{x^{4n+1}}{(2n+1)!}.$$

Note that $f(x) = (\sin(x^2))/x$ is not defined at $x = 0$ but does have a limit (namely 0) as x approaches 0. If we define $f(0) = 0$ (the continuous extension of $f(x)$ to $x = 0$), then the series converges to $f(x)$ for all x.

(c) We use a trigonometric identity to express $\sin^2 x$ in terms of $\cos 2x$ and then use the Maclaurin series for $\cos x$ with x replaced by $2x$.

$$
\begin{aligned}
\sin^2 x &= \frac{1 - \cos 2x}{2} = \frac{1}{2} - \frac{1}{2}\left(1 - \frac{(2x)^2}{2!} + \frac{(2x)^4}{4!} - \cdots\right) \\
&= \frac{1}{2}\left(\frac{(2x)^2}{2!} - \frac{(2x)^4}{4!} + \frac{(2x)^6}{6!} - \cdots\right) \\
&= \sum_{n=0}^{\infty} (-1)^n \frac{2^{2n+1}}{(2n+2)!} x^{2n+2} \qquad \text{(for all real } x\text{).}
\end{aligned}
$$

Taylor series about points other than 0 can often be obtained from known Maclaurin series by a change of variable.

Example 4 Find the Taylor series for $\ln x$ in powers of $x - 2$. Where does the series converge to $\ln x$?

Solution Note that if $t = (x - 2)/2$, then

$$
\ln x = \ln(2 + (x - 2)) = \ln\left[2\left(1 + \frac{x-2}{2}\right)\right] = \ln 2 + \ln(1 + t).
$$

We use the known Maclaurin series for $\ln(1 + t)$:

$$
\begin{aligned}
\ln x &= \ln 2 + \ln(1 + t) \\
&= \ln 2 + t - \frac{t^2}{2} + \frac{t^3}{3} - \frac{t^4}{4} - \cdots \\
&= \ln 2 + \frac{x-2}{2} - \frac{(x-2)^2}{2 \times 2^2} + \frac{(x-2)^3}{3 \times 2^3} - \frac{(x-2)^4}{4 \times 2^4} + \cdots \\
&= \ln 2 + \sum_{n=1}^{\infty} \frac{(-1)^{n-1}}{n\, 2^n} (x - 2)^n.
\end{aligned}
$$

Since the series for $\ln(1 + t)$ is valid for $-1 < t \le 1$, this series for $\ln x$ is valid for $-1 < (x - 2)/2 \le 1$, that is, for $0 < x \le 4$.

Example 5 Find the Taylor series for $\cos x$ about $\pi/3$. Where is the series valid?

Solution We use the addition formula for cosine:

$$
\begin{aligned}
\cos x &= \cos\left(x - \frac{\pi}{3} + \frac{\pi}{3}\right) = \cos\left(x - \frac{\pi}{3}\right)\cos\frac{\pi}{3} - \sin\left(x - \frac{\pi}{3}\right)\sin\frac{\pi}{3} \\
&= \frac{1}{2}\left[1 - \frac{1}{2!}\left(x - \frac{\pi}{3}\right)^2 + \frac{1}{4!}\left(x - \frac{\pi}{3}\right)^4 - \cdots\right] \\
&\quad - \frac{\sqrt{3}}{2}\left[\left(x - \frac{\pi}{3}\right) - \frac{1}{3!}\left(x - \frac{\pi}{3}\right)^3 + \cdots\right] \\
&= \frac{1}{2} - \frac{\sqrt{3}}{2}\left(x - \frac{\pi}{3}\right) - \frac{1}{2}\frac{1}{2!}\left(x - \frac{\pi}{3}\right)^2 + \frac{\sqrt{3}}{2}\frac{1}{3!}\left(x - \frac{\pi}{3}\right)^3 \\
&\quad + \frac{1}{2}\frac{1}{4!}\left(x - \frac{\pi}{3}\right)^4 - \cdots.
\end{aligned}
$$

This series representation is valid for all x. A similar calculation would enable us to expand $\cos x$ or $\sin x$ in powers of $x - c$ for any real c; both functions are analytic at every point of the real line.

Sometimes it is quite difficult, if not impossible, to find a formula for the general term of a Maclaurin or Taylor series. In such cases it is usually possible to obtain the first few terms before the calculations get too cumbersome. Had we attempted to solve Example 3(c) by multiplying the series for $\sin x$ by itself we might have found ourselves in this bind. Other examples occur when it is necessary to substitute one series into another or to divide one by another.

Example 6 Obtain the first three nonzero terms of the Maclaurin series for
(a) $\tan x$ and (b) $\ln \cos x$.

Solution

(a) $\tan x = (\sin x)/(\cos x)$. We can obtain the first three terms of the Maclaurin series for $\tan x$ by long division of the series for $\cos x$ into that for $\sin x$:

$$
\begin{array}{r}
x \;+\; \dfrac{x^3}{3} \;+\; \dfrac{2}{15}x^5 \;+\; \cdots \\[4pt]
\hline
\end{array}
$$

$$
1 \;-\; \frac{x^2}{2} \;+\; \frac{x^4}{24} \;\Big)\; x \;-\; \frac{x^3}{6} \;+\; \frac{x^5}{120} \;-\; \cdots
$$

$$
\begin{array}{r}
x \;-\; \dfrac{x^3}{2} \;+\; \dfrac{x^5}{24} \;-\; \cdots \\[4pt]
\hline
\dfrac{x^3}{3} \;-\; \dfrac{x^5}{30} \;+\; \cdots \\[4pt]
\dfrac{x^3}{3} \;-\; \dfrac{x^5}{6} \;+\; \cdots \\[4pt]
\hline
\dfrac{2x^5}{15} \;-\; \cdots \\[4pt]
\dfrac{2x^5}{15} \;-\; \cdots \\[4pt]
\hline
\end{array}
$$

Thus $\tan x = x + \dfrac{1}{3}x^3 + \dfrac{2}{15}x^5 + \cdots$.

We cannot easily find all the terms of the series; only with considerable computational effort can we find many more terms than we have already found. This Maclaurin series for $\tan x$ converges for $|x| < \pi/2$, but we cannot demonstrate this fact by the techniques we have at our disposal now. (It is true because the complex number $z = x + iy$ closest to 0 where the "denominator" of $\tan z$, that is $\cos z$, is zero, is, in fact, the real value $z = \pi/2$.)

(b) $\ln \cos x = \ln \left(1 + \left(-\dfrac{x^2}{2!} + \dfrac{x^4}{4!} - \dfrac{x^6}{6!} + \cdots \right) \right)$

$\qquad = \left(-\dfrac{x^2}{2!} + \dfrac{x^4}{4!} - \dfrac{x^6}{6!} + \cdots \right) - \dfrac{1}{2}\left(-\dfrac{x^2}{2!} + \dfrac{x^4}{4!} - \dfrac{x^6}{6!} + \cdots \right)^2$

$\qquad\quad + \dfrac{1}{3}\left(-\dfrac{x^2}{2!} + \dfrac{x^4}{4!} - \dfrac{x^6}{6!} + \cdots \right)^3 - \cdots$

$\qquad = -\dfrac{x^2}{2} + \dfrac{x^4}{24} - \dfrac{x^6}{720} + \cdots - \dfrac{1}{2}\left(\dfrac{x^4}{4} - \dfrac{x^6}{24} + \cdots \right)$

$\qquad\quad + \dfrac{1}{3}\left(-\dfrac{x^6}{8} + \cdots \right) - \cdots$

$$= -\frac{x^2}{2} - \frac{x^4}{12} - \frac{x^6}{45} - \cdots.$$

Note that at each stage of the calculation we kept only enough terms to ensure that we could get all the terms with powers up to x^6. Being an even function, $\ln \cos x$ has only even powers in its Maclaurin series. Again, we cannot find the general term of this series. We could try to calculate terms by using the formula $a_k = f^{(k)}(0)/k!$ but even this becomes difficult after the first few values of k.

Observe that the series for $\tan x$ can also have been derived from that of $\ln \cos x$ because we have $\tan x = -\dfrac{d}{dx} \ln \cos x$.

∎

Taylor's Formula Revisited

In the examples above we have used a variety of techniques to obtain Taylor series for functions and verify that functions are analytic. As shown in Section 4.8, Taylor's Theorem provides a means for estimating the size of the error $E_n(x) = f(x) - P_n(x)$ involved when the Taylor polynomial

$$P_n(x) = \sum_{k=0}^{n} \frac{f^{(k)}(c)}{k!}(x-c)^k$$

is used to approximate the value of $f(x)$ for $x \neq c$. Since the Taylor polynomials are partial sums of the Taylor series for f at c (if the latter exists), another technique for verifying the convergence of a Taylor series is to use the formula for $E_n(x)$ provided by Taylor's Theorem to show, at least for an interval of values of x containing c, that $\lim_{n \to \infty} E_n(x) = 0$. This implies that $\lim_{n \to \infty} P_n(x) = f(x)$ so that f is indeed the sum of its Taylor series about c on that inverval, and f is analytic at c. Here is a somewhat more general version of Taylor's theroem.

THEOREM 22

Taylor's Theorem

If the $(n+1)$st derivative of f exists on an interval containing c and x, and if $P_n(x)$ is the Taylor polynomial of degree n for f about the point $x = c$, then

$$f(x) = P_n(x) + E_n(x) \qquad \textbf{Taylor's Formula}$$

holds, where the error term $E_n(x)$ is given by *either* of the following formulas:

Lagrange remainder $\qquad E_n(x) = \dfrac{f^{(n+1)}(s)}{(n+1)!}(x-c)^{n+1},$

for some s between c and x

Integral remainder $\qquad E_n(x) = \dfrac{1}{n!}\displaystyle\int_c^x (x-t)^n f^{(n+1)}(t)\, dt.$

Taylor's Theorem with Lagrange remainder was proved in Section 4.8 (Theorem 10) by using the Mean-Value Theorem and induction on n. The Integral remainder version is also proved by induction on n. See Exercise 42 for hints on how to carry out the proof. We will not make any use of the Integral form of the remainder here.

Our final example in this section reestablishes the Maclaurin series for e^x by finding the limit of the Lagrange remainder as suggested above.

Example 7 Use Taylor's Theorem to find the Maclaurin series for $f(x) = e^x$. Where does the series converge to $f(x)$?

Solution Since e^x is positive and increasing, $e^s \leq e^{|x|}$ for any $s \leq |x|$. Since $f^{(k)}(x) = e^x$ for any k we have, taking $c = 0$ in the Lagrange remainder in Taylor's Formula,

$$|E_n(x)| = \left| \frac{f^{(n+1)}(s)}{(n+1)!} x^{n+1} \right| \quad \text{for some } s \text{ between } 0 \text{ and } x$$

$$= \frac{e^s}{(n+1)!} |x|^{n+1} \leq e^{|x|} \frac{|x|^{n+1}}{(n+1)!} \to 0 \text{ as } n \to \infty$$

for any real x, as shown in Theorem 3(b) of Section 9.1. Thus $\lim_{n\to\infty} E_n(x) = 0$. Since the nth-order Maclaurin polynomial for e^x is $\sum_{k=0}^{n}(x^k/k!)$,

$$e^x = \lim_{n\to\infty} \left(\sum_{k=0}^{n} \frac{x^k}{k!} + E_n(x) \right) = \sum_{k=0}^{\infty} \frac{x^k}{k!} = 1 + x + \frac{x^2}{2!} + \frac{x^3}{3!} + \cdots,$$

and the series converges to e^x for all real numbers x.

Exercises 9.6

Find Maclaurin series representations for the functions in Exercises 1–14. For what values of x is each representation valid?

1. e^{3x+1}

2. $\cos(2x^3)$

3. $\sin(x - \pi/4)$

4. $\cos(2x - \pi)$

5. $x^2 \sin(x/3)$

6. $\cos^2(x/2)$

7. $\sin x \cos x$

8. $\tan^{-1}(5x^2)$

9. $\dfrac{1 + x^3}{1 + x^2}$

10. $\ln(2 + x^2)$

11. $\ln \dfrac{1 + x}{1 - x}$

12. $(e^{2x^2} - 1)/x^2$

13. $\cosh x - \cos x$

14. $\sinh x - \sin x$

Find the required Taylor series representations of the functions in Exercises 15–26. Where is each series representation valid?

15. $f(x) = e^{-2x}$ about -1

16. $f(x) = \sin x$ about $\pi/2$

17. $f(x) = \cos x$ in powers of $x - \pi$

18. $f(x) = \ln x$ in powers of $x - 3$

19. $f(x) = \ln(2 + x)$ in powers of $x - 2$

20. $f(x) = e^{2x+3}$ in powers of $x + 1$

21. $f(x) = \sin x - \cos x$ about $\dfrac{\pi}{4}$

22. $f(x) = \cos^2 x$ about $\dfrac{\pi}{8}$

23. $f(x) = 1/x^2$ in powers of $x + 2$

24. $f(x) = \dfrac{x}{1 + x}$ in powers of $x - 1$

25. $f(x) = x \ln x$ in powers of $x - 1$

26. $f(x) = xe^x$ in powers of $x + 2$

Find the first three nonzero terms in the Maclaurin series for the functions in Exercises 27–30.

27. $\sec x$

28. $\sec x \tan x$

29. $\tan^{-1}(e^x - 1)$

30. $e^{\tan^{-1} x} - 1$

∗31. Use the fact that $(\sqrt{1 + x})^2 = 1 + x$ to find the first three nonzero terms of the Maclaurin series for $\sqrt{1 + x}$.

32. Does $\csc x$ have a Maclaurin series? Why? Find the first three nonzero terms of the Taylor series for $\csc x$ about the point $x = \pi/2$.

Find the sums of the series in Exercises 33–36.

33. $1 + x^2 + \dfrac{x^4}{2!} + \dfrac{x^6}{3!} + \dfrac{x^8}{4!} + \cdots$

∗34. $x^3 - \dfrac{x^9}{3! \times 4} + \dfrac{x^{15}}{5! \times 16} - \dfrac{x^{21}}{7! \times 64} + \dfrac{x^{27}}{9! \times 256} - \cdots$

35. $1 + \dfrac{x^2}{3!} + \dfrac{x^4}{5!} + \dfrac{x^6}{7!} + \dfrac{x^8}{9!} + \cdots$

∗36. $1 + \dfrac{1}{2 \times 2!} + \dfrac{1}{4 \times 3!} + \dfrac{1}{8 \times 4!} + \cdots$

37. Let $P(x) = 1 + x + x^2$. Find (a) the Maclaurin series for $P(x)$ and (b) the Taylor series for $P(x)$ about 1.

∗38. Verify by direct calculation that $f(x) = 1/x$ is analytic at a for every $a \neq 0$.

∗39. Verify by direct calculation that $\ln x$ is analytic at a for every $a > 0$.

∗40. Review Exercise 41 of Section 4.3. It shows that the function

$$f(x) = \begin{cases} e^{-1/x^2} & \text{if } x \neq 0 \\ 0 & \text{if } x = 0 \end{cases}$$

has derivatives of all orders at every point of the real line, and $f^{(k)}(0) = 0$ for every positive integer k. What is the Maclaurin series for $f(x)$? What is the interval of convergence of this Maclaurin series? On what interval does the series converge to $f(x)$? Is f analytic at 0?

∗41. By direct multiplication of the Maclaurin series for e^x and e^y show that $e^x e^y = e^{x+y}$.

*42. **(Taylor's Formula with integral remainder)** Verify that if $f^{(n+1)}$ exists on an interval containing c and x, and if $P_n(x)$ is the nth-order Taylor polynomial for f about c, then $f(x) = P_n(x) + E_n(x)$, where

$$E_n(x) = \frac{1}{n!} \int_c^x (x-t)^n f^{(n+1)}(t)\, dt.$$

Proceed as follows:

(a) First observe that the case $n = 0$ is just the Fundamental Theorem of Calculus:

$$f(x) = f(c) + \int_c^x f'(t)\, dt.$$

Now integrate by parts in this formula, taking $U = f'(t)$ and $dV = dt$. Contrary to our usual policy of not including a constant of integration in V, here write $V = -(x-t)$ rather than just $V = t$. Observe that the result of the integration by parts is the case $n = 1$ of the formula.

(b) Use induction argument (and integration by parts again) to show that if the formula is valid for $n = k$, then it is also valid for $n = k + 1$.

*43. Use Taylor's formula with integral remainder to reprove that the Maclaurin series for $\ln(1+x)$ converges to $\ln(1+x)$ for $-1 < x \le 1$.

*44. **(Stirling's Formula)** The limit

$$\lim_{n \to \infty} \frac{n!}{\sqrt{2\pi}\, n^{n+1/2} e^{-n}} = 1$$

says that the *relative error* in the approximation

$$n! \approx \sqrt{2\pi}\, n^{n+1/2} e^{-n}$$

approaches zero as n increases. That is, $n!$ grows at a rate comparable to $\sqrt{2\pi}\, n^{n+1/2} e^{-n}$. This result, known as Stirling's Formula, is often very useful in applied mathematics and statistics. Prove it by carrying out the following steps.

(a) Use the identity $\ln(n!) = \sum_{j=1}^{n} \ln j$ and the increasing nature of $\ln$ to show that if $n \ge 1$,

$$\int_0^n \ln x\, dx < \ln(n!) < \int_1^{n+1} \ln x\, dx$$

and hence that

$$n \ln n - n < \ln(n!) < (n+1)\ln(n+1) - n.$$

(b) If $c_n = \ln(n!) - \left(n + \frac{1}{2}\right)\ln n + n$, show that

$$c_n - c_{n+1} = \left(n + \frac{1}{2}\right) \ln \frac{n+1}{n} - 1$$

$$= \left(n + \frac{1}{2}\right) \ln \frac{1 + 1/(2n+1)}{1 - 1/(2n+1)} - 1.$$

(c) Use the Maclaurin series for $\ln \dfrac{1+t}{1-t}$ (see Exercise 11) to show that

$$0 < c_n - c_{n+1} < \frac{1}{3}\left(\frac{1}{(2n+1)^2} + \frac{1}{(2n+1)^4} + \cdots\right)$$

$$= \frac{1}{12}\left(\frac{1}{n} - \frac{1}{n+1}\right),$$

and therefore that $\{c_n\}$ is decreasing and $\left\{c_n - \frac{1}{12n}\right\}$ is increasing. Hence conclude that $\lim_{n \to \infty} c_n = c$ exists, and that

$$\lim_{n \to \infty} \frac{n!}{n^{n+1/2} e^{-n}} = \lim_{n \to \infty} e^{c_n} = e^c.$$

(d) Now use the Wallis Product from Exercise 38 of Section 6.1 to show that

$$\lim_{n \to \infty} \frac{(2^n n!)^2}{(2n)!\sqrt{2n}} = \sqrt{\frac{\pi}{2}},$$

and hence deduce that $e^c = \sqrt{2\pi}$, which completes the proof.

9.7 Applications of Taylor and Maclaurin Series

Approximating the Values of Functions

We saw in Section 4.8 how Taylor and Maclaurin polynomials (the partial sums of Taylor and Maclaurin series) can be used as polynomial approximations to more complicated functions. In Example 5 of that section we used the Lagrange remainder in Taylor's Formula to determine how many terms of the Maclaurin series for e^x are needed to calculate $e^1 = e$ correct to 3 decimal places. For comparison, we obtained the same result in Example 7 in Section 9.3 by using a geometric series to bound the tail of the series for e.

The following example shows how the error bound associated with the alternating series test (see Theorem 15 in Section 9.4) can also be used for such approximations. When the terms a_n of a series (i) alternate in sign, (ii) decrease steadily in size, and (iii) approach zero as $n \to \infty$, then the error involved in using a partial sum of the series as an approximation to the sum of the series has the same sign as, and is no greater in absolute value than, the first omitted term.

Example 1 Find $\cos 43°$ with error less than $1/10,000$.

Solution We give two alternative solutions:

METHOD I. We can use the Maclaurin series for cosine:

$$\cos 43° = \cos \frac{43\pi}{180} = 1 - \frac{1}{2!}\left(\frac{43\pi}{180}\right)^2 + \frac{1}{4!}\left(\frac{43\pi}{180}\right)^4 - \cdots.$$

Now $43\pi/180 \approx 0.750\,49\cdots < 1$, so the series above must satisfy the conditions (i)–(iii) mentioned above. If we truncate the series after the nth term

$$(-1)^{n-1}\frac{1}{(2n-2)!}\left(\frac{43\pi}{180}\right)^{2n-2},$$

then the error E will be bounded by the size of the first omitted term:

$$|E| \le \frac{1}{(2n)!}\left(\frac{43\pi}{180}\right)^{2n} < \frac{1}{(2n)!}.$$

The error will not exceed $1/10,000$ if $(2n)! > 10,000$, so $n = 4$ will do ($8! = 40,320$).

$$\cos 43° \approx 1 - \frac{1}{2!}\left(\frac{43\pi}{180}\right)^2 + \frac{1}{4!}\left(\frac{43\pi}{180}\right)^4 - \frac{1}{6!}\left(\frac{43\pi}{180}\right)^6 \approx 0.731\,35\cdots$$

METHOD II. Since $43°$ is close to $45° = \pi/4$ rad, we can do a bit better by using the Taylor series about $\pi/4$ instead of the Maclaurin series:

$$\begin{aligned}
\cos 43° &= \cos\left(\frac{\pi}{4} - \frac{\pi}{90}\right) \\
&= \cos\frac{\pi}{4}\cos\frac{\pi}{90} + \sin\frac{\pi}{4}\sin\frac{\pi}{90} \\
&= \frac{1}{\sqrt{2}}\left[\left(1 - \frac{1}{2!}\left(\frac{\pi}{90}\right)^2 + \frac{1}{4!}\left(\frac{\pi}{90}\right)^4 - \cdots\right) \right. \\
&\qquad \left. + \left(\frac{\pi}{90} - \frac{1}{3!}\left(\frac{\pi}{90}\right)^3 + \cdots\right)\right].
\end{aligned}$$

Since

$$\frac{1}{4!}\left(\frac{\pi}{90}\right)^4 < \frac{1}{3!}\left(\frac{\pi}{90}\right)^3 < \frac{1}{20,000},$$

we need only the first two terms of the first series and the first term of the second series:

$$\cos 43° \approx \frac{1}{\sqrt{2}}\left(1 + \frac{\pi}{90} - \frac{1}{2}\left(\frac{\pi}{90}\right)^2\right) \approx 0.731\,358\cdots.$$

(In fact, $\cos 43° = 0.731\,353\,7\cdots$.)

When finding approximate values of functions, it is best, whenever possible, to use a power series about a point as close as possible to the point where the approximation is desired.

Functions Defined by Integrals

Many functions that can be expressed as simple combinations of elementary functions cannot be antidifferentiated by elementary techniques; their antiderivatives are not simple combinations of elementary functions. We can, however, often find the Taylor series for the antiderivatives of such functions and hence approximate their definite integrals.

Example 2 Find the Maclaurin series for

$$E(x) = \int_0^x e^{-t^2}\, dt,$$

and use it to evaluate $E(1)$ correct to 3 decimal places.

Solution   The Maclaurin series for $E(x)$ is given by

$$
\begin{aligned}
E(x) &= \int_0^x \left(1 - t^2 + \frac{t^4}{2!} - \frac{t^6}{3!} + \frac{t^8}{4!} - \cdots \right) dt \\
&= \left(t - \frac{t^3}{3} + \frac{t^5}{5 \times 2!} - \frac{t^7}{7 \times 3!} + \frac{t^9}{9 \times 4!} - \cdots \right) \Bigg|_0^x \\
&= x - \frac{x^3}{3} + \frac{x^5}{5 \times 2!} - \frac{x^7}{7 \times 3!} + \frac{x^9}{9 \times 4!} - \cdots = \sum_{n=0}^{\infty} (-1)^n \frac{x^{2n+1}}{(2n+1)n!},
\end{aligned}
$$

and is valid for all x because the series for e^{-t^2} is valid for all t. Therefore,

$$
\begin{aligned}
E(1) &= 1 - \frac{1}{3} + \frac{1}{5 \times 2!} - \frac{1}{7 \times 3!} + \cdots \\
&\approx 1 - \frac{1}{3} + \frac{1}{5 \times 2!} - \frac{1}{7 \times 3!} + \cdots + \frac{(-1)^{n-1}}{(2n-1)(n-1)!}.
\end{aligned}
$$

We stopped with the nth term. Again, the alternating series test assures us that the error in this approximation does not exceed the first omitted term, so it will be less than 0.0005, provided $(2n+1)n! > 2{,}000$. Since $13 \times 6! = 9{,}360$, $n = 6$ will do. Thus,

$$E(1) \approx 1 - \frac{1}{3} + \frac{1}{10} - \frac{1}{42} + \frac{1}{216} - \frac{1}{1{,}320} \approx 0.747,$$

rounded to 3 decimal places. ∎

Indeterminate Forms

Examples 1 and 2 of Section 4.9 showed how Maclaurin polynomials could be used for evaluating the limits of indeterminate forms. Here are two more examples, this time using the series directly and keeping enough terms to allow cancellation of the [0/0] factors.

Example 3   Evaluate   (a) $\displaystyle \lim_{x \to 0} \frac{x - \sin x}{x^3}$ and   (b) $\displaystyle \lim_{x \to 0} \frac{(e^{2x} - 1)\ln(1 + x^3)}{(1 - \cos 3x)^2}.$

Solution

(a) $\displaystyle\lim_{x\to 0}\frac{x-\sin x}{x^3}$ $\quad\begin{bmatrix}0\\0\end{bmatrix}$

$$= \lim_{x\to 0}\frac{x-\left(x-\dfrac{x^3}{3!}+\dfrac{x^5}{5!}-\cdots\right)}{x^3}$$

$$= \lim_{x\to 0}\frac{\dfrac{x^3}{3!}-\dfrac{x^5}{5!}+\cdots}{x^3}$$

$$= \lim_{x\to 0}\left(\frac{1}{3!}-\frac{x^2}{5!}+\cdots\right)=\frac{1}{3!}=\frac{1}{6}.$$

(b) $\displaystyle\lim_{x\to 0}\frac{(e^{2x}-1)\ln(1+x^3)}{(1-\cos 3x)^2}$ $\quad\begin{bmatrix}0\\0\end{bmatrix}$

$$= \lim_{x\to 0}\frac{\left(1+(2x)+\dfrac{(2x)^2}{2!}+\dfrac{(2x)^3}{3!}+\cdots-1\right)\left(x^3-\dfrac{x^6}{2}+\cdots\right)}{\left(1-\left(1-\dfrac{(3x)^2}{2!}+\dfrac{(3x)^4}{4!}-\cdots\right)\right)^2}$$

$$= \lim_{x\to 0}\frac{2x^4+2x^5+\cdots}{\left(\dfrac{9}{2}x^2-\dfrac{3^4}{4!}x^4+\cdots\right)^2}$$

$$= \lim_{x\to 0}\frac{2+2x+\cdots}{\left(\dfrac{9}{2}-\dfrac{3^4}{4!}x^2+\cdots\right)^2}=\frac{2}{\left(\dfrac{9}{2}\right)^2}=\frac{8}{81}.$$

You can check that the second of these examples is much more difficult if attempted using l'Hôpital's Rule.

Exercises 9.7

1. Estimate the error if the Maclaurin polynomial of degree 5 for $\sin x$ is used to approximate $\sin(0.2)$.

2. Estimate the error if the Taylor polynomial of degree 4 for $\ln x$ in powers of $x-2$ is used to approximate $\ln(1.95)$.

Use Maclaurin or Taylor series to calculate the function values indicated in Exercises 3–14, with error less than 5×10^{-5} in absolute value.

3. $e^{0.2}$
4. $1/e$
5. $e^{1.2}$
6. $\sin(0.1)$
7. $\cos 5°$
8. $\ln(6/5)$
9. $\ln(0.9)$
10. $\sin 80°$
11. $\cos 65°$
12. $\tan^{-1} 0.2$
13. $\cosh(1)$
14. $\ln(3/2)$

Find Maclaurin series for the functions in Exercises 15–19.

15. $I(x)=\displaystyle\int_0^x \frac{\sin t}{t}\,dt$

16. $J(x)=\displaystyle\int_0^x \frac{e^t-1}{t}\,dt$

17. $K(x)=\displaystyle\int_1^{1+x} \frac{\ln t}{t-1}\,dt$

18. $L(x)=\displaystyle\int_0^x \cos(t^2)\,dt$

19. $M(x)=\displaystyle\int_0^x \frac{\tan^{-1} t^2}{t^2}\,dt$

20. Find $L(0.5)$ correct to 3 decimal places, with L defined as in Exercise 16.

21. Find $I(1)$ correct to 3 decimal places, with I defined as in Exercise 13.

Evaluate the limits in Exercises 22–27.

22. $\displaystyle\lim_{x\to 0}\frac{\sin(x^2)}{\sinh x}$

23. $\displaystyle\lim_{x\to 0}\frac{1-\cos(x^2)}{(1-\cos x)^2}$

24. $\displaystyle\lim_{x\to 0}\frac{(e^x-1-x)^2}{x^2-\ln(1+x^2)}$

25. $\displaystyle\lim_{x\to 0}\frac{2\sin 3x-3\sin 2x}{5x-\tan^{-1} 5x}$

26. $\displaystyle\lim_{x\to 0}\frac{\sin(\sin x)-x}{x(\cos(\sin x)-1)}$

27. $\displaystyle\lim_{x\to 0}\frac{\sinh x-\sin x}{\cosh x-\cos x}$

9.8 The Binomial Theorem and Binomial Series

Example 1   Use Taylor's Formula to prove the Binomial Theorem: if n is a positive integer, then

$$(a + x)^n = a^n + n\,a^{n-1}x + \frac{n(n-1)}{2!}\,a^{n-2}x^2 + \cdots + n\,ax^{n-1} + x^n$$

$$= \sum_{k=0}^{n} \binom{n}{k}\,a^{n-k}x^k,$$

where $\binom{n}{k} = \dfrac{n!}{(n-k)!\,k!}$.

Solution   Let $f(x) = (a+x)^n$. Then

$$f'(x) = n(a+x)^{n-1} = \frac{n!}{(n-1)!}\,(a+x)^{n-1}$$

$$f''(x) = \frac{n!}{(n-1)!}\,(n-1)(a+x)^{n-2} = \frac{n!}{(n-2)!}\,(a+x)^{n-2}$$

$$\vdots$$

$$f^{(k)}(x) = \frac{n!}{(n-k)!}\,(a+x)^{n-k} \qquad (0 \le k \le n).$$

In particular, $f^{(n)}(x) = \dfrac{n!}{0!}\,(a+x)^{n-n} = n!$, a constant, and so

$$f^{(k)}(x) = 0 \qquad \text{for all } x, \text{ if } k > n.$$

For $0 \le k \le n$ we have $f^{(k)}(0) = \dfrac{n!}{(n-k)!}\,a^{n-k}$. Thus, by Taylor's Theorem with Lagrange remainder, for some s between a and x,

$$(a + x)^n = f(x) = \sum_{k=0}^{n} \frac{f^{(k)}(0)}{k!}\,x^k + \frac{f^{(n+1)}(s)}{(n+1)!}\,x^{n+1}$$

$$= \sum_{k=0}^{n} \frac{n!}{(n-k)!\,k!}\,a^{n-k}x^k + 0 = \sum_{k=0}^{n} \binom{n}{k}\,a^{n-k}x^k.$$

This is, in fact, the Maclaurin *series* for $(a+x)^n$, not just the Maclaurin polynomial of degree n. Since all higher-degree terms are zero, the series has only finitely many nonzero terms and so converges for all x.

Remark   If $f(x) = (a+x)^r$, where $a > 0$ and r is any real number, then calculations similar to those above show that the Maclaurin polynomial of degree n for f is

$$P_n(x) = a^r + \sum_{k=1}^{n} \frac{r(r-1)(r-2)\cdots(r-k+1)}{k!}\,a^{r-k}x^k.$$

However, if r is not a positive integer, then there will be no positive integer n for which the remainder $E_n(x) = f(x) - P_n(x)$ vanishes identically, and the corresponding Maclaurin series will not be a polynomial.

The Binomial Series

To simplify the discussion of the function $(a + x)^r$ when r is not a positive integer, we take $a = 1$ and consider the function $(1 + x)^r$. Results for the general case follow via the identity

$$(a + x)^r = a^r \left(1 + \frac{x}{a}\right)^r,$$

valid for any $a > 0$.

If r is any real number and $x > -1$, then the kth derivative of $(1 + x)^r$ is

$$r(r - 1)(r - 2) \cdots (r - k + 1)(1 + x)^{r-k}, \qquad (k = 1, 2, \ldots).$$

Thus, the Maclaurin series for $(1 + x)^r$ is

$$1 + \sum_{k=1}^{\infty} \frac{r(r - 1)(r - 2) \cdots (r - k + 1)}{k!} x^k,$$

which is called the **binomial series**. The following theorem shows that the binomial series does, in fact, converge to $(1 + x)^r$ if $|x| < 1$. We could accomplish this by writing Taylor's Formula for $(1 + x)^r$ with $c = 0$ and showing that the remainder $E_n(x) \to 0$ as $n \to \infty$. (We would need to use the integral form of the remainder to prove this for all $|x| < 1$.) However, we will use an easier method, similar to the one used for the exponential and trigonometric functions in Section 9.6.

THEOREM 23

The binomial series

If $|x| < 1$, then

$$(1 + x)^r = 1 + rx + \frac{r(r - 1)}{2!} x^2 + \frac{r(r - 1)(r - 2)}{3!} x^3 + \cdots$$

$$= 1 + \sum_{n=1}^{\infty} \frac{r(r - 1)(r - 2) \cdots (r - n + 1)}{n!} x^n \quad (-1 < x < 1).$$

PROOF   If $|x| < 1$, then the series

$$f(x) = 1 + \sum_{n=1}^{\infty} \frac{r(r - 1)(r - 2) \cdots (r - n + 1)}{n!} x^n$$

converges by the ratio test, since

$$\rho = \lim_{n \to \infty} \left| \frac{\dfrac{r(r - 1)(r - 2) \cdots (r - n + 1)(r - n)}{(n + 1)!} x^{n+1}}{\dfrac{r(r - 1)(r - 2) \cdots (r - n + 1)}{n!} x^n} \right|$$

$$= \lim_{n \to \infty} \left| \frac{r - n}{n + 1} \right| |x| = |x| < 1.$$

Note that $f(0) = 1$. We need to show that $f(x) = (1 + x)^r$ for $|x| < 1$.

By Theorem 19, we can differentiate the series for $f(x)$ termwise on $|x| < 1$ to obtain

$$f'(x) = \sum_{n=1}^{\infty} \frac{r(r - 1)(r - 2) \cdots (r - n + 1)}{(n - 1)!} x^{n-1}$$

$$= \sum_{n=0}^{\infty} \frac{r(r - 1)(r - 2) \cdots (r - n)}{n!} x^n.$$

We have replaced n with $n + 1$ to get the second version of the sum from the first version. Adding the second version to x times the first version, we get

$$(1 + x)f'(x) = \sum_{n=0}^{\infty} \frac{r(r - 1)(r - 2) \cdots (r - n)}{n!} x^n$$

$$+ \sum_{n=1}^{\infty} \frac{r(r - 1)(r - 2) \cdots (r - n + 1)}{(n - 1)!} x^n$$

$$= r + \sum_{n=1}^{\infty} \frac{r(r - 1)(r - 2) \cdots (r - n + 1)}{n!} x^n \left[(r - n) + n\right]$$

$$= r f(x).$$

The differential equation $(1 + x)f'(x) = rf(x)$ implies that

$$\frac{d}{dx} \frac{f(x)}{(1 + x)^r} = \frac{(1 + x)^r f'(x) - r(1 + x)^{r-1} f(x)}{(1 + x)^{2r}} = 0$$

for all x satisfying $|x| < 1$. Thus, $f(x)/(1 + x)^r$ is constant on that interval, and since $f(0) = 1$, the constant must be 1. Thus $f(x) = (1 + x)^r$.

Remark For some values of r the binomial series may converge at the endpoints $x = 1$ or $x = -1$. As observed above, if r is a positive integer, the series has only finitely many nonzero terms, and so converges for all x.

Example 2 Find the Maclaurin series for $\dfrac{1}{\sqrt{1 + x}}$.

Solution Here $r = -(1/2)$:

$$\frac{1}{\sqrt{1 + x}} = (1 + x)^{-1/2}$$

$$= 1 - \frac{1}{2} x + \frac{1}{2!} \left(-\frac{1}{2}\right) \left(-\frac{3}{2}\right) x^2 + \frac{1}{3!} \left(-\frac{1}{2}\right) \left(-\frac{3}{2}\right) \left(-\frac{5}{2}\right) x^3 + \cdots$$

$$= 1 - \frac{1}{2} x + \frac{1 \times 3}{2^2 2!} x^2 - \frac{1 \times 3 \times 5}{2^3 3!} x^3 + \cdots$$

$$= 1 + \sum_{n=1}^{\infty} (-1)^n \frac{1 \times 3 \times 5 \times \cdots \times (2n - 1)}{2^n n!} x^n.$$

This series converges for $-1 < x \le 1$. (Use the alternating series test to get the endpoint $x = 1$.)

Example 3 Find the Maclaurin series for $\sin^{-1} x$.

Solution Replace x with $-t^2$ in the series obtained in the previous example to get

$$\frac{1}{\sqrt{1 - t^2}} = 1 + \sum_{n=1}^{\infty} \frac{1 \times 3 \times 5 \times \cdots \times (2n - 1)}{2^n n!} t^{2n} \qquad (-1 < t < 1).$$

Now integrate t from 0 to x:

$$\sin^{-1} x = \int_0^x \frac{dt}{\sqrt{1 - t^2}} = \int_0^x \left(1 + \sum_{n=1}^{\infty} \frac{1 \times 3 \times 5 \times \cdots \times (2n - 1)}{2^n n!} t^{2n}\right) dt$$

$$= x + \sum_{n=1}^{\infty} \frac{1 \times 3 \times 5 \times \cdots \times (2n - 1)}{2^n n! (2n + 1)} x^{2n+1}$$

$$= x + \frac{x^3}{6} + \frac{3}{40} x^5 + \cdots \qquad (-1 < x < 1).$$

▌▌▌▌▌▌▌ Exercises 9.8 ▌▌▌▌▌▌▌

Find Maclaurin series representations for the functions in Exercises 1–6. Use the binomial series to calculate the answers.

1. $\sqrt{1+x}$

2. $x\sqrt{1-x}$

3. $\sqrt{4+x}$

4. $\dfrac{1}{\sqrt{4+x^2}}$

5. $(1-x)^{-2}$

6. $(1+x)^{-3}$

7. (Binomial coefficients) Show that the binomial coefficients $\dbinom{n}{k} = \dfrac{n!}{k!\,(n-k)!}$ satisfy

(i) $\dbinom{n}{0} = \dbinom{n}{n} = 1$ for every n, and

(ii) if $0 \le k \le n$, then $\dbinom{n}{k-1} + \dbinom{n}{k} = \dbinom{n+1}{k}$.

It follows that, for fixed $n \ge 1$, the binomial coefficients $\dbinom{n}{0}$, $\dbinom{n}{1}$, $\dbinom{n}{2}$, $\ldots$, $\dbinom{n}{n}$ are the elements of the nth row of **Pascal's triangle** below, where each element with value > 1 is the sum of the two diagonally above it.

$$
\begin{array}{ccccccccccc}
 & & & & & 1 & & 1 & & & \\
 & & & & 1 & & 2 & & 1 & & \\
 & & & 1 & & 3 & & 3 & & 1 & \\
 & & 1 & & 4 & & 6 & & 4 & & 1 \\
 & 1 & & 5 & & 10 & & 10 & & 5 & & 1
\end{array}
$$

8. (An inductive proof of the Binomial Theorem) Use mathematical induction and the results of Exercise 7 to prove the Binomial Theorem:

$$
(a+b)^n = \sum_{k=0}^{n} \binom{n}{k} a^{n-k} b^k
$$

$$
= a^n + na^{n-1}b + \binom{n}{2}a^{n-2}b^2 + \binom{n}{3}a^{n-3}b^3 + \cdots + b^n.
$$

9. (The Leibniz Rule) Use mathematical induction, the Product Rule, and Exercise 7 to verify the Leibniz Rule for the nth derivative of a product of two functions:

$$
(fg)^{(n)} = \sum_{k=0}^{n} \binom{n}{k} f^{(n-k)} g^{(k)}
$$

$$
= f^{(n)}g + nf^{(n-1)}g' + \binom{n}{2}f^{(n-2)}g''
$$

$$
+ \binom{n}{3}f^{(n-3)}g^{(3)} + \cdots + fg^{(n)}.
$$

9.9 Fourier Series

As we have seen, power series representations of functions make it possible to approximate those functions as closely as we want in intervals near a particular point of interest by using partial sums of the series, that is, polynomials. However, in many important applications of mathematics, the functions involved are required to be periodic. For example, much of electrical engineering is concerned with the analysis and manipulation of *waveforms*, which are periodic functions of time. Polynomials are not periodic functions, and for this reason power series are not well suited to representing such functions.

Much more appropriate for the representations of periodic functions over extended intervals are certain infinite series of periodic functions called Fourier series.

Periodic Functions

Recall that a function f defined on the real line is **periodic** with period T if

$$
f(t+T) = f(t) \quad \text{for all real } t. \tag{$*$}
$$

This implies that $f(t+mT) = f(t)$ for any integer m, so that if T is a period of f, then so is any multiple mT of T. The smallest positive number T for which $(*)$ holds is called the **fundamental period**, or simply **the period** of f.

The entire graph of a function with period T can be obtained by shifting the part of the graph in any half-open interval of length T (e.g., the interval $[0, T)$) to the left or right by integer multiples of the period T. Figure 9.6 shows the graph of a function of period 2.

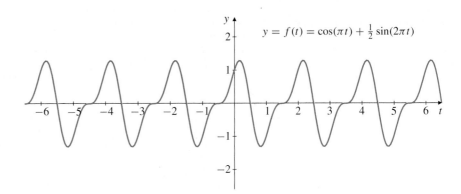

Figure 9.6   This function has period 2. Observe how the graph repeats the part in the interval $[0, 2)$ over and over to the left and right

Example 1   The functions $g(t) = \cos(\pi t)$ and $h(t) = \sin(\pi t)$ are both periodic with period 2:

$$g(t + 2) = \cos(\pi t + 2\pi) = \cos(\pi t) = g(t).$$

The function $k(t) = \sin(2\pi t)$ also has period 2, but this is not its fundamental period. The fundamental period is 1:

$$k(t + 1) = \sin(2\pi t + 2\pi) = \sin(2\pi t) = k(t).$$

The sum $f(t) = g(t) + \frac{1}{2}k(t) = \cos(\pi t) + \frac{1}{2}\sin(2\pi t)$, graphed in Figure 9.6, has period 2, the least common multiple of the periods of its two terms.

Example 2   For any positive integer n, the functions

$$f_n(t) = \cos(n\omega t) \qquad \text{and} \qquad g_n(t) = \sin(n\omega t)$$

both have fundamental period $T = 2\pi/(n\omega)$. The collection of all such functions corresponding to all positive integers n have common period $T = 2\pi/\omega$, the fundamental period of f_1 and g_1. T is an integer multiple of the fundamental periods of all the functions f_n and g_n. The subject of Fourier series is concerned with expressing general functions with period T as series whose terms are real multiples of these functions.

Fourier Series

It can be shown (but we won't do it here) that if $f(t)$ is periodic with fundamental period T, is continuous, and has a piecewise continuous derivative on the real line, then $f(t)$ is everywhere the sum of a series of the form

$$f(t) = \frac{a_0}{2} + \sum_{n=1}^{\infty} \bigl(a_n \cos(n\omega t) + b_n \sin(n\omega t)\bigr), \qquad (**)$$

called the **Fourier series** of f, where $\omega = 2\pi/T$ and the sequences $\{a_n\}_{n=0}^{\infty}$ and $\{b_n\}_{n=1}^{\infty}$ are the **Fourier coefficients** of f. Determining the values of these coefficients for a given such function f is made possible by the following identities, valid for integers m and n, which are easily proved by using the addition formulas for sine and cosine. (See Exercises 49–51 in Section 5.6.)

$$\int_0^T \cos(n\omega t)\,dt = \begin{cases} 0 & \text{if } n \neq 0 \\ T & \text{if } n = 0 \end{cases}$$

$$\int_0^T \sin(n\omega t)\,dt = 0$$

$$\int_0^T \cos(m\omega t)\cos(n\omega t)\,dt = \begin{cases} 0 & \text{if } m \neq n \\ T/2 & \text{if } m = n \end{cases}$$

$$\int_0^T \sin(m\omega t)\sin(n\omega t)\,dt = \begin{cases} 0 & \text{if } m \neq n \\ T/2 & \text{if } m = n \end{cases}$$

$$\int_0^T \cos(m\omega t)\sin(n\omega t)\,dt = 0.$$

If we multiply equation ($**$) by $\cos(m\omega t)$ (or by $\sin(m\omega t)$) and integrate the resulting equation over $[0, T]$ term by term, all the terms on the right except the one involving a_m (or b_m) will be 0. (The term-by-term integration requires justification, but we won't try to do that here either.) The integration results in

$$\int_0^T f(t)\cos(m\omega t)\,dt = \frac{1}{2}T a_m$$

$$\int_0^T f(t)\sin(m\omega t)\,dt = \frac{1}{2}T b_m.$$

(Note that the first of these formulas is even valid for $m = 0$ because we chose to call the constant term in the Fourier series $a_0/2$ instead of a_0.) Since the integrands are all periodic with period T, the integrals can be taken over any interval of length T; it is often convenient to use $[-T/2, T/2]$ instead of $[0, T]$. The Fourier coefficients of f are therefore given by

$$a_n = \frac{2}{T}\int_{-T/2}^{T/2} f(t)\cos(n\omega t)\,dt \quad (n = 0,\ 1,\ 2,\ \dots)$$

$$b_n = \frac{2}{T}\int_{-T/2}^{T/2} f(t)\sin(n\omega t)\,dt \quad (n = 1,\ 2,\ 3,\ \dots),$$

where $\omega = 2\pi/T$.

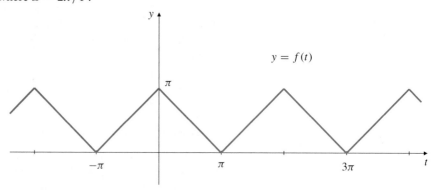

Figure 9.7   A sawtooth function of period 2π

Example 3   Find the Fourier series of the sawtooth function $f(t)$ of period 2π whose values in the interval $[-\pi, \pi]$ are given by $f(t) = \pi - |t|$. (See Figure 9.7.)

Solution   Here $T = 2\pi$ and $\omega = 2\pi/(2\pi) = 1$. Since $f(t)$ is an even function, $f(t)\sin(nt)$ is odd, so all the Fourier sine coefficients b_n are zero:

$$b_n = \frac{2}{2\pi}\int_{-\pi}^{\pi} f(t)\sin(nt)\,dt = 0.$$

Also, $f(t)\cos(nt)$ is an even function, so

$$
\begin{aligned}
a_n &= \frac{2}{2\pi}\int_{-\pi}^{\pi} f(t)\cos(nt)\,dt = \frac{4}{2\pi}\int_{0}^{\pi} f(t)\cos(nt)\,dt \\
&= \frac{2}{\pi}\int_{0}^{\pi} (\pi - t)\cos(nt)\,dt \\
&= \begin{cases}
\pi & \text{if } n = 0 \\
0 & \text{if } n \neq 0 \text{ and } n \text{ is even} \\
4/(\pi n^2) & \text{if } n \text{ is odd}
\end{cases}
\end{aligned}
$$

Since odd positive integers n are of the form $n = 2k - 1$, where k is a positive integer, the Fourier series of f is given by

$$
f(t) = \frac{\pi}{2} + \sum_{k=1}^{\infty} \frac{4}{\pi(2k-1)^2}\cos\big((2k-1)t\big).
$$

Convergence of Fourier Series

The partial sums of a Fourier series are called Fourier polynomials because they can be expressed as polynomials in $\sin(\omega t)$ and $\cos(\omega t)$, although we will not actually try to write them that way. The Fourier polynomial of order m of the periodic function f having period T is

$$
f_m(t) = \frac{a_0}{2} + \sum_{n=1}^{m}\big(a_n\cos(n\omega t) + b_n\sin(n\omega t)\big),
$$

where $\omega = 2\pi/T$ and the coefficients a_n $(0 \leq n \leq m)$ and b_n $(1 \leq n \leq m)$ are given by the integral formulas developed earlier.

Example 4   The Fourier polynomial of order 3 of the sawtooth function of Example 3 is

$$
f_3(t) = \frac{\pi}{2} + \frac{4}{\pi}\cos t + \frac{4}{9\pi}\cos(3t).
$$

The graph of this function is shown in Figure 9.8. Observe that it appears to be a reasonable approximation to the graph of f in Figure 9.7, but, being a finite sum of differentiable functions, $f_3(t)$ is itself differentiable everywhere, even at the integer multiples of π where f is not differentiable.

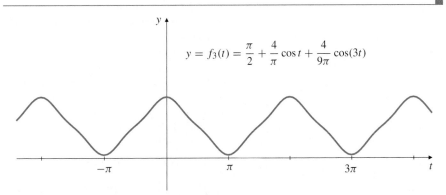

Figure 9.8   The Fourier polynomial approximation $f_3(t)$ to the sawtooth function of Example 3

As noted earlier, the Fourier series of a function $f(t)$ that is periodic, continuous, and has a piecewise continuous derivative on the real line converges to $f(t)$ at each real number t. However, the Fourier coefficients (and hence the Fourier series) can be calculated (by the formulas given above) for periodic functions with piecewise continuous derivative even if the functions are not themselves continuous, but only piecewise continuous.

Recall that $f(t)$ is piecewise continuous on the interval $[a, b]$ if there exists a partition $\{a = x_0 < x_1 < x_2 < \cdots < x_k = b\}$ of $[a, b]$ and functions $F_1, F_2, \ldots, F_k$, such that

(i) F_i is continuous on $[x_{i-1}, x_i]$, and

(ii) $f(t) = F_i(t)$ on (x_{i-1}, x_i).

The integral of such a function f is the sum of integrals of the functions F_i:

$$\int_a^b f(t)\, dt = \sum_{i=1}^k \int_{x_{i-1}}^{x_i} F_i(t)\, dt.$$

Since $f(t)\cos(n\omega t)$ and $f(t)\sin(n\omega t)$ are piecewise continuous if f is, the Fourier coefficients of a piecewise continuous, periodic function can be calculated by the same formulas given for a continuous periodic function. The question of where and to what the Fourier series converges in this case is answered by the following theorem, proved in textbooks on Fourier analysis.

THEOREM 24

The Fourier series of a piecewise continuous, periodic function f with piecewise continuous derivative converges to that function at every point t where f is continuous. Moreover, if f is discontinuous at $t = c$, then f has different, but finite, left and right limits at c:

$$\lim_{t \to c-} f(t) = f(c-), \qquad \text{and} \qquad \lim_{t \to c+} f(t) = f(c+).$$

The Fourier series of f converges at $t = c$ to the average of these left and right limits:

$$\frac{a_0}{2} + \sum_{n=1}^\infty \big(a_n \cos(n\omega c) + b_n \sin(n\omega c)\big) = \frac{f(c-) + f(c+)}{2},$$

where $\omega = 2\pi/T$.

Example 5   Calculate the Fourier series for the periodic function f with period 2 satisfying

$$f(t) = \begin{cases} -1 & \text{if } -1 < x < 0 \\ 1 & \text{if } 0 < x < 1. \end{cases}$$

Where does f fail to be continuous? To what does the Fourier series of f converge at these points?

Solution   Here $T = 2$ and $\omega = 2\pi/2 = \pi$. Since f is an odd function, its cosine coefficients are all zero:

$$a_n = \int_{-1}^1 f(t)\cos(n\pi t)\, dt = 0. \qquad \text{(The integrand is odd.)}$$

The same symmetry implies that

$$b_n = \int_{-1}^1 f(t)\sin(n\pi t)\, dt$$

$$= 2\int_0^1 \sin(n\pi t)\, dt = -\left.\frac{2\cos(n\pi t)}{n\pi}\right|_0^1$$

$$= -\frac{2}{n\pi}\big((-1)^n - 1\big) = \begin{cases} 4/(n\pi) & \text{if } n \text{ is odd} \\ 0 & \text{if } n \text{ is even.} \end{cases}$$

Odd integers n are of the form $n = 2k - 1$ for $k = 1, 2, 3, \ldots$. Therefore, the Fourier series of f is

$$\frac{4}{\pi} \sum_{k=1}^{\infty} \frac{1}{2k-1} \sin\big((2k-1)\pi t\big)$$

$$= \frac{4}{\pi} \left(\sin(\pi t) + \frac{1}{3} \sin(3\pi t) + \frac{1}{5} \sin(5\pi t) + \cdots \right).$$

Note that f is continuous except at the points where t is an integer. At each of these points f jumps from -1 to 1 or from 1 to -1, so the average of the left and right limits of f at these points is 0. Observe that the sum of the Fourier series is 0 at integer values of t, in accordance with Theorem 24. See Figure 9.9.

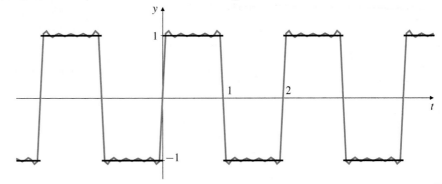

Figure 9.9 The piecewise continuous function f (black) of Example 5 and its Fourier polynomial f_{15} (colour)

$$f_{15}(t) = \sum_{k=1}^{8} \frac{4 \sin\big((2k-1)\pi t\big)}{(2k-1)\pi}$$

Fourier Cosine and Sine Series

As observed in Example 3 and Example 5, even functions have no sine terms in their Fourier series, and odd functions have no cosine terms (including the constant term $a_0/2$). It is often necessary in applications to find a Fourier series representation of a given function defined on a finite interval $[0, a]$ having either no sine terms (a **Fourier cosine series**) or no cosine terms (a **Fourier sine series**). This is accomplished by extending the domain of f to $[-a, 0)$ so as to make f either even or odd on $[-a, a]$,

$$f(-t) = f(t) \text{ if } -a \le t < 0 \text{ for the even extension}$$
$$f(-t) = -f(t) \text{ if } -a \le t < 0 \text{ for the odd extension},$$

and then calculating its Fourier series considering the extended f to have period $2a$. (If we want the odd extension, we may have to redefine $f(0)$ to be 0.)

Example 6   Find the Fourier cosine series of $g(t) = \pi - t$ defined on $[0, \pi]$.

Solution   The even extension of $g(t)$ to $[-\pi, \pi]$ is the function f of Example 3. Thus the Fourier cosine series of g is

$$\frac{\pi}{2} + \sum_{k=1}^{\infty} \frac{4}{\pi(2k-1)^2} \cos\big((2k-1)t\big).$$

Example 7   Find the Fourier sine series of $h(t) = 1$ defined on $[0, 1]$.

Solution   If we redefine $h(0) = 0$, then the odd extension of h to $[-1, 1]$ coincides with the function $f(t)$ of Example 5 except that the latter function is undefined at $t = 0$. The Fourier sine series of h is the series obtained in Example 5, namely,

$$\frac{4}{\pi} \sum_{k=1}^{\infty} \frac{1}{2k-1} \sin\big((2k-1)\pi t\big).$$

Remark   Fourier cosine and sine series are treated from a different perspective in Section 13.4.

Exercises 9.9

In Exercises 1–4, what is the fundamental period of the given function?

1. $f(t) = \sin(3t)$

2. $g(t) = \cos(3 + \pi t)$

3. $h(t) = \cos^2 t$

4. $k(t) = \sin(2t) + \cos(3t)$

In Exercises 5–8, find the Fourier series of the given function.

5. $f(t) = t$, $-\pi < t \le \pi$, f has period 2π.

6. $f(t) = \begin{cases} 0 & \text{if } 0 \le t < 1 \\ 1 & \text{if } 1 \le t < 2, \end{cases}$ f has period 2.

7. $f(t) = \begin{cases} 0 & \text{if } -1 \le t < 0 \\ t & \text{if } 0 \le t < 1, \end{cases}$ f has period 2.

8. $f(t) = \begin{cases} t & \text{if } 0 \le t < 1 \\ 1 & \text{if } 1 \le t < 2 \\ 3 - t & \text{if } 2 \le t < 3, \end{cases}$ f has period 3.

9. What is the Fourier cosine series of the function $h(t)$ of Example 7?

10. Calculate the Fourier sine series of the function $g(t)$ of Example 6.

11. Find the Fourier sine series of $f(t) = t$ on $[0, 1]$.

12. Find the Fourier cosine series of $f(t) = t$ on $[0, 1]$.

13. Use the result of Example 3 to evaluate

$$\sum_{n=1}^{\infty} \frac{1}{(2n-1)^2} = 1 + \frac{1}{3^2} + \frac{1}{5^2} + \cdots.$$

14. Verify that if f is an even function of period T, then the Fourier sine coefficients b_n of f are all zero and the Fourier cosine coefficients a_n of f are given by

$$a_n = \frac{4}{T} \int_0^{T/2} f(t) \cos(n\omega t)\, dt, \qquad n = 0, 1, 2, \ldots,$$

where $\omega = 2\pi/T$. State and verify the corresponding result for odd functions f.

Chapter Review

Key Ideas

- **What does it mean to say that the sequence $\{a_n\}$**
  - ◇ is bounded above?
  - ◇ is ultimately positive?
  - ◇ is alternating?
  - ◇ is increasing?
  - ◇ converges?
  - ◇ diverges to infinity?
- **What does it mean to say that the series $\sum_{n=1}^{\infty} a_n$**
  - ◇ converges?
  - ◇ diverges?
  - ◇ is geometric?
  - ◇ is telescoping?
  - ◇ is a p-series?
  - ◇ is positive?
  - ◇ converges absolutely?
  - ◇ converges conditionally?
- **State the following convergence tests for series.**
  - ◇ the integral test
  - ◇ the comparison test
  - ◇ the limit comparison test
  - ◇ the ratio test
  - ◇ the alternating series test
- **How can you find bounds for the tail of a series?**
- **What is a bound for the tail of an alternating series?**
- **What do the following terms and phrases mean?**
  - ◇ a power series
  - ◇ interval of convergence
  - ◇ radius of convergence
  - ◇ centre of convergence
  - ◇ a Taylor series
  - ◇ a Maclaurin series
  - ◇ a Taylor polynomial
  - ◇ a binomial series
  - ◇ an analytic function
- **Where is the sum of a power series differentiable?**
- **Where does the integral of a power series converge?**
- **Where is the sum of a power series continuous?**
- **State Taylor's Theorem with Lagrange remainder.**
- **State Taylor's Theorem with integral remainder.**

- **What is the binomial theorem?**
- **What is a Fourier series?**
- **What is a Fourier cosine series? a Fourier sine series?**

Review Exercises

In Exercises 1–4, determine whether the given sequence converges, and find its limit if it does converge.

1. $\left\{ \dfrac{(-1)^n e^n}{n!} \right\}$

2. $\left\{ \dfrac{n^{100} + 2^n \pi}{2^n} \right\}$

3. $\left\{ \dfrac{\ln n}{\tan^{-1} n} \right\}$

4. $\left\{ \dfrac{(-1)^n n^2}{\pi n(n - \pi)} \right\}$

5. Let $a_1 > \sqrt{2}$, and let

$$a_{n+1} = \frac{a_n}{2} + \frac{1}{a_n} \quad \text{for} \quad n = 1, 2, 3, \ldots$$

Show that $\{a_n\}$ is decreasing and that $a_n > \sqrt{2}$ for $n \ge 1$. Why must $\{a_n\}$ converge? Find $\lim_{n\to\infty} a_n$.

6. Find the limit of the sequence $\{\ln \ln(n + 1) - \ln \ln n\}$.

Evaluate the sums of the series in Exercises 7–10.

7. $\displaystyle\sum_{n=1}^{\infty} 2^{-(n-5)/2}$

8. $\displaystyle\sum_{n=0}^{\infty} \frac{4^{n-1}}{(\pi - 1)^{2n}}$

9. $\displaystyle\sum_{n=1}^{\infty} \frac{1}{n^2 - \frac{1}{4}}$

10. $\displaystyle\sum_{n=1}^{\infty} \frac{1}{n^2 - \frac{9}{4}}$

Determine whether the series in Exercises 11–16 converge or diverge. Give reasons for your answers.

11. $\displaystyle\sum_{n=1}^{\infty} \frac{n-1}{n^3}$

12. $\displaystyle\sum_{n=1}^{\infty} \frac{n + 2^n}{1 + 3^n}$

13. $\displaystyle\sum_{n=1}^{\infty} \frac{n}{(1+n)(1+n\sqrt{n})}$ **14.** $\displaystyle\sum_{n=1}^{\infty} \frac{n^2}{(1+2^n)(1+n\sqrt{n})}$

15. $\displaystyle\sum_{n=1}^{\infty} \frac{3^{2n+1}}{n!}$ **16.** $\displaystyle\sum_{n=1}^{\infty} \frac{n!}{(n+2)!+1}$

Do the series in Exercises 17–20 converge absolutely, converge conditionally, or diverge?

17. $\displaystyle\sum_{n=1}^{\infty} \frac{(-1)^{n-1}}{1+n^3}$ **18.** $\displaystyle\sum_{n=1}^{\infty} \frac{(-1)^n}{2^n - n}$

19. $\displaystyle\sum_{n=10}^{\infty} \frac{(-1)^{n-1}}{\ln\ln n}$ **20.** $\displaystyle\sum_{n=1}^{\infty} \frac{n^2 \cos(n\pi)}{1+n^3}$

For what values of x do the series in Exercises 21–22 converge absolutely? converge conditionally? diverge?

21. $\displaystyle\sum_{n=1}^{\infty} \frac{(x-2)^n}{3^n \sqrt{n}}$ **22.** $\displaystyle\sum_{n=1}^{\infty} \frac{(5-2x)^n}{n}$

Determine the sums of the series in Exercises 23–24 to within 0.001.

23. $\displaystyle\sum_{n=1}^{\infty} \frac{1}{n^3}$ **24.** $\displaystyle\sum_{n=1}^{\infty} \frac{1}{4+n^2}$

In Exercises 25–32, find Maclaurin series for the given functions. State where each series converges to the function.

25. $\dfrac{1}{3-x}$ **26.** $\dfrac{x}{3-x^2}$

27. $\ln(e+x^2)$ **28.** $\dfrac{1-e^{-2x}}{x}$

29. $x\cos^2 x$ **30.** $\sin(x+(\pi/3))$

31. $(8+x)^{-1/3}$ **32.** $(1+x)^{1/3}$

Find Taylor series for the functions in Exercises 33–34 about the indicated points $x = c$.

33. $1/x$, $c = \pi$ **34.** $\sin x + \cos x$, $c = \pi/4$

Find the Maclaurin polynomial of the indicated degree for the functions in Exercises 35–38.

35. e^{x^2+2x}, degree 3   **36.** $\sin(1+x)$, degree 3

37. $\cos(\sin x)$, degree 4   **38.** $\sqrt{1+\sin x}$, degree 4

39. What function has Maclaurin series

$$1 - \frac{x}{2!} + \frac{x^2}{4!} - \cdots = \sum_{n=0}^{\infty} \frac{(-1)^n x^n}{(2n)!}?$$

40. A function $f(x)$ has Maclaurin series

$$1 + x^2 + \frac{x^4}{2^2} + \frac{x^6}{3^2} + \cdots = 1 + \sum_{n=1}^{\infty} \frac{x^{2n}}{n^2}.$$

Find $f^{(k)}(0)$ for all positive integers k.

Find the sums of the series in Exercises 41–44.

41. $\displaystyle\sum_{n=0}^{\infty} \frac{n+1}{\pi^n}$ ***42.** $\displaystyle\sum_{n=0}^{\infty} \frac{n^2}{\pi^n}$

43. $\displaystyle\sum_{n=1}^{\infty} \frac{1}{ne^n}$ ***44.** $\displaystyle\sum_{n=2}^{\infty} \frac{(-1)^n \pi^{2n-4}}{(2n-1)!}$

45. If $S(x) = \displaystyle\int_0^x \sin(t^2)\, dt$, find $\displaystyle\lim_{x\to 0} \frac{x^3 - 3S(x)}{x^7}$.

46. Use series to evaluate $\displaystyle\lim_{x\to 0} \frac{(x - \tan^{-1}x)(e^{2x} - 1)}{2x^2 - 1 + \cos(2x)}$.

47. How many nonzero terms in the Maclaurin series for e^{-x^4} are needed to evaluate $\int_0^{1/2} e^{-x^4}\, dx$ correct to 5 decimal places? Evaluate the integral to that accuracy.

48. Estimate the size of the error if the Taylor polynomial of degree 4 about $x = \pi/2$ for $f(x) = \ln\sin x$ is used to approximate $\ln\sin(1.5)$.

49. Find the Fourier sine series for $f(t) = \pi - t$ on $[0, \pi]$.

50. Find the Fourier series for $f(t) = \begin{cases} 1 & \text{if } -\pi < t \leq 0 \\ t & \text{if } 0 < t \leq \pi. \end{cases}$

Challenging Problems

1. (A refinement of the ratio test) Suppose $a_n > 0$ and $a_{n+1}/a_n \geq n/n+1$ for all n. Show that $\sum_{n=1}^{\infty} a_n$ diverges. *Hint:* $a_n \geq K/n$ for some constant K.

* 2. (Summation by parts) Let $\{u_n\}$ and $\{v_n\}$ be two sequences, and let $s_n = \sum_{k=1}^{n} v_k$.

(a) Show that $\sum_{k=1}^{n} u_k v_k = u_{n+1}s_n + \sum_{k=1}^{n}(u_k - u_{k+1})s_n$. (*Hint:* Write $v_n = s_n - s_{n-1}$, with $s_0 = 0$, and rearrange the sum.)

(b) If $\{u_n\}$ is positive, decreasing, and convergent to 0, and if $\{v_n\}$ has bounded partial sums, $|s_n| \leq K$ for all n, where K is a constant, show that $\sum_{n=1}^{\infty} u_n v_n$ converges. (*Hint:* Show that the series $\sum_{n=1}^{\infty}(u_n - u_{n+1})s_n$ converges by comparing it to the telescoping series $\sum_{n=1}^{\infty}(u_n - u_{n+1})$.)

* 3. Show that $\sum_{n=1}^{\infty}(1/n)\sin(nx)$ converges for every x. *Hint:* If x is an integer multiple of π, all the terms in the series are 0 so there is nothing to prove. Otherwise, $\sin(x/2) \neq 0$. In this case show that

$$\sum_{n=1}^{N} \sin(nx) = \frac{\cos(x/2) - \cos((N+1/2)x)}{2\sin(x/2)}$$

using the identity

$$\sin a \sin b = \frac{\cos(a-b) - \cos(a+b)}{2}$$

to make the sum telescope. Then apply the result of Problem 2(b) with $u_n = 1/n$ and $v_n = \sin(nx)$.

4. Let $a_1, a_2, a_3, \ldots$ be those positive integers that do not contain the digit 0 in their decimal representations. Thus $a_1 = 1$, $a_2 = 2, \ldots, a_9 = 9, a_{10} = 11, \ldots, a_{18} = 19, a_{19} = 21,$ $\ldots, a_{90} = 99, a_{91} = 111$, etc. Show that the series $\displaystyle\sum_{n=1}^{\infty} \frac{1}{a_n}$ converges and that the sum is less than 90. (*Hint:* How many of these integers have m digits? Each term $1/a_n$, where a_n has m digits, is less than 10^{-m+1}.)

* 5. (Using an integral to improve convergence) Recall the error formula for the Midpoint Rule, according to which

$$\int_{k-1/2}^{k+1/2} f(x)\, dx - f(k) = \frac{f''(c)}{24},$$

where $k - (1/2) \le c \le k + (1/2)$.

(a) If $f''(x)$ is a decreasing function of x, show that

$$f'(k + \tfrac{3}{2}) - f'(k + \tfrac{1}{2}) \le f''(c) \le f'(k - \tfrac{1}{2}) - f'(k - \tfrac{3}{2}).$$

(b) If (i) $f''(x)$ is a decreasing function of x,
(ii) $\int_{N+1/2}^{\infty} f(x)\, dx$ converges, and (iii) $f'(x) \to 0$ as $x \to \infty$, show that

$$\frac{f'(N - \tfrac{1}{2})}{24} \le \sum_{n=N+1}^{\infty} f(n) - \int_{N+1/2}^{\infty} f(x)\, dx \le \frac{f'(N + \tfrac{3}{2})}{24}.$$

(c) Use the result of part (b) to approximate $\sum_{n=1}^{\infty} 1/n^2$ to within 0.001.

* **6. (The number e is irrational)** Start with $e = \sum_{n=0}^{\infty} 1/n!$.

(a) Use the technique of Example 7 in Section 9.3 to show that for any $n > 0$,

$$0 < e - \sum_{j=0}^{n} \frac{1}{j!} < \frac{1}{n!n}.$$

(Note that the sum here has $n + 1$ terms, not n terms.)

(b) Suppose that e is a rational number, say $e = M/N$ for certain positive integers M and N. Show that $N!\left(e - \sum_{j=0}^{N}(1/j!)\right)$ is an integer.

(c) Combine parts (a) and (b) to show that there is an integer between 0 and $1/N$. Why is this not possible? Conclude that e cannot be a rational number.

7. Let

$$f(x) = \sum_{k=0}^{\infty} \frac{2^{2k} k!}{(2k + 1)!} x^{2k+1}$$

$$= x + \frac{2}{3}x^3 + \frac{4}{3 \times 5}x^5 + \frac{8}{3 \times 5 \times 7}x^7 + \cdots.$$

(a) Find the radius of convergence of this power series.

(b) Show that $f'(x) = 1 + 2xf(x)$.

(c) What is $\dfrac{d}{dx}\left(e^{-x^2} f(x)\right)$?

(d) Express $f(x)$ in terms of an integral.

* **8. (The number π is irrational)** Problem 6 above shows how to prove that e is irrational by assuming the contrary and deducing a contradiction. In this problem you will show that π is also irrational. The proof for π is also by contradiction but is rather more complicated, so it will be broken down into several parts.

(a) Let $f(x)$ be a polynomial, and let

$$g(x) = f(x) - f''(x) + f^{(4)}(x) - f^{(6)}(x) + \cdots$$

$$= \sum_{j=0}^{\infty} (-1)^j f^{(2j)}(x).$$

(Since f is a polynomial, all but a finite number of terms in the above sum are identically zero, so there are no convergence problems.) Verify that

$$\frac{d}{dx}\left(g'(x) \sin x - g(x) \cos x\right) = f(x) \sin x,$$

and hence that $\int_0^{\pi} f(x) \sin x\, dx = g(\pi) + g(0)$.

(b) Suppose that π is rational, say, $\pi = m/n$, where m and n are positive integers. You will show that this leads to a contradiction and thus cannot be true. Choose a positive integer k such that $(\pi m)^k / k! < 1/2$. (Why is this possible?) Consider the polynomial

$$f(x) = \frac{x^k (m - nx)^k}{k!} = \frac{1}{k!} \sum_{j=0}^{k} \binom{k}{j} m^{k-j} (-n)^j x^{j+k}.$$

Show that $0 < f(x) < 1/2$ for $0 < x < \pi$, and hence that

$$0 < \int_0^{\pi} f(x) \sin x\, dx < 1.$$ Thus, $0 < g(\pi) + g(0) < 1$, where $g(x)$ is defined as in part (a).

(c) Show that the ith derivative of $f(x)$ is given by

$$f^{(i)}(x) = \frac{1}{k!} \sum_{j=0}^{k} \binom{k}{j} m^{k-j} (-n)^j \frac{(j+k)!}{(j+k-i)!} x^{j+k-i}.$$

(d) Show that $f^{(i)}(0)$ is an integer for $i = 0, 1, 2, \ldots$. (*Hint:* Observe for $i < k$ that $f^{(i)}(0) = 0$, and for $i > 2k$ that $f^{(i)}(x) = 0$ for all x. For $k \le i \le 2k$, show that only one term in the sum for $f^{(i)}(0)$ is not 0, and that this term is an integer. You will need the fact that the binomial coefficients $\binom{k}{j}$ are integers.)

(e) Show that $f(\pi - x) = f(x)$ for all x, and hence that $f^{(i)}(\pi)$ is also an integer for each $i = 0, 1, 2, \ldots$. Therefore, if $g(x)$ is defined as in (a), then $g(\pi) + g(0)$ is an integer. This contradicts the conclusion of part (b) and so shows that π cannot be rational.

* **9. (An asymptotic series)** Integrate by parts to show that

$$\int_0^x e^{-1/t}\, dt = e^{-1/x} \sum_{n=2}^{N} (-1)^n (n - 1)! x^n$$

$$+ (-1)^{N+1} N! \int_0^x t^{N-1} e^{-1/t}\, dt.$$

Why can't you just use a Maclaurin series to approximate this integral? Using $N = 5$, find an approximate value for $\int_0^{0.1} e^{-1/t}\, dt$, and estimate the error. Estimate the error for $N = 10$ and $N = 20$.

Note that the series $\sum_{n=2}^{\infty} (-1)^n (n - 1)! x^n$ **diverges** for any $x \neq 0$. This is an example of what is called an **asymptotic series**. Even though it diverges, a properly chosen partial sum gives a good approximation to our function when x is small.

CHAPTER 10

Vectors and Coordinate Geometry in 3-Space

Lord Ronald said nothing; he flung himself from the room, flung himself upon his horse and rode madly off in all directions. ...

And who is this tall young man who draws nearer to Gertrude with every revolution of the horse? ...

The two were destined to meet. Nearer and nearer they came. And then still nearer. Then for one brief moment they met. As they passed Gertrude raised her head and directed towards the young nobleman two eyes so eye-like in their expression as to be absolutely circular, while Lord Ronald directed towards the occupant of the dogcart a gaze so gaze-like that nothing but a gazelle, or a gas-pipe, could have emulated its intensity.

Stephen Leacock   1869–1944
from *Gertrude the Governess: or, Simple Seventeen*

Introduction   A complete real-variable calculus program involves the study of

(i) real-valued functions of a single real variable,

(ii) vector-valued functions of a single real variable,

(iii) real-valued functions of a real vector variable,

(iv) vector-valued functions of a real vector variable.

Chapters 1–9 are concerned with item (i). The remaining chapters deal with items (ii), (iii), and (iv). Specifically, Chapter 11 deals with vector-valued functions of a single real variable. Chapters 12–14 are concerned with the differentiation and integration of real-valued functions of several real variables, that is, of a real vector variable. Chapters 15 and 16 present aspects of the calculus of functions whose domains and ranges both have dimension greater than one, that is, vector-valued functions of a vector variable. Most of the time we will limit our attention to vector functions with domains and ranges in the plane, or in 3-dimensional space.

In this chapter we will lay the foundation for multivariable and vector calculus by extending the concepts of analytic geometry to three or more dimensions and by introducing vectors as a convenient way of dealing with several variables as a single entity. We also introduce matrices, because these will prove useful for formulating some of the concepts of calculus. This chapter is not intended to be a course in linear algebra. We develop only those aspects that we will use in later chapters and omit most proofs.

10.1 Analytic Geometry in Three Dimensions

We say that the physical world in which we live is three-dimensional because through any point there can pass three, and no more, straight lines that are **mutually perpendicular**, that is, each of them is perpendicular to the other two. This is equivalent to the fact that we require three numbers to locate a point in space with respect to some reference point (the **origin**). One way to use three numbers to locate a point is by having them represent (signed) distances from the origin, measured in the directions of three mutually perpendicular lines passing through the origin. We call such a set of lines a Cartesian coordinate system, and each of the lines is called a coordinate axis. We usually call these axes the x-axis, the y-axis, and the z-axis, regarding the x- and y-axes as lying in a horizontal plane and the z-axis as vertical. Moreover, the coordinate system should have a **right-handed orientation**. This means that the thumb, forefinger, and middle finger of the right hand can be extended so as to point, respectively, in the directions of the positive x-axis, the positive y-axis, and the positive z-axis. For the more mechanically minded, a right-handed screw will advance in the positive z direction if twisted in the direction of rotation from the positive x-axis toward the positive y-axis. (See Figure 10.1(a).)

Figure 10.1

(a) The screw moves upward when twisted counterclockwise as seen from above

(b) The three coordinates of a point in 3-space

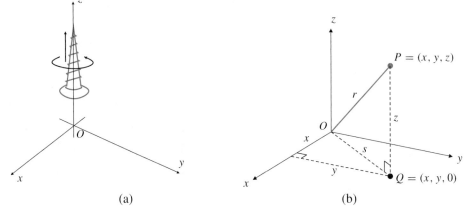

(a) (b)

With respect to such a Cartesian coordinate system, the **coordinates** of a point P in 3-space constitute an ordered triple of real numbers, (x, y, z). The numbers x, y, and z are, respectively, the signed distances of P from the origin, measured in the directions of the x-axis, the y-axis, and the z-axis. (See Figure 10.1(b).)

Let Q be the point with coordinates $(x, y, 0)$. Then Q lies in the xy-plane (the plane containing the x- and y-axes) directly under (or over) P. We say that Q is the vertical projection of P onto the xy-plane. If r is the distance from the origin O to P and s is the distance from O to Q, then, using two right-angled triangles, we have

$$s^2 = x^2 + y^2 \quad \text{and} \quad r^2 = s^2 + z^2 = x^2 + y^2 + z^2.$$

Thus, the distance from P to the origin is given by

$$r = \sqrt{x^2 + y^2 + z^2}.$$

Similarly, the distance r between points $P_1 = (x_1, y_1, z_1)$ and $P_2 = (x_2, y_2, z_2)$ (see Figure 10.2) is

$$r = \sqrt{(x_2 - x_1)^2 + (y_2 - y_1)^2 + (z_2 - z_1)^2}.$$

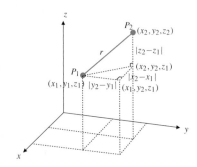

Figure 10.2

Example 1 Show that the triangle with vertices $A = (1, -1, 2)$, $B = (3, 3, 8)$, and $C = (2, 0, 1)$ has a right angle.

Solution   We calculate the lengths of the three sides of the triangle:

$$a = |BC| = \sqrt{(2-3)^2 + (0-3)^2 + (1-8)^2} = \sqrt{59}$$
$$b = |AC| = \sqrt{(2-1)^2 + (0+1)^2 + (1-2)^2} = \sqrt{3}$$
$$c = |AB| = \sqrt{(3-1)^2 + (3+1)^2 + (8-2)^2} = \sqrt{56}$$

By the cosine law, $a^2 = b^2 + c^2 - 2bc \cos A$. In this case $a^2 = 59 = 3 + 56 = b^2 + c^2$, so that $2bc \cos A$ must be 0. Therefore $\cos A = 0$ and $A = 90°$.

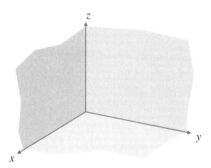

Figure 10.3   The first octant

Just as the x- and y-axes divide the xy-plane into four quadrants, so also the three **coordinate planes** in 3-space (the xy-plane, the xz-plane, and the yz-plane) divide 3-space into eight **octants**. We call the octant in which $x \geq 0$, $y \geq 0$, and $z \geq 0$ the **first octant**. When drawing graphs in 3-space it is sometimes easier to draw only the part lying in the first octant (Figure 10.3).

An equation or inequality involving the three variables x, y, and z defines a subset of points in 3-space whose coordinates satisfy the equation or inequality. A single equation usually represents a surface (a two-dimensional object) in 3-space.

Example 2   (Some equations and the surfaces they represent)

(a) The equation $z = 0$ represents all points with coordinates $(x, y, 0)$, that is, the xy-plane. The equation $z = -2$ represents all points with coordinates $(x, y, -2)$, that is, the horizontal plane passing through the point $(0, 0, -2)$ on the z-axis.

(b) The equation $x = y$ represents all points with coordinates (x, x, z). This is a vertical plane containing the straight line with equation $x = y$ in the xy-plane. The plane also contains the z-axis. (See Figure 10.4.)

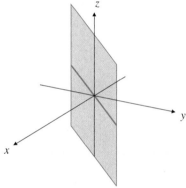

Figure 10.4   Equation $x = y$ defines a vertical plane

(c) The equation $x + y + z = 1$ represents all points the sum of whose coordinates is 1. This set is a plane that passes through the three points $(1, 0, 0)$, $(0, 1, 0)$, and $(0, 0, 1)$. These points are not collinear (they do not lie on a straight line), so there is only one plane passing through all three. (See Figure 10.5.) The equation $x + y + z = 0$ represents a plane parallel to the one with equation $x + y + z = 1$ but passing through the origin.

(d) The equation $x^2 + y^2 = 4$ represents all points on the vertical circular cylinder containing the circle with equation $x^2 + y^2 = 4$ in the xy-plane. This cylinder has radius 2 and axis along the z-axis. (See Figure 10.6.)

(e) The equation $z = x^2$ represents all points with coordinates (x, y, x^2). This surface is a parabolic cylinder tangent to the xy-plane along the y-axis. (See Figure 10.7.)

(f) The equation $x^2 + y^2 + z^2 = 25$ represents all points (x, y, z) at distance 5 from the origin. This set of points is a *sphere* of radius 5 centred at the origin.

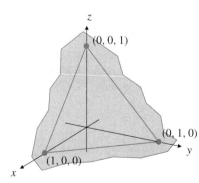

Figure 10.5   The plane with equation $x + y + z = 1$

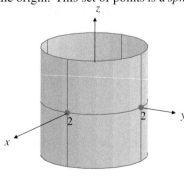

Figure 10.6   The circular cylinder with equation $x^2 + y^2 = 4$

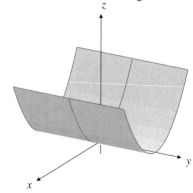

Figure 10.7   The parabolic cylinder with equation $z = x^2$

Observe that equations in x, y, and z need not involve each variable explicitly. When one of the variables is missing from the equation, the equation represents a surface *parallel to* the axis of the missing variable. Such a surface may be a plane or a cylinder. For example, if z is absent from the equation, the equation represents in 3-space a vertical (i.e., parallel to the z-axis) surface containing the curve with the same equation in the xy-plane.

Occasionally a single equation may not represent a two-dimensional object (a surface). It can represent a one-dimensional object (a line or curve), a zero-dimensional object (one or more points), or even nothing at all.

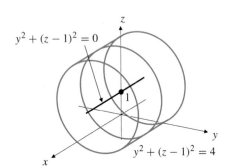

Figure 10.8   The cylinder $y^2 + (z-1)^2 = 4$ and its axial line $y^2 + (z-1)^2 = 0$

Example 3   Identify the graphs of the equations: (a) $y^2 + (z-1)^2 = 4$, (b) $y^2 + (z-1)^2 = 0$,   (c) $x^2 + y^2 + z^2 = 0$, and (d) $x^2 + y^2 + z^2 = -1$.

Solution

(a) Since x is absent, the equation $y^2 + (z-1)^2 = 4$ represents an object parallel to the x-axis. In the yz-plane the equation represents a circle of radius 2 centred at $(y, z) = (0, 1)$. In 3-space it represents a horizontal circular cylinder, parallel to the x-axis, with axis one unit above the x-axis. (See Figure 10.8.)

(b) Since squares cannot be negative, the equation $y^2 + (z-1)^2 = 0$ implies that $y = 0$ and $z = 1$, so it represents points $(x, 0, 1)$. All these points lie on the line parallel to the x-axis and one unit above it. (See Figure 10.8.)

(c) As in part (b), $x^2 + y^2 + z^2 = 0$ implies that $x = 0$, $y = 0$, and $z = 0$. The equation represents only one point, the origin.

(d) The equation $x^2 + y^2 + z^2 = -1$ is not satisfied by any real numbers x, y, and z, so it represents no points at all.   ∎

A single inequality in x, y, and z typically represents points lying on one side of the surface represented by the corresponding equation (together with points on the surface if the inequality is not strict).

Example 4

(a) The inequality $z > 0$ represents all points above the xy-plane.

(b) The inequality $x^2 + y^2 \geq 4$ says that the square of the distance from (x, y, z) to the nearest point $(0, 0, z)$ on the z-axis is at least 4. This inequality represents all points lying on or outside the cylinder of Example 2(d).

(c) The inequality $x^2 + y^2 + z^2 \leq 25$ says that the square of the distance from (x, y, z) to the origin is no greater than 25. It represents the solid ball of radius 5 centred at the origin, which consists of all points lying inside or on the sphere of Example 2(f).   ∎

Two equations in x, y, and z normally represent a one-dimensional object, the line or curve along which the two surfaces represented by the two equations intersect. Any point whose coordinates satisfy both equations must lie on both the surfaces, so must lie on their intersection.

Example 5   What sets of points in 3-space are represented by the following pairs of equations?

(a) $\begin{cases} x + y + z = 1 \\ y - 2x = 0 \end{cases}$ (b) $\begin{cases} x^2 + y^2 + z^2 = 1 \\ x + y = 1 \end{cases}$

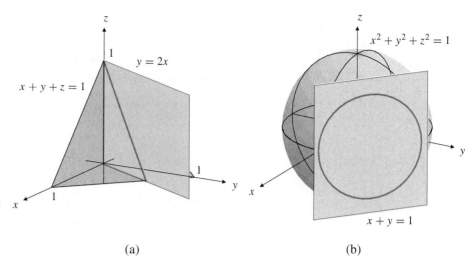

Figure 10.9

(a) The two planes intersect in a straight line

(b) The plane intersects the sphere in a circle

(a)   (b)

Solution

(a) The equation $x + y + z = 1$ represents the oblique plane of Example 2(c), and the equation $y - 2x = 0$ represents a vertical plane through the origin and the point $(1, 2, 0)$. Together these two equations represent the line of intersection of the two planes. This line passes through, for example, the points $(0, 0, 1)$ and $(\frac{1}{3}, \frac{2}{3}, 0)$. (See Figure 10.9(a).)

(b) The equation $x^2 + y^2 + z^2 = 1$ represents a sphere of radius 1 with centre at the origin, and $x + y = 1$ represents a vertical plane through the points $(1, 0, 0)$ and $(0, 1, 0)$. The two surfaces intersect in a circle, as shown in Figure 10.9(b). The line from $(1, 0, 0)$ to $(0, 1, 0)$ is a diameter of the circle, so the centre of the circle is $(\frac{1}{2}, \frac{1}{2}, 0)$, and its radius is $\sqrt{2}/2$.

In Sections 10.4 and 10.5 we will see many more examples of geometric objects in 3-space represented by simple equations.

Euclidean *n*-Space

Mathematicians and users of mathematics frequently need to consider **n-dimensional space**, where n is greater than 3 and may even be infinite. It is difficult to visualize a space of dimension 4 or higher geometrically. The secret to dealing with these spaces is to regard the points in n-space as *being* ordered n-tuples of real numbers; that is, $(x_1, x_2, \ldots, x_n)$ is a point in n-space instead of just being the coordinates of such a point. We stop thinking of points as existing in physical space and start thinking of them as algebraic objects. We usually denote n-space by the symbol $\mathbb{R}^n$ to show that its points are n-tuples of *real* numbers. Thus $\mathbb{R}^2$ and $\mathbb{R}^3$ denote the plane and 3-space, respectively. Note that in passing from $\mathbb{R}^3$ to $\mathbb{R}^n$ we have altered the notation a bit: in $\mathbb{R}^3$ we called the coordinates x, y, and z, while in $\mathbb{R}^n$ we called them $x_1, x_2, \ldots$ and x_n so as not to run out of letters. We could, of course, talk about coordinates (x_1, x_2, x_3) in $\mathbb{R}^3$ and (x_1, x_2) in the plane $\mathbb{R}^2$, but (x, y, z) and (x, y) are traditionally used there.

Although we think of points in $\mathbb{R}^n$ as n-tuples rather than geometric objects, we do not want to lose all sight of the underlying geometry. By analogy with the two- and three-dimensional cases, we still consider the quantity

$$\sqrt{(y_1 - x_1)^2 + (y_2 - x_2)^2 + \cdots + (y_n - x_n)^2}$$

as representing the *distance* between the points with coordinates $(x_1, x_2, \ldots, x_n)$ and $(y_1, y_2, \ldots, y_n)$. Also, we call the $(n - 1)$-dimensional set of points in $\mathbb{R}^n$ that satisfy the equation $x_n = 0$ a **hyperplane**, by analogy with the plane $z = 0$ in $\mathbb{R}^3$.

Describing Sets in the Plane, 3-Space, and *n*-Space

We conclude this section by collecting some definitions of terms used to describe sets of points in $\mathbb{R}^n$ for $n \geq 2$. These terms belong to the branch of mathematics called

topology, and they generalize the notions of open and closed intervals and endpoints used to describe sets on the real line $\mathbb{R}$. We state the definitions for $\mathbb{R}^n$, but we are most interested in the cases where $n = 2$ or $n = 3$.

A **neighbourhood** of a point P in $\mathbb{R}^n$ is a set of the form

$$B_r(P) = \{Q \in \mathbb{R}^n : \text{distance from } Q \text{ to } P < r\}$$

for some $r > 0$.

For $n = 1$, if $p \in \mathbb{R}$, then $B_r(p)$ is the **open interval** $(p - r, p + r)$ centred at p.

For $n = 2$, $B_r(P)$ is the **open disk** of radius r centred at point P.

For $n = 3$, $B_r(P)$ is the **open ball** of radius r centred at point P.

A set S is **open** in $\mathbb{R}^n$ if every point of S has a neighbourhood contained in S. Every neighbourhood is itself an open set. Other examples of open sets in $\mathbb{R}^2$ include the sets of points (x, y) such that $x > 0$, or such that $y > x^2$, or even such that $y \neq x^2$. Typically, sets defined by strict inequalities (using $>$ and $<$) are open. Examples in $\mathbb{R}^3$ include the sets of points (x, y, z) satisfying $x + y + z > 2$, or $1 < x < 3$.

The whole space $\mathbb{R}^n$ is an open set in itself. For technical reasons, the empty set (containing no points) is also considered to be open. (No point in the empty set fails to have a neighbourhood contained in the empty set.)

The **complement**, S^c, of a set S in $\mathbb{R}^n$ is the set of all points in $\mathbb{R}^n$ that do not belong to S. For example, the complement of the set of points (x, y) in $\mathbb{R}^2$ such that $x > 0$ is the set of points for which $x \leq 0$. A set is said to be **closed** if its complement is open. Typically, sets defined by nonstrict inequalities (using $\geq$ and $\leq$) are closed. Closed intervals are closed sets in $\mathbb{R}$. Since the whole space and the empty set are both open in $\mathbb{R}^n$ and are complements of each other, they are also both closed. They are the only sets that are both open and closed.

A point P is called a **boundary point** of a set S if every neighbourhood of P contains both points in S and points in S^c. The **boundary**, bdry(S), of a set S is the set of all boundary points of S. For example, the boundary of the closed disk $x^2 + y^2 \leq 1$ in $\mathbb{R}^2$ is the circle $x^2 + y^2 = 1$. A closed set contains all its boundary points. An open set contains none of its boundary points.

A point P is an **interior point** of a set S if it belongs to S but not to the boundary of S. P is an **exterior point** of S if it belongs to the complement of S but not to the boundary of S. The **interior**, int(S), and **exterior**, ext(S), of S consist of all the interior points and exterior points of S, respectively. Both int(S) and ext(S) are open sets. If S is open, then int$(S) = S$. If S is closed, then ext$(S) = S^c$. See Figure 10.10.

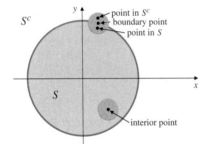

Figure 10.10 The closed disk S consisting of points $(x, y) \in \mathbb{R}^2$ that satisfy $x^2 + y^2 \leq 1$. Note the shaded neighbourhoods of the boundary point and the interior point.
bdry(S) is the circle $x^2 + y^2 = 1$
int(S) is the open disk $x^2 + y^2 < 1$
ext(S) is the open set $x^2 + y^2 > 1$

‖‖‖‖‖ Exercises 10.1 ‖‖‖‖‖‖

Find the distance between the pairs of points in Exercises 1–4.

1. $(0, 0, 0)$ and $(2, -1, -2)$ **2.** $(-1, -1, -1)$ and $(1, 1, 1)$

3. $(1, 1, 0)$ and $(0, 2, -2)$ **4.** $(3, 8, -1)$ and $(-2, 3, -6)$

5. What is the shortest distance from the point (x, y, z) to
(a) the xy-plane? (b) the x-axis?

6. Show that the triangle with vertices $(1, 2, 3)$, $(4, 0, 5)$, and $(3, 6, 4)$ has a right angle.

7. Find the angle A in the triangle with vertices $A = (2, -1, -1)$, $B = (0, 1, -2)$, and $C = (1, -3, 1)$.

8. Show that the triangle with vertices $(1, 2, 3)$, $(1, 3, 4)$, and $(0, 3, 3)$ is equilateral.

9. Find the area of the triangle with vertices $(1, 1, 0)$, $(1, 0, 1)$, and $(0, 1, 1)$.

10. What is the distance from the origin to the point $(1, 1, \ldots, 1)$ in $\mathbb{R}^n$?

11. What is the distance from the point $(1, 1, \ldots, 1)$ in n-space to the closest point on the x_1-axis?

In Exercises 12–23, describe (and sketch if possible) the set of points in $\mathbb{R}^3$ that satisfy the given equation or inequality.

12. $z = 2$ **13.** $y \geq -1$

14. $z = x$ **15.** $x + y = 1$

16. $x^2 + y^2 + z^2 = 4$

17. $(x - 1)^2 + (y + 2)^2 + (z - 3)^2 = 4$

18. $x^2 + y^2 + z^2 = 2z$ **19.** $y^2 + z^2 \leq 4$

20. $x^2 + z^2 = 4$ **21.** $z = y^2$

22. $z \geq \sqrt{x^2 + y^2}$ **23.** $x + 2y + 3z = 6$

In Exercises 24–32, describe (and sketch if possible) the set of points in $\mathbb{R}^3$ that satisfy the given pair of equations or inequalities.

24. $\begin{cases} x = 1 \\ y = 2 \end{cases}$ **25.** $\begin{cases} x = 1 \\ y = z \end{cases}$

26. $\begin{cases} x^2 + y^2 + z^2 = 4 \\ z = 1 \end{cases}$ **27.** $\begin{cases} x^2 + y^2 + z^2 = 4 \\ x^2 + y^2 + z^2 = 4x \end{cases}$

28. $\begin{cases} x^2 + y^2 + z^2 = 4 \\ x^2 + z^2 = 1 \end{cases}$ **29.** $\begin{cases} x^2 + y^2 = 1 \\ z = x \end{cases}$

30. $\begin{cases} y \geq x \\ z \leq y \end{cases}$ **31.** $\begin{cases} x^2 + y^2 \leq 1 \\ z \geq y \end{cases}$

32. $\begin{cases} x^2 + y^2 + z^2 \leq 1 \\ \sqrt{x^2 + y^2} \leq z \end{cases}$

In Exercises 33–36, specify the boundary and the interior of the plane sets S whose points (x, y) satisfy the given conditions. Is S open, closed, or neither?

33. $0 < x^2 + y^2 < 1$ **34.** $x \geq 0, \quad y < 0$

35. $x + y = 1$ **36.** $|x| + |y| \leq 1$

In Exercises 37–40, specify the boundary and the interior of the sets S in 3-space whose points (x, y, z) satisfy the given conditions. Is S open, closed, or neither?

37. $1 \leq x^2 + y^2 + z^2 \leq 4$ **38.** $x \geq 0, \quad y > 1, \quad z < 2$

39. $(x - z)^2 + (y - z)^2 = 0$ **40.** $x^2 + y^2 < 1, \, y + z > 2$

10.2 Vectors

A **vector** is a quantity that involves both **magnitude** (size or length) and **direction**. For instance, the *velocity* of a moving object involves its speed and direction of motion, so is a vector. Such quantities are represented geometrically by arrows (directed line segments) and are often actually identified with these arrows. For instance, the vector $\overrightarrow{AB}$ is an arrow with tail at the point A and head at the point B. In print, such a vector is usually denoted by a single letter in boldface type,

$$\mathbf{v} = \overrightarrow{AB}.$$

Figure 10.11   The vector $\mathbf{v} = \overrightarrow{AB}$

(See Figure 10.11.) In handwriting, an arrow over a letter ($\overrightarrow{v} = \overrightarrow{AB}$) can be used to denote a vector. The *magnitude* of the vector $\mathbf{v}$ is the length of the arrow and is denoted $|\mathbf{v}|$ or $|\overrightarrow{AB}|$.

While vectors have magnitude and direction, they do not generally have *position*, that is, they are not regarded as being in a particular place. Two vectors, $\mathbf{u}$ and $\mathbf{v}$, are considered *equal* if they have *the same length and the same direction*, even if their representative arrows do not coincide. The arrows must be parallel, have the same length, and point in the same direction. In Figure 10.12, for example, if $ABYX$ is a parallelogram, then $\overrightarrow{AB} = \overrightarrow{XY}$.

Figure 10.12 $\overrightarrow{AB} = \overrightarrow{XY}$

For the moment, we consider plane vectors, that is, vectors whose representative arrows lie in a plane. If we introduce a Cartesian coordinate system into the plane, we can talk about the x and y components of any vector. If $A = (a, b)$ and $P = (p, q)$, as shown in Figure 10.13, then the x and y components of $\overrightarrow{AP}$ are, respectively, $p - a$ and $q - b$. Note that if O is the origin and X is the point $(p - a, q - b)$, then

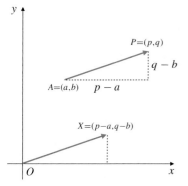
Figure 10.13   Components of a vector

$$|\overrightarrow{AP}| = \sqrt{(p - a)^2 + (q - b)^2} = |\overrightarrow{OX}|$$

$$\text{slope of } \overrightarrow{AP} = \frac{q - b}{p - a} = \text{slope of } \overrightarrow{OX}.$$

Hence $\overrightarrow{AP} = \overrightarrow{OX}$. In general, two vectors are equal if and only if they have the same x components and y components.

There are two important algebraic operations defined for vectors: addition and scalar multiplication.

DEFINITION **1**

Vector addition

Given two vectors **u** and **v**, their **sum u + v** is defined as follows. If an arrow representing **v** is placed with its tail at the head of an arrow representing **u**, then an arrow from the tail of **u** to the head of **v** represents **u + v**. Equivalently, if **u** and **v** have tails at the same point, then **u + v** is represented by an arrow with its tail at that point and its head at the opposite vertex of the parallelogram spanned by **u** and **v**. This is shown in Figure 10.14(a).

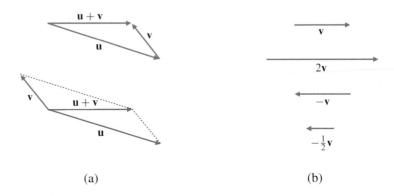

Figure 10.14

(a) Vector addition

(b) Scalar multiplication

(a)   (b)

DEFINITION **2**

Scalar multiplication

If **v** is a vector and t is a real number (also called a **scalar**), then the **scalar multiple** $t\mathbf{v}$ is a vector with magnitude $|t|$ times that of **v** and direction the same as **v** if $t > 0$, or opposite to that of **v** if $t < 0$. See Figure 10.14(b). If $t = 0$, then $t\mathbf{v}$ has zero length and therefore no particular direction. It is the **zero vector**, denoted **0**.

Suppose that **u** has components a and b and that **v** has components x and y. Then the components of **u + v** are $a + x$ and $b + y$, and those of $t\mathbf{v}$ are tx and ty. See Figure 10.15.

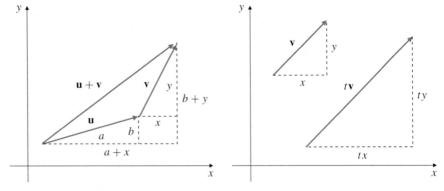

Figure 10.15   The components of a sum of vectors or a scalar multiple of a vector is the same sum or multiple of the corresponding components of the vectors

In $\mathbb{R}^2$ we single out two particular vectors for special attention. They are

(i) the vector **i** from the origin to the point $(1, 0)$, and

(ii) the vector **j** from the origin to the point $(0, 1)$.

Thus, **i** has components 1 and 0, and **j** has components 0 and 1. These vectors are called the **standard basis vectors** in the plane. The vector **r** from the origin to the point (x, y) has components x and y and can be expressed in the form

$$\mathbf{r} = \langle x, y \rangle = x\mathbf{i} + y\mathbf{j}.$$

In the first form we specify the vector by listing its components between angle brackets; in the second we write **r** as a **linear combination** of the standard basis vectors **i** and **j**. (See Figure 10.16.) The vector **r** is called the **position vector** of the point (x, y). A position vector has its tail at the origin and its head at the point whose position it is specifying. The length of **r** is $|\mathbf{r}| = \sqrt{x^2 + y^2}$.

More generally, the vector $\overrightarrow{AP}$ from $A = (a, b)$ to $P = (p, q)$ in Figure 10.13 can also be written as a list of components or as a linear combination of the standard basis vectors:

$$\overrightarrow{AP} = \langle p - a, q - b \rangle = (p - a)\mathbf{i} + (q - b)\mathbf{j}.$$

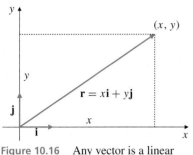

Figure 10.16   Any vector is a linear combination of the basis vectors

Sums and scalar multiples of vectors are easily expressed in terms of components. If $\mathbf{u} = u_1\mathbf{i} + u_2\mathbf{j}$ and $\mathbf{v} = v_1\mathbf{i} + v_2\mathbf{j}$, and if t is a scalar (i.e., a real number), then

$$\mathbf{u} + \mathbf{v} = (u_1 + v_1)\mathbf{i} + (u_2 + v_2)\mathbf{j},$$
$$t\mathbf{u} = (tu_1)\mathbf{i} + (tu_2)\mathbf{j}.$$

The zero vector is $\mathbf{0} = 0\mathbf{i} + 0\mathbf{j}$. It has length zero and no specific direction. For any vector **u** we have $0\mathbf{u} = \mathbf{0}$. A **unit vector** is a vector of length 1. The standard basis vectors **i** and **j** are unit vectors. Given any nonzero vector **v**, we can form a unit vector $\hat{\mathbf{v}}$ in the same direction as **v** by multiplying **v** by the reciprocal of its length (a scalar):

$$\hat{\mathbf{v}} = \left(\frac{1}{|\mathbf{v}|}\right)\mathbf{v}.$$

Example 1   If $A = (2, -1)$, $B = (-1, 3)$, and $C = (0, 1)$, express each of the following vectors as a linear combination of the standard basis vectors:

(a) $\overrightarrow{AB}$ (b) $\overrightarrow{BC}$ (c) $\overrightarrow{AC}$ (d) $\overrightarrow{AB} + \overrightarrow{BC}$ (e) $2\overrightarrow{AC} - 3\overrightarrow{CB}$

(f) a unit vector in the direction of $\overrightarrow{AB}$.

Solution

(a) $\overrightarrow{AB} = (-1 - 2)\mathbf{i} + (3 - (-1))\mathbf{j} = -3\mathbf{i} + 4\mathbf{j}$

(b) $\overrightarrow{BC} = (0 - (-1))\mathbf{i} + (1 - 3)\mathbf{j} = \mathbf{i} - 2\mathbf{j}$

(c) $\overrightarrow{AC} = (0 - 2)\mathbf{i} + (1 - (-1))\mathbf{j} = -2\mathbf{i} + 2\mathbf{j}$

(d) $\overrightarrow{AB} + \overrightarrow{BC} = \overrightarrow{AC} = -2\mathbf{i} + 2\mathbf{j}$

(e) $2\overrightarrow{AC} - 3\overrightarrow{CB} = 2(-2\mathbf{i} + 2\mathbf{j}) - 3(-\mathbf{i} + 2\mathbf{j}) = -\mathbf{i} - 2\mathbf{j}$

(f) A unit vector in the direction of $\overrightarrow{AB}$ is $\dfrac{\overrightarrow{AB}}{|\overrightarrow{AB}|} = -\dfrac{3}{5}\mathbf{i} + \dfrac{4}{5}\mathbf{j}$.

Implicit in the above example is the fact that the operations of addition and scalar multiplication obey appropriate algebraic rules, such as

$$\mathbf{u} + \mathbf{v} = \mathbf{v} + \mathbf{u},$$
$$(\mathbf{u} + \mathbf{v}) + \mathbf{w} = \mathbf{u} + (\mathbf{v} + \mathbf{w}),$$
$$\mathbf{u} - \mathbf{v} = \mathbf{u} + (-1)\mathbf{v},$$
$$t(\mathbf{u} + \mathbf{v}) = t\mathbf{u} + t\mathbf{v}.$$

Vectors in 3-Space

The algebra and geometry of vectors described here extends to spaces of any number of dimensions; we can still think of vectors as represented by arrows, and sums and scalar multiples are formed just as for plane vectors.

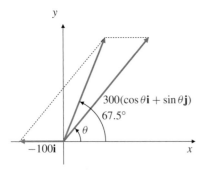

Figure 10.17   The standard basis vectors **i**, **j**, and **k**

Given a Cartesian coordinate system in 3-space, we define three **standard basis vectors**, **i**, **j**, and **k**, represented by arrows from the origin to the points $(1, 0, 0)$, $(0, 1, 0)$, and $(0, 0, 1)$, respectively. (See Figure 10.17.) Any vector in 3-space can be written as a *linear combination* of these basis vectors; for instance, the position vector of the point (x, y, z) is given by

$$\mathbf{r} = x\mathbf{i} + y\mathbf{j} + z\mathbf{k}.$$

We say that **r** has **components** x, y, and z. The length of **r** is

$$|\mathbf{r}| = \sqrt{x^2 + y^2 + z^2}.$$

If $P_1 = (x_1, y_1, z_1)$ and $P_2 = (x_2, y_2, z_2)$ are two points in 3-space, then the vector $\mathbf{v} = \overrightarrow{P_1 P_2}$ from P_1 to P_2 has components $x_2 - x_1$, $y_2 - y_1$, and $z_2 - z_1$ and is therefore represented in terms of the standard basis vectors by

$$\mathbf{v} = \overrightarrow{P_1 P_2} = (x_2 - x_1)\mathbf{i} + (y_2 - y_1)\mathbf{j} + (z_2 - z_1)\mathbf{k}.$$

Example 2   If $\mathbf{u} = 2\mathbf{i} + \mathbf{j} - 2\mathbf{k}$ and $\mathbf{v} = 3\mathbf{i} - 2\mathbf{j} - \mathbf{k}$, find $\mathbf{u} + \mathbf{v}$, $\mathbf{u} - \mathbf{v}$, $3\mathbf{u} - 2\mathbf{v}$, $|\mathbf{u}|$, $|\mathbf{v}|$, and a unit vector $\hat{\mathbf{u}}$ in the direction of **u**.

Solution

$$\mathbf{u} + \mathbf{v} = (2 + 3)\mathbf{i} + (1 - 2)\mathbf{j} + (-2 - 1)\mathbf{k} = 5\mathbf{i} - \mathbf{j} - 3\mathbf{k}$$
$$\mathbf{u} - \mathbf{v} = (2 - 3)\mathbf{i} + (1 + 2)\mathbf{j} + (-2 + 1)\mathbf{k} = -\mathbf{i} + 3\mathbf{j} - \mathbf{k}$$
$$3\mathbf{u} - 2\mathbf{v} = (6 - 6)\mathbf{i} + (3 + 4)\mathbf{j} + (-6 + 2)\mathbf{k} = 7\mathbf{j} - 4\mathbf{k}$$
$$|\mathbf{u}| = \sqrt{4 + 1 + 4} = 3, \qquad |\mathbf{v}| = \sqrt{9 + 4 + 1} = \sqrt{14}$$
$$\hat{\mathbf{u}} = \left(\frac{1}{|\mathbf{u}|}\right)\mathbf{u} = \frac{2}{3}\mathbf{i} + \frac{1}{3}\mathbf{j} - \frac{2}{3}\mathbf{k}.$$

The following example illustrates the way vectors can be used to solve problems involving relative velocities. If A moves with velocity $\mathbf{v}_{A\,\text{rel}\,B}$ relative to B, and B moves with velocity $\mathbf{v}_{B\,\text{rel}\,C}$ relative to C, then A moves with velocity $\mathbf{v}_{A\,\text{rel}\,C}$ relative to C, where

$$\mathbf{v}_{A\,\text{rel}\,C} = \mathbf{v}_{A\,\text{rel}\,B} + \mathbf{v}_{B\,\text{rel}\,C}.$$

Example 3   An aircraft cruises at a speed of 300 km/h in still air. If the wind is blowing from the east at 100 km/h, in what direction should the aircraft head in order to fly in a straight line from city P to city Q, 400 km north northeast of P? How long will the trip take?

Solution   The problem is two-dimensional, so we use plane vectors. Let us choose our coordinate system so that the x- and y-axes point east and north, respectively. Figure 10.18 illustrates the three velocities that must be considered. The velocity of the air relative to the ground is

$$\mathbf{v}_{\text{air rel ground}} = -100\,\mathbf{i}.$$

If the aircraft heads in a direction making angle θ with the positive direction of the x-axis, then the velocity of the aircraft relative to the air is

$$\mathbf{v}_{\text{aircraft rel air}} = 300\cos\theta\,\mathbf{i} + 300\sin\theta\,\mathbf{j}.$$

Thus, the velocity of the aircraft relative to the ground is

$$\mathbf{v}_{\text{aircraft rel ground}} = \mathbf{v}_{\text{aircraft rel air}} + \mathbf{v}_{\text{air rel ground}}$$
$$= (300\cos\theta - 100)\,\mathbf{i} + 300\sin\theta\,\mathbf{j}.$$

Figure 10.18   Velocity diagram for the aircraft in Example 3

We want this latter velocity to be in a north-northeasterly direction, that is, in the direction making angle $3\pi/8 = 67.5°$ with the positive direction of the x-axis. Thus we will have

$$\mathbf{v}_{\text{aircraft rel ground}} = v\left[\left(\cos 67.5°\right)\mathbf{i} + \left(\sin 67.5°\right)\mathbf{j}\right],$$

where v is the actual groundspeed of the aircraft. Comparing the two expressions for $\mathbf{v}_{\text{aircraft rel ground}}$ we obtain

$$300\cos\theta - 100 = v\,\cos 67.5°$$
$$300\sin\theta = v\,\sin 67.5°.$$

Eliminating v between these two equations we get

$$300\,\cos\theta\,\sin 67.5° - 300\,\sin\theta\,\cos 67.5° = 100\,\sin 67.5°,$$

or

$$3\,\sin(67.5° - \theta) = \sin 67.5°.$$

Therefore, the aircraft should head in direction θ given by

$$\theta = 67.5° - \arcsin\left(\frac{1}{3}\sin 67.5°\right) \approx 49.56°,$$

that is, $49.56°$ north of east. The groundspeed is now seen to be

$$v = 300\sin\theta / \sin 67.5° \approx 247.15 \text{ km/h}.$$

Thus, the 400 km trip will take about $400/247.15 \approx 1.618$ hours, or about 1 hour and 37 minutes.

Hanging Cables and Chains

When it is suspended from both ends and allowed to hang under gravity, a heavy cable or chain assumes the shape of a **catenary** curve, which is the graph of the hyperbolic cosine function. We will demonstrate this now, using vectors to keep track of the various forces acting on the cable.

Suppose that the cable has line density δ (units of mass per unit length) and hangs as shown in Figure 10.19. Let us choose a coordinate system so that the lowest point L on the cable is at $(0, y_0)$; we will specify the value of y_0 later. If $P = (x, y)$ is another point on the cable, there are three forces acting on the arc LP of the cable between L and P. These are all forces that we can represent using horizontal and vertical components.

(i) The horizontal tension $\mathbf{H} = -H\mathbf{i}$ at L. This is the force that the part of the cable to the left of L exerts on the arc LP at L.

(ii) The tangential tension $\mathbf{T} = T_h\mathbf{i} + T_v\mathbf{j}$. This is the force the part of the cable to the right of P exerts on arc LP at P.

(iii) The weight $\mathbf{W} = -\delta gs\mathbf{j}$ of arc LP, where g is the acceleration of gravity and s is the length of the arc LP.

Since the cable is not moving, these three forces must balance; their vector sum must be zero:

$$\mathbf{T} + \mathbf{H} + \mathbf{W} = \mathbf{0}$$
$$(T_h - H)\mathbf{i} + (T_v - \delta gs)\mathbf{j} = \mathbf{0}$$

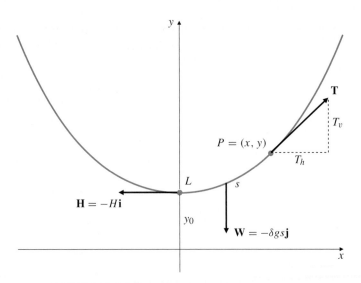

Figure 10.19 A hanging cable and the forces acting on arc LP

Thus $T_h = H$ and $T_v = \delta gs$. Since $\mathbf{T}$ is tangent to the cable at P, the slope of the cable there is

$$\frac{dy}{dx} = \frac{T_v}{T_h} = \frac{\delta gs}{H} = as,$$

where $a = \delta g/H$ is a constant for the given cable. Differentiating with respect to x and using the fact, from our study of arc length, that

$$\frac{ds}{dx} = \sqrt{1 + \left(\frac{dy}{dx}\right)^2},$$

we obtain a second-order differential equation,

$$\frac{d^2y}{dx^2} = a\frac{ds}{dx} = a\sqrt{1 + \left(\frac{dy}{dx}\right)^2},$$

to be solved for the equation of the curve along which the hanging cable lies. The appropriate initial conditions are $y = y_0$ and $dy/dx = 0$ at $x = 0$.

Since the differential equation depends on dy/dx rather than y, we substitute $m(x) = dy/dx$ and obtain a first-order equation for m:

$$\frac{dm}{dx} = a\sqrt{1 + m^2}.$$

This equation is separable; we integrate it using the substitution $m = \sinh u$:

$$\int \frac{1}{\sqrt{1 + m^2}}\, dm = \int a\, dx$$

$$\int du = \int \frac{\cosh u}{\sqrt{1 + \sinh^2 u}}\, du = ax + C_1$$

$$\sinh^{-1} m = u = ax + C_1$$

$$m = \sinh(ax + C_1).$$

Since $m = dy/dx = 0$ at $x = 0$, we have $0 = \sinh C_1$, so $C_1 = 0$ and

$$\frac{dy}{dx} = m = \sinh(ax).$$

This equation is easily integrated to find y. (Had we used a tangent substitution instead of the hyperbolic sine substitution for m we would have had more trouble here.)

$$y = \frac{1}{a}\cosh(ax) + C_2.$$

If we choose $y_0 = y(0) = 1/a$, then, substituting $x = 0$ we will get $C_2 = 0$. With this choice of y_0, we therefore find that the equation of the curve along which the hanging cable lies is the catenary

$$y = \frac{1}{a}\cosh(ax).$$

Remark If a hanging cable bears loads other than its own weight, it will assume a different shape. For example, a cable supporting a level suspension bridge whose weight per unit length is much greater than that of the cable will assume the shape of a parabola. See Exercise 34 below.

The Dot Product and Projections

There is another operation on vectors in any dimension by which two vectors are combined to produce a number called their *dot product*.

DEFINITION 3

> **The dot product of two vectors**
>
> Given two vectors, $\mathbf{u} = u_1\mathbf{i} + u_2\mathbf{j}$ and $\mathbf{v} = v_1\mathbf{i} + v_2\mathbf{j}$ in $\mathbb{R}^2$, we define their **dot product $\mathbf{u} \bullet \mathbf{v}$** to be the sum of the products of their corresponding components:
>
> $$\mathbf{u} \bullet \mathbf{v} = u_1 v_1 + u_2 v_2.$$
>
> The terms **scalar product** and **inner product** are also used in place of dot product. Similarly, for vectors $\mathbf{u} = u_1\mathbf{i} + u_2\mathbf{j} + u_3\mathbf{k}$ and $\mathbf{v} = v_1\mathbf{i} + v_2\mathbf{j} + v_3\mathbf{k}$ in $\mathbb{R}^3$,
>
> $$\mathbf{u} \bullet \mathbf{v} = u_1 v_1 + u_2 v_2 + u_3 v_3.$$

The dot product has the following algebraic properties, easily checked using the definition above:

$$\mathbf{u} \bullet \mathbf{v} = \mathbf{v} \bullet \mathbf{u} \qquad \text{(commutative law)},$$
$$\mathbf{u} \bullet (\mathbf{v} + \mathbf{w}) = \mathbf{u} \bullet \mathbf{v} + \mathbf{u} \bullet \mathbf{w} \qquad \text{(distributive law)},$$
$$(t\mathbf{u}) \bullet \mathbf{v} = \mathbf{u} \bullet (t\mathbf{v}) = t(\mathbf{u} \bullet \mathbf{v}) \qquad \text{(for real } t\text{)},$$
$$\mathbf{u} \bullet \mathbf{u} = |\mathbf{u}|^2.$$

The real significance of the dot product is shown by the following result, which could have been used as the definition of dot product:

THEOREM 1

If θ is the angle between the directions of $\mathbf{u}$ and $\mathbf{v}$ ($0 \leq \theta \leq \pi$), then

$$\mathbf{u} \bullet \mathbf{v} = |\mathbf{u}||\mathbf{v}|\cos\theta.$$

In particular, $\mathbf{u} \bullet \mathbf{v} = 0$ if and only if $\mathbf{u}$ and $\mathbf{v}$ are perpendicular. (Of course, the zero vector is perpendicular to every vector.)

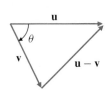

PROOF   Refer to Figure 10.20 and apply the Cosine Law to the triangle with the arrows $\mathbf{u}$, $\mathbf{v}$, and $\mathbf{u} - \mathbf{v}$ as sides:

$$|\mathbf{u}|^2 + |\mathbf{v}|^2 - 2|\mathbf{u}||\mathbf{v}|\cos\theta = |\mathbf{u} - \mathbf{v}|^2 = (\mathbf{u} - \mathbf{v}) \bullet (\mathbf{u} - \mathbf{v})$$
$$= \mathbf{u} \bullet (\mathbf{u} - \mathbf{v}) - \mathbf{v} \bullet (\mathbf{u} - \mathbf{v})$$
$$= \mathbf{u} \bullet \mathbf{u} - \mathbf{u} \bullet \mathbf{v} - \mathbf{v} \bullet \mathbf{u} + \mathbf{v} \bullet \mathbf{v}$$
$$= |\mathbf{u}|^2 + |\mathbf{v}|^2 - 2\mathbf{u} \bullet \mathbf{v}$$

Hence $|\mathbf{u}||\mathbf{v}|\cos\theta = \mathbf{u} \bullet \mathbf{v}$, as claimed.

Figure 10.20

Example 4   Find the angle θ between the vectors $\mathbf{u} = 2\mathbf{i} + \mathbf{j} - 2\mathbf{k}$ and $\mathbf{v} = 3\mathbf{i} - 2\mathbf{j} - \mathbf{k}$.

Solution   Solving the formula $\mathbf{u} \bullet \mathbf{v} = |\mathbf{u}||\mathbf{v}|\cos\theta$ for θ, we obtain

$$\theta = \cos^{-1}\frac{\mathbf{u} \bullet \mathbf{v}}{|\mathbf{u}||\mathbf{v}|} = \cos^{-1}\left(\frac{(2)(3) + (1)(-2) + (-2)(-1)}{3\sqrt{14}}\right)$$
$$= \cos^{-1}\frac{2}{\sqrt{14}} \approx 57.69°.$$

It is sometimes useful to project one vector along another. We define both scalar and vector projections of $\mathbf{u}$ in the direction of $\mathbf{v}$:

DEFINITION   4

Scalar and vector projections

The **scalar projection** s of any vector $\mathbf{u}$ in the direction of a nonzero vector $\mathbf{v}$ is the dot product of $\mathbf{u}$ with a unit vector in the direction of $\mathbf{v}$. Thus, it is the *number*

$$s = \frac{\mathbf{u} \bullet \mathbf{v}}{|\mathbf{v}|} = |\mathbf{u}|\cos\theta,$$

where θ is the angle between $\mathbf{u}$ and $\mathbf{v}$.

The **vector projection**, $\mathbf{u_v}$, of $\mathbf{u}$ in the direction of $\mathbf{v}$ (see Figure 10.21) is the scalar multiple of a unit vector $\hat{\mathbf{v}}$ in the direction of $\mathbf{v}$, by the scalar projection of $\mathbf{u}$ in the direction of $\mathbf{v}$, that is,

$$\text{vector projection of } \mathbf{u} \text{ along } \mathbf{v} = \mathbf{u_v} = \frac{\mathbf{u} \bullet \mathbf{v}}{|\mathbf{v}|}\hat{\mathbf{v}} = \frac{\mathbf{u} \bullet \mathbf{v}}{|\mathbf{v}|^2}\mathbf{v}.$$

Note that $|s|$ is the length of the line segment along the line of $\mathbf{v}$ obtained by dropping perpendiculars to that line from the tail and head of $\mathbf{u}$. (See Figure 10.21.) Also, s is negative if $\theta > 90°$.

It is often necessary to express a vector as a sum of two other vectors parallel and perpendicular to a given direction.

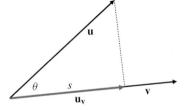

Figure 10.21   The scalar projection s and the vector projection $\mathbf{u_v}$ of vector $\mathbf{u}$ along vector $\mathbf{v}$

Example 5   Express the vector $3\mathbf{i} + \mathbf{j}$ as a sum of vectors $\mathbf{u} + \mathbf{v}$, where $\mathbf{u}$ is parallel to the vector $\mathbf{i} + \mathbf{j}$ and $\mathbf{v}$ is perpendicular to $\mathbf{u}$.

Solution

METHOD I (Using vector projection) Note that $\mathbf{u}$ must be the vector projection of $3\mathbf{i} + \mathbf{j}$ in the direction of $\mathbf{i} + \mathbf{j}$. Thus,

$$\mathbf{u} = \frac{(3\mathbf{i} + \mathbf{j}) \bullet (\mathbf{i} + \mathbf{j})}{|\mathbf{i} + \mathbf{j}|^2}(\mathbf{i} + \mathbf{j}) = \frac{4}{2}(\mathbf{i} + \mathbf{j}) = 2\mathbf{i} + 2\mathbf{j}$$
$$\mathbf{v} = 3\mathbf{i} + \mathbf{j} - \mathbf{u} = \mathbf{i} - \mathbf{j}.$$

METHOD II (From basic principles) Since $\mathbf{u}$ is parallel to $\mathbf{i} + \mathbf{j}$ and $\mathbf{v}$ is perpendicular to $\mathbf{u}$, we have

$$\mathbf{u} = t(\mathbf{i} + \mathbf{j}) \qquad \text{and} \qquad \mathbf{v} \bullet (\mathbf{i} + \mathbf{j}) = 0,$$

for some scalar t. We want $\mathbf{u} + \mathbf{v} = 3\mathbf{i} + \mathbf{j}$. Take the dot product of this equation with $\mathbf{i} + \mathbf{j}$:

$$\mathbf{u} \bullet (\mathbf{i} + \mathbf{j}) + \mathbf{v} \bullet (\mathbf{i} + \mathbf{j}) = (3\mathbf{i} + \mathbf{j}) \bullet (\mathbf{i} + \mathbf{j})$$
$$t(\mathbf{i} + \mathbf{j}) \bullet (\mathbf{i} + \mathbf{j}) + 0 = 4.$$

Thus $2t = 4$, so $t = 2$. Therefore,

$$\mathbf{u} = 2\mathbf{i} + 2\mathbf{j} \qquad \text{and} \qquad \mathbf{v} = 3\mathbf{i} + \mathbf{j} - \mathbf{u} = \mathbf{i} - \mathbf{j}.$$

Vectors in *n*-Space

All the above ideas make sense for vectors in spaces of any dimension. Vectors in $\mathbb{R}^n$ can be expressed as linear combinations of the n unit vectors

$\mathbf{e_1}$ from the origin to the point $(1, 0, 0, \ldots, 0)$

$\mathbf{e_2}$ from the origin to the point $(0, 1, 0, \ldots, 0)$

$\vdots$

$\mathbf{e_n}$ from the origin to the point $(0, 0, 0, \ldots, 1)$.

These vectors constitute a *standard basis* in $\mathbb{R}^n$. The n-vector $\mathbf{x}$ with components $x_1, x_2, \ldots, x_n$ is expressed in the form

$$\mathbf{x} = x_1\mathbf{e_1} + x_2\mathbf{e_2} + \cdots + x_n\mathbf{e_n}.$$

The length of $\mathbf{x}$ is $|\mathbf{x}| = \sqrt{x_1^2 + x_2^2 + \cdots + x_n^2}$. The angle between two vectors $\mathbf{x}$ and $\mathbf{y}$ is

$$\theta = \cos^{-1} \frac{\mathbf{x} \bullet \mathbf{y}}{|\mathbf{x}||\mathbf{y}|},$$

where

$$\mathbf{x} \bullet \mathbf{y} = x_1 y_1 + x_2 y_2 + \cdots + x_n y_n.$$

We will not make much use of n-vectors for $n > 3$ but you should be aware that everything said up until now for 2-vectors or 3-vectors extends to n-vectors.

Exercises 10.2

1. Let $A = (-1, 2)$, $B = (2, 0)$, $C = (1, -3)$, $D = (0, 4)$. Express each of the following vectors as a linear combination of the standard basis vectors $\mathbf{i}$ and $\mathbf{j}$ in $\mathbb{R}^2$.

(a) $\overrightarrow{AB}$,   (b) $\overrightarrow{BA}$,   (c) $\overrightarrow{AC}$,   (d) $\overrightarrow{BD}$,   (e) $\overrightarrow{DA}$,

(f) $\overrightarrow{AB} - \overrightarrow{BC}$,   (g) $\overrightarrow{AC} - 2\overrightarrow{AB} + 3\overrightarrow{CD}$,

(h) $\dfrac{\overrightarrow{AB} + \overrightarrow{AC} + \overrightarrow{AD}}{3}$.

In Exercises 2–3, calculate the following for the given vectors $\mathbf{u}$ and $\mathbf{v}$:

(a) $\mathbf{u} + \mathbf{v}$, $\mathbf{u} - \mathbf{v}$, $2\mathbf{u} - 3\mathbf{v}$,

(b) the lengths $|\mathbf{u}|$ and $|\mathbf{v}|$,

(c) unit vectors $\hat{\mathbf{u}}$ and $\hat{\mathbf{v}}$ in the directions of $\mathbf{u}$ and $\mathbf{v}$, respectively,

   (d) the dot product $\mathbf{u} \bullet \mathbf{v}$,

   (e) the angle between $\mathbf{u}$ and $\mathbf{v}$,

   (f) the scalar projection of $\mathbf{u}$ in the direction of $\mathbf{v}$,

   (g) the vector projection of $\mathbf{v}$ along $\mathbf{u}$.

2. $\mathbf{u} = \mathbf{i} - \mathbf{j}$ and $\mathbf{v} = \mathbf{j} + 2\mathbf{k}$

3. $\mathbf{u} = 3\mathbf{i} + 4\mathbf{j} - 5\mathbf{k}$ and $\mathbf{v} = 3\mathbf{i} - 4\mathbf{j} - 5\mathbf{k}$

4. Use vectors to show that the triangle with vertices $(-1, 1)$, $(2, 5)$, and $(10, -1)$ is right-angled.

In Exercises 5–8, prove the stated geometric result using vectors.

5. The line segment joining the midpoints of two sides of a triangle is parallel to and half as long as the third side.

6. If P, Q, R, and S are midpoints of sides AB, BC, CD, and DA, respectively, of quadrilateral $ABCD$, then $PQRS$ is a parallelogram.

$*$ **7.** The diagonals of any parallelogram bisect each other.

$*$ **8.** The medians of any triangle meet in a common point. (A median is a line joining one vertex to the midpoint of the opposite side. The common point is the *centroid* of the triangle.)

9. A weather vane mounted on the top of a car moving due north at 50 km/h indicates that the wind is coming from the west. When the car doubles its speed, the weather vane indicates that the wind is coming from the northwest. From what direction is the wind coming, and what is its speed?

10. A straight river 500 m wide flows due east at a constant speed of 3 km/h. If you can row your boat at a speed of 5 km/h in still water, in what direction should you head if you wish to row from point A on the south shore to point B on the north shore directly north of A? How long will the trip take?

$*$**11.** In what direction should you head to cross the river in Exercise 10 if you can only row at 2 km/h, and you wish to row from A to point C on the north shore, k km downstream from B? For what values of k is the trip not possible?

12. A certain aircraft flies with an airspeed of 750 km/h. In what direction should it head in order to make progress in a true easterly direction if the wind is from the northeast at 100 km/h? How long will it take to complete a trip to a city 1,500 km from its starting point?

13. For what value of t is the vector $2t\mathbf{i} + 4\mathbf{j} - (10 + t)\mathbf{k}$ perpendicular to the vector $\mathbf{i} + t\mathbf{j} + \mathbf{k}$?

14. Find the angle between a diagonal of a cube and one of the edges of the cube.

15. Find the angle between a diagonal of a cube and a diagonal of one of the faces of the cube. Give all possible answers.

16. **(Direction cosines)** If a vector $\mathbf{u}$ in $\mathbb{R}^3$ makes angles α, β, and γ with the coordinate axes, show that

$$\hat{\mathbf{u}} = \cos\alpha\,\mathbf{i} + \cos\beta\,\mathbf{j} + \cos\gamma\,\mathbf{k}$$

is a unit vector in the direction of $\mathbf{u}$, so $\cos^2\alpha + \cos^2\beta + \cos^2\gamma = 1$. The numbers $\cos\alpha$, $\cos\beta$, and $\cos\gamma$ are called the *direction cosines* of $\mathbf{u}$.

17. Find a unit vector that makes equal angles with the three coordinate axes.

18. Find the three angles of the triangle with vertices $(1, 0, 0)$, $(0, 2, 0)$, and $(0, 0, 3)$.

19. If $\mathbf{r}_1$ and $\mathbf{r}_2$ are the position vectors of two points, P_1 and P_2, and λ is a real number, show that

$$\mathbf{r} = (1 - \lambda)\mathbf{r}_1 + \lambda\mathbf{r}_2$$

is the position vector of a point P on the straight line joining P_1 and P_2. Where is P if $\lambda = 1/2$? if $\lambda = 2/3$? if $\lambda = -1$? if $\lambda = 2$?

20. Let $\mathbf{a}$ be a nonzero vector. Describe the set of all points in 3-space whose position vectors $\mathbf{r}$ satisfy $\mathbf{a} \bullet \mathbf{r} = 0$.

21. Let $\mathbf{a}$ be a nonzero vector, and let b be any real number. Describe the set of all points in 3-space whose position vectors $\mathbf{r}$ satisfy $\mathbf{a} \bullet \mathbf{r} = b$.

In Exercises 22–24, $\mathbf{u} = 2\mathbf{i} + \mathbf{j} - 2\mathbf{k}$, $\mathbf{v} = \mathbf{i} + 2\mathbf{j} - 2\mathbf{k}$, and $\mathbf{w} = 2\mathbf{i} - 2\mathbf{j} + \mathbf{k}$.

22. Find two unit vectors each of which is perpendicular to both $\mathbf{u}$ and $\mathbf{v}$.

23. Find a vector $\mathbf{x}$ satisfying the system of equations $\mathbf{x} \bullet \mathbf{u} = 9$, $\mathbf{x} \bullet \mathbf{v} = 4$, $\mathbf{x} \bullet \mathbf{w} = 6$.

24. Find two unit vectors each of which makes equal angles with $\mathbf{u}$, $\mathbf{v}$, and $\mathbf{w}$.

25. Find a unit vector that bisects the angle between any two nonzero vectors $\mathbf{u}$ and $\mathbf{v}$.

26. Given two nonparallel vectors $\mathbf{u}$ and $\mathbf{v}$, describe the set of all points whose position vectors $\mathbf{r}$ are of the form $\mathbf{r} = \lambda\mathbf{u} + \mu\mathbf{v}$, where λ and μ are arbitrary real numbers.

27. **(The triangle inequality)** Let $\mathbf{u}$ and $\mathbf{v}$ be two vectors.

   (a) Show that $|\mathbf{u} + \mathbf{v}|^2 = |\mathbf{u}|^2 + 2\mathbf{u} \bullet \mathbf{v} + |\mathbf{v}|^2$.

   (b) Show that $\mathbf{u} \bullet \mathbf{v} \le |\mathbf{u}||\mathbf{v}|$.

   (c) Deduce from (a) and (b) that $|\mathbf{u} + \mathbf{v}| \le |\mathbf{u}| + |\mathbf{v}|$.

28. (a) Why is the inequality in Exercise 27(c) called a triangle inequality?

   (b) What conditions on $\mathbf{u}$ and $\mathbf{v}$ imply that $|\mathbf{u} + \mathbf{v}| = |\mathbf{u}| + |\mathbf{v}|$?

29. **(Orthonormal bases)** Let $\mathbf{u} = \frac{3}{5}\mathbf{i} + \frac{4}{5}\mathbf{j}$, $\mathbf{v} = \frac{4}{5}\mathbf{i} - \frac{3}{5}\mathbf{j}$, and $\mathbf{w} = \mathbf{k}$.

   (a) Show that $|\mathbf{u}| = |\mathbf{v}| = |\mathbf{w}| = 1$ and $\mathbf{u} \bullet \mathbf{v} = \mathbf{u} \bullet \mathbf{w} = \mathbf{v} \bullet \mathbf{w} = 0$. The vectors $\mathbf{u}$, $\mathbf{v}$, and $\mathbf{w}$ are mutually perpendicular unit vectors and as such are said to constitute an **orthonormal basis** for $\mathbb{R}^3$.

   (b) If $\mathbf{r} = x\mathbf{i} + y\mathbf{j} + z\mathbf{k}$, show by direct calculation that

$$\mathbf{r} = (\mathbf{r} \bullet \mathbf{u})\mathbf{u} + (\mathbf{r} \bullet \mathbf{v})\mathbf{v} + (\mathbf{r} \bullet \mathbf{w})\mathbf{w}.$$

30. Show that if $\mathbf{u}$, $\mathbf{v}$, and $\mathbf{w}$ are any three mutually perpendicular unit vectors in $\mathbb{R}^3$ and $\mathbf{r} = a\mathbf{u} + b\mathbf{v} + c\mathbf{w}$, then $a = \mathbf{r} \bullet \mathbf{u}$, $b = \mathbf{r} \bullet \mathbf{v}$, and $c = \mathbf{r} \bullet \mathbf{w}$.

31. **(Resolving a vector in perpendicular directions)** If $\mathbf{a}$ is a nonzero vector and $\mathbf{w}$ is any vector, find vectors $\mathbf{u}$ and $\mathbf{v}$ such that $\mathbf{w} = \mathbf{u} + \mathbf{v}$, $\mathbf{u}$ is parallel to $\mathbf{a}$, and $\mathbf{v}$ is perpendicular to $\mathbf{a}$.

32. **(Expressing a vector as a linear combination of two other vectors with which it is coplanar)** Suppose that $\mathbf{u}$, $\mathbf{v}$, and $\mathbf{r}$ are position vectors of points U, V, and P, respectively, that $\mathbf{u}$ is not parallel to $\mathbf{v}$, and that P lies in the plane containing the origin, U, and V. Show that there exist numbers λ and μ such that $\mathbf{r} = \lambda\mathbf{u} + \mu\mathbf{v}$. *Hint:* Resolve both $\mathbf{v}$ and $\mathbf{r}$ as sums of vectors parallel and perpendicular to $\mathbf{u}$ as suggested in Exercise 31.

***33.** Given constants r, s, and t, with $r \neq 0$ and $s \neq 0$, and given a vector $\mathbf{a}$ satisfying $|\mathbf{a}|^2 > 4rst$, solve the system of equations

$$\begin{cases} r\mathbf{x} + s\mathbf{y} = \mathbf{a} \\ \mathbf{x} \bullet \mathbf{y} = t \end{cases}$$

for the unknown vectors $\mathbf{x}$ and $\mathbf{y}$.

Hanging cables

34. (A suspension bridge) If a hanging cable is supporting weight with constant horizontal line density (so that the weight supported by the arc LP in Figure 10.19 is δgx rather than δgs, show that the cable assumes the shape of a parabola rather than a catenary. Such is likely to be the case for the cables of a suspension bridge.

35. At a point P, 10 m away horizontally from its lowest point L, a cable makes an angle 55° with the horizontal. Find the length of the cable between L and P.

36. Calculate the length s of the arc LP of the hanging cable in Figure 10.19 using the equation $y = (1/a)\cosh(ax)$ obtained for the cable. Hence, verify that the magnitude $T = |\mathbf{T}|$ of the tension in the cable at any point $P = (x, y)$ is $T = \delta gy$.

37. A cable 100 m long hangs between two towers 90 m apart so that its ends are attached at the same height on the two towers. How far below that height is the lowest point on the cable?

10.3 The Cross Product in 3-Space

There is defined, *in 3-space only*, another kind of product of two vectors called a *cross product* or *vector product*, and denoted $\mathbf{u} \times \mathbf{v}$.

DEFINITION 5

For any vectors $\mathbf{u}$ and $\mathbf{v}$ in $\mathbb{R}^3$, the **cross product** $\mathbf{u} \times \mathbf{v}$ is the unique vector satisfying the following three conditions:

(i) $(\mathbf{u} \times \mathbf{v}) \bullet \mathbf{u} = 0$ and $(\mathbf{u} \times \mathbf{v}) \bullet \mathbf{v} = 0$,

(ii) $|\mathbf{u} \times \mathbf{v}| = |\mathbf{u}||\mathbf{v}| \sin\theta$, where θ is the angle between $\mathbf{u}$ and $\mathbf{v}$, and

(iii) $\mathbf{u}$, $\mathbf{v}$, and $\mathbf{u} \times \mathbf{v}$ form a right-handed triad.

If $\mathbf{u}$ and $\mathbf{v}$ are parallel, condition (ii) says that $\mathbf{u} \times \mathbf{v} = \mathbf{0}$, the zero vector. Otherwise, through any point in $\mathbb{R}^3$ there is a unique straight line that is perpendicular to both $\mathbf{u}$ and $\mathbf{v}$. Condition (i) says that $\mathbf{u} \times \mathbf{v}$ is parallel to this line. Condition (iii) determines which of the two directions along this line is the direction of $\mathbf{u} \times \mathbf{v}$; a right-handed screw advances in the direction of $\mathbf{u} \times \mathbf{v}$ if rotated in the direction from $\mathbf{u}$ toward $\mathbf{v}$. (This is equivalent to saying that the thumb, forefinger, and middle finger of the right hand can be made to point in the directions of $\mathbf{u}$, $\mathbf{v}$, and $\mathbf{u} \times \mathbf{v}$, respectively.)

If $\mathbf{u}$ and $\mathbf{v}$ have their tails at the point P, then $\mathbf{u} \times \mathbf{v}$ is normal (i.e., perpendicular) to the plane through P in which $\mathbf{u}$ and $\mathbf{v}$ lie and, by condition (ii), $\mathbf{u} \times \mathbf{v}$ has length equal to the area of the parallelogram spanned by $\mathbf{u}$ and $\mathbf{v}$. (See Figure 10.22.) These properties make the cross product very useful for the description of tangent planes and normal lines to surfaces in $\mathbb{R}^3$.

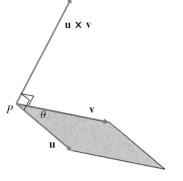

Figure 10.22 $\mathbf{u} \times \mathbf{v}$ is perpendicular to both $\mathbf{u}$ and $\mathbf{v}$ and has length equal to the area of the shaded parallelogram

The definition of cross product given above does not involve any coordinate system and therefore does not directly show the components of the cross product with respect to the standard basis. These components are provided by the following theorem:

THEOREM 2

Components of the cross product

If $\mathbf{u} = u_1\mathbf{i} + u_2\mathbf{j} + u_3\mathbf{k}$ and $\mathbf{v} = v_1\mathbf{i} + v_2\mathbf{j} + v_3\mathbf{k}$, then

$$\mathbf{u} \times \mathbf{v} = (u_2 v_3 - u_3 v_2)\mathbf{i} + (u_3 v_1 - u_1 v_3)\mathbf{j} + (u_1 v_2 - u_2 v_1)\mathbf{k}.$$

PROOF First, we observe that the vector

$$\mathbf{w} = (u_2 v_3 - u_3 v_2)\mathbf{i} + (u_3 v_1 - u_1 v_3)\mathbf{j} + (u_1 v_2 - u_2 v_1)\mathbf{k}$$

is perpendicular to both $\mathbf{u}$ and $\mathbf{v}$ since

$$\mathbf{u} \bullet \mathbf{w} = u_1(u_2 v_3 - u_3 v_2) + u_2(u_3 v_1 - u_1 v_3) + u_3(u_1 v_2 - u_2 v_1) = 0,$$

and similarly $\mathbf{v} \bullet \mathbf{w} = 0$. Thus $\mathbf{u} \times \mathbf{v}$ is parallel to $\mathbf{w}$. Next, we show that $\mathbf{w}$ and $\mathbf{u} \times \mathbf{v}$ have the same length. In fact,

$$\begin{aligned}
|\mathbf{w}|^2 &= (u_2v_3 - u_3v_2)^2 + (u_3v_1 - u_1v_3)^2 + (u_1v_2 - u_2v_1)^2 \\
&= u_2^2v_3^2 + u_3^2v_2^2 - 2u_2v_3u_3v_2 + u_3^2v_1^2 + u_1^2v_3^2 \\
&\quad - 2u_3v_1u_1v_3 + u_1^2v_2^2 + u_2^2v_1^2 - 2u_1v_2u_2v_1,
\end{aligned}$$

while

$$\begin{aligned}
|\mathbf{u} \times \mathbf{v}|^2 &= |\mathbf{u}|^2|\mathbf{v}|^2 \sin^2\theta \\
&= |\mathbf{u}|^2|\mathbf{v}|^2 (1 - \cos^2\theta) \\
&= |\mathbf{u}|^2|\mathbf{v}|^2 - (\mathbf{u} \bullet \mathbf{v})^2 \\
&= (u_1^2 + u_2^2 + u_3^2)(v_1^2 + v_2^2 + v_3^2) - (u_1v_1 + u_2v_2 + u_3v_3)^2 \\
&= u_1^2v_1^2 + u_1^2v_2^2 + u_1^2v_3^2 + u_2^2v_1^2 + u_2^2v_2^2 + u_2^2v_3^2 + u_3^2v_1^2 + u_3^2v_2^2 + u_3^2v_3^2 \\
&\quad - u_1^2v_1^2 - u_2^2v_2^2 - u_3^2v_3^2 - 2u_1v_1u_2v_2 - 2u_1v_1u_3v_3 - 2u_2v_2u_3v_3 \\
&= |\mathbf{w}|^2.
\end{aligned}$$

Since $\mathbf{w}$ is parallel to, and has the same length as, $\mathbf{u} \times \mathbf{v}$, we must have either $\mathbf{u} \times \mathbf{v} = \mathbf{w}$ or $\mathbf{u} \times \mathbf{v} = -\mathbf{w}$. It remains to be shown that the first of these is the correct choice. To see this, suppose that the triad of vectors $\mathbf{u}$, $\mathbf{v}$, and $\mathbf{w}$ is rigidly rotated in 3-space so that $\mathbf{u}$ points in the direction of the positive x-axis and $\mathbf{v}$ lies in the upper half of the xy-plane. Then $\mathbf{u} = u_1\mathbf{i}$, and $\mathbf{v} = v_1\mathbf{i} + v_2\mathbf{j}$, where $u_1 > 0$ and $v_2 > 0$. By the "right-hand rule" $\mathbf{u} \times \mathbf{v}$ must point in the direction of the positive z-axis. But $\mathbf{w} = u_1v_2\mathbf{k}$ does point in that direction, so $\mathbf{u} \times \mathbf{v} = \mathbf{w}$, as asserted.

The formula for the cross product in terms of components may seem awkward and asymmetric. As we shall see, however, it can be written more easily in terms of a determinant. We introduce determinants later in this section.

Example 1 **(Calculating cross products)**

(a)
$$\begin{array}{lll}
\mathbf{i} \times \mathbf{i} = \mathbf{0}, & \mathbf{i} \times \mathbf{j} = \mathbf{k}, & \mathbf{j} \times \mathbf{i} = -\mathbf{k}, \\
\mathbf{j} \times \mathbf{j} = \mathbf{0}, & \mathbf{j} \times \mathbf{k} = \mathbf{i}, & \mathbf{k} \times \mathbf{j} = -\mathbf{i}, \\
\mathbf{k} \times \mathbf{k} = \mathbf{0}, & \mathbf{k} \times \mathbf{i} = \mathbf{j}, & \mathbf{i} \times \mathbf{k} = -\mathbf{j}.
\end{array}$$

(b) $(2\mathbf{i} + \mathbf{j} - 3\mathbf{k}) \times (-2\mathbf{j} + 5\mathbf{k})$
$$\begin{aligned}
&= \big((1)(5) - (-2)(-3)\big)\mathbf{i} + \big((-3)(0) - (2)(5)\big)\mathbf{j} + \big((2)(-2) - (1)(0)\big)\mathbf{k} \\
&= -\mathbf{i} - 10\mathbf{j} - 4\mathbf{k}.
\end{aligned}$$

The cross product has some but not all of the properties we usually ascribe to products. We summarize its algebraic properties as follows:

Properties of the cross product

If $\mathbf{u}$, $\mathbf{v}$, and $\mathbf{w}$ are any vectors in $\mathbb{R}^3$, and t is a real number (a scalar), then

(i) $\mathbf{u} \times \mathbf{u} = \mathbf{0}$,

(ii) $\mathbf{u} \times \mathbf{v} = -\mathbf{v} \times \mathbf{u}$,   (The cross product is **anticommutative**.)

(iii) $(\mathbf{u} + \mathbf{v}) \times \mathbf{w} = \mathbf{u} \times \mathbf{w} + \mathbf{v} \times \mathbf{w}$,

(iv) $\mathbf{u} \times (\mathbf{v} + \mathbf{w}) = \mathbf{u} \times \mathbf{v} + \mathbf{u} \times \mathbf{w}$,

(v) $(t\mathbf{u}) \times \mathbf{v} = \mathbf{u} \times (t\mathbf{v}) = t(\mathbf{u} \times \mathbf{v})$,

(vi) $\mathbf{u} \bullet (\mathbf{u} \times \mathbf{v}) = \mathbf{v} \bullet (\mathbf{u} \times \mathbf{v}) = 0$.

These identities are all easily verified using the components or the definition of the cross product or by using properties of determinants discussed below. They are left as exercises for the reader. Note the absence of an associative law. The cross product is not associative. (See Exercise 21 at the end of this section.) In general,

$$\mathbf{u} \times (\mathbf{v} \times \mathbf{w}) \neq (\mathbf{u} \times \mathbf{v}) \times \mathbf{w}.$$

Determinants

In order to simplify certain formulas such as the component representation of the cross product, we introduce 2×2 and 3×3 **determinants**. General $n \times n$ determinants are normally studied in courses on linear algebra; we will encounter them in Section 10.6. In this section we will outline enough of the properties of determinants to enable us to use them as shorthand in some otherwise complicated formulas.

A determinant is an expression that involves the elements of a square array (matrix) of numbers. The determinant of the 2×2 array of numbers

$$\begin{matrix} a & b \\ c & d \end{matrix}$$

is denoted by enclosing the array between vertical bars, and its value is the number $ad - bc$:

$$\begin{vmatrix} a & b \\ c & d \end{vmatrix} = ad - bc.$$

Figure 10.23

This is the product of elements in the *downward diagonal* of the array minus the product of elements in the *upward diagonal* as shown in Figure 10.23. For example,

$$\begin{vmatrix} 1 & 2 \\ 3 & 4 \end{vmatrix} = (1)(4) - (2)(3) = -2.$$

Similarly, the determinant of a 3×3 array of numbers is defined by

$$\begin{vmatrix} a & b & c \\ d & e & f \\ g & h & i \end{vmatrix} = aei + bfg + cdh - gec - hfa - idb.$$

Observe that each of the six products in the value of the determinant involves exactly one element from each row and exactly one from each column of the array. As such, each term is the product of elements in a *diagonal* of an *extended* array obtained by repeating the first two columns of the array to the right of the third column, as shown in Figure 10.24. The value of the determinant is the sum of products corresponding to the three complete *downward* diagonals minus the sum corresponding to the three *upward* diagonals. With practice you will be able to form these diagonal products without having to write the extended array.

If we group the terms in the expansion of the determinant to factor out the elements of the first row, we obtain

Figure 10.24   WARNING: This method does not work for 4×4 or higher-order determinants!

$$\begin{vmatrix} a & b & c \\ d & e & f \\ g & h & i \end{vmatrix} = a(ei - fh) - b(di - fg) + c(dh - eg)$$

$$= a \begin{vmatrix} e & f \\ h & i \end{vmatrix} - b \begin{vmatrix} d & f \\ g & i \end{vmatrix} + c \begin{vmatrix} d & e \\ g & h \end{vmatrix}.$$

The 2×2 determinants appearing here (called *minors* of the given 3×3 determinant) are obtained by deleting the row and column containing the corresponding element from the original 3×3 determinant. This process is called *expanding* the 3×3 determinant *in minors* about the first row.

Such expansions in minors can be carried out about any row or column. Note that a minus sign appears in any term whose minor is obtained by deleting the ith row and jth column, where $i + j$ is an *odd* number. For example, we can expand the above determinant in minors about the second column as follows:

$$\begin{vmatrix} a & b & c \\ d & e & f \\ g & h & i \end{vmatrix} = -b \begin{vmatrix} d & f \\ g & i \end{vmatrix} + e \begin{vmatrix} a & c \\ g & i \end{vmatrix} - h \begin{vmatrix} a & c \\ d & f \end{vmatrix}$$

$$= -bdi + bfg + eai - ecg - haf + hcd.$$

(Of course, this is the same value as the one obtained previously.)

Example 2

$$\begin{vmatrix} 1 & 4 & -2 \\ -3 & 1 & 0 \\ 2 & 2 & -3 \end{vmatrix} = 3 \begin{vmatrix} 4 & -2 \\ 2 & -3 \end{vmatrix} + 1 \begin{vmatrix} 1 & -2 \\ 2 & -3 \end{vmatrix}$$

$$= 3(-8) + 1 = -23.$$

We expanded about the second row; the third column would also have been a good choice. (Why?)

Any row (or column) of a determinant may be regarded as the components of a vector. Then the determinant is a *linear function* of that vector. For example,

$$\begin{vmatrix} a & b & c \\ d & e & f \\ sx + tl & sy + tm & sz + tn \end{vmatrix} = s \begin{vmatrix} a & b & c \\ d & e & f \\ x & y & z \end{vmatrix} + t \begin{vmatrix} a & b & c \\ d & e & f \\ l & m & n \end{vmatrix}$$

because the determinant is a linear function of its third row. This and other properties of determinants follow directly from the definition. Some other properties are summarized below. These are stated for rows and for 3×3 determinants, but similar statements can be made for columns and for determinants of any order.

Properties of determinants

(i) If two rows of a determinant are interchanged, then the determinant changes sign:

$$\begin{vmatrix} d & e & f \\ a & b & c \\ g & h & i \end{vmatrix} = - \begin{vmatrix} a & b & c \\ d & e & f \\ g & h & i \end{vmatrix}.$$

(ii) If two rows of a determinant are equal, the determinant has value 0:

$$\begin{vmatrix} a & b & c \\ a & b & c \\ g & h & i \end{vmatrix} = 0.$$

(iii) If a multiple of one row of a determinant is added to another row, the value of the determinant remains unchanged:

$$\begin{vmatrix} a & b & c \\ d + ta & e + tb & f + tc \\ g & h & i \end{vmatrix} = \begin{vmatrix} a & b & c \\ d & e & f \\ g & h & i \end{vmatrix}.$$

The Cross Product as a Determinant

The elements of a determinant are usually numbers because they have to be multiplied to get the value of the determinant. However, it is possible to use vectors as the elements of *one row* (or column) of a determinant. When expanding in minors about that row (or column), the minor for each vector element is a number that determines the scalar multiple of the vector. The formula for the cross product of

$$\mathbf{u} = u_1\mathbf{i} + u_2\mathbf{j} + u_3\mathbf{k} \quad \text{and} \quad \mathbf{v} = v_1\mathbf{i} + v_2\mathbf{j} + v_3\mathbf{k}$$

presented in Theorem 2 can be expressed symbolically as a determinant with the standard basis vectors as the elements of the first row:

$$\mathbf{u} \times \mathbf{v} = \begin{vmatrix} \mathbf{i} & \mathbf{j} & \mathbf{k} \\ u_1 & u_2 & u_3 \\ v_1 & v_2 & v_3 \end{vmatrix} = \begin{vmatrix} u_2 & u_3 \\ v_2 & v_3 \end{vmatrix}\mathbf{i} - \begin{vmatrix} u_1 & u_3 \\ v_1 & v_3 \end{vmatrix}\mathbf{j} + \begin{vmatrix} u_1 & u_2 \\ v_1 & v_2 \end{vmatrix}\mathbf{k}.$$

The formula for the cross product given in that theorem is just the expansion of this determinant in minors about the first row.

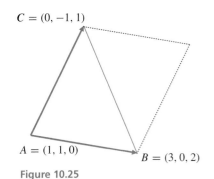

$C = (0, -1, 1)$

$A = (1, 1, 0)$

$B = (3, 0, 2)$

Figure 10.25

Example 3　Find the area of the triangle with vertices at the three points $A = (1, 1, 0)$, $B = (3, 0, 2)$, and $C = (0, -1, 1)$.

Solution　Two sides of the triangle (Figure 10.25) are given by the vectors:

$$\overrightarrow{AB} = 2\mathbf{i} - \mathbf{j} + 2\mathbf{k} \quad \text{and} \quad \overrightarrow{AC} = -\mathbf{i} - 2\mathbf{j} + \mathbf{k}.$$

The area of the triangle is half the area of the parallelogram spanned by $\overrightarrow{AB}$ and $\overrightarrow{AC}$. By the definition of cross product, the area of the triangle must therefore be

$$\frac{1}{2}|\overrightarrow{AB} \times \overrightarrow{AC}| = \frac{1}{2}\left| \begin{vmatrix} \mathbf{i} & \mathbf{j} & \mathbf{k} \\ 2 & -1 & 2 \\ -1 & -2 & 1 \end{vmatrix} \right|$$

$$= \frac{1}{2}|3\mathbf{i} - 4\mathbf{j} - 5\mathbf{k}| = \frac{1}{2}\sqrt{9 + 16 + 25} = \frac{5}{2}\sqrt{2} \text{ square units.}$$

A **parallelepiped** is the three-dimensional analogue of a parallelogram. It is a solid with three pairs of parallel planar faces. Each face is in the shape of a parallelogram. A rectangular brick is a special case of a parallelepiped in which nonparallel faces intersect at right angles. We say that a parallelepiped is **spanned** by three vectors coinciding with three of its edges that meet at one vertex. (See Figure 10.26.)

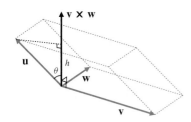

$\mathbf{v} \times \mathbf{w}$

$\mathbf{u}$

θ　h

$\mathbf{w}$

$\mathbf{v}$

Figure 10.26

Example 4　Find the volume of the parallelepiped spanned by $\mathbf{u}$, $\mathbf{v}$, and $\mathbf{w}$.

Solution　The volume of the parallelepiped is equal to the area of one of its faces, say, the face spanned by $\mathbf{v}$ and $\mathbf{w}$, multiplied by the height of the parallelepiped measured in a direction perpendicular to that face. The area of the face is $|\mathbf{v} \times \mathbf{w}|$. Since $\mathbf{v} \times \mathbf{w}$ is perpendicular to the face, the height h of the parallelepiped will be the absolute value of the scalar projection of $\mathbf{u}$ along $\mathbf{v} \times \mathbf{w}$. If θ is the angle between $\mathbf{u}$ and $\mathbf{v} \times \mathbf{w}$, then the volume of the parallelepiped is given by

$$\text{Volume} = |\mathbf{u}||\mathbf{v} \times \mathbf{w}||\cos\theta| = |\mathbf{u} \bullet (\mathbf{v} \times \mathbf{w})| \text{ cubic units.}$$

DEFINITION　6

The quantity $\mathbf{u} \bullet (\mathbf{v} \times \mathbf{w})$ is called the **scalar triple product** of the vectors $\mathbf{u}$, $\mathbf{v}$, and $\mathbf{w}$.

The scalar triple product is easily expressed in terms of a determinant. If $\mathbf{u} = u_1\mathbf{i} + u_2\mathbf{j} + u_3\mathbf{k}$, and similar representations hold for $\mathbf{v}$ and $\mathbf{w}$, then

$$\mathbf{u} \bullet (\mathbf{v} \times \mathbf{w}) = u_1 \begin{vmatrix} v_2 & v_3 \\ w_2 & w_3 \end{vmatrix} - u_2 \begin{vmatrix} v_1 & v_3 \\ w_1 & w_3 \end{vmatrix} + u_3 \begin{vmatrix} v_1 & v_2 \\ w_1 & w_2 \end{vmatrix}$$

$$= \begin{vmatrix} u_1 & u_2 & u_3 \\ v_1 & v_2 & v_3 \\ w_1 & w_2 & w_3 \end{vmatrix}.$$

The volume of the parallelepiped spanned by $\mathbf{u}$, $\mathbf{v}$, and $\mathbf{w}$ is the absolute value of this determinant.

Using the properties of the determinant, it is easily verified that

$$\mathbf{u} \bullet (\mathbf{v} \times \mathbf{w}) = \mathbf{v} \bullet (\mathbf{w} \times \mathbf{u}) = \mathbf{w} \bullet (\mathbf{u} \times \mathbf{v}).$$

(See Exercise 18 below.) Note that $\mathbf{u}$, $\mathbf{v}$, and $\mathbf{w}$ remain in the same *cyclic order* in these three expressions. Reversing the order would introduce a factor -1:

$$\mathbf{u} \bullet (\mathbf{v} \times \mathbf{w}) = -\mathbf{u} \bullet (\mathbf{w} \times \mathbf{v}).$$

Three vectors in 3-space are said to be **coplanar** if the parallelepiped they span has zero volume; if their tails coincide, three such vectors must lie in the same plane.

$$\mathbf{u}, \mathbf{v}, \text{ and } \mathbf{w} \text{ are coplanar} \iff \mathbf{u} \bullet (\mathbf{v} \times \mathbf{w}) = 0$$

$$\iff \begin{vmatrix} u_1 & u_2 & u_3 \\ v_1 & v_2 & v_3 \\ w_1 & w_2 & w_3 \end{vmatrix} = 0.$$

Three vectors are certainly coplanar if any of them is $\mathbf{0}$, or if any pair of them is parallel. If neither of these degenerate conditions apply, they are only coplanar if any one of them can be expressed as a linear combination of the other two. (See Exercise 20 below.)

Applications of Cross Products

Cross products are of considerable importance in mechanics and electromagnetic theory, as well as in the study of motion in general. For example:

(a) The linear velocity $\mathbf{v}$ of a particle located at position $\mathbf{r}$ in a body rotating with angular velocity $\boldsymbol{\Omega}$ about the origin is given by $\mathbf{v} = \boldsymbol{\Omega} \times \mathbf{r}$. (See Section 11.2 for more details.)

(b) The angular momentum of a planet of mass m moving with velocity $\mathbf{v}$ in its orbit around the sun is given by $\mathbf{h} = \mathbf{r} \times m\mathbf{v}$, where $\mathbf{r}$ is the position vector of the planet relative to the sun as origin. (See Section 11.6.)

(c) If a particle of electric charge q is travelling with velocity $\mathbf{v}$ through a magnetic field whose strength and direction are given by vector $\mathbf{B}$, then the force that the field exerts on the particle is given by $\mathbf{F} = q\mathbf{v} \times \mathbf{B}$. The electron beam in a television tube is controlled by magnetic fields using this principle.

(d) The torque $\mathbf{T}$ of a force $\mathbf{F}$ applied at the point P with position vector $\mathbf{r}$ about another point P_0 with position vector $\mathbf{r}_0$ is defined to be

$$\mathbf{T} = \overrightarrow{P_0 P} \times \mathbf{F} = (\mathbf{r} - \mathbf{r}_0) \times \mathbf{F}.$$

This torque measures the effectiveness of the force $\mathbf{F}$ in causing rotation about P_0. The direction of $\mathbf{T}$ is along the axis through P_0 about which $\mathbf{F}$ acts to rotate P.

Example 5 An automobile wheel has centre at the origin and axle along the y-axis. One of the retaining nuts holding the wheel is at position $P_0 = (0, 0, 10)$. (Distances are measured in centimetres.) A bent tire wrench with arm 25 cm long and inclined at an angle of 60° to the direction of its handle is fitted to the nut in an upright direction, as shown in Figure 10.27. If a horizontal force $\mathbf{F} = 500\mathbf{i}$ newtons (N) is applied to the handle of the wrench, what is its torque on the nut? What part (component) of this torque is effective in trying to rotate the nut about its horizontal axis? What is the effective torque trying to rotate the wheel?

Solution The nut is at position $\mathbf{r}_0 = 10\mathbf{k}$, and the handle of the wrench is at position

$$\mathbf{r} = 25\cos 60°\mathbf{j} + (10 + 25\sin 60°)\mathbf{k} \approx 12.5\mathbf{j} + 31.65\mathbf{k}.$$

The torque of the force $\mathbf{F}$ on the nut is

$$\mathbf{T} = (\mathbf{r} - \mathbf{r}_0) \times \mathbf{F}$$
$$\approx (12.5\mathbf{j} + 21.65\mathbf{k}) \times 500\mathbf{i} \approx 10{,}825\mathbf{j} - 6{,}250\mathbf{k},$$

which is at right angles to $\mathbf{F}$ and to the arm of the wrench. Only the horizontal component of this torque is effective in turning the nut. This component is 10,825 N·cm or 108.25 N·m in magnitude. For the effective torque on the wheel itself, we have to replace $\mathbf{r}_0$ by $\mathbf{0}$, the position of the centre of the wheel. In this case the horizontal torque is

$$31.65\mathbf{k} \times 500\mathbf{i} \approx 15{,}825\mathbf{j},$$

that is, about 158.25 N·m.

Figure 10.27 The force on the handle is 500 N in a direction directly toward you

Exercises 10.3

1. Calculate $\mathbf{u} \times \mathbf{v}$ if $\mathbf{u} = \mathbf{i} - 2\mathbf{j} + 3\mathbf{k}$ and $\mathbf{v} = 3\mathbf{i} + \mathbf{j} - 4\mathbf{k}$.

2. Calculate $\mathbf{u} \times \mathbf{v}$ if $\mathbf{u} = \mathbf{j} + 2\mathbf{k}$ and $\mathbf{v} = -\mathbf{i} - \mathbf{j} + \mathbf{k}$.

3. Find the area of the triangle with vertices $(1, 2, 0)$, $(1, 0, 2)$, and $(0, 3, 1)$.

4. Find a unit vector perpendicular to the plane containing the points $(a, 0, 0)$, $(0, b, 0)$, and $(0, 0, c)$. What is the area of the triangle with these vertices?

5. Find a unit vector perpendicular to the vectors $\mathbf{i} + \mathbf{j}$ and $\mathbf{j} + 2\mathbf{k}$.

6. Find a unit vector with positive $\mathbf{k}$ component that is perpendicular to both $2\mathbf{i} - \mathbf{j} - 2\mathbf{k}$ and $2\mathbf{i} - 3\mathbf{j} + \mathbf{k}$.

Verify the identities in Exercises 7–11, either by using the definition of cross product or the properties of determinants.

7. $\mathbf{u} \times \mathbf{u} = \mathbf{0}$

8. $\mathbf{u} \times \mathbf{v} = -\mathbf{v} \times \mathbf{u}$

9. $(\mathbf{u} + \mathbf{v}) \times \mathbf{w} = \mathbf{u} \times \mathbf{w} + \mathbf{v} \times \mathbf{w}$

10. $(t\mathbf{u}) \times \mathbf{v} = \mathbf{u} \times (t\mathbf{v}) = t(\mathbf{u} \times \mathbf{v})$

11. $\mathbf{u} \bullet (\mathbf{u} \times \mathbf{v}) = \mathbf{v} \bullet (\mathbf{u} \times \mathbf{v}) = 0$

12. Obtain the addition formula

$$\sin(\alpha - \beta) = \sin\alpha\cos\beta - \cos\alpha\sin\beta$$

by examining the cross product of the two unit vectors $\mathbf{u} = \cos\beta\mathbf{i} + \sin\beta\mathbf{j}$ and $\mathbf{v} = \cos\alpha\mathbf{i} + \sin\alpha\mathbf{j}$. Assume $0 \le \alpha - \beta \le \pi$. *Hint:* Regard $\mathbf{u}$ and $\mathbf{v}$ as position vectors. What is the area of the parallelogram they span?

13. If $\mathbf{u} + \mathbf{v} + \mathbf{w} = \mathbf{0}$, show that $\mathbf{u} \times \mathbf{v} = \mathbf{v} \times \mathbf{w} = \mathbf{w} \times \mathbf{u}$.

14. **(Volume of a tetrahedron)** A **tetrahedron** is a pyramid with a triangular base and three other triangular faces. It has four vertices and six edges. Like any pyramid or cone, its volume is equal to $\frac{1}{3}Ah$, where A is the area of the base and h is the height measured perpendicular to the base. If $\mathbf{u}$, $\mathbf{v}$, and $\mathbf{w}$ are vectors coinciding with the three edges of a tetrahedron that meet at one vertex, show that the tetrahedron has volume given by

$$\text{Volume} = \frac{1}{6}|\mathbf{u} \bullet (\mathbf{v} \times \mathbf{w})| = \frac{1}{6}\left|\begin{matrix} u_1 & u_2 & u_3 \\ v_1 & v_2 & v_3 \\ w_1 & w_2 & w_3 \end{matrix}\right|.$$

Thus, the volume of a tetrahedron spanned by three vectors is one-sixth of the volume of the parallelepiped spanned by the same vectors.

15. Find the volume of the tetrahedron with vertices $(1, 0, 0)$, $(1, 2, 0)$, $(2, 2, 2)$, and $(0, 3, 2)$.

16. Find the volume of the parallelepiped spanned by the diagonals of the three faces of a cube of side a that meet at one vertex of the cube.

17. For what value of k do the four points $(1, 1, -1)$, $(0, 3, -2)$, $(-2, 1, 0)$, and $(k, 0, 2)$ all lie in a plane?

18. (The scalar triple product) Verify the identities

$$\mathbf{u} \bullet (\mathbf{v} \times \mathbf{w}) = \mathbf{v} \bullet (\mathbf{w} \times \mathbf{u}) = \mathbf{w} \bullet (\mathbf{u} \times \mathbf{v}).$$

19. If $\mathbf{u} \bullet (\mathbf{v} \times \mathbf{w}) \neq 0$ and $\mathbf{x}$ is an arbitrary 3-vector, find the numbers λ, μ, and ν such that

$$\mathbf{x} = \lambda \mathbf{u} + \mu \mathbf{v} + \nu \mathbf{w}.$$

20. If $\mathbf{u} \bullet (\mathbf{v} \times \mathbf{w}) = 0$ but $\mathbf{v} \times \mathbf{w} \neq \mathbf{0}$, show that there are constants λ and μ such that

$$\mathbf{u} = \lambda \mathbf{v} + \mu \mathbf{w}.$$

Hint: Use the result of Exercise 19 with $\mathbf{u}$ in place of $\mathbf{x}$ and $\mathbf{v} \times \mathbf{w}$ in place of $\mathbf{u}$.

21. Calculate $\mathbf{u} \times (\mathbf{v} \times \mathbf{w})$ and $(\mathbf{u} \times \mathbf{v}) \times \mathbf{w}$, given that $\mathbf{u} = \mathbf{i} + 2\mathbf{j} + 3\mathbf{k}$, $\mathbf{v} = 2\mathbf{i} - 3\mathbf{j}$, and $\mathbf{w} = \mathbf{j} - \mathbf{k}$. Why would you not expect these to be equal?

22. Does the notation $\mathbf{u} \bullet \mathbf{v} \times \mathbf{w}$ make sense? Why? How about the notation $\mathbf{u} \times \mathbf{v} \times \mathbf{w}$?

23. (The vector triple product) The product $\mathbf{u} \times (\mathbf{v} \times \mathbf{w})$ is called a **vector triple product**. Since it is perpendicular to $\mathbf{v} \times \mathbf{w}$, it must lie in the plane of $\mathbf{v}$ and $\mathbf{w}$. Show that

$$\mathbf{u} \times (\mathbf{v} \times \mathbf{w}) = (\mathbf{u} \bullet \mathbf{w})\mathbf{v} - (\mathbf{u} \bullet \mathbf{v})\mathbf{w}.$$

Hint: This can be done by direct calculation of the components of both sides of the equation, but the job is much easier if you choose coordinate axes so that $\mathbf{v}$ lies along the x-axis and $\mathbf{w}$ lies in the xy-plane.

24. If $\mathbf{u}$, $\mathbf{v}$, and $\mathbf{w}$ are mutually perpendicular vectors, show that $\mathbf{u} \times (\mathbf{v} \times \mathbf{w}) = \mathbf{0}$. What is $\mathbf{u} \bullet (\mathbf{v} \times \mathbf{w})$ in this case?

25. Show that $\mathbf{u} \times (\mathbf{v} \times \mathbf{w}) + \mathbf{v} \times (\mathbf{w} \times \mathbf{u}) + \mathbf{w} \times (\mathbf{u} \times \mathbf{v}) = \mathbf{0}$.

26. Find all vectors $\mathbf{x}$ that satisfy the equation

$$(-\mathbf{i} + 2\mathbf{j} + 3\mathbf{k}) \times \mathbf{x} = \mathbf{i} + 5\mathbf{j} - 3\mathbf{k}.$$

27. Show that the equation

$$(-\mathbf{i} + 2\mathbf{j} + 3\mathbf{k}) \times \mathbf{x} = \mathbf{i} + 5\mathbf{j}$$

has no solutions for the unknown vector $\mathbf{x}$.

28. What condition must be satisfied by the nonzero vectors $\mathbf{a}$ and $\mathbf{b}$ to guarantee that the equation $\mathbf{a} \times \mathbf{x} = \mathbf{b}$ has a solution for $\mathbf{x}$? Is the solution unique?

10.4 Planes and Lines

A single equation in the three variables, x, y, and z, constitutes a single constraint on the freedom of the point $P = (x, y, z)$ to lie anywhere in 3-space. Such a constraint usually results in the loss of exactly one *degree of freedom* and so forces P to lie on a two-dimensional surface. For example, the equation

$$x^2 + y^2 + z^2 = 4$$

states that the point (x, y, z) is at distance 2 from the origin. All points satisfying this condition lie on a **sphere** (i.e., the surface of a ball) of radius 2 centred at the origin. The equation above therefore represents that sphere, and the sphere is the graph of the equation. In this section we will investigate the graphs of linear equations in three variables.

Planes in 3-Space

Let $P_0 = (x_0, y_0, z_0)$ be a point in $\mathbb{R}^3$ with position vector

$$\mathbf{r}_0 = x_0\mathbf{i} + y_0\mathbf{j} + z_0\mathbf{k}.$$

If $\mathbf{n} = A\mathbf{i} + B\mathbf{j} + C\mathbf{k}$ is any given *nonzero* vector, then there exists exactly one **plane** (flat surface) passing through P_0 and perpendicular to $\mathbf{n}$. We say that $\mathbf{n}$ is a **normal vector** to the plane. The plane is the set of all points P for which $\overrightarrow{P_0P}$ is perpendicular to $\mathbf{n}$. (See Figure 10.28.)

If $P = (x, y, z)$ has position vector $\mathbf{r}$, then $\overrightarrow{P_0P} = \mathbf{r} - \mathbf{r}_0$. This vector is perpendicular to $\mathbf{n}$ if and only if $\mathbf{n} \bullet (\mathbf{r} - \mathbf{r}_0) = 0$. This is the equation of the plane in vector form. We can rewrite it in terms of coordinates to obtain the corresponding scalar equation.

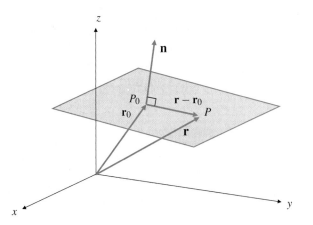

Figure 10.28   The plane through P_0 with normal $\mathbf{n}$ contains all points P for which $\overrightarrow{P_0P}$ is perpendicular to $\mathbf{n}$

The point-normal equation of a plane

The plane having nonzero normal vector $\mathbf{n} = A\mathbf{i} + B\mathbf{j} + C\mathbf{k}$, and passing through the point $P_0 = (x_0, y_0, z_0)$ with position vector $\mathbf{r}_0$, has equation

$$\mathbf{n} \bullet (\mathbf{r} - \mathbf{r}_0) = 0$$

in vector form, or, equivalently,

$$A(x - x_0) + B(y - y_0) + C(z - z_0) = 0$$

in scalar form.

The scalar form can be written more simply in the **standard form** $Ax + By + Cz = D$, where $D = Ax_0 + By_0 + Cz_0$.

If at least one of the constants A, B, and C is not zero, then the *linear equation* $Ax + By + Cz = D$ always represents a plane in $\mathbb{R}^3$. For example, if $A \neq 0$, it represents the plane through $(D/A, 0, 0)$ with normal vector $\mathbf{n} = A\mathbf{i} + B\mathbf{j} + C\mathbf{k}$. A vector normal to a plane can always be determined from the coefficients of x, y, and z. If the constant term $D = 0$, then the plane must pass through the origin.

Example 1   (Recognizing and writing the equations of planes)

(a) The equation $2x - 3y - 4z = 0$ represents a plane that passes through the origin and is normal (perpendicular) to the vector $\mathbf{n} = 2\mathbf{i} - 3\mathbf{j} - 4\mathbf{k}$.

(b) The plane that passes through the point $(2, 0, 1)$ and is perpendicular to the straight line passing through the points $(1, 1, 0)$ and $(4, -1, -2)$ has normal vector $\mathbf{n} = (4 - 1)\mathbf{i} + (-1 - 1)\mathbf{j} + (-2 - 0)\mathbf{k} = 3\mathbf{i} - 2\mathbf{j} - 2\mathbf{k}$. Therefore, its equation is $3(x - 2) - 2(y - 0) - 2(z - 1) = 0$, or, more simply, $3x - 2y - 2z = 4$.

(c) The plane with equation $2x - y = 1$ has a normal $2\mathbf{i} - \mathbf{j}$ that is perpendicular to the z-axis. The plane is therefore parallel to the z-axis. Note that the equation is independent of z. In the xy-plane, the equation $2x - y = 1$ represents a straight line; in 3-space it represents a plane containing that line and parallel to the z-axis. What does the equation $y = z$ represent in $\mathbb{R}^3$? the equation $y = -2$?

(d) The equation $2x + y + 3z = 6$ represents a plane with normal $\mathbf{n} = 2\mathbf{i} + \mathbf{j} + 3\mathbf{k}$. In this case we cannot directly read from the equation the coordinates of a particular point on the plane, but it is not difficult to discover some points. For instance, if we put $y = z = 0$ in the equation we get $x = 3$, so $(3, 0, 0)$ is a point on the plane. We say that the **x-intercept** of the plane is 3 since $(3, 0, 0)$ is the point where the plane intersects the x-axis. Similarly, the y-intercept is 6 and the z-intercept is 2 because the plane intersects the y- and z-axes at $(0, 6, 0)$ and $(0, 0, 2)$, respectively.

(e) In general, if a, b, and c are all nonzero, the plane with intercepts a, b, and c on

$$\frac{x}{a} + \frac{y}{b} + \frac{z}{c} = 1$$

Figure 10.29   The plane with intercepts a, b, and c on the coordinate axes

the coordinate axes has equation

$$\frac{x}{a} + \frac{y}{b} + \frac{z}{c} = 1,$$

called the **intercept form** of the equation of the plane. (See Figure 10.29.)

Example 2   Find an equation of the plane that passes through the three points $P = (1, 1, 0)$, $Q = (0, 2, 1)$, and $R = (3, 2, -1)$.

Solution   We need to find a vector, $\mathbf{n}$, normal to the plane. Such a vector will be perpendicular to the vectors $\overrightarrow{PQ} = -\mathbf{i} + \mathbf{j} + \mathbf{k}$ and $\overrightarrow{PR} = 2\mathbf{i} + \mathbf{j} - \mathbf{k}$. Therefore, we can use

$$\mathbf{n} = \overrightarrow{PQ} \times \overrightarrow{PR} = \begin{vmatrix} \mathbf{i} & \mathbf{j} & \mathbf{k} \\ -1 & 1 & 1 \\ 2 & 1 & -1 \end{vmatrix} = -2\mathbf{i} + \mathbf{j} - 3\mathbf{k}.$$

We can use this normal vector together with the coordinates of any one of the three given points to write the equation of the plane. Using point P leads to the equation $-2(x - 1) + 1(y - 1) - 3(z - 0) = 0$, or

$$2x - y + 3z = 1.$$

You can check that using either Q or R leads to the same equation. (If the cross product $\overrightarrow{PQ} \times \overrightarrow{PR}$ had been the zero vector, what would have been true about the three points P, Q, and R? Would they have determined a unique plane?)

Example 3   Show that the two planes

$$x - y = 3 \qquad \text{and} \qquad x + y + z = 0$$

intersect, and find a vector, $\mathbf{v}$, parallel to their line of intersection.

Solution   The two planes have normal vectors

$$\mathbf{n}_1 = \mathbf{i} - \mathbf{j} \qquad \text{and} \qquad \mathbf{n}_2 = \mathbf{i} + \mathbf{j} + \mathbf{k},$$

respectively. Since these vectors are not parallel, the planes are not parallel, and they intersect in a straight line perpendicular to both $\mathbf{n}_1$ and $\mathbf{n}_2$. This line must therefore be parallel to

$$\mathbf{v} = \mathbf{n}_1 \times \mathbf{n}_2 = \begin{vmatrix} \mathbf{i} & \mathbf{j} & \mathbf{k} \\ 1 & -1 & 0 \\ 1 & 1 & 1 \end{vmatrix} = -\mathbf{i} - \mathbf{j} + 2\mathbf{k}.$$

Figure 10.30   A pencil of planes

A family of planes intersecting in a straight line is called a **pencil of planes**. (See Figure 10.30.) Such a pencil of planes is determined by any two nonparallel planes in it, since these have a unique line of intersection. If the two nonparallel planes have equations

$$A_1 x + B_1 y + C_1 z = D_1 \qquad \text{and} \qquad A_2 x + B_2 y + C_2 z = D_2,$$

then, for any value of the real number λ, the equation

$$A_1 x + B_1 y + C_1 z - D_1 + \lambda(A_2 x + B_2 y + C_2 z - D_2) = 0$$

represents a plane in the pencil. To see this, observe that the equation is linear, and so represents a plane, and that any point (x, y, z) satisfying the equations of both given planes also satisfies this equation for any value of λ. Any plane in the pencil except the second defining plane, $A_2 x + B_2 y + C_2 z = D_2$, can be obtained by suitably choosing the value of λ.

Example 4   Find an equation of the plane passing through the line of intersection of the two planes

$$x + y - 2z = 6 \quad \text{and} \quad 2x - y + z = 2$$

and also passing through the point $(-2, 0, 1)$.

Solution   For any constant λ, the equation

$$x + y - 2z - 6 + \lambda(2x - y + z - 2) = 0$$

represents a plane and is satisfied by the coordinates of all points on the line of intersection of the given planes. This plane passes through the point $(-2, 0, 1)$ if $-2 - 2 - 6 + \lambda(-4 + 1 - 2) = 0$, that is, if $\lambda = -2$. The equation of the required plane therefore simplifies to $3x - 3y + 4z + 2 = 0$. (This solution would not have worked if the given point had been on the second plane, $2x - y + z = 2$. Why?)

Lines in 3-Space

As we observed above, any two nonparallel planes in $\mathbb{R}^3$ determine a unique (straight) line of intersection, and a vector parallel to this line can be obtained by taking the cross product of normal vectors to the two planes.

Suppose that $\mathbf{r}_0 = x_0\mathbf{i} + y_0\mathbf{j} + z_0\mathbf{k}$ is the position vector of point P_0 and $\mathbf{v} = a\mathbf{i} + b\mathbf{j} + c\mathbf{k}$ is a nonzero vector. There is a unique line passing through P_0 parallel to $\mathbf{v}$. If $\mathbf{r} = x\mathbf{i} + y\mathbf{j} + z\mathbf{k}$ is the position vector of any other point P on the line, then $\mathbf{r} - \mathbf{r}_0$ lies along the line and so is parallel to $\mathbf{v}$. (See Figure 10.31.) Thus, $\mathbf{r} - \mathbf{r}_0 = t\mathbf{v}$ for some real number t. This equation, usually rewritten in the form

$$\mathbf{r} = \mathbf{r}_0 + t\mathbf{v},$$

is called the **vector parametric equation of the straight line**. All points on the line can be obtained as the parameter t ranges from $-\infty$ to ∞. The vector $\mathbf{v}$ is called a **direction vector** of the line.

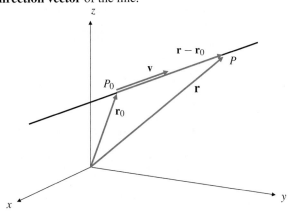

Figure 10.31   The line through P_0 parallel to $\mathbf{v}$

Breaking the vector parametric equation down into its components yields the **scalar parametric** equations of the line:

$$\begin{cases} x = x_0 + at \\ y = y_0 + bt \qquad (-\infty < t < \infty) \\ z = z_0 + ct. \end{cases}$$

These appear to be *three* linear equations, but the parameter t can be eliminated to give *two* linear equations in x, y, and z. If $a \neq 0$, $b \neq 0$, and $c \neq 0$, then we can solve each of the scalar equations for t and so obtain

$$\frac{x - x_0}{a} = \frac{y - y_0}{b} = \frac{z - z_0}{c},$$

which is called the **standard form** for the equations of the straight line through (x_0, y_0, z_0) parallel to $\mathbf{v}$. The standard form must be modified if any component of $\mathbf{v}$ vanishes. For example, if $c = 0$, the equations are

$$\frac{x - x_0}{a} = \frac{y - y_0}{b}, \quad z = z_0.$$

Note that none of the above equations for straight lines is unique; each depends on the particular choice of the point (x_0, y_0, z_0) on the line. In general, you can always use the equations of two nonparallel planes to represent their line of intersection.

Example 5 (Equations of straight lines)

(a) The equations

$$\begin{cases} x = 2 + t \\ y = 3 \\ z = -4t \end{cases}$$

represent the straight line through $(2, 3, 0)$ parallel to the vector $\mathbf{i} - 4\mathbf{k}$.

(b) The straight line through $(1, -2, 3)$ perpendicular to the plane $x - 2y + 4z = 5$ is parallel to the normal vector $\mathbf{i} - 2\mathbf{j} + 4\mathbf{k}$ of the plane. Therefore, the line has vector parametric equation

$$\mathbf{r} = \mathbf{i} - 2\mathbf{j} + 3\mathbf{k} + t(\mathbf{i} - 2\mathbf{j} + 4\mathbf{k}),$$

or scalar parametric equations

$$\begin{cases} x = 1 + t \\ y = -2 - 2t \\ z = 3 + 4t. \end{cases}$$

Its standard form equations are

$$\frac{x - 1}{1} = \frac{y + 2}{-2} = \frac{z - 3}{4}.$$

Example 6 Find a direction vector for the line of intersection of the two planes

$$x + y - z = 0 \quad \text{and} \quad y + 2z = 6,$$

and find a set of equations for the line in standard form.

Solution The two planes have respective normals $\mathbf{n}_1 = \mathbf{i} + \mathbf{j} - \mathbf{k}$ and $\mathbf{n}_2 = \mathbf{j} + 2\mathbf{k}$. Thus, a direction vector of their line of intersection is

$$\mathbf{v} = \mathbf{n}_1 \times \mathbf{n}_2 = 3\mathbf{i} - 2\mathbf{j} + \mathbf{k}.$$

We need to know one point on the line in order to write equations in standard form. We can find a point by assigning a value to one coordinate and calculating the other two from the given equations. For instance, taking $z = 0$ in the two equations we are led to $y = 6$ and $x = -6$, so $(-6, 6, 0)$ is one point on the line. Thus, the line has standard form equations

$$\frac{x + 6}{3} = \frac{y - 6}{-2} = z.$$

This answer is not unique; the coordinates of any other point on the line could be used in place of $(-6, 6, 0)$. You could even find a direction vector $\mathbf{v}$ by subtracting the position vectors of two different points on the line.

Distances

The **distance** between two geometric objects always means the minimum distance between two points, one in each object. In the case of *flat* objects like lines and planes defined by linear equations, such minimum distances can usually be determined by geometric arguments without having to use calculus.

Example 7 (Distance from a point to a plane)

(a) Find the distance from the point $P_0 = (x_0, y_0, z_0)$ to the plane $\mathcal{P}$ having equation $Ax + By + Cz = D$.

(b) What is the distance from $(2, -1, 3)$ to the plane $2x - 2y - z = 9$?

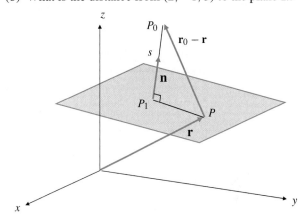

Figure 10.32 The distance from P_0 to the plane $\mathcal{P}$ is the length of the vector projection of $P P_0$ along the normal $\mathbf{n}$ to $\mathcal{P}$, where P is any point on $\mathcal{P}$

Solution

(a) Let $\mathbf{r}_0$ be the position vector of P_0, and let $\mathbf{n} = A\mathbf{i} + B\mathbf{j} + C\mathbf{k}$ be the normal to $\mathcal{P}$. Let P_1 be the point on $\mathcal{P}$ that is closest to P_0. Then $\overrightarrow{P_1 P_0}$ is perpendicular to $\mathcal{P}$ and so is parallel to $\mathbf{n}$. The distance from P_0 to $\mathcal{P}$ is $s = |\overrightarrow{P_1 P_0}|$. If P, having position vector $\mathbf{r}$, is any point on $\mathcal{P}$, then s is the length of the projection of $\overrightarrow{P P_0} = \mathbf{r}_0 - \mathbf{r}$ in the direction of $\mathbf{n}$. (See Figure 10.32.) Thus

$$s = \left| \frac{\overrightarrow{P P_0} \bullet \mathbf{n}}{|\mathbf{n}|} \right| = \frac{|(\mathbf{r}_0 - \mathbf{r}) \bullet \mathbf{n}|}{|\mathbf{n}|} = \frac{|\mathbf{r}_0 \bullet \mathbf{n} - \mathbf{r} \bullet \mathbf{n}|}{|\mathbf{n}|}.$$

Since $P = (x, y, z)$ lies on $\mathcal{P}$, we have $\mathbf{r} \bullet \mathbf{n} = Ax + By + Cz = D$. In terms of the coordinates (x_0, y_0, z_0) of P_0, we can therefore represent the distance from P_0 to $\mathcal{P}$ as

$$s = \frac{|Ax_0 + By_0 + Cz_0 - D|}{\sqrt{A^2 + B^2 + C^2}}.$$

(b) The distance from $(2, -1, 3)$ to the plane $2x - 2y - z = 9$ is

$$s = \frac{|2(2) - 2(-1) - 1(3) - 9|}{\sqrt{2^2 + (-2)^2 + (-1)^2}} = \frac{|-6|}{3} = 2 \text{ units.}$$

Example 8 (Distance from a point to a line)

(a) Find the distance from the point P_0 to the straight line $\mathcal{L}$ through P_1 parallel to the nonzero vector $\mathbf{v}$.

(b) What is the distance from $(2, 0, -3)$ to the line $\mathbf{r} = \mathbf{i} + (1 + 3t)\mathbf{j} - (3 - 4t)\mathbf{k}$?

Solution

(a) Let $\mathbf{r}_0$ and $\mathbf{r}_1$ be the position vectors of P_0 and P_1, respectively. The point P_2 on $\mathcal{L}$ that is closest to P_0 is such that $P_2 P_0$ is perpendicular to $\mathcal{L}$. The distance from P_0 to $\mathcal{L}$ is

$$s = |P_2 P_0| = |P_1 P_0| \sin \theta = |\mathbf{r}_0 - \mathbf{r}_1| \sin \theta,$$

where θ is the angle between $\mathbf{r}_0 - \mathbf{r}_1$ and $\mathbf{v}$. (See Figure 10.33(a).) Since

$$|(\mathbf{r}_0 - \mathbf{r}_1) \times \mathbf{v}| = |\mathbf{r}_0 - \mathbf{r}_1|\,|\mathbf{v}| \sin \theta,$$

we have

$$s = \frac{|(\mathbf{r}_0 - \mathbf{r}_1) \times \mathbf{v}|}{|\mathbf{v}|}.$$

(b) The line $\mathbf{r} = \mathbf{i} + (1 + 3t)\mathbf{j} - (3 - 4t)\mathbf{k}$ passes through $P_1 = (1, 1, -3)$ and is parallel to $\mathbf{v} = 3\mathbf{j} + 4\mathbf{k}$. The distance from $P_0 = (2, 0, -3)$ to this line is

$$s = \frac{\left|\big((2 - 1)\mathbf{i} + (0 - 1)\mathbf{j} + (-3 + 3)\mathbf{k}\big) \times (3\mathbf{j} + 4\mathbf{k})\right|}{\sqrt{3^2 + 4^2}}$$

$$= \frac{|(\mathbf{i} - \mathbf{j}) \times (3\mathbf{j} + 4\mathbf{k})|}{5} = \frac{|-4\mathbf{i} - 4\mathbf{j} + 3\mathbf{k}|}{5} = \frac{\sqrt{41}}{5} \text{ units.}$$

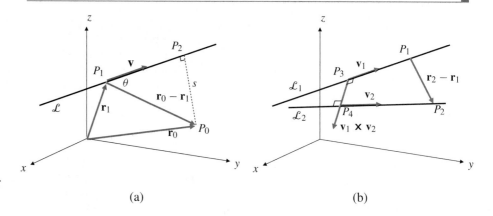

Figure 10.33

(a) The distance from P_0 to the line $\mathcal{L}$ is $s = |P_0 P_1| \sin \theta$

(b) The distance between the lines $\mathcal{L}_1$ and $\mathcal{L}_2$ is the length of the projection of $P_1 P_2$ along the vector $\mathbf{v}_1 \times \mathbf{v}_2$

(a)

(b)

Example 9 **(The distance between two lines)** Find the distance between the two lines $\mathcal{L}_1$ through point P_1 parallel to vector $\mathbf{v}_1$ and $\mathcal{L}_2$ through point P_2 parallel to vector $\mathbf{v}_2$.

Solution Let $\mathbf{r}_1$ and $\mathbf{r}_2$ be the position vectors of points P_1 and P_2, respectively. If P_3 and P_4 (with position vectors $\mathbf{r}_3$ and $\mathbf{r}_4$) are the points on $\mathcal{L}_1$ and $\mathcal{L}_2$, respectively, that are closest to one another, then $\overrightarrow{P_3 P_4}$ is perpendicular to both lines and is therefore parallel to $\mathbf{v}_1 \times \mathbf{v}_2$. (See Figure 10.33(b).) $\overrightarrow{P_3 P_4}$ is the vector projection of $\overrightarrow{P_1 P_2} = \mathbf{r}_2 - \mathbf{r}_1$ along $\mathbf{v}_1 \times \mathbf{v}_2$. Therefore, the distance $s = |\overrightarrow{P_3 P_4}|$ between the lines is given by

$$s = |\mathbf{r}_4 - \mathbf{r}_3| = \frac{|(\mathbf{r}_2 - \mathbf{r}_1) \bullet (\mathbf{v}_1 \times \mathbf{v}_2)|}{|\mathbf{v}_1 \times \mathbf{v}_2|}.$$

▌▌▌▌▌▌▌ Exercises 10.4 ▌▌▌▌▌▌▌▌

1. A single equation involving the coordinates (x, y, z) need not always represent a two-dimensional "surface" in $\mathbb{R}^3$. For example, $x^2 + y^2 + z^2 = 0$ represents the single point $(0, 0, 0)$, which has dimension zero. Give examples of single equations in x, y, and z that represent

   (a) a (one-dimensional) straight line,

   (b) the whole of $\mathbb{R}^3$,

   (c) no points at all (i.e., the empty set).

In Exercises 2–9, find equations of the planes satisfying the given conditions.

2. Passing through $(0, 2, -3)$ and normal to the vector $4\mathbf{i} - \mathbf{j} - 2\mathbf{k}$

3. Passing through the origin and having normal $\mathbf{i} - \mathbf{j} + 2\mathbf{k}$

4. Passing through $(1, 2, 3)$ and parallel to the plane $3x + y - 2z = 15$

5. Passing through the three points $(1, 1, 0)$, $(2, 0, 2)$, and $(0, 3, 3)$

6. Passing through the three points $(-2, 0, 0)$, $(0, 3, 0)$, and $(0, 0, 4)$

7. Passing through $(1, 1, 1)$ and $(2, 0, 3)$ and perpendicular to the plane $x + 2y - 3z = 0$

8. Passing through the line of intersection of the planes $2x + 3y - z = 0$ and $x - 4y + 2z = -5$, and passing through the point $(-2, 0, -1)$

9. Passing through the line $x + y = 2$, $y - z = 3$, and perpendicular to the plane $2x + 3y + 4z = 5$

10. Under what geometric condition will three distinct points in $\mathbb{R}^3$ not determine a unique plane passing through them? How can this condition be expressed algebraically in terms of the position vectors, $\mathbf{r}_1$, $\mathbf{r}_2$, and $\mathbf{r}_3$, of the three points?

11. Give a condition on the position vectors of four points that guarantees that the four points are *coplanar*, that is, all lie on one plane.

Describe geometrically the one-parameter families of planes in Exercises 12–14. (λ is a real parameter.)

12. $x + y + z = \lambda$.   * 13. $x + \lambda y + \lambda z = \lambda$.

* 14. $\lambda x + \sqrt{1 - \lambda^2}\, y = 1$.

In Exercises 15–19, find equations of the line specified in vector and scalar parametric forms and in standard form.

15. Through the point $(1, 2, 3)$ and parallel to $2\mathbf{i} - 3\mathbf{j} - 4\mathbf{k}$

16. Through $(-1, 0, 1)$ and perpendicular to the plane $2x - y + 7z = 12$

17. Through the origin and parallel to the line of intersection of the planes $x + 2y - z = 2$ and $2x - y + 4z = 5$

18. Through $(2, -1, -1)$ and parallel to each of the two planes $x + y = 0$ and $x - y + 2z = 0$

19. Through $(1, 2, -1)$ and making equal angles with the positive directions of the coordinate axes

In Exercises 20–22, find the equations of the given line in standard form.

20. $\mathbf{r} = (1 - 2t)\mathbf{i} + (4 + 3t)\mathbf{j} + (9 - 4t)\mathbf{k}$.

21. $\begin{cases} x = 4 - 5t \\ y = 3t \\ z = 7 \end{cases}$ 22. $\begin{cases} x - 2y + 3z = 0 \\ 2x + 3y - 4z = 4 \end{cases}$

23. If $P_1 = (x_1, y_1, z_1)$ and $P_2 = (x_2, y_2, z_2)$, show that the equations

$$\begin{cases} x = x_1 + t(x_2 - x_1) \\ y = y_1 + t(y_2 - y_1) \\ z = z_1 + t(z_2 - z_1) \end{cases}$$

   represent a line through P_1 and P_2.

24. What points on the line in Exercise 23 correspond to the parameter values $t = -1$, $t = 1/2$, and $t = 2$? Describe their locations.

25. Under what conditions on the position vectors of four distinct points P_1, P_2, P_3, and P_4 will the straight line through P_1 and P_2 intersect the straight line through P_3 and P_4 at a unique point?

Find the required distances in Exercises 26–29.

26. From the origin to the plane $x + 2y + 3z = 4$

27. From $(1, 2, 0)$ to the plane $3x - 4y - 5z = 2$

28. From the origin to the line $x + y + z = 0$, $2x - y - 5z = 1$

29. Between the lines

$$\begin{cases} x + 2y = 3 \\ y + 2z = 3 \end{cases} \quad \text{and} \quad \begin{cases} x + y + z = 6 \\ x - 2z = -5 \end{cases}$$

30. Show that the line $x - 2 = \dfrac{y + 3}{2} = \dfrac{z - 1}{4}$ is parallel to the plane $2y - z = 1$. What is the distance between the line and the plane?

In Exercises 31–32, describe the one-parameter families of straight lines represented by the given equations. (λ is a real parameter.)

* 31. $(1 - \lambda)(x - x_0) = \lambda(y - y_0)$, $z = z_0$.

* 32. $\dfrac{x - x_0}{\sqrt{1 - \lambda^2}} = \dfrac{y - y_0}{\lambda} = z - z_0$.

33. Why does the factored second-degree equation

$$(A_1 x + B_1 y + C_1 z - D_1)(A_2 x + B_2 y + C_2 z - D_2) = 0$$

   represent a pair of planes rather than a single straight line?

10.5 Quadric Surfaces

The most general second-degree equation in three variables is

$$Ax^2 + By^2 + Cz^2 + Dxy + Exz + Fyz + Gx + Hy + Iz = J.$$

We will not attempt the (rather difficult) task of classifying all the surfaces that can be represented by such an equation, but will examine some interesting special cases. Let us observe at the outset that if the above equation can be factored in the form

$$(A_1x + B_1y + C_1z - D_1)(A_2x + B_2y + C_2z - D_2) = 0,$$

then the graph is, in fact, a pair of planes,

$$A_1x + B_1y + C_1z = D_1 \qquad \text{and} \qquad A_2x + B_2y + C_2z = D_2,$$

or one plane if the two linear equations represent the same plane. This is considered a degenerate case. Where such factorization is not possible, the surface (called a **quadric surface**) will not be flat, although there may still be straight lines that lie on the surface. Nondegenerate quadric surfaces fall into the following six categories.

Spheres. The equation $x^2 + y^2 + z^2 = a^2$ represents a sphere of radius a centred at the origin. More generally,

$$(x - x_0)^2 + (y - y_0)^2 + (z - z_0)^2 = a^2$$

represents a sphere of radius a centred at the point (x_0, y_0, z_0). If a quadratic equation in x, y, and z has equal coefficients for the x^2, y^2, and z^2 terms and has no other second-degree terms, then it will represent, if any surface at all, a sphere. The centre can be found by completing the squares as for circles in the plane.

Cylinders. The equation $x^2 + y^2 = a^2$, being independent of z, represents a **right-circular cylinder** of radius a and axis along the z-axis. (See Figure 10.34(a).) The intersection of the cylinder with the horizontal plane $z = k$ is the circle with equations

$$\begin{cases} x^2 + y^2 = a^2 \\ z = k. \end{cases}$$

Quadric cylinders also come in other shapes: elliptic, parabolic, and hyperbolic. For instance, $z = x^2$ represents a parabolic cylinder with vertex line along the y-axis. (See Figure 10.34(b).) In general, an equation in two variables only will represent a cylinder in 3-space.

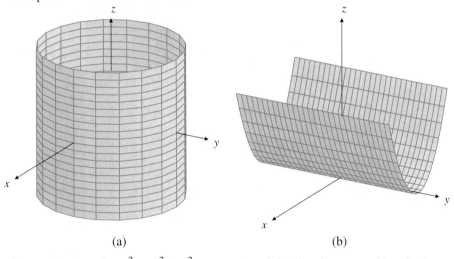

Figure 10.34

(a) The circular cylinder
$x^2 + y^2 = a^2$

(b) The parabolic cylinder $z = x^2$

(a)   (b)

Cones. The equation $z^2 = x^2 + y^2$ represents a **right-circular cone** with axis along the z-axis. The surface is generated by rotating about the z-axis the line $z = y$ in the yz-plane. This *generator* makes an angle of $45°$ with the axis of the cone. Cross-sections of the cone in planes parallel to the xy-plane are circles. (See Figure 10.35(a).) The equation $x^2 + y^2 = a^2z^2$ also represents a right-circular cone with vertex at the origin and axis along the z-axis but having semi-vertical angle $\alpha = \tan^{-1} a$. A circular cone has plane cross-sections that are elliptical, parabolic, and hyperbolic. Conversely, any nondegenerate quadric cone has a direction perpendicular to which the cross-sections of the cone are circular. In that sense, every quadric cone is a circular cone, although it may be *oblique* rather than right-circular in that the line joining the centres of the circular cross-sections need not be perpendicular to those cross-sections. (See Exercise 24.)

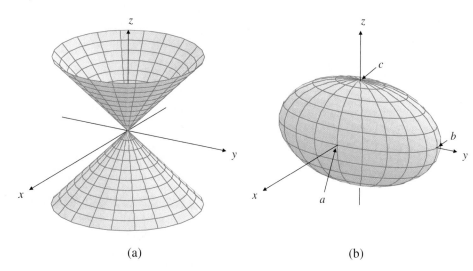

Figure 10.35

(a) The circular cone $a^2 z^2 = x^2 + y^2$

(b) The ellipsoid $\dfrac{x^2}{a^2} + \dfrac{y^2}{b^2} + \dfrac{z^2}{c^2} = 1$

(a)   (b)

Ellipsoids. The equation

$$\frac{x^2}{a^2} + \frac{y^2}{b^2} + \frac{z^2}{c^2} = 1$$

represents an **ellipsoid** with *semi-axes* a, b, and c. (See Figure 10.35(b).) The surface is oval, and it is enclosed inside the rectangular parallelepiped $-a \le x \le a$, $-b \le y \le b$, $-c \le z \le c$. If $a = b = c$, the ellipsoid is a sphere. In general, all plane cross-sections of ellipsoids are ellipses. This is easy to see for cross-sections parallel to coordinate planes, but somewhat harder to see for other planes.

Paraboloids. The equations

$$z = \frac{x^2}{a^2} + \frac{y^2}{b^2} \qquad \text{and} \qquad z = \frac{x^2}{a^2} - \frac{y^2}{b^2}$$

represent, respectively, an **elliptic paraboloid** and a **hyperbolic paraboloid**. (See Figure 10.36(a) and (b).) Cross-sections in planes $z = k$ (k being a positive constant) are ellipses (circles if $a = b$) and hyperbolas, respectively. Parabolic reflective mirrors have the shape of circular paraboloids. The hyperbolic paraboloid is a **ruled surface**. (A ruled surface is one through every point of which there passes a straight line lying wholly on the surface. Cones and cylinders are also examples of ruled surfaces.) There are two one-parameter families of straight lines that lie on the hyperbolic paraboloid:

$$\begin{cases} \lambda z = \dfrac{x}{a} - \dfrac{y}{b} \\[2mm] \dfrac{1}{\lambda} = \dfrac{x}{a} + \dfrac{y}{b} \end{cases} \qquad \text{and} \qquad \begin{cases} \mu z = \dfrac{x}{a} + \dfrac{y}{b} \\[2mm] \dfrac{1}{\mu} = \dfrac{x}{a} - \dfrac{y}{b} \end{cases}$$

where λ and μ are real parameters. Every point on the hyperbolic paraboloid lies on one line of each family.

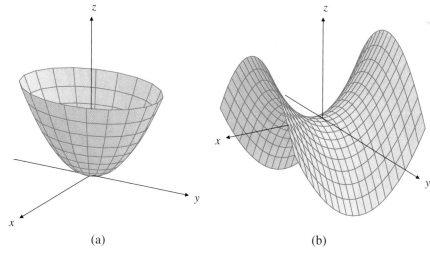

Figure 10.36

(a) The elliptic paraboloid $z = \dfrac{x^2}{a^2} + \dfrac{y^2}{b^2}$

(b) The hyperbolic paraboloid

$z = \dfrac{x^2}{a^2} - \dfrac{y^2}{b^2}$

(a)   (b)

Hyperboloids. The equation

$$\frac{x^2}{a^2} + \frac{y^2}{b^2} - \frac{z^2}{c^2} = 1$$

represents a surface called a **hyperboloid of one sheet**. (See Figure 10.37(a).) The equation

$$\frac{x^2}{a^2} + \frac{y^2}{b^2} - \frac{z^2}{c^2} = -1$$

represents a **hyperboloid of two sheets**. (See Figure 10.37(b).) Both surfaces

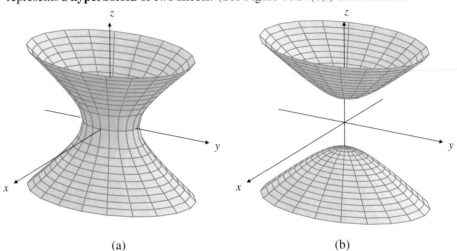

Figure 10.37

(a) The hyperboloid of one sheet
$$\frac{x^2}{a^2} + \frac{y^2}{b^2} - \frac{z^2}{c^2} = 1$$

(b) The hyperboloid of two sheets
$$\frac{x^2}{a^2} + \frac{y^2}{b^2} - \frac{z^2}{c^2} = -1$$

(a)   (b)

have elliptical cross-sections in horizontal planes and hyperbolic cross-sections in vertical planes. Both are *asymptotic* to the elliptic cone with equation

$$\frac{x^2}{a^2} + \frac{y^2}{b^2} = \frac{z^2}{c^2};$$

they approach arbitrarily close to the cone as they recede arbitrarily far away from the origin. Like the hyperbolic paraboloid, the hyperboloid of one sheet is a ruled surface.

Exercises 10.5

Identify the surfaces represented by the equations in Exercises 1–16 and sketch their graphs.

1. $x^2 + 4y^2 + 9z^2 = 36$ **2.** $x^2 + y^2 + 4z^2 = 4$

3. $2x^2 + 2y^2 + 2z^2 - 4x + 8y - 12z + 27 = 0$

4. $x^2 + 4y^2 + 9z^2 + 4x - 8y = 8$

5. $z = x^2 + 2y^2$ **6.** $z = x^2 - 2y^2$

7. $x^2 - y^2 - z^2 = 4$ **8.** $-x^2 + y^2 + z^2 = 4$

9. $z = xy$ **10.** $x^2 + 4z^2 = 4$

11. $x^2 - 4z^2 = 4$ **12.** $y = z^2$

13. $x = z^2 + z$ **14.** $x^2 = y^2 + 2z^2$

15. $(z - 1)^2 = (x - 2)^2 + (y - 3)^2$

16. $(z - 1)^2 = (x - 2)^2 + (y - 3)^2 + 4$

Describe and sketch the geometric objects represented by the systems of equations in Exercises 17–20.

17. $\begin{cases} x^2 + y^2 + z^2 = 4 \\ x + y + z = 1 \end{cases}$ **18.** $\begin{cases} x^2 + y^2 = 1 \\ z = x + y \end{cases}$

19. $\begin{cases} z^2 = x^2 + y^2 \\ z = 1 + x \end{cases}$ **20.** $\begin{cases} x^2 + 2y^2 + 3z^2 = 6 \\ y = 1 \end{cases}$

21. Find two one-parameter families of straight lines that lie on the hyperboloid of one sheet

$$\frac{x^2}{a^2} + \frac{y^2}{b^2} - \frac{z^2}{c^2} = 1.$$

22. Find two one-parameter families of straight lines that lie on the hyperbolic paraboloid $z = xy$.

23. The equation $2x^2 + y^2 = 1$ represents a cylinder with elliptical cross-sections in planes perpendicular to the z-axis. Find a vector **a** perpendicular to which the cylinder has circular cross-sections.

***24.** The equation $z^2 = 2x^2 + y^2$ represents a cone with elliptical cross-sections in planes perpendicular to the z-axis. Find a vector **a** perpendicular to which the cone has circular cross-sections. *Hint:* Do Exercise 23 first and use its result.

10.6 A Little Linear Algebra

Differential calculus is essentially the study of linear approximations to functions. The tangent line to the graph $y = f(x)$ at $x = x_0$ provides the "best linear approximation" to $f(x)$ near x_0. Differentiation of functions of several variables can also be viewed as a process of finding *best linear approximations*. Therefore the language of linear algebra can be very useful for expressing certain concepts in the calculus of several variables.

Linear algebra is a vast subject and is usually studied independently of calculus. This is unfortunate because understanding the relationship between the two subjects can greatly enhance your understanding and appreciation of each of them. Knowledge of linear algebra, and therefore familiarity with the material covered in this section, is *not essential* for fruitful study of the rest of this book. However, we shall occasionally comment on the significance of the subject at hand from the point of view of linear algebra. To this end we need only a little of the terminology and content of this subject, especially that part pertaining to matrix manipulation and systems of linear equations. In the rest of this section we present an outline of this material. Some students will already be familiar with it; others will encounter it later. We make no attempt at completeness here and refer interested students to standard linear algebra texts for proofs of some assertions. Students proceeding beyond this book to further study of advanced calculus and differential equations will certainly need a much more extensive background in linear algebra.

Matrices

An $m \times n$ **matrix** $\mathcal{A}$ is a rectangular array of mn numbers arranged in m rows and n columns. If a_{ij} is the element in the ith row and the jth column, then

$$\mathcal{A} = \begin{pmatrix} a_{11} & a_{12} & \cdots & a_{1n} \\ a_{21} & a_{22} & \cdots & a_{2n} \\ \vdots & \vdots & & \vdots \\ a_{m1} & a_{m2} & \cdots & a_{mn} \end{pmatrix}.$$

Sometimes, as a shorthand notation, we write $\mathcal{A} = (a_{ij})$. In this case i is assumed to range from 1 to m and j from 1 to n. If $m = n$, we say that $\mathcal{A}$ is a square matrix. The elements a_{ij} of the matrices we use in this book will always be real numbers.

The **transpose** of an $m \times n$ matrix $\mathcal{A}$ is the $n \times m$ matrix $\mathcal{A}^T$ whose rows are the columns of $\mathcal{A}$:

$$\mathcal{A}^T = \begin{pmatrix} a_{11} & a_{21} & \cdots & a_{m1} \\ a_{12} & a_{22} & \cdots & a_{m2} \\ \vdots & \vdots & & \vdots \\ a_{1n} & a_{2n} & \cdots & a_{mn} \end{pmatrix}.$$

Matrix $\mathcal{A}$ is called **symmetric** if $\mathcal{A}^T = \mathcal{A}$. Symmetric matrices are necessarily square. Observe that $(\mathcal{A}^T)^T = \mathcal{A}$ for *every* matrix $\mathcal{A}$. Frequently we want to consider an n-vector **x** as an $n \times 1$ matrix having n rows and one column:

$$\mathbf{x} = \begin{pmatrix} x_1 \\ x_2 \\ \vdots \\ x_n \end{pmatrix}.$$

As such, $\mathbf{x}$ is called a **column vector**. $\mathbf{x}^T$ then has one row and n columns and is called a **row vector**:

$$\mathbf{x}^T = (x_1 \, x_2 \, \cdots \, x_n).$$

Note that $\mathbf{x}$ and $\mathbf{x}^T$ have the same components, so they are identical as vectors even though they appear differently as matrices.

Most of the usefulness of matrices depends on the following definition of matrix multiplication, which enables two arrays to be combined into a single one in a manner that preserves linear relationships.

DEFINITION **7**

Multiplying matrices

If $\mathbf{A} = (a_{ij})$ is an $m \times n$ matrix and $\mathbf{B} = (b_{ij})$ is an $n \times p$ matrix, then the product $\mathbf{AB}$ is the $m \times p$ matrix $\mathbf{C} = (c_{ij})$ with elements given by

$$c_{ij} = \sum_{k=1}^{n} a_{ik}b_{kj}, \qquad i = 1, \ldots, m, \quad j = 1, \ldots, p.$$

That is, c_{ij} is the *dot product* of the ith row of $\mathbf{A}$ and the jth column of $\mathbf{B}$ (both of which are n-vectors).

Note that only *some* pairs of matrices can be multiplied. The product $\mathbf{AB}$ is only defined if the number of columns of $\mathbf{A}$ is equal to the number of rows of $\mathbf{B}$.

Example 1

$$\begin{pmatrix} 1 & 0 & 3 \\ 2 & 1 & -1 \end{pmatrix} \begin{pmatrix} 2 & 1 & 1 & 0 \\ 0 & -1 & 3 & 1 \\ 1 & 0 & 4 & 5 \end{pmatrix} = \begin{pmatrix} 5 & 1 & 13 & 15 \\ 3 & 1 & 1 & -4 \end{pmatrix}$$

The left factor has 2 rows and 3 columns, and the right factor has 3 rows and 4 columns. Therefore, the product has 2 rows and 4 columns. The element in the first row and third column of the product, 13, is the dot product of the first row, $(1, 0, 3)$, of the left factor and the third column, $(1, 3, 4)$, of the second factor:

$$1 \times 1 + 0 \times 3 + 3 \times 4 = 13.$$

With a little practice you can easily calculate the elements of a matrix product by simultaneously running your left index finger across rows of the left factor and your right index finger down columns of the right factor while taking the dot products.

Example 2

$$\begin{pmatrix} 1 & 2 & 3 \\ 0 & 1 & -1 \\ -2 & 3 & 0 \end{pmatrix} \begin{pmatrix} x \\ y \\ z \end{pmatrix} = \begin{pmatrix} x + 2y + 3z \\ y - z \\ -2x + 3y \end{pmatrix}$$

The product of a 3×3 matrix with a column 3-vector is a column 3-vector.

Matrix multiplication is *associative*. This means that

$$\mathbf{A}(\mathbf{BC}) = (\mathbf{AB})\mathbf{C}$$

(provided $\mathbf{A}$, $\mathbf{B}$, and $\mathbf{C}$ have dimensions compatible with the formation of the various products); therefore, it makes sense to write $\mathbf{ABC}$. However, matrix multiplication is *not commutative*. Indeed, if $\mathbf{A}$ is an $m \times n$ matrix and $\mathbf{B}$ is an $n \times p$ matrix, then the product $\mathbf{AB}$ is defined, but the product $\mathbf{BA}$ is not defined unless $m = p$. Even if $\mathbf{A}$ and $\mathbf{B}$ are square matrices of the same size, it is not necessarily true that $\mathbf{AB} = \mathbf{BA}$.

Example 3

$$\begin{pmatrix} 1 & 2 \\ 3 & 0 \end{pmatrix} \begin{pmatrix} 1 & -1 \\ 1 & 1 \end{pmatrix} = \begin{pmatrix} 3 & 1 \\ 3 & -3 \end{pmatrix} \quad \text{but} \quad \begin{pmatrix} 1 & -1 \\ 1 & 1 \end{pmatrix} \begin{pmatrix} 1 & 2 \\ 3 & 0 \end{pmatrix} = \begin{pmatrix} -2 & 2 \\ 4 & 2 \end{pmatrix}$$

The reader should verify that if the product $\mathcal{A}\mathcal{B}$ is defined, then the transpose of the product is the product of the transposes *in the reverse order*:

$$(\mathcal{A}\mathcal{B})^T = \mathcal{B}^T \mathcal{A}^T.$$

Determinants and Matrix Inverses

In Section 10.3 we introduced 2×2 and 3×3 determinants as certain algebraic expressions associated with 2×2 and 3×3 square arrays of numbers. In general, it is possible to define the determinant $\det(\mathcal{A})$ for any square matrix. For an $n \times n$ matrix $\mathcal{A}$ we continue to denote

$$\det(\mathcal{A}) = \begin{vmatrix} a_{11} & a_{12} & \cdots & a_{1n} \\ a_{21} & a_{22} & \cdots & a_{2n} \\ \vdots & \vdots & \ddots & \vdots \\ a_{n1} & a_{n2} & \cdots & a_{nn} \end{vmatrix}.$$

We will not attempt to give a formal definition of the determinant here but will note that the properties of determinants stated for the 3×3 case in Section 10.3 continue to be true. In particular, an $n \times n$ determinant can be expanded in minors about any row or column and so expressed as a sum of multiples of $(n - 1) \times (n - 1)$ determinants. Continuing this process, we can eventually reduce the evaluation of any $n \times n$ determinant to the evaluation of (perhaps many) 2×2 or 3×3 determinants. It is important to realize that the "diagonal" method for evaluating 2×2 or 3×3 determinants does not extend to 4×4 or higher-order determinants.

Example 4

$$\begin{vmatrix} 2 & 1 & 0 & 1 \\ 1 & 0 & 1 & 1 \\ 3 & 0 & 0 & 2 \\ -1 & 1 & 1 & 0 \end{vmatrix} = - \begin{vmatrix} 2 & 1 & 1 \\ 3 & 0 & 2 \\ -1 & 1 & 0 \end{vmatrix} - \begin{vmatrix} 2 & 1 & 1 \\ 1 & 0 & 1 \\ 3 & 0 & 2 \end{vmatrix}$$

$$= - \left(-3 \begin{vmatrix} 1 & 1 \\ 1 & 0 \end{vmatrix} - 2 \begin{vmatrix} 2 & 1 \\ -1 & 1 \end{vmatrix} \right) - \left(-1 \begin{vmatrix} 1 & 1 \\ 3 & 2 \end{vmatrix} \right)$$

$$= 3(0 - 1) + 2(2 + 1) + 1(2 - 3) = 2.$$

We expanded the 4×4 determinant in minors about the third column to obtain the two 3×3 determinants. The first of these was expanded about the second row, the other about the second column.

In addition to the properties stated in Section 10.3, determinants have two other very important properties, which are stated in the following theorem.

THEOREM 3

If $\mathcal{A}$ and $\mathcal{B}$ are $n \times n$ matrices, then

(a) $\det(\mathcal{A}^T) = \det(\mathcal{A})$ and

(b) $\det(\mathcal{A}\mathcal{B}) = \det(\mathcal{A})\det(\mathcal{B})$.

We will not attempt any proof of this or other theorems in this section. The reader is referred to texts on linear algebra. Part (a) is not very difficult to prove, even in the case of general n. Part (b) cannot really be proved in general without a formal definition of determinant. However, the reader should verify (b) for 2×2 matrices by direct calculation.

We say that the square matrix $\mathcal{A}$ is **singular** if $\det(\mathcal{A}) = 0$. If $\det(\mathcal{A}) \neq 0$, we say that $\mathcal{A}$ is **nonsingular** or **invertible**.

Remark If A is a 3×3 matrix, then $\det(A)$ is the scalar triple product of the rows of A, and its absolute value is the volume of the parallelepiped spanned by those rows. Therefore, A is nonsingular if and only if its rows span a parallelepiped of positive volume; the row vectors cannot all lie in the same plane. The same may be said of the columns of A.

In general, an $n \times n$ matrix is singular if its rows (or columns), considered as vectors, satisfy one or more linear equations of the form

$$c_1 \mathbf{x}_1 + c_2 \mathbf{x}_2 + \cdots + c_n \mathbf{x}_n = \mathbf{0},$$

with at least one nonzero coefficient c_i. A set of vectors satisfying such a linear equation is called **linearly dependent** because one of the vectors can always be expressed as a linear combination of the others; if $c_1 \neq 0$, then

$$\mathbf{x}_1 = -\frac{c_2}{c_1} \mathbf{x}_2 - \frac{c_3}{c_1} \mathbf{x}_3 - \cdots - \frac{c_n}{c_1} \mathbf{x}_n.$$

All linear combinations of the vectors in a linearly dependent set of n vectors in $\mathbb{R}^n$ must lie in a **subspace** of dimension lower than n.

The $n \times n$ **identity matrix** is the matrix

$$I = \begin{pmatrix} 1 & 0 & \cdots & 0 \\ 0 & 1 & \cdots & 0 \\ \vdots & \vdots & \ddots & \vdots \\ 0 & 0 & \cdots & 1 \end{pmatrix}$$

with "1" in every position on the **main diagonal** and "0" in every other position. Evidently I commutes with every $n \times n$ matrix: $IA = AI = A$. Also $\det(I) = 1$. The identity matrix plays the same role in matrix algebra that the number 1 plays in arithmetic.

Any nonzero number x has a reciprocal x^{-1} such that $x x^{-1} = x^{-1} x = 1$. A similar situation holds for square matrices. The **inverse** of a *nonsingular* square matrix A is a nonsingular square matrix A^{-1} satisfying

$$A A^{-1} = A^{-1} A = I.$$

THEOREM 4

Every nonsingular square matrix A has a *unique* inverse A^{-1}. Moreover, the inverse satisfies

(a) $\det(A^{-1}) = \dfrac{1}{\det(A)},$

(b) $(A^{-1})^T = (A^T)^{-1}.$

We will not have much cause to calculate inverses, but we note that it can be done by solving systems of linear equations, as the following simple example illustrates.

Example 5 Show that the matrix $A = \begin{pmatrix} 1 & -1 \\ 1 & 1 \end{pmatrix}$ is nonsingular and find its inverse.

Solution $\det(A) = \begin{vmatrix} 1 & -1 \\ 1 & 1 \end{vmatrix} = 1 + 1 = 2.$ Therefore, A is nonsingular and invertible. Let $A^{-1} = \begin{pmatrix} a & b \\ c & d \end{pmatrix}$. Then $A A^{-1} = I$, that is,

$$\begin{pmatrix} 1 & 0 \\ 0 & 1 \end{pmatrix} = \begin{pmatrix} 1 & -1 \\ 1 & 1 \end{pmatrix} \begin{pmatrix} a & b \\ c & d \end{pmatrix} = \begin{pmatrix} a - c & b - d \\ a + c & b + d \end{pmatrix},$$

so a, b, c, and d must satisfy the systems of equations

$$\begin{cases} a - c = 1 \\ a + c = 0 \end{cases} \qquad \begin{cases} b - d = 0 \\ b + d = 1. \end{cases}$$

Evidently $a = b = d = 1/2$, $c = -1/2$, and

$$\mathcal{A}^{-1} = \begin{pmatrix} \frac{1}{2} & \frac{1}{2} \\ -\frac{1}{2} & \frac{1}{2} \end{pmatrix}.$$

Generally, matrix inversion is not carried out by the method of the above example but rather by an orderly process of performing operations on the rows of the matrix to transform it into the identity. When the same operations are performed on the rows of the identity matrix, the inverse of the original matrix results. See a text on linear algebra for a description of the method. A singular matrix has no inverse.

Linear Transformations

A function $\mathbf{F}$ whose domain is the m-dimensional space $\mathbb{R}^m$ and whose range is contained in the n-dimensional space $\mathbb{R}^n$ is called a **linear transformation from $\mathbb{R}^m$ to $\mathbb{R}^n$** if it satisfies

$$\mathbf{F}(\lambda \mathbf{x} + \mu \mathbf{y}) = \lambda \mathbf{F}(\mathbf{x}) + \mu \mathbf{F}(\mathbf{y})$$

for all points $\mathbf{x}$ and $\mathbf{y}$ in $\mathbb{R}^m$ and all real numbers λ and μ. To such a linear transformation $\mathbf{F}$ there corresponds an $n \times m$ matrix $\mathcal{F}$ such that for all $\mathbf{x}$ in $\mathbb{R}^m$,

$$\mathbf{F}(\mathbf{x}) = \mathcal{F}\mathbf{x},$$

or, expressed in terms of the components of $\mathbf{x}$,

$$\mathbf{F}(x_1, x_2, \cdots, x_m) = \mathcal{F} \begin{pmatrix} x_1 \\ x_2 \\ \vdots \\ x_m \end{pmatrix}.$$

We say that $\mathcal{F}$ is a **matrix representation** of the linear transformation $\mathbf{F}$. If $m = n$ so that $\mathbf{F}$ maps $\mathbb{R}^m$ into itself, then $\mathcal{F}$ is a square matrix. In this case $\mathcal{F}$ is nonsingular if and only if $\mathbf{F}$ is one-to-one and has the whole of $\mathbb{R}^m$ as range.

A composition of linear transformations is still a linear transformation and will have a matrix representation. The real motivation lying behind the definition of matrix multiplication is that the matrix representation of a *composition* of linear transformations is the *product* of the individual matrix representations of the transformations being composed.

THEOREM 5

If $\mathbf{F}$ is a linear transformation from $\mathbb{R}^m$ to $\mathbb{R}^n$ represented by the $n \times m$ matrix $\mathcal{F}$, and if $\mathbf{G}$ is a linear transformation from $\mathbb{R}^n$ to $\mathbb{R}^p$ represented by the $p \times n$ matrix $\mathcal{G}$, then the composition $\mathbf{G} \circ \mathbf{F}$ defined by

$$\mathbf{G} \circ \mathbf{F}(x_1, x_2, \ldots, x_m) = \mathbf{G}\Big(\mathbf{F}(x_1, x_2, \ldots, x_m)\Big)$$

is itself a linear transformation from $\mathbb{R}^m$ to $\mathbb{R}^p$ represented by the $p \times m$ matrix $\mathcal{G}\mathcal{F}$. That is,

$$\mathbf{G}\Big(\mathbf{F}(\mathbf{x})\Big) = \mathcal{G}\mathcal{F}\mathbf{x}.$$

Linear Equations

A system of n linear equations in n unknowns:

$$a_{11}x_1 + a_{12}x_2 + \cdots + a_{1n}x_n = b_1$$
$$a_{21}x_1 + a_{22}x_2 + \cdots + a_{2n}x_n = b_2$$
$$\vdots$$
$$a_{n1}x_1 + a_{n2}x_2 + \cdots + a_{nn}x_n = b_n$$

can be written compactly as a single matrix equation,

$$\mathcal{A}\mathbf{x} = \mathbf{b},$$

where

$$\mathcal{A} = \begin{pmatrix} a_{11} & a_{12} & \cdots & a_{1n} \\ a_{21} & a_{22} & \cdots & a_{2n} \\ \vdots & \vdots & \ddots & \vdots \\ a_{n1} & a_{n2} & \cdots & a_{nn} \end{pmatrix}, \qquad \mathbf{x} = \begin{pmatrix} x_1 \\ x_2 \\ \vdots \\ x_n \end{pmatrix}, \quad \text{and} \quad \mathbf{b} = \begin{pmatrix} b_1 \\ b_2 \\ \vdots \\ b_n \end{pmatrix}.$$

Compare the equation $\mathcal{A}\mathbf{x} = \mathbf{b}$ with the equation $ax = b$ for a single unknown x. The equation $ax = b$ has the unique solution $x = a^{-1}b$ provided $a \neq 0$. By analogy, the linear system $\mathcal{A}\mathbf{x} = \mathbf{b}$ has a unique solution given by

$$\mathbf{x} = \mathcal{A}^{-1}\mathbf{b},$$

provided $\mathcal{A}$ is nonsingular. To see this, just multiply both sides of the equation $\mathcal{A}\mathbf{x} = \mathbf{b}$ on the left by $\mathcal{A}^{-1}$; $\mathbf{x} = I\mathbf{x} = \mathcal{A}^{-1}\mathcal{A}\mathbf{x} = \mathcal{A}^{-1}\mathbf{b}$.

If $\mathcal{A}$ is singular, then the system $\mathcal{A}\mathbf{x} = \mathbf{b}$ may or may not have a solution, and if a solution exists it will not be unique. Consider the case $\mathbf{b} = \mathbf{0}$ (the zero vector). Then any vector $\mathbf{x}$ perpendicular to all the rows of $\mathcal{A}$ will satisfy the system. Since the rows of $\mathcal{A}$ lie in a space of dimension less than n (because $\det(\mathcal{A}) = 0$), there will be at least a line of such vectors $\mathbf{x}$. Thus, solutions of $\mathcal{A}\mathbf{x} = \mathbf{0}$ are not unique if $\mathcal{A}$ is singular. The same must be true of the system $\mathcal{A}^T\mathbf{y} = \mathbf{0}$; there will be nonzero vectors $\mathbf{y}$ satisfying it if $\mathcal{A}$ is singular. But then, if the system $\mathcal{A}\mathbf{x} = \mathbf{b}$ has any solution $\mathbf{x}$, we must have

$$(\mathbf{y} \bullet \mathbf{b}) = \mathbf{y}^T\mathbf{b} = \mathbf{y}^T\mathcal{A}\mathbf{x} = (\mathbf{x}^T\mathcal{A}^T\mathbf{y})^T = (\mathbf{x}^T\mathbf{0})^T = (0).$$

Hence, $\mathcal{A}\mathbf{x} = \mathbf{b}$ can only have solutions for those vectors $\mathbf{b}$ that are perpendicular to *every* solution $\mathbf{y}$ of $\mathcal{A}^T\mathbf{y} = \mathbf{0}$.

A system of m linear equations in n unknowns may or may not have any solutions if $n < m$. It will have solutions if some $m - n$ of the equations are *linear combinations* (sums of multiples) of the other n equations. If $n > m$, then we can try to solve the m equations for m of the variables, allowing the solutions to depend on the other $n - m$ variables. Such a solution exists if the determinant of the coefficients of the m variables for which we want to solve is not zero. This is a special case of the **Implicit Function Theorem**, which we will examine in Section 12.8.

Example 6 Solve $\begin{cases} 2x + y - 3z = 4 \\ x + 2y + 6z = 5 \end{cases}$ for x and y in terms of z.

Solution The system can be expressed in the form

$$\mathcal{A}\begin{pmatrix} x \\ y \end{pmatrix} = \begin{pmatrix} 4 + 3z \\ 5 - 6z \end{pmatrix}, \qquad \text{where} \quad \mathcal{A} = \begin{pmatrix} 2 & 1 \\ 1 & 2 \end{pmatrix}.$$

$\mathcal{A}$ has determinant 3 and inverse $\mathcal{A}^{-1} = \begin{pmatrix} 2/3 & -1/3 \\ -1/3 & 2/3 \end{pmatrix}$. Thus

$$\begin{pmatrix} x \\ y \end{pmatrix} = \mathcal{A}^{-1} \begin{pmatrix} 4 + 3z \\ 5 - 6z \end{pmatrix} = \begin{pmatrix} 2/3 & -1/3 \\ -1/3 & 2/3 \end{pmatrix} \begin{pmatrix} 4 + 3z \\ 5 - 6z \end{pmatrix} = \begin{pmatrix} 1 + 4z \\ 2 - 5z \end{pmatrix}$$

The solution is $x = 1 + 4z$, $y = 2 - 5z$. (Of course, this solution could have been found by elimination of x or y from the given equations without using matrix methods.) ∎

The following theorem states a result of some theoretical importance expressing the solution of the system $\mathcal{A}\mathbf{x} = \mathbf{b}$ for nonsingular $\mathcal{A}$ in terms of determinants.

THEOREM 6

Cramer's Rule

Let $\mathcal{A}$ be a nonsingular $n \times n$ matrix. Then the solution $\mathbf{x}$ of the system

$$\mathcal{A}\mathbf{x} = \mathbf{b}$$

has components given by

$$x_1 = \frac{\det(\mathcal{A}_1)}{\det(\mathcal{A})}, \quad x_2 = \frac{\det(\mathcal{A}_2)}{\det(\mathcal{A})}, \quad \cdots, \quad x_n = \frac{\det(\mathcal{A}_n)}{\det(\mathcal{A})},$$

where $\mathcal{A}_j$ is the matrix $\mathcal{A}$ with its jth column replaced by the column vector $\mathbf{b}$. That is,

$$\det(\mathcal{A}_j) = \begin{vmatrix} a_{11} & \cdots & a_{1(j-1)} & b_1 & a_{1(j+1)} & \cdots & a_{1n} \\ a_{21} & \cdots & a_{2(j-1)} & b_2 & a_{2(j+1)} & \cdots & a_{2n} \\ \vdots & & \vdots & \vdots & \vdots & & \vdots \\ a_{n1} & \cdots & a_{n(j-1)} & b_n & a_{n(j+1)} & \cdots & a_{nn} \end{vmatrix}.$$

The following example provides a concrete illustration of the use of Cramer's Rule to solve a specific linear system. However, Cramer's Rule is primarily used in a more general (theoretical) context; it is not efficient to use determinants to calculate solutions of linear systems.

Example 7 Find the point of intersection of the three planes

$$\begin{aligned} x + y + 2z &= 1 \\ 3x + 6y - z &= 0 \\ x - y - 4z &= 3. \end{aligned}$$

Solution The solution of the linear system above provides the coordinates of the intersection point. The determinant of the coefficient matrix of this system is

$$\det(\mathcal{A}) = \begin{vmatrix} 1 & 1 & 2 \\ 3 & 6 & -1 \\ 1 & -1 & -4 \end{vmatrix} = -32,$$

so the system does have a unique solution. We have

$$x = \frac{1}{-32} \begin{vmatrix} 1 & 1 & 2 \\ 0 & 6 & -1 \\ 3 & -1 & -4 \end{vmatrix} = \frac{-64}{-32} = 2,$$

$$y = \frac{1}{-32} \begin{vmatrix} 1 & 1 & 2 \\ 3 & 0 & -1 \\ 1 & 3 & -4 \end{vmatrix} = \frac{32}{-32} = -1,$$

$$z = \frac{1}{-32} \begin{vmatrix} 1 & 1 & 1 \\ 3 & 6 & 0 \\ 1 & -1 & 3 \end{vmatrix} = \frac{0}{-32} = 0.$$

The intersection point is $(2, -1, 0)$. ∎

Quadratic Forms, Eigenvalues, and Eigenvectors

If $\mathbf{x}$ is a column vector in $\mathbb{R}^n$ and $\mathcal{A} = (a_{ij})$ is an $n \times n$, real, symmetric matrix (i.e., $a_{ij} = a_{ji}$ for $1 \leq i, j \leq n$), then the expression

$$Q(\mathbf{x}) = \mathbf{x}^T \mathcal{A} \mathbf{x} = \sum_{i,j=1}^{n} a_{ij} x_i x_j$$

is called a **quadratic form** on $\mathbb{R}^n$ corresponding to the matrix $\mathcal{A}$. Observe that $Q(\mathbf{x})$ is a real number for every n-vector $\mathbf{x}$.

We say that $\mathcal{A}$ is **positive definite** if $Q(\mathbf{x}) > 0$ for every nonzero vector $\mathbf{x}$. Similarly, $\mathcal{A}$ is **negative definite** if $Q(\mathbf{x}) < 0$ for every nonzero vector $\mathbf{x}$. We say that $\mathcal{A}$ is **positive semidefinite** (or **negative semidefinite**) if $Q(\mathbf{x}) \geq 0$ (or $Q(\mathbf{x}) \leq 0$) for every nonzero vector $\mathbf{x}$.

If $Q(\mathbf{x}) > 0$ for some nonzero vectors $\mathbf{x}$ while $Q(\mathbf{x}) < 0$ for other such $\mathbf{x}$ (i.e., if $\mathcal{A}$ is neither positive semidefinite nor negative semidefinite), then we will say that $\mathcal{A}$ is **indefinite**.

Example 8   The expression $Q(x, y, z) = 3x^2 + 2y^2 + 5z^2 - 2xy + 4xz + 2yz$ is a quadratic form on $\mathbb{R}^3$ corresponding to the symmetric matrix

$$\mathcal{A} = \begin{pmatrix} 3 & -1 & 2 \\ -1 & 2 & 1 \\ 2 & 1 & 5 \end{pmatrix}.$$

Observe how the elements of the matrix are obtained from the coefficients of Q; the coefficients of x^2, y^2, and z^2 form the main diagonal elements, while the coefficients of the product terms are cut in half and half is put in each of the two corresponding symmetric off-diagonal positions.

The matrix $\mathcal{A}$ is positive definite since $Q(x, y, z)$ can be rewritten in the form

$$Q(x, y, z) = x^2 + (x - y)^2 + (x + 2z)^2 + (y + z)^2,$$

from which it is apparent that $Q(x, y, z) \geq 0$ for all (x, y, z) and $Q(x, y, z) = 0$ only if $x = y = z = 0$.

In Section 13.1 we will use the positive or negative definiteness of certain matrices to classify critical points of functions of several variables as local maxima and minima. Useful criteria for definiteness can be expressed in terms of the *eigenvalues* of the matrix $\mathcal{A}$.

We say that λ is an **eigenvalue** of the $n \times n$ square matrix $\mathcal{A} = (a_{ij})$ if there exists a *nonzero* column vector $\mathbf{x}$ such that $\mathcal{A}\mathbf{x} = \lambda \mathbf{x}$, or, equivalently,

$$(\mathcal{A} - \lambda I)\mathbf{x} = \mathbf{0},$$

where I is the $n \times n$ identity matrix. The nonzero vector $\mathbf{x}$ is called an **eigenvector** of $\mathcal{A}$ corresponding to the eigenvalue λ and can exist only if $\mathcal{A} - \lambda I$ is a singular matrix, that is, if

$$\det(\mathcal{A} - \lambda I) = \begin{vmatrix} a_{11} - \lambda & a_{12} & \cdots & a_{1n} \\ a_{21} & a_{22} - \lambda & \cdots & a_{2n} \\ \vdots & \vdots & \ddots & \vdots \\ a_{n1} & a_{n2} & \cdots & a_{nn} - \lambda \end{vmatrix} = 0.$$

The eigenvalues of $\mathcal{A}$ must satisfy this nth-degree polynomial equation, so they can be either real or complex. The following theorems are proved in standard linear algebra texts.

THEOREM 7   If $\mathcal{A} = (a_{ij})_{i,j=1}^{n}$ is a real, symmetric matrix, then

(a) all the eigenvalues of $\mathcal{A}$ are real,

(b) all the eigenvalues of $\mathcal{A}$ are nonzero if $\det(\mathcal{A}) \neq 0$,

(c) $\mathcal{A}$ is positive definite if all its eigenvalues are positive,

(d) $\mathcal{A}$ is negative definite if all its eigenvalues are negative,

(e) $\mathcal{A}$ is positive semidefinite if all its eigenvalues are nonnegative,

(f) $\mathcal{A}$ is negative semidefinite if all its eigenvalues are nonpositive,

(g) $\mathcal{A}$ is indefinite if it has at least one positive eigenvalue and at least one negative eigenvalue.

THEOREM 8   Let $\mathcal{A} = (a_{ij})_{i,j=1}^{n}$ be a real symmetric matrix and consider the determinants

$$D_i = \begin{vmatrix} a_{11} & a_{12} & \cdots & a_{1i} \\ a_{21} & a_{22} & \cdots & a_{2i} \\ \vdots & \vdots & \ddots & \vdots \\ a_{i1} & a_{i2} & \cdots & a_{ii} \end{vmatrix} \qquad \text{for } 1 \leq i \leq n.$$

Thus, $D_1 = a_{11}$, $D_2 = \begin{vmatrix} a_{11} & a_{12} \\ a_{21} & a_{22} \end{vmatrix} = a_{11}a_{22} - a_{12}a_{21} = a_{11}a_{22} - a_{12}^2$, etc.

(a) If $D_i > 0$ for $1 \leq i \leq n$, then $\mathcal{A}$ is positive definite.

(b) If $D_i > 0$ for even numbers i in $\{1, 2, \ldots, n\}$, and $D_i < 0$ for odd numbers i in $\{1, 2, \ldots, n\}$, then $\mathcal{A}$ is negative definite.

(c) If $\det(\mathcal{A}) = D_n \neq 0$ but neither of the above conditions hold, then $Q(\mathbf{x})$ is indefinite.

(d) If $\det(\mathcal{A}) = 0$, then $\mathcal{A}$ is not positive or negative definite and may be semidefinite or indefinite.

Example 9   For the matrix $\mathcal{A}$ of Example 8, we have

$$D_1 = 3 > 0, \quad D_2 = \begin{vmatrix} 3 & -1 \\ -1 & 2 \end{vmatrix} = 5 > 0, \quad D_3 = \begin{vmatrix} 3 & -1 & 2 \\ -1 & 2 & 1 \\ 2 & 1 & 5 \end{vmatrix} = 10 > 0,$$

which reconfirms that the quadratic form of that exercise is positive definite.

Exercises 10.6

Evaluate the matrix products in Exercises 1–4.

1. $\begin{pmatrix} 3 & 0 & -2 \\ 1 & 1 & 2 \\ -1 & 1 & -1 \end{pmatrix} \begin{pmatrix} 2 & 1 \\ 3 & 0 \\ 0 & -2 \end{pmatrix}$

2. $\begin{pmatrix} 1 & 1 & 1 \\ 0 & 1 & 1 \\ 0 & 0 & 1 \end{pmatrix} \begin{pmatrix} 1 & 1 & 1 \\ 0 & 1 & 1 \\ 0 & 0 & 1 \end{pmatrix}$

3. $\begin{pmatrix} a & b \\ c & d \end{pmatrix} \begin{pmatrix} w & x \\ y & z \end{pmatrix}$ **4.** $\begin{pmatrix} w & x \\ y & z \end{pmatrix} \begin{pmatrix} a & b \\ c & d \end{pmatrix}$

5. Evaluate $\mathcal{A}\mathcal{A}^T$ and $\mathcal{A}^2 = \mathcal{A}\mathcal{A}$, where

$$\mathcal{A} = \begin{pmatrix} 1 & 1 & 1 & 1 \\ 0 & 1 & 1 & 1 \\ 0 & 0 & 1 & 1 \\ 0 & 0 & 0 & 1 \end{pmatrix}$$

6. Evaluate $\mathbf{x}\mathbf{x}^T$, $\mathbf{x}^T\mathbf{x}$, and $\mathbf{x}^T\mathcal{A}\mathbf{x}$, where

$$\mathbf{x} = \begin{pmatrix} x \\ y \\ z \end{pmatrix} \quad \text{and} \quad \mathcal{A} = \begin{pmatrix} a & p & q \\ p & b & r \\ q & r & c \end{pmatrix}.$$

Evaluate the determinants in Exercises 7–8.

7. $\begin{vmatrix} 2 & 3 & -1 & 0 \\ 4 & 0 & 2 & 1 \\ 1 & 0 & -1 & 1 \\ -2 & 0 & 0 & 1 \end{vmatrix}$ **8.** $\begin{vmatrix} 1 & 1 & 1 & 1 \\ 1 & 2 & 3 & 4 \\ -2 & 0 & 2 & 4 \\ 3 & -3 & 2 & -2 \end{vmatrix}$

9. Show that if $\mathcal{A} = (a_{ij})$ is an $n \times n$ matrix for which $a_{ij} = 0$ whenever $i > j$, then $\det(\mathcal{A}) = \prod_{k=1}^{n} a_{kk}$, the product of the elements on the main diagonal of $\mathcal{A}$.

10. Show that $\begin{vmatrix} 1 & 1 \\ x & y \end{vmatrix} = y - x$, and

$$\begin{vmatrix} 1 & 1 & 1 \\ x & y & z \\ x^2 & y^2 & z^2 \end{vmatrix} = (y - x)(z - x)(z - y).$$

Try to generalize this result to the $n \times n$ case.

11. Verify the associative law $(\mathcal{A}\mathcal{B})\mathcal{C} = \mathcal{A}(\mathcal{B}\mathcal{C})$ by direct calculation for three arbitrary 2×2 matrices.

12. Show that $\det(\mathcal{A}^T) = \det(\mathcal{A})$ for $n \times n$ matrices by induction on n. Start with the 2×2 case.

13. Verify by direct calculation that $\det(\mathcal{A}\mathcal{B}) = \det(\mathcal{A})\det(\mathcal{B})$ holds for two arbitrary 2×2 matrices.

14. Let $\mathcal{A}_\theta = \begin{pmatrix} \cos\theta & \sin\theta \\ -\sin\theta & \cos\theta \end{pmatrix}$. Show that $(\mathcal{A}_\theta)^T = (\mathcal{A}_\theta)^{-1} = \mathcal{A}_{-\theta}$.

Find the inverses of the matrices in Exercises 15–16.

15. $\begin{pmatrix} 1 & 1 & 1 \\ 0 & 1 & 1 \\ 0 & 0 & 1 \end{pmatrix}$ **16.** $\begin{pmatrix} 1 & 0 & -1 \\ -1 & 1 & 0 \\ 2 & 1 & 3 \end{pmatrix}$

17. Use your result from Exercise 16 to solve the linear system

$$\begin{cases} x - z = -2 \\ -x + y = 1 \\ 2x + y + 3z = 13. \end{cases}$$

18. Solve the system of Exercise 17 by using Cramer's Rule.

19. Solve the system $\begin{cases} x_1 + x_2 + x_3 + x_4 = 0 \\ x_1 + x_2 + x_3 - x_4 = 4 \\ x_1 + x_2 - x_3 - x_4 = 6 \\ x_1 - x_2 - x_3 - x_4 = 2. \end{cases}$

20. Verify Theorem 5 for the special case where **F** and **G** are linear transformations from $\mathbb{R}^2$ to $\mathbb{R}^2$.

In Exercises 21–26, classify the given symmetric matrices as positive or negative definite, positive or negative semidefinite, or indefinite.

21. $\begin{pmatrix} -1 & 1 \\ 1 & -2 \end{pmatrix}$ **22.** $\begin{pmatrix} 1 & 2 & 0 \\ 2 & 1 & 0 \\ 0 & 0 & 1 \end{pmatrix}$

23. $\begin{pmatrix} 2 & 1 & 1 \\ 1 & 2 & 1 \\ 1 & 1 & 2 \end{pmatrix}$ **24.** $\begin{pmatrix} 1 & 1 & 0 \\ 1 & 1 & 0 \\ 0 & 0 & 1 \end{pmatrix}$

25. $\begin{pmatrix} 1 & 0 & 1 \\ 0 & 1 & -1 \\ 1 & -1 & 1 \end{pmatrix}$ **26.** $\begin{pmatrix} 2 & 0 & 1 \\ 0 & 4 & -1 \\ 1 & -1 & 1 \end{pmatrix}$

$\boxed{\textbf{10.7}}$ Using Maple for Vector and Matrix Calculations

The use of a computer algebra system can free us from much of the tedious calculation needed to do calculus. This is especially true of calculations in multivariable and vector calculus, where the calculations can quickly become unmanageable as the number of variables increases. This author's colleague, Dr. Robert Israel, has written an excellent book, *Calculus, the Maple Way*, to show how Maple can be used effectively for doing calculus involving both single-variable and multivariable functions.

In this book we will occasionally call on the power of Maple to carry out calculations involving functions of several variables and vector-valued functions of one or more variables. This section illustrates some of the most basic techniques for calculating with vectors and matrices. The examples here were calculated using Maple 9, but Maple 6 or later should give similar output.

Most of Maple's capability to deal with vectors and matrices is not in its kernel but is written into a package of procedures called **LinearAlgebra**. Therefore, it is customary to load this package at the beginning of a session where it will be needed:

```
>   with(LinearAlgebra):
```

One usually completes a Maple command with a semicolon rather than a colon. You can use a colon to suppress output. Had we used a semicolon to complete the command the result would have produced a list of all the procedures defined in the LinearAlgebra package.

Maple also includes a second linear algebra package called **linalg**, but it is inferior to LinearAlgebra, especially for heavy-duty numerical calculations using large matrices; it is also somewhat more difficult to use. However, the linalg package was present in releases of Maple earlier than release 6, and is still present in release 9. We will not make any use of linalg here, but it was used instead of LinearAlgebra in the fifth edition of this book.

Vectors

There are several ways to define vectors in Maple; the easiest are to use the `Vector([,])` or `<,>` constructions, where a comma-separated list of the components of the vector is placed in the square or angle brackets. Both of these constructions produce column vectors:

```
>  Uc := Vector([1,2,3]); Vc := <a,b,c>;
```

$$Uc := \begin{bmatrix} 1 \\ 2 \\ 3 \end{bmatrix}$$

$$Vc := \begin{bmatrix} a \\ b \\ c \end{bmatrix}$$

You can use `Vector[row]([,])` to produce a row vector; alternatively, you can define a row vector using angle brackets with "|" to separate the components:

```
>  Ur := Vector[row]([1,2,3]); Vr := <a|b|c>;
```

$$Ur := [1, 2, 3]$$

$$Vr := [a, b, c]$$

Vectors can be of any dimension; simply include the appropriate number of commas or | separated components. You can also use the `Vector()` construct with two arguments, the first a positive integer giving the dimension of the vector and the second either a square-bracket-enclosed list of components or an assignment rule giving the value of the ith component:

```
>  <5|-2|3|x>; W := Vector[row](5, i -> i^2);
```

$$[5, -2, 3, x]$$

$$W := [1, 4, 9, 16, 25]$$

We can also construct a vector with arbitrary components like this:

```
>  X := Vector(2, symbol=x);
   Y := Vector[row](4, symbol=y);
```

$$X := \begin{bmatrix} x_1 \\ x_2 \end{bmatrix}$$

$$Y := [y_1, y_2, y_3, y_4]$$

The components of a vector can be referenced by appending the index of the component, enclosed in square brackets, to the name or constructor of the vector. The fourth component of vector W above is `W[4]`:

```
>  W[4]; Vector(16, i -> 3*i - 1)[10]; X[2]+Y[3];
```

$$16$$
$$29$$
$$x_2 + y_3$$

Vectors of the same dimension and type (row or column) can be added, subtracted, and multiplied by scalars using the ordinary operators $+$, $-$, and $*$:

```
>  Uc + Vc; Vc - 3*Uc;
```

$$\begin{bmatrix} 1+a \\ 2+b \\ 3+c \end{bmatrix}$$

$$\begin{bmatrix} a-3 \\ b-6 \\ c-9 \end{bmatrix}$$

For most vector calculations it doesn't matter whether you think of vectors as row or column vectors, but it does make a difference for some LinearAlgebra operators; if you try to add a row vector to a column vector, or two vectors of different dimensions, you will get an error message.

The LinearAlgebra package also defines the product functions `DotProduct` and `CrossProduct`, each of which takes two vector arguments. For DotProduct, the arguments must be of the same but arbitrary dimension. For CrossProduct, both arguments must have dimension 3. However, neither requires both arguments to be of the same type (row or column). The cross product will be a column vector unless both its arguments are row vectors.

As defined in the LinearAlgebra package, DotProduct can produce some strange results. Consider the following:

```
>   DotProduct(Uc,Vc); DotProduct(Vc,Uc);
    DotProduct(Ur,Vr);
```

$$a + 2b + 3c$$
$$\bar{a} + 2\bar{b} + 3\bar{c}$$
$$\bar{a} + 2\bar{b} + 3\bar{c}$$

What is going on here? The bars on the unknown quantities a, b, and c denote complex conjugates of these quantities; The LinearAlgebra package is designed to meet the needs of a great many users of linear algebra, not just calculus students for whom all vectors are assumed to have real components. In fact, `DotProduct(U,V)` sums the products of the complex conjugates of the components of `U` and the unconjugated components of `V` if both vectors are column vectors, and vice versa if both are row vectors. In the first example above, the components of `Uc` are real numbers so no conjugates appeared over them; in the other two cases it is the components of `Vc` or `Vr` that require conjugation, and since Maple doesn't know that these are real, it puts on the bars. To avoid this difficulty when using real vectors, include "`conjugate=false`" as a third argument when using DotProduct from the LinearAlgebra package:

```
>   DotProduct(Ur,Vr, conjugate=false);
```

$$a + 2b + 3c$$

It is also possible to use a dot "." as a binary operator to calculate a dot product. However, dot also represents matrix multiplication, so you must use a row vector to the left of the dot and a column vector to the right to be sure of getting a dot product.

```
>   <a|b|c>.<x,y,z>; <a,b,c>.<x|y|z>;
```

$$a + 2b + 3c$$

$$\begin{bmatrix} a & b & c \\ 2a & 2b & 2c \\ 3a & 3b & 3c \end{bmatrix}$$

LinearAlgebra also has a CrossProduct function, which applies only to 3-vectors. It does not matter whether either of the arguments is a row or column vector. This function can be called using either `CrossProduct(U,V)` or `U &x V`.

```
>   CrossProduct(Uc,Vc); Ur &x Vr;
```

$$\begin{bmatrix} 2c - 3b \\ 3a - c \\ b - 2a \end{bmatrix}$$

$$[2c-3b, 3a - c, b - 2a]$$

LinearAlgebra has a function `Norm()` for calculating the length of a vector. Unfortunately, Maple knows many different definitions for the length of a vector. The one we use is the Euclidean length. The Euclidean length of a vector `V` is calculated by `Norm(V,Euclidean)` or `Norm(V,2)`. (In the latter case the 2 stands for the fact that we use the *square root* of the sum of the *squares* of the components to find the length.)

```
>  Norm(Ur,Euclidean); Norm(<1,-1,2,-3,1>,2);
```

$$\sqrt{14}$$

$$4$$

You can use `Normalize(U,Euclidean)` of `Normalize(U,2)` to find a unit vector in the same direction as `U`. Of course, you could always just multiply `U` by the scalar which is the reciprocal of its length:

```
>  Normalize(<2|-2|1>,2); (1/Norm(Uc,2))*Uc;
```

$$\begin{bmatrix} \dfrac{2}{3}, \dfrac{-2}{3}, \dfrac{1}{3} \end{bmatrix}$$

$$\begin{bmatrix} \dfrac{1}{14}\sqrt{14} \\ \dfrac{1}{7}\sqrt{14} \\ \dfrac{3}{14}\sqrt{14} \end{bmatrix}$$

LinearAlgebra has a function `VectorAngle` to give the angle between two vectors. It doesn't matter whether either vector is a row or column. The result will be in radian measure so you will have to multiply it by $180/\pi$ to get the angle in degrees.

```
>  VectorAngle(<2,2,1>,<1,-2,2>);
```

$$\frac{1}{2}\pi$$

To further illustrate these ideas, let us get Maple to calculate an equation of the plane through $(2, 1, -1)$ perpendicular to the line of intersection of the two planes $2x + 3y + z = 5$ and $3x - 2y - 4z = 1$.

```
>  (<2|3|1> &x <3|-2|-4>) .  (<x,y,z>-<2,1,-1>) = 0;
```

$$-10x - 4 + 11y - 13z = 0$$

or, as we would write it, $10x - 11y + 13z = -4$. Note how we used the cross product of two row vectors (which is itself, therefore, a row vector) to the left of the "." and a difference of two column vectors (which is itself a column vector) to the right of the "." for calculating the dot product.

Finally, let us use Maple to verify the identity

$$(\mathbf{U} \times \mathbf{V}) \times \mathbf{W} = (\mathbf{W} \bullet \mathbf{U})\mathbf{V} - (\mathbf{W} \bullet \mathbf{V})\mathbf{U}.$$

First, we define $\mathbf{U}$, $\mathbf{V}$, and $\mathbf{W}$ to be vectors with arbitrary components. In view of the two dot products on the right-hand side of the identity, we make $\mathbf{W}$ a row vector and the other two column vectors:

```
> U := Vector(3,symbol=u);
  V := Vector(3, symbol=v);
  W := Vector[row](3, symbol=w);
```

$$U := \begin{bmatrix} u_1 \\ u_2 \\ u_3 \end{bmatrix}$$

$$V := \begin{bmatrix} v_1 \\ v_2 \\ v_3 \end{bmatrix}$$

$$W := [w_1, w_2, w_3]$$

Now we only need to subtract the right side of the identity from the left side and simplify the result:

```
> simplify((U &x V) &x W - (W . U)*V + (W . V)*U);
```

$$\begin{bmatrix} 0 \\ 0 \\ 0 \end{bmatrix}$$

The result is the zero vector, thus confirming the identity.

Remark   Maple 8 and later releases have a new package called VectorCalculus, which provides greater functionality than the LinearAlgebra package for dealing with vector-valued functions and functions of vector variables. We will be illustrating the use of this package in later chapters, but note here that it also defines the vector operations considered above but not all of the matrix functions considered below. VectorCalculus reports vectors as linear combinations of basis vectors rather than as row or column matrices. The default bases it uses consist of vectors e_x, e_y, e_z (rather than **i**, **j**, **k**) for vectors of dimension up to 3, but $e_{x1}, e_{x2}, \ldots$ for dimensions higher than 3. Nevertheless, although it is not apparent from the way VectorCalculus displays vectors, it still maintains the distinction between row and column vectors and won't let you add a row vector to a column vector. A big advantage of the VectorCalculus package over LinearAlgebra is that VectorCalculus uses the usual definition of dot product (even when using the "." notation), so that the order of factors in a dot product is irrelevant and no complex conjugation is used. If you want to use the VectorCalculus package and still have access to all the matrix operations provided by LinearAlgebra, load the VectorCalculus package *after* the LinearAlgebra package, so that its new definitions of vector operations will replace those of the LinearAlgebra package.

```
> with(LinearAlgebra):
  with(VectorCalculus):
```

Even with output suppressed, the second `with` above produces a few lines of "warnings" mainly about the changed definitions of some vector operations.

```
> V1 := <2,-3,4>; V2 := <a|b|c>; V3 := <2,-3,4,-5,6>;
```

$$V1 := 2e_x - 3e_y + 4e_z$$

$$V2 := ae_x + be_y + ce_z$$

$$V3 := 2e_{x1} - 3e_{x2} + 4e_{x3} - 5e_{x4} + 6e_{x5}$$

```
> V1.V2; V2.V1;
```

$$2a - 3b + 4c$$
$$2a - 3b + 4c$$

Because `V1` is a column vector and `V2` is a row vector, any attempt to calculate a linear combination of these vectors will generate an error, as will attempts to calculate `M.V2` or `V1.M` if M is a 3×3 matrix. Of course, `M.V1` will work fine, as will `V2.M`, although the result will be a one-row matrix rather than a vector. We will examine VectorCalculus further in later chapters.

Matrices

The LinearAlgebra package also provides a variety of ways to define and manipulate matrices. We can define a matrix as a column vector whose elements are row vectors, or as a row vector whole elements are column vectors:

```
>   <<1|1|1>,<2|1|3>>; <<1,2>|<1,1>|<1,3>>;
```

$$\begin{bmatrix} 1 & 1 & 1 \\ 2 & 1 & 3 \end{bmatrix}$$

$$\begin{bmatrix} 1 & 1 & 1 \\ 2 & 1 & 3 \end{bmatrix}$$

You can also use the `Matrix` function to define a matrix. This function can either be supplied with a list of lists specifying the rows of the matrix, or two positive integers (the number of rows and columns, respectively) and a rule for calculating the element in the ith row and jth column.

```
>   L := Matrix([[1,1,1],[2,1,3]]);
    M := Matrix(3,3, (i,j) -> i-j);
```

$$L := \begin{bmatrix} 1 & 1 & 1 \\ 2 & 1 & 3 \end{bmatrix}$$

$$M := \begin{bmatrix} 0 & -1 & -2 \\ 1 & 0 & -1 \\ 2 & 1 & 0 \end{bmatrix}$$

A matrix P with 2 rows and four columns having arbitrary elements $p_{i,j}$ can be constructed as follows:

```
>   P := Matrix(2,4,symbol=p);
```

$$P := \begin{bmatrix} p_{1,1} & p_{1,2} & p_{1,3} & p_{1,4} \\ p_{2,1} & p_{2,2} & p_{2,3} & p_{2,4} \end{bmatrix}$$

As with vectors, particular elements in a matrix can be accessed by including the row and column indices in square brackets following the name of the matrix.

```
>   P[1,2] := Pi; P[1,4]+P[2,4]; P;
```

$$P_{1,2} := \pi$$

$$p_{1,4} + p_{2,4}$$

$$\begin{bmatrix} p_{1,1} & \pi & p_{1,3} & p_{1,4} \\ p_{2,1} & p_{2,2} & p_{2,3} & p_{2,4} \end{bmatrix}$$

There are also shorthand constructs for special kinds of matrices, such as ones with all zero entries, identity (square) matrices, and diagonal matrices:

```
>   Matrix(2,3); IdentityMatrix(3);
    DiagonalMatrix([a,b,c]);
```

$$\begin{bmatrix} 0 & 0 & 0 \\ 0 & 0 & 0 \end{bmatrix}$$

$$\begin{bmatrix} 1 & 0 & 0 \\ 0 & 1 & 0 \\ 0 & 0 & 1 \end{bmatrix}$$

$$\begin{bmatrix} a & 0 & 0 \\ 0 & b & 0 \\ 0 & 0 & c \end{bmatrix}$$

The transpose T of the matrix is obtained by using the `Transpose` function.

```
>   T := Transpose(L);
```

$$T := \begin{bmatrix} 1 & 2 \\ 1 & 1 \\ 1 & 3 \end{bmatrix}$$

The product, AB, of two matrices A and B is calculated using the binary operator `.`; that is, we calculate `A.B`. Of course, the number of columns of A must be equal to the number of rows of B.

```
>   L.T; T.L;
```

$$\begin{bmatrix} 3 & 6 \\ 6 & 14 \end{bmatrix}$$

$$\begin{bmatrix} 5 & 3 & 7 \\ 3 & 2 & 4 \\ 7 & 4 & 10 \end{bmatrix}$$

The determinant and inverse of a square matrix are calculated with the `Determinant` and `MatrixInverse` functions.

```
>   A := <<1|1|1>,<2|1|3>,<1|1|2>>;
    DetA := Determinant(A); Ainv := MatrixInverse(A);
```

$$A := \begin{bmatrix} 1 & 1 & 1 \\ 2 & 1 & 3 \\ 1 & 1 & 2 \end{bmatrix}$$

$$DetA := -1$$

$$Ainv := \begin{bmatrix} 1 & 1 & -2 \\ 1 & -1 & 1 \\ -1 & 0 & 1 \end{bmatrix}$$

```
>   A.Ainv = Ainv.A;
```

$$\begin{bmatrix} 1 & 0 & 0 \\ 0 & 1 & 0 \\ 0 & 0 & 1 \end{bmatrix} = \begin{bmatrix} 1 & 0 & 0 \\ 0 & 1 & 0 \\ 0 & 0 & 1 \end{bmatrix}$$

Linear Equations

A set of n linear equations in n variables can be written in the form $AX = B$, where A is an $n \times n$ matrix and X and B are column n-vectors. Thus, the solution can be calculated as $X = A^{-1}B$. For example, the system

$$x + y + z = 2, \qquad 2x + y + 3z = 9, \qquad x + y + 2z = 1$$

has the matrix A defined above as its coefficient matrix, and B the column vector `<2,9,1>`. The solution of the system is:

```
>   X : = Ainv.<2,9,1>;
```

$$X := \begin{bmatrix} 9 \\ -6 \\ -1 \end{bmatrix}$$

that is, $x = 9$, $y = -6$, $z = -1$. LinearAlgebra provides a simpler way of solving the system $AX = B$; we just need to use the function `LinearSolve(A,B)`:

```
>   X := linsolve(A, [2,9,1]);
```

$$X := \begin{bmatrix} 9 \\ -6 \\ -1 \end{bmatrix}$$

`LinearSolve` is better at solving linear systems than is matrix inversion, since it can solve some systems for which the matrix is singular or not square. Consider the two systems

$$\begin{array}{ccc} x + y = 1 & & x + y = 1 \\ 2x + 2y = 2 & \text{and} & 2x + 2y = 1 \end{array}$$

The first system has a one-parameter family of solutions $x = 1 - t$, $y = t$ for arbitrary t. The second system is inconsistent and has no solutions.

```
>   L := Matrix([[1,1],[2,2]]); B1 := <1,2>; B2 := <1,1>;
```

$$L := \begin{bmatrix} 1 & 1 \\ 2 & 2 \end{bmatrix}$$

$$B1 := \begin{bmatrix} 1 \\ 2 \end{bmatrix}$$

$$B2 := \begin{bmatrix} 1 \\ 1 \end{bmatrix}$$

```
>   X := LinearSolve(L,B1,free=t);
```

$$X := \begin{bmatrix} 1 - t_2 \\ t_2 \end{bmatrix}$$

The extra argument `free=t` was included to force LinearSolve to use subscripted t variables for any parameters. It is always safe to include an argument of this type; omitting it can cause output that looks somewhat strange. (Try it and see.) If the system has a unique solution, the `free=t` parameter will just be ignored.

```
>   X := LinearSolve(L,B2,free=t);
```

```
Error, (in LinearSolve) inconsistent system
```

Eigenvectors and Eigenfunctions

The LinearAlgebra package has procedures for finding the eigenvalues and eigenvectors of matrices. For a real symmetric matrix, the eigenvalues are always real.

```
>   K := Matrix([[3,1,-1],[1,4,1],[-1,1,3]]);
```

$$K := \begin{bmatrix} 3 & 1 & -1 \\ 1 & 4 & 1 \\ -1 & 1 & 3 \end{bmatrix}$$

```
>   Eigenvalues(K);
```

$$\begin{bmatrix} 4 \\ 3 + \sqrt{3} \\ 3 - \sqrt{3} \end{bmatrix}$$

The `Eigenvalues` function produces a column vector of the eigenvalues of the square matrix that is its argument. In this example all three eigenvalues are positive, so K is a positive definite matrix. Our main use for eigenvalues will be the classification of critical points of functions of several variables. This use does not require knowledge of the corresponding eigenvectors, but if we did need to know them, we could have used the function `Eigenvectors(K)` instead. The output would then have consisted of two items separated by a comma. The first item would be the column vector of eigenvalues of K; the second would be a square matrix whose columns are the eigenvectors corresponding to those eigenvalues. (Corresponding to an eigenvalue having multiplicity m there would be m linearly independent columns in the matrix.)

```
>   Eigenvectors(K);
```

$$\begin{bmatrix} 4 \\ 3+\sqrt{3} \\ 3-\sqrt{3} \end{bmatrix}, \quad \begin{bmatrix} -1 & -\dfrac{(-2+\sqrt{3})\sqrt{3}}{-3+2\sqrt{3}} & -\dfrac{(-2-\sqrt{3})\sqrt{3}}{-3-2\sqrt{3}} \\ 0 & -\dfrac{-3+\sqrt{3}}{-3+2\sqrt{3}} & -\dfrac{-3-\sqrt{3}}{-3-2\sqrt{3}} \\ 1 & 1 & 1 \end{bmatrix}$$

Maple isn't always good at spotting simplifications. If you follow the above Maple command with `simplify(%[2])`, you will see that the top row in the matrix of eigenvectors is, in fact, much simpler than it looks.

Remark All the matrices and vectors used in the examples of this section were of very small dimension. The LinearAlgebra package is capable of dealing with large matrices with hundreds of rows and columns, but for such matrices it is best to avoid simple expressions like `2*M-3*N` and `M.N` for linear combinations and products of matrices, and use instead `MatrixAdd(M,N,2,-3)` and `MatrixMatrixMultiply(M,N)`, which are much more efficient in their calculations. Similarly, use `MatrixVectorMultiply(M,X)` rather than `M.X` if X is a column vector and `ScalarMultiply(M,c)` rather than `c*M` if c is a number.

Exercises 10.7

Use Maple to calculate the quantities in Exercises 1–2.

1. The distance between the line through $(3, 0, 2)$ parallel to the vector $2\mathbf{i} + \mathbf{j} - 2\mathbf{k}$ and the line through $(1, 2, 4)$ parallel to $\mathbf{i} + 3\mathbf{j} + 4\mathbf{k}$

2. The angle (in degrees) between the vector $\mathbf{i} - \mathbf{j} + 2\mathbf{k}$ and the plane through the origin containing the vectors $\mathbf{i} - 2\mathbf{j} - 3\mathbf{k}$ and $2\mathbf{i} + 3\mathbf{j} + 4\mathbf{k}$

Use Maple to verify the identities in Exercises 3–4.

3. $\mathbf{U} \bullet (\mathbf{V} \times \mathbf{W}) = \mathbf{V} \bullet (\mathbf{W} \times \mathbf{U}) = \mathbf{W} \bullet (\mathbf{U} \times \mathbf{V})$

4. $(\mathbf{U} \times \mathbf{V}) \times (\mathbf{U} \times \mathbf{W}) = (\mathbf{U} \bullet (\mathbf{V} \times \mathbf{W}))\mathbf{U}$

In Exercises 5–10, define Maple functions to produce the indicated results. You may use functions already defined in **LinearAlgebra**.

5. A function `sp(U,V)` that gives the scalar projection of vector $\mathbf{U}$ along the nonzero vector $\mathbf{V}$

6. A function `vp(U,V)` that gives the vector projection of vector $\mathbf{U}$ along the nonzero vector $\mathbf{V}$

7. A function `ang(U,V)` that gives the angle between the nonzero vectors $\mathbf{U}$ and $\mathbf{V}$ in degrees as a decimal number

8. A function `unitn(U,V)` that gives a unit vector normal to the two nonparallel vectors $\mathbf{U}$ and $\mathbf{V}$ in 3-space

9. A function `VolT(U,V,W)` that gives the volume of the tetrahedron in 3-space that is spanned by the vectors $\mathbf{U}$, $\mathbf{V}$, and $\mathbf{W}$

10. A function `dist(A,B)` giving the distance between two points having position vectors $\mathbf{A}$ and $\mathbf{B}$. Use your function to find the distance between $[1, 1, 1, 1]$ and $[3, -1, 2, 5]$

In Exercises 11–12, use `LinearSolve` to solve the given systems.

11. $\begin{cases} u + 2v + 3x + 4y + 5z = 20 \\ 6u - v + 6x + 2y - 3z = 0 \\ 2u + 8v - 8x - 2y + z = 6 \\ u + v + x + y + z = 5 \\ 10u - 3v + 3x - 2y + 2z = 5 \end{cases}$

12. $\begin{cases} u + v + x + y = 10 \\ u + y + z = 10 \\ u + x + y = 8 \\ u + v + x + z = 11 \\ v + y - z = 1 \end{cases}$

13. Evaluate the determinant of the coefficient matrix for the system in Exercise 11.

14. Find the eigenvalues of the coefficient matrix for the system in Exercise 12. Quote your answers as decimal numbers (use `evalf`) to 5 decimal places. Do you think any of them are really complex?

15. Find the inverse of the matrix

$$A = \begin{bmatrix} 1 & 1/2 & 1/3 \\ 1/2 & 1/3 & 1/4 \\ 1/3 & 1/4 & 1/5 \end{bmatrix}.$$

16. Find, in decimal form (using `evalf(Eigenvals(A))`), the eigenvalues of the matrix A of Exercise 15 and the Eigenvalues of its inverse. Use `Digits := 10`. How do you account for the fact that some of the eigenvalues appear to be complex? What relationship appears to exist between the eigenvalues of A and those of its inverse?

Chapter Review

Key Ideas

- **What is each of the following?**
  - ◇ a neighbourhood ◇ an open set ◇ a closed set
  - ◇ the boundary of a set ◇ the interior of a set
  - ◇ a vector in 3-space ◇ the dot product of vectors
  - ◇ the cross product of two vectors in $\mathbb{R}^3$
  - ◇ a scalar triple product ◇ a vector triple product
  - ◇ a matrix ◇ a determinant
  - ◇ a plane ◇ a straight line ◇ a cone
  - ◇ a cylinder ◇ an ellipsoid ◇ a paraboloid
  - ◇ a hyperboloid of 1 sheet ◇ a hyperboloid of 2 sheets
  - ◇ the transpose of a matrix ◇ the inverse of a matrix
  - ◇ a linear transformation ◇ an eigenvalue of a matrix

- **What is the angle between the vectors u and v?**
- **How do you calculate u × v, given the components of u and v?**
- **What is an equation of the plane through P_0 having normal vector N?**
- **What is an equation of the straight line through P_0 parallel to a?**
- **Given two 3 × 3 matrices A and B, how do you calculate AB?**
- **What is the distance from P_0 to the plane $Ax + By + Cz + D = 0$?**
- **What is Cramer's Rule, and how is it used?**

Review Exercises

Describe the sets of points in 3-space that satisfy the given equations or inequalities in Exercises 1–18.

1. $x + 3z = 3$
2. $y - z \geq 1$
3. $x + y + z \geq 0$
4. $x - 2y - 4z = 8$
5. $y = 1 + x^2 + z^2$
6. $y = z^2$
7. $x = y^2 - z^2$
8. $z = xy$
9. $x^2 + y^2 + 4z^2 < 4$
10. $x^2 + y^2 - 4z^2 = 4$
11. $x^2 - y^2 - 4z^2 = 0$
12. $x^2 - y^2 - 4z^2 = 4$
*13. $(x - z)^2 + y^2 = 1$
*14. $(x - z)^2 + y^2 = z^2$
15. $\begin{cases} x + 2y = 0 \\ z = 3 \end{cases}$
16. $\begin{cases} x + y + 2z = 1 \\ x + y + z = 0 \end{cases}$
17. $\begin{cases} x^2 + y^2 + z^2 = 4 \\ x + y + z = 3 \end{cases}$
18. $\begin{cases} x^2 + z^2 \leq 1 \\ x - y \geq 0 \end{cases}$

Find equations of the planes and lines specified in Exercises 19–28.

19. The plane through the origin perpendicular to the line

$$\frac{x - 1}{2} = \frac{y + 3}{-1} = \frac{z + 2}{3}$$

20. The plane through $(2, -1, 1)$ and $(1, 0, -1)$ parallel to the line in Exercise 19

21. The plane through $(2, -1, 1)$ perpendicular to the planes $x - y + z = 0$ and $2x + y - 3z = 2$

22. The plane through $(-1, 1, 0)$, $(0, 4, -1)$, and $(2, 0, 0)$

23. The plane containing the line of intersection of the planes $x + y + z = 0$ and $2x + y - 3z = 2$, and passing through the point $(2, 0, 1)$

24. The plane containing the line of intersection of the planes $x + y + z = 0$ and $2x + y - 3z = 2$, and perpendicular to the plane $x - 2y - 5z = 17$

25. The vector parametric equation of the line through $(2, 1, -1)$ and $(-1, 0, 1)$

26. Standard form equations of the line through $(1, 0, -1)$ parallel to each of the planes $x - y = 3$ and $x + 2y + z = 1$

27. Scalar parametric equations of the line through the origin perpendicular to the plane $3x - 2y + 4z = 5$

28. The vector parametric equation of the line that joins points on the two lines

$$\mathbf{r} = (1 + t)\mathbf{i} - t\mathbf{j} - (2 + 2t)\mathbf{k}$$
$$\mathbf{r} = 2t\mathbf{i} + (t - 2)\mathbf{j} - (1 + 3t)\mathbf{k}$$

and is perpendicular to both those lines

Express the given conditions or quantities in Exercises 29–30 in terms of dot and cross products.

29. The three points with position vectors $\mathbf{r}_1$, $\mathbf{r}_2$, and $\mathbf{r}_3$ all lie on a straight line.

30. The four points with position vectors $\mathbf{r}_1$, $\mathbf{r}_2$, $\mathbf{r}_3$, and $\mathbf{r}_4$ do not all lie on a plane.

31. Find the area of the triangle with vertices $(1, 2, 1)$, $(4, -1, 1)$, and $(3, 4, -2)$.

32. Find the volume of the tetrahedron with vertices $(1, 2, 1)$, $(4, -1, 1)$, $(3, 4, -2)$, and $(2, 2, 2)$.

33. Show that the matrix

$$\mathcal{A} = \begin{pmatrix} 1 & 0 & 0 & 0 \\ 2 & 1 & 0 & 0 \\ 3 & 2 & 1 & 0 \\ 4 & 3 & 2 & 1 \end{pmatrix}$$

has an inverse, and find the inverse $\mathcal{A}^{-1}$.

34. Let $\mathcal{A} = \begin{pmatrix} 1 & 1 & 1 \\ 2 & 1 & 0 \\ 1 & 0 & -1 \end{pmatrix}$. What condition must the vector $\mathbf{b}$ satisfy in order that the equation $\mathcal{A}\mathbf{x} = \mathbf{b}$ has solutions $\mathbf{x}$? What are the solutions $\mathbf{x}$ if $\mathbf{b}$ satisfies the condition?

35. Is the matrix $\begin{pmatrix} 3 & -1 & 1 \\ -1 & 1 & -1 \\ 1 & -1 & 2 \end{pmatrix}$ positive or negative definite or neither?

Challenging Problems

1. Show that the distance d from point P to the line AB can be expressed in terms of the position vectors of P, A, and B by

$$d = \frac{|(\mathbf{r}_A - \mathbf{r}_P) \times (\mathbf{r}_B - \mathbf{r}_P)|}{|\mathbf{r}_A - \mathbf{r}_B|}$$

2. For any vectors $\mathbf{u}$, $\mathbf{v}$, $\mathbf{w}$, and $\mathbf{x}$, show that

$$(\mathbf{u} \times \mathbf{v}) \times (\mathbf{w} \times \mathbf{x}) = \Big((\mathbf{u} \times \mathbf{v}) \bullet \mathbf{x}\Big)\mathbf{w} - \Big((\mathbf{u} \times \mathbf{v}) \bullet \mathbf{w}\Big)\mathbf{x}$$

$$= \Big((\mathbf{w} \times \mathbf{x}) \bullet \mathbf{u}\Big)\mathbf{v} - \Big((\mathbf{w} \times \mathbf{x}) \bullet \mathbf{v}\Big)\mathbf{u}.$$

In particular, show that

$$(\mathbf{u} \times \mathbf{v}) \times (\mathbf{u} \times \mathbf{w}) = \Big((\mathbf{u} \times \mathbf{v}) \bullet \mathbf{w}\Big)\mathbf{u}.$$

3. Show that the area A of a triangle with vertices $(x_1, y_1, 0)$, $(x_2, y_2, 0)$, and $(x_3, y_3, 0)$ in the xy-plane is given by

$$A = \frac{1}{2} \left| \begin{vmatrix} x_1 & y_1 & 1 \\ x_2 & y_2 & 1 \\ x_3 & y_3 & 1 \end{vmatrix} \right|.$$

4. (a) If L_1 and L_2 are two **skew** (i.e., nonparallel and nonintersecting) lines, show that there is a pair of parallel planes P_1 and P_2 such that L_1 lies in P_1 and L_2 lies in P_2.

   (b) Find parallel planes containing the following two lines: L_1 through points $(1, 1, 0)$ and $(2, 0, 1)$ and L_2 through points $(0, 1, 1)$ and $(1, 2, 2)$.

5. What condition must the vectors $\mathbf{a}$ and $\mathbf{b}$ satisfy to ensure that the equation $\mathbf{a} \times \mathbf{x} = \mathbf{b}$ has solutions? If this condition is satisfied, find all solutions of the equation. Describe the set of solutions.

CHAPTER 11

Vector Functions and Curves

Philosophy is written in this grand book — I mean the universe — which stands continually open to our gaze, but it cannot be understood unless one first learns to comprehend the language and interpret the characters in which it is written. It is written in the language of mathematics, and its characters are triangles, circles, and other geometrical figures, without which it is humanly impossible to understand a single word of it; without these, one is wandering about in a dark labyrinth.

Galileo Galilei   1564–1642

Introduction   This chapter is concerned with functions of a single real variable that have *vector* values. Such functions can be thought of as parametric representations of curves, and we will examine them from both a *kinematic* point of view (involving position, velocity, and acceleration of a moving particle) and a *geometric* point of view (involving tangents, normals, curvature, and torsion). Finally, we will work through a simple derivation of Kepler's laws of planetary motion.

11.1 Vector Functions of One Variable

In this section we will examine several aspects of differential and integral calculus as applied to **vector-valued functions** of a single real variable. Such functions can be used to represent curves parametrically. It is natural to interpret a vector-valued function of the real variable t as giving the position, at time t, of a point or "particle" moving around in space. Derivatives of this *position vector* are then other vector-valued functions giving the velocity and acceleration of the particle. To motivate the study of vector functions, we will consider such a vectorial description of motion in 3-space. Some of our examples will involve motion in the plane; in this case the third components of the vectors will be 0 and will be omitted.

If a particle moves around in 3-space, its motion can be described by giving the three coordinates of its position as functions of time t:

$$x = x(t), \qquad y = y(t), \qquad \text{and} \qquad z = z(t).$$

It is more convenient, however, to replace these three equations by a single vector equation,

$$\mathbf{r} = \mathbf{r}(t),$$

giving the position vector of the moving particle as a function of t. (Recall that the position vector of a point is the vector from the origin to that point.) In terms of the standard basis vectors $\mathbf{i}$, $\mathbf{j}$, and $\mathbf{k}$, the position of the particle at time t is

position: $\mathbf{r} = \mathbf{r}(t) = x(t)\,\mathbf{i} + y(t)\,\mathbf{j} + z(t)\,\mathbf{k}.$

As t increases, the particle moves along a *path*, a curve $\mathcal{C}$ in 3-space. If $z(t) = 0$, then $\mathcal{C}$ is a plane curve in the xy-plane. We assume that $\mathcal{C}$ is a *continuous curve*; the particle cannot instantaneously jump from one point to a distant point. This is equivalent to requiring that the component functions $x(t)$, $y(t)$, and $z(t)$ are continuous functions of t, and we therefore say that $\mathbf{r}(t)$ is a continuous vector function of t.

In the time interval from t to $t + \Delta t$ the particle moves from position $\mathbf{r}(t)$ to position $\mathbf{r}(t + \Delta t)$. Therefore, its **average velocity** is

$$\frac{\mathbf{r}(t + \Delta t) - \mathbf{r}(t)}{\Delta t},$$

which is a vector parallel to the secant vector from $\mathbf{r}(t)$ to $\mathbf{r}(t + \Delta t)$. If the average velocity has a limit as $\Delta t \to 0$, then we say that $\mathbf{r}$ is **differentiable** at t, and we call the limit the (instantaneous) **velocity** of the particle at time t. We denote the velocity vector by $\mathbf{v}(t)$:

velocity: $\mathbf{v}(t) = \lim\limits_{\Delta t \to 0} \dfrac{\mathbf{r}(t + \Delta t) - \mathbf{r}(t)}{\Delta t} = \dfrac{d}{dt}\mathbf{r}(t).$

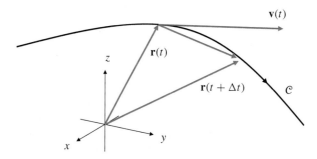

Figure 11.1   The velocity $\mathbf{v}(t)$ is the derivative of the position $\mathbf{r}(t)$ and is tangent to the path of motion at the point with position vector $\mathbf{r}(t)$

This velocity vector has direction tangent to the path $\mathcal{C}$ at the point $\mathbf{r}(t)$ (see Figure 11.1), and it points in the direction of motion. The length of the velocity vector, $v(t) = |\mathbf{v}(t)|$, is called the **speed** of the particle:

speed: $v(t) = |\mathbf{v}(t)|.$

Wherever the velocity vector exists, is continuous, and does not vanish, the path $\mathcal{C}$ is a **smooth** curve, that is, it has a continuously turning tangent line. The path may not be smooth at points where the velocity is zero, even if the components of the velocity vector are smooth functions of t.

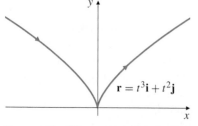

Figure 11.2   The components of $\mathbf{r}(t)$ are smooth functions of t, but the curve fails to be smooth at the origin, where $\mathbf{v} = \mathbf{0}$

Example 1   Consider the plane curve $\mathbf{r} = t^3\mathbf{i} + t^2\mathbf{j}$. Its component functions t^3 and t^2 have continuous derivatives of all orders. However, the curve is not smooth at the origin ($t = 0$), where its velocity $\mathbf{v} = 3t^2\mathbf{i} + 2t\mathbf{j} = \mathbf{0}$. (See Figure 11.2.) The curve is smooth at all other points where $\mathbf{v}(t) \neq \mathbf{0}$.

The rules for addition and scalar multiplication of vectors imply that

$$\mathbf{v} = \frac{d\mathbf{r}}{dt}$$
$$= \lim_{\Delta t \to 0} \left(\frac{x(t + \Delta t) - x(t)}{\Delta t}\mathbf{i} + \frac{y(t + \Delta t) - y(t)}{\Delta t}\mathbf{j} + \frac{z(t + \Delta t) - z(t)}{\Delta t}\mathbf{k} \right)$$
$$= \frac{dx}{dt}\mathbf{i} + \frac{dy}{dt}\mathbf{j} + \frac{dz}{dt}\mathbf{k}.$$

Thus, the vector function $\mathbf{r}$ is differentiable at t if and only if its three scalar components, x, y, and z, are differentiable at t. In general, vector functions can be differentiated (or integrated) by differentiating (or integrating) their component functions, provided that the basis vectors with respect to which the components are taken are fixed in space and not changing with time.

Continuing our analysis of the moving particle, we define the **acceleration** of the particle to be the time derivative of the velocity:

acceleration: $\quad \mathbf{a}(t) = \dfrac{d\mathbf{v}}{dt} = \dfrac{d^2\mathbf{r}}{dt^2}.$

Newton's Second Law of Motion asserts that this acceleration is proportional to, and *in the same direction as*, the force $\mathbf{F}$ causing the motion: if the particle has mass m, then the law is expressed by the *vector equation* $\mathbf{F} = m\mathbf{a}$.

Example 2 Describe the curve $\mathbf{r} = t\mathbf{i} + t^2\mathbf{j} + t^3\mathbf{k}$. Find the velocity and acceleration vectors for this curve at $(1, 1, 1)$.

Solution Since the scalar parametric equations for the curve are

$$x = t, \qquad y = t^2, \qquad \text{and} \qquad z = t^3,$$

which satisfy $y = x^2$ and $z = x^3$, the curve is the curve of intersection of the two cylinders $y = x^2$ and $z = x^3$. At any time t the velocity and acceleration vectors are given by

$$\mathbf{v} = \frac{d\mathbf{r}}{dt} = \mathbf{i} + 2t\mathbf{j} + 3t^2\mathbf{k},$$
$$\mathbf{a} = \frac{d\mathbf{v}}{dt} = 2\mathbf{j} + 6t\mathbf{k}.$$

The point $(1, 1, 1)$ on the curve corresponds to $t = 1$, so the velocity and acceleration at that point are $\mathbf{v} = \mathbf{i} + 2\mathbf{j} + 3\mathbf{k}$ and $\mathbf{a} = 2\mathbf{j} + 6\mathbf{k}$, respectively.

Example 3 Find the velocity, speed, and acceleration, and describe the motion of a particle whose position at time t is

$$\mathbf{r} = 3 \cos \omega t \, \mathbf{i} + 4 \cos \omega t \, \mathbf{j} + 5 \sin \omega t \, \mathbf{k}.$$

Solution The velocity, speed, and acceleration are readily calculated:

$$\mathbf{v} = \frac{d\mathbf{r}}{dt} = -3\omega \sin \omega t \, \mathbf{i} - 4\omega \sin \omega t \, \mathbf{j} + 5\omega \cos \omega t \, \mathbf{k}$$
$$v = |\mathbf{v}| = 5\omega$$
$$\mathbf{a} = \frac{d\mathbf{v}}{dt} = -3\omega^2 \cos \omega t \, \mathbf{i} - 4\omega^2 \cos \omega t \, \mathbf{j} - 5\omega^2 \sin \omega t \, \mathbf{k} = -\omega^2 \mathbf{r}.$$

Observe that $|\mathbf{r}| = 5$. Therefore, the path of the particle lies on the sphere with equation $x^2 + y^2 + z^2 = 25$. Since $x = 3 \cos \omega t$ and $y = 4 \cos \omega t$, the path also lies on the vertical plane $4x = 3y$. Hence, the particle moves around a circle of radius 5 centred at the origin and lying in the plane $4x = 3y$. Observe also that $\mathbf{r}$ is periodic with period $2\pi/\omega$. Therefore, the particle makes one revolution around the circle in time $2\pi/\omega$. The acceleration is always in the direction of $-\mathbf{r}$, that is, toward the origin. The term **centripetal acceleration** is used to describe such a "centre-seeking" acceleration.

Example 4   (**The projectile problem**) Describe the path followed by a particle experiencing a constant downward acceleration, $-g\mathbf{k}$, caused by gravity. Assume that at time $t = 0$ the position of the particle is $\mathbf{r}_0$ and its velocity is $\mathbf{v}_0$.

Solution   If the position of the particle at time t is $\mathbf{r}(t)$, then its acceleration is $d^2\mathbf{r}/dt^2$. The position of the particle can be found by solving the *initial-value problem*

$$\frac{d^2\mathbf{r}}{dt^2} = -g\mathbf{k}, \qquad \frac{d\mathbf{r}}{dt}\bigg|_{t=0} = \mathbf{v}_0, \qquad \mathbf{r}(0) = \mathbf{r}_0.$$

We integrate the differential equation twice. Each integration introduces a *vector* constant of integration that we can determine from the given data by evaluating at $t = 0$:

$$\frac{d\mathbf{r}}{dt} = -gt\mathbf{k} + \mathbf{v}_0$$

$$\mathbf{r} = -\frac{gt^2}{2}\mathbf{k} + \mathbf{v}_0 t + \mathbf{r}_0.$$

The latter equation represents a parabola in the vertical plane passing through the point with position vector $\mathbf{r}_0$ and containing the vector $\mathbf{v}_0$. (See Figure 11.3.) The parabola has scalar parametric equations

$$x = u_0 t + x_0,$$
$$y = v_0 t + y_0,$$
$$z = -\frac{gt^2}{2} + w_0 t + z_0,$$

where $\mathbf{r}_0 = x_0\mathbf{i} + y_0\mathbf{j} + z_0\mathbf{k}$ and $\mathbf{v}_0 = u_0\mathbf{i} + v_0\mathbf{j} + w_0\mathbf{k}$.

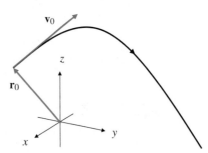

Figure 11.3   The path of a projectile fired from position $\mathbf{r}_0$ with velocity $\mathbf{v}_0$

Example 5   An object moves to the right along the plane curve $y = x^2$ with constant speed $v = 5$. Find the velocity and acceleration of the object when it is at the point $(1, 1)$.

Solution   The position of the object at time t is

$$\mathbf{r} = x\mathbf{i} + x^2\mathbf{j},$$

where x, the x-coordinate of the object's position, is a function of t. The object's velocity, speed, and acceleration at time t are given by

$$\mathbf{v} = \frac{d\mathbf{r}}{dt} = \frac{dx}{dt}\mathbf{i} + 2x\frac{dx}{dt}\mathbf{j} = \frac{dx}{dt}(\mathbf{i} + 2x\mathbf{j}),$$

$$v = |\mathbf{v}| = \left|\frac{dx}{dt}\right|\sqrt{1 + (2x)^2} = \frac{dx}{dt}\sqrt{1 + 4x^2},$$

$$\mathbf{a} = \frac{d\mathbf{v}}{dt} = \frac{d^2x}{dt^2}(\mathbf{i} + 2x\mathbf{j}) + 2\left(\frac{dx}{dt}\right)^2\mathbf{j}.$$

(In the speed calculation we used $|dx/dt| = dx/dt$ because the object is moving to the right.) We are given that the speed is constant; $v = 5$. Therefore,

$$\frac{dx}{dt} = \frac{5}{\sqrt{1 + 4x^2}}.$$

When $x = 1$, we have $dx/dt = 5/\sqrt{1+4} = \sqrt{5}$, so the velocity of the object at that point is $\mathbf{v} = \sqrt{5}\mathbf{i} + 2\sqrt{5}\mathbf{j}$. Now we can calculate

$$\frac{d^2x}{dt^2} = \frac{d}{dt}\frac{5}{\sqrt{1 + 4x^2}} = \left(\frac{d}{dx}\frac{5}{\sqrt{1 + 4x^2}}\right)\frac{dx}{dt}$$

$$= -\frac{5}{2(1 + 4x^2)^{3/2}}(8x)\frac{5}{\sqrt{1 + 4x^2}} = -\frac{100x}{(1 + 4x^2)^2}.$$

At $x = 1$, we have $d^2x/dt^2 = -4$. Thus, the acceleration at that point is $\mathbf{a} = -4(\mathbf{i} + 2\mathbf{j}) + 10\mathbf{j} = -4\mathbf{i} + 2\mathbf{j}$.

Remark   Note that we used x as the parameter for the curve in the above example, so we could use t for time. If you want to analyze motion along a curve $\mathbf{r} = \mathbf{r}(t)$, where t is just a parameter, not necessarily time, then you will have to use a different symbol, say τ (Greek "tau"), for time. The physical *velocity* and *acceleration* of a particle moving along the curve are then

$$\mathbf{v} = \frac{d\mathbf{r}}{d\tau} = \frac{dt}{d\tau}\frac{d\mathbf{r}}{dt} \qquad \text{and} \qquad \mathbf{a} = \frac{d\mathbf{v}}{d\tau} = \frac{d^2t}{d\tau^2}\frac{d\mathbf{r}}{dt} + \left(\frac{dt}{d\tau}\right)^2\frac{d^2\mathbf{r}}{dt^2}.$$

Be careful how you interpret t in a problem where time is meaningful.

Differentiating Combinations of Vectors

Vectors and scalars can be combined in a variety of ways to form other vectors or scalars. Vectors can be added and multiplied by scalars and can be factors in dot and cross products. Appropriate differentiation rules apply to all such combinations of vector and scalar functions; we summarize them in the following theorem.

THEOREM **1**

Differentiation rules for vector functions

Let $\mathbf{u}(t)$ and $\mathbf{v}(t)$ be differentiable vector-valued functions, and let $\lambda(t)$ be a differentiable scalar-valued function. Then $\mathbf{u}(t) + \mathbf{v}(t)$, $\lambda(t)\mathbf{u}(t)$, $\mathbf{u}(t) \bullet \mathbf{v}(t)$, $\mathbf{u}(t) \times \mathbf{v}(t)$, and $\mathbf{u}(\lambda(t))$ are differentiable, and

(a) $\quad \dfrac{d}{dt}\Big(\mathbf{u}(t) + \mathbf{v}(t)\Big) = \mathbf{u}'(t) + \mathbf{v}'(t)$

(b) $\quad \dfrac{d}{dt}\Big(\lambda(t)\mathbf{u}(t)\Big) = \lambda'(t)\mathbf{u}(t) + \lambda(t)\mathbf{u}'(t)$

(c) $\quad \dfrac{d}{dt}\Big(\mathbf{u}(t) \bullet \mathbf{v}(t)\Big) = \mathbf{u}'(t) \bullet \mathbf{v}(t) + \mathbf{u}(t) \bullet \mathbf{v}'(t)$

(d) $\quad \dfrac{d}{dt}\Big(\mathbf{u}(t) \times \mathbf{v}(t)\Big) = \mathbf{u}'(t) \times \mathbf{v}(t) + \mathbf{u}(t) \times \mathbf{v}'(t)$

(e) $\quad \dfrac{d}{dt}\Big(\mathbf{u}\big(\lambda(t)\big)\Big) = \lambda'(t)\mathbf{u}'\big(\lambda(t)\big).$

Also, at any point where $\mathbf{u}(t) \neq 0$,

(f) $\quad \dfrac{d}{dt}|\mathbf{u}(t)| = \dfrac{\mathbf{u}(t) \bullet \mathbf{u}'(t)}{|\mathbf{u}(t)|}.$

Remark Formulas (b), (c), and (d) are versions of the Product Rule. Formula (e) is a version of the Chain Rule. Formula (f) is also a case of the Chain Rule applied to $|\mathbf{u}| = \sqrt{\mathbf{u} \bullet \mathbf{u}}$. All have the obvious form. Note that the order of the factors is the same in the terms on both sides of the cross product formula (d). It is essential that the order be preserved because, unlike the dot product or the product of a vector with a scalar, the cross product is *not commutative*.

Remark The formula for the derivative of a cross product is a special case of that for the derivative of a 3×3 determinant. (See Section 10.3.) Since every term in the expansion of a determinant of any order is a product involving one element from each row (or column), the general Product Rule implies that the derivative of an $n \times n$ determinant whose elements are functions will be the sum of n such $n \times n$ determinants, each with the elements of one of the rows (or columns) differentiated. For the 3×3 case we have

$$\frac{d}{dt}\begin{vmatrix} a_{11}(t) & a_{12}(t) & a_{13}(t) \\ a_{21}(t) & a_{22}(t) & a_{23}(t) \\ a_{31}(t) & a_{32}(t) & a_{33}(t) \end{vmatrix} = \begin{vmatrix} a'_{11}(t) & a'_{12}(t) & a'_{13}(t) \\ a_{21}(t) & a_{22}(t) & a_{23}(t) \\ a_{31}(t) & a_{32}(t) & a_{33}(t) \end{vmatrix}$$

$$+ \begin{vmatrix} a_{11}(t) & a_{12}(t) & a_{13}(t) \\ a'_{21}(t) & a'_{22}(t) & a'_{23}(t) \\ a_{31}(t) & a_{32}(t) & a_{33}(t) \end{vmatrix} + \begin{vmatrix} a_{11}(t) & a_{12}(t) & a_{13}(t) \\ a_{21}(t) & a_{22}(t) & a_{23}(t) \\ a'_{31}(t) & a'_{32}(t) & a'_{33}(t) \end{vmatrix}.$$

Example 6 Show that the speed of a moving particle remains constant over an interval of time if and only if the acceleration is perpendicular to the velocity throughout that interval.

Solution Since $\big(v(t)\big)^2 = \mathbf{v}(t) \bullet \mathbf{v}(t)$, we have

$$2v(t)\frac{dv}{dt} = \frac{d}{dt}\big(v(t)\big)^2 = \frac{d}{dt}\Big(\mathbf{v}(t) \bullet \mathbf{v}(t)\Big)$$
$$= \mathbf{a}(t) \bullet \mathbf{v}(t) + \mathbf{v}(t) \bullet \mathbf{a}(t) = 2\mathbf{v}(t) \bullet \mathbf{a}(t).$$

If we assume that $v(t) \neq 0$, it follows that $dv/dt = 0$ if and only if $\mathbf{v} \bullet \mathbf{a} = 0$. The speed is constant if and only if the velocity is perpendicular to the acceleration.

Example 7   If $\mathbf{u}$ is three times differentiable, calculate and simplify the triple product derivative

$$\frac{d}{dt}\left(\mathbf{u}\bullet\left(\frac{d\mathbf{u}}{dt}\times\frac{d^2\mathbf{u}}{dt^2}\right)\right).$$

Solution   Using various versions of the Product Rule, we calculate

$$\frac{d}{dt}\left(\mathbf{u}\bullet\left(\frac{d\mathbf{u}}{dt}\times\frac{d^2\mathbf{u}}{dt^2}\right)\right)$$

$$=\frac{d\mathbf{u}}{dt}\bullet\left(\frac{d\mathbf{u}}{dt}\times\frac{d^2\mathbf{u}}{dt^2}\right)+\mathbf{u}\bullet\left(\frac{d^2\mathbf{u}}{dt^2}\times\frac{d^2\mathbf{u}}{dt^2}\right)+\mathbf{u}\bullet\left(\frac{d\mathbf{u}}{dt}\times\frac{d^3\mathbf{u}}{dt^3}\right)$$

$$=0+0+\mathbf{u}\bullet\left(\frac{d\mathbf{u}}{dt}\times\frac{d^3\mathbf{u}}{dt^3}\right)=\mathbf{u}\bullet\left(\frac{d\mathbf{u}}{dt}\times\frac{d^3\mathbf{u}}{dt^3}\right).$$

The first term vanishes because $d\mathbf{u}/dt$ is perpendicular to its cross product with another vector; the second term vanishes because of the cross product of identical vectors.

Exercises 11.1

In Exercises 1–14, find the velocity, speed, and acceleration at time t of the particle whose position is $\mathbf{r}(t)$. Describe the path of the particle.

1. $\mathbf{r}=\mathbf{i}+t\mathbf{j}$
2. $\mathbf{r}=t^2\mathbf{i}+\mathbf{k}$
3. $\mathbf{r}=t^2\mathbf{j}+t\mathbf{k}$
4. $\mathbf{r}=\mathbf{i}+t\mathbf{j}+t\mathbf{k}$
5. $\mathbf{r}=t^2\mathbf{i}-t^2\mathbf{j}+\mathbf{k}$
6. $\mathbf{r}=t\mathbf{i}+t^2\mathbf{j}+t^2\mathbf{k}$
7. $\mathbf{r}=a\cos t\,\mathbf{i}+a\sin t\,\mathbf{j}+ct\mathbf{k}$
8. $\mathbf{r}=a\cos\omega t\,\mathbf{i}+b\mathbf{j}+a\sin\omega t\,\mathbf{k}$
9. $\mathbf{r}=3\cos t\,\mathbf{i}+4\cos t\,\mathbf{j}+5\sin t\,\mathbf{k}$
10. $\mathbf{r}=3\cos t\,\mathbf{i}+4\sin t\,\mathbf{j}+t\mathbf{k}$
11. $\mathbf{r}=ae^t\mathbf{i}+be^t\mathbf{j}+ce^t\mathbf{k}$
12. $\mathbf{r}=at\cos\omega t\,\mathbf{i}+at\sin\omega t\,\mathbf{j}+b\ln t\,\mathbf{k}$
13. $\mathbf{r}=e^{-t}\cos(e^t)\mathbf{i}+e^{-t}\sin(e^t)\mathbf{j}-e^t\mathbf{k}$
14. $\mathbf{r}=a\cos t\sin t\,\mathbf{i}+a\sin^2 t\,\mathbf{j}+a\cos t\,\mathbf{k}$

15. A particle moves around the circle $x^2+y^2=25$ at constant speed, making one revolution in 2 s. Find its acceleration when it is at $(3, 4)$.

16. A particle moves to the right along the curve $y=3/x$. If its speed is 10 when it passes through the point $(2,\frac{3}{2})$, what is its velocity at that time?

17. A point P moves along the curve of intersection of the cylinder $z=x^2$ and the plane $x+y=2$ in the direction of increasing y with constant speed $v=3$. Find the velocity of P when it is at $(1, 1, 1)$.

18. An object moves along the curve $y=x^2$, $z=x^3$, with constant vertical speed $dz/dt=3$. Find the velocity and acceleration of the object when it is at the point $(2, 4, 8)$.

19. A particle moves along the curve $\mathbf{r}=3u\mathbf{i}+3u^2\mathbf{j}+2u^3\mathbf{k}$ in the direction corresponding to increasing u and with a constant speed of 6. Find the velocity and acceleration of the particle when it is at the point $(3, 3, 2)$.

20. A particle moves along the curve of intersection of the cylinders $y=-x^2$ and $z=x^2$ in the direction in which x increases. (All distances are in centimetres.) At the instant when the particle is at the point $(1, -1, 1)$, its speed is 9 cm/s, and that speed is increasing at a rate of 3 cm/s^2. Find the velocity and acceleration of the particle at that instant.

21. Show that if the dot product of the velocity and acceleration of a moving particle is positive (or negative), then the speed of the particle is increasing (or decreasing).

22. Verify the formula for the derivative of a dot product given in Theorem 1(c).

23. Verify the formula for the derivative of a 3×3 determinant in the second remark following Theorem 1. Use this formula to verify the formula for the derivative of the cross product in Theorem 1.

24. If the position and velocity vectors of a moving particle are always perpendicular, show that the path of the particle lies on a sphere.

25. Generalize Exercise 24 to the case where the velocity of the particle is always perpendicular to the line joining the particle to a fixed point P_0.

26. What can be said about the motion of a particle at a time when its position and velocity satisfy $\mathbf{r}\bullet\mathbf{v}>0$? What can be said when $\mathbf{r}\bullet\mathbf{v}<0$?

In Exercises 27–32, assume that the vector functions encountered have continuous derivatives of all required orders.

27. Show that $\dfrac{d}{dt}\left(\dfrac{d\mathbf{u}}{dt}\times\dfrac{d^2\mathbf{u}}{dt^2}\right)=\dfrac{d\mathbf{u}}{dt}\times\dfrac{d^3\mathbf{u}}{dt^3}$.

28. Write the Product Rule for $\dfrac{d}{dt}\Big(\mathbf{u}\bullet(\mathbf{v}\times\mathbf{w})\Big)$.

29. Write the Product Rule for $\dfrac{d}{dt}\Big(\mathbf{u}\times(\mathbf{v}\times\mathbf{w})\Big)$.

30. Expand and simplify: $\dfrac{d}{dt}\left(\mathbf{u}\times\left(\dfrac{d\mathbf{u}}{dt}\times\dfrac{d^2\mathbf{u}}{dt^2}\right)\right)$.

31. Expand and simplify: $\dfrac{d}{dt}\Big((\mathbf{u}+\mathbf{u}'')\bullet(\mathbf{u}\times\mathbf{u}')\Big)$.

32. Expand and simplify: $\dfrac{d}{dt}\Big((\mathbf{u}\times\mathbf{u}')\bullet(\mathbf{u}'\times\mathbf{u}'')\Big)$.

33. If at all times t the position and velocity vectors of a moving particle satisfy $\mathbf{v}(t)=2\mathbf{r}(t)$, and if $\mathbf{r}(0)=\mathbf{r}_0$, find $\mathbf{r}(t)$ and the acceleration $\mathbf{a}(t)$. What is the path of motion?

⊛34. Verify that $\mathbf{r}=\mathbf{r}_0\cos(\omega t)+(\mathbf{v}_0/\omega)\sin(\omega t)$ satisfies the initial-value problem

$$\frac{d^2\mathbf{r}}{dt^2}=-\omega^2\mathbf{r}, \qquad \mathbf{r}'(0)=\mathbf{v}_0, \qquad \mathbf{r}(0)=\mathbf{r}_0.$$

(It is the unique solution.) Describe the path $\mathbf{r}(t)$. What is the path if $\mathbf{r}_0$ is perpendicular to $\mathbf{v}_0$?

⊛35. (Free fall with air resistance) A projectile falling under gravity and slowed by air resistance proportional to its speed has position satisfying

$$\frac{d^2\mathbf{r}}{dt^2}=-g\mathbf{k}-c\frac{d\mathbf{r}}{dt},$$

where c is a positive constant. If $\mathbf{r}=\mathbf{r}_0$ and $d\mathbf{r}/dt=\mathbf{v}_0$ at time $t=0$, find $\mathbf{r}(t)$. (*Hint:* Let $\mathbf{w}=e^{ct}(d\mathbf{r}/dt)$.) Show that the solution approaches that of the projectile problem given in this section as $c\to 0$.

11.2 Some Applications of Vector Differentiation

Many interesting problems in mechanics involve the differentiation of vector functions. This section is devoted to a brief discussion of a few of these.

Motion Involving Varying Mass

The **momentum p** of a moving object is the product of its (scalar) mass m and its (vector) velocity $\mathbf{v}$; $\mathbf{p}=m\mathbf{v}$. Newton's Second Law of Motion states that the rate of change of *momentum* is equal to the external force acting on the object:

$$\mathbf{F}=\frac{d\mathbf{p}}{dt}=\frac{d}{dt}\big(m\mathbf{v}\big).$$

It is only when the mass of the object remains constant that this law reduces to the more familiar $\mathbf{F}=m\mathbf{a}$. When mass is changing you must deal with momentum rather than acceleration.

Example 1   (**The changing velocity of a rocket**) A rocket accelerates by burning its onboard fuel. If the exhaust gases are ejected with constant velocity $\mathbf{v}_e$ *relative to the rocket*, and if the rocket ejects $p\%$ of its initial mass while its engines are firing, by what amount will the velocity of the rocket change? Assume the rocket is in deep space so that gravitational and other external forces acting on it can be neglected.

Solution   Since the rocket is not acted on by any external forces (i.e., $\mathbf{F}=\mathbf{0}$), Newton's law implies that the total momentum of the rocket and its exhaust gases will remain constant. At time t the rocket has mass $m(t)$ and velocity $\mathbf{v}(t)$. At time $t+\Delta t$ the rocket's mass is $m+\Delta m$ (where $\Delta m<0$), its velocity is $\mathbf{v}+\Delta\mathbf{v}$, and the mass $-\Delta m$ of exhaust gases has escaped with velocity $\mathbf{v}+\mathbf{v}_e$ (relative to a coordinate system fixed in space). Equating total momenta at t and $t+\Delta t$ we obtain

$$(m+\Delta m)(\mathbf{v}+\Delta\mathbf{v})+(-\Delta m)(\mathbf{v}+\mathbf{v}_e)=m\mathbf{v}.$$

Simplifying this equation and dividing by Δt gives

$$(m+\Delta m)\frac{\Delta\mathbf{v}}{\Delta t}=\frac{\Delta m}{\Delta t}\mathbf{v}_e,$$

and, on taking the limit as $\Delta t\to 0$,

$$m\frac{d\mathbf{v}}{dt}=\frac{dm}{dt}\mathbf{v}_e.$$

Suppose that the engine fires from $t = 0$ to $t = T$. By the Fundamental Theorem of Calculus, the velocity of the rocket will change by

$$\mathbf{v}(T) - \mathbf{v}(0) = \int_0^T \frac{d\mathbf{v}}{dt}\, dt = \left(\int_0^T \frac{1}{m} \frac{dm}{dt}\, dt \right) \mathbf{v}_e$$

$$= \left(\ln m(T) - \ln m(0) \right) \mathbf{v}_e = -\ln\left(\frac{m(0)}{m(T)} \right) \mathbf{v}_e.$$

Since $m(0) > m(T)$, we have $\ln\big(m(0)/m(T)\big) > 0$ and, as was to be expected, the change in velocity of the rocket is in the opposite direction to the exhaust velocity $\mathbf{v}_e$. If $p\%$ of the mass of the rocket is ejected during the burn, then the velocity of the rocket will change by the amount $-\mathbf{v}_e \ln(100/(100 - p))$.

Remark It is interesting that this model places no restriction on how great a velocity the rocket can achieve, provided that a sufficiently large percentage of its initial mass is fuel. See Exercise 1 at the end of the section.

Circular Motion

The angular speed Ω of a rotating body is its rate of rotation measured in radians per unit time. For instance, a lighthouse lamp rotating at a rate of three revolutions per minute has an angular speed of $\Omega = 6\pi$ radians per minute. It is useful to represent the rate of rotation of a rigid body about an axis in terms of an **angular velocity** vector rather than just the scalar angular speed. The angular velocity vector, $\mathbf{\Omega}$, has magnitude equal to the angular speed, Ω, and direction along the axis of rotation such that if the extended right thumb points in the direction of $\mathbf{\Omega}$, then the fingers surround the axis in the direction of rotation.

If the origin of the coordinate system is on the axis of rotation, and $\mathbf{r} = \mathbf{r}(t)$ is the position vector at time t of a point P in the rotating body, then P moves around a circle of radius $D = |\mathbf{r}(t)| \sin\theta$, where θ is the (constant) angle between $\mathbf{\Omega}$ and $\mathbf{r}(t)$. (See Figure 11.4.) Thus, P travels a distance $2\pi D$ in time $2\pi/\Omega$, and its linear speed is

$$\frac{\text{distance}}{\text{time}} = \frac{2\pi D}{2\pi/\Omega} = \Omega D = |\mathbf{\Omega}||\mathbf{r}(t)| \sin\theta = |\mathbf{\Omega} \times \mathbf{r}(t)|.$$

Since the direction of $\mathbf{\Omega}$ was defined so that $\mathbf{\Omega} \times \mathbf{r}(t)$ would point in the direction of motion of P, the linear velocity of P at time t is given by

$$\frac{d\mathbf{r}}{dt} = \mathbf{v}(t) = \mathbf{\Omega} \times \mathbf{r}(t).$$

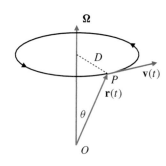

Figure 11.4 Rotation with angular velocity $\mathbf{\Omega}$: $\mathbf{v} = \mathbf{\Omega} \times \mathbf{r}$

Example 2 The position vector $\mathbf{r}(t)$ of a moving particle P satisfies the initial-value problem

$$\begin{cases} \dfrac{d\mathbf{r}}{dt} = 2\mathbf{i} \times \mathbf{r}(t) \\ \mathbf{r}(0) = \mathbf{i} + 3\mathbf{j}. \end{cases}$$

Find $\mathbf{r}(t)$ and describe the motion of P.

Solution There are two ways to solve this problem. We will do it both ways.

METHOD I. By the discussion above, the given differential equation is consistent with rotation about the x-axis with angular velocity $2\mathbf{i}$, so that the angular speed is 2, and the motion is counterclockwise as seen from far out on the positive x-axis. Therefore, the particle P moves on a circle in a plane $x =$ constant and centred on the x-axis. Since P is at $(1, 3, 0)$ at time $t = 0$, the plane of motion is $x = 1$, and the radius of the circle is 3. Therefore, the circle has a parametric equation of the form

$$\mathbf{r} = \mathbf{i} + 3\cos(\lambda t)\mathbf{j} + 3\sin(\lambda t)\mathbf{k}.$$

P travels once around this circle (2π radians) in time $t = 2\pi/\lambda$, so the angular speed is λ. Therefore, $\lambda = 2$ and the motion of the particle is given by

$$\mathbf{r} = \mathbf{i} + 3\cos(2t)\mathbf{j} + 3\sin(2t)\mathbf{k}.$$

METHOD II. Break the given vector differential equation into components:

$$\frac{dx}{dt}\mathbf{i} + \frac{dy}{dt}\mathbf{j} + \frac{dz}{dt}\mathbf{k} = 2\mathbf{i} \times (x\mathbf{i} + y\mathbf{j} + z\mathbf{k}) = -2z\mathbf{j} + 2y\mathbf{k}$$

$$\frac{dx}{dt} = 0, \qquad \frac{dy}{dt} = -2z, \qquad \frac{dz}{dt} = 2y.$$

The first equation implies that $x =$ constant. Since $x(0) = 1$, we have $x(t) = 1$ for all t. Differentiate the second equation with respect to t and substitute the third equation. This leads to the equation of simple harmonic motion for y,

$$\frac{d^2y}{dt^2} = -2\frac{dz}{dt} = -4y,$$

for which a general solution is

$$y = A\cos(2t) + B\sin(2t).$$

Thus, $z = -\frac{1}{2}(dy/dt) = A\sin(2t) - B\cos(2t)$. Since $y(0) = 3$ and $z(0) = 0$, we have $A = 3$ and $B = 0$. Thus, the particle P travels counterclockwise around the circular path

$$\mathbf{r} = \mathbf{i} + 3\cos(2t)\mathbf{j} + 3\sin(2t)\mathbf{k}$$

in the plane $x = 1$ with angular speed 2.

Remark Newton's Second Law states that $\mathbf{F} = (d/dt)(m\mathbf{v}) = d\mathbf{p}/dt$, where $\mathbf{p} = m\mathbf{v}$ is the (linear) momentum of a particle of mass m moving under the influence of a force $\mathbf{F}$. This law may be reformulated in a manner appropriate for describing rotational motion as follows. If $\mathbf{r}(t)$ is the position of the particle at time t, then, since $\mathbf{v} \times \mathbf{v} = \mathbf{0}$,

$$\frac{d}{dt}(\mathbf{r} \times \mathbf{p}) = \frac{d}{dt}\Big(\mathbf{r} \times (m\mathbf{v})\Big) = \mathbf{v} \times (m\mathbf{v}) + \mathbf{r} \times \frac{d}{dt}(m\mathbf{v}) = \mathbf{r} \times \mathbf{F}.$$

The quantities $\mathbf{H} = \mathbf{r} \times (m\mathbf{v})$ and $\mathbf{T} = \mathbf{r} \times \mathbf{F}$ are, respectively, the **angular momentum** of the particle about the origin and the **torque** of $\mathbf{F}$ about the origin. We have shown that

$$\mathbf{T} = \frac{d\mathbf{H}}{dt};$$

the torque of the external forces is equal to the rate of change of the angular momentum of the particle. This is the analogue for rotational motion of $\mathbf{F} = d\mathbf{p}/dt$.

Rotating Frames and the Coriolis Effect

The procedure of differentiating a vector function by differentiating its components is valid only if the basis vectors themselves do not depend on the variable of differentiation. In some situations in mechanics this is not the case. For instance, in modelling large-scale weather phenomena the analysis is affected by the fact that a coordinate system fixed with respect to the earth is, in fact, rotating (along with the earth) relative to directions fixed in space.

In order to understand the effect that the rotation of the coordinate system has on representations of velocity and acceleration, let us consider two Cartesian coordinate frames (i.e., systems of axes with corresponding unit basis vectors), a "fixed" frame with basis {**I**, **J**, **K**}, not rotating with the earth, and a rotating frame with basis {**i**, **j**, **k**} attached to the earth and therefore rotating with the same angular speed as the earth, namely, $\pi/12$ radians/hour. Let us take the origin of the fixed frame to be at the centre of the earth, with **K** pointing north. Then the angular velocity of the earth is $\boldsymbol{\Omega} = (\pi/12)\mathbf{K}$. The fixed frame is being carried along with the earth in its orbit around the sun, but it is not rotating with the earth, and, since the earth's orbital rotation around the sun has angular speed only 1/365th of the angular speed of its rotation about its axis, we can ignore the much smaller effect of the motion of the earth along its orbit.

Let us take the origin of the rotating frame to be at the location of an observer on the surface of the earth, say, at point P_0 with position vector $\mathbf{R}_0$ with respect to the fixed frame.[1] Assume that P_0 has colatitude ϕ (the angle between $\mathbf{R}_0$ and **K**) satisfying $0 < \phi < \pi$, so that P_0 is not at either the north pole or the south pole. Let us assume that **i** and **j** point, respectively, due east and north at P_0. Thus, **k** must point directly upward there. (See Figure 11.5.)

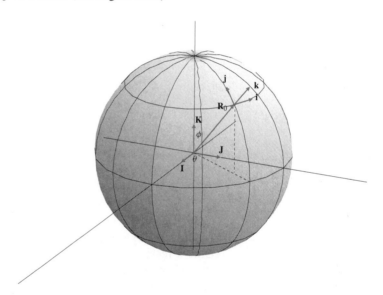

Figure 11.5   The fixed and local frames

Since each of the vectors **i**, **j**, **k**, and $\mathbf{R}_0$ is rotating with the earth (with angular velocity $\boldsymbol{\Omega}$), we have, as shown earlier in this section,

$$\frac{d\,\mathbf{i}}{dt} = \boldsymbol{\Omega} \times \mathbf{i}, \quad \frac{d\,\mathbf{j}}{dt} = \boldsymbol{\Omega} \times \mathbf{j}, \quad \frac{d\,\mathbf{k}}{dt} = \boldsymbol{\Omega} \times \mathbf{k}, \quad \text{and} \quad \frac{d\,\mathbf{R}_0}{dt} = \boldsymbol{\Omega} \times \mathbf{R}_0.$$

Any vector function can be expressed in terms of either basis. Let us denote by $\mathbf{R}(t)$, $\mathbf{V}(t)$, and $\mathbf{A}(t)$ the position, velocity, and acceleration of a moving object with respect to the fixed frame, and by $\mathbf{r}(t)$, $\mathbf{v}(t)$, and $\mathbf{a}(t)$ the same quantities with respect to the rotating frame. Thus,

$$\mathbf{R} = X\mathbf{I} + Y\mathbf{J} + Z\mathbf{K} \qquad\qquad \mathbf{r} = x\mathbf{i} + y\mathbf{j} + z\mathbf{k}$$

$$\mathbf{V} = \frac{dX}{dt}\mathbf{I} + \frac{dY}{dt}\mathbf{J} + \frac{dZ}{dt}\mathbf{K} \qquad \mathbf{v} = \frac{dx}{dt}\mathbf{i} + \frac{dy}{dt}\mathbf{j} + \frac{dz}{dt}\mathbf{k}$$

$$\mathbf{A} = \frac{d^2X}{dt^2}\mathbf{I} + \frac{d^2Y}{dt^2}\mathbf{J} + \frac{d^2Z}{dt^2}\mathbf{K} \qquad \mathbf{a} = \frac{d^2x}{dt^2}\mathbf{i} + \frac{d^2y}{dt^2}\mathbf{j} + \frac{d^2z}{dt^2}\mathbf{k}$$

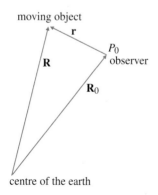

Figure 11.6   Position vectors relative to the fixed and rotating frames

How are the rotating-frame values of these vectors related to the fixed-frame values? Since the origin of the rotating frame is at $\mathbf{R}_0$, we have (see Figure 11.6)

$$\mathbf{R} = \mathbf{R}_0 + \mathbf{r}.$$

[1]   The author is grateful to his colleague, Professor Lon Rosen, for suggesting this approach to the analysis of the rotating frame.

When we differentiate with respect to time, we must remember that $\mathbf{R}_0$, $\mathbf{i}$, $\mathbf{j}$, and $\mathbf{k}$ all depend on time. Therefore,

$$\mathbf{V} = \frac{d\mathbf{R}}{dt} = \frac{d\mathbf{R}_0}{dt} + \frac{dx}{dt}\mathbf{i} + x\frac{d\mathbf{i}}{dt} + \frac{dy}{dt}\mathbf{j} + y\frac{d\mathbf{j}}{dt} + \frac{dz}{dt}\mathbf{k} + z\frac{d\mathbf{k}}{dt}$$

$$= \mathbf{v} + \mathbf{\Omega} \times \mathbf{R}_0 + x\mathbf{\Omega} \times \mathbf{i} + y\mathbf{\Omega} \times \mathbf{j} + z\mathbf{\Omega} \times \mathbf{k}$$

$$= \mathbf{v} + \mathbf{\Omega} \times \mathbf{R}_0 + \mathbf{\Omega} \times \mathbf{r}$$

$$= \mathbf{v} + \mathbf{\Omega} \times \mathbf{R}.$$

Similarly,

$$\mathbf{A} = \frac{d\mathbf{V}}{dt} = \frac{d}{dt}(\mathbf{v} + \mathbf{\Omega} \times \mathbf{R})$$

$$= \frac{d^2x}{dt^2}\mathbf{i} + \frac{dx}{dt}\frac{d\mathbf{i}}{dt} + \frac{d^2y}{dt^2}\mathbf{j} + \frac{dy}{dt}\frac{d\mathbf{j}}{dt} + \frac{d^2z}{dt^2}\mathbf{k} + \frac{dz}{dt}\frac{d\mathbf{k}}{dt} + \mathbf{\Omega} \times \frac{d\mathbf{R}}{dt}$$

$$= \mathbf{a} + \mathbf{\Omega} \times \mathbf{v} + \mathbf{\Omega} \times (\mathbf{V})$$

$$= \mathbf{a} + 2\mathbf{\Omega} \times \mathbf{v} + \mathbf{\Omega} \times (\mathbf{\Omega} \times \mathbf{R}).$$

The term $2\mathbf{\Omega} \times \mathbf{v}$ is called the **Coriolis acceleration**, and the term $\mathbf{\Omega} \times (\mathbf{\Omega} \times \mathbf{R})$ is called the **centripetal acceleration**.

Suppose our moving object has mass m and is acted on by an external force $\mathbf{F}$. By Newton's Second Law,

$$\mathbf{F} = m\mathbf{A} = m\mathbf{a} + 2m\mathbf{\Omega} \times \mathbf{v} + m\mathbf{\Omega} \times (\mathbf{\Omega} \times \mathbf{R}),$$

or, equivalently,

$$\mathbf{a} = \frac{\mathbf{F}}{m} - 2\mathbf{\Omega} \times \mathbf{v} - \mathbf{\Omega} \times (\mathbf{\Omega} \times \mathbf{R}).$$

To the observer on the rotating earth, the object appears to be subject to $\mathbf{F}$ and to two other forces, a **Coriolis force**, whose value per unit mass is $-2\mathbf{\Omega} \times \mathbf{v}$, and a **centrifugal force**, whose value per unit mass is $-\mathbf{\Omega} \times (\mathbf{\Omega} \times \mathbf{R})$. The centrifugal and Coriolis forces are not "real" forces acting on the object. They are fictitious forces that compensate for the fact that we are measuring acceleration with respect to a frame that we are regarding as fixed, although it is really rotating and hence accelerating.

Observe that the centrifugal force points directly away from the polar axis of the earth. It represents the effect that the moving object wants to continue moving in a straight line and "fly off" from the earth rather than continuing to rotate along with the observer. This force is greatest at the equator (where $\mathbf{\Omega}$ is perpendicular to $\mathbf{R}$), but it is of very small magnitude: $|\mathbf{\Omega}|^2|\mathbf{R}_0| \approx 0.003g$.

The Coriolis force is quite different in nature from the centrifugal force. In particular, it is zero if the observer perceives the object to be at rest. It is perpendicular to both the velocity of the object and the polar axis of the earth, and its magnitude can be as large as $2|\mathbf{\Omega}||\mathbf{v}|$; and, in particular, it can be larger than that of the centrifugal force if $|\mathbf{v}|$ is sufficiently large.

Figure 11.7

(a) Tangential and normal components of the angular velocity of the earth in the northern and southern hemispheres

(b) In the northern hemisphere the tangential Coriolis force deflects winds to the right of the path toward the low-pressure area L so the winds move counterclockwise around the centre of L

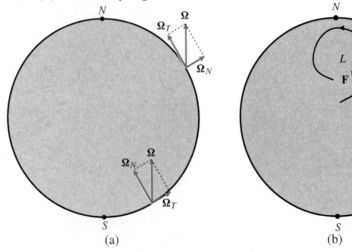

(a)

(b)

Example 3   (**Winds around the eye of a storm**) The circulation of winds around a storm centre is an example of the Coriolis effect. The eye of a storm is an area of low pressure sucking air toward it. The direction of rotation of the earth is such that the angular velocity $\boldsymbol{\Omega}$ points north and is parallel to the earth's axis of rotation. At any point P on the surface of the earth we can express $\boldsymbol{\Omega}$ as a sum of tangential (to the earth's surface) and normal components (see Figure 11.7(a)),

$$\boldsymbol{\Omega}(P) = \boldsymbol{\Omega}_T(P) + \boldsymbol{\Omega}_N(P).$$

If P is in the northern hemisphere, $\boldsymbol{\Omega}_N(P)$ points upward (away from the centre of the earth). At such a point the Coriolis "force" $\mathbf{C} = -2\boldsymbol{\Omega}(P) \times \mathbf{v}$ on a particle of air moving with horizontal velocity $\mathbf{v}$ would itself have horizontal and normal components

$$\mathbf{C} = -2\boldsymbol{\Omega}_T \times \mathbf{v} - 2\boldsymbol{\Omega}_N \times \mathbf{v} = \mathbf{C}_N + \mathbf{C}_T.$$

The normal component of the Coriolis force has negligible effect, since air is not free to travel great distances vertically. However, the tangential component of the Coriolis force, $\mathbf{C}_T = -2\boldsymbol{\Omega}_N \times \mathbf{v}$, is 90° to the right of $\mathbf{v}$ (i.e., clockwise from $\mathbf{v}$). Therefore, particles of air that are being sucked toward the eye of the storm experience Coriolis deflection to the right and so actually spiral into the eye in a counterclockwise direction. The opposite is true in the southern hemisphere, where the normal component $\boldsymbol{\Omega}_N$ is downward (into the earth). The suction force $\mathbf{F}$, the velocity $\mathbf{v}$, and the component of the Coriolis force tangential to the earth's surface, $\mathbf{C}_T$, are shown at two positions on the path of an air particle spiralling around a low-pressure area in the northern hemisphere in Figure 11.7(b). ∎

Remark   Strong winds spiralling inward around low-pressure areas are called **cyclones**. Strong winds spiralling outward around high-pressure areas are called **anticyclones**. The latter spiral counterclockwise in the southern hemisphere and clockwise in the northern hemisphere. The Coriolis effect also accounts for the high-velocity eastward-flowing jet streams in the upper atmosphere at midlatitudes in both hemispheres, the energy being supplied by the rising of warm tropical air and its subsequent moving toward the poles.

The relationships between the basis vectors in the fixed and rotating frames can be used to analyze many phenomena. Recall that $\mathbf{R}_0$ makes angle ϕ with $\mathbf{K}$. Suppose the projection of $\mathbf{R}_0$ onto the equatorial plane (containing $\mathbf{I}$ and $\mathbf{J}$) makes angle θ with $\mathbf{I}$ as shown in Figure 11.5. Careful consideration of that figure should convince you that

$$\mathbf{i} = -\sin\theta\,\mathbf{I} + \cos\theta\,\mathbf{J}$$
$$\mathbf{j} = -\cos\phi\cos\theta\,\mathbf{I} - \cos\phi\sin\theta\,\mathbf{J} + \sin\phi\,\mathbf{K}$$
$$\mathbf{k} = \sin\phi\cos\theta\,\mathbf{I} + \sin\phi\sin\theta\,\mathbf{J} + \cos\phi\,\mathbf{K}.$$

Similarly, or by solving the above equations for $\mathbf{I}$, $\mathbf{J}$, and $\mathbf{K}$,

$$\mathbf{I} = -\sin\theta\,\mathbf{i} - \cos\phi\cos\theta\,\mathbf{j} + \sin\phi\cos\theta\,\mathbf{k}$$
$$\mathbf{J} = \cos\theta\,\mathbf{i} - \cos\phi\sin\theta\,\mathbf{j} + \sin\phi\sin\theta\,\mathbf{k}$$
$$\mathbf{K} = \sin\phi\,\mathbf{j} + \cos\phi\,\mathbf{k}.$$

Note that as the earth rotates on its axis, ϕ remains constant while θ increases at $(\pi/12)$ radians/hour.

Example 4   Suppose that the direction to the sun lies in the plane of $\mathbf{I}$ and $\mathbf{K}$, and makes angle σ with $\mathbf{I}$. Thus, the sun lies in the direction of the vector

$$\mathbf{S} = \cos\sigma\,\mathbf{I} + \sin\sigma\,\mathbf{K}.$$

($\sigma = 0$ at the March and September equinoxes, and $\sigma \approx 23.5°$ and $-23.5°$ at the June and December solstices.) Find the length of the day (the time between sunrise and sunset) for an observer at colatitude ϕ.

Solution   The sun will be "up" for the observer if the angle between $\mathbf{S}$ and $\mathbf{k}$ does not exceed $\pi/2$, that is, if $\mathbf{S} \bullet \mathbf{k} \geq 0$. Thus, daytime corresponds to

$$\cos \sigma \sin \phi \cos \theta + \sin \sigma \cos \phi \geq 0,$$

or, equivalently, $\cos \theta \geq -\dfrac{\tan \sigma}{\tan \phi}$. Sunup and sundown occur where equality occurs, namely, when

$$\theta = \theta_0 = \pm \cos^{-1} \left(-\frac{\tan \sigma}{\tan \phi} \right)$$

if such values exist. (They will exist if $\phi \geq \sigma \geq 0$ or if $\pi - \phi \geq -\sigma \geq 0$.) In this case, daytime for the observer lasts

$$\frac{2\theta_0}{2\pi} \times 24 = \frac{24}{\pi} \cos^{-1} \left(-\frac{\tan \sigma}{\tan \phi} \right) \text{ hours.}$$

For instance, on June 21st at the Arctic Circle (so $\phi = \sigma$), daytime lasts $(24/\pi) \cos^{-1}(-1) = 24$ hours. ∎

Exercises 11.2

1. What fraction of its total initial mass would the rocket considered in Example 1 have to burn as fuel in order to accelerate in a straight line from rest to the speed of its own exhaust gases? to twice that speed?

*   2. When run at maximum power output, the motor in a self-propelled tank car can accelerate the full car (mass M kg) along a horizontal track at a m/s^2. The tank is full at time zero, but the contents pour out of a hole in the bottom at rate k kg/s thereafter. If the car is at rest at time zero and full forward power is turned on at that time, how fast will it be moving at any time t before the tank is empty?

◈   3. Solve the initial-value problem

$$\frac{d\mathbf{r}}{dt} = \mathbf{k} \times \mathbf{r}, \qquad \mathbf{r}(0) = \mathbf{i} + \mathbf{k}.$$

Describe the curve $\mathbf{r} = \mathbf{r}(t)$.

◈   4. An object moves so that its position vector $\mathbf{r}(t)$ satisfies

$$\frac{d\mathbf{r}}{dt} = \mathbf{a} \times \Big(\mathbf{r}(t) - \mathbf{b}\Big)$$

and $\mathbf{r}(0) = \mathbf{r}_0$. Here, $\mathbf{a}$, $\mathbf{b}$, and $\mathbf{r}_0$ are given constant vectors with $\mathbf{a} \neq \mathbf{0}$. Describe the path along which the object moves.

The Coriolis effect

*   5. A satellite is in a low, circular, polar orbit around the earth (i.e., passing over the north and south poles). It makes one revolution every two hours. An observer standing on the earth at the equator sees the satellite pass directly overhead.

In what direction does it seem to the observer to be moving? From the observer's point of view, what is the approximate value of the Coriolis force acting on the satellite?

*   6. Repeat Exercise 5 for an observer at a latitude of 45° in the northern hemisphere.

*   7. Describe the tangential and normal components of the Coriolis force on a particle moving with horizontal velocity $\mathbf{v}$ at (a) the north pole, (b) the south pole, and (c) the equator. In general, what is the effect of the normal component of the Coriolis force near the eye of a storm?

*   8. **(The location of sunrise and sunset)** Extend the argument in Example 4 to determine where on the horizon of the observer at P_0 the sun will rise and set. Specifically, if μ is the angle between $\mathbf{j}$ and $\mathbf{S}$ (the direction to the sun) at sunrise or sunset, show that

$$\cos \mu = \frac{\sin \sigma}{\sin \phi}.$$

For example, if $\sigma = 0$ (the equinoxes), then $\mu = \pi/2$ at all colatitudes ϕ; the sun rises due east and sets due west on those days.

9. Vancouver, Canada, has latitude 49.2° N, so its colatitude is 40.8°. How long is the sun visible at Vancouver on June 21st? Or rather, how long would it be visible if it weren't raining and if there were not so many mountains around? At what angle away from north would the sun rise and set?

10. Repeat Exercise 9 for Umeå, Sweden (latitude 63.5° N).

11.3 Curves and Parametrizations

In this section we will consider curves as geometric objects rather than as paths of moving particles. Everyone has an intuitive idea of what a curve is, but it is difficult to give a formal definition of a curve as a geometric object (i.e., as a certain kind of set of points) without involving the concept of parametric representation. We will avoid this difficulty by continuing to regard a curve in 3-space as the set of points whose positions are given by the position vector function

$$\mathbf{r} = \mathbf{r}(t) = x(t)\mathbf{i} + y(t)\mathbf{j} + z(t)\mathbf{k}, \qquad a \le t \le b.$$

However, the parameter t need no longer represent time or any other specific physical quantity.

Curves can be very pathological. For instance, there exist continuous curves that pass through every point in a cube. It is difficult to think of such a curve as a one-dimensional object. In order to avoid such strange objects we *assume* hereafter that the defining function $\mathbf{r}(t)$ has a *continuous* first derivative, $d\mathbf{r}/dt$, which we will continue to call "velocity" and denote by $\mathbf{v}(t)$ by analogy with the physical case where t is time. (We also continue to call $v(t) = |\mathbf{v}(t)|$ the "speed.") As we will see later, this implies that the curve has an *arc length* between any two points corresponding to parameter values t_1 and t_2; if $t_1 < t_2$, this arc length is

$$\int_{t_1}^{t_2} v(t)\, dt = \int_{t_1}^{t_2} |\mathbf{v}(t)|\, dt = \int_{t_1}^{t_2} \left|\frac{d\mathbf{r}}{dt}\right| dt.$$

Frequently we will want $\mathbf{r}(t)$ to have continuous derivatives of higher order. Whenever needed, we will assume that the "acceleration," $\mathbf{a}(t) = d^2\mathbf{r}/dt^2$, and even the third derivative, $d^3\mathbf{r}/dt^3$, are continuous. Of course, most of the curves we encounter in practice have parametrizations with continuous derivatives of all orders.

It must be recalled, however, that no assumptions on the continuity of derivatives of the function $\mathbf{r}(t)$ are sufficient to guarantee that the curve $\mathbf{r} = \mathbf{r}(t)$ is a "smooth" curve. It may fail to be smooth at a point where $\mathbf{v} = \mathbf{0}$. (See Example 1 in Section 11.1.) We will show in the next section that if, besides being continuous, the velocity vector $\mathbf{v}(t)$ is *never the zero vector*, then the curve $\mathbf{r} = \mathbf{r}(t)$ is **smooth** in the sense that it has a continuously turning tangent line.

Although we have said that a curve is a set of points given by a parametric equation $\mathbf{r} = \mathbf{r}(t)$, there is no *unique* way of representing a given curve parametrically. Just as two cars can travel the same highway at different speeds, stopping and starting at different places, so too can the same curve be defined by different parametrizations; a given curve can have infinitely many different parametrizations.

Example 1 Show that each of the vector functions

$$\mathbf{r}_1(t) = \sin t\, \mathbf{i} + \cos t\, \mathbf{j}, \qquad (-\pi/2 \le t \le \pi/2),$$
$$\mathbf{r}_2(t) = (t-1)\mathbf{i} + \sqrt{2t - t^2}\, \mathbf{j}, \qquad (0 \le t \le 2), \quad \text{and}$$
$$\mathbf{r}_3(t) = t\sqrt{2 - t^2}\, \mathbf{i} + (1 - t^2)\mathbf{j}, \qquad (-1 \le t \le 1)$$

all represent the same curve. Describe the curve.

Solution All three functions represent points in the xy-plane. The function $\mathbf{r}_1(t)$ starts at the point $(-1, 0)$ with position vector $\mathbf{r}_1(-\pi/2) = -\mathbf{i}$ and ends at the point $(1, 0)$ with position vector $\mathbf{i}$. It lies in the half of the xy-plane where $y \ge 0$ (because $\cos t \ge 0$ for $(-\pi/2 \le t \le \pi/2)$). Finally, all points on the curve are at distance 1 from the origin:

$$|\mathbf{r}_1(t)| = \sqrt{(\sin t)^2 + (\cos t)^2} = 1.$$

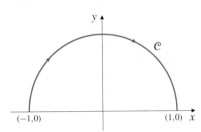

Figure 11.8 Three parametrizations of the semicircle $\mathcal{C}$ are given in Example 1

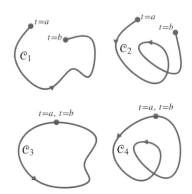

Figure 11.9 Curves $\mathcal{C}_1$ and $\mathcal{C}_3$ are non-self-intersecting

Curves $\mathcal{C}_2$ and $\mathcal{C}_4$ intersect themselves

Curves $\mathcal{C}_1$ and $\mathcal{C}_2$ are not closed

Curves $\mathcal{C}_3$ and $\mathcal{C}_4$ are closed

Curve $\mathcal{C}_3$ is a simple closed curve

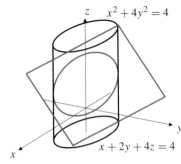

Figure 11.10 The curve of intersection of an oblique plane and an elliptic cylinder

Therefore, $\mathbf{r}_1(t)$ represents the semicircle $y = \sqrt{1 - x^2}$ in the xy-plane traversed from left to right.

The other two functions have the same properties: both graphs lie in $y \geq 0$,

$$\mathbf{r}_2(0) = -\mathbf{i}, \qquad \mathbf{r}_2(2) = \mathbf{i}, \qquad |\mathbf{r}_2(t)| = \sqrt{(t-1)^2 + 2t - t^2} = 1,$$
$$\mathbf{r}_3(-1) = -\mathbf{i}, \qquad \mathbf{r}_3(1) = \mathbf{i}, \qquad |\mathbf{r}_3(t)| = \sqrt{t^2(2 - t^2) + (1 - t^2)^2} = 1.$$

Thus all three functions represent the same semicircle (see Figure 11.8). Of course, the three parametrizations trace out the curve with different velocities.

The curve $\mathbf{r} = \mathbf{r}(t)$, $(a \leq t \leq b)$, is called a **closed curve** if $\mathbf{r}(a) = \mathbf{r}(b)$, that is, if the curve begins and ends at the same point. The curve $\mathcal{C}$ is **non-self-intersecting** if there exists some parametrization $\mathbf{r} = \mathbf{r}(t)$, $(a \leq t \leq b)$, of $\mathcal{C}$ that is one-to-one except that the endpoints could be the same:

$$\mathbf{r}(t_1) = \mathbf{r}(t_2) \quad a \leq t_1 < t_2 \leq b \quad \Longrightarrow \quad t_1 = a \quad \text{and} \quad t_2 = b.$$

Such a curve can be closed, but otherwise does not intersect itself; it is then called a **simple closed curve**. Circles and ellipses are examples of simple closed curves. Every parametrization of a particular curve determines one of two possible **orientations** corresponding to the direction along the curve in which the parameter is increasing. Figure 11.9 illustrates these concepts. All three parametrizations of the semicircle in Example 1 orient the semicircle clockwise as viewed from a point above the xy-plane. This orientation is shown by the arrowheads on the curve in Figure 11.8. The same semicircle could be given the opposite orientation by, for example, the parametrization

$$\mathbf{r}(t) = \cos t\, \mathbf{i} + \sin t\, \mathbf{j}, \qquad 0 \leq t \leq \pi.$$

Parametrizing the Curve of Intersection of Two Surfaces

Frequently a curve is specified as the intersection of two surfaces with given Cartesian equations. We may want to represent the curve by parametric equations. There is no unique way to do this, but if one of the given surfaces is a cylinder parallel to a coordinate axis (so its equation is independent of one of the variables), we can begin by parametrizing that surface. The following examples clarify the method.

Example 2 Parametrize the curve of intersection of the plane $x + 2y + 4z = 4$ and the elliptic cylinder $x^2 + 4y^2 = 4$.

Solution We begin with the equation $x^2 + 4y^2 = 4$, which is independent of z. It can be parametrized in many ways; one convenient way is

$$x = 2\cos t, \qquad y = \sin t, \qquad (0 \leq t \leq 2\pi).$$

The equation of the plane can then be solved for z, so that z can be expressed in terms of t:

$$z = \frac{1}{4}(4 - x - 2y) = 1 - \frac{1}{2}(\cos t + \sin t).$$

Thus, the given surfaces intersect in the curve (see Figure 11.10)

$$\mathbf{r} = 2\cos t\, \mathbf{i} + \sin t\, \mathbf{j} + \left(1 - \frac{\cos t + \sin t}{2}\right)\mathbf{k}, \qquad (0 \leq t \leq 2\pi).$$

Example 3   Find a parametric representation of the curve of intersection of the two surfaces

$$x^2 + y + z = 2 \quad \text{and} \quad xy + z = 1.$$

Solution   Here, neither given equation is independent of a variable, but we can obtain a third equation representing a surface containing the curve of intersection of the two given surfaces by subtracting the two given equations to eliminate z:

$$x^2 + y - xy = 1.$$

This equation is readily parametrized. If, for example, we let $x = t$, then

$$t^2 + y(1 - t) = 1, \quad \text{so} \quad y = \frac{1 - t^2}{1 - t} = 1 + t.$$

Either of the given equations can then be used to express z in terms of t:

$$z = 1 - xy = 1 - t(1 + t) = 1 - t - t^2.$$

Thus, a possible parametrization of the curve is

$$\mathbf{r} = t\mathbf{i} + (1 + t)\mathbf{j} + (1 - t - t^2)\mathbf{k}.$$

Of course, this answer is not unique. Many other parametrizations can be found for the curve, providing orientations in either direction.

Arc Length

We now consider how to define and calculate the length of a curve. Let $\mathcal{C}$ be a bounded, continuous curve specified by

$$\mathbf{r} = \mathbf{r}(t), \quad a \le t \le b.$$

Subdivide the closed interval $[a, b]$ into n subintervals by points

$$a = t_0 < t_1 < t_2 < \cdots < t_{n-1} < t_n = b.$$

The points $\mathbf{r}_i = \mathbf{r}(t_i)$, $(0 \le i \le n)$, subdivide $\mathcal{C}$ into n arcs. If we use the chord length $|\mathbf{r}_i - \mathbf{r}_{i-1}|$ as an approximation to the arc length between $\mathbf{r}_{i-1}$ and $\mathbf{r}_i$, then the sum

$$s_n = \sum_{i=1}^{n} |\mathbf{r}_i - \mathbf{r}_{i-1}|$$

approximates the length of $\mathcal{C}$ by the length of a polygonal line. (See Figure 11.11.) Evidently, any such approximation is less than or equal to the actual length of $\mathcal{C}$. We say that $\mathcal{C}$ is **rectifiable** if there exists a constant K such that $s_n \le K$ for every n and every choice of the points t_i. In this case, the completeness axiom of the real number system assures us that there will be a smallest such number K. We call this smallest K the **length** of $\mathcal{C}$ and denote it by s.

Let $\Delta t_i = t_i - t_{i-1}$ and $\Delta \mathbf{r}_i = \mathbf{r}_i - \mathbf{r}_{i-1}$. Then s_n can be written in the form

$$s_n = \sum_{i=1}^{n} \left| \frac{\Delta \mathbf{r}_i}{\Delta t_i} \right| \Delta t_i.$$

Figure 11.11   A polygonal approximation to a curve $\mathcal{C}$. The length of the polygonal line cannot exceed the length of the curve. In this figure the points on the curve are labelled with their position vectors, but the origin and these vectors are not themselves shown

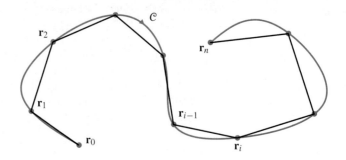

If $\mathbf{r}(t)$ has a continuous derivative $\mathbf{v}(t)$, then

$$s = \lim_{\substack{n \to \infty \\ \max \Delta t_i \to 0}} s_n = \int_a^b \left| \frac{d\mathbf{r}}{dt} \right| dt = \int_a^b |\mathbf{v}(t)| \, dt = \int_a^b v(t) \, dt.$$

In kinematic terms, this formula states that the distance travelled by a moving particle is the integral of its speed.

Remark   Although the above formula is expressed in terms of the parameter t, the arc length, as defined above, is a strictly geometric property of the curve $\mathcal{C}$. It is independent of the particular parametrization used to represent $\mathcal{C}$. See Exercise 27 below.

If $s(t)$ denotes the arc length of that part of $\mathcal{C}$ corresponding to parameter values in $[a, t]$, then

$$\frac{ds}{dt} = \frac{d}{dt} \int_a^t v(\tau) \, d\tau = v(t),$$

so that the **arc length element** for $\mathcal{C}$ is given by

$$ds = v(t) \, dt = \left| \frac{d}{dt} \mathbf{r}(t) \right| dt.$$

The length of $\mathcal{C}$ is the integral of these arc length elements; we write

$$\int_{\mathcal{C}} ds = \text{length of } \mathcal{C} = \int_a^b v(t) \, dt.$$

Several familiar formulas for arc length follow from the above formula by using specific parametrizations of curves. For instance, the arc length element ds for the Cartesian plane curve $y = f(x)$ on $[a, b]$ is obtained by using x as parameter; here, $\mathbf{r} = x\mathbf{i} + f(x)\mathbf{j}$, so $\mathbf{v} = \mathbf{i} + f'(x)\mathbf{j}$ and

$$ds = \sqrt{1 + \left(f'(x) \right)^2} \, dx.$$

Similarly, the arc length element ds for a plane polar curve $r = g(\theta)$ can be calculated from the parametrization

$$\mathbf{r}(\theta) = g(\theta) \cos\theta \mathbf{i} + g(\theta) \sin\theta \mathbf{j}.$$

It is

$$ds = \sqrt{\left(g(\theta) \right)^2 + \left(g'(\theta) \right)^2} \, d\theta.$$

Figure 11.12   The helix

$x = a \cos t$

$y = a \sin t$

$z = bt$

Example 4   Find the length s of that part of the **circular helix**

$$\mathbf{r} = a \cos t\, \mathbf{i} + a\, \sin t\, \mathbf{j} + bt\, \mathbf{k}$$

between the points $(a, 0, 0)$ and $(a, 0, 2\pi b)$.

Solution   This curve spirals around the z-axis, rising as it turns. (See Figure 11.12.) It lies on the surface of the circular cylinder $x^2 + y^2 = a^2$. We have

$$\mathbf{v} = \frac{d\mathbf{r}}{dt} = -a\, \sin t\, \mathbf{i} + a \cos t\, \mathbf{j} + b\mathbf{k}$$

$$v = \sqrt{a^2 + b^2},$$

so that in terms of the parameter t the helix is traced out at constant speed. The required length s corresponds to parameter interval $[0, 2\pi]$. Thus,

$$s = \int_0^{2\pi} v(t)\, dt = \int_0^{2\pi} \sqrt{a^2 + b^2}\, dt = 2\pi\sqrt{a^2 + b^2}.$$

∎

Piecewise Smooth Curves

As observed earlier, a parametric curve $\mathcal{C}$ given by $\mathbf{r} = \mathbf{r}(t)$ can fail to be smooth at points where $d\mathbf{r}/dt = \mathbf{0}$. If there are finitely many such points, we will say that the curve is piecewise smooth.

In general, a **piecewise smooth curve** $\mathcal{C}$ consists of a finite number of smooth arcs, $\mathcal{C}_1, \mathcal{C}_2, \ldots, \mathcal{C}_k$, as shown in Figure 11.13.

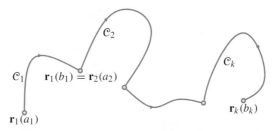

Figure 11.13   A piecewise smooth curve

In this case we express $\mathcal{C}$ as the sum of the individual arcs:

$$\mathcal{C} = \mathcal{C}_1 + \mathcal{C}_2 + \cdots + \mathcal{C}_k.$$

Each arc $\mathcal{C}_i$ can have its own parametrization

$$\mathbf{r} = \mathbf{r}_i(t), \qquad (a_i \leq t \leq b_i),$$

where $\mathbf{v}_i = d\mathbf{r}_i/dt \neq \mathbf{0}$ for $a_i < t < b_i$. The fact that $\mathcal{C}_{i+1}$ must begin at the point where $\mathcal{C}_i$ ends requires the conditions

$$\mathbf{r}_{i+1}(a_{i+1}) = \mathbf{r}_i(b_i) \qquad \text{for } 1 \leq i \leq k - 1.$$

If also $\mathbf{r}_k(b_k) = \mathbf{r}_1(a_1)$, then $\mathcal{C}$ is a closed piecewise smooth curve.

The length of a piecewise smooth curve $\mathcal{C} = \mathcal{C}_1 + \mathcal{C}_2 + \cdots + \mathcal{C}_k$ is the sum of the lengths of its component arcs:

$$\text{length of } \mathcal{C} = \sum_{i=1}^{k} \int_{a_i}^{b_i} \left| \frac{d\mathbf{r}_i}{dt} \right| dt.$$

The Arc-Length Parametrization

The selection of a particular parameter in terms of which to specify a given curve will usually depend on the problem in which the curve arises; there is no one "right way" to parametrize a curve. However, there is one parameter that is "natural" in that it arises from the geometry (shape and size) of the curve itself and not from any particular coordinate system in which the equation of the curve is to be expressed. This parameter is the *arc length* measured from some particular point (the *initial point*) on the curve. The position vector of an arbitrary point P on the curve can be specified as a function of the arc length s along the curve from the initial point P_0 to P,

$$\mathbf{r} = \mathbf{r}(s).$$

This equation is called an **arc-length parametrization** or **intrinsic parametrization** of the curve. Since $ds = v(t)\,dt$ for any parametrization $\mathbf{r} = \mathbf{r}(t)$, for the arc-length parametrization we have $ds = v(s)\,ds$. Thus $v(s) = 1$, identically; *a curve parametrized in terms of arc length is traced at unit speed.* Although it is seldom easy (and usually not possible) to find $\mathbf{r}(s)$ explicitly when the curve is given in terms of some other parameter, smooth curves always have such parametrizations (see Exercise 28 below), and they will prove useful when we develop the fundamentals of the *differential geometry* for 3-space curves in the next section.

Suppose that a curve is specified in terms of an arbitrary parameter t. If the arc length over a parameter interval $[t_0, t]$,

$$s = s(t) = \int_{t_0}^{t} \left| \frac{d}{d\tau} \mathbf{r}(\tau) \right| d\tau,$$

can be evaluated explicitly, and if the equation $s = s(t)$ can be explicitly solved for t as a function of s ($t = t(s)$), then the curve can be reparametrized in terms of arc length by substituting for t in the original parametrization:

$$\mathbf{r} = \mathbf{r}(t(s)).$$

Example 5 Parametrize the circular helix

$$\mathbf{r} = a \cos t\mathbf{i} + a \sin t\mathbf{j} + bt\mathbf{k}$$

in terms of the arc length measured from the point $(a, 0, 0)$ in the direction of increasing t. (See Figure 11.12.)

Solution The initial point corresponds to $t = 0$. As shown in Example 4, we have $ds/dt = \sqrt{a^2 + b^2}$, so

$$s = s(t) = \int_{0}^{t} \sqrt{a^2 + b^2} \, d\tau = \sqrt{a^2 + b^2}\, t.$$

Therefore, $t = s/\sqrt{a^2 + b^2}$, and the arc-length parametrization is

$$\mathbf{r}(s) = a \cos\left(\frac{s}{\sqrt{a^2 + b^2}} \right)\mathbf{i} + a \sin\left(\frac{s}{\sqrt{a^2 + b^2}} \right)\mathbf{j} + \frac{bs}{\sqrt{a^2 + b^2}}\mathbf{k}.$$

▌▌▌▌▌▌▌▌ Exercises 11.3 ▌▌▌▌▌▌▌▌

In Exercises 1–4, find the required parametrization of the first quadrant part of the circular arc $x^2 + y^2 = a^2$.

1. In terms of the y-coordinate, oriented counterclockwise

2. In terms of the x-coordinate, oriented clockwise

3. In terms of the angle between the tangent line and the positive x-axis, oriented counterclockwise

4. In terms of arc length measured from $(0, a)$, oriented clockwise

5. The cylinders $z = x^2$ and $z = 4y^2$ intersect in two curves, one of which passes through the point $(2, -1, 4)$. Find a parametrization of that curve using $t = y$ as parameter.

6. The plane $x + y + z = 1$ intersects the cylinder $z = x^2$ in a parabola. Parametrize the parabola using $t = x$ as parameter.

In Exercises 7–10, parametrize the curve of intersection of the given surfaces. *Note*: the answers are not unique.

7. $x^2 + y^2 = 9$ and $z = x + y$

8. $z = \sqrt{1 - x^2 - y^2}$ and $x + y = 1$

9. $z = x^2 + y^2$ and $2x - 4y - z - 1 = 0$

10. $yz + x = 1$ and $xz - x = 1$

11. The plane $z = 1 + x$ intersects the cone $z^2 = x^2 + y^2$ in a parabola. Try to parametrize the parabola using as parameter: (a) $t = x$, (b) $t = y$, and (c) $t = z$. Which of these choices for t leads to a single parametrization that represents the whole parabola? What is that parametrization? What happens with the other two choices?

∗12. The plane $x + y + z = 1$ intersects the sphere $x^2 + y^2 + z^2 = 1$ in a circle $\mathcal{C}$. Find the centre $\mathbf{r}_0$ and radius r of $\mathcal{C}$. Also find two perpendicular unit vectors $\hat{\mathbf{v}}_1$ and $\hat{\mathbf{v}}_2$ parallel to the plane of $\mathcal{C}$. (*Hint:* To be specific, show that $\hat{\mathbf{v}}_1 = (\mathbf{i} - \mathbf{j})/\sqrt{2}$ is one such vector; then find a second that is perpendicular to $\hat{\mathbf{v}}_1$.) Use your results to construct a parametrization of $\mathcal{C}$.

13. Find the length of the curve $\mathbf{r} = t^2\mathbf{i} + t^2\mathbf{j} + t^3\mathbf{k}$ from $t = 0$ to $t = 1$.

14. For what values of the parameter λ is the length $s(T)$ of the curve $\mathbf{r} = t\mathbf{i} + \lambda t^2\mathbf{j} + t^3\mathbf{k}$, $(0 \le t \le T)$ given by $s(T) = T + T^3$?

15. Express the length of the curve $\mathbf{r} = at^2\mathbf{i} + bt\mathbf{j} + c\ln t\,\mathbf{k}$, $(1 \le t \le T)$, as a definite integral. Evaluate the integral if $b^2 = 4ac$.

16. Describe the parametric curve $\mathcal{C}$ given by
$$x = a\cos t \sin t, \qquad y = a\sin^2 t, \qquad z = bt.$$
What is the length of $\mathcal{C}$ between $t = 0$ and $t = T > 0$?

17. Find the length of the conical helix $\mathbf{r} = t\cos t\,\mathbf{i} + t\sin t\,\mathbf{j} + t\,\mathbf{k}$, $(0 \le t \le 2\pi)$. Why is the curve called a conical helix?

18. Describe the intersection of the sphere $x^2 + y^2 + z^2 = 1$ and the elliptic cylinder $x^2 + 2z^2 = 1$. Find the total length of this intersection curve.

19. Let $\mathcal{C}$ be the curve $x = e^t \cos t$, $y = e^t \sin t$, $z = t$ between $t = 0$ and $t = 2\pi$. Find the length of $\mathcal{C}$.

20. Find the length of the piecewise smooth curve $\mathbf{r} = t^3\mathbf{i} + t^2\mathbf{j}$, $(-1 \le t \le 2)$.

21. Describe the piecewise smooth curve $\mathcal{C} = \mathcal{C}_1 + \mathcal{C}_2$, where $\mathbf{r}_1(t) = t\mathbf{i} + t\mathbf{j}$, $(0 \le t \le 1)$, and $\mathbf{r}_2(t) = (1 - t)\mathbf{i} + (1 + t)\mathbf{j}$, $(0 \le t \le 1)$.

∗22. A cable of length L and circular cross-section of radius a is wound around a cylindrical spool of radius b with no overlapping and with the adjacent windings touching one another. What length of the spool is covered by the cable?

In Exercises 23–26, reparametrize the given curve in the same orientation in terms of arc length measured from the point where $t = 0$.

23. $\mathbf{r} = At\mathbf{i} + Bt\mathbf{j} + Ct\mathbf{k}$, $\qquad (A^2 + B^2 + C^2 > 0)$

24. $\mathbf{r} = e^t\mathbf{i} + \sqrt{2}t\mathbf{j} - e^{-t}\mathbf{k}$

∗25. $\mathbf{r} = a\cos^3 t\,\mathbf{i} + a\sin^3 t\,\mathbf{j} + b\cos 2t\,\mathbf{k}$, $\qquad (0 \le t \le \frac{\pi}{2})$

∗26. $\mathbf{r} = 3t\cos t\,\mathbf{i} + 3t\sin t\,\mathbf{j} + 2\sqrt{2}t^{3/2}\mathbf{k}$

∗27. Let $\mathbf{r} = \mathbf{r}_1(t)$, $(a \le t \le b)$, and $\mathbf{r} = \mathbf{r}_2(u)$, $(c \le u \le d)$, be two parametrizations of the same curve $\mathcal{C}$, each one-to-one on its domain and each giving $\mathcal{C}$ the same orientation (so that $\mathbf{r}_1(a) = \mathbf{r}_2(c)$ and $\mathbf{r}_1(b) = \mathbf{r}_2(d)$). Then for each t in $[a, b]$ there is a unique $u = u(t)$ such that $\mathbf{r}_2(u(t)) = \mathbf{r}_1(t)$. Show that
$$\int_a^b \left| \frac{d}{dt}\mathbf{r}_1(t) \right| dt = \int_c^d \left| \frac{d}{du}\mathbf{r}_2(u) \right| du,$$
and thus that the length of $\mathcal{C}$ is independent of parametrization.

∗28. If the curve $\mathbf{r} = \mathbf{r}(t)$ has continuous, nonvanishing velocity $\mathbf{v}(t)$ on the interval $[a, b]$, and if t_0 is some point in $[a, b]$, show that the function
$$s = g(t) = \int_{t_0}^t |\mathbf{v}(u)|\, du$$
is an increasing function on $[a, b]$ and so has an inverse:
$$t = g^{-1}(s) \iff s = g(t).$$
Hence, show that the curve can be parametrized in terms of arc length measured from $\mathbf{r}(t_0)$.

11.4   Curvature, Torsion, and the Frenet Frame

In this section and the next we develop the fundamentals of differential geometry of curves in 3-space. We will introduce several new scalar and vector functions associated with a curve $\mathcal{C}$. The most important of these are the curvature and torsion of the curve and a right-handed triad of mutually perpendicular unit vectors forming a basis at any point on the curve, and called the Frenet frame. The curvature measures the rate at which a curve is turning (away from its tangent line) at any point. The torsion measures the rate at which the curve is twisting (out of the plane in which it is turning) at any point.

The Unit Tangent Vector

The velocity vector $\mathbf{v}(t) = d\mathbf{r}/dt$ is tangent to the parametric curve $\mathbf{r} = \mathbf{r}(t)$ at the point $\mathbf{r}(t)$ and points in the direction of the orientation of the curve there. Since we are assuming that $\mathbf{v}(t) \neq \mathbf{0}$, we can find a **unit tangent vector**, $\hat{\mathbf{T}}(t)$, at $\mathbf{r}(t)$ by dividing $\mathbf{v}(t)$ by its length:

$$\hat{\mathbf{T}}(t) = \frac{\mathbf{v}(t)}{v(t)} = \frac{d\mathbf{r}}{dt} \Big/ \left| \frac{d\mathbf{r}}{dt} \right|.$$

Recall that a curve parametrized in terms of arc length, $\mathbf{r} = \mathbf{r}(s)$, is traced at unit speed; $v(s) = 1$. In terms of arc-length parametrization, the unit tangent vector is

$$\hat{\mathbf{T}}(s) = \frac{d\mathbf{r}}{ds}.$$

Example 1   Find the unit tangent vector, $\hat{\mathbf{T}}$, for the circular helix of Example 4 of Section 11.3, in terms of both t and the arc-length parameter s.

Solution   In terms of t we have

$$\mathbf{r} = a\cos t\,\mathbf{i} + a\sin t\,\mathbf{j} + bt\,\mathbf{k}$$
$$\mathbf{v}(t) = -a\sin t\,\mathbf{i} + a\cos t\,\mathbf{j} + b\,\mathbf{k}$$
$$v(t) = \sqrt{a^2\sin^2 t + a^2\cos^2 t + b^2} = \sqrt{a^2 + b^2}$$
$$\hat{\mathbf{T}}(t) = -\frac{a}{\sqrt{a^2+b^2}}\sin t\,\mathbf{i} + \frac{a}{\sqrt{a^2+b^2}}\cos t\,\mathbf{j} + \frac{b}{\sqrt{a^2+b^2}}\mathbf{k}.$$

In terms of the arc-length parameter (see Example 5 of Section 11.3)

$$\mathbf{r}(s) = a\cos\left(\frac{s}{\sqrt{a^2+b^2}}\right)\mathbf{i} + a\sin\left(\frac{s}{\sqrt{a^2+b^2}}\right)\mathbf{j} + \frac{bs}{\sqrt{a^2+b^2}}\mathbf{k}$$

$$\hat{\mathbf{T}}(s) = \frac{d\mathbf{r}}{ds} = -\frac{a}{\sqrt{a^2+b^2}}\sin\left(\frac{s}{\sqrt{a^2+b^2}}\right)\mathbf{i} + \frac{a}{\sqrt{a^2+b^2}}\cos\left(\frac{s}{\sqrt{a^2+b^2}}\right)\mathbf{j}$$
$$+ \frac{b}{\sqrt{a^2+b^2}}\mathbf{k}.$$

Remark   If the curve $\mathbf{r} = \mathbf{r}(t)$ has a continuous, nonvanishing velocity $\mathbf{v}(t)$, then the unit tangent vector $\hat{\mathbf{T}}(t)$ is a continuous function of t. The angle $\theta(t)$ between $\hat{\mathbf{T}}(t)$ and any fixed unit vector $\hat{\mathbf{u}}$ is also continuous in t:

$$\theta(t) = \cos^{-1}(\hat{\mathbf{T}}(t) \bullet \hat{\mathbf{u}}).$$

Thus, as asserted previously, the curve is *smooth* in the sense that it has a continuously turning tangent line. The rate of this turning is quantified by the curvature, which we introduce now.

Curvature and the Unit Normal

In the rest of this section we will deal abstractly with a curve $\mathcal{C}$ parametrized in terms of arc length measured from some point on it:

$$\mathbf{r} = \mathbf{r}(s).$$

In the next section we return to curves with arbitrary parametrizations and apply the principles developed in this section to specific problems. Throughout we assume that the parametric equations of curves have continuous derivatives up to third order on the intervals where they are defined.

Having unit length, the tangent vector $\hat{\mathbf{T}}(s) = d\mathbf{r}/ds$ satisfies $\hat{\mathbf{T}}(s) \bullet \hat{\mathbf{T}}(s) = 1$. Differentiating this equation with respect to s we get

$$2\hat{\mathbf{T}}(s) \bullet \frac{d\hat{\mathbf{T}}}{ds} = 0,$$

so that $d\hat{\mathbf{T}}/ds$ is perpendicular to $\hat{\mathbf{T}}(s)$.

DEFINITION 1

> **Curvature and radius of curvature**
>
> The **curvature** of $\mathcal{C}$ at the point $\mathbf{r}(s)$ is the length of $d\hat{\mathbf{T}}/ds$ there. It is denoted by κ, the Greek letter "kappa:"
>
> $$\kappa(s) = \left| \frac{d\hat{\mathbf{T}}}{ds} \right|.$$
>
> The **radius of curvature**, denoted ρ, the Greek letter "rho," is the reciprocal of the curvature:
>
> $$\rho(s) = \frac{1}{\kappa(s)}.$$

As we will see below, the curvature of $\mathcal{C}$ at $\mathbf{r}(s)$ measures the rate of turning of the tangent line to the curve there. The radius of curvature is the radius of the circle through $\mathbf{r}(s)$ that most closely approximates the curve $\mathcal{C}$ near that point.

According to its definition, $\kappa(s) \geq 0$ everywhere on $\mathcal{C}$. If $\kappa(s) \neq 0$, we can divide $d\hat{\mathbf{T}}/ds$ by its length, $\kappa(s)$, and obtain a unit vector $\hat{\mathbf{N}}(s)$ in the same direction. This unit vector is called the **unit principal normal** to $\mathcal{C}$ at $\mathbf{r}(s)$, or, more commonly, just the **unit normal**:

$$\hat{\mathbf{N}}(s) = \frac{1}{\kappa(s)} \frac{d\hat{\mathbf{T}}}{ds} = \frac{d\hat{\mathbf{T}}}{ds} \bigg/ \left| \frac{d\hat{\mathbf{T}}}{ds} \right|.$$

Note that $\hat{\mathbf{N}}(s)$ is perpendicular to $\mathcal{C}$ at $\mathbf{r}(s)$ and points in the direction that $\hat{\mathbf{T}}$, and therefore $\mathcal{C}$, is turning. The principal normal is not defined at points where the curvature $\kappa(s)$ is zero. For instance, a straight line has no principal normal. Figure 11.14(a) shows $\hat{\mathbf{T}}$ and $\hat{\mathbf{N}}$ at a point on a typical curve.

Example 2 Let $a > 0$. Show that the curve $\mathcal{C}$ given by

$$\mathbf{r} = a\cos\left(\frac{s}{a}\right)\mathbf{i} + a\sin\left(\frac{s}{a}\right)\mathbf{j}$$

is a circle in the xy-plane having radius a and centre at the origin and that it is parametrized in terms of arc length. Find the curvature, the radius of curvature, and the unit tangent and principal normal vectors at any point on $\mathcal{C}$.

Figure 11.14

(a) The unit tangent and principal normal vectors for a curve

(b) The unit tangent and principal normal vectors for a circle

(a)   (b)

Solution   Since

$$|\mathbf{r}(s)| = a\sqrt{\left(\cos\left(\frac{s}{a}\right)\right)^2 + \left(\sin\left(\frac{s}{a}\right)\right)^2} = a,$$

$\mathcal{C}$ is indeed a circle of radius a centred at the origin in the xy-plane. Since the speed

$$\left|\frac{d\mathbf{r}}{ds}\right| = \left|-\sin\left(\frac{s}{a}\right)\mathbf{i} + \cos\left(\frac{s}{a}\right)\mathbf{j}\right| = 1,$$

the parameter s must represent arc length; hence the unit tangent vector is

$$\hat{\mathbf{T}}(s) = -\sin\left(\frac{s}{a}\right)\mathbf{i} + \cos\left(\frac{s}{a}\right)\mathbf{j}.$$

Therefore,

$$\frac{d\hat{\mathbf{T}}}{ds} = -\frac{1}{a}\cos\left(\frac{s}{a}\right)\mathbf{i} - \frac{1}{a}\sin\left(\frac{s}{a}\right)\mathbf{j}$$

and the curvature and radius of curvature at $\mathbf{r}(s)$ are

$$\kappa(s) = \left|\frac{d\hat{\mathbf{T}}}{ds}\right| = \frac{1}{a}, \qquad \rho(s) = \frac{1}{\kappa(s)} = a.$$

Finally, the unit principal normal is

$$\hat{\mathbf{N}}(s) = -\cos\left(\frac{s}{a}\right)\mathbf{i} - \sin\left(\frac{s}{a}\right)\mathbf{j} = -\frac{1}{a}\mathbf{r}(s).$$

Note that the curvature and radius of curvature are constant; the latter is in fact the radius of the circle. The circle and its unit tangent and normal vectors at a typical point are sketched in Figure 11.14(b). Note that $\hat{\mathbf{N}}$ points toward the centre of the circle. ∎

Remark   Another observation can be made about the above example. The position vector $\mathbf{r}(s)$ makes angle $\theta = s/a$ with the positive x-axis; therefore, $\hat{\mathbf{T}}(s)$ makes the same angle with the positive y-axis. Therefore, the rate of rotation of $\hat{\mathbf{T}}$ with respect to s is

$$\frac{d\theta}{ds} = \frac{1}{a} = \kappa.$$

That is, κ is the rate at which $\hat{\mathbf{T}}$ is turning (measured with respect to arc length). This observation extends to a general smooth curve.

THEOREM **2**

Curvature is the rate of turning of the unit tangent

Let $\kappa > 0$ on an interval containing s, and let $\Delta\theta$ be the angle between $\hat{\mathbf{T}}(s + \Delta s)$ and $\hat{\mathbf{T}}(s)$, the unit tangent vectors at neighbouring points on the curve. Then

$$\kappa(s) = \lim_{\Delta s \to 0} \left| \frac{\Delta\theta}{\Delta s} \right|.$$

PROOF   Let $\Delta\hat{\mathbf{T}} = \hat{\mathbf{T}}(s + \Delta s) - \hat{\mathbf{T}}(s)$. Because both $\hat{\mathbf{T}}(s)$ and $\hat{\mathbf{T}}(s + \Delta s)$ are unit vectors, $|\Delta\hat{\mathbf{T}}/\Delta\theta|$ is the ratio of the length of a chord to the length of the corresponding arc on a circle of radius 1. (See Figure 11.15.) Thus

$$\lim_{\Delta s \to 0} \left| \frac{\Delta\hat{\mathbf{T}}}{\Delta\theta} \right| = 1 \qquad \text{and}$$

$$\kappa(s) = \lim_{\Delta s \to 0} \left| \frac{\Delta\hat{\mathbf{T}}}{\Delta s} \right| = \lim_{\Delta s \to 0} \left| \frac{\Delta\hat{\mathbf{T}}}{\Delta\theta} \right| \left| \frac{\Delta\theta}{\Delta s} \right| = \lim_{\Delta s \to 0} \left| \frac{\Delta\theta}{\Delta s} \right|.$$

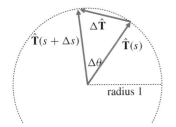

Figure 11.15 $|\Delta\hat{\mathbf{T}}| \approx |\Delta\theta|$ for small values of $|\Delta s|$

The unit tangent $\hat{\mathbf{T}}$ and unit normal $\hat{\mathbf{N}}$ at a point $\mathbf{r}(s)$ on a curve $\mathcal{C}$ are regarded as having their tails at that point. They are perpendicular, and $\hat{\mathbf{N}}$ points in the direction toward which $\hat{\mathbf{T}}(s)$ turns as s increases. The plane passing through $\mathbf{r}(s)$ and containing the vectors $\hat{\mathbf{T}}(s)$ and $\hat{\mathbf{N}}(s)$ is called the **osculating plane** of $\mathcal{C}$ at $\mathbf{r}(s)$ (from the Latin *osculum*, meaning *kiss*). For a *plane curve*, such as the circle in Example 2, the osculating plane is just the plane containing the curve. For more general three-dimensional curves the osculating plane varies from point to point; at any point it is the plane that comes closest to containing the part of the curve near that point. The osculating plane is not properly defined at a point where $\kappa(s) = 0$, although if such points are isolated, it can sometimes be defined as a limit of osculating planes for neighbouring points.

Still assuming that $\kappa(s) \neq 0$, let

$$\mathbf{r}_c(s) = \mathbf{r}(s) + \rho(s)\hat{\mathbf{N}}(s).$$

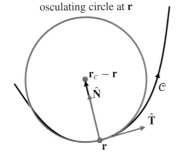

osculating circle at **r**

Figure 11.16   An osculating circle

For each s the point with position vector $\mathbf{r}_c(s)$ lies in the osculating plane of $\mathcal{C}$ at $\mathbf{r}(s)$, on the concave side of $\mathcal{C}$ and at distance $\rho(s)$ from $\mathbf{r}(s)$. It is called the **centre of curvature** of $\mathcal{C}$ for the point $\mathbf{r}(s)$. The circle in the osculating plane having centre at the centre of curvature and radius equal to the radius of curvature $\rho(s)$ is called the **osculating circle** for $\mathcal{C}$ at $\mathbf{r}(s)$. Among all circles that pass through the point $\mathbf{r}(s)$, the osculating circle is the one that best describes the behaviour of $\mathcal{C}$ near that point. Of course, the osculating circle of a circle at any point is the same circle. A typical example of an osculating circle is shown in Figure 11.16.

Torsion and Binormal, the Frenet–Serret Formulas

At any point $\mathbf{r}(s)$ on the curve $\mathcal{C}$ where $\hat{\mathbf{T}}$ and $\hat{\mathbf{N}}$ are defined, a third unit vector, the **unit binormal** $\hat{\mathbf{B}}$, is defined by the formula

$$\hat{\mathbf{B}} = \hat{\mathbf{T}} \times \hat{\mathbf{N}}.$$

Note that $\hat{\mathbf{B}}(s)$ is normal to the osculating plane of $\mathcal{C}$ at $\mathbf{r}(s)$; if $\mathcal{C}$ is a plane curve, then $\hat{\mathbf{B}}$ is a constant vector, independent of s on any interval where $\kappa(s) \neq 0$. At each point $\mathbf{r}(s)$ on $\mathcal{C}$, the three vectors $\{\hat{\mathbf{T}}, \hat{\mathbf{N}}, \hat{\mathbf{B}}\}$ constitute a right-handed basis of mutually perpendicular unit vectors like the standard basis $\{\mathbf{i}, \mathbf{j}, \mathbf{k}\}$. (See Figure 11.17.) This basis is called the **Frenet frame** for $\mathcal{C}$ at the point $\mathbf{r}(s)$. Note that

$$\hat{\mathbf{B}} \times \hat{\mathbf{T}} = \hat{\mathbf{N}} \quad \text{and} \quad \hat{\mathbf{N}} \times \hat{\mathbf{B}} = \hat{\mathbf{T}}.$$

Figure 11.17 The Frenet frame $\{\hat{\mathbf{T}}, \hat{\mathbf{N}}, \hat{\mathbf{B}}\}$ at some points on $\mathcal{C}$

Since $1 = \hat{\mathbf{B}}(s) \bullet \hat{\mathbf{B}}(s)$, then $\hat{\mathbf{B}}(s) \bullet (d\hat{\mathbf{B}}/ds) = 0$, and $d\hat{\mathbf{B}}/ds$ is perpendicular to $\hat{\mathbf{B}}(s)$. Also, differentiating $\hat{\mathbf{B}} = \hat{\mathbf{T}} \times \hat{\mathbf{N}}$ we obtain

$$\frac{d\hat{\mathbf{B}}}{ds} = \frac{d\hat{\mathbf{T}}}{ds} \times \hat{\mathbf{N}} + \hat{\mathbf{T}} \times \frac{d\hat{\mathbf{N}}}{ds} = \kappa \hat{\mathbf{N}} \times \hat{\mathbf{N}} + \hat{\mathbf{T}} \times \frac{d\hat{\mathbf{N}}}{ds} = \hat{\mathbf{T}} \times \frac{d\hat{\mathbf{N}}}{ds}.$$

Therefore, $d\hat{\mathbf{B}}/ds$ is also perpendicular to $\hat{\mathbf{T}}$. Being perpendicular to both $\hat{\mathbf{T}}$ and $\hat{\mathbf{B}}$, $d\hat{\mathbf{B}}/ds$ must be parallel to $\hat{\mathbf{N}}$. This fact is the basis for our definition of torsion.

DEFINITION 2

> **Torsion**
>
> On any interval where $\kappa(s) \neq 0$ there exists a function $\tau(s)$ such that
>
> $$\frac{d\hat{\mathbf{B}}}{ds} = -\tau(s)\hat{\mathbf{N}}(s).$$
>
> The number $\tau(s)$ is called the **torsion** of $\mathcal{C}$ at $\mathbf{r}(s)$.

The torsion measures the degree of twisting that the curve exhibits near a point, that is, the extent to which the curve fails to be planar. It may be positive or negative, depending on the right-handedness or left-handedness of the twisting. We will present an example later in this section.

Theorem 2 has an analogue for torsion, for which the proof is similar. It states that the absolute value of the torsion, $|\tau(s)|$, at point $\mathbf{r}(s)$ on the curve $\mathcal{C}$ is the rate of turning of the unit binormal:

$$\lim_{\Delta s \to 0} \left| \frac{\Delta \psi}{\Delta s} \right| = |\tau(s)|,$$

where $\Delta \psi$ is the angle between $\hat{\mathbf{B}}(s + \Delta s)$ and $\hat{\mathbf{B}}(s)$.

Example 3 (**The circular helix**) As observed in Example 5 of Section 11.3, the parametric equation

$$\mathbf{r}(s) = a \cos(cs)\mathbf{i} + a \sin(cs)\mathbf{j} + bcs\mathbf{k}, \qquad \text{where } c = \frac{1}{\sqrt{a^2 + b^2}},$$

represents a circular helix wound on the cylinder $x^2 + y^2 = a^2$ and parametrized in terms of arc length. Assume $a > 0$. Find the curvature and torsion functions $\kappa(s)$ and $\tau(s)$ for this helix and also the unit vectors comprising the Frenet frame at any point $\mathbf{r}(s)$ on the helix.

Solution   In Example 1 we calculated the unit tangent vector to be

$$\hat{\mathbf{T}}(s) = -ac\sin(cs)\mathbf{i} + ac\cos(cs)\mathbf{j} + bc\mathbf{k}.$$

Differentiating again leads to

$$\frac{d\hat{\mathbf{T}}}{ds} = -ac^2\cos(cs)\mathbf{i} - ac^2\sin(cs)\mathbf{j},$$

so that the curvature of the helix is

$$\kappa(s) = \left|\frac{d\hat{\mathbf{T}}}{ds}\right| = ac^2 = \frac{a}{a^2 + b^2},$$

and the unit normal vector is

$$\hat{\mathbf{N}}(s) = \frac{1}{\kappa(s)}\frac{d\hat{\mathbf{T}}}{ds} = -\cos(cs)\mathbf{i} - \sin(cs)\mathbf{j}.$$

Now we have

$$\hat{\mathbf{B}}(s) = \hat{\mathbf{T}}(s) \times \hat{\mathbf{N}}(s) = \begin{vmatrix} \mathbf{i} & \mathbf{j} & \mathbf{k} \\ -ac\sin(cs) & ac\cos(cs) & bc \\ -\cos(cs) & -\sin(cs) & 0 \end{vmatrix}$$

$$= bc\sin(cs)\mathbf{i} - bc\cos(cs)\mathbf{j} + ac\mathbf{k}.$$

Differentiating this formula leads to

$$\frac{d\hat{\mathbf{B}}}{ds} = bc^2\cos(cs)\mathbf{i} + bc^2\sin(cs)\mathbf{j} = -bc^2\hat{\mathbf{N}}(s).$$

Therefore, the torsion is given by

$$\tau(s) = -(-bc^2) = \frac{b}{a^2 + b^2}.$$

Remark   Observe that the curvature $\kappa(s)$ and the torsion $\tau(s)$ are both constant (i.e., independent of s) for a circular helix. In the above example, $\tau > 0$ (assuming that $b > 0$). This corresponds to the fact that the helix is *right-handed*. (See Figure 11.12 in the previous section.) If you grasp the helix with your right hand so your fingers surround it in the direction of increasing s (counterclockwise, looking down from the positive z-axis), then your thumb also points in the axial direction corresponding to increasing s (the upward direction). Had we started with a left-handed helix, such as

$$\mathbf{r} = a\sin t\,\mathbf{i} + a\cos t\,\mathbf{j} + bt\mathbf{k}, \qquad (a, b > 0),$$

we would have obtained $\tau = -b/(a^2 + b^2)$.

Making use of the formulas $d\hat{\mathbf{T}}/ds = \kappa\hat{\mathbf{N}}$ and $d\hat{\mathbf{B}}/ds = -\tau\hat{\mathbf{N}}$ we can calculate $d\hat{\mathbf{N}}/ds$ as well:

$$\frac{d\hat{\mathbf{N}}}{ds} = \frac{d}{ds}(\hat{\mathbf{B}} \times \hat{\mathbf{T}}) = \frac{d\hat{\mathbf{B}}}{ds} \times \hat{\mathbf{T}} + \hat{\mathbf{B}} \times \frac{d\hat{\mathbf{T}}}{ds}$$

$$= -\tau\hat{\mathbf{N}} \times \hat{\mathbf{T}} + \kappa\hat{\mathbf{B}} \times \hat{\mathbf{N}} = -\kappa\hat{\mathbf{T}} + \tau\hat{\mathbf{B}}.$$

Together, the three formulas

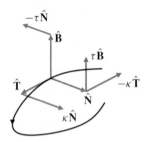

Figure 11.18 $\hat{\mathbf{T}}$, $\hat{\mathbf{N}}$, and $\hat{\mathbf{B}}$, and their directions of change

$$\frac{d\hat{\mathbf{T}}}{ds} = \kappa \hat{\mathbf{N}}$$

$$\frac{d\hat{\mathbf{N}}}{ds} = -\kappa \hat{\mathbf{T}} + \tau \hat{\mathbf{B}}$$

$$\frac{d\hat{\mathbf{B}}}{ds} = -\tau \hat{\mathbf{N}}$$

are known as the **Frenet–Serret formulas**. (See Figure 11.18.) They are of fundamental importance in the theory of curves in 3-space. The Frenet–Serret formulas can be written in matrix form as follows:

$$\frac{d}{ds} \begin{pmatrix} \hat{\mathbf{T}} \\ \hat{\mathbf{N}} \\ \hat{\mathbf{B}} \end{pmatrix} = \begin{pmatrix} 0 & \kappa & 0 \\ -\kappa & 0 & \tau \\ 0 & -\tau & 0 \end{pmatrix} \begin{pmatrix} \hat{\mathbf{T}} \\ \hat{\mathbf{N}} \\ \hat{\mathbf{B}} \end{pmatrix}.$$

Using the Frenet–Serret formulas we can show that the shape of a curve with nonvanishing curvature is completely determined by the curvature and torsion functions $\kappa(s)$ and $\tau(s)$.

THEOREM 3

The Fundamental Theorem of Space Curves

Let $\mathcal{C}_1$ and $\mathcal{C}_2$ be two curves, both of which have the same nonvanishing curvature function $\kappa(s)$ and the same torsion function $\tau(s)$. Then the curves are congruent. That is, one can be moved rigidly (translated and rotated) so as to coincide exactly with the other.

PROOF   We require $\kappa \neq 0$ because $\hat{\mathbf{N}}$ and $\hat{\mathbf{B}}$ are not defined where $\kappa = 0$. Move $\mathcal{C}_2$ rigidly so that its initial point coincides with the initial point of $\mathcal{C}_1$ and so that the Frenet frames of both curves coincide at that point. Let $\hat{\mathbf{T}}_1$, $\hat{\mathbf{T}}_2$, $\hat{\mathbf{N}}_1$, $\hat{\mathbf{N}}_2$, $\hat{\mathbf{B}}_1$, and $\hat{\mathbf{B}}_2$ be the unit tangents, normals, and binormals for the two curves. Let

$$f(s) = \hat{\mathbf{T}}_1(s) \bullet \hat{\mathbf{T}}_2(s) + \hat{\mathbf{N}}_1(s) \bullet \hat{\mathbf{N}}_2(s) + \hat{\mathbf{B}}_1(s) \bullet \hat{\mathbf{B}}_2(s).$$

We calculate the derivative of $f(s)$ using the Product Rule and the Frenet–Serret formulas:

$$f'(s) = \hat{\mathbf{T}}_1' \bullet \hat{\mathbf{T}}_2 + \hat{\mathbf{T}}_1 \bullet \hat{\mathbf{T}}_2' + \hat{\mathbf{N}}_1' \bullet \hat{\mathbf{N}}_2 + \hat{\mathbf{N}}_1 \bullet \hat{\mathbf{N}}_2' + \hat{\mathbf{B}}_1' \bullet \hat{\mathbf{B}}_2 + \hat{\mathbf{B}}_1 \bullet \hat{\mathbf{B}}_2'$$

$$= \kappa \hat{\mathbf{N}}_1 \bullet \hat{\mathbf{T}}_2 + \kappa \hat{\mathbf{T}}_1 \bullet \hat{\mathbf{N}}_2 - \kappa \hat{\mathbf{T}}_1 \bullet \hat{\mathbf{N}}_2 + \tau \hat{\mathbf{B}}_1 \bullet \hat{\mathbf{N}}_2 - \kappa \hat{\mathbf{N}}_1 \bullet \hat{\mathbf{T}}_2$$

$$+ \tau \hat{\mathbf{N}}_1 \bullet \hat{\mathbf{B}}_2 - \tau \hat{\mathbf{N}}_1 \bullet \hat{\mathbf{B}}_2 - \tau \hat{\mathbf{B}}_1 \bullet \hat{\mathbf{N}}_2$$

$$= 0.$$

Therefore $f(s)$ is constant. Since the frames coincide at $s = 0$, the constant must, in fact, be 3:

$$\hat{\mathbf{T}}_1(s) \bullet \hat{\mathbf{T}}_2(s) + \hat{\mathbf{N}}_1(s) \bullet \hat{\mathbf{N}}_2(s) + \hat{\mathbf{B}}_1(s) \bullet \hat{\mathbf{B}}_2(s) = 3.$$

However, each dot product cannot exceed 1 since the factors are unit vectors. Therefore, each dot product must be equal to 1. In particular, $\hat{\mathbf{T}}_1(s) \bullet \hat{\mathbf{T}}_2(s) = 1$ for all s; hence

$$\frac{d\mathbf{r}_1}{ds} = \hat{\mathbf{T}}_1(s) = \hat{\mathbf{T}}_2(s) = \frac{d\mathbf{r}_2}{ds}.$$

Integrating with respect to s and using the fact that both curves start from the same point when $s = 0$, we obtain $\mathbf{r}_1(s) = \mathbf{r}_2(s)$ for all s, which is what we wanted to show.

Remark   It is a consequence of the above theorem that any curve having nonzero constant curvature and constant torsion must, in fact, be a circle (if the torsion is zero) or a circular helix (if the torsion is nonzero). See Exercises 7 and 8 below.

Exercises 11.4

Find the unit tangent vector $\hat{\mathbf{T}}(t)$ for the curves in Exercises 1–4.

1. $\mathbf{r} = t\mathbf{i} - 2t^2\mathbf{j} + 3t^3\mathbf{k}$

2. $\mathbf{r} = a\sin\omega t\,\mathbf{i} + a\cos\omega t\,\mathbf{k}$

3. $\mathbf{r} = \cos t\sin t\,\mathbf{i} + \sin^2 t\,\mathbf{j} + \cos t\,\mathbf{k}$

4. $\mathbf{r} = a\cos t\,\mathbf{i} + b\sin t\,\mathbf{j} + t\mathbf{k}$

5. Show that if $\kappa(s) = 0$ for all s, then the curve $\mathbf{r} = \mathbf{r}(s)$ is a straight line.

6. Show that if $\tau(s) = 0$ for all s, then the curve $\mathbf{r} = \mathbf{r}(s)$ is a plane curve. *Hint:* Show that $\mathbf{r}(s)$ lies in the plane through $\mathbf{r}(0)$ with normal $\hat{\mathbf{B}}(0)$.

7. Show that if $\kappa(s) = C$ is a positive constant and $\tau(s) = 0$ for all s, then the curve $\mathbf{r} = \mathbf{r}(s)$ is a circle. *Hint:* Find a circle having the given constant curvature. Then use Theorem 3.

8. Show that if the curvature $\kappa(s)$ and the torsion $\tau(s)$ are both nonzero constants, then the curve $\mathbf{r} = \mathbf{r}(s)$ is a circular helix. *Hint:* Find a helix having the given curvature and torsion.

11.5 Curvature and Torsion for General Parametrizations

The formulas developed above for curvature and torsion as well as for the unit normal and binormal vectors are not very useful if the curve we want to analyze is not expressed in terms of the arc-length parameter. We will now consider how to find these quantities in terms of a general parametrization $\mathbf{r} = \mathbf{r}(t)$. We will express them all in terms of the velocity, $\mathbf{v}(t)$, the speed, $v(t) = |\mathbf{v}(t)|$, and the acceleration, $\mathbf{a}(t)$. First observe that

$$\mathbf{v} = \frac{d\mathbf{r}}{dt} = \frac{d\mathbf{r}}{ds}\frac{ds}{dt} = v\hat{\mathbf{T}}$$

$$\mathbf{a} = \frac{d\mathbf{v}}{dt} = \frac{dv}{dt}\hat{\mathbf{T}} + v\frac{d\hat{\mathbf{T}}}{dt}$$

$$= \frac{dv}{dt}\hat{\mathbf{T}} + v\frac{d\hat{\mathbf{T}}}{ds}\frac{ds}{dt} = \frac{dv}{dt}\hat{\mathbf{T}} + v^2\kappa\hat{\mathbf{N}}$$

$$\mathbf{v}\times\mathbf{a} = v\frac{dv}{dt}\hat{\mathbf{T}}\times\hat{\mathbf{T}} + v^3\kappa\hat{\mathbf{T}}\times\hat{\mathbf{N}} = v^3\kappa\hat{\mathbf{B}}.$$

Note that $\hat{\mathbf{B}}$ is in the direction of $\mathbf{v}\times\mathbf{a}$. From these formulas we obtain useful formulas for $\hat{\mathbf{T}}$, $\hat{\mathbf{B}}$, and κ:

$$\hat{\mathbf{T}} = \frac{\mathbf{v}}{v}, \qquad \hat{\mathbf{B}} = \frac{\mathbf{v}\times\mathbf{a}}{|\mathbf{v}\times\mathbf{a}|}, \qquad \kappa = \frac{|\mathbf{v}\times\mathbf{a}|}{v^3}.$$

There are several ways to calculate $\hat{\mathbf{N}}$. Perhaps the easiest is

$$\hat{\mathbf{N}} = \hat{\mathbf{B}}\times\hat{\mathbf{T}}.$$

Sometimes it may be easier to use $\dfrac{d\hat{\mathbf{T}}}{dt} = \dfrac{d\hat{\mathbf{T}}}{ds}\dfrac{ds}{dt} = v\dfrac{d\hat{\mathbf{T}}}{ds} = v\kappa\hat{\mathbf{N}}$ to calculate

$$\hat{\mathbf{N}} = \frac{1}{v\kappa}\frac{d\hat{\mathbf{T}}}{dt} = \frac{\rho}{v}\frac{d\hat{\mathbf{T}}}{dt} = \frac{d\hat{\mathbf{T}}}{dt}\bigg/\left|\frac{d\hat{\mathbf{T}}}{dt}\right|.$$

The torsion remains to be calculated. Observe that

$$\frac{d\mathbf{a}}{dt} = \frac{d}{dt}\left(\frac{dv}{dt}\hat{\mathbf{T}} + v^2\kappa\hat{\mathbf{N}}\right).$$

This differentiation will produce several terms. The only one that involves $\hat{\mathbf{B}}$ is the one that comes from evaluating $v^2\kappa(d\hat{\mathbf{N}}/dt) = v^3\kappa(d\hat{\mathbf{N}}/ds) = v^3\kappa(\tau\hat{\mathbf{B}} - \kappa\hat{\mathbf{T}})$. Therefore,

$$\frac{d\mathbf{a}}{dt} = \lambda\hat{\mathbf{T}} + \mu\hat{\mathbf{N}} + v^3\kappa\tau\hat{\mathbf{B}},$$

for certain scalars λ and μ. Since $\mathbf{v} \times \mathbf{a} = v^3 \kappa \hat{\mathbf{B}}$, it follows that

$$(\mathbf{v} \times \mathbf{a}) \bullet \frac{d\mathbf{a}}{dt} = (v^3 \kappa)^2 \tau = |\mathbf{v} \times \mathbf{a}|^2 \tau.$$

Hence

$$\tau = \frac{(\mathbf{v} \times \mathbf{a}) \bullet (d\mathbf{a}/dt)}{|\mathbf{v} \times \mathbf{a}|^2}.$$

Example 1 Find the curvature, the torsion, and the Frenet frame at a general point on the curve

$$\mathbf{r} = (t + \cos t)\mathbf{i} + (t - \cos t)\mathbf{j} + \sqrt{2} \sin t \mathbf{k}.$$

Describe this curve.

Solution We calculate the various quantities using the recipe given above. First the preliminaries:

$$\mathbf{v} = (1 - \sin t)\mathbf{i} + (1 + \sin t)\mathbf{j} + \sqrt{2} \cos t \mathbf{k}$$

$$\mathbf{a} = -\cos t \mathbf{i} + \cos t \mathbf{j} - \sqrt{2} \sin t \mathbf{k}$$

$$\frac{d\mathbf{a}}{dt} = \sin t \mathbf{i} - \sin t \mathbf{j} - \sqrt{2} \cos t \mathbf{k}$$

$$\mathbf{v} \times \mathbf{a} = \begin{vmatrix} \mathbf{i} & \mathbf{j} & \mathbf{k} \\ 1 - \sin t & 1 + \sin t & \sqrt{2} \cos t \\ -\cos t & \cos t & -\sqrt{2} \sin t \end{vmatrix}$$

$$= -\sqrt{2}(1 + \sin t)\mathbf{i} - \sqrt{2}(1 - \sin t)\mathbf{j} + 2 \cos t \mathbf{k}$$

$$(\mathbf{v} \times \mathbf{a}) \bullet \frac{d\mathbf{a}}{dt} = -\sqrt{2} \sin t (1 + \sin t) + \sqrt{2} \sin t (1 - \sin t) - 2\sqrt{2} \cos^2 t$$

$$= -2\sqrt{2}$$

$$v = |\mathbf{v}| = \sqrt{2 + 2 \sin^2 t + 2 \cos^2 t} = 2$$

$$|\mathbf{v} \times \mathbf{a}| = \sqrt{2(2 + 2 \sin^2 t) + 4 \cos^2 t} = \sqrt{8} = 2\sqrt{2}.$$

Thus we have

$$\kappa = \frac{|\mathbf{v} \times \mathbf{a}|}{v^3} = \frac{2\sqrt{2}}{8} = \frac{1}{2\sqrt{2}}$$

$$\tau = \frac{(\mathbf{v} \times \mathbf{a}) \bullet (d\mathbf{a}/dt)}{|\mathbf{v} \times \mathbf{a}|^2} = \frac{-2\sqrt{2}}{(2\sqrt{2})^2} = -\frac{1}{2\sqrt{2}}$$

$$\hat{\mathbf{T}} = \frac{\mathbf{v}}{v} = \frac{1 - \sin t}{2}\mathbf{i} + \frac{1 + \sin t}{2}\mathbf{j} + \frac{1}{\sqrt{2}} \cos t \mathbf{k}$$

$$\hat{\mathbf{B}} = \frac{\mathbf{v} \times \mathbf{a}}{|\mathbf{v} \times \mathbf{a}|} = -\frac{1 + \sin t}{2}\mathbf{i} - \frac{1 - \sin t}{2}\mathbf{j} + \frac{1}{\sqrt{2}} \cos t \mathbf{k}$$

$$\hat{\mathbf{N}} = \hat{\mathbf{B}} \times \hat{\mathbf{T}} = -\frac{1}{\sqrt{2}} \cos t \mathbf{i} + \frac{1}{\sqrt{2}} \cos t \mathbf{j} - \sin t \mathbf{k}.$$

Since the curvature and torsion are both constant (they are therefore constant when expressed in terms of any parametrization), the curve must be a circular helix by Theorem 3. It is left-handed, since $\tau < 0$. By Example 3 in Section 11.4, it is congruent to the helix

$$\mathbf{r} = a \cos t \mathbf{i} + a \sin t \mathbf{j} + bt \mathbf{k},$$

provided that $a/(a^2 + b^2) = 1/(2\sqrt{2}) = -b/(a^2 + b^2)$. Solving these equations gives $a = \sqrt{2}$ and $b = -\sqrt{2}$, so the helix is wound on a cylinder of radius $\sqrt{2}$. The axis of this cylinder is the line $x = y$, $z = 0$, as can be seen by inspecting the components of $\mathbf{r}(t)$.

Example 2 (**Curvature of the graph of a function of one variable**) Find the curvature of the plane curve with equation $y = f(x)$ at an arbitrary point $(x, f(x))$ on the curve.

Solution The graph can be parametrized: $\mathbf{r} = x\mathbf{i} + f(x)\mathbf{j}$. Thus,

$$\mathbf{v} = \mathbf{i} + f'(x)\mathbf{j},$$
$$\mathbf{a} = f''(x)\mathbf{j},$$
$$\mathbf{v} \times \mathbf{a} = f''(x)\mathbf{k}.$$

Therefore the curvature is

$$\kappa(x) = \frac{|\mathbf{v} \times \mathbf{a}|}{v^3} = \frac{|f''(x)|}{\left(1 + (f'(x))^2\right)^{3/2}}.$$

Tangential and Normal Acceleration

In the formula obtained earlier for the acceleration in terms of the unit tangent and normal,

$$\mathbf{a} = \frac{dv}{dt}\hat{\mathbf{T}} + v^2 \kappa \hat{\mathbf{N}},$$

the term $(dv/dt)\hat{\mathbf{T}}$ is called the **tangential** acceleration, and the term $v^2\kappa\hat{\mathbf{N}}$ is called the **normal** or **centripetal** acceleration. This latter component is directed toward the centre of curvature and its magnitude is $v^2\kappa = v^2/\rho$. Highway, railway, and roller coaster designers attempt to bank curves in such a way that the resultant of the corresponding centrifugal force, $-m(v^2/\rho)\hat{\mathbf{N}}$, and the weight, $-mg\mathbf{k}$, of the vehicle will be normal to the surface at a desired speed.

Example 3 **Banking a curve.** A level, curved road lies along the curve $y = x^2$ in the horizontal xy-plane. Find, as a function of x, the angle at which the road should be banked (i.e., the angle between the vertical and the normal to the surface of the road) so that the resultant of the centrifugal and gravitational $(-mg\mathbf{k})$ forces acting on the vehicle travelling at constant speed v_0 along the road is always normal to the surface of the road.

Solution By Example 2 the path of the road, $y = x^2$, has curvature

$$\kappa = \frac{|d^2y/dx^2|}{\left(1 + (dy/dx)^2\right)^{3/2}} = \frac{2}{(1 + 4x^2)^{3/2}}.$$

The normal component of the acceleration of a vehicle travelling at speed v_0 along the road is

$$a_N = v_0^2\kappa = \frac{2v_0^2}{(1 + 4x^2)^{3/2}}.$$

If the road is banked at angle θ (see Figure 11.19), then the resultant of the centrifugal force $-ma_N\hat{\mathbf{N}}$ and the gravitational force $-mg\mathbf{k}$ is normal to the roadway provided

$$\tan\theta = \frac{ma_N}{mg}, \qquad \text{that is,} \qquad \theta = \tan^{-1}\frac{2v_0^2}{g(1 + 4x^2)^{3/2}}.$$

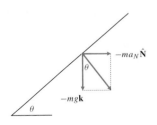

Figure 11.19 Banking a curve on a roadway

Remark The definition of centripetal acceleration given above is consistent with the one that arose in the discussion of rotating frames in Section 11.2. If $\mathbf{r}(t)$ is the position of a moving particle at time t, we can regard the motion at any instant as being a rotation about the centre of curvature, so that the angular velocity must be $\mathbf{\Omega} = \Omega\hat{\mathbf{B}}$. The linear velocity is $\mathbf{v} = \mathbf{\Omega} \times (\mathbf{r} - \mathbf{r}_c) = v\hat{\mathbf{T}}$, so the speed is $v = \Omega\rho$, and $\mathbf{\Omega} = (v/\rho)\hat{\mathbf{B}}$. As developed in Section 11.2, the centripetal acceleration is

$$\mathbf{\Omega} \times (\mathbf{\Omega} \times (\mathbf{r} - \mathbf{r}_c)) = \mathbf{\Omega} \times \mathbf{v} = \frac{v^2}{\rho}\hat{\mathbf{B}} \times \hat{\mathbf{T}} = \frac{v^2}{\rho}\hat{\mathbf{N}}.$$

Evolutes

The centre of curvature $\mathbf{r}_c(t)$ of a given curve can itself trace out another curve as t varies. This curve is called the **evolute** of the given curve $\mathbf{r}(t)$.

Example 4 Find the evolute of the exponential spiral

$$\mathbf{r} = ae^{-t}\cos t\,\mathbf{i} + ae^{-t}\sin t\,\mathbf{j}.$$

Solution The curve is a plane curve so $\tau = 0$. We will take a shortcut to the curvature and the unit normal without calculating $\mathbf{v} \times \mathbf{a}$. First we calculate

$$\mathbf{v} = ae^{-t}\Big(-(\cos t + \sin t)\mathbf{i} - (\sin t - \cos t)\mathbf{j}\Big)$$

$$\frac{ds}{dt} = v = \sqrt{2}ae^{-t}$$

$$\hat{\mathbf{T}}(t) = \frac{1}{\sqrt{2}}\Big(-(\cos t + \sin t)\mathbf{i} - (\sin t - \cos t)\mathbf{j}\Big)$$

$$\frac{d\hat{\mathbf{T}}}{ds} = \frac{1}{(ds/dt)}\frac{d\hat{\mathbf{T}}}{dt} = \frac{1}{2ae^{-t}}\Big((\sin t - \cos t)\mathbf{i} - (\cos t + \sin t)\mathbf{j}\Big)$$

$$\kappa(t) = \left|\frac{d\hat{\mathbf{T}}}{ds}\right| = \frac{1}{\sqrt{2}ae^{-t}}.$$

It follows that the radius of curvature is $\rho(t) = \sqrt{2}ae^{-t}$. Since $d\hat{\mathbf{T}}/ds = \kappa\hat{\mathbf{N}}$, we have $\hat{\mathbf{N}} = \rho(d\hat{\mathbf{T}}/ds)$. The centre of curvature is

$$\mathbf{r}_c(t) = \mathbf{r}(t) + \rho(t)\hat{\mathbf{N}}(t)$$

$$= \mathbf{r}(t) + \rho^2\frac{d\hat{\mathbf{T}}}{ds}$$

$$= ae^{-t}\Big(\cos t\,\mathbf{i} + \sin t\,\mathbf{j}\Big)$$

$$\quad + 2a^2e^{-2t}\frac{1}{2ae^{-t}}\Big((\sin t - \cos t)\mathbf{i} - (\cos t + \sin t)\mathbf{j}\Big)$$

$$= ae^{-t}\Big(\sin t\,\mathbf{i} - \cos t\,\mathbf{j}\Big)$$

$$= ae^{-t}\Big(\cos(t - \frac{\pi}{2})\mathbf{i} + \sin(t - \frac{\pi}{2})\mathbf{j}\Big).$$

Thus, interestingly, the evolute of the exponential spiral is the same exponential spiral rotated 90° clockwise in the plane. (See Figure 11.20(a).)

An Application to Track (or Road) Design

Model trains frequently come with two kinds of track sections: straight and curved. The curved sections are arcs of a circle of radius R, and the track is intended to be laid out in the shape shown in Figure 11.20(b); AB and CD are straight, and BC and DA are semicircles. The track looks smooth, but is it smooth enough?

Figure 11.20

(a) The evolute of an exponential spiral is another exponential spiral

(b) The shape of a model train track

The track is held together by friction, and occasionally it can come apart as the train is racing around. It is especially likely to come apart at the points A, B, C, and D. To see why, assume that the train is travelling at constant speed v. Then the tangential acceleration, $(dv/dt)\hat{\mathbf{T}}$, is zero, and the total acceleration is just the centripetal acceleration, $\mathbf{a} = (v^2/\rho)\hat{\mathbf{N}}$. Therefore, $|\mathbf{a}| = 0$ along the straight sections, and $|\mathbf{a}| = v^2\kappa = v^2/R$ on the semicircular sections. The acceleration is *discontinuous* at the points A, B, C, and D, and the reactive force exerted by the train on the track is also discontinuous at these points. There is a "shock" or "jolt" as the train enters or leaves a curved part of the track. In order to avoid such stress points, tracks should be designed so that the curvature varies continuously from point to point.

Example 5   Existing track along the negative x-axis and along the ray $y = x - 1$, $x \geq 2$, is to be joined smoothly by track along the transition curve $y = f(x)$, $0 \leq x \leq 2$, where $f(x)$ is a polynomial of degree as small as possible. Find $f(x)$ so that a train moving along the track will not experience discontinuous acceleration at the joins.

Solution   The situation is shown in Figure 11.21. The polynomial $f(x)$ must be chosen so that the track is continuous, has continuous slope, and has continuous curvature at $x = 0$ and $x = 2$. Since the curvature of $y = f(x)$ is

$$\kappa = |f''(x)|\left(1 + (f'(x))^2\right)^{-3/2},$$

we need only arrange that f, f', and f'' take the same values at $x = 0$ and $x = 2$ that the straight sections do there:

$$f(0) = 0, \qquad f'(0) = 0, \qquad f''(0) = 0,$$
$$f(2) = 1, \qquad f'(2) = 1, \qquad f''(2) = 0.$$

These six independent conditions suggest we should try a polynomial of degree 5 involving six arbitrary coefficients:

$$f(x) = A + Bx + Cx^2 + Dx^3 + Ex^4 + Fx^5$$
$$f'(x) = B + 2Cx + 3Dx^2 + 4Ex^3 + 5Fx^4$$
$$f''(x) = 2C + 6Dx + 12Ex^2 + 20Fx^3.$$

The three conditions at $x = 0$ imply that $A = B = C = 0$. Those at $x = 2$ imply that

$$8D + 16E + 32F = f(2) = 1$$
$$12D + 32E + 80F = f'(2) = 1$$
$$12D + 48E + 160F = f''(2) = 0.$$

This system has solution $D = 1/4$, $E = -1/16$, and $F = 0$, so we should use $f(x) = (x^3/4) - (x^4/16)$. ∎

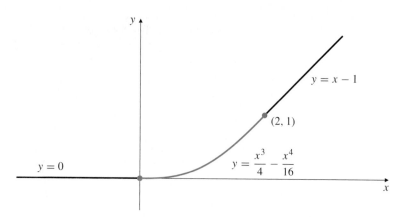

Figure 11.21   Joining two straight tracks with a curved track

Remark   Road and railroad builders do not usually use polynomial graphs as transition curves. Other kinds of curves called **clothoids** and **lemniscates** are usually used. (See Exercise 7 in the Review Exercises at the end of this chapter.)

Maple Calculations

Calculations of the sort done in this section for fairly simple curves can become quite oppressive for more complicated curves. As usual, Maple can come to the rescue. Before the advent of the VectorCalculus package in Maple 8, which introduces a true vector data structure, defining a vector-valued function was a bit tricky: one had to use something like

```
>   R := [t -> f(t), t -> g(t), t -> h(t)];
```

rather than the more obvious

```
>   R := t -> [f(t), g(t), h(t)];
```

because in the latter the function R was not considered to be a vector, even though its values are vectors.

In the following we assume the LinearAlgebra and VectorCalculus packages have been loaded (in that order):

```
>   with(LinearAlgebra):   with(VectorCalculus):
```

Here is how we might define a vector-valued function representing a circular helix:

```
>   R := t -> <a*cos(t), a*sin(t), b*t>;
```

The output from this definition (which we omit here) may appear a bit cryptic at first; it asserts that R is defined as a procedure "VectorCalculus-<,>" whose arguments are three "VectorCalculus-*" procedures for the products that represent the three components of R. Calling the function generates the expected results.

```
>   R(t); R(Pi);
```

$$a \, \cos(t) \, e_x + a \, \sin(t) \, e_y + b \, t \, e_z$$
$$-a \, e_x + b \, \pi \, e_z$$

Velocity, acceleration, and speed functions can now be defined in the obvious way and the results used to find these quantities at any point:

```
>   V := t -> diff(R(t),t):   A := t -> diff(V(t),t):
>   v := t -> Norm(V(t),2):
>   V(t); A(t); v(t);
```

$$-a \, \sin(t) \, e_x + a \, \cos(t) \, e_y + b \, e_z$$
$$-a \, \cos(t) \, e_x - a \, \sin(t) \, e_y$$

$$\sqrt{|a \, \sin(t)|^2 + |a \, \cos(t)|^2 + |b|^2}$$

No attempt to simplify the last expression has much effect unless we tell Maple that a and b are real numbers. In fact, it is useful for purposes of simplification to tell Maple that a, b, and t are all real, and to suppress Maple's urge to beat us over the head with that fact by subsequently placing a tilde ($\tilde{\ }$) after each of these variables in all subsequent output. We can accomplish this with

```
>  assume(a::real, b::real, t::real);
>  interface(showassumed=0);
>  simplify(v(t));
```

$$\sqrt{b^2 + a^2}$$

The VectorCalculus package has a function called TNBFrame whose output is a list of functions generating the unit tangent, principal normal, and binormal vectors $\hat{\mathbf{T}}(t)$, $\hat{\mathbf{N}}(t)$, and $\hat{\mathbf{B}}(t)$. We can use the components of this list to define each vector:

```
>  T := TNBFrame(R,t)[1]:  T(t);
```

$$-\frac{a\,\sin(t)}{\sqrt{b^2 + a^2}}\,e_x + \frac{a\,\cos(t)}{\sqrt{b^2 + a^2}}\,e_y + \frac{b}{\sqrt{b^2 + a^2}}\,e_z$$

```
>  N := TNBFrame(R,t)[2]:  N(t);
```

$$-\frac{a\,\cos(t)}{|a|}\,e_x - \frac{a\,\sin(t)}{|a|}\,e_y$$

```
>  B := TNBFrame(R,t)[3]:  B(t);
```

$$\frac{b\,a\,\sin(t)}{\sqrt{b^2 + a^2}\,|a|}\,e_x - \frac{b\,a\,\cos(t)}{\sqrt{b^2 + a^2}\,|a|}\,e_y + \left(\frac{a^2\,\sin(t)^2}{\sqrt{b^2 + a^2}\,|a|} + \frac{a^2\,\cos(t)^2}{\sqrt{b^2 + a^2}\,|a|}\right)\,e_z$$

```
>  simplify(%);
```

$$\frac{b\,a\,\sin(t)}{\sqrt{b^2 + a^2}\,|a|}\,e_x - \frac{b\,a\,\cos(t)}{\sqrt{b^2 + a^2}\,|a|}\,e_y + \frac{|a|}{\sqrt{b^2 + a^2}}\,e_z$$

VectorCalculus also defines Curvature and Torsion functions that can be invoked as follows:

```
>  simplify(Curvature(R,t)(t));
```

$$\frac{|a|}{\sqrt{b^2 + a^2}}$$

```
>  simplify(Torsion(R,t)(t));
```

$$\frac{|b|}{\sqrt{b^2 + a^2}}$$

The torsion is in error if $b < 0$; the torsion should be negative in this case. In fact, the result above would make the third Serret-Frenet formula false as we can see from

```
>  simplify(diff(B(t),t) + tau*N(t));
```

$$\frac{a\,\cos(t)\left(b - \tau\sqrt{b^2 + a^2}\right)}{\sqrt{b^2 + a^2}\,|a|}\,e_x + \frac{a\,\sin(t)\left(b - \tau\sqrt{b^2 + a^2}\right)}{\sqrt{b^2 + a^2}\,|a|}\,e_y$$

which must be **0**, but will be only if $\tau = b/\sqrt{b^2 + a^2}$. This error has been reported to MapleSoft and is corrected in Maple 10.

Exercises 11.5

Find the radius of curvature of the curves in Exercises 1–4 at the points indicated.

1. $y = x^2$ at $x = 0$ and at $x = \sqrt{2}$

2. $y = \cos x$ at $x = 0$ and at $x = \pi/2$

3. $\mathbf{r} = 2t\mathbf{i} + (1/t)\mathbf{j} - 2t\mathbf{k}$ at $(2, 1, -2)$

4. $\mathbf{r} = t^3\mathbf{i} + t^2\mathbf{j} + t\mathbf{k}$ at the point where $t = 1$

Find the Frenet frames $\{\hat{\mathbf{T}}, \hat{\mathbf{N}}, \hat{\mathbf{B}}\}$ for the curves in Exercises 5–6 at the points indicated.

5. $\mathbf{r} = t\mathbf{i} + t^2\mathbf{j} + 2\mathbf{k}$ at $(1, 1, 2)$

6. $\mathbf{r} = t\mathbf{i} + t^2\mathbf{j} + t\mathbf{k}$ at $(1, 1, 1)$

In Exercises 7–8, find the unit tangent, normal, and binormal vectors and the curvature and torsion at a general point on the given curve.

7. $\mathbf{r} = t\mathbf{i} + \dfrac{t^2}{2}\mathbf{j} + \dfrac{t^3}{3}\mathbf{k}$ **8.** $\mathbf{r} = e^t(\cos t\,\mathbf{i} + \sin t\,\mathbf{j} + \mathbf{k})$

9. Find the curvature and torsion of the parametric curve

$$x = 2 + \sqrt{2}\cos t, \quad y = 1 - \sin t, \quad z = 3 + \sin t$$

at an arbitrary point t. What is the curve?

10. A particle moves along the plane curve $y = \sin x$ in the direction of increasing x with constant horizontal speed $dx/dt = k$. Find the tangential and normal components of the acceleration of the particle when it is at position x.

11. Find the unit tangent, normal and binormal, and the curvature and torsion for the curve

$$\mathbf{r} = \sin t \cos t\,\mathbf{i} + \sin^2 t\,\mathbf{j} + \cos t\,\mathbf{k}$$

at the points   (a) $t = 0$ and   (b) $t = \pi/4$.

12. A particle moves on an elliptical path in the xy-plane so that its position at time t is $\mathbf{r} = a\cos t\,\mathbf{i} + b\sin t\,\mathbf{j}$. Find the tangential and normal components of its acceleration at time t. At what points is the tangential acceleration zero?

13. Find the maximum and minimum values for the curvature of the ellipse $x = a\cos t$, $y = b\sin t$, where $a > b > 0$.

14. A bead of mass m slides without friction down a wire bent in the shape of the curve $y = x^2$, under the influence of the gravitational force $-mg\mathbf{j}$. The speed of the bead is v as it passes through the point $(1, 1)$. Find, at that instant, the magnitude of the normal acceleration of the bead and the rate of change of its speed.

15. Find the curvature of the plane curve $y = e^x$ at x. Find the equation of the evolute of this curve.

16. Show that the curvature of the plane polar graph $r = f(\theta)$ at a general point θ is

$$\kappa(\theta) = \frac{\left|2\left(f'(\theta)\right)^2 + \left(f(\theta)\right)^2 - f(\theta)f''(\theta)\right|}{\left[\left(f'(\theta)\right)^2 + \left(f(\theta)\right)^2\right]^{3/2}}.$$

17. Find the curvature of the cardioid $r = a(1 - \cos\theta)$.

∗18. Find the curve $\mathbf{r} = \mathbf{r}(t)$ for which $\kappa(t) = 1$ and $\tau(t) = 1$ for all t, and $\mathbf{r}(0) = \hat{\mathbf{T}}(0) = \mathbf{i}$, $\hat{\mathbf{N}}(0) = \mathbf{j}$, and $\hat{\mathbf{B}}(0) = \mathbf{k}$.

19. Suppose the curve $\mathbf{r} = \mathbf{r}(t)$ satisfies $\dfrac{d\mathbf{r}}{dt} = \mathbf{c} \times \mathbf{r}(t)$, where $\mathbf{c}$ is a constant vector. Show that the curve is the circle in which the plane through $\mathbf{r}(0)$ normal to $\mathbf{c}$ intersects a sphere with radius $|\mathbf{r}(0)|$ centred at the origin.

20. Find the evolute of the circular helix $\mathbf{r} = a\cos t\,\mathbf{i} + a\sin t\,\mathbf{j} + bt\mathbf{k}$.

21. Find the evolute of the parabola $y = x^2$.

22. Find the evolute of the ellipse $x = 2\cos t$, $y = \sin t$.

23. Find the polynomial $f(x)$ of lowest degree so that track along $y = f(x)$ from $x = -1$ to $x = 1$ joins with existing straight tracks $y = -1$, $x \le -1$ and $y = 1$, $x \ge 1$ sufficiently smoothly that a train moving at constant speed will not experience discontinuous acceleration at the joins.

∗24. Help out model train manufacturers. Design a track segment $y = f(x)$, $-1 \le x \le 0$, to provide a jolt-free link between a straight track section $y = 1$, $x \le -1$, and a semicircular arc section $x^2 + y^2 = 1$, $x \ge 0$.

∗25. If the position $\mathbf{r}$, velocity $\mathbf{v}$, and acceleration $\mathbf{a}$ of a moving particle satisfy $\mathbf{a}(t) = \lambda(t)\mathbf{r}(t) + \mu(t)\mathbf{v}(t)$, where $\lambda(t)$ and $\mu(t)$ are scalar functions of time t, and if $\mathbf{v} \times \mathbf{a} \ne \mathbf{0}$, show that the path of the particle lies in a plane.

Use Maple in Exercises 26–31. Make sure to load the LinearAlgebra and VectorCalculus packages.

In Exercises 26–29, determine the curvature and torsion functions for the given curves. Because of the problem with the `Torsion` function in some versions of the VectorCalculus package (as mentioned at the end of this section), you may want to use the formulas derived from the derivatives of position to determine it, and probably the curvature as well. Try to describe the curve.

26. $\mathbf{r}(t) = \cos(t)\mathbf{i} + 2\sin(t)\mathbf{j} + \cos(t)\mathbf{k}$. Why should you not be surprised at the value of the torsion? What are the maximum and minimum curvatures? Describe the curve.

27. $\mathbf{r}(t) = (t - \sin t)\mathbf{i} + (1 - \cos t)\mathbf{j} + t\mathbf{k}$. Are the curvature and torsion continuous for all t?

28. $\mathbf{r}(t) = \cos(t)\cos(2t)\mathbf{i} + \cos(t)\sin(2t)\mathbf{j} + \sin(t)\mathbf{k}$. Show that the curve lies on the sphere $x^2 + y^2 + z^2 = 1$. What is the minimum value of its curvature?

29. $\mathbf{r}(t) = (t + \cos t)\mathbf{i} + (t + \sin t)\mathbf{j} + (1 + t - \cos t)\mathbf{k}$.

In Exercises 30–31, define new Maple functions to calculate the requested items. Assume the LinearAlgebra and VectorCalculus packages are loaded.

30. The `evolute(R)(t)` whose value at R is the function whose value at t is the position vector of the centre of curvature of the curve R for the point R(t).

31. A function `tanline(R)(t,u)` whose value at **R** is the function whose value at (t,u) is the position vector of the point on the tangent line to the curve R at t at distance u from R(t) in the direction of increasing t.

11.6 Kepler's Laws of Planetary Motion

The German mathematician and astronomer Johannes Kepler (1571–1630) was a student and colleague of Danish astronomer Tycho Brahe (1546–1601). Over a lifetime of observing the positions of planets without the aid of a telescope, Brahe compiled a vast amount of data, which Kepler analyzed. Although Polish astronomer Nicolaus Copernicus (1473–1543) had postulated that the earth and other planets moved around the sun, the religious and philosophical climate in Europe at the end of the sixteenth century still favoured explaining the motion of heavenly bodies in terms of circular orbits around the earth. It was known that planets such as Mars could not move on circular orbits centred at the earth, but models were proposed in which they moved on other circles (epicycles) whose centres moved on circles centred at the earth.

Brahe's observations of Mars were sufficiently detailed that Kepler realized that no simple model based on circles could be made to conform very closely with the actual orbit. He was, however, able to fit a more general quadratic curve, an ellipse with one focus at the sun. Based on this success and on Brahe's data on other planets, he formulated the following three laws of planetary motion:

Kepler's Laws

1. The planets move on elliptical orbits with the sun at one focus.
2. The radial line from the sun to a planet sweeps out equal areas in equal times.
3. The squares of the periods of revolution of the planets around the sun are proportional to the cubes of the major axes of their orbits.

Kepler's statement of the third law actually says that the squares of the periods of revolution of the planets are proportional to the cubes of their mean distances from the sun. The mean distance of points on an ellipse from a focus of the ellipse is equal to the semi-major axis. (See Exercise 17 below.) Therefore, the two statements are equivalent.

The choice of ellipses was reasonable once it became clear that circles would not work. The properties of the conic sections were well understood, having been developed by the Greek mathematician Apollonius of Perga around 200 BC. Nevertheless, based, as it was, on observations rather than theory, Kepler's formulation of his laws without any causal explanation was a truly remarkable feat. The theoretical underpinnings came later when Newton, with the aid of his newly created calculus, showed that Kepler's laws implied an inverse square gravitational force. (See Review Exercises 14–16 at the end of this chapter.) Newton believed that his universal gravitational law also implied Kepler's laws, but his writings fail to provide a proof that is convincing by today's standards.[1]

Later in this section we will derive Kepler's laws from the gravitational law by an elegant method that exploits vector differentiation to the fullest. First, however, we need to attend to some preliminaries.

Ellipses in Polar Coordinates

The polar coordinates $[r, \theta]$ of a point in the plane whose distance r from the origin is ε times its distance $p - r \cos \theta$ from the line $x = p$ (see Figure 11.22) satisfy the equation $r = \varepsilon(p - r \cos \theta)$, or, solving for r,

$$r = \frac{\ell}{1 + \varepsilon \cos \theta},$$

where $\ell = \varepsilon p$. As observed in Sections 8.1 and 8.5, for $0 \le \varepsilon < 1$ this equation represents an ellipse having **eccentricity** ε. (It is a circle if $\varepsilon = 0$.) To see this, let us transform the equation to Cartesian coordinates:

$$x^2 + y^2 = r^2 = \varepsilon^2(p - r \cos \theta)^2 = \varepsilon^2(p - x)^2 = \varepsilon^2(p^2 - 2px + x^2).$$

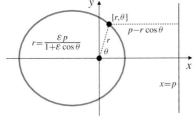

Figure 11.22 An ellipse with focus at the origin, directrix $x = p$, and eccentricity ε

---

[1] There are interesting articles debating the historical significance of Newton's work by Robert Weinstock, Curtis Wilson, and others in *The College Mathematics Journal*, vol. 25, No. 3, 1994.

With some algebraic manipulation, this equation can be juggled into the form

$$\frac{\left(x + \dfrac{\varepsilon\ell}{1 - \varepsilon^2}\right)^2}{\left(\dfrac{\ell}{1 - \varepsilon^2}\right)^2} + \frac{y^2}{\left(\dfrac{\ell}{\sqrt{1 - \varepsilon^2}}\right)^2} = 1,$$

which can be recognized as an ellipse with **centre** at the point $C = (-c, 0)$, where $c = \varepsilon\ell/(1 - \varepsilon^2)$, and semi-axes a and b given by

$$a = \frac{\ell}{1 - \varepsilon^2} \qquad \textbf{(semi-major axis)},$$

$$b = \frac{\ell}{\sqrt{1 - \varepsilon^2}} \qquad \textbf{(semi-minor axis)}.$$

The Cartesian equation of the ellipse shows that the curve is symmetric about the lines $x = -c$ and $y = 0$ and so has a second focus at $F = (-2c, 0)$ and a second directrix with equation $x = -2c - p$. (See Figure 11.23.) The ends of the major axis are $A = (a - c, 0)$ and $A' = (-a - c, 0)$, and the ends of the minor axis are $B = (-c, b)$ and $B' = (-c, -b)$.

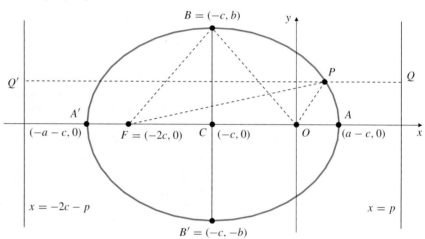

Figure 11.23 The sum of the distances from any point P on the ellipse to the two foci O and F is constant, ε times the distance between the directrices

If P is any point on the ellipse, then the distance OP is ε times the distance PQ from P to the right directrix. Similarly, the distance FP is ε times the distance $Q'P$ from P to the left directrix. Thus, the sum of the focal radii $OP + FP$ is the constant $\varepsilon Q'Q = \varepsilon(2c + 2p)$, regardless of where P is on the ellipse. Letting P be A or B we get for this sum

$$2a = (a - c) + (a + c) = OA + FA = OB + FB = 2\sqrt{b^2 + c^2}.$$

It follows that

$$a^2 = b^2 + c^2, \qquad c = \sqrt{a^2 - b^2} = \frac{\ell\varepsilon}{1 - \varepsilon^2} = \varepsilon a.$$

The number ℓ is called the **semi-latus rectum** of the ellipse; the latus rectum is the width measured along the line through a focus, perpendicular to the major axis. (See Figure 11.24.)

Remark The polar equation $r = \ell/(1 + \varepsilon\cos\theta)$ represents a *bounded* curve only if $\varepsilon < 1$; in this case we have $\ell/(1 + \varepsilon) \le r \le \ell/(1 - \varepsilon)$ for all directions θ. If $\varepsilon = 1$, the equation represents a parabola, and if $\varepsilon > 1$, a hyperbola. It is possible for objects to travel on parabolic or hyperbolic orbits, but they will approach the sun only once, rather than continue to loop around it. Some comets have hyperbolic orbits.

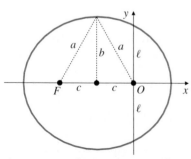

Figure 11.24 Some parameters of an ellipse

Polar Components of Velocity and Acceleration

Let $\mathbf{r}(t)$ be the position vector at time t of a particle P moving in the xy-plane. We construct two unit vectors at P, the vector $\hat{\mathbf{r}}$ points in the direction of the position vector $\mathbf{r}$, and the vector $\hat{\boldsymbol{\theta}}$ is rotated $90°$ counterclockwise from $\hat{\mathbf{r}}$. (See Figure 11.25.) If P has polar coordinates $[r, \theta]$, then $\hat{\mathbf{r}}$ points in the direction of increasing r at P, and $\hat{\boldsymbol{\theta}}$ points in the direction of increasing θ. Evidently

$$\hat{\mathbf{r}} = \cos\theta\,\mathbf{i} + \sin\theta\,\mathbf{j}$$
$$\hat{\boldsymbol{\theta}} = -\sin\theta\,\mathbf{i} + \cos\theta\,\mathbf{j}.$$

Note that $\hat{\mathbf{r}}$ and $\hat{\boldsymbol{\theta}}$ do not depend on r but only on θ:

$$\frac{d\hat{\mathbf{r}}}{d\theta} = \hat{\boldsymbol{\theta}} \qquad \text{and} \qquad \frac{d\hat{\boldsymbol{\theta}}}{d\theta} = -\hat{\mathbf{r}}.$$

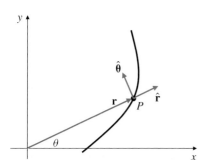

Figure 11.25 Basis vectors in the direction of increasing r and θ

The pair $\{\hat{\mathbf{r}}, \hat{\boldsymbol{\theta}}\}$ forms a reference frame (a basis) at P so that vectors in the plane can be expressed in terms of these two unit vectors. The $\hat{\mathbf{r}}$ component of a vector is called the **radial component**, and the $\hat{\boldsymbol{\theta}}$ component is called the **transverse component**. The frame varies from point to point, so we must remember that $\hat{\mathbf{r}}$ and $\hat{\boldsymbol{\theta}}$ are both functions of t. In terms of this moving frame, the position $\mathbf{r}(t)$ of P can be expressed very simply:

$$\mathbf{r} = r\hat{\mathbf{r}},$$

where $r = r(t) = |\mathbf{r}(t)|$ is the distance from P to the origin at time t.

We are going to differentiate this equation with respect to t in order to express the velocity and acceleration of P in terms of the moving frame. Along the path of motion, $\mathbf{r}$ can be regarded as a function of either θ or t; θ is itself a function of t. To avoid confusion, let us adopt a notation that is used extensively in mechanics and that resembles the notation originally used by Newton in his calculus.

> A dot over a quantity denotes the time derivative of that quantity. Two dots denote the second derivative with respect to time. Thus
>
> $$\dot{u} = du/dt \qquad \text{and} \qquad \ddot{u} = d^2u/dt^2.$$

First, let us record the time derivatives of the vectors $\hat{\mathbf{r}}$ and $\hat{\boldsymbol{\theta}}$. By the Chain Rule, we have

$$\dot{\hat{\mathbf{r}}} = \frac{d\hat{\mathbf{r}}}{d\theta}\frac{d\theta}{dt} = \dot{\theta}\hat{\boldsymbol{\theta}},$$

$$\dot{\hat{\boldsymbol{\theta}}} = \frac{d\hat{\boldsymbol{\theta}}}{d\theta}\frac{d\theta}{dt} = -\dot{\theta}\hat{\mathbf{r}}.$$

Now the velocity of P is

$$\mathbf{v} = \dot{\mathbf{r}} = \frac{d}{dt}(r\hat{\mathbf{r}}) = \dot{r}\hat{\mathbf{r}} + r\dot{\theta}\hat{\boldsymbol{\theta}}.$$

> **Polar components of velocity:**
>
> The **radial component of velocity** is $\dot{r}$.
> The **transverse component of velocity** is $r\dot{\theta}$.

Since $\hat{\mathbf{r}}$ and $\hat{\boldsymbol{\theta}}$ are perpendicular unit vectors, the speed of P is given by

$$v = |\mathbf{v}| = \sqrt{\dot{r}^2 + r^2\dot{\theta}^2}.$$

Similarly, the acceleration of P can be expressed in terms of radial and transverse components:

$$\mathbf{a} = \dot{\mathbf{v}} = \ddot{\mathbf{r}} = \frac{d}{dt}(\dot{r}\hat{\mathbf{r}} + r\dot{\theta}\hat{\boldsymbol{\theta}})$$

$$= \ddot{r}\hat{\mathbf{r}} + \dot{r}\dot{\theta}\hat{\boldsymbol{\theta}} + \dot{r}\dot{\theta}\hat{\boldsymbol{\theta}} + r\ddot{\theta}\hat{\boldsymbol{\theta}} - r\dot{\theta}^2\hat{\mathbf{r}}$$

$$= (\ddot{r} - r\dot{\theta}^2)\hat{\mathbf{r}} + (r\ddot{\theta} + 2\dot{r}\dot{\theta})\hat{\boldsymbol{\theta}}.$$

> **Polar components of acceleration:**
>
> The **radial component of acceleration** is $\ddot{r} - r\dot{\theta}^2$.
> The **transverse component of acceleration** is $r\ddot{\theta} + 2\dot{r}\dot{\theta}$.

Central Forces and Kepler's Second Law

Polar coordinates are most appropriate for analyzing motion due to a **central force** that is always directed toward (or away from) a single point, the origin: $\mathbf{F} = \lambda(\mathbf{r})\mathbf{r}$, where the scalar $\lambda(\mathbf{r})$ depends on the position $\mathbf{r}$ of the object. If the velocity and acceleration of the object are $\mathbf{v} = \dot{\mathbf{r}}$ and $\mathbf{a} = \dot{\mathbf{v}}$, then Newton's Second Law of Motion ($\mathbf{F} = m\mathbf{a}$) says that $\mathbf{a}$ is parallel to $\mathbf{r}$. Therefore,

$$\frac{d}{dt}(\mathbf{r} \times \mathbf{v}) = \dot{\mathbf{r}} \times \mathbf{v} + \mathbf{r} \times \dot{\mathbf{v}} = \mathbf{v} \times \mathbf{v} + \mathbf{r} \times \mathbf{a} = \mathbf{0} + \mathbf{0} = \mathbf{0},$$

and $\mathbf{r} \times \mathbf{v} = \mathbf{h}$, a constant vector representing the object's angular momentum per unit mass about the origin. This says that $\mathbf{r}$ is always perpendicular to $\mathbf{h}$, so motion due to a central force always takes place in a *plane* through the origin having normal $\mathbf{h}$.

If we choose the z-axis to be in the direction of $\mathbf{h}$ and let $|\mathbf{h}| = h$, then $\mathbf{h} = h\mathbf{k}$, and the path of the object is in the xy-plane. In this case the position and velocity of the object satisfy

$$\mathbf{r} = r\hat{\mathbf{r}} \qquad \text{and} \qquad \mathbf{v} = \dot{r}\hat{\mathbf{r}} + r\dot{\theta}\hat{\boldsymbol{\theta}}.$$

Since $\hat{\mathbf{r}} \times \hat{\boldsymbol{\theta}} = \mathbf{k}$, we have

$$h\mathbf{k} = \mathbf{r} \times \mathbf{v} = r\dot{r}\hat{\mathbf{r}} \times \hat{\mathbf{r}} + r^2\dot{\theta}\hat{\mathbf{r}} \times \hat{\boldsymbol{\theta}} = r^2\dot{\theta}\mathbf{k}.$$

Hence, for any motion under a central force,

> $$r^2\dot{\theta} = h \qquad \text{(a constant for the path of motion)}.$$

This formula is equivalent to Kepler's Second Law; if $A(t)$ is the area in the plane of motion bounded by the orbit and radial lines $\theta = \theta_0$ and $\theta = \theta(t)$, then

$$A(t) = \frac{1}{2}\int_{\theta_0}^{\theta(t)} r^2\, d\theta,$$

so that

$$\frac{dA}{dt} = \frac{dA}{d\theta}\frac{d\theta}{dt} = \frac{1}{2}r^2\dot{\theta} = \frac{h}{2}.$$

Thus, area is being swept out at the constant rate $h/2$, and equal areas are swept out in equal times. Note that this law does not depend on the magnitude or direction of the force on the moving object other than the fact that it is *central*. You can also derive the equation $r^2\dot{\theta} = h$ (constant) directly from the fact that the transverse acceleration is zero:

$$\frac{d}{dt}(r^2\dot{\theta}) = 2r\dot{r}\dot{\theta} + r^2\ddot{\theta} = r(2\dot{r}\dot{\theta} + r\ddot{\theta}) = 0.$$

Example 1 An object moves along the polar curve $r = 1/\theta$ under the influence of a force attracting it toward the origin. If the speed of the object is v_0 at the instant when $\theta = 1$, find the magnitude of the acceleration of the object at any point on its path as a function of its distance r from the origin.

Solution Since the force is central, we know that the transverse acceleration is zero and that $r^2\dot\theta = h$ is constant. Differentiating the equation of the path with respect to time and expressing the result in terms of r, we obtain

$$\dot r = -\frac{1}{\theta^2}\dot\theta = -r^2\frac{h}{r^2} = -h.$$

Hence, the radial component of acceleration is

$$a_r = \ddot r - r(\dot\theta)^2 = 0 - r\frac{h^2}{r^4} = -\frac{h^2}{r^3}.$$

At $\theta = 1$ we have $r = 1$, so $\dot\theta = h$. At that instant the square of the speed is

$$v_0^2 = \dot r^2 + r^2\dot\theta^2 = h^2 + h^2 = 2h^2.$$

Hence, $h^2 = v_0^2/2$, and, at any point of its path, the magnitude of the acceleration of the object is

$$|a_r| = \frac{v_0^2}{2r^3}.$$

■

Derivation of Kepler's First and Third Laws

The planets and the sun move around their common centre of mass. Since the sun is vastly more massive than the planets, that centre of mass is quite close to the centre of the sun. For example, the joint centre of mass of the sun and the earth lies inside the sun. For the following derivation we will take the sun and a planet as *point masses* and consider the sun to be fixed at the origin. We will specify the directions of the coordinate axes later, when the need arises.

According to Newton's law of gravitation, the force that the sun exerts on a planet of mass m whose position vector is **r** is

$$\mathbf{F} = -\frac{km}{r^2}\,\hat{\mathbf{r}} = -\frac{km}{r^3}\mathbf{r},$$

where k is a positive constant depending on the mass of the sun, and $\hat{\mathbf{r}} = \mathbf{r}/r$.

As observed above, the fact that the force on the planet is always directed toward the origin implies that $\mathbf{r} \times \mathbf{v}$ is constant. We choose the direction of the z-axis so that $\mathbf{r} \times \mathbf{v} = h\mathbf{k}$, so the motion will be in the xy-plane and $r^2\dot\theta = h$. We have not yet specified the directions of the x- and y-axes but will do so shortly. Using polar coordinates in the xy-plane, we calculate

$$\frac{d\mathbf{v}}{d\theta} = \frac{\dot{\mathbf{v}}}{\dot\theta} = \frac{-\dfrac{k}{r^2}\hat{\mathbf{r}}}{\dfrac{h}{r^2}} = -\frac{k}{h}\hat{\mathbf{r}}.$$

Since $d\hat{\boldsymbol{\theta}}/d\theta = -\hat{\mathbf{r}}$, we can integrate the differential equation above to find **v**:

$$\mathbf{v} = -\frac{k}{h}\int \hat{\mathbf{r}}\,d\theta = \frac{k}{h}\hat{\boldsymbol{\theta}} + \mathbf{C},$$

where **C** is a vector constant of integration. Therefore, we have shown that

$$|\mathbf{v} - \mathbf{C}| = \frac{k}{h}.$$

This result, known as **Hamilton's Theorem**, says that as a planet moves around its orbit, its velocity vector (when positioned with its tail at the origin) traces out a *circle* with centre at point *C* having position vector **C**. It is perhaps surprising that there is a circle associated with the orbit of a planet after all. Only it is not the *position* vector that moves on a circle but the *velocity* vector. (See Figure 11.26.)

Recall that so far we have specified only the position of the origin and the direction of the *z*-axis. Therefore, the *xy*-plane is determined but not the directions of the *x*-axis or the *y*-axis. Let us choose these axes in the *xy*-plane so that **C** is in the direction of the *y*-axis; say $\mathbf{C} = (\varepsilon k / h)\mathbf{j}$, where ε is a positive constant. We therefore have

$$\mathbf{v} = \frac{k}{h}(\hat{\boldsymbol{\theta}} + \varepsilon\mathbf{j}).$$

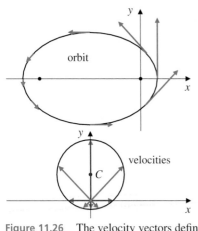

Figure 11.26 The velocity vectors define a circle

The position of the *x*-axis is now determined by the fact that the three vectors **i**, **j**, and **k** are mutually perpendicular and form a right-handed basis. We calculate **r** × **v** again. Remember that $\mathbf{r} = r\cos\theta\mathbf{i} + r\sin\theta\mathbf{j}$, and also $\mathbf{r} = r\hat{\mathbf{r}}$:

$$h\mathbf{k} = \mathbf{r} \times \mathbf{v} = \frac{k}{h}(r\hat{\mathbf{r}} \times \hat{\boldsymbol{\theta}} + r\varepsilon\cos\theta\mathbf{i} \times \mathbf{j} + r\varepsilon\sin\theta\mathbf{j} \times \mathbf{j})$$

$$= \frac{k}{h}r(1 + \varepsilon\cos\theta)\mathbf{k}.$$

Thus $h = \dfrac{kr}{h}(1 + \varepsilon\cos\theta)$, or, solving for r,

$$r = \frac{h^2/k}{1 + \varepsilon\cos\theta}.$$

This is the polar equation of the orbit. If $\varepsilon < 1$, it is an ellipse with one focus at the origin (the sun) and with parameters given by

Semi-latus rectum:	$\ell = \dfrac{h^2}{k}$
Semi-major axis:	$a = \dfrac{h^2}{k(1 - \varepsilon^2)} = \dfrac{\ell}{1 - \varepsilon^2}$
Semi-minor axis:	$b = \dfrac{h^2}{k\sqrt{1 - \varepsilon^2}} = \dfrac{\ell}{\sqrt{1 - \varepsilon^2}}$
Semi-focal separation:	$c = \sqrt{a^2 - b^2} = \dfrac{\varepsilon\ell}{1 - \varepsilon^2}.$

We have deduced Kepler's First Law! The choices we made for the coordinate axes result in **perihelion** (the point on the orbit that is closest to the sun) being on the positive *x*-axis ($\theta = 0$).

Example 2 A planet's orbit has eccentricity ε (where $0 < \varepsilon < 1$), and its speed at perihelion is v_P. Find its speed v_A at **aphelion** (the point on its orbit farthest from the sun).

Solution At perihelion and aphelion the planet's radial velocity $\dot{r}$ is zero (since r is minimum or maximum), so the velocity is entirely transverse. Thus $v_P = r_P\dot{\theta}_P$ and $v_A = r_A\dot{\theta}_A$. Since $r^2\dot{\theta} = h$ has the same value at all points of the orbit, we have

$$r_P v_P = r_P^2\dot{\theta}_P = h = r_A^2\dot{\theta}_A = r_A v_A.$$

The planet's orbit has equation

$$r = \frac{\ell}{1 + \varepsilon \cos \theta},$$

so perihelion corresponds to $\theta = 0$ and aphelion to $\theta = \pi$:

$$r_P = \frac{\ell}{1 + \varepsilon} \qquad \text{and} \qquad r_A = \frac{\ell}{1 - \varepsilon}.$$

Therefore, $v_A = \dfrac{r_P}{r_A} v_P = \dfrac{1 - \varepsilon}{1 + \varepsilon} v_P.$

We can obtain Kepler's Third Law from the other two as follows. Since the radial line from the sun to a planet sweeps out area at a constant rate $h/2$, the total area A enclosed by the orbit is $A = (h/2)T$, where T is the period of revolution. The area of an ellipse with semi-axes a and b is $A = \pi a b$. Since $b^2 = \ell a = h^2 a/k$, we have

$$T^2 = \frac{4}{h^2} A^2 = \frac{4}{h^2} \pi^2 a^2 b^2 = \frac{4\pi^2}{k} a^3.$$

Note how the final expression for T^2 does not depend on h, which is a constant for the orbit of any one planet, but varies from planet to planet. The constant $4\pi^2/k$ does not depend on the particular planet. (k depends on the mass of the sun and a universal gravitational constant.) Thus,

$$T^2 = \frac{4\pi^2}{k} a^3$$

says that the square of the period of a planet is proportional to the cube of the length, $2a$, of the major axis of its orbit, the proportionality extending over all the planets. This is Kepler's Third Law. Modern astronomical data show that T^2/a^3 varies by only about three-tenths of one percent over the nine known planets.

Conservation of Energy

Solving the second-order differential equation of motion $\mathbf{F} = m\ddot{\mathbf{r}}$ to find the orbit of a planet requires two integrations. In the above derivation we exploited properties of the cross product to make these integrations easy. More traditional derivations of Kepler's laws usually begin with separating the radial and transverse components in the equation of motion:

$$\ddot{r} - r\dot{\theta}^2 = -\frac{k}{r^2}, \qquad r\ddot{\theta} + 2\dot{r}\dot{\theta} = 0.$$

As observed earlier, the second equation above implies that $r^2\dot{\theta} = h = \text{constant}$, which is Kepler's Second Law. This can be used to eliminate θ from the first equation to give

$$\ddot{r} - \frac{h^2}{r^3} = -\frac{k}{r^2}.$$

Therefore,

$$\frac{d}{dt}\left(\frac{\dot{r}^2}{2} + \frac{h^2}{2r^2}\right) = \dot{r}\left(\ddot{r} - \frac{h^2}{r^3}\right) = -\frac{k}{r^2}\dot{r}.$$

If we integrate this equation, we obtain

$$\frac{1}{2}\left(\dot{r}^2 + \frac{h^2}{r^2}\right) - \frac{k}{r} = E.$$

This is a **conservation of energy** law. The first term on the left is $v^2/2$, the kinetic energy (per unit mass) of the planet. The term $-k/r$ is the potential energy per unit mass. It is difficult to integrate this equation and to find r as a function of t. In any event, we really want r as a function of θ so that we can recognize that we have an ellipse. Another way to obtain this is suggested in Exercise 18 below.

Remark The procedure used above to demonstrate Kepler's laws in fact shows that if any object moves under the influence of a force that attracts it toward the origin (or repels it away from the origin) and has magnitude proportional to the reciprocal of the square of distance from the origin, then the object must move in a plane orbit whose shape is a conic section. If the total energy E defined above is negative, then the orbit is *bounded* and must therefore be an ellipse. If $E = 0$, the orbit is a parabola. If $E > 0$, the orbit is a hyperbola. Hyperbolic orbits are typical for repulsive forces but may also occur for attractions if the object has high enough velocity (exceeding the *escape velocity*). See Exercise 22 below for an example.

Exercises 11.6

1. **(Polar ellipses)** Fill in the details of the calculation suggested in the text to transform the polar equation of an ellipse, $r = \ell/(1 + \varepsilon \cos\theta)$, where $0 < \varepsilon < 1$, to Cartesian coordinates in a form showing the centre and semi-axes explicitly.

Polar components of velocity and acceleration

2. A particle moves on the circle with polar equation $r = k$, $(k > 0)$. What are the radial and transverse components of its velocity and acceleration? Show that the transverse component of the acceleration is equal to the rate of change of the speed of the particle.

3. Find the radial and transverse components of velocity and acceleration of a particle moving at unit speed along the exponential spiral $r = e^\theta$. Express your answers in terms of the angle θ.

4. If a particle moves along the polar curve $r = \theta$ under the influence of a central force attracting it to the origin, find the magnitude of the acceleration as a function of r and the speed of the particle.

5. An object moves along the polar curve $r = \theta^{-2}$ under the influence of a force attracting it toward the origin. If the speed of the object is v_0 at the instant when $\theta = 1$, find the magnitude of the acceleration of the object at any point on its path as a function of its distance r from the origin.

Deductions from Kepler's laws

6. The mean distance from the earth to the sun is approximately 150 million km. Comet Halley approaches perihelion (comes closest to the sun) in its elliptical orbit approximately every 76 years. Estimate the major axis of the orbit of Comet Halley.

7. The mean distance from the moon to the earth is about 385,000 km, and its period of revolution around the earth is about 27 days (the sidereal month). At approximately what distance from the centre of the earth, and in what plane, should a communications satellite be inserted into circular orbit if it must remain directly above the same position on the earth at all times?

8. An asteroid is in a circular orbit around the sun. If its period of revolution is T, find the radius of its orbit.

9. If the asteroid in Exercise 8 is instantaneously stopped in its orbit, it will fall toward the sun. How long will it take to get there? *Hint:* You can do this question easily if instead you regard the asteroid as *almost* stopped, so that it goes into a highly eccentric elliptical orbit whose major axis is a bit greater than the radius of the original circular orbit.

10. Find the eccentricity of an asteroid's orbit if the asteroid's speed at perihelion is twice its speed at aphelion.

11. Show that the orbital speed of a planet is constant if and only if the orbit is circular. *Hint:* Use the conservation of energy identity.

12. A planet's distance from the sun at perihelion is 80% of its distance at aphelion. Find the ratio of its speeds at perihelion and aphelion and the eccentricity of its orbit.

*13. As a result of a collision, an asteroid originally in a circular orbit about the sun suddenly has its velocity cut in half, so that it falls into an elliptical orbit with maximum distance from the sun equal to the radius of the original circular orbit. Find the eccentricity of its new orbit.

14. If the speeds of a planet at perihelion and aphelion are v_P and v_A, respectively, what is its speed when it is at the ends of the minor axis of its orbit?

15. What fraction of its "year" (i.e., the period of its orbit) does a planet spend traversing the half of its orbit that is closest to the sun? Give your answer in terms of the eccentricity ε of the planet's orbit.

*16. Suppose that a planet is travelling at speed v_0 at an instant when it is at distance r_0 from the sun. Show that the period of the planet's orbit is

$$T = \frac{2\pi}{\sqrt{k}} \left(\frac{2}{r_0} - \frac{v_0^2}{k} \right)^{-3/2}.$$

Hint: The quantity $\dfrac{k}{r} - \dfrac{1}{2}v^2$ is constant at all points of the orbit, as shown in the discussion of conservation of energy. Find the value of this expression at perihelion in terms of the semi-major axis, a.

*17. The sum of the distances from a point P on an ellipse $\mathcal{E}$ to the foci of $\mathcal{E}$ is the constant $2a$, the length of the major axis of the ellipse. Use this fact in a *geometric* argument to show that the mean distance from points P to one focus of $\mathcal{E}$ is a. That is, show that

$$\frac{1}{c(\mathcal{E})} \int_{\mathcal{E}} r \, ds = a,$$

where $c(\mathcal{E})$ is the circumference of $\mathcal{E}$, and r is the distance from a point on $\mathcal{E}$ to one focus.

*18. **(A direct approach to Kepler's First Law)** The result
of eliminating θ between the equations for the radial and
transverse components of acceleration for a planet is

$$\ddot{r} - \frac{h^2}{r^3} = -\frac{k}{r^2}.$$

Show that the change of dependent and independent
variables:

$$r(t) = \frac{1}{u(\theta)}, \qquad \theta = \theta(t),$$

transforms this equation to the simpler equation

$$\frac{d^2u}{d\theta^2} + u = \frac{k}{h^2}.$$

Show that the solution of this equation is

$$u = \frac{k}{h^2}\Big(1 + \varepsilon \cos(\theta - \theta_0)\Big),$$

where ε and θ_0 are constants. Hence, show that the orbit is
elliptical if $|\varepsilon| < 1$.

*19. **(What if gravitation were an inverse cube law?)** Use
the technique of Exercise 18 to find the trajectory of an
object of unit mass attracted to the origin by a force of
magnitude $f(r) = k/r^3$. Are there any orbits that do not
approach infinity or the origin as $t \to \infty$?

20. Use the conservation of energy formula to show that if
$E < 0$ the orbit must be bounded; that is, it cannot get
arbitrarily far away from the origin.

*21. **(Polar hyperbolas)** If $\varepsilon > 1$, then the equation

$$r = \frac{\ell}{1 + \varepsilon \cos \theta}$$

represents a hyperbola rather than an ellipse. Sketch the
hyperbola, find its centre and the directions of its asymptotes,
and determine its semi-transverse axis, its semi-conjugate
axis, and semi-focal separation in terms of ℓ and ε.

*22. **(Hyperbolic orbits)** A meteor travels from infinity on a
hyperbolic orbit passing near the sun. At a very large
distance from the sun it has speed v_∞. The asymptotes of its
orbit pass at perpendicular distance D from the sun. (See
Figure 11.27.) Show that the angle δ through which the
meteor's path is deflected by the gravitational attraction of
the sun is given by

$$\cot\left(\frac{\delta}{2}\right) = \frac{Dv_\infty^2}{k}.$$

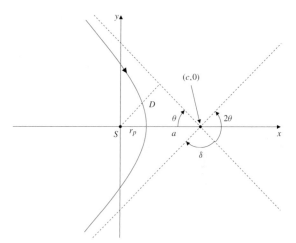

Figure 11.27   Path of a meteor

(*Hint:* You will need the result of Exercise 21.) The same
analysis and results hold for electrostatic attraction or
repulsion; $f(r) = \pm k/r^2$ in that case also. The constant k
depends on the charges of two particles, and r is the distance
between them.

Chapter Review

Key Ideas

- **What is a vector function of a real variable, and why
does it represent a curve?**

- **State the Product Rule for the derivative of
$u(t) \bullet \big(v(t) \times w(t)\big)$.**

- **What do the following terms mean?**
  - angular velocity
  - angular momentum
  - centripetal acceleration
  - Coriolis acceleration
  - arc-length parametrization
  - central force

- **Find the following quantities associated with a
parametric curve $\mathcal{C}$ with parametrization
$r = r(t)$, $(a \le t \le b)$.**
  - the velocity $v(t)$
  - the speed $v(t)$
  - the arc length
  - the acceleration $a(t)$
  - the unit tangent $\hat{T}(t)$
  - the unit normal $\hat{N}(t)$
  - the curvature $\kappa(t)$
  - the radius of curvature $\rho(t)$
  - the osculating plane
  - the osculating circle
  - the unit binormal $\hat{B}(t)$
  - the torsion $\tau(t)$
  - the tangential acceleration
  - the normal acceleration
  - the evolute

- **State the Frenet–Serret formulas.**

- **State Kepler's laws of planetary motion.**

- **What are the radial and transverse components of velocity and acceleration?**

Review Exercises

1. If $\mathbf{r}(t)$, $\mathbf{v}(t)$, and $\mathbf{a}(t)$ represent the position, velocity, and acceleration at time t of a particle moving in 3-space, and if, at every time t, the $\mathbf{a}$ is perpendicular to both $\mathbf{r}$ and $\mathbf{v}$, show that the vector $\mathbf{r}(t) - t\mathbf{v}(t)$ has constant length.

2. Describe the parametric curve

$$\mathbf{r} = t\cos t\,\mathbf{i} + t\sin t\,\mathbf{j} + (2\pi - t)\mathbf{k},$$

$(0 \le t \le 2\pi)$, and find its length.

3. A particle moves along the curve of intersection of the surfaces $y = x^2$ and $z = 2x^3/3$ with constant speed $v = 6$. It is moving in the direction of increasing x. Find its velocity and acceleration when it is at the point $(1, 1, 2/3)$.

4. A particle moves along the curve $y = x^2$ in the xy-plane so that at time t its speed is $v = t$. Find its acceleration at time $t = 3$ if it is at the point $(\sqrt{2}, 2)$ at that time.

5. Find the curvature and torsion at a general point of the curve $\mathbf{r} = e^t\mathbf{i} + \sqrt{2}t\mathbf{j} + e^{-t}\mathbf{k}$.

6. A particle moves on the curve of Exercise 5 so that it is at position $\mathbf{r}(t)$ at time t. Find its normal acceleration and tangential acceleration at any time t. What is its minimum speed?

7. **(A clothoid curve)** The plane curve $\mathcal{C}$ in Figure 11.28 has parametric equations

$$x(s) = \int_0^s \cos\frac{kt^2}{2}\,dt \quad \text{and} \quad y(s) = \int_0^s \sin\frac{kt^2}{2}\,dt.$$

Verify that s is, in fact, the arc length along $\mathcal{C}$ measured from $(0, 0)$ and that the curvature of $\mathcal{C}$ is given by $\kappa(s) = ks$. Because the curvature changes linearly with distance along the curve, such curves, called *clothoids*, are useful for joining track sections of different curvatures.

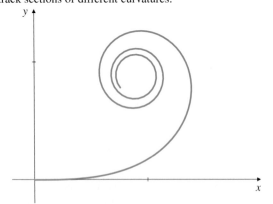

Figure 11.28 A clothoid curve

8. A particle moves along the polar curve $r = e^{-\theta}$ with constant angular speed $\dot\theta = k$. Express its velocity and acceleration in terms of radial and transverse components depending only on the distance r from the origin.

Some properties of cycloids

Exercises 9–12 all deal with the cycloid

$$\mathbf{r} = a(t - \sin t)\mathbf{i} + a(1 - \cos t)\mathbf{j}.$$

Recall that this curve is the path of a point on the circumference of a circle of radius a rolling along the x-axis.

9. Find the arc length $s = s(T)$ of the part of the cycloid from $t = 0$ to $t = T \le 2\pi$.

10. Find the arc-length parametrization $\mathbf{r} = \mathbf{r}(s)$ of the arch $0 \le t \le 2\pi$ of the cycloid, with s measured from the point $(0, 0)$.

11. Find the *evolute* of the cycloid, that is, find parametric equations of the centre of curvature $\mathbf{r} = \mathbf{r}_c(t)$ of the cycloid. Show that the evolute is the same cycloid translated πa units to the right and $2a$ units downward.

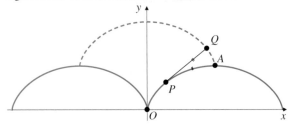

Figure 11.29

12. A string of length $4a$ has one end fixed at the origin and is wound along the arch of the cycloid to the right of the origin. Since that arch has total length $8a$, the free end of the string lies at the highest point A of the arch. Find the path followed by the free end Q of the string as it is unwound from the cycloid and is held taught during the unwinding. (See Figure 11.29.) If the string leaves the cycloid at P, then

$$(\text{arc } OP) + PQ = 4a.$$

The path of Q is called the *involute* of the cycloid. Show that, like the evolute, the involute is also a translate of the original cycloid. In fact, the cycloid is the evolute of its involute.

13. Let P be a point in 3-space with spherical coordinates (ρ, ϕ, θ). Suppose that P is not on the z-axis. Find a triad of mutually perpendicular unit vectors, $\{\hat{\rho}, \hat{\phi}, \hat{\theta}\}$, at P in the directions of increasing ρ, ϕ, and θ, respectively. Is the triad right- or left-handed?

Kepler's laws imply Newton's law of gravitation

In Exercises 14–16, it is *assumed* that a planet of mass m moves in an elliptical orbit $r = \ell/(1 + \varepsilon\cos\theta)$, with focus at the origin (the sun), under the influence of a force $\mathbf{F} = \mathbf{F}(\mathbf{r})$ that depends only on the position of the planet.

14. Assuming Kepler's Second Law, show that $\mathbf{r} \times \mathbf{v} = \mathbf{h}$ is constant and, hence, that $r^2\dot\theta = h$ is constant.

15. Use Newton's Second Law of Motion ($\mathbf{F} = m\ddot{\mathbf{r}}$) to show that $\mathbf{r} \times \mathbf{F}(\mathbf{r}) = \mathbf{0}$. Therefore $\mathbf{F}(\mathbf{r})$ is parallel to $\mathbf{r}$: $\mathbf{F}(\mathbf{r}) = -f(\mathbf{r})\,\hat{\mathbf{r}}$, for some scalar-valued function $f(\mathbf{r})$, and the transverse component of $\mathbf{F}(\mathbf{r})$ is zero.

16. By direct calculation of the radial acceleration of the planet, show that $f(\mathbf{r}) = mh^2/(\ell r^2)$, where $r = |\mathbf{r}|$. Thus, $\mathbf{F}$ is an attraction to the origin, proportional to the mass of the planet, and inversely proportional to the square of its distance from the sun.

Challenging Problems

1. Let P be a point on the surface of the earth at $45°$ north latitude. Use a coordinate system with origin at P and basis vectors $\mathbf{i}$ and $\mathbf{j}$ pointing east and north, respectively, so that $\mathbf{k}$ points vertically upward.

(a) Express the angular velocity $\boldsymbol{\Omega}$ of the earth in terms of the basis vectors at P. What is the magnitude Ω of $\boldsymbol{\Omega}$ in radians per second?

(b) Find the Coriolis acceleration $\mathbf{a}_C = 2\boldsymbol{\Omega} \times \mathbf{v}$ of an object falling vertically with speed v above P.

(c) If the object in (b) drops from rest from a height of 100 m above P, approximately where will it strike the ground? Ignore air resistance but not the Coriolis acceleration. Since the Coriolis acceleration is much smaller than the gravitational acceleration in magnitude, you can use the vertical velocity as a good approximation to the actual velocity of the object at any time during its fall.

2. (The spin of a baseball) When a ball is thrown with spin about an axis that is not parallel to its velocity, it experiences a lateral acceleration due to differences in friction along its sides. This spin acceleration is given by $\mathbf{a}_s = k\mathbf{S} \times \mathbf{v}$, where $\mathbf{v}$ is the velocity of the ball, $\mathbf{S}$ is the angular velocity of its spin, and k is a positive constant depending on the surface of the ball. Suppose that a ball for which $k = 0.001$ is thrown horizontally along the x-axis with an initial speed of 70 ft/s and a spin of 1,000 radians/s about a vertical axis. Its velocity $\mathbf{v}$ must satisfy

$$\begin{cases} \dfrac{d\mathbf{v}}{dt} = (0.001)(1{,}000\mathbf{k}) \times \mathbf{v} - 32\mathbf{k} = \mathbf{k} \times \mathbf{v} - 32\mathbf{k} \\ \mathbf{v}(0) = 70\mathbf{i}, \end{cases}$$

since the acceleration of gravity is 32 ft/s^2.

(a) Show that the components of $\mathbf{v} = v_1\mathbf{i} + v_2\mathbf{j} + v_3\mathbf{k}$ satisfy

$$\begin{cases} \dfrac{dv_1}{dt} = -v_2 \\ v_1(0) = 70 \end{cases} \quad \begin{cases} \dfrac{dv_2}{dt} = v_1 \\ v_2(0) = 0 \end{cases} \quad \begin{cases} \dfrac{dv_3}{dt} = -32 \\ v_3(0) = 0. \end{cases}$$

(b) Solve these equations, and find the position of the ball t s after it is thrown. Assume that it is thrown from the origin at time $t = 0$.

(c) At $t = 1/5$ s, how far, and in what direction, has the ball deviated from the parabolic path it would have followed if it had been thrown without spin?

3. (Charged particles moving in magnetic fields) Magnetic fields exert forces on moving charged particles. If a particle of mass m and charge q is moving with velocity $\mathbf{v}$ in a magnetic field $\mathbf{B}$, then it experiences a force $\mathbf{F} = q\mathbf{v} \times \mathbf{B}$, and hence its velocity is governed by the equation

$$m \frac{d\mathbf{v}}{dt} = q\mathbf{v} \times \mathbf{B}.$$

For this exercise, suppose that the magnetic field is constant and vertical, say, $\mathbf{B} = B\mathbf{k}$ (as, e.g., in a cathode-ray tube). If the moving particle has initial velocity $\mathbf{v}_0$, then its velocity at time t is determined by

$$\begin{cases} \dfrac{d\mathbf{v}}{dt} = \omega\mathbf{v} \times \mathbf{k}, \quad \text{where } \omega = \dfrac{qB}{m} \\ \mathbf{v}(0) = \mathbf{v}_0. \end{cases}$$

(a) Show that $\mathbf{v} \bullet \mathbf{k} = \mathbf{v}_0 \bullet \mathbf{k}$ and $|\mathbf{v}| = |\mathbf{v}_0|$ for all t.

(b) Let $\mathbf{w}(t) = \mathbf{v}(t) - (\mathbf{v}_0 \bullet \mathbf{k})\mathbf{k}$, so that $\mathbf{w}$ is perpendicular to $\mathbf{k}$ for all t. Show that $\mathbf{w}$ satisfies

$$\begin{cases} \dfrac{d^2\mathbf{w}}{dt^2} = -\omega^2\mathbf{w} \\ \mathbf{w}(0) = \mathbf{v}_0 - (\mathbf{v}_0 \bullet \mathbf{k})\mathbf{k} \\ \mathbf{w}'(0) = \omega\mathbf{v}_0 \times \mathbf{k}. \end{cases}$$

(c) Solve the initial-value problem in (b) for $\mathbf{w}(t)$, and hence find $\mathbf{v}(t)$.

(d) Find the position vector $\mathbf{r}(t)$ of the particle at time t if it is at the origin at time $t = 0$. Verify that the path of the particle is, in general, a circular helix. Under what circumstances is the path a straight line? a circle?

4. (The tautochrone) The parametric equations

$$x = a(\theta - \sin\theta) \quad \text{and} \quad y = a(\cos\theta - 1)$$

(for $0 \le \theta \le 2\pi$), describe an arch of the cycloid followed by a point on a circle of radius a rolling along the underside of the x-axis. Suppose the curve is made of wire along which a bead can slide without friction. (See Figure 11.30.) If the bead slides from rest under gravity, starting at a point having parameter value θ_0, show that the time it takes for the bead to fall to the lowest point on the arch (corresponding to $\theta = \pi$) is a *constant*, independent of the starting position θ_0. Thus, two such beads released simultaneously from different positions along the wire will always collide at the lowest point. For this reason, the cycloid is sometimes called the *tautochrone*, from the Greek for "constant time." *Hint:* When the bead has fallen from height $y(\theta_0)$ to height $y(\theta)$, its speed is $v = \sqrt{2g\left(y(\theta_0) - y(\theta)\right)}$. (Why?) The time for the bead to fall to the bottom is

$$T = \int_{\theta=\theta_0}^{\theta=\pi} \frac{1}{v}\,ds,$$

where ds is the arc length element along the cycloid.

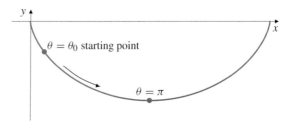

Figure 11.30

* **5. (The Drop of Doom)** An amusement park ride at the West Edmonton Mall in Alberta, Canada, gives thrill seekers a taste of free-fall. It consists of a car moving along a track consisting of straight vertical and horizontal sections joined by a smooth curve. The car drops from the top and falls vertically under gravity for $10 - 2\sqrt{2} \approx 7.2$ m before entering the curved section at B. (See Figure 11.31.) It falls another $2\sqrt{2} \approx 2.8$ m as it whips around the curve and into the horizontal section DE at ground level, where brakes are applied to stop it. (Thus, the total vertical drop from A to D or E is 10 m, a figure, like the others in this problem, chosen for mathematical convenience rather than engineering precision.) For purposes of this problem it is helpful to take the coordinate axes at a $45°$ angle to the vertical, so that the two straight sections of the track lie along the graph $y = |x|$. The curved section then goes from $(-2, 2)$ to $(2, 2)$ and can be taken to be symmetric about the y-axis. With this coordinate system, the gravitational acceleration is in the direction of $\mathbf{i} - \mathbf{j}$.

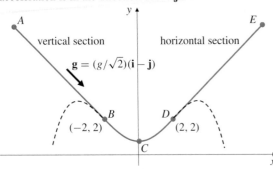

Figure 11.31

(a) Find a fourth-degree polynomial whose graph can be used to link the two straight sections of track without producing discontinuous accelerations for the falling car. (Why is fourth degree adequate?)

(b) Ignoring friction and air resistance, how fast is the car moving when it enters the curve at B? at the midpoint C of the curve? and when it leaves the curve at D?

(c) Find the magnitude of the normal acceleration and of the total acceleration of the car as it passes through C.

* **6. (A chase problem)** A fox and a hare are running in the xy-plane. Both are running at the same speed v. The hare is running up the y-axis; at time $t = 0$ it is at the origin. The fox is always running straight toward the hare. At time $t = 0$ the fox is at the point $(a, 0)$, where $a > 0$. Let the fox's position at time t be $\left(x(t), y(t)\right)$.

(a) Verify that the tangent to the fox's path at time t has slope

$$\frac{dy}{dx} = \frac{y(t) - vt}{x(t)}.$$

(b) Show that the equation of the path of the fox satisfies the equation

$$x \frac{d^2y}{dx^2} = \sqrt{1 + \left(\frac{dy}{dx}\right)^2}.$$

Hint: Differentiate the equation in (a) with respect to t. On the left side note that $(d/dt) = (dx/dt)(d/dx)$.

(c) Solve the equation in (b) by substituting $u(x) = dy/dx$ and separating variables. Note that $y = 0$ and $u = 0$ when $x = a$.

7. Suppose the earth is a perfect sphere of radius a. You set out from the point on the equator whose spherical coordinates are $(\rho, \phi, \theta) = (a, \pi/2, 0)$ and travel on the surface of the earth at constant speed v, always moving toward the northeast ($45°$ east of north).

(a) Will you ever get to the north pole? If so, how long will it take to get there?

(b) Find the functions $\phi(t)$ and $\theta(t)$ that are the angular spherical coordinates of your position at time $t > 0$.

(c) How many times does your path cross the meridian $\theta = 0$?

CHAPTER 12
Partial Differentiation

I have a very wide command of matters mathematical,
I understand equations both the simple and quadratical.
About binomial theorem I'm teeming with a lot of news,
And many cheerful facts about the square on the hypotenuse.

William Schwenck Gilbert   **1836–1911**
from *The Pirates of Penzance*

Introduction   This chapter is concerned with extending the idea of the derivative to real functions of a vector variable, that is, to functions depending on several real variables. Although differentiation is carried out one variable at a time, the relationship between derivatives with respect to different variables makes the analysis of such functions much more complicated and subtle than in the single-variable case.

12.1 Functions of Several Variables

The notation $y = f(x)$ is used to indicate that the variable y depends on the single real variable x, that is, that y is a function of x. The domain of such a function f is a set of real numbers. Many quantities can be regarded as depending on more than one real variable and thus to be functions of more than one variable. For example, the volume of a circular cylinder of radius r and height h is given by $V = \pi r^2 h$; we say that V is a function of the two variables r and h. If we choose to denote this function by f, then we would write $V = f(r, h)$ where

$$f(r, h) = \pi r^2 h, \qquad (r \geq 0, \quad h \geq 0).$$

Thus, f is a function of two variables having as *domain* the set of points in the rh-plane with coordinates (r, h) satisfying $r \geq 0$ and $h \geq 0$. Similarly, the relationship $w = f(x, y, z) = x + 2y - 3z$ defines w as a function of the three variables x, y, and z, with domain the whole of $\mathbb{R}^3$, or, if we state explicitly, some particular subset of $\mathbb{R}^3$.

By analogy with the corresponding definition for functions of one variable, we define a function of n variables as follows:

DEFINITION   **1**

A **function** f of n real variables is a rule that assigns a *unique* real number $f(x_1, x_2, \ldots, x_n)$ to each point $(x_1, x_2, \ldots, x_n)$ in some subset $\mathcal{D}(f)$ of $\mathbb{R}^n$. $\mathcal{D}(f)$ is called the **domain** of f. The set of real numbers $f(x_1, x_2, \ldots, x_n)$ obtained from points in the domain is called the **range** of f.

As for functions of one variable, the **domain convention** specifies that the domain of a function of n variables is the largest set of points $(x_1, x_2, \ldots, x_n)$ for which $f(x_1, x_2, \ldots, x_n)$ makes sense as a real number, unless that domain is explicitly stated to be a smaller set.

Most of the examples we consider hereafter will be functions of two or three independent variables. When a function f depends on two variables, we will usually call these independent variables x and y, and we will use z to denote the dependent variable that represents the value of the function; that is, $z = f(x, y)$. We will normally use x, y, and z as the independent variables of a function of three variables and w as the value of the function: $w = f(x, y, z)$. Some definitions will be given, and some theorems will be stated (and proved) only for the two-variable case, but extensions to three or more variables will usually be obvious.

Graphical Representations

The graph of a *function f* of one variable (i.e., the graph of the *equation $y = f(x)$*) is the set of points in the xy-plane having coordinates $(x, f(x))$, where x is in the domain of f. Similarly, the graph of a *function f* of two variables (i.e., the graph of the *equation $z = f(x, y)$*) is the set of points in 3-space having coordinates $(x, y, f(x, y))$, where (x, y) belongs to the domain of f. This graph is a surface in $\mathbb{R}^3$ lying above (if $f(x, y) > 0$) or below (if $f(x, y) < 0$) the domain of f in the xy-plane. (See Figure 12.1.) The graph of a function of three variables is a three-dimensional *hypersurface* in 4-space, $\mathbb{R}^4$. In general, the graph of a function of n variables is an n-dimensional *surface* in $\mathbb{R}^{n+1}$. We will not attempt to *draw* graphs of functions of more than two variables!

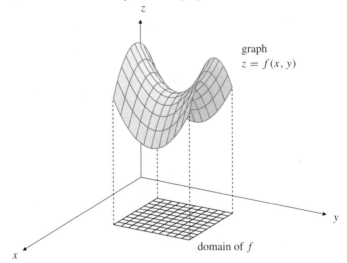

Figure 12.1   The graph of $f(x, y)$ is the surface with equation $z = f(x, y)$ defined for points (x, y) in the domain of f

Figure 12.2

Example 1   Consider the function

$$f(x, y) = 3\left(1 - \frac{x}{2} - \frac{y}{4}\right), \qquad (0 \le x \le 2, \quad 0 \le y \le 4 - 2x).$$

The graph of f is the plane triangular surface with vertices at $(2, 0, 0)$, $(0, 4, 0)$, and $(0, 0, 3)$. (See Figure 12.2.) If the domain of f had not been explicitly stated to be a particular set in the xy-plane, the graph would have been the whole plane through these three points.

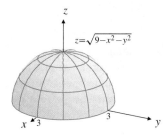

Figure 12.3

Example 2 Consider $f(x, y) = \sqrt{9 - x^2 - y^2}$. The expression under the square root cannot be negative, so the domain is the disk $x^2 + y^2 \leq 9$ in the xy-plane.

If we square the equation $z = \sqrt{9 - x^2 - y^2}$, we can rewrite the result in the form $x^2 + y^2 + z^2 = 9$. This is a sphere of radius 3 centred at the origin. However, the graph of f is only the upper hemisphere where $z \geq 0$. (See Figure 12.3.) ∎

Since it is necessary to project the surface $z = f(x, y)$ onto a two-dimensional page, most such graphs are difficult to sketch without considerable artistic talent and training. Nevertheless, you should always try to visualize such a graph and sketch it as best you can. Sometimes it is convenient to sketch only part of a graph, for instance, the part lying in the first octant. It is also helpful to determine (and sketch) the intersections of the graph with various planes, especially the coordinate planes, and planes parallel to the coordinate planes. (See Figure 12.1.)

Some mathematical software packages will produce plots of three-dimensional graphs to help you get a feeling for how the corresponding functions behave. Figure 12.1 is an example of such a computer-drawn graph, as is Figure 12.4 below. Along with most of the other mathematical graphics in this book, both were produced using the mathematical graphics software package **MG**. Later in this section we discuss how to use Maple to produce such graphs.

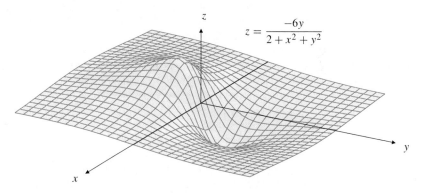

$$z = \frac{-6y}{2 + x^2 + y^2}$$

Figure 12.4 The graph of $z = \dfrac{-6y}{2 + x^2 + y^2}$

Another way to represent the function $f(x, y)$ graphically is to produce a two-dimensional *topographic map* of the surface $z = f(x, y)$. In the xy-plane we sketch the curves $f(x, y) = C$ for various values of the constant C. These curves are called **level curves** of f because they are the vertical projections onto the xy-plane of the curves in which the graph $z = f(x, y)$ intersects the horizontal (level) planes $z = C$. The graph and some level curves of the function $f(x, y) = x^2 + y^2$ are shown in Figure 12.5. The graph is a circular paraboloid in 3-space; the level curves are circles centred at the origin in the xy-plane.

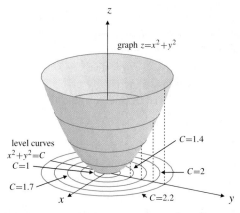

Figure 12.5 The graph of $f(x, y) = x^2 + y^2$ and some level curves of f

Figure 12.6 Level curves (contours) representing elevation in a topographic map

The contour curves in the topographic map in Figure 12.6 show the elevations, in 100 m increments above sea level, on part of Nelson Island on the British Columbia coast. Since these contours are drawn for equally spaced values of C, the spacing of the contours themselves conveys information about the relative steepness at various places on the mountains; the land is steepest where the contour lines are closest together. Observe also that the streams shown cross the contours at right angles. They take the route of steepest descent. Isotherms (curves of constant temperature) and isobars (curves of constant pressure) on weather maps are also examples of level curves.

Example 3   The level curves of the function $f(x, y) = 3\left(1 - \dfrac{x}{2} - \dfrac{y}{4}\right)$ of Example 1 are the segments of the straight lines

$$3\left(1 - \frac{x}{2} - \frac{y}{4}\right) = C \qquad \text{or} \qquad \frac{x}{2} + \frac{y}{4} = 1 - \frac{C}{3}, \qquad (0 \le C \le 3),$$

which lie in the first quadrant. Several such level curves are shown in Figure 12.7(a). They correspond to equally spaced values of C, and their equal spacing indicates the uniform steepness of the graph of f in Figure 12.2.

Figure 12.7

(a) Level curves of $3\left(1 - \dfrac{x}{2} - \dfrac{y}{4}\right)$

(b) Level curves of $\sqrt{9 - x^2 - y^2}$

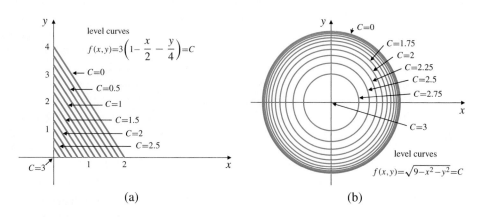

(a)

(b)

Example 4   The level curves of the function $f(x, y) = \sqrt{9 - x^2 - y^2}$ of Example 2 are the concentric circles

$$\sqrt{9 - x^2 - y^2} = C \qquad \text{or} \qquad x^2 + y^2 = 9 - C^2, \qquad (0 \le C \le 3).$$

Observe the spacing of these circles in Figure 12.7(b); they are plotted for several equally spaced values of C. The bunching of the circles as $C \to 0+$ indicates the steepness of the hemispherical surface that is the graph of f. (See Figure 12.3.)

A function determines its level curves with any given spacing between consecutive values of C. However, level curves only determine the function if *all of them* are known.

Example 5   The level curves of the function $f(x, y) = x^2 - y^2$ are the curves $x^2 - y^2 = C$. For $C = 0$ the level "curve" is the pair of straight lines $x = y$ and $x = -y$. For other values of C the level curves are rectangular hyperbolas with these lines as asymptotes. (See Figure 12.8(a).) The graph of f is the saddle-like hyperbolic paraboloid in Figure 12.8(b).

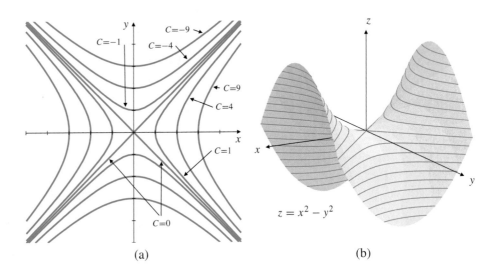

Figure 12.8

(a) Level curves of $x^2 - y^2$

(b) The graph of $x^2 - y^2$

Example 6 Describe and sketch some level curves of the function $z = g(x, y)$ defined by $z \geq 0$, $x^2 + (y - z)^2 = 2z^2$. Also sketch the graph of g.

Solution The level curve $z = g(x, y) = C$ (where C is a positive constant) has equation $x^2 + (y - C)^2 = 2C^2$ and is, therefore, a circle of radius $\sqrt{2}C$ centred at $(0, C)$. Level curves for C in increments of 0.1 from 0 to 1 are shown in Figure 12.9(a). These level curves intersect rays from the origin at equal spacing (the spacing is different for different rays) indicating that the surface $z = g(x, y)$ is an oblique cone. See Figure 12.9(b).

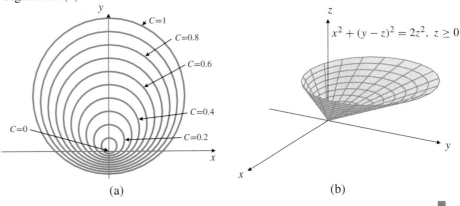

Figure 12.9

(a) Level curves of $z = g(x, y)$ for Example 6

(b) The graph of $z = g(x, y)$

Although the *graph* of a function $f(x, y, z)$ of three variables cannot easily be drawn (it is a three-dimensional *hypersurface* in 4-space), such a function has **level surfaces** in 3-space that can, perhaps, be drawn. These level surfaces have equations $f(x, y, z) = C$ for various choices of the constant C. For instance, the level surfaces of the function $f(x, y, z) = x^2 + y^2 + z^2$ are concentric spheres centred at the origin. Figure 12.10 shows a few level surfaces of the function $f(x, y, z) = x^2 - z$. They are parabolic cylinders.

Using Maple Graphics

Like many mathematical software packages, Maple has several plotting routines to help you visualize the behaviour of functions of two and three variables. We mention only a few of them here; there are many more. Most of the plotting routines are in the **plots** package, so you should begin any Maple session where you want to use them with the input

```
>  with(plots):
```

To save space, we won't show any of the plot output here. You will need to play with modifications to the various plot commands to obtain the kind of output you desire.

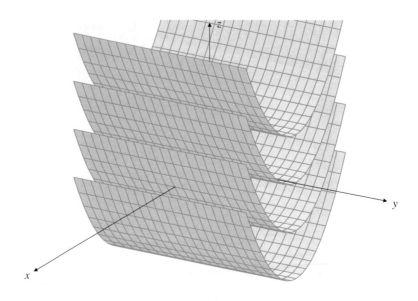

Figure 12.10   Level surfaces of $f(x, y, z) = x^2 - z$

The graph of a function $f(x, y)$ of two variables (or an expression in x and y) can be plotted over a rectangle in the xy-plane with a call to the **plot3d** routine. For example,

```
>  f := -6*y/(2+x^2+y^2);
>  plot3d(f, x=-6..6, y=-6..6);
```

will plot a surface similar to the one in Figure 12.4 but without axes and viewed from a steeper angle. You can add many kinds of options to the command to change the output. For instance,

```
>  plot3d(f, x=-6..6, y=-6..6, axes=boxed,
     orientation=[30,70]);
```

will plot the same surface within a three-dimensional rectangular box with scales on three of its edges indicating the coordinate values. (If we had said axes=normal instead, we would have gotten the usual coordinate axes through the origin, but they tend to be more difficult to see against the background of the surface, so axes=boxed is usually preferable. The option orientation=[30,70] results in the plot's being viewed from the direction making angle 70° with the z-axis and lying in a plane containing the z-axis making an angle 30° with the xz-plane. (The default value of the orientation is [45, 45] if the option is not specified.) By default, the surface plotted by plot3d is ruled by two families of curves, representing its intersection with vertical planes $x = a$ and $y = b$ for several equally spaced values of a and b, and it is coloured opaquely so that hidden parts do not show.

Instead of plot3d, you can use **contourplot3d** to get a plot of the surface ruled by contours on which the value of the function is constant. If you don't get enough contours by default, you can include a contours=n option to specify the number you want.

```
>  contourplot3d(f, x=-6..6, y=-6..6, axes=boxed,
     contours=24);
```

The contours are the projections of the level curves onto the graph of the surface. Alternatively, you can get a two-dimensional plot of the level curves themselves using **contourplot**:

```
>  contourplot(f, x=-6..6, y=-6..6, axes=normal,
     contours=24);
```

Other options you may want to include with plot3d or contourplot3d are

(a) view=zmin..zmax to specify the range of values of the function (i.e., z) to show in the plot.

(b) `grid=[m,n]` to specify the number of x and y values at which to evaluate the function. If your plot doesn't look smooth enough, try $m = n = 20$ or 30 or even higher values.

The graph of an equation, $f(x, y) = 0$, in the xy-plane can be generated without solving the equation for x or y first, by using **implicitplot**.

```
>   implicitplot(x^3-y^2-5*x*y-x-5, x=-6..7, y=-5..6);
```

will produce the graph of $x^3 - y^2 - 5xy - x - 5 = 0$ on the rectangle $-6 \le x \le 7$, $-5 \le y \le 6$. There is also an **implicitplot3d** routine to plot the surface in 3-space having an equation of the form $f(x, y, z) = 0$. For this routine you must specify ranges for all three variables;

```
>   implicitplot3d(x^2+y^2-z^2-1, x=-4..4, y=-4..4,
       z=-3..3, axes=boxed);
```

plots the hyperboloid $z^2 = x^2 + y^2 - 1$.

Finally, we observe that Maple is no more capable than we are of drawing graphs of functions of three or more variables, since it doesn't have four-dimensional plot capability. The best we can do is plot a set of level surfaces for such a function:

```
>   implicitplot3d({z-x^2-2,z-x^2,z-x^2+2},x=-2..2,
       y=-2..2, z=-2..5, axes=boxed);
```

It is possible to construct a sequence of *plot structures* and assign them to, say, the elements of a list variable, without actually plotting them. Then all the plots can be plotted simultaneously using the **display** function.

```
>   for c from -1 to 1 do
       p[c] := implicitplot3d(z^2-x^2-y^2-2*c, x=-3..3,
       y=-3..3, z=0..2, color=COLOR(RGB,(1+c)/2,(1-c)/2,1))
    od:
>   display([seq(p[c],c=-1..1)], axes=boxed,
       orientation=[30,40]);
```

Note that the command creating the plots is terminated with a colon rather than the usual semicolon. If you don't suppress the output in this way, you will get vast amounts of meaningless numerical output as the plots are constructed. The `color=...` option is an attempt to give the three plots a different colour so they can be distinguished from each other.

Exercises 12.1

Specify the domains of the functions in Exercises 1–10.

1. $f(x, y) = \dfrac{x + y}{x - y}$

2. $f(x, y) = \sqrt{xy}$

3. $f(x, y) = \dfrac{x}{x^2 + y^2}$

4. $f(x, y) = \dfrac{xy}{x^2 - y^2}$

5. $f(x, y) = \sqrt{4x^2 + 9y^2 - 36}$

6. $f(x, y) = \dfrac{1}{\sqrt{x^2 - y^2}}$

7. $f(x, y) = \ln(1 + xy)$

8. $f(x, y) = \sin^{-1}(x + y)$

9. $f(x, y, z) = \dfrac{xyz}{x^2 + y^2 + z^2}$

10. $f(x, y, z) = \dfrac{e^{xyz}}{\sqrt{xyz}}$

Sketch the graphs of the functions in Exercises 11–18.

11. $f(x, y) = x$, $\quad (0 \le x \le 2, \quad 0 \le y \le 3)$

12. $f(x, y) = \sin x$, $\quad (0 \le x \le 2\pi, \quad 0 \le y \le 1)$

13. $f(x, y) = y^2$, $\quad (-1 \le x \le 1, \quad -1 \le y \le 1)$

14. $f(x, y) = 4 - x^2 - y^2$, $\quad (x^2 + y^2 \le 4, \ x \ge 0, \ y \ge 0)$

15. $f(x, y) = \sqrt{x^2 + y^2}$

16. $f(x, y) = 4 - x^2$

17. $f(x, y) = |x| + |y|$

18. $f(x, y) = 6 - x - 2y$

Sketch some of the level curves of the functions in Exercises 19–26.

19. $f(x, y) = x - y$

20. $f(x, y) = x^2 + 2y^2$

21. $f(x, y) = xy$

22. $f(x, y) = \dfrac{x^2}{y}$

23. $f(x, y) = \dfrac{x - y}{x + y}$

24. $f(x, y) = \dfrac{y}{x^2 + y^2}$

25. $f(x, y) = xe^{-y}$

26. $f(x, y) = \sqrt{\dfrac{1}{y} - x^2}$

Exercises 27–28 refer to Figure 12.11, which shows contours of a hilly region with heights given in metres.

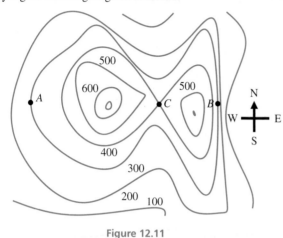

Figure 12.11

27. At which of the points A or B is the landscape steeper? How do you know?

28. Describe the topography of the region near point C.

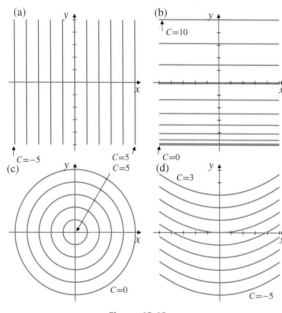

Figure 12.12

Describe the graphs of the functions $f(x, y)$ for which families of level curves $f(x, y) = C$ are shown in the figures referred to in Exercises 29–32. Assume that each family corresponds to equally spaced values of C and that the behaviour of the family is representative of all such families for the function.

29. See Figure 12.12(a).   **30.** See Figure 12.12(b).

31. See Figure 12.12(c).   **32.** See Figure 12.12(d).

33. Are the curves $y = (x - C)^2$ level curves of a function $f(x, y)$? What property must a family of curves in a region of the xy-plane have to be the family of level curves of a function defined in the region?

34. If we assume $z \geq 0$, the equation $4z^2 = (x - z)^2 + (y - z)^2$ defines z as a function of x and y. Sketch some level curves of this function. Describe its graph.

35. Find $f(x, y)$ if each level curve $f(x, y) = C$ is a circle centred at the origin and having radius
(a) C (b) C^2 (c) $\sqrt{C}$ (d) $\ln C$.

36. Find $f(x, y, z)$ if for each constant C the level surface $f(x, y, z) = C$ is a plane having intercepts C^3, $2C^3$, and $3C^3$ on the x-axis, the y-axis, and the z-axis, respectively.

Describe the level surfaces of the functions specified in Exercises 37–41.

37. $f(x, y, z) = x^2 + y^2 + z^2$

38. $f(x, y, z) = x + 2y + 3z$

39. $f(x, y, z) = x^2 + y^2$ **40.** $f(x, y, z) = \dfrac{x^2 + y^2}{z^2}$

41. $f(x, y, z) = |x| + |y| + |z|$

42. Describe the "level hypersurfaces" of the function

$$f(x, y, z, t) = x^2 + y^2 + z^2 + t^2.$$

Use Maple or other computer graphing software to plot the graphs and the level curves of the functions in Exercises 43–48.

43. $\dfrac{1}{1 + x^2 + y^2}$ **44.** $\dfrac{\cos x}{1 + y^2}$

45. $\dfrac{y}{1 + x^2 + y^2}$ **46.** $\dfrac{x}{(x^2 - 1)^2 + y^2}$

47. xy **48.** $\dfrac{1}{xy}$

12.2 Limits and Continuity

Before reading this section you should review the concepts of neighbourhood, open and closed sets, and boundary and interior points introduced in Section 10.1.

The concept of the limit of a function of several variables is similar to that for functions of one variable. For clarity we present the definition for functions of two variables only; the general case is similar.

We might say that $f(x, y)$ approaches the limit L as the point (x, y) approaches the point (a, b), and write

$$\lim_{(x,y)\to(a,b)} f(x, y) = L,$$

if all points of a neighbourhood of (a, b), except possibly the point (a, b) itself, belong to the domain of f, and if $f(x, y)$ approaches L as (x, y) approaches (a, b). However, it is more convenient to define the limit in such a way that (a, b) can be a boundary point of the domain of f. Thus, our formal definition will generalize the one-dimensional notion of one-sided limit as well.

DEFINITION **2**

> **Definition of Limit**
>
> We say that $\lim\limits_{(x,y)\to(a,b)} f(x, y) = L$, provided that
>
> (i) every neighbourhood of (a, b) contains points of the domain of f different from (a, b), and
>
> (ii) for every positive number ϵ there exists a positive number $\delta = \delta(\epsilon)$ such that $|f(x, y) - L| < \epsilon$ holds whenever (x, y) is in the domain of f and satisfies $0 < \sqrt{(x-a)^2 + (y-b)^2} < \delta$.

Condition (i) is included in Definition 2 because it is not appropriate to consider limits at *isolated* points of the domain of f, that is, points with neighbourhoods that contain no other points of the domain.

If a limit exists it is unique. For a single-variable function f, the existence of $\lim_{x\to a} f(x)$ implies that $f(x)$ approaches the same finite number as x approaches a from either the right or the left. Similarly, for a function of two variables, we can have $\lim_{(x,y)\to(a,b)} f(x, y) = L$ only if $f(x, y)$ approaches the same number L no matter how (x, y) approaches (a, b) in the domain of f. In particular, (x, y) can approach (a, b) along any curve that lies in $\mathcal{D}(f)$. It is not necessary that $L = f(a, b)$ even if $f(a, b)$ is defined. The examples below illustrate these assertions.

All the usual laws of limits extend to functions of several variables in the obvious way. For example, if $\lim_{(x,y)\to(a,b)} f(x, y) = L$, $\lim_{(x,y)\to(a,b)} g(x, y) = M$, and every neighbourhood of (a, b) contains points in $\mathcal{D}(f) \cap \mathcal{D}(g)$ other than (a, b), then

$$\lim_{(x,y)\to(a,b)} \big(f(x, y) \pm g(x, y)\big) = L \pm M,$$

$$\lim_{(x,y)\to(a,b)} f(x, y)\, g(x, y) = LM,$$

$$\lim_{(x,y)\to(a,b)} \frac{f(x, y)}{g(x, y)} = \frac{L}{M}, \qquad \text{provided } M \neq 0.$$

Also, if $F(t)$ is continuous at $t = L$, then

$$\lim_{(x,y)\to(a,b)} F\big(f(x, y)\big) = F(L).$$

Example 1 (a) $\quad \lim\limits_{(x,y)\to(2,3)} 2x - y^2 = 4 - 9 = -5,$

(b) $\quad \lim\limits_{(x,y)\to(a,b)} x^2 y = a^2 b,$

(c) $\quad \lim\limits_{(x,y)\to(\pi/3,2)} y \sin\left(\dfrac{x}{y}\right) = 2\sin\left(\dfrac{\pi}{6}\right) = 1.$

Example 2 The function $f(x, y) = \sqrt{1 - x^2 - y^2}$ has limit $f(a, b)$ at all points (a, b) of its domain, the *closed* disk $x^2 + y^2 \leq 1$, and is therefore considered to be *continuous* on its domain. Of course, (x, y) can approach points of the bounding circle $x^2 + y^2 = 1$ only from within the disk.

The following examples show that the requirement that $f(x, y)$ approach the same limit *no matter how* (x, y) approaches (a, b) can be very restrictive, and makes limits in two or more variables much more subtle than in the single-variable case.

Example 3   Investigate the limiting behaviour of $f(x, y) = \dfrac{2xy}{x^2 + y^2}$ as (x, y) approaches $(0, 0)$.

Solution   Note that $f(x, y)$ is defined at all points of the xy-plane except the origin $(0, 0)$. We can still ask whether $\lim_{(x,y)\to(0,0)} f(x, y)$ exists. If we let (x, y) approach $(0, 0)$ along the x-axis ($y = 0$), then $f(x, y) = f(x, 0) \to 0$ (because $f(x, 0) = 0$ identically). Thus, $\lim_{(x,y)\to(0,0)} f(x, y)$ must be 0 if it exists at all. Similarly, at all points of the y-axis we have $f(x, y) = f(0, y) = 0$. However, at points of the line $x = y$, f has a different constant value; $f(x, x) = 1$. Since the limit of $f(x, y)$ is 1 as (x, y) approaches $(0, 0)$ along this line, it follows that $f(x, y)$ cannot have a unique limit at the origin. That is,

$$\lim_{(x,y)\to(0,0)} \frac{2xy}{x^2 + y^2} \qquad \text{does not exist.}$$

Observe that $f(x, y)$ has a constant value on any ray from the origin (on the ray $y = kx$ the value is $2k/(1 + k^2)$), but these values differ on different rays. The level curves of f are the rays from the origin (with the origin itself removed). It is difficult to sketch the graph of f near the origin. The first octant part of the graph is the "hood-shaped" surface in Figure 12.13(a).

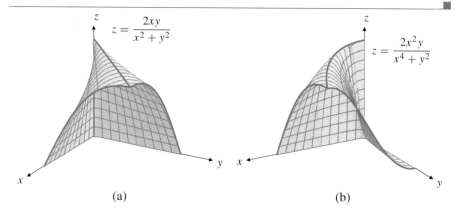

Figure 12.13

(a) $f(x, y)$ has different limits as $(x, y) \to (0, 0)$ along different straight lines

(b) $f(x, y)$ has the same limit 0 as $(x, y) \to (0, 0)$ along any straight line but has limit 1 as $(x, y) \to (0, 0)$ along $y = x^2$

Example 4   Investigate the limiting behaviour of $f(x, y) = \dfrac{2x^2 y}{x^4 + y^2}$ as (x, y) approaches $(0, 0)$.

Solution   As in Example 3, $f(x, y)$ vanishes identically on the coordinate axes, so $\lim_{(x,y)\to(0,0)} f(x, y)$ must be 0 if it exists at all. If we examine $f(x, y)$ at points of the ray $y = kx$, we obtain

$$f(x, kx) = \frac{2kx^3}{x^4 + k^2 x^2} = \frac{2kx}{x^2 + k^2} \to 0, \qquad \text{as} \quad x \to 0 \quad (k \neq 0).$$

Thus, $f(x, y) \to 0$ as $(x, y) \to (0, 0)$ along *any* straight line through the origin. We might be tempted to conclude, therefore, that $\lim_{(x,y)\to(0,0)} f(x, y) = 0$, but this is incorrect. Observe the behaviour of $f(x, y)$ along the curve $y = x^2$:

$$f(x, x^2) = \frac{2x^4}{x^4 + x^4} = 1.$$

Thus, $f(x, y)$ does not approach 0 as (x, y) approaches the origin along this curve, so $\lim_{(x,y)\to(0,0)} f(x, y)$ does not exist. The level curves of f are pairs of parabolas of the form $y = kx^2$, $y = x^2/k$ with the origin removed. See Figure 12.13(b) for the first octant part of the graph of f.

$$z = \frac{x^2 y}{x^2 + y^2}$$

Figure 12.14 $\displaystyle\lim_{(x,y)\to(0,0)} \frac{x^2 y}{x^2 + y^2} = 0$

Example 5   Show that the function $f(x, y) = \dfrac{x^2 y}{x^2 + y^2}$ does have a limit at the origin; specifically,

$$\lim_{(x,y)\to(0,0)} \frac{x^2 y}{x^2 + y^2} = 0.$$

Solution   This function is also defined everywhere except at the origin. Observe that since $x^2 \le x^2 + y^2$, we have

$$|f(x, y) - 0| = \left| \frac{x^2 y}{x^2 + y^2} \right| \le |y| \le \sqrt{x^2 + y^2},$$

which approaches zero as $(x, y) \to (0, 0)$. (See Figure 12.14.) Formally, if $\epsilon > 0$ is given and we take $\delta = \epsilon$, then $|f(x, y) - 0| < \epsilon$ whenever $0 < \sqrt{x^2 + y^2} < \delta$, so $f(x, y)$ has limit 0 as $(x, y) \to (0, 0)$ by Definition 2. ∎

As for functions of one variable, continuity of a function f at a point of its domain is defined directly in terms of the limit. (See, for instance, Example 2.)

DEFINITION 3

> The function $f(x, y)$ is continuous at the point (a, b) if
>
> $$\lim_{(x,y)\to(a,b)} f(x, y) = f(a, b).$$

It remains true that sums, differences, products, quotients, and compositions of continuous functions are continuous. The functions of Examples 3 and 4 above are continuous wherever they are defined, that is, at all points except the origin. There is no way to define $f(0, 0)$ so that these functions become continuous at the origin. They show that the continuity of the single-variable functions $f(x, b)$ at $x = a$ and $f(a, y)$ at $y = b$ does *not* imply that $f(x, y)$ is continuous at (a, b). In fact, even if $f(x, y)$ is continuous along every straight line through (a, b), it still need not be continuous at (a, b). (See Exercises 16–17 below.) Note, however, that the function $f(x, y)$ of Example 5, although not defined at the origin, has a continuous extension to that point. If we extend the domain of f by defining $f(0, 0) = \lim_{(x,y)\to(0,0)} f(x, y) = 0$, then f is continuous on the whole xy-plane.

As for functions of one variable, the existence of a limit of a function at a point does not imply that the function is continuous at that point. The function

$$f(x, y) = \begin{cases} 0 & \text{if } (x, y) \ne (0, 0) \\ 1 & \text{if } (x, y) = (0, 0) \end{cases}$$

satisfies $\lim_{(x,y)\to(0,0)} f(x,y) = 0$, which is not equal to $f(0,0)$, so f is not continuous at $(0,0)$. Of course, we can *make* f continuous at $(0,0)$ by redefining its value at that point to be 0.

Exercises 12.2

In Exercises 1–12, evaluate the indicated limit or explain why it does not exist.

1. $\lim_{(x,y)\to(2,-1)} xy + x^2$

2. $\lim_{(x,y)\to(0,0)} \sqrt{x^2 + y^2}$

3. $\lim_{(x,y)\to(0,0)} \dfrac{x^2 + y^2}{y}$

4. $\lim_{(x,y)\to(0,0)} \dfrac{x}{x^2 + y^2}$

5. $\lim_{(x,y)\to(1,\pi)} \dfrac{\cos(xy)}{1 - x - \cos y}$

6. $\lim_{(x,y)\to(0,1)} \dfrac{x^2(y-1)^2}{x^2 + (y-1)^2}$

7. $\lim_{(x,y)\to(0,0)} \dfrac{y^3}{x^2 + y^2}$

8. $\lim_{(x,y)\to(0,0)} \dfrac{\sin(x-y)}{\cos(x+y)}$

9. $\lim_{(x,y)\to(0,0)} \dfrac{\sin(xy)}{x^2 + y^2}$

10. $\lim_{(x,y)\to(1,2)} \dfrac{2x^2 - xy}{4x^2 - y^2}$

11. $\lim_{(x,y)\to(0,0)} \dfrac{x^2y^2}{x^2 + y^4}$

12. $\lim_{(x,y)\to(0,0)} \dfrac{x^2y^2}{2x^4 + y^4}$

13. How can the function

$$f(x,y) = \dfrac{x^2 + y^2 - x^3y^3}{x^2 + y^2}, \qquad (x,y) \neq (0,0),$$

be defined at the origin so that it becomes continuous at all points of the xy-plane?

14. How can the function

$$f(x,y) = \dfrac{x^3 - y^3}{x - y}, \qquad (x \neq y),$$

be defined along the line $x = y$ so that the resulting function is continuous on the whole xy-plane?

15. What is the domain of

$$f(x,y) = \dfrac{x - y}{x^2 - y^2}?$$

Does $f(x,y)$ have a limit as $(x,y) \to (1,1)$? Can the domain of f be extended so that the resulting function is continuous at $(1,1)$? Can the domain be extended so that the resulting function is continuous everywhere in the xy-plane?

****16.** Given a function $f(x,y)$ and a point (a,b) in its domain, define single-variable functions g and h as follows:

$$g(x) = f(x,b), \qquad h(y) = f(a,y).$$

If g is continuous at $x = a$ and h is continuous at $y = b$, does it follow that f is continuous at (a,b)? Conversely, does the continuity of f at (a,b) guarantee the continuity of g at a and the continuity of h at b? Justify your answers.

****17.** Let $\mathbf{u} = u\mathbf{i} + v\mathbf{j}$ be a unit vector, and let

$$f_{\mathbf{u}}(t) = f(a + tu, b + tv)$$

be the single-variable function obtained by restricting the domain of $f(x,y)$ to points of the straight line through (a,b) parallel to $\mathbf{u}$. If $f_{\mathbf{u}}(t)$ is continuous at $t = 0$ for every unit vector $\mathbf{u}$, does it follow that f is continuous at (a,b)? Conversely, does the continuity of f at (a,b) guarantee the continuity of $f_{\mathbf{u}}(t)$ at $t = 0$? Justify your answers.

****18.** What condition must the nonnegative integers m, n, and p satisfy to guarantee that $\lim_{(x,y)\to(0,0)} x^m y^n / (x^2 + y^2)^p$ exists? Prove your answer.

****19.** What condition must the constants a, b, and c satisfy to guarantee that $\lim_{(x,y)\to(0,0)} xy/(ax^2 + bxy + cy^2)$ exists? Prove your answer.

****20.** Can the function $f(x,y) = \dfrac{\sin x \sin^3 y}{1 - \cos(x^2 + y^2)}$ be defined at $(0,0)$ in such a way that it becomes continuous there? If so, how?

21. Use two- and three-dimensional mathematical graphing software to examine the graph and level curves of the function $f(x,y)$ of Example 3 on the region $-1 \leq x \leq 1$, $-1 \leq y \leq 1$, $(x,y) \neq (0,0)$. How would you describe the behaviour of the graph near $(x,y) = (0,0)$?

22. Use two- and three-dimensional mathematical graphing software to examine the graph and level curves of the function $f(x,y)$ of Example 4 on the region $-1 \leq x \leq 1$, $-1 \leq y \leq 1$, $(x,y) \neq (0,0)$. How would you describe the behaviour of the graph near $(x,y) = (0,0)$?

23. The graph of a single-variable function $f(x)$ that is continuous on an interval is a curve that has no *breaks* in it there and that intersects any vertical line through a point in the interval exactly once. What analogous statement can you make about the graph of a bivariate function $f(x,y)$ that is continuous on a region of the xy-plane?

12.3 Partial Derivatives

In this section we begin the process of extending the concepts and techniques of single-variable calculus to functions of more than one variable. It is convenient to begin by considering the rate of change of such functions with respect to one variable at a time.

Thus, a function of n variables has n *first-order partial derivatives*, one with respect to each of its independent variables. For a function of two variables, we make this precise in the following definition:

DEFINITION **4**

> The **first partial derivatives** of the function $f(x, y)$ **with respect to the variables** x and y are the functions $f_1(x, y)$ and $f_2(x, y)$ given by
>
> $$f_1(x, y) = \lim_{h \to 0} \frac{f(x + h, y) - f(x, y)}{h},$$
> $$f_2(x, y) = \lim_{k \to 0} \frac{f(x, y + k) - f(x, y)}{k},$$
>
> provided these limits exist.

Each of the two partial derivatives is the limit of a Newton quotient in one of the variables. Observe that $f_1(x, y)$ is just the ordinary first derivative of $f(x, y)$ considered as a function of x only, regarding y as a constant parameter. Similarly, $f_2(x, y)$ is the first derivative of $f(x, y)$ considered as a function of y alone, with x held fixed.

Example 1 If $f(x, y) = x^2 \sin y$, then

$$f_1(x, y) = 2x \sin y \qquad \text{and} \qquad f_2(x, y) = x^2 \cos y.$$

The subscripts 1 and 2 in the notations for the partial derivatives refer to the first and second variables of f. For functions of one variable we use the notation f' for the derivative; the *prime* (′) denotes differentiation with respect to the only variable on which f depends. For functions f of two variables we use f_1 or f_2 to show the variable of differentiation. Do not confuse these subscripts with subscripts used for other purposes, for example, to denote the components of vectors.

The partial derivative $f_1(a, b)$ measures the rate of change of $f(x, y)$ with respect to x at $x = a$ while y is held fixed at b. In graphical terms, the surface $z = f(x, y)$ intersects the vertical plane $y = b$ in a curve. If we take horizontal and vertical lines through the point $(0, b, 0)$ as coordinate axes in the plane $y = b$, then the curve has equation $z = f(x, b)$, and its slope at $x = a$ is $f_1(a, b)$. (See Figure 12.15.) Similarly, $f_2(a, b)$ represents the rate of change of f with respect to y at $y = b$ with x held fixed

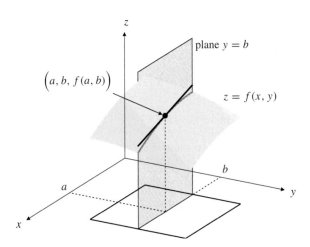

Figure 12.15 $f_1(a, b)$ is the slope of the curve of intersection of $z = f(x, y)$ and the vertical plane $y = b$ at $x = a$

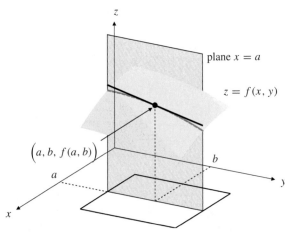

Figure 12.16 $f_2(a, b)$ is the slope of the curve of intersection of $z = f(x, y)$ and the vertical plane $x = a$ at $y = b$

at a. The surface $z = f(x, y)$ intersects the vertical plane $x = a$ in a curve $z = f(a, y)$ whose slope at $y = b$ is $f_2(a, b)$. (See Figure 12.16.)

Various notations can be used to denote the partial derivatives of $z = f(x, y)$ considered as functions of x and y:

> **Notations for first partial derivatives**
>
> $$\frac{\partial z}{\partial x} = \frac{\partial}{\partial x} f(x, y) = f_1(x, y) = D_1 f(x, y)$$
>
> $$\frac{\partial z}{\partial y} = \frac{\partial}{\partial y} f(x, y) = f_2(x, y) = D_2 f(x, y)$$

The symbol $\partial/\partial x$ should be read as "partial with respect to x" so $\partial z/\partial x$ is "partial z with respect to x." The reason for distinguishing ∂ from the d of ordinary derivatives of single-variable functions will be made clear later.

Similar notations can be used to denote the values of partial derivatives at a particular point (a, b):

> **Values of partial derivatives**
>
> $$\left.\frac{\partial z}{\partial x}\right|_{(a,b)} = \left(\frac{\partial}{\partial x} f(x, y)\right)\bigg|_{(a,b)} = f_1(a, b) = D_1 f(a, b)$$
>
> $$\left.\frac{\partial z}{\partial y}\right|_{(a,b)} = \left(\frac{\partial}{\partial y} f(x, y)\right)\bigg|_{(a,b)} = f_2(a, b) = D_2 f(a, b)$$

BEWARE

The argument in the paragraph at the right is a bit subtle. It tries to explain why, at least for the time being, we are using subscripts 1 and 2 instead of subscripts x and y for the partial derivatives of $f(x, y)$. Later on, and especially when we are discussing partial differential equations or dealing with vector-valued functions for which numerical subscripts normally represent components, we will prefer to use letter subscripts for partial derivatives.

Some authors prefer to use f_x, $D_x f$, or $\partial f/\partial x$, and f_y, $D_y f$, or $\partial f/\partial y$, instead of f_1 and f_2. However, this can lead to problems of ambiguity when compositions of functions arise. For instance, suppose $f(x, y) = x^2 y$. By $f_1(x^2, xy)$ we clearly mean

$$\left(\frac{\partial}{\partial u} f(u, v)\right)\bigg|_{u=x^2, v=xy} = 2uv\bigg|_{u=x^2, v=xy} = (2)(x^2)(xy) = 2x^3 y.$$

But does $f_x(x^2, xy)$ mean the same thing? One could argue that $f_x(x^2, xy)$ should mean

$$\frac{\partial}{\partial x}\left(f(x^2, xy)\right) = \frac{\partial}{\partial x}\left((x^2)^2(xy)\right) = \frac{\partial}{\partial x}(x^5 y) = 5x^4 y.$$

In order to avoid such ambiguities we usually prefer to use f_1 and f_2 instead of f_x and f_y. (However, in some situations where no confusion is likely to occur we may still use the notations f_x and f_y, and also $D_x f$, $D_y f$, $\partial f/\partial x$ and $\partial f/\partial y$.)

All the standard differentiation rules for sums, products, reciprocals, and quotients continue to apply to partial derivatives.

Example 2 Find $\partial z/\partial x$ and $\partial z/\partial y$ if $z = x^3 y^2 + x^4 y + y^4$.

Solution $\partial z/\partial x = 3x^2 y^2 + 4x^3 y$ and $\partial z/\partial y = 2x^3 y + x^4 + 4y^3$. ∎

Example 3 Find $f_1(0, \pi)$ if $f(x, y) = e^{xy} \cos(x + y)$.

Solution

$$f_1(x, y) = y\, e^{xy} \cos(x + y) - e^{xy} \sin(x + y),$$
$$f_1(0, \pi) = \pi\, e^0 \cos(\pi) - e^0 \sin(\pi) = -\pi.$$

∎

The single-variable version of the Chain Rule also continues to apply to, say, $f\big(g(x, y)\big)$, where f is a function of only one variable having derivative f':

$$\frac{\partial}{\partial x} f\big(g(x, y)\big) = f'\big(g(x, y)\big) g_1(x, y), \quad \frac{\partial}{\partial y} f\big(g(x, y)\big) = f'\big(g(x, y)\big) g_2(x, y).$$

We will develop versions of the Chain Rule for more complicated compositions of multivariate functions in Section 12.5.

Example 4   If f is an everywhere differentiable function of one variable, show that $z = f(x/y)$ satisfies the *partial differential equation*

$$x \frac{\partial z}{\partial x} + y \frac{\partial z}{\partial y} = 0.$$

Solution   By the (single-variable) Chain Rule,

$$\frac{\partial z}{\partial x} = f'\left(\frac{x}{y}\right)\left(\frac{1}{y}\right) \quad \text{and} \quad \frac{\partial z}{\partial y} = f'\left(\frac{x}{y}\right)\left(\frac{-x}{y^2}\right).$$

Hence,

$$x \frac{\partial z}{\partial x} + y \frac{\partial z}{\partial y} = f'\left(\frac{x}{y}\right)\left(x \times \frac{1}{y} + y \times \frac{-x}{y^2}\right) = 0.$$

Definition 4 can be extended in the obvious way to cover functions of more than two variables. If f is a function of n variables $x_1, x_2, \ldots, x_n$, then f has n first partial derivatives, $f_1(x_1, x_2, \ldots, x_n)$, $f_2(x_1, x_2, \ldots, x_n)$, $\ldots$, $f_n(x_1, x_2, \ldots, x_n)$, one with respect to each variable.

Example 5

$$\frac{\partial}{\partial z}\left(\frac{2xy}{1 + xz + yz}\right) = -\frac{2xy}{(1 + xz + yz)^2}(x + y).$$

Again, all the standard differentiation rules are applied to calculate partial derivatives.

Remark   If a single-variable function $f(x)$ has a derivative $f'(a)$ at $x = a$, then f is necessarily continuous at $x = a$. This property does *not* extend to partial derivatives. Even if all the first partial derivatives of a function of several variables exist at a point, the function may still fail to be continuous at that point. See Exercise 36 below.

Tangent Planes and Normal Lines

If the graph $z = f(x, y)$ is a "smooth" surface near the point P with coordinates $\big(a, b, f(a, b)\big)$, then that graph will have a **tangent plane** and a **normal line** at P. The normal line is the line through P that is perpendicular to the surface; for instance, a line joining a point on a sphere to the centre of the sphere is normal to the sphere. Any nonzero vector that is parallel to the normal line at P is called a normal vector to the surface at P. The tangent plane to the surface $z = f(x, y)$ at P is the plane through P that is perpendicular to the normal line at P.

Let us assume that the surface $z = f(x, y)$ has a *nonvertical* tangent plane (and therefore a *nonhorizontal* normal line) at point P. (Later in this chapter we will state precise conditions that guarantee that the graph of a function has a nonvertical tangent plane at a point.) The tangent plane intersects the vertical plane $y = b$ in a straight line that is tangent at P to the curve of intersection of the surface $z = f(x, y)$ and the plane $y = b$. (See Figures 12.15 and 12.17.) This line has slope $f_1(a, b)$, so it is parallel to the vector $\mathbf{T}_1 = \mathbf{i} + f_1(a, b)\mathbf{k}$. Similarly, the tangent plane intersects the vertical plane $x = a$ in a straight line having slope $f_2(a, b)$. This line is therefore parallel to the vector $\mathbf{T}_2 = \mathbf{j} + f_2(a, b)\mathbf{k}$. It follows that the tangent plane, and therefore the surface $z = f(x, y)$ itself, has normal vector

$$\mathbf{n} = \mathbf{T}_2 \times \mathbf{T}_1 = \begin{vmatrix} \mathbf{i} & \mathbf{j} & \mathbf{k} \\ 0 & 1 & f_2(a, b) \\ 1 & 0 & f_1(a, b) \end{vmatrix} = f_1(a, b)\mathbf{i} + f_2(a, b)\mathbf{j} - \mathbf{k}.$$

A normal vector to $z = f(x, y)$ at $\big(a, b, f(a, b)\big)$ is

$$\mathbf{n} = f_1(a, b)\mathbf{i} + f_2(a, b)\mathbf{j} - \mathbf{k}.$$

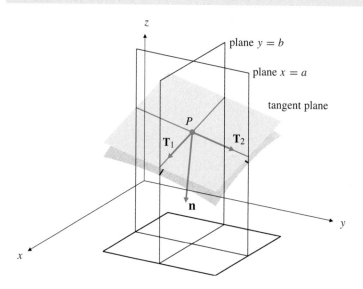

Figure 12.17 The tangent plane and a normal vector to $z = f(x, y)$ at $P = \big(a, b, f(a, b)\big)$

Since the tangent plane passes through $P = (a, b, f(a, b))$, it has equation

$$f_1(a, b)(x - a) + f_2(a, b)(y - b) - (z - f(a, b)) = 0,$$

or, equivalently,

An equation of the tangent plane to $z = f(x, y)$ at $\big(a, b, f(a, b)\big)$ is

$$z = f(a, b) + f_1(a, b)(x - a) + f_2(a, b)(y - b).$$

We shall obtain this result by a different method in Section 12.7.

The normal line to $z = f(x, y)$ at $\big(a, b, f(a, b)\big)$ has direction vector $f_1(a, b)\mathbf{i} + f_2(a, b)\mathbf{j} - \mathbf{k}$ and so has equations

$$\frac{x - a}{f_1(a, b)} = \frac{y - b}{f_2(a, b)} = \frac{z - f(a, b)}{-1}$$

with suitable modifications if either $f_1(a, b) = 0$ or $f_2(a, b) = 0$.

Example 6   Find a normal vector and equations of the tangent plane and normal line to the graph $z = \sin(xy)$ at the point where $x = \pi/3$ and $y = -1$.

Solution   The point on the graph has coordinates $(\pi/3, -1, -\sqrt{3}/2)$. Now

$$\frac{\partial z}{\partial x} = y\,\cos(xy) \qquad \text{and} \qquad \frac{\partial z}{\partial y} = x\,\cos(xy).$$

At $(\pi/3, -1)$ we have $\partial z/\partial x = -1/2$ and $\partial z/\partial y = \pi/6$. Therefore, the surface has normal vector $\mathbf{n} = -(1/2)\mathbf{i} + (\pi/6)\mathbf{j} - \mathbf{k}$ and tangent plane

$$z = \frac{-\sqrt{3}}{2} - \frac{1}{2}\left(x - \frac{\pi}{3}\right) + \frac{\pi}{6}(y + 1),$$

or, more simply, $3x - \pi y + 6z = 2\pi - 3\sqrt{3}$. The normal line has equation

$$\frac{x - \dfrac{\pi}{3}}{-\dfrac{1}{2}} = \frac{y + 1}{\dfrac{\pi}{6}} = \frac{z + \dfrac{\sqrt{3}}{2}}{-1} \qquad \text{or} \qquad \frac{6x - 2\pi}{-3} = \frac{6y + 6}{\pi} = \frac{6z + 3\sqrt{3}}{-6}.$$

Example 7   What horizontal plane is tangent to the surface

$$z = x^2 - 4xy - 2y^2 + 12x - 12y - 1,$$

and what is the point of tangency?

Solution   A plane is horizontal only if its equation is of the form $z = k$, that is, it is independent of x and y. Therefore, we must have $\partial z/\partial x = \partial z/\partial y = 0$ at the point of tangency. The equations

$$\frac{\partial z}{\partial x} = 2x - 4y + 12 = 0$$

$$\frac{\partial z}{\partial y} = -4x - 4y - 12 = 0$$

have solution $x = -4$, $y = 1$. For these values we have $z = -31$, so the required tangent plane has equation $z = -31$ and the point of tangency is $(-4, 1, -31)$.

Distance from a Point to a Surface: A Geometric Example

Example 8   Find the distance from the point $(3, 0, 0)$ to the hyperbolic paraboloid with equation $z = x^2 - y^2$.

Solution   This is an optimization problem of a sort we will deal with in a more systematic way in the next chapter. However, such problems involving minimizing distances from points to surfaces can frequently be solved using geometric methods.

If $Q = (X, Y, Z)$ is the point on the surface $z = x^2 - y^2$ that is closest to $P = (3, 0, 0)$, then the vector $\vec{PQ} = (X - 3)\mathbf{i} + Y\mathbf{j} + Z\mathbf{k}$ must be normal to the surface at Q. (See Figure 12.18(a).) Using the partial derivatives of $z = x^2 - y^2$, we know that the vector $\mathbf{n} = 2X\mathbf{i} - 2Y\mathbf{j} - \mathbf{k}$ is normal to the surface at Q. Thus, $\vec{PQ}$ must be parallel to $\mathbf{n}$, and $\vec{PQ} = t\mathbf{n}$ for some scalar t. Separated into components, this vector equation states that

$$X - 3 = 2Xt, \qquad Y = -2Yt, \qquad \text{and} \qquad Z = -t.$$

The middle equation implies that either $Y = 0$ or $t = -\frac{1}{2}$. We must consider both of these possibilities.

CASE I   If $Y = 0$, then

$$X = \frac{3}{1 - 2t} \qquad \text{and} \qquad Z = -t.$$

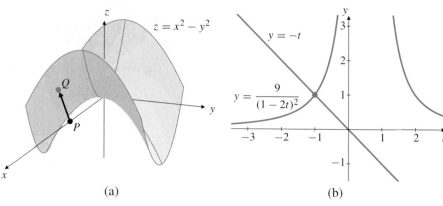

Figure 12.18

(a) If Q is the point on $z = x^2 - y^2$ closest to P, then $\overrightarrow{PQ}$ is normal to the surface

(b) Equation $-t = \dfrac{9}{(1-2t)^2}$ has only one real root, $t = -1$

But $Z = X^2 - Y^2$, so we must have

$$-t = \frac{9}{(1-2t)^2}.$$

This is a cubic equation in t, so we might expect to have to solve it numerically, for instance, by using Newton's Method. However, if we try small integer values of t, we will quickly discover that $t = -1$ is a solution. The graphs of both sides of the equation are shown in Figure 12.18(b). They show that $t = -1$ is the only real solution. Calculating the corresponding values of X and Z, we obtain $(1, 0, 1)$ as a candidate for Q. The distance from this point to P is $\sqrt{5}$.

CASE II   If $t = -1/2$, then $X = 3/2$, $Z = 1/2$, and $Y = \pm\sqrt{X^2 - Z} = \pm\sqrt{7}/2$, and the distance from these points to P is $\sqrt{17}/2$.

Since $\frac{17}{4} < 5$, the points $(3/2, \pm\sqrt{7}/2, 1/2)$ are the points on $z = x^2 - y^2$ closest to $(3, 0, 0)$, and the distance from $(3, 0, 0)$ to the surface is $\sqrt{17}/2$ units.

Exercises 12.3

In Exercises 1–10, find all the first partial derivatives of the function specified, and evaluate them at the given point.

1. $f(x, y) = x - y + 2$, $(3, 2)$

2. $f(x, y) = xy + x^2$, $(2, 0)$

3. $f(x, y, z) = x^3 y^4 z^5$, $(0, -1, -1)$

4. $g(x, y, z) = \dfrac{xz}{y + z}$, $(1, 1, 1)$

5. $z = \tan^{-1}\left(\dfrac{y}{x}\right)$, $(-1, 1)$

6. $w = \ln(1 + e^{xyz})$, $(2, 0, -1)$

7. $f(x, y) = \sin(x\sqrt{y})$, $\left(\dfrac{\pi}{3}, 4\right)$

8. $f(x, y) = \dfrac{1}{\sqrt{x^2 + y^2}}$, $(-3, 4)$

9. $w = x^{(y \ln z)}$, $(e, 2, e)$

10. $g(x_1, x_2, x_3, x_4) = \dfrac{x_1 - x_2^2}{x_3 + x_4^2}$, $(3, 1, -1, -2)$

In Exercises 11–12, calculate the first partial derivatives of the given functions at $(0, 0)$. You will have to use Definition 4.

11. $f(x, y) = \begin{cases} \dfrac{2x^3 - y^3}{x^2 + 3y^2}, & \text{if } (x, y) \neq (0, 0) \\ 0, & \text{if } (x, y) = (0, 0). \end{cases}$

12. $f(x, y) = \begin{cases} \dfrac{x^2 - 2y^2}{x - y}, & \text{if } x \neq y \\ 0, & \text{if } x = y. \end{cases}$

In Exercises 13–22, find equations of the tangent plane and normal line to the graph of the given function at the point with specified values of x and y.

13. $f(x, y) = x^2 - y^2$ at $(-2, 1)$

14. $f(x, y) = \dfrac{x - y}{x + y}$ at $(1, 1)$

15. $f(x, y) = \cos(x/y)$ at $(\pi, 4)$

16. $f(x, y) = e^{xy}$ at $(2, 0)$

17. $f(x, y) = \dfrac{x}{x^2 + y^2}$ at $(1, 2)$

18. $f(x, y) = y\, e^{-x^2}$ at $(0, 1)$

19. $f(x, y) = \ln(x^2 + y^2)$ at $(1, -2)$

20. $f(x, y) = \dfrac{2xy}{x^2 + y^2}$ at $(0, 2)$

21. $f(x, y) = \tan^{-1}(y/x)$ at $(1, -1)$

22. $f(x, y) = \sqrt{1 + x^3 y^2}$ at $(2, 1)$

23. Find the coordinates of all points on the surface with equation $z = x^4 - 4xy^3 + 6y^2 - 2$ where the surface has a horizontal tangent plane.

24. Find all horizontal planes that are tangent to the surface with equation $z = xye^{-(x^2+y^2)/2}$. At what points are they tangent?

In Exercises 25–31, show that the given function satisfies the given partial differential equation.

◆**25.** $z = x\,e^y$, $\quad x\dfrac{\partial z}{\partial x} = \dfrac{\partial z}{\partial y}$

◆**26.** $z = \dfrac{x+y}{x-y}$, $\quad x\dfrac{\partial z}{\partial x} + y\dfrac{\partial z}{\partial y} = 0$

◆**27.** $z = \sqrt{x^2+y^2}$, $\quad x\dfrac{\partial z}{\partial x} + y\dfrac{\partial z}{\partial y} = z$

◆**28.** $w = x^2 + yz$, $\quad x\dfrac{\partial w}{\partial x} + y\dfrac{\partial w}{\partial y} + z\dfrac{\partial w}{\partial z} = 2w$

◆**29.** $w = \dfrac{1}{x^2+y^2+z^2}$, $\quad x\dfrac{\partial w}{\partial x} + y\dfrac{\partial w}{\partial y} + z\dfrac{\partial w}{\partial z} = -2w$

◆**30.** $z = f(x^2 + y^2)$, where f is any differentiable function of one variable,

$$y\dfrac{\partial z}{\partial x} - x\dfrac{\partial z}{\partial y} = 0.$$

◆**31.** $z = f(x^2 - y^2)$, where f is any differentiable function of one variable,

$$y\dfrac{\partial z}{\partial x} + x\dfrac{\partial z}{\partial y} = 0.$$

32. Give a formal definition of the three first partial derivatives of the function $f(x, y, z)$.

33. What is an equation of the "tangent hyperplane" to the graph $w = f(x, y, z)$ at $\big(a, b, c, f(a, b, c)\big)$?

∗**34.** Find the distance from the point $(1, 1, 0)$ to the circular paraboloid with equation $z = x^2 + y^2$.

∗**35.** Find the distance from the point $(0, 0, 1)$ to the elliptic paraboloid having equation $z = x^2 + 2y^2$.

∗**36.** Let $f(x, y) = \begin{cases} \dfrac{2xy}{x^2+y^2}, & \text{if } (x, y) \neq (0, 0) \\ 0, & \text{if } (x, y) = (0, 0). \end{cases}$

Note that f is not continuous at $(0, 0)$. (See Example 3 of Section 12.2.) Therefore its graph is not smooth there. Show, however, that $f_1(0, 0)$ and $f_2(0, 0)$ both exist. Hence, the existence of partial derivatives does not imply that a function of several variables is continuous. This is in contrast to the single-variable case.

37. Determine $f_1(0, 0)$ and $f_2(0, 0)$ if they exist, where

$$f(x, y) = \begin{cases} (x^3 + y)\sin\dfrac{1}{x^2+y^2}, & \text{if } (x, y) \neq (0, 0) \\ 0, & \text{if } (x, y) = (0, 0). \end{cases}$$

38. Calculate $f_1(x, y)$ for the function in Exercise 37. Is $f_1(x, y)$ continuous at $(0, 0)$?

∗**39.** Let $f(x, y) = \begin{cases} \dfrac{x^3 - y^3}{x^2+y^2}, & \text{if } (x, y) \neq (0, 0) \\ 0, & \text{if } (x, y) = (0, 0). \end{cases}$

Calculate $f_1(x, y)$ and $f_2(x, y)$ at all points (x, y) in the plane. Is f continuous at $(0, 0)$? Are f_1 and f_2 continuous at $(0, 0)$?

∗**40.** Let $f(x, y, z) = \begin{cases} \dfrac{xy^2z}{x^4+y^4+z^4}, & \text{if } (x, y, z) \neq (0, 0, 0) \\ 0, & \text{if } (x, y, z) = (0, 0, 0). \end{cases}$

Find $f_1(0, 0, 0)$, $f_2(0, 0, 0)$, and $f_3(0, 0, 0)$. Is f continuous at $(0, 0, 0)$? Are f_1, f_2, and f_3 continuous at $(0, 0, 0)$?

12.4 Higher-Order Derivatives

Partial derivatives of second and higher orders are calculated by taking partial derivatives of already calculated partial derivatives. The order in which the differentiations are performed is indicated in the notations used. If $z = f(x, y)$, we can calculate *four* partial derivatives of second order, namely, two **pure** second partial derivatives with respect to x or y,

$$\dfrac{\partial^2 z}{\partial x^2} = \dfrac{\partial}{\partial x}\dfrac{\partial z}{\partial x} = f_{11}(x, y) = f_{xx}(x, y),$$

$$\dfrac{\partial^2 z}{\partial y^2} = \dfrac{\partial}{\partial y}\dfrac{\partial z}{\partial y} = f_{22}(x, y) = f_{yy}(x, y),$$

and two **mixed** second partial derivatives with respect to x and y,

$$\frac{\partial^2 z}{\partial x \partial y} = \frac{\partial}{\partial x}\frac{\partial z}{\partial y} = f_{21}(x, y) = f_{yx}(x, y),$$

$$\frac{\partial^2 z}{\partial y \partial x} = \frac{\partial}{\partial y}\frac{\partial z}{\partial x} = f_{12}(x, y) = f_{xy}(x, y).$$

Again, we remark that the notations f_{11}, f_{12}, f_{21}, and f_{22} are usually preferable to f_{xx}, f_{xy}, f_{yx}, and f_{yy}, although the latter are often used in partial differential equations. Note that f_{12} indicates differentiation of f *first* with respect to its first variable and *then* with respect to its second variable; f_{21} indicates the opposite order of differentiation. The subscript closest to f indicates which differentiation occurs first.

Similarly, if $w = f(x, y, z)$, then

$$\frac{\partial^5 w}{\partial y \partial x \partial y^2 \partial z} = \frac{\partial}{\partial y}\frac{\partial}{\partial x}\frac{\partial}{\partial y}\frac{\partial}{\partial y}\frac{\partial w}{\partial z} = f_{32212}(x, y, z) = f_{zyyxy}(x, y, z).$$

Example 1   Find the four second partial derivatives of $f(x, y) = x^3 y^4$.

Solution

$$f_1(x, y) = 3x^2 y^4, \qquad\qquad f_2(x, y) = 4x^3 y^3,$$

$$f_{11}(x, y) = \frac{\partial}{\partial x}(3x^2 y^4) = 6xy^4, \qquad f_{21}(x, y) = \frac{\partial}{\partial x}(4x^3 y^3) = 12x^2 y^3,$$

$$f_{12}(x, y) = \frac{\partial}{\partial y}(3x^2 y^4) = 12x^2 y^3, \qquad f_{22}(x, y) = \frac{\partial}{\partial y}(4x^3 y^3) = 12x^3 y^2.$$

Example 2   Calculate $f_{223}(x, y, z)$, $f_{232}(x, y, z)$, and $f_{322}(x, y, z)$ for the function $f(x, y, z) = e^{x - 2y + 3z}$.

Solution

$$f_{223}(x, y, z) = \frac{\partial}{\partial z}\frac{\partial}{\partial y}\frac{\partial}{\partial y} e^{x - 2y + 3z}$$

$$= \frac{\partial}{\partial z}\frac{\partial}{\partial y}\left(-2e^{x - 2y + 3z}\right)$$

$$= \frac{\partial}{\partial z}\left(4e^{x - 2y + 3z}\right) = 12\, e^{x - 2y + 3z},$$

$$f_{232}(x, y, z) = \frac{\partial}{\partial y}\frac{\partial}{\partial z}\frac{\partial}{\partial y} e^{x - 2y + 3z}$$

$$= \frac{\partial}{\partial y}\frac{\partial}{\partial z}\left(-2e^{x - 2y + 3z}\right)$$

$$= \frac{\partial}{\partial y}\left(-6e^{x - 2y + 3z}\right) = 12\, e^{x - 2y + 3z},$$

$$f_{322}(x, y, z) = \frac{\partial}{\partial y}\frac{\partial}{\partial y}\frac{\partial}{\partial z} e^{x - 2y + 3z}$$

$$= \frac{\partial}{\partial y}\frac{\partial}{\partial y}\left(3e^{x - 2y + 3z}\right)$$

$$= \frac{\partial}{\partial y}\left(-6e^{x - 2y + 3z}\right) = 12\, e^{x - 2y + 3z}.$$

In both of the examples above observe that the mixed partial derivatives taken with respect to the same variables but in different orders turned out to be equal. This is not a coincidence. It will always occur for sufficiently smooth functions. In particular, the mixed partial derivatives involved are required to be *continuous*. The following theorem presents a more precise statement of this important phenomenon.

THEOREM **1**

Equality of mixed partials

Suppose that two mixed nth-order partial derivatives of a function f involve the same differentiations but in different orders. If those partials are continuous at a point P, and if f and all partials of f of order less than n are continuous in a neighbourhood of P, then the two mixed partials are equal at the point P.

PROOF   We shall prove only a representative special case, showing the equality of $f_{12}(a, b)$ and $f_{21}(a, b)$ for a function f of two variables, provided f_{12} and f_{21} are defined and f_1, f_2, and f are continuous throughout a disk of positive radius centred at (a, b), and f_{12} and f_{21} are continuous at (a, b). Let h and k have sufficiently small absolute values that the point $(a + h, b + k)$ lies in this disk. Then so do all points of the rectangle with sides parallel to the coordinate axes and diagonally opposite corners at (a, b) and $(a + h, b + k)$. (See Figure 12.19.)

Let $Q = f(a + h, b + k) - f(a + h, b) - f(a, b + k) + f(a, b)$ and define single-variable functions $u(x)$ and $v(y)$ by

$$u(x) = f(x, b + k) - f(x, b) \qquad \text{and} \qquad v(y) = f(a + h, y) - f(a, y).$$

Evidently $Q = u(a + h) - u(a)$ and also $Q = v(b + k) - v(b)$. By the (single-variable) Mean-Value Theorem, there exists a number θ_1 satisfying $0 < \theta_1 < 1$ (so that $a + \theta_1 h$ lies between a and $a + h$) such that

$$Q = u(a + h) - u(a) = h\, u'(a + \theta_1 h) = h\left[f_1(a + \theta_1 h, b + k) - f_1(a + \theta_1 h, b) \right].$$

Now we apply the Mean-Value Theorem again, this time to f_1 considered as a function of its second variable, and obtain another number θ_2 satisfying $0 < \theta_2 < 1$ such that

$$f_1(a + \theta_1 h, b + k) - f_1(a + \theta_1 h, b) = k\, f_{12}(a + \theta_1 h, b + \theta_2 k).$$

Thus $Q = hk\, f_{12}(a + \theta_1 h, b + \theta_2 k)$. Two similar applications of the Mean-Value Theorem to $Q = v(b + k) - v(b)$ lead to $Q = hk\, f_{21}(a + \theta_3 h, b + \theta_4 k)$, where θ_3 and θ_4 are two numbers each between 0 and 1. Equating these two expressions for Q and cancelling the common factor hk, we obtain

$$f_{12}(a + \theta_1 h, b + \theta_2 k) = f_{21}(a + \theta_3 h, b + \theta_4 k).$$

Since f_{12} and f_{21} are continuous at (a, b), we can let h and k approach zero to obtain $f_{12}(a, b) = f_{21}(a, b)$, as required.

Figure 12.19   A rectangle contained in the disk where f and certain partials are continuous

points in figure: $(a+h, b+k)$, $(a, b+k)$, (a, b), $(a+h, b)$

⚠ BEWARE

The Mean-Value Theorem is used four times in this proof, each time to write a difference of the form $g(p + m) - g(p)$ in the form $g'(c)m$, where c is some number between p and $p + m$. It is convenient to write c in the form $p + \theta m$, where θ is some number between 0 and 1.

Exercise 16 below develops an example of a function for which f_{12} and f_{21} exist but are not continuous at $(0, 0)$, and for which $f_{12}(0, 0) \neq f_{21}(0, 0)$.

Remark   **Partial Derivatives in Maple**   When you use the Maple function **diff** to calculate a derivative, you must include the name of the variable of differentiation. For example, `diff(x^2+y^3, x)` gives the result $2x$. It doesn't matter that the function being differentiated depends on more than one variable since you are telling Maple to differentiate with respect to x. If you wanted the derivative with respect to y, you would input `diff(x^2+y^3, y)` and the output would be $3y^2$. In this context, there is no distinction between ordinary and partial derivatives. There is, however, a difference when you want to apply a *differential operator* to a function f. If f is a function of one variable, you can denote its derivative f' in Maple by `D(f)`. For example,

```
>   f := x -> sin(2*x); fprime := D(f);
```

$$f := x \rightarrow \sin(2\,x)$$

$$fprime := x \rightarrow 2\,\cos(2\,x)$$

The input `fprime(Pi/6)` will now give the output 1, as expected.

If f is a function of two (or more) variables, then $D(f)$ no longer makes sense; do we mean f_1 or f_2? We distinguish the two (or more) first partials by using subscripts with D.

```
>   f := (x,y) -> exp(3*y)*sin(2*x);
```

$$f := (x, y) \rightarrow e^{(3y)} * \sin(2\,x)$$

```
>   fone := D[1](f); ftwo := D[2](f);
```

$$fone := (x, y) \rightarrow 2e^{(3y)} * \cos(2\,x)$$

$$ftwo := (x, y) \rightarrow 3e^{(3y)} * \sin(2\,x)$$

Higher-order partials are denoted with multiple subscripts (within one set of square brackets).

```
>   D[1,1,2](f)(Pi/4, 0);
```

$$-12$$

You don't need to worry about the order of the subscripts in a mixed partial. Maple assumes the partials are continuous, even if it doesn't know what the function is. Even if g has not been assigned any meaning during the current Maple session, the input `D[1,2](g)(x,y)-D[2,1](g)(x,y);` produces the output 0.

The Laplace and Wave Equations

Many important and interesting phenomena are modelled by functions of several variables that satisfy certain *partial differential equations*. In the following examples we encounter two particular partial differential equations that arise frequently in mathematics and the physical sciences. Exercises 17–19 below introduce another such equation with important applications.

Example 3   Show that for any real number k the functions

$$z = e^{kx}\,\cos(ky) \qquad \text{and} \qquad z = e^{kx}\,\sin(ky)$$

satisfy the partial differential equation

$$\frac{\partial^2 z}{\partial x^2} + \frac{\partial^2 z}{\partial y^2} = 0$$

at every point in the xy-plane.

Solution   For $z = e^{kx}\,\cos(ky)$ we have

$$\frac{\partial z}{\partial x} = k\,e^{kx}\,\cos(ky), \qquad \frac{\partial z}{\partial y} = -k\,e^{kx}\,\sin(ky),$$

$$\frac{\partial^2 z}{\partial x^2} = k^2\,e^{kx}\,\cos(ky), \qquad \frac{\partial^2 z}{\partial y^2} = -k^2\,e^{kx}\,\cos(ky).$$

Thus,

$$\frac{\partial^2 z}{\partial x^2} + \frac{\partial^2 z}{\partial y^2} = k^2\,e^{kx}\,\cos(ky) - k^2\,e^{kx}\,\cos(ky) = 0.$$

The calculation for $z = e^{kx}\,\sin(ky)$ is similar.

Remark The partial differential equation in the above example is called the (two-dimensional) **Laplace equation**. A function of two variables having continuous second partial derivatives in a region of the plane is said to be **harmonic** there if it satisfies Laplace's equation. Such functions play a critical role in the theory of differentiable functions of a *complex variable* (see Appendix II) and are used to model various physical quantities such as steady-state temperature distributions, fluid flows, and electric and magnetic potential fields. Harmonic functions have many interesting properties. They have derivatives of all orders, and they are *analytic*, that is, they are the sums of their (multivariable) Taylor series. Moreover, a harmonic function can achieve maximum and minimum values only on the boundary of its domain. Laplace's equation, and therefore harmonic functions, can be considered in any number of dimensions. (See Exercises 13 and 14 below.)

Example 4 If f and g are any twice-differentiable functions of one variable, show that

$$w = f(x - ct) + g(x + ct)$$

satisfies the partial differential equation

$$\frac{\partial^2 w}{\partial t^2} = c^2 \frac{\partial^2 w}{\partial x^2}.$$

Solution Using the Chain Rule for functions of one variable we obtain

$$\frac{\partial w}{\partial t} = -c\, f'(x - ct) + c\, g'(x + ct), \qquad \frac{\partial w}{\partial x} = f'(x - ct) + g'(x + ct),$$

$$\frac{\partial^2 w}{\partial t^2} = c^2\, f''(x - ct) + c^2\, g''(x + ct), \qquad \frac{\partial^2 w}{\partial x^2} = f''(x - ct) + g''(x + ct).$$

Thus w satisfies the given differential equation.

Remark The partial differential equation in the above example is called the (one-dimensional) **wave equation**. If t measures time, then $f(x - ct)$ represents a waveform travelling to the right along the x-axis with speed c. (See Figure 12.20.) Similarly, $g(x + ct)$ represents a waveform travelling to the left with speed c. Unlike the solutions of Laplace's equation that must be infinitely differentiable, solutions of the wave equation need only have enough derivatives to satisfy the differential equation. The functions f and g are otherwise arbitrary.

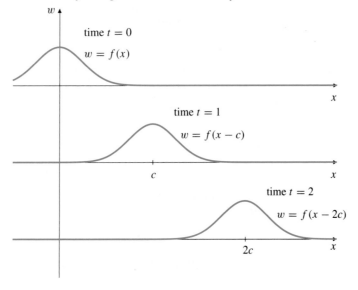

Figure 12.20 $w = f(x - ct)$ represents a waveform moving to the right with speed c

||||||||| **Exercises 12.4** |||||||||

In Exercises 1–6, find all the second partial derivatives of the given function.

1. $z = x^2(1 + y^2)$ **2.** $f(x, y) = x^2 + y^2$

3. $w = x^3 y^3 z^3$ **4.** $z = \sqrt{3x^2 + y^2}$

5. $z = x e^y - y e^x$

6. $f(x, y) = \ln\left(1 + \sin(xy)\right)$

* **7.** How many mixed partial derivatives of order 3 can a function of three variables have? If they are all continuous, how many different values can they have at one point? Find the mixed partials of order 3 for $f(x, y, z) = x e^{xy} \cos(xz)$ that involve two differentiations with respect to z and one with respect to x.

Show that the functions in Exercises 8–12 are harmonic in the plane regions indicated.

8. $f(x, y) = A(x^2 - y^2) + Bxy$ in the whole plane (A and B are constants.)

9. $f(x, y) = 3x^2 y - y^3$ in the whole plane (Can you think of another polynomial of degree 3 in x and y that is also harmonic?)

10. $f(x, y) = \dfrac{x}{x^2 + y^2}$ everywhere except at the origin

11. $f(x, y) = \ln(x^2 + y^2)$ everywhere except at the origin

12. $\tan^{-1}(y/x)$ except at points on the y-axis

13. Show that $w = e^{3x+4y} \sin(5z)$ is harmonic in all of $\mathbb{R}^3$, that is, it satisfies everywhere the 3-dimensional Laplace equation

$$\frac{\partial^2 w}{\partial x^2} + \frac{\partial^2 w}{\partial y^2} + \frac{\partial^2 w}{\partial z^2} = 0.$$

14. Assume that $f(x, y)$ is harmonic in the xy-plane. Show that each of the functions $z f(x, y)$, $x f(y, z)$, and $y f(z, x)$ is harmonic in the whole of $\mathbb{R}^3$. What condition should the constants a, b, and c satisfy to ensure that $f(ax + by, cz)$ is harmonic in $\mathbb{R}^3$?

15. Let the functions $u(x, y)$ and $v(x, y)$ have continuous second partial derivatives and satisfy the **Cauchy–Riemann equations**

$$\frac{\partial u}{\partial x} = \frac{\partial v}{\partial y} \qquad \text{and} \qquad \frac{\partial v}{\partial x} = -\frac{\partial u}{\partial y}.$$

Show that u and v are both harmonic.

* **16.** Let $F(x, y) = \begin{cases} \dfrac{2xy(x^2 - y^2)}{x^2 + y^2}, & \text{if } (x, y) \neq (0, 0) \\ 0, & \text{if } (x, y) = (0, 0) \end{cases}$

Calculate $F_1(x, y)$, $F_2(x, y)$, $F_{12}(x, y)$, and $F_{21}(x, y)$ at points $(x, y) \neq (0, 0)$. Also calculate these derivatives at $(0, 0)$. Observe that $F_{21}(0, 0) = 2$ and $F_{12}(0, 0) = -2$. Does this result contradict Theorem 1? Explain why.

The heat (diffusion) equation

17. Show that the function $u(x, t) = t^{-1/2} e^{-x^2/4t}$ satisfies the partial differential equation

$$\frac{\partial u}{\partial t} = \frac{\partial^2 u}{\partial x^2}.$$

This equation is called the **one-dimensional heat equation** because it models heat diffusion in an insulated rod (with $u(x, t)$ representing the temperature at position x at time t) and other similar phenomena.

18. Show that the function $u(x, y, t) = t^{-1} e^{-(x^2+y^2)/4t}$ satisfies the **two-dimensional heat equation**

$$\frac{\partial u}{\partial t} = \frac{\partial^2 u}{\partial x^2} + \frac{\partial^2 u}{\partial y^2}.$$

19. By comparing the results of Exercises 17 and 18, guess a solution to the **three-dimensional heat equation**

$$\frac{\partial u}{\partial t} = \frac{\partial^2 u}{\partial x^2} + \frac{\partial^2 u}{\partial y^2} + \frac{\partial^2 u}{\partial z^2}.$$

Verify your guess. (If you're feeling lazy, use Maple.)

Biharmonic functions

A function $u(x, y)$ with continuous partials of fourth order is **biharmonic** if $\dfrac{\partial^2 u}{\partial x^2} + \dfrac{\partial^2 u}{\partial y^2}$ is a harmonic function.

20. Show that $u(x, y)$ is biharmonic if and only if it satisfies the biharmonic equation

$$\frac{\partial^4 u}{\partial x^4} + 2\frac{\partial^4 u}{\partial x^2 \partial y^2} + \frac{\partial^4 u}{\partial y^4} = 0$$

21. Verify that $u(x, y) = x^4 - 3x^2 y^2$ is biharmonic.

22. Show that if $u(x, y)$ is harmonic, then $v(x, y) = xu(x, y)$ and $w(x, y) = yu(x, y)$ are biharmonic.

Use the result of Exercise 22 to show that the functions in Exercises 23–25 are biharmonic.

23. $x e^x \sin y$ **24.** $y \ln(x^2 + y^2)$

25. $\dfrac{xy}{x^2 + y^2}$

26. Propose a definition of a biharmonic function of three variables, and prove results analogous to those of Exercises 20 and 22 for biharmonic functions $u(x, y, z)$.

27. Use Maple to verify directly that the function of Exercise 25 is biharmonic.

12.5 The Chain Rule

The Chain Rule for functions of one variable is a formula that gives the derivative of a composition $f\big(g(x)\big)$ of two functions f and g:

$$\frac{d}{dx} f\big(g(x)\big) = f'\big(g(x)\big)g'(x).$$

The situation for several variables is more complicated. If f depends on more than one variable, and any of those variables can be functions of one or more other variables, we cannot expect a simple formula for partial derivatives of the composition to cover all possible cases. We must come to think of the Chain Rule as a *procedure for differentiating compositions* rather than as a formula for their derivatives. In order to motivate a formulation of the Chain Rule for functions of two variables, we begin with a concrete example.

Example 1 Suppose you are hiking in a mountainous region for which you have a map. Let (x, y) be the coordinates of your position on the map (i.e., the horizontal coordinates of your actual position in the region). Let $z = f(x, y)$ denote the height of land (above sea level, say) at position (x, y). Suppose you are walking along a trail so that your position at time t is given by $x = u(t)$ and $y = v(t)$. (These are parametric equations of the trail on the map.) At time t your altitude above sea level is given by the composite function

$$z = f\big(u(t), v(t)\big) = g(t),$$

a function of only one variable. How fast is your altitude changing with respect to time at time t?

Solution The answer is the derivative of $g(t)$:

$$g'(t) = \lim_{h\to 0} \frac{g(t+h) - g(t)}{h} = \lim_{h\to 0} \frac{f\big(u(t+h), v(t+h)\big) - f\big(u(t), v(t)\big)}{h}$$

$$= \lim_{h\to 0} \frac{f\big(u(t+h), v(t+h)\big) - f\big(u(t), v(t+h)\big)}{h}$$

$$+ \lim_{h\to 0} \frac{f\big(u(t), v(t+h)\big) - f\big(u(t), v(t)\big)}{h}.$$

We added 0 to the numerator of the Newton quotient in a creative way so as to separate the quotient into the sum of two quotients, in the first of which the difference of values of f involves only the first variable of f, and in the second of which the difference involves only the second variable of f. The single-variable Chain Rule suggests that the sum of the two limits above is

$$g'(t) = f_1\big(u(t), v(t)\big)u'(t) + f_2\big(u(t), v(t)\big)v'(t).$$

The above formula is the Chain Rule for $\dfrac{d}{dt} f\big(u(t), v(t)\big)$. In terms of Leibniz notation we have

A version of the Chain Rule

If z is a function of x and y with continuous first partial derivatives, and if x and y are differentiable functions of t, then

$$\frac{dz}{dt} = \frac{\partial z}{\partial x} \frac{dx}{dt} + \frac{\partial z}{\partial y} \frac{dy}{dt}.$$

Note that there are two terms in the expression for dz/dt (or $g'(t)$), one arising from each variable of f that depends on t.

Now consider a function f of two variables, x and y, each of which is in turn a function of two other variables, s and t:

$$z = f(x, y), \qquad \text{where} \qquad x = u(s, t) \quad \text{and} \quad y = v(s, t).$$

We can form the composite function

$$z = f\big(u(s, t), v(s, t)\big) = g(s, t).$$

For instance, if $f(x, y) = x^2 + 3y$, where $u(s, t) = st^2$ and $v(s, t) = s - t$, then $g(s, t) = s^2 t^4 + 3(s - t)$.

Let us assume that f, u, and v have first partial derivatives with respect to their respective variables and that those of f are continuous. Then g has first partial derivatives given by

$$g_1(s, t) = f_1\big(u(s, t), v(s, t)\big)u_1(s, t) + f_2\big(u(s, t), v(s, t)\big)v_1(s, t),$$
$$g_2(s, t) = f_1\big(u(s, t), v(s, t)\big)u_2(s, t) + f_2\big(u(s, t), v(s, t)\big)v_2(s, t).$$

These formulas can be expressed more simply using Leibniz notation:

Another version of the Chain Rule

If z is a function of x and y with continuous first partial derivatives, and if x and y depend on s and t, then

$$\frac{\partial z}{\partial s} = \frac{\partial z}{\partial x}\frac{\partial x}{\partial s} + \frac{\partial z}{\partial y}\frac{\partial y}{\partial s},$$
$$\frac{\partial z}{\partial t} = \frac{\partial z}{\partial x}\frac{\partial x}{\partial t} + \frac{\partial z}{\partial y}\frac{\partial y}{\partial t}.$$

This can be deduced from the version obtained in Example 1 by allowing u and v there to depend on two variables, but holding one of them fixed while we differentiate with respect to the other. A more formal proof of this simple but representative case of the Chain Rule will be given in the next section.

The two equations in the box above can be combined into a single matrix equation:

$$\begin{pmatrix} \dfrac{\partial z}{\partial s} & \dfrac{\partial z}{\partial t} \end{pmatrix} = \begin{pmatrix} \dfrac{\partial z}{\partial x} & \dfrac{\partial z}{\partial y} \end{pmatrix} \begin{pmatrix} \dfrac{\partial x}{\partial s} & \dfrac{\partial x}{\partial t} \\ \dfrac{\partial y}{\partial s} & \dfrac{\partial y}{\partial t} \end{pmatrix}.$$

We will comment on the significance of this matrix form at the end of the next section.

In general, if z is a function of several "primary" variables, and each of these depends on some "secondary" variables, then the partial derivative of z with respect to one of the secondary variables will have several terms, one for the contribution to the derivative arising from each of the primary variables on which z depends.

Remark Note the significance of the various subscripts denoting partial derivatives in the functional form of the Chain Rule:

$$g_1(s, t) = f_1\big(u(s, t), v(s, t)\big)u_1(s, t) + f_2\big(u(s, t), v(s, t)\big)v_1(s, t).$$

The "1" in $g_1(s, t)$ refers to differentiation with respect to s, the first variable on which g depends. By contrast, the "1" in $f_1(u(s, t), v(s, t))$ refers to differentiation with respect to x, the first variable on which f depends. (This derivative is then evaluated at $x = u(s, t)$, $y = v(s, t)$.)

Example 2 If $z = \sin(x^2 y)$, where $x = st^2$ and $y = s^2 + \dfrac{1}{t}$, find $\partial z/\partial s$ and $\partial z/\partial t$

(a) by direct substitution and the single-variable form of the Chain Rule, and

(b) by using the (two-variable) Chain Rule.

Solution

(a) By direct substitution:

$$z = \sin\left((st^2)^2\left(s^2 + \frac{1}{t}\right)\right) = \sin(s^4 t^4 + s^2 t^3),$$

$$\frac{\partial z}{\partial s} = (4s^3 t^4 + 2st^3)\cos(s^4 t^4 + s^2 t^3),$$

$$\frac{\partial z}{\partial t} = (4s^4 t^3 + 3s^2 t^2)\cos(s^4 t^4 + s^2 t^3).$$

(b) Using the Chain Rule:

$$\frac{\partial z}{\partial s} = \frac{\partial z}{\partial x}\frac{\partial x}{\partial s} + \frac{\partial z}{\partial y}\frac{\partial y}{\partial s}$$

$$= \left(2xy\cos(x^2 y)\right)t^2 + \left(x^2\cos(x^2 y)\right)2s$$

$$= \left(2st^2\left(s^2 + \frac{1}{t}\right)t^2 + 2s^3 t^4\right)\cos(s^4 t^4 + s^2 t^3)$$

$$= (4s^3 t^4 + 2st^3)\cos(s^4 t^4 + s^2 t^3),$$

$$\frac{\partial z}{\partial t} = \frac{\partial z}{\partial x}\frac{\partial x}{\partial t} + \frac{\partial z}{\partial y}\frac{\partial y}{\partial t}$$

$$= \left(2xy\cos(x^2 y)\right)2st + \left(x^2\cos(x^2 y)\right)\left(\frac{-1}{t^2}\right)$$

$$= \left(2st^2(s^2 + \frac{1}{t})2st + s^2 t^4\left(\frac{-1}{t^2}\right)\right)\cos(s^4 t^4 + s^2 t^3)$$

$$= (4s^4 t^3 + 3s^2 t^2)\cos(s^4 t^4 + s^2 t^3).$$

Note that we still had to use direct substitution on the derivatives obtained in (b) in order to show that the values were the same as those obtained in (a). ∎

Example 3 Find $\dfrac{\partial}{\partial x} f(x^2 y, x + 2y)$ and $\dfrac{\partial}{\partial y} f(x^2 y, x + 2y)$ in terms of the partial derivatives of f, assuming that these partial derivatives are continuous.

Solution We have

$$\frac{\partial}{\partial x} f(x^2 y, x + 2y) = f_1(x^2 y, x + 2y)\frac{\partial}{\partial x}(x^2 y) + f_2(x^2 y, x + 2y)\frac{\partial}{\partial x}(x + 2y)$$

$$= 2xyf_1(x^2 y, x + 2y) + f_2(x^2 y, x + 2y),$$

$$\frac{\partial}{\partial y} f(x^2 y, x + 2y) = f_1(x^2 y, x + 2y)\frac{\partial}{\partial y}(x^2 y) + f_2(x^2 y, x + 2y)\frac{\partial}{\partial y}(x + 2y)$$

$$= x^2 f_1(x^2 y, x + 2y) + 2f_2(x^2 y, x + 2y). \quad ∎$$

Example 4 Express the partial derivatives of $z = h(s, t) = f\big(g(s, t)\big)$ in terms of the derivative f' of f and the partial derivatives of g.

Solution The partial derivatives of h can be calculated using the single-variable version of the Chain Rule: if $x = g(s, t)$, then $z = f(x)$ and

$$h_1(s, t) = \frac{\partial z}{\partial s} = \frac{dz}{dx}\frac{\partial x}{\partial s} = f'\big(g(s, t)\big)g_1(s, t),$$

$$h_2(s, t) = \frac{\partial z}{\partial t} = \frac{dz}{dx}\frac{\partial x}{\partial t} = f'\big(g(s, t)\big)g_2(s, t).$$

The following example involves a hybrid application of the Chain Rule to a function that depends both directly and indirectly on the variable of differentiation.

Example 5 Find dz/dt, where $z = f(x, y, t)$, $x = g(t)$, and $y = h(t)$. (Assume that f, g, and h all have continuous derivatives.)

Solution Since z depends on t through each of the three variables of f, there will be three terms in the appropriate Chain Rule:

$$\frac{dz}{dt} = \frac{\partial z}{\partial x}\frac{dx}{dt} + \frac{\partial z}{\partial y}\frac{dy}{dt} + \frac{\partial z}{\partial t}$$

$$= f_1(x, y, t)g'(t) + f_2(x, y, t)h'(t) + f_3(x, y, t).$$

Remark In the above example we can easily distinguish between the meanings of the symbols dz/dt and $\partial z/\partial t$. If, however, we had been dealing with the situation

$$z = f(x, y, s, t), \qquad \text{where} \quad x = g(s, t) \quad \text{and} \quad y = h(s, t),$$

then the meaning of the symbol $\partial z/\partial t$ would be unclear; it could refer to the simple partial derivative of f with respect to its fourth primary variable (i.e., $f_4(x, y, s, t)$), or it could refer to the derivative of the composite function $f(g(s, t), h(s, t), s, t)$. Three of the four primary variables of f depend on t and, therefore, contribute to the rate of change of z with respect to t. The partial derivative $f_4(x, y, s, t)$ denotes the contribution of only one of these three variables. It is conventional to use $\partial z/\partial t$ to denote the whole derivative of the *composite* function with respect to the secondary variable t:

$$\frac{\partial z}{\partial t} = \frac{\partial}{\partial t}f(g(s, t), h(s, t), s, t)$$

$$= f_1(x, y, s, t)g_2(s, t) + f_2(x, y, s, t)h_2(s, t) + f_4(x, y, s, t).$$

When it is necessary, we can denote the contribution coming from the primary variable t by

$$\left(\frac{\partial z}{\partial t}\right)_{x,y,s} = \frac{\partial}{\partial t}f(x, y, s, t) = f_4(x, y, s, t).$$

Here, the subscripts denote those primary variables of f being *held fixed*, that is, whose contributions to the rate of change of z with respect to t are being *ignored*. Of course, in the situation described above, $(\partial z/\partial t)_s$ means the same as $\partial z/\partial t$.

In applications, the variables that contribute to a particular partial derivative will usually be clear from the context. The following example contains such an application. This is an example of a procedure called *differentiation following the motion*.

Example 6 Atmospheric temperature depends on position and time. If we denote position by three spatial coordinates x, y, and z (measured in kilometres) and time by t (measured in hours), then the temperature $T \, °C$ is a function of four variables, $T(x, y, z, t)$.

(a) If a thermometer is attached to a weather balloon that moves through the atmosphere on a path with parametric equations $x = f(t)$, $y = g(t)$, and $z = h(t)$, what is the rate of change at time t of the temperature T recorded by the thermometer?

(b) Find the rate of change of the recorded temperature at time $t = 1$ if

$$T(x, y, z, t) = \frac{xy}{1+z}(1+t),$$

and if the balloon moves along the curve

$$x = t, \qquad y = 2t, \qquad z = t - t^2.$$

Solution

(a) Here, the rate of change of the thermometer reading depends on the change in position of the thermometer as well as increasing time. Thus, none of the four variables of T can be ignored in the differentiation. The rate is given by

$$\frac{dT}{dt} = \frac{\partial T}{\partial x}\frac{dx}{dt} + \frac{\partial T}{\partial y}\frac{dy}{dt} + \frac{\partial T}{\partial z}\frac{dz}{dt} + \frac{\partial T}{\partial t}.$$

The term $\partial T/\partial t$ refers only to the rate of change of the temperature with respect to time at a fixed position in the atmosphere. The other three terms arise from the motion of the balloon.

(b) The values of the three coordinates and their derivatives at $t = 1$ are $x = 1$, $y = 2$, $z = 0$, $dx/dt = 1$, $dy/dt = 2$, and $dz/dt = -1$. Also, at $t = 1$,

$$\frac{\partial T}{\partial x} = \frac{y}{1+z}(1+t) = 4, \qquad \frac{\partial T}{\partial z} = \frac{-xy}{(1+z)^2}(1+t) = -4,$$

$$\frac{\partial T}{\partial y} = \frac{x}{1+z}(1+t) = 2, \qquad \frac{\partial T}{\partial t} = \frac{xy}{1+z} = 2.$$

Thus,

$$\left.\frac{dT}{dt}\right|_{t=1} = (4)(1) + (2)(2) + (-4)(-1) + 2 = 14.$$

The recorded temperature is increasing at a rate of $14\,°$C/h at time $t = 1$.

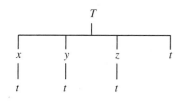

Figure 12.21   Chart showing the dependence of T on t in Example 6

The discussion and examples above show that the Chain Rule for functions of several variables can take different forms depending on the numbers of variables of the various functions being composed. As an aid in determining the correct form of the Chain Rule in a given situation you can construct a chart showing which variables depend on which. Figure 12.21 shows such a chart for the temperature function of Example 6. The Chain Rule for dT/dt involves a term for every route from T to t in the chart. The route from T through x to t produces the term $\dfrac{\partial T}{\partial x}\dfrac{dx}{dt}$ and so on.

Example 7   Write the appropriate Chain Rule for $\partial z/\partial x$, where z depends on u, v, and r; u and v depend on x, y, and r; and r depends on x and y.

Solution   The appropriate chart is shown in Figure 12.22. There are five routes from z to x:

$$\frac{\partial z}{\partial x} = \frac{\partial z}{\partial u}\frac{\partial u}{\partial x} + \frac{\partial z}{\partial u}\frac{\partial u}{\partial r}\frac{\partial r}{\partial x} + \frac{\partial z}{\partial v}\frac{\partial v}{\partial x} + \frac{\partial z}{\partial v}\frac{\partial v}{\partial r}\frac{\partial r}{\partial x} + \frac{\partial z}{\partial r}\frac{\partial r}{\partial x}.$$

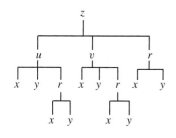

Figure 12.22   Dependence chart for Example 7

Homogeneous Functions

A function $f(x_1, \ldots, x_n)$ is said to be **positively homogeneous of degree** k if, for every point $(x_1, x_2, \ldots, x_n)$ in its domain and every real number $t > 0$, we have

$$f(tx_1, tx_2, \ldots, tx_n) = t^k f(x_1, \ldots, x_n).$$

For example,

$$f(x, y) = x^2 + xy - y^2 \qquad \text{is positively homogeneous of degree 2,}$$
$$f(x, y) = \sqrt{x^2 + y^2} \qquad \text{is positively homogeneous of degree 1,}$$
$$f(x, y) = \frac{2xy}{x^2 + y^2} \qquad \text{is positively homogeneous of degree 0,}$$
$$f(x, y, z) = \frac{x - y + 5z}{yz - z^2} \qquad \text{is positively homogeneous of degree } -1,$$
$$f(x, y) = x^2 + y \qquad \text{is not positively homogeneous.}$$

Observe that a positively homogeneous function of degree 0 remains constant along rays from the origin. More generally, along such rays a positively homogeneous function of degree k grows or decays proportionally to the kth power of distance from the origin.

THEOREM 2

Euler's Theorem

If $f(x_1, \ldots, x_n)$ has continuous first partial derivatives and is positively homogeneous of degree k, then

$$\sum_{i=1}^{n} x_i f_i(x_1, \ldots, x_n) = kf(x_1, \ldots, x_n).$$

PROOF   Differentiate the equation $f(tx_1, tx_2, \ldots, tx_n) = t^k f(x_1, \ldots, x_n)$ with respect to t to get

$$x_1 f_1(tx_1, \ldots, tx_n) + x_2 f_2(tx_1, \ldots, tx_n) + \ldots + x_n f_n(tx_1, \ldots, tx_n)$$
$$= k\, t^{k-1} f(x_1, \ldots, x_n).$$

Now substitute $t = 1$ to get the desired result.

Note that Exercises 26–29 in Section 12.3 illustrate this theorem.

Higher-Order Derivatives

Applications of the Chain Rule to higher-order derivatives can become quite complicated. It is important to keep in mind at each stage which variables are independent of one another.

Example 8   Calculate $\dfrac{\partial^2}{\partial x \partial y} f(x^2 - y^2, xy)$ in terms of partial derivatives of the function f. Assume that the second-order partials of f are continuous.

Solution   In this problem symbols for the primary variables on which f depends are not stated explicitly. Let them be u and v. The problem therefore asks us to find

$$\frac{\partial^2}{\partial x \partial y} f(u, v), \qquad \text{where} \quad u = x^2 - y^2 \quad \text{and} \quad v = xy.$$

First differentiate with respect to y:

$$\frac{\partial}{\partial y} f(u, v) = -2y f_1(u, v) + x f_2(u, v).$$

Now differentiate this result with respect to x. Note that the second term on the right is a product of two functions of x so we need to use the Product Rule:

$$\frac{\partial^2}{\partial x \partial y} f(u, v) = -2y\Big(2xf_{11}(u, v) + yf_{12}(u, v)\Big)$$

$$+ f_2(u, v) + x\Big(2xf_{21}(u, v) + yf_{22}(u, v)\Big)$$

$$= f_2(u, v) - 4xyf_{11}(u, v) + 2(x^2 - y^2)f_{12}(u, v) + xyf_{22}(u, v).$$

In the last step we have used the fact that the mixed partials of f are continuous so we could equate f_{12} and f_{21}.

Review the above calculation very carefully and make sure you understand what is being done at each step. Note that all the derivatives of f that appear are evaluated at $(u, v) = (x^2 - y^2, xy)$, not at (x, y), because x and y are not themselves the primary variables on which f depends.

Remark   The kind of calculation done in the above example (and the following ones) is easily carried out by a computer algebra system. In Maple:

```
>   g := (x,y) -> f(x^2 - y^2, x*y):
      simplify(D[1,2](g)(x,y));
```

$$-4y\,D_{1,1}(f)(x^2 - y^2, xy)x - 2 * D_{1,2}(f)(x^2 - y^2, xy)y^2$$

$$+ 2D_{1,2}(f)(x^2 - y^2, xy)x^2$$

$$+ x\,D_{2,2}(f)(x^2 - y^2, xy)y + D_2(f)(x^2 - y^2, xy)$$

which, on close inspection, is the same answer we calculated in the example.

Example 9   If $f(x, y)$ is harmonic, show that $f(x^2 - y^2, 2xy)$ is also harmonic.

Solution   Let $u = x^2 - y^2$ and $v = 2xy$. If $z = f(u, v)$, then

$$\frac{\partial z}{\partial x} = 2xf_1(u, v) + 2yf_2(u, v),$$

$$\frac{\partial z}{\partial y} = -2yf_1(u, v) + 2xf_2(u, v),$$

$$\frac{\partial^2 z}{\partial x^2} = 2f_1(u, v) + 2x\Big(2xf_{11}(u, v) + 2yf_{12}(u, v)\Big)$$

$$+ 2y\Big(2xf_{21}(u, v) + 2yf_{22}(u, v)\Big)$$

$$= 2f_1(u, v) + 4x^2 f_{11}(u, v) + 8xyf_{12}(u, v) + 4y^2 f_{22}(u, v),$$

$$\frac{\partial^2 z}{\partial y^2} = -2f_1(u, v) - 2y\Big(-2yf_{11}(u, v) + 2xf_{12}(u, v)\Big)$$

$$+ 2x\Big(-2yf_{21}(u, v) + 2xf_{22}(u, v)\Big)$$

$$= -2f_1(u, v) + 4y^2 f_{11}(u, v) - 8xyf_{12}(u, v) + 4x^2 f_{22}(u, v).$$

Therefore,

$$\frac{\partial^2 z}{\partial x^2} + \frac{\partial^2 z}{\partial y^2} = 4(x^2 + y^2)\Big(f_{11}(u, v) + f_{22}(u, v)\Big) = 0$$

because f is harmonic. Thus, $z = f(x^2 - y^2, 2xy)$ is a harmonic function of x and y.

In the following example we show that the two-dimensional Laplace differential equation (see Example 3 in Section 12.4) takes the form

$$\frac{\partial^2 z}{\partial r^2} + \frac{1}{r}\frac{\partial z}{\partial r} + \frac{1}{r^2}\frac{\partial^2 z}{\partial \theta^2} = 0$$

when stated for a function z expressed in terms of polar coordinates r and θ.

Example 10 (**Laplace's equation in polar coordinates**) If $z = f(x, y)$ has continuous partial derivatives of second order, and if $x = r\cos\theta$ and $y = r\sin\theta$, show that

$$\frac{\partial^2 z}{\partial r^2} + \frac{1}{r}\frac{\partial z}{\partial r} + \frac{1}{r^2}\frac{\partial^2 z}{\partial \theta^2} = \frac{\partial^2 z}{\partial x^2} + \frac{\partial^2 z}{\partial y^2}.$$

Solution It is possible to do this in two different ways; we can start with either side and use the Chain Rule to show that it is equal to the other side. Here, we will calculate the partial derivatives with respect to r and θ that appear on the left side and express them in terms of partial derivatives with respect to x and y. The other approach, involving expressing partial derivatives with respect to x and y in terms of partial derivatives with respect to r and θ, is a little more difficult. (See Exercise 24 below.) However, we would have to do it that way if we were not given the form of the differential equation in polar coordinates and had to find it.

First note that

$$\frac{\partial x}{\partial r} = \cos\theta, \qquad \frac{\partial x}{\partial \theta} = -r\sin\theta, \qquad \frac{\partial y}{\partial r} = \sin\theta, \qquad \frac{\partial y}{\partial \theta} = r\cos\theta.$$

Thus,

$$\frac{\partial z}{\partial r} = \frac{\partial z}{\partial x}\frac{\partial x}{\partial r} + \frac{\partial z}{\partial y}\frac{\partial y}{\partial r} = \cos\theta\,\frac{\partial z}{\partial x} + \sin\theta\,\frac{\partial z}{\partial y}.$$

> **⚠ BEWARE**
>
> This is a difficult but important example. Examine each step carefully to make sure you understand what is being done.

Now differentiate with respect to r again. Remember that r and θ are independent variables, so the factors $\cos\theta$ and $\sin\theta$ can be regarded as constants. However, $\partial z/\partial x$ and $\partial z/\partial y$ depend on x and y and, therefore, on r and θ.

$$\frac{\partial^2 z}{\partial r^2} = \cos\theta\,\frac{\partial}{\partial r}\frac{\partial z}{\partial x} + \sin\theta\,\frac{\partial}{\partial r}\frac{\partial z}{\partial y}$$

$$= \cos\theta\left(\cos\theta\,\frac{\partial^2 z}{\partial x^2} + \sin\theta\,\frac{\partial^2 z}{\partial y\partial x}\right) + \sin\theta\left(\cos\theta\,\frac{\partial^2 z}{\partial x\partial y} + \sin\theta\,\frac{\partial^2 z}{\partial y^2}\right)$$

$$= \cos^2\theta\,\frac{\partial^2 z}{\partial x^2} + 2\cos\theta\sin\theta\,\frac{\partial^2 z}{\partial x\partial y} + \sin^2\theta\,\frac{\partial^2 z}{\partial y^2}.$$

We have used the equality of mixed partials in the last line. Similarly,

$$\frac{\partial z}{\partial \theta} = -r\sin\theta\,\frac{\partial z}{\partial x} + r\cos\theta\,\frac{\partial z}{\partial y}.$$

When we differentiate a second time with respect to θ, we can regard r as constant, but each term above is still a product of two functions that depend on θ. Thus,

$$\frac{\partial^2 z}{\partial \theta^2} = -r\left(\cos\theta\,\frac{\partial z}{\partial x} + \sin\theta\,\frac{\partial}{\partial \theta}\frac{\partial z}{\partial x}\right) + r\left(-\sin\theta\,\frac{\partial z}{\partial y} + \cos\theta\,\frac{\partial}{\partial \theta}\frac{\partial z}{\partial y}\right)$$

$$= -r\,\frac{\partial z}{\partial r} - r\sin\theta\left(-r\sin\theta\,\frac{\partial^2 z}{\partial x^2} + r\cos\theta\,\frac{\partial^2 z}{\partial y\partial x}\right)$$

$$\qquad + r\cos\theta\left(-r\sin\theta\,\frac{\partial^2 z}{\partial x\partial y} + r\cos\theta\,\frac{\partial^2 z}{\partial y^2}\right)$$

$$= -r\,\frac{\partial z}{\partial r} + r^2\left(\sin^2\theta\,\frac{\partial^2 z}{\partial x^2} - 2\sin\theta\cos\theta\,\frac{\partial^2 z}{\partial x\partial y} + \cos^2\theta\,\frac{\partial^2 z}{\partial y^2}\right).$$

Combining these results, we obtain the desired formula:

$$\frac{\partial^2 z}{\partial r^2} + \frac{1}{r}\frac{\partial z}{\partial r} + \frac{1}{r^2}\frac{\partial^2 z}{\partial \theta^2} = \frac{\partial^2 z}{\partial x^2} + \frac{\partial^2 z}{\partial y^2}.$$

▌▌▌▌▌▌▌▌ Exercises 12.5 ▌▌▌▌▌▌▌▌▌

In Exercises 1–4, write appropriate versions of the Chain Rule for the indicated derivatives.

1. $\partial w / \partial t$ if $w = f(x, y, z)$, where $x = g(s, t)$, $y = h(s, t)$, and $z = k(s, t)$

2. $\partial w / \partial t$ if $w = f(x, y, z)$, where $x = g(s)$, $y = h(s, t)$, and $z = k(t)$

3. $\partial z / \partial u$ if $z = g(x, y)$, where $y = f(x)$ and $x = h(u, v)$

4. dw / dt if $w = f(x, y)$, $x = g(r, s)$, $y = h(r, t)$, $r = k(s, t)$, and $s = m(t)$

5. If $w = f(x, y, z)$, where $x = g(y, z)$ and $y = h(z)$, state appropriate versions of the Chain Rule for $\dfrac{dw}{dz}$, $\left(\dfrac{\partial w}{\partial z}\right)_x$, and $\left(\dfrac{\partial w}{\partial z}\right)_{x,y}$.

6. Use two different methods to calculate $\partial u / \partial t$ if $u = \sqrt{x^2 + y^2}$, $x = e^{st}$, and $y = 1 + s^2 \cos t$.

7. Use two different methods to calculate $\partial z / \partial x$ if $z = \tan^{-1}(u/v)$, $u = 2x + y$, and $v = 3x - y$.

8. Use two methods to calculate dz / dt given that $z = txy^2$, $x = t + \ln(y + t^2)$, and $y = e^t$.

In Exercises 9–12, find the indicated derivatives, assuming that the function $f(x, y)$ has continuous first partial derivatives.

9. $\dfrac{\partial}{\partial x} f(2x, 3y)$ **10.** $\dfrac{\partial}{\partial x} f(2y, 3x)$

11. $\dfrac{\partial}{\partial x} f(y^2, x^2)$ **12.** $\dfrac{\partial}{\partial y} f\left(yf(x, t), f(y, t)\right)$

13. Suppose that the temperature T in a certain liquid varies with depth z and time t according to the formula $T = e^{-t}z$. Find the rate of change of temperature with respect to time at a point that is moving through the liquid so that at time t its depth is $f(t)$. What is this rate if $f(t) = e^t$? What is happening in this case?

14. Suppose the strength E of an electric field in space varies with position (x, y, z) and time t according to the formula $E = f(x, y, z, t)$. Find the rate of change with respect to time of the electric field strength measured by an instrument moving through space along the helix $x = \sin t$, $y = \cos t$, $z = t$.

In Exercises 15–20, assume that f has continuous partial derivatives of all orders.

15. If $z = f(x, y)$, where $x = 2s + 3t$ and $y = 3s - 2t$, find

(a) $\dfrac{\partial^2 z}{\partial s^2}$, (b) $\dfrac{\partial^2 z}{\partial s \partial t}$, and (c) $\dfrac{\partial^2 z}{\partial t^2}$.

16. If $f(x, y)$ is harmonic, show that $f\left(\dfrac{x}{x^2 + y^2}, \dfrac{-y}{x^2 + y^2}\right)$ is also harmonic.

17. If $x = t \sin s$ and $y = t \cos s$, find $\dfrac{\partial^2}{\partial s \partial t} f(x, y)$.

18. Find $\dfrac{\partial^3}{\partial x \partial y^2} f(2x + 3y, xy)$ in terms of partial derivatives of the function f.

19. Find $\dfrac{\partial^2}{\partial y \partial x} f(y^2, xy, -x^2)$ in terms of partial derivatives of

the function f.

20. Find $\dfrac{\partial^3}{\partial t^2 \partial s} f(s^2 - t, s + t^2)$ in terms of partial derivatives of the function f.

21. Suppose that $u(x, y)$ and $v(x, y)$ have continuous second partial derivatives and satisfy the Cauchy–Riemann equations

$$\frac{\partial u}{\partial x} = \frac{\partial v}{\partial y} \quad \text{and} \quad \frac{\partial v}{\partial x} = -\frac{\partial u}{\partial y}.$$

Suppose also that $f(u, v)$ is a harmonic function of u and v. Show that $f\left(u(x, y), v(x, y)\right)$ is a harmonic function of x and y. *Hint:* u and v are harmonic functions by Exercise 15 in Section 12.4.

22. If $r^2 = x^2 + y^2 + z^2$, verify that $u(x, y, z) = 1/r$ is harmonic throughout $\mathbb{R}^3$ except at the origin.

∗23. If $x = e^s \cos t$, $y = e^s \sin t$, and $z = u(x, y) = v(s, t)$, show that

$$\frac{\partial^2 z}{\partial s^2} + \frac{\partial^2 z}{\partial t^2} = (x^2 + y^2)\left(\frac{\partial^2 z}{\partial x^2} + \frac{\partial^2 z}{\partial y^2}\right).$$

∗24. **(Converting Laplace's equation to polar coordinates)** The transformation to polar coordinates, $x = r \cos \theta$, $y = r \sin \theta$, implies that $r^2 = x^2 + y^2$ and $\tan \theta = y/x$. Use these equations to show that

$$\frac{\partial r}{\partial x} = \cos \theta \qquad \frac{\partial r}{\partial y} = \sin \theta$$

$$\frac{\partial \theta}{\partial x} = -\frac{\sin \theta}{r} \qquad \frac{\partial \theta}{\partial y} = \frac{\cos \theta}{r}.$$

Use these formulas to help you express $\dfrac{\partial^2 u}{\partial x^2} + \dfrac{\partial^2 u}{\partial y^2}$ in terms of partials of u with respect to r and θ, and hence reprove the formula for the Laplace differential equation in polar coordinates given in Example 10.

25. If $u(x, y) = r^2 \ln r$, where $r^2 = x^2 + y^2$, verify that u is a biharmonic function by showing that

$$\left(\frac{\partial^2}{\partial x^2} + \frac{\partial^2}{\partial y^2}\right)\left(\frac{\partial^2 u}{\partial x^2} + \frac{\partial^2 u}{\partial y^2}\right) = 0.$$

26. If $f(x, y)$ is positively homogeneous of degree k and has continuous partial derivatives of second order, show that

$$x^2 f_{11}(x, y) + 2xy f_{12}(x, y) + y^2 f_{22}(x, y)$$
$$= k(k - 1) f(x, y).$$

∗27. Generalize the result of Exercise 26 to functions of n variables.

∗28. Generalize the results of Exercises 26 and 27 to expressions involving mth-order partial derivatives of the function f.

Exercises 29–30 revisit Exercise 16 of Section 12.4. Let

$$F(x, y) = \begin{cases} \dfrac{2xy(x^2 - y^2)}{x^2 + y^2}, & \text{if } (x, y) \neq (0, 0) \\ 0, & \text{if } (x, y) = (0, 0). \end{cases}$$

29. (a) Show that $F(x, y) = -F(y, x)$ for all (x, y).

(b) Show that $F_1(x, y) = -F_2(y, x)$ and
$F_{12}(x, y) = -F_{21}(y, x)$ for $(x, y) \neq (0, 0)$.

(c) Show that $F_1(0, y) = -2y$ for all y and, hence, that
$F_{12}(0, 0) = 2$.

(d) Deduce that $F_2(x, 0) = 2x$ and $F_{21}(0, 0) = 2$.

30. (a) Use Exercise 29(b) to find $F_{12}(x, x)$ for $x \neq 0$.

(b) Is $F_{12}(x, y)$ continuous at $(0, 0)$? Why?

◆31. Use the change of variables $\xi = x + ct$, $\eta = x$ to transform
the partial differential equation

$$\frac{\partial u}{\partial t} = c \frac{\partial u}{\partial x}, \qquad (c = \text{constant}),$$

into the simpler equation $\partial v / \partial \eta = 0$, where
$v(\xi, \eta) = v(x + ct, x) = u(x, t)$. This equation says that
$v(\xi, \eta)$ does not depend on η, so $v = f(\xi)$ for some arbitrary
differentiable function f. What is the corresponding
"general solution" $u(x, t)$ of the original partial differential
equation?

◆32. Having considered Exercise 31, guess a "general solution"
$w(r, s)$ of the second-order partial differential equation

$$\frac{\partial^2}{\partial r \partial s} w(r, s) = 0.$$

Your answer should involve two arbitrary functions.

◆33. Use the change of variables $r = x + ct$, $s = x - ct$,
$w(r, s) = u(x, t)$ to transform the one-dimensional wave
equation

$$\frac{\partial^2 u}{\partial t^2} = c^2 \frac{\partial^2 u}{\partial x^2}$$

to a simpler form. Now use the result of Exercise 32 to find
the *general solution* of this wave equation in the form given
in Example 4 in Section 12.4.

◆34. Show that the initial-value problem for the one-dimensional
wave equation

$$\begin{cases} u_{tt}(x, t) = c^2 u_{xx}(x, t) \\ u(x, 0) = p(x) \\ u_t(x, 0) = q(x) \end{cases}$$

has the solution

$$u(x, t) = \frac{1}{2}\left[p(x - ct) + p(x + ct) \right] + \frac{1}{2c} \int_{x-ct}^{x+ct} q(s)\, ds.$$

(Note that we have used subscripts x and t instead of 1 and 2
to denote the partial derivatives here. This is common usage
in dealing with partial differential equations.)

Remark   The initial-value problem in Exercise 34 gives the
small lateral displacement $u(x, t)$ at position x at time t of a
vibrating string held under tension along the x-axis. The function
$p(x)$ gives the *initial* displacement at position x, that is, the
displacement at time $t = 0$. Similarly, $q(x)$ gives the initial
velocity at position x. Observe that the position at time t depends
only on values of these initial data at points no further than ct
units away. This is consistent with the previous observation that
the solutions of the wave equation represent waves travelling
with speed c.

Redo the examples and exercises listed in Exercises 35–40 using
Maple to do the calculations.

35. Example 10   **36.** Exercise 16

37. Exercise 19   **38.** Exercise 20

39. Exercise 23   **40.** Exercise 34

12.6 Linear Approximations, Differentiability, and Differentials

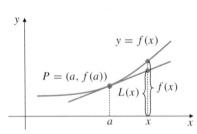

Figure 12.23   The linearization of f at
$x = a$

The tangent line to the graph $y = f(x)$ at $x = a$ provides a convenient approximation
for values of $f(x)$ for x near a (see Figure 12.23):

$$f(x) \approx L(x) = f(a) + f'(a)(x - a).$$

Here, $L(x)$ is the **linearization** of f at a; its graph is the tangent line to $y = f(x)$ there.
The mere existence of $f'(a)$ is sufficient to guarantee that the error in the approximation
(the vertical distance between the curve and tangent at x) is small compared with the
distance $h = x - a$ between a and x, that is,

$$\lim_{h \to 0} \frac{f(a+h) - L(a+h)}{h} = \lim_{h \to 0} \frac{f(a+h) - f(a) - f'(a)h}{h}$$
$$= \lim_{h \to 0} \frac{f(a+h) - f(a)}{h} - f'(a)$$
$$= f'(a) - f'(a) = 0.$$

Similarly, the tangent plane to the graph of $z = f(x, y)$ at (a, b) is $z = L(x, y)$, where

$$L(x, y) = f(a, b) + f_1(a, b)(x - a) + f_2(a, b)(y - b)$$

is the **linearization** of f at (a, b). We can use $L(x, y)$ to approximate values of $f(x, y)$ near (a, b):

$$f(x, y) \approx L(x, y) = f(a, b) + f_1(a, b)(x - a) + f_2(a, b)(y - b).$$

Example 1   Find an approximate value for $f(x, y) = \sqrt{2x^2 + e^{2y}}$ at $(2.2, -0.2)$.

Solution   It is convenient to use the linearization at $(2, 0)$, where the values of f and its partials are easily evaluated:

$$f(2, 0) = 3,$$

$$f_1(x, y) = \frac{2x}{\sqrt{2x^2 + e^{2y}}}, \qquad f_1(2, 0) = \frac{4}{3},$$

$$f_2(x, y) = \frac{e^{2y}}{\sqrt{2x^2 + e^{2y}}}, \qquad f_2(2, 0) = \frac{1}{3}.$$

Thus, $L(x, y) = 3 + \dfrac{4}{3}(x - 2) + \dfrac{1}{3}(y - 0)$, and

$$f(2.2, -0.2) \approx L(2.2, -0.2) = 3 + \frac{4}{3}(2.2 - 2) + \frac{1}{3}(-0.2 - 0) = 3.2.$$

(For the sake of comparison, $f(2.2, -0.2) \approx 3.2172$ to 4 decimal places.)

Unlike the single-variable case, the mere existence of the partial derivatives $f_1(a, b)$ and $f_2(a, b)$ does not even imply that f is continuous at (a, b), let alone that the error in the linearization is small compared with the distance $\sqrt{(x - a)^2 + (y - b)^2}$ between (a, b) and (x, y). We adopt this latter condition as our definition of what it means for a function of two variables to be *differentiable* at a point.

DEFINITION 5

We say that the function $f(x, y)$ is **differentiable** at the point (a, b) if

$$\lim_{(h,k) \to (0,0)} \frac{f(a + h, b + k) - f(a, b) - h\, f_1(a, b) - k f_2(a, b)}{\sqrt{h^2 + k^2}} = 0.$$

This definition and the following theorems can be generalized to functions of any number of variables in the obvious way. For the sake of simplicity, we state them for the two-variable case only.

The function $f(x, y)$ is differentiable at the point (a, b) if and only if the surface $z = f(x, y)$ has a *nonvertical tangent plane* at (a, b). This implies that $f_1(a, b)$ and $f_2(a, b)$ must exist and that f must be continuous at (a, b). (Recall, however, that the existence of the partial derivatives does *not* even imply that f is continuous, let alone differentiable.) In particular, the function is *continuous* wherever it is differentiable. We will prove a two-variable version of the Mean-Value Theorem and use it to show that functions are differentiable wherever they have *continuous* first partial derivatives.

THEOREM 3   **A Mean-Value Theorem**

If $f_1(x, y)$ and $f_2(x, y)$ are continuous in a neighbourhood of the point (a, b), and if the absolute values of h and k are sufficiently small, then there exist numbers θ_1 and θ_2, each between 0 and 1, such that

$$f(a + h, b + k) - f(a, b) = h f_1(a + \theta_1 h, b + k) + k f_2(a, b + \theta_2 k).$$

PROOF   The proof of this theorem is very similar to that of Theorem 1 in Section 12.4, so we give only a sketch here. The reader can fill in the details. Write

$$f(a+h, b+k) - f(a,b) = \big(f(a+h, b+k) - f(a, b+k)\big) + \big(f(a, b+k) - f(a,b)\big),$$

and then apply the single-variable Mean-Value Theorem separately to $f(x, b+k)$ on the interval between a and $a+h$, and to $f(a, y)$ on the interval between b and $b+k$ to get the desired result.

THEOREM   4

If f_1 and f_2 are continuous in a neighbourhood of the point (a, b), then f is differentiable at (a, b).

PROOF   Using Theorem 3 and the facts that

$$\left| \frac{h}{\sqrt{h^2 + k^2}} \right| \le 1 \quad \text{and} \quad \left| \frac{k}{\sqrt{h^2 + k^2}} \right| \le 1,$$

we estimate

$$\left| \frac{f(a+h, b+k) - f(a,b) - hf_1(a,b) - kf_2(a,b)}{\sqrt{h^2 + k^2}} \right|$$

$$= \left| \frac{h}{\sqrt{h^2 + k^2}} \Big(f_1(a + \theta_1 h, b+k) - f_1(a,b) \Big) \right.$$

$$\left. + \frac{k}{\sqrt{h^2 + k^2}} \Big(f_2(a, b + \theta_2 k) - f_2(a,b) \Big) \right|$$

$$\le \big| f_1(a + \theta_1 h, b+k) - f_1(a,b) \big| + \big| f_2(a, b + \theta_2 k) - f_2(a,b) \big|.$$

Since f_1 and f_2 are continuous at (a, b), each of these latter terms approaches 0 as h and k approach 0. This is what we needed to prove.

We illustrate differentiability with an example where we can calculate directly the error in the tangent plane approximation.

Example 2   Calculate $f(x + h, y + k) - f(x, y) - f_1(x, y)h - f_2(x, y)k$ if $f(x, y) = x^3 + xy^2$.

Solution   Since $f_1(x, y) = 3x^2 + y^2$ and $f_2(x, y) = 2xy$, we have

$$f(x + h, y + k) - f(x, y) - f_1(x, y)h - f_2(x, y)k$$

$$= (x + h)^3 + (x + h)(y + k)^2 - x^3 - xy^2 - (3x^2 + y^2)h - 2xyk$$

$$= 3xh^2 + h^3 + 2yhk + hk^2 + xk^2.$$

Observe that the result above is a polynomial in h and k with no term of degree less than 2 in these variables. Therefore, this difference approaches zero like the *square* of the distance $\sqrt{h^2 + k^2}$ from (x, y) to $(x+h, y+k)$ as $(h, k) \to (0, 0)$, so the condition for differentiability is certainly satisfied:

$$\lim_{(h,k) \to (0,0)} \frac{3xh^2 + h^3 + 2yhk + hk^2 + xk^2}{\sqrt{h^2 + k^2}} = 0.$$

This quadratic behaviour is the case for any function f with continuous *second* partial derivatives. (See Exercise 19 below.)

Proof of the Chain Rule

We are now able to give a formal statement and proof of a simple but representative case of the Chain Rule for multivariate functions.

THEOREM **5**

A Chain Rule

Let $z = f(x, y)$, where $x = u(s, t)$ and $y = v(s, t)$. Suppose that

(i) $u(a, b) = p$ and $v(a, b) = q$,

(ii) the first partial derivatives of u and v exist at the point (a, b), and

(iii) f is differentiable at the point (p, q).

Then $z = w(s, t) = f(u(s, t), v(s, t))$ has first partial derivatives with respect to s and t at (a, b), and

$$w_1(a, b) = f_1(p, q)u_1(a, b) + f_2(p, q)v_1(a, b),$$
$$w_2(a, b) = f_1(p, q)u_2(a, b) + f_2(p, q)v_2(a, b),$$

that is,

$$\frac{\partial z}{\partial s} = \frac{\partial z}{\partial x}\frac{\partial x}{\partial s} + \frac{\partial z}{\partial y}\frac{\partial y}{\partial s} \qquad \text{and} \qquad \frac{\partial z}{\partial t} = \frac{\partial z}{\partial x}\frac{\partial x}{\partial t} + \frac{\partial z}{\partial y}\frac{\partial y}{\partial t}.$$

PROOF Define a function E of two variables as follows: $E(0, 0) = 0$, and if $(h, k) \neq (0, 0)$, then

$$E(h, k) = \frac{f(p + h, q + k) - f(p, q) - hf_1(p, q) - kf_2(p, q)}{\sqrt{h^2 + k^2}}.$$

Observe that $E(h, k)$ is continuous at $(0, 0)$ because f is differentiable at (p, q). Now,

$$f(p + h, q + k) - f(p, q) = hf_1(p, q) + kf_2(p, q) + \sqrt{h^2 + k^2}\, E(h, k).$$

In this formula put $h = u(a + \sigma, b) - u(a, b)$ and $k = v(a + \sigma, b) - v(a, b)$ and divide by σ to obtain

$$\frac{w(a + \sigma, b) - w(a, b)}{\sigma} = \frac{f(u(a + \sigma, b), v(a + \sigma, b)) - f(u(a, b), v(a, b))}{\sigma}$$
$$= \frac{f(p + h, q + k) - f(p, q)}{\sigma}$$
$$= f_1(p, q)\frac{h}{\sigma} + f_2(p, q)\frac{k}{\sigma} + \sqrt{\left(\frac{h}{\sigma}\right)^2 + \left(\frac{k}{\sigma}\right)^2}\, E(h, k).$$

We want to let σ approach 0 in this formula. Note that

$$\lim_{\sigma \to 0} \frac{h}{\sigma} = \lim_{\sigma \to 0} \frac{u(a + \sigma, b) - u(a, b)}{\sigma} = u_1(a, b),$$

and, similarly, $\lim_{\sigma \to 0}(k/\sigma) = v_1(a, b)$. Since $(h, k) \to (0, 0)$ if $\sigma \to 0$, we have

$$w_1(a, b) = f_1(p, q)u_1(a, b) + f_2(p, q)v_1(a, b).$$

The proof for w_2 is similar.

Differentials

If the first partial derivatives of a function $z = f(x_1, \ldots, x_n)$ exist at a point, we may construct a **differential** dz or df of the function at that point in a manner similar to that used for functions of one variable:

$$
dz = df = \frac{\partial z}{\partial x_1}\, dx_1 + \frac{\partial z}{\partial x_2}\, dx_2 + \cdots + \frac{\partial z}{\partial x_n}\, dx_n
$$
$$
= f_1(x_1, \ldots, x_n)\, dx_1 + \cdots + f_n(x_1, \ldots, x_n)\, dx_n.
$$

Here, the differential dz is considered to be a function of the $2n$ independent variables $x_1, x_2, \ldots, x_n, dx_1, dx_2, \ldots, dx_n$.

For a *differentiable* function f, the differential df is an approximation to the change Δf in value of the function given by

$$
\Delta f = f(x_1 + dx_1, \ldots, x_n + dx_n) - f(x_1, \ldots, x_n).
$$

The error in this approximation is small compared with the distance between the two points in the domain of f, that is,

$$
\frac{\Delta f - df}{\sqrt{(dx_1)^2 + \cdots + (dx_n)^2}} \to 0 \quad \text{if all } dx_i \to 0, \qquad (1 \le i \le n).
$$

In this sense, differentials are just another way of looking at linearization.

Example 3   Estimate the percentage change in the period

$$
T = 2\pi \sqrt{\frac{L}{g}}
$$

of a simple pendulum if the length L of the pendulum increases by 2% and the acceleration of gravity g decreases by 0.6%.

Solution   We calculate the differential of T:

$$
dT = \frac{\partial T}{\partial L}\, dL + \frac{\partial T}{\partial g}\, dg
$$
$$
= \frac{2\pi}{2\sqrt{Lg}}\, dL - \frac{2\pi\sqrt{L}}{2g^{3/2}}\, dg.
$$

We are given that $dL = \dfrac{2}{100}\, L$ and $dg = -\dfrac{6}{1{,}000}\, g$. Thus

$$
dT = \frac{1}{100}\, 2\pi \sqrt{\frac{L}{g}} - \left(-\frac{6}{1{,}000}\right) \frac{2\pi}{2} \sqrt{\frac{L}{g}} = \frac{13}{1{,}000}\, T.
$$

Therefore, the period T of the pendulum increases by 1.3%. ∎

Functions from *n*-space to *m*-space

(*This is an optional topic.*)   A vector $\mathbf{f} = (f_1, f_2, \ldots, f_m)$ of m functions, each depending on n variables $(x_1, x_2, \ldots, x_n)$, defines a *transformation* (i.e., a function) from $\mathbb{R}^n$ to $\mathbb{R}^m$; specifically, if $\mathbf{x} = (x_1, x_2, \ldots, x_n)$ is a point in $\mathbb{R}^n$, and

$$
y_1 = f_1(x_1, x_2, \ldots, x_n)
$$
$$
y_2 = f_2(x_1, x_2, \ldots, x_n)
$$
$$
\vdots
$$
$$
y_m = f_m(x_1, x_2, \ldots, x_n),
$$

then $\mathbf{y} = (y_1, y_2, \ldots, y_m)$ is the point in $\mathbb{R}^m$ that corresponds to $\mathbf{x}$ under the transformation $\mathbf{f}$. We can write these equations more compactly as

$$\mathbf{y} = \mathbf{f}(\mathbf{x}).$$

Information about the rate of change of $\mathbf{y}$ with respect to $\mathbf{x}$ is contained in the various partial derivatives $\partial y_i / \partial x_j$, $(1 \le i \le m, \ 1 \le j \le n)$, and is conveniently organized into an $m \times n$ matrix, $D\mathbf{f}(\mathbf{x})$, called the **Jacobian matrix** of the transformation $\mathbf{f}$:

$$D\mathbf{f}(\mathbf{x}) = \begin{pmatrix} \dfrac{\partial y_1}{\partial x_1} & \dfrac{\partial y_1}{\partial x_2} & \cdots & \dfrac{\partial y_1}{\partial x_n} \\[2mm] \dfrac{\partial y_2}{\partial x_1} & \dfrac{\partial y_2}{\partial x_2} & \cdots & \dfrac{\partial y_2}{\partial x_n} \\[2mm] \vdots & \vdots & & \vdots \\[2mm] \dfrac{\partial y_m}{\partial x_1} & \dfrac{\partial y_m}{\partial x_2} & \cdots & \dfrac{\partial y_m}{\partial x_n} \end{pmatrix}$$

The linear transformation (see Section 10.6) represented by the Jacobian matrix is called **the derivative** of the transformation $\mathbf{f}$.

Remark   We can regard the scalar-valued function of two variables, $f(x, y)$ say, as a transformation from $\mathbb{R}^2$ to $\mathbb{R}$. Its derivative is then the linear transformation with matrix

$$Df(x, y) = \big(f_1(x, y), \ f_2(x, y)\big).$$

It is not our purpose to enter into a study of such *vector-valued functions of a vector variable* at this point, but we can observe here that the Jacobian matrix of the composition of two such transformations is the matrix product of their Jacobian matrices.

To see this, let $\mathbf{y} = \mathbf{f}(\mathbf{x})$ be a transformation from $\mathbb{R}^n$ to $\mathbb{R}^m$ as described above, and let $\mathbf{z} = \mathbf{g}(\mathbf{y})$ be another such transformation from $\mathbb{R}^m$ to $\mathbb{R}^k$ given by

$$z_1 = g_1(y_1, y_2, \ldots, y_m)$$
$$z_2 = g_2(y_1, y_2, \ldots, y_m)$$
$$\vdots$$
$$z_k = g_k(y_1, y_2, \ldots, y_m),$$

which has the $k \times m$ Jacobian matrix

$$D\mathbf{g}(\mathbf{y}) = \begin{pmatrix} \dfrac{\partial z_1}{\partial y_1} & \dfrac{\partial z_1}{\partial y_2} & \cdots & \dfrac{\partial z_1}{\partial y_m} \\[2mm] \dfrac{\partial z_2}{\partial y_1} & \dfrac{\partial z_2}{\partial y_2} & \cdots & \dfrac{\partial z_2}{\partial y_m} \\[2mm] \vdots & \vdots & & \vdots \\[2mm] \dfrac{\partial z_k}{\partial y_1} & \dfrac{\partial z_k}{\partial y_2} & \cdots & \dfrac{\partial z_k}{\partial y_m} \end{pmatrix}.$$

Then the composition $\mathbf{z} = \mathbf{g} \circ \mathbf{f}(\mathbf{x}) = \mathbf{g}\big(\mathbf{f}(\mathbf{x})\big)$ given by

$$z_1 = g_1\big(f_1(x_1, \ldots, x_n), \ldots, f_m(x_1, \ldots, x_n)\big)$$
$$z_2 = g_2\big(f_1(x_1, \ldots, x_n), \ldots, f_m(x_1, \ldots, x_n)\big)$$
$$\vdots$$
$$z_k = g_k\big(f_1(x_1, \ldots, x_n), \ldots, f_m(x_1, \ldots, x_n)\big)$$

has, according to the Chain Rule, the $k \times n$ Jacobian matrix

$$
\begin{pmatrix}
\frac{\partial z_1}{\partial x_1} & \frac{\partial z_1}{\partial x_2} & \cdots & \frac{\partial z_1}{\partial x_n} \\
\frac{\partial z_2}{\partial x_1} & \frac{\partial z_2}{\partial x_2} & \cdots & \frac{\partial z_2}{\partial x_n} \\
\vdots & \vdots & & \vdots \\
\frac{\partial z_k}{\partial x_1} & \frac{\partial z_k}{\partial x_2} & \cdots & \frac{\partial z_k}{\partial x_n}
\end{pmatrix}
=
\begin{pmatrix}
\frac{\partial z_1}{\partial y_1} & \frac{\partial z_1}{\partial y_2} & \cdots & \frac{\partial z_1}{\partial y_m} \\
\frac{\partial z_2}{\partial y_1} & \frac{\partial z_2}{\partial y_2} & \cdots & \frac{\partial z_2}{\partial y_m} \\
\vdots & \vdots & & \vdots \\
\frac{\partial z_k}{\partial y_1} & \frac{\partial z_k}{\partial y_2} & \cdots & \frac{\partial z_k}{\partial y_m}
\end{pmatrix}
\begin{pmatrix}
\frac{\partial y_1}{\partial x_1} & \frac{\partial y_1}{\partial x_2} & \cdots & \frac{\partial y_1}{\partial x_n} \\
\frac{\partial y_2}{\partial x_1} & \frac{\partial y_2}{\partial x_2} & \cdots & \frac{\partial y_2}{\partial x_n} \\
\vdots & \vdots & & \vdots \\
\frac{\partial y_m}{\partial x_1} & \frac{\partial y_m}{\partial x_2} & \cdots & \frac{\partial y_m}{\partial x_n}
\end{pmatrix}
$$

This is, in fact, the Chain Rule for compositions of transformations:

$$D(\mathbf{g} \circ \mathbf{f})(\mathbf{x}) = D\mathbf{g}\big(\mathbf{f}(\mathbf{x})\big)D\mathbf{f}(\mathbf{x}),$$

and exactly mimics the one-variable Chain Rule $D(g \circ f)(x) = Dg\big(f(x)\big)Df(x)$.

The transformation $\mathbf{y} = \mathbf{f}(\mathbf{x})$ also defines a vector $d\mathbf{y}$ of differentials of the variables y_i in terms of the vector $d\mathbf{x}$ of differentials of the variables x_j. Writing $d\mathbf{y}$ and $d\mathbf{x}$ as column vectors we have

$$
d\mathbf{y} =
\begin{pmatrix}
dy_1 \\ dy_2 \\ \vdots \\ dy_m
\end{pmatrix}
=
\begin{pmatrix}
\frac{\partial y_1}{\partial x_1} & \frac{\partial y_1}{\partial x_2} & \cdots & \frac{\partial y_1}{\partial x_n} \\
\frac{\partial y_2}{\partial x_1} & \frac{\partial y_2}{\partial x_2} & \cdots & \frac{\partial y_2}{\partial x_n} \\
\vdots & \vdots & & \vdots \\
\frac{\partial y_m}{\partial x_1} & \frac{\partial y_m}{\partial x_2} & \cdots & \frac{\partial y_m}{\partial x_n}
\end{pmatrix}
\begin{pmatrix}
dx_1 \\ dx_2 \\ \vdots \\ dx_n
\end{pmatrix}
= D\mathbf{f}(\mathbf{x})d\mathbf{x}.
$$

Example 4   Find the Jacobian matrix $D\mathbf{f}(1, 0)$ for the transformation from $\mathbb{R}^2$ to $\mathbb{R}^3$ given by

$$\mathbf{f}(x, y) = \big(xe^y + \cos(\pi y), x^2, x - e^y\big)$$

and use it to find an approximate value for $\mathbf{f}(1.02, 0.01)$.

Solution $D\mathbf{f}(x, y)$ is the 3×2 matrix whose jth row consists of the partial derivatives of the jth component of $\mathbf{f}$ with respect to x and y. Thus,

$$
D\mathbf{f}(1, 0) =
\begin{pmatrix}
e^y & xe^y - \pi \sin(\pi y) \\
2x & 0 \\
1 & -e^y
\end{pmatrix}\bigg|_{(1,0)}
=
\begin{pmatrix}
1 & 1 \\
2 & 0 \\
1 & -1
\end{pmatrix}.
$$

Since $\mathbf{f}(1, 0) = (2, 1, 0)$ and $d\mathbf{x} = \begin{pmatrix} 0.02 \\ 0.01 \end{pmatrix}$, we have

$$
d\mathbf{f} = D\mathbf{f}(1, 0)\, d\mathbf{x} =
\begin{pmatrix}
1 & 1 \\
2 & 0 \\
1 & -1
\end{pmatrix}
\begin{pmatrix} 0.02 \\ 0.01 \end{pmatrix}
=
\begin{pmatrix} 0.03 \\ 0.04 \\ 0.01 \end{pmatrix}.
$$

Therefore, $\mathbf{f}(1.02, 0.01) \approx (2.03, 1.04, 0.01)$.

For transformations between spaces of the same dimension (say from $\mathbb{R}^n$ to $\mathbb{R}^n$), the corresponding Jacobian matrices are square and have determinants. These Jacobian determinants will play an important role in our consideration of implicit functions and inverse functions in Section 12.8 and in changes of variables in multiple integrals in Chapter 14.

Maple's **VectorCalculus** package has a function **Jacobian** that takes two inputs, a list (or vector) of expressions and a list of variables, and produces the Jacobian matrix of the partial derivatives of those expressions with respect to the variables. For example,

```
> with(VectorCalculus):
> Jacobian([x*y*exp(z), (x+2*y)*cos(z)],[x,y,z]);
```

$$\begin{bmatrix} ye^z & xe^z & xye^z \\ \cos(z) & 2\cos(z) & -(x+2y)\sin(z) \end{bmatrix}$$

VectorCalculus has only been included since Maple 8. If you have an earlier release use `linalg` instead, and the function `jacobian`.

▌▌▌▌▌▌▌ Exercises 12.6 ▌▌▌▌▌▌▌▌

In Exercises 1–6, use suitable linearizations to find approximate values for the given functions at the points indicated.

1. $f(x, y) = x^2 y^3$ at $(3.1, 0.9)$

2. $f(x, y) = \tan^{-1}\left(\dfrac{y}{x}\right)$ at $(3.01, 2.99)$

3. $f(x, y) = \sin(\pi xy + \ln y)$ at $(0.01, 1.05)$

4. $f(x, y) = \dfrac{24}{x^2 + xy + y^2}$ at $(2.1, 1.8)$

5. $f(x, y, z) = \sqrt{x + 2y + 3z}$ at $(1.9, 1.8, 1.1)$

6. $f(x, y) = x\, e^{y+x^2}$ at $(2.05, -3.92)$

7. The edges of a rectangular box are each measured to within an accuracy of 1% of their values. What is the approximate maximum percentage error in

(a) the calculated volume of the box,

(b) the calculated area of one of the faces of the box, and

(c) the calculated length of a diagonal of the box?

8. The radius and height of a right-circular conical tank are measured to be 25 ft and 21 ft, respectively. Each measurement is accurate to within 0.5 in. By about how much can the calculated volume of the tank be in error?

9. By approximately how much can the calculated area of the conical surface of the tank in Exercise 8 be in error?

10. Two sides and the contained angle of a triangular plot of land are measured to be 224 m, 158 m, and 64°, respectively. The length measurements were accurate to within 0.4 m and the angle measurement to within 2°. What is the approximate maximum percentage error if the area of the plot is calculated from these measurements?

11. The angle of elevation of the top of a tower is measured at two points A and B on the ground in the same direction from the base of the tower. The angles are 50° at A and 35° at B, each measured to within 1°. The distance AB is measured to be 100 m with error at most 0.1%. What is the calculated height of the building, and by about how much can it be in error? To which of the three measurements is the calculated height most sensitive?

12. By approximately what percentage will the value of

$$w = \frac{x^2 y^3}{z^4}$$

increase or decrease if x increases by 1%, y increases by 2%, and z increases by 3%?

13. Find the Jacobian matrix for the transformation
$\mathbf{f}(r, \theta) = (x, y)$, where

$$x = r\cos\theta \quad \text{and} \quad y = r\sin\theta.$$

(Although (r, θ) can be regarded as *polar coordinates* in the xy-plane, they are Cartesian coordinates in their own $r\theta$-plane.)

14. Find the Jacobian matrix for the transformation
$\mathbf{f}(\rho, \phi, \theta) = (x, y, z)$, where

$$x = \rho\sin\phi\cos\theta, \quad y = \rho\sin\phi\sin\theta, \quad z = \rho\cos\phi.$$

Here, (ρ, ϕ, θ) are *spherical coordinates* in xyz-space. They will be formally introduced in Section 14.5.

15. Find the Jacobian matrix $D\mathbf{f}(x, y, z)$ for the transformation of $\mathbb{R}^3$ to $\mathbb{R}^2$ given by

$$\mathbf{f}(x, y, z) = (x^2 + yz, y^2 - x\ln z).$$

Use $D\mathbf{f}(2, 2, 1)$ to help you find an approximate value for $\mathbf{f}(1.98, 2.01, 1.03)$.

16. Find the Jacobian matrix $D\mathbf{g}(1, 3, 3)$ for the transformation of $\mathbb{R}^3$ to $\mathbb{R}^3$ given by

$$\mathbf{g}(r, s, t) = (r^2 s, r^2 t, s^2 - t^2)$$

and use the result to find an approximate value for $\mathbf{g}(0.99, 3.02, 2.97)$.

17. Prove that if $f(x, y)$ is differentiable at (a, b), then $f(x, y)$ is continuous at (a, b).

∗18. Prove the following version of the Mean-Value Theorem: If $f(x, y)$ has first partial derivatives continuous near every point of the straight line segment joining the points (a, b) and $(a + h, b + k)$, then there exists a number θ satisfying $0 < \theta < 1$ such that

$$f(a + h, b + k) = f(a, b) + hf_1(a + \theta h, b + \theta k) \\ + kf_2(a + \theta h, b + \theta k).$$

(*Hint:* Apply the single-variable Mean-Value Theorem to $g(t) = f(a + th, b + tk)$.) Why could we not have used this result in place of Theorem 3 to prove Theorem 4 and hence the version of the Chain Rule given in this section?

∗19. Generalize Exercise 18 as follows: show that, if $f(x, y)$ has continuous partial derivatives of second order near the point (a, b), then there exists a number θ satisfying $0 < \theta < 1$ such that, for h and k sufficiently small in absolute value,

$$f(a + h, b + k) = f(a, b) + hf_1(a, b) + kf_2(a, b) \\ + h^2 f_{11}(a + \theta h, b + \theta k) \\ + 2hk f_{12}(a + \theta h, b + \theta k) \\ + k^2 f_{22}(a + \theta h, b + \theta k).$$

Hence, show that there is a constant K such that for all values of h and k that are sufficiently small in absolute value,

$$\left| f(a+h, b+k) - f(a, b) - hf_1(a, b) - kf_2(a, b) \right|$$
$$\leq K(h^2 + k^2).$$

12.7 Gradients and Directional Derivatives

A first partial derivative of a function of several variables gives the rate of change of that function with respect to distance measured in the direction of one of the coordinate axes. In this section we will develop a method for finding the rate of change of such a function with respect to distance measured in *any direction* in the domain of the function.

To begin, it is useful to combine the first partial derivatives of a function into a single *vector function* called a **gradient**. For simplicity, we will develop and interpret the gradient for functions of two variables. Extension to functions of three or more variables is straightforward and will be discussed later in this section.

DEFINITION 6

> At any point (x, y) where the first partial derivatives of the function $f(x, y)$ exist, we define the **gradient vector** $\nabla f(x, y) = \mathbf{grad}\, f(x, y)$ by
>
> $$\nabla f(x, y) = \mathbf{grad}\, f(x, y) = f_1(x, y)\mathbf{i} + f_2(x, y)\mathbf{j}.$$

Recall that $\mathbf{i}$ and $\mathbf{j}$ denote the unit basis vectors from the origin to the points $(1, 0)$ and $(0, 1)$, respectively. The symbol ∇, called *del* or *nabla*, is a *vector differential operator*:

$$\nabla = \mathbf{i}\frac{\partial}{\partial x} + \mathbf{j}\frac{\partial}{\partial y}.$$

We can *apply* this operator to a function $f(x, y)$ by writing the operator to the left of the function. The result is the gradient of the function

$$\nabla f(x, y) = \left(\mathbf{i}\frac{\partial}{\partial x} + \mathbf{j}\frac{\partial}{\partial y}\right) f(x, y) = f_1(x, y)\mathbf{i} + f_2(x, y)\mathbf{j}.$$

We will make extensive use of the del operator in Chapter 16.

Example 1 If $f(x, y) = x^2 + y^2$, then $\nabla f(x, y) = 2x\mathbf{i} + 2y\mathbf{j}$. In particular, $\nabla f(1, 2) = 2\mathbf{i} + 4\mathbf{j}$. Observe that this vector is perpendicular to the tangent line $x + 2y = 5$ to the circle $x^2 + y^2 = 5$ at $(1, 2)$. This circle is the level curve of f that passes through the point $(1, 2)$. (See Figure 12.24.) As the following theorem shows, this perpendicularity is not a coincidence.

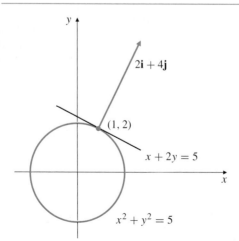

Figure 12.24 The gradient of $f(x, y) = x^2 + y^2$ at $(1, 2)$ is normal to the level curve of f through $(1, 2)$

THEOREM 6

If $f(x, y)$ is differentiable at the point (a, b) and $\nabla f(a, b) \neq \mathbf{0}$, then $\nabla f(a, b)$ is a normal vector to the level curve of f that passes through (a, b).

PROOF   Let $\mathbf{r} = \mathbf{r}(t) = x(t)\mathbf{i} + y(t)\mathbf{j}$ be a parametrization of the level curve of f such that $x(0) = a$ and $y(0) = b$. Then for all t near 0, $f\big(x(t), y(t)\big) = f(a, b)$. Differentiating this equation with respect to t using the Chain Rule, we obtain

$$f_1\big(x(t), y(t)\big) \frac{dx}{dt} + f_2\big(x(t), y(t)\big) \frac{dy}{dt} = 0.$$

At $t = 0$ this says that $\nabla f(a, b) \bullet \left.\frac{d\mathbf{r}}{dt}\right|_{t=0} = 0$, that is, ∇f is perpendicular to the tangent vector $d\mathbf{r}/dt$ to the level curve at (a, b).

Directional Derivatives

The first partial derivatives $f_1(a, b)$ and $f_2(a, b)$ give the rates of change of $f(x, y)$ at (a, b) measured in the directions of the positive x- and y-axes, respectively. If we want to know how fast $f(x, y)$ changes value as we move through the domain of f at (a, b) in some other direction, we require a more general **directional derivative**. We can specify the direction by means of a nonzero vector. It is most convenient to use a *unit vector*.

DEFINITION 7

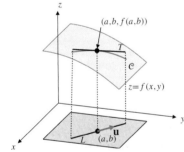

Figure 12.25   Unit vector $\mathbf{u}$ determines a line L through (a, b) in the domain of f. The vertical plane containing L intersects the graph of f in a curve $\mathcal{C}$ whose tangent T at $(a, b, f(a, b))$ has slope $D_{\mathbf{u}} f(a, b)$

Let $\mathbf{u} = u\mathbf{i} + v\mathbf{j}$ be a unit vector, so that $u^2 + v^2 = 1$. The **directional derivative** of $f(x, y)$ at (a, b) in the direction of $\mathbf{u}$ is the rate of change of $f(x, y)$ with respect to distance measured at (a, b) along a ray in the direction of $\mathbf{u}$ in the xy-plane. (See Figure 12.25.) This directional derivative is given by

$$D_{\mathbf{u}} f(a, b) = \lim_{h \to 0+} \frac{f(a + hu, b + hv) - f(a, b)}{h}.$$

It is also given by

$$D_{\mathbf{u}} f(a, b) = \left.\frac{d}{dt} f(a + tu, b + tv)\right|_{t=0}$$

if the derivative on the right side exists.

Observe that directional derivatives in directions parallel to the coordinate axes are given directly by first partials: $D_{\mathbf{i}} f(a, b) = f_1(a, b)$, $D_{\mathbf{j}} f(a, b) = f_2(a, b)$, $D_{-\mathbf{i}} f(a, b) = -f_1(a, b)$, and $D_{-\mathbf{j}} f(a, b) = -f_2(a, b)$. The following theorem shows how the gradient can be used to calculate any directional derivative.

THEOREM 7

Using the gradient to find directional derivatives

If f is differentiable at (a, b) and $\mathbf{u} = u\mathbf{i} + v\mathbf{j}$ is a unit vector, then the directional derivative of f at (a, b) in the direction of $\mathbf{u}$ is given by

$$D_{\mathbf{u}} f(a, b) = \mathbf{u} \bullet \nabla f(a, b).$$

PROOF   By the Chain Rule:

$$D_{\mathbf{u}} f(a, b) = \left.\frac{d}{dt} f(a + tu, b + tv)\right|_{t=0}$$
$$= u f_1(a, b) + v f_2(a, b) = \mathbf{u} \bullet \nabla f(a, b).$$

We already know that having partial derivatives at a point does not imply that a function is continuous there, let alone that it is differentiable. The same can be said about directional derivatives. It is possible for a function to have a directional derivative in every direction at a given point and still not be continuous at that point. See Exercise 37 below for an example of such a function.

Given any nonzero vector $\mathbf{v}$, we can always obtain a unit vector in the same direction by dividing $\mathbf{v}$ by its length. The directional derivative of f at (a, b) in the direction of $\mathbf{v}$ is therefore given by

$$D_{\mathbf{v}/|\mathbf{v}|} f(a, b) = \frac{\mathbf{v}}{|\mathbf{v}|} \bullet \nabla f(a, b).$$

Example 2   Find the rate of change of $f(x, y) = y^4 + 2xy^3 + x^2y^2$ at $(0, 1)$ measured in each of the following directions:

(a) $\mathbf{i} + 2\mathbf{j}$,   (b) $\mathbf{j} - 2\mathbf{i}$,   (c) $3\mathbf{i}$,   (d) $\mathbf{i} + \mathbf{j}$.

Solution   We calculate

$$\nabla f(x, y) = (2y^3 + 2xy^2)\mathbf{i} + (4y^3 + 6xy^2 + 2x^2y)\mathbf{j},$$
$$\nabla f(0, 1) = 2\mathbf{i} + 4\mathbf{j}.$$

(a) The directional derivative of f at $(0, 1)$ in the direction of $\mathbf{i} + 2\mathbf{j}$ is

$$\frac{\mathbf{i} + 2\mathbf{j}}{|\mathbf{i} + 2\mathbf{j}|} \bullet (2\mathbf{i} + 4\mathbf{j}) = \frac{2 + 8}{\sqrt{5}} = 2\sqrt{5}.$$

Observe that $\mathbf{i} + 2\mathbf{j}$ points in the same direction as $\nabla f(0, 1)$ so the directional derivative is positive and equal to the length of $\nabla f(0, 1)$.

(b) The directional derivative of f at $(0, 1)$ in the direction of $\mathbf{j} - 2\mathbf{i}$ is

$$\frac{-2\mathbf{i} + \mathbf{j}}{|-2\mathbf{i} + \mathbf{j}|} \bullet (2\mathbf{i} + 4\mathbf{j}) = \frac{-4 + 4}{\sqrt{5}} = 0.$$

Since $\mathbf{j} - 2\mathbf{i}$ is perpendicular to $\nabla f(0, 1)$, it is tangent to the level curve of f through $(0, 1)$, so the directional derivative in that direction is zero.

(c) The directional derivative of f at $(0, 1)$ in the direction of $3\mathbf{i}$ is

$$\mathbf{i} \bullet (2\mathbf{i} + 4\mathbf{j}) = 2.$$

As noted previously, the directional derivative of f in the direction of the positive x-axis is just $f_1(0, 1)$.

(d) The directional derivative of f at $(0, 1)$ in the direction of $\mathbf{i} + \mathbf{j}$ is

$$\frac{\mathbf{i} + \mathbf{j}}{|\mathbf{i} + \mathbf{j}|} \bullet (2\mathbf{i} + 4\mathbf{j}) = \frac{2 + 4}{\sqrt{2}} = 3\sqrt{2}.$$

If we move along the surface $z = f(x, y)$ through the point $(0, 1, 1)$ in a direction making horizontal angles of $45°$ with the positive directions of the x- and y-axes, we would be rising at a rate of $3\sqrt{2}$ vertical units per horizontal unit moved.

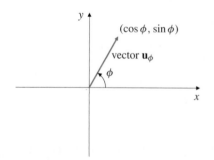

Figure 12.26   The unit vector specified by a polar angle ϕ

Remark   A direction in the plane can be specified by a polar angle. The direction making angle ϕ with the positive direction of the x-axis corresponds to the unit vector (see Figure 12.26)

$$\mathbf{u}_\phi = \cos\phi\,\mathbf{i} + \sin\phi\,\mathbf{j},$$

so the directional derivative of f at (x, y) in that direction is

$$D_\phi f(x, y) = D_{\mathbf{u}_\phi} f(x, y) = \mathbf{u}_\phi \bullet \nabla f(x, y) = f_1(x, y) \cos \phi + f_2(x, y) \sin \phi.$$

Note the use of the symbol $D_\phi f(x, y)$ to denote a derivative of f with respect to *distance* measured in the direction ϕ.

As observed in the previous example, Theorem 7 provides a useful interpretation for the gradient vector. For any unit vector $\mathbf{u}$ we have

$$D_{\mathbf{u}} f(a, b) = \mathbf{u} \bullet \nabla f(a, b) = |\nabla f(a, b)| \cos \theta,$$

where θ is the angle between the vectors $\mathbf{u}$ and $\nabla f(a, b)$. Since $\cos \theta$ only takes on values between -1 and 1, $D_{\mathbf{u}} f(a, b)$ only takes on values between $-|\nabla f(a, b)|$ and $|\nabla f(a, b)|$. Moreover, $D_{\mathbf{u}} f(a, b) = -|\nabla f(a, b)|$ if and only if $\mathbf{u}$ points in the opposite direction to $\nabla f(a, b)$ (so that $\cos \theta = -1$), and $D_{\mathbf{u}} f(a, b) = |\nabla f(a, b)|$ if and only if $\mathbf{u}$ points in the same direction as $\nabla f(a, b)$ (so that $\cos \theta = 1$). The directional derivative is zero in the direction $\theta = \pi/2$; this is the direction of the (tangent line to the) level curve of f through (a, b).

We summarize these properties of the gradient as follows:

Geometric properties of the gradient vector

(i) At (a, b), $f(x, y)$ increases most rapidly in the direction of the gradient vector $\nabla f(a, b)$. The maximum rate of increase is $|\nabla f(a, b)|$.

(ii) At (a, b), $f(x, y)$ decreases most rapidly in the direction of $-\nabla f(a, b)$. The maximum rate of decrease is $|\nabla f(a, b)|$.

(iii) The rate of change of $f(x, y)$ at (a, b) is zero in directions tangent to the level curve of f that passes through (a, b).

Look again at the topographic map in Figure 12.6 in Section 12.1. The streams on the map flow in the direction of steepest descent, that is, in the direction of $-\nabla f$, where f measures the elevation of land. The streams therefore cross the contours (the level curves of f) at right angles. Like the stream, an experienced skier might choose a downhill path close to the direction of the negative gradient, while a novice skier would prefer to stay closer to the level curves.

Example 3   The temperature at position (x, y) in a region of the xy-plane is $T\,°\text{C}$, where

$$T(x, y) = x^2 e^{-y}.$$

In what direction at the point $(2, 1)$ does the temperature increase most rapidly? What is the rate of increase of f in that direction?

Solution   We have

$$\nabla T(x, y) = 2x\,e^{-y}\mathbf{i} - x^2 e^{-y}\mathbf{j},$$

$$\nabla T(2, 1) = \frac{4}{e}\mathbf{i} - \frac{4}{e}\mathbf{j} = \frac{4}{e}(\mathbf{i} - \mathbf{j}).$$

At $(2, 1)$, $T(x, y)$ increases most rapidly in the direction of the vector $\mathbf{i} - \mathbf{j}$. The rate of increase in this direction is $|\nabla T(2, 1)| = 4\sqrt{2}/e$ °C/unit distance.

Example 4   A hiker is standing beside a stream on the side of a mountain, examining her map of the region. The height of land (in metres) at any point (x, y) is given by the function

$$h(x, y) = \frac{20{,}000}{3 + x^2 + 2y^2},$$

where x and y (in kilometres) denote the coordinates of the point on the hiker's map. The hiker is at the point $(3, 2)$.

(a) What is the direction of flow of the stream at $(3, 2)$ on the hiker's map? How fast is the stream descending at her location?

(b) Find the equation of the path of the stream on the hiker's map.

(c) At what angle to the path of the stream (on the map) should the hiker set out if she wishes to climb at a $15°$ inclination to the horizontal?

(d) Make a sketch of the hiker's map, showing some curves of constant elevation, and showing the stream.

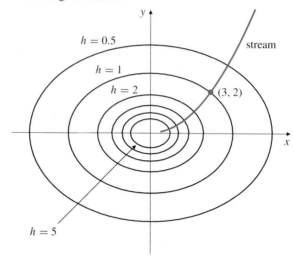

Figure 12.27 The hiker's map. Unlike most mountains, this one has perfectly elliptical contours

Solution

(a) We begin by calculating the gradient of h and its length at $(3, 2)$:

$$\nabla h(x, y) = -\frac{20{,}000}{(3 + x^2 + 2y^2)^2}(2x\mathbf{i} + 4y\mathbf{j}),$$

$$\nabla h(3, 2) = -100(3\mathbf{i} + 4\mathbf{j}),$$

$$|\nabla h(3, 2)| = 500.$$

The stream is flowing in the direction whose horizontal projection at $(3, 2)$ is $-\nabla h(3, 2)$, that is, in the horizontal direction of the vector $3\mathbf{i} + 4\mathbf{j}$. The stream is descending at a rate of 500 m/km, that is, 0.5 m per horizontal metre travelled.

(b) Coordinates on the map are the coordinates (x, y) in the domain of the height function h. We can find an equation of the path of the stream on a map of the region by setting up a differential equation for a change of position along the path. If the vector $d\mathbf{r} = dx\,\mathbf{i} + dy\,\mathbf{j}$ is tangent to the path of the stream at point (x, y) on the map, then $d\mathbf{r}$ is parallel to $\nabla h(x, y)$. Hence, the components of these two vectors are proportional:

$$\frac{dx}{2x} = \frac{dy}{4y} \quad \text{or} \quad \frac{dy}{y} = \frac{2dx}{x}.$$

Integrating both sides of this equation, we get $\ln y = 2 \ln x + \ln C$, or $y = Cx^2$. Since the path of the stream passes through $(3, 2)$, we have $C = 2/9$ and the equation is $9y = 2x^2$.

(c) Suppose the hiker moves away from $(3, 2)$ in the direction of the unit vector $\mathbf{u}$. She will be ascending at an inclination of $15°$ if the directional derivative of h in the direction of $\mathbf{u}$ is $1{,}000 \tan 15° \approx 268$. (The $1{,}000$ compensates for the fact that the vertical units are metres while the horizontal units are kilometres.) If θ is the angle between $\mathbf{u}$ and the upstream direction, then

$$500 \cos \theta = |\nabla h(3, 2)| \cos \theta = D_{\mathbf{u}}h(3, 2) \approx 268.$$

Hence, $\cos\theta \approx 0.536$ and $\theta \approx 57.6°$. She should set out in a direction making a horizontal angle of about 58° with the upstream direction.

(d) A suitable sketch of the map is given in Figure 12.27. ∎

Example 5 Find the second directional derivative of $f(x, y)$ in the direction making angle ϕ with the positive x-axis.

Solution As observed earlier, the first directional derivative is

$$D_\phi f(x, y) = (\cos\phi\,\mathbf{i} + \sin\phi\,\mathbf{j}) \bullet \nabla f(x, y) = f_1(x, y)\cos\phi + f_2(x, y)\sin\phi.$$

The second directional derivative is therefore

$$\begin{aligned}
D_\phi^2 f(x, y) &= D_\phi\left(D_\phi f(x, y)\right) \\
&= (\cos\phi\,\mathbf{i} + \sin\phi\,\mathbf{j}) \bullet \nabla\left(f_1(x, y)\cos\phi + f_2(x, y)\sin\phi\right) \\
&= \left(f_{11}(x, y)\cos\phi + f_{21}(x, y)\sin\phi\right)\cos\phi \\
&\quad + \left(f_{12}(x, y)\cos\phi + f_{22}(x, y)\sin\phi\right)\sin\phi \\
&= f_{11}(x, y)\cos^2\phi + 2f_{12}(x, y)\cos\phi\sin\phi + f_{22}(x, y)\sin^2\phi.
\end{aligned}$$

Note that if $\phi = 0$ or $\phi = \pi$ (so the directional derivative is in a direction parallel to the x-axis), then $D_\phi^2 f(x, y) = f_{11}(x, y)$. Similarly, $D_\phi^2 f(x, y) = f_{22}(x, y)$ if $\phi = \pi/2$ or $3\pi/2$. ∎

Rates Perceived by a Moving Observer

Suppose that an observer is moving around in the xy-plane measuring the value of a function $f(x, y)$ defined in the plane as he passes through each point (x, y). (For instance, $f(x, y)$ might be the temperature at (x, y).) If the observer is moving with velocity $\mathbf{v}$ at the instant when he passes through the point (a, b), how fast would he observe $f(x, y)$ to be changing at that moment?

At the moment in question the observer is moving in the direction of the unit vector $\mathbf{v}/|\mathbf{v}|$. The rate of change of $f(x, y)$ at (a, b) in that direction is

$$D_{\mathbf{v}/|\mathbf{v}|} f(a, b) = \frac{\mathbf{v}}{|\mathbf{v}|} \bullet \nabla f(a, b)$$

measured in units of f per unit distance in the xy-plane. To convert this rate to units of f per unit time, we must multiply by the speed of the observer, $|\mathbf{v}|$ units of distance per unit time. Thus, the time rate of change of $f(x, y)$ as measured by the observer passing through (a, b) is

$$|\mathbf{v}|\,\frac{\mathbf{v}}{|\mathbf{v}|} \bullet \nabla f(a, b) = \mathbf{v} \bullet \nabla f(a, b).$$

It is natural to extend our use of the symbol $D_{\mathbf{v}} f(a, b)$ to represent this rate even though $\mathbf{v}$ is not (necessarily) a unit vector. Thus

The rate of change of $f(x, y)$ at (a, b) as measured by an observer moving through (a, b) with velocity $\mathbf{v}$ is

$$D_{\mathbf{v}} f(a, b) = \mathbf{v} \bullet \nabla f(a, b)$$

units of f per unit time.

If the hiker in Example 4 moves away from $(3, 2)$ with horizontal velocity $\mathbf{v} = -\mathbf{i} - \mathbf{j}$ km/h, then she will be rising at a rate of

$$\mathbf{v} \bullet \nabla h(3, 2) = (-\mathbf{i} - \mathbf{j}) \bullet \left(-\frac{1}{10}(3\mathbf{i} + 4\mathbf{j}) \right) = \frac{7}{10} \text{ km/h}.$$

As defined here, $D_{\mathbf{v}} f$ is the spatial component of the derivative of f following the motion. See Example 6 in Section 12.5. The rate of change of the reading on the moving thermometer in that example can be expressed as

$$\frac{dT}{dt} = D_{\mathbf{v}} T(x, y, z, t) + \frac{\partial T}{\partial t},$$

where $\mathbf{v}$ is the velocity of the moving thermometer and $D_{\mathbf{v}} T = \mathbf{v} \bullet \nabla T$. The gradient is being taken with respect to the *three spatial variables* only. (See below for the gradient in 3-space.)

The Gradient in Three and More Dimensions

By analogy with the two-dimensional case, a function $f(x_1, x_2, \ldots, x_n)$ of n variables possessing first partial derivatives has gradient given by

$$\nabla f(x_1, x_2, \ldots, x_n) = \frac{\partial f}{\partial x_1}\mathbf{e_1} + \frac{\partial f}{\partial x_2}\mathbf{e_2} + \cdots + \frac{\partial f}{\partial x_n}\mathbf{e_n},$$

where $\mathbf{e}_j$ is the unit vector from the origin to the unit point on the jth coordinate axis. In particular, for a function of three variables,

$$\nabla f(x, y, z) = \frac{\partial f}{\partial x}\mathbf{i} + \frac{\partial f}{\partial y}\mathbf{j} + \frac{\partial f}{\partial z}\mathbf{k}.$$

The level surface of $f(x, y, z)$ passing through (a, b, c) has a tangent plane there if f is differentiable at (a, b, c) and $\nabla f(a, b, c) \neq \mathbf{0}$.

For functions of any number of variables, the vector $\nabla f(P_0)$ is normal to the "level surface" of f passing through the point P_0 (i.e., the (hyper)surface with equation $f(x_1, \ldots, x_n) = f(P_0)$), and, if f is differentiable at P_0, the rate of change of f at P_0 in the direction of the unit vector $\mathbf{u}$ is given by $\mathbf{u} \bullet \nabla f(P_0)$. Equations of tangent planes to surfaces in 3-space can be found easily with the aid of gradients.

Example 6 Let $f(x, y, z) = x^2 + y^2 + z^2$.

(a) Find $\nabla f(x, y, z)$ and $\nabla f(1, -1, 2)$.

(b) Find an equation of the tangent plane to the sphere $x^2 + y^2 + z^2 = 6$ at the point $(1, -1, 2)$.

(c) What is the maximum rate of increase of f at $(1, -1, 2)$?

(d) What is the rate of change with respect to distance of f at $(1, -1, 2)$ measured in the direction from that point toward the point $(3, 1, 1)$?

Solution

(a) $\nabla f(x, y, z) = 2x\mathbf{i} + 2y\mathbf{j} + 2z\mathbf{k}$, so $\nabla f(1, -1, 2) = 2\mathbf{i} - 2\mathbf{j} + 4\mathbf{k}$.

(b) The required tangent plane has $\nabla f(1, -1, 2)$ as normal. (See Figure 12.28(a).) Therefore, its equation is given by $2(x - 1) - 2(y + 1) + 4(z - 2) = 0$ or, more simply, $x - y + 2z = 6$.

(c) The maximum rate of increase of f at $(1, -1, 2)$ is $|\nabla f(1, -1, 2)| = 2\sqrt{6}$, and it occurs in the direction of the vector $\mathbf{i} - \mathbf{j} + 2\mathbf{k}$.

(d) The direction from $(1, -1, 2)$ toward $(3, 1, 1)$ is specified by $2\mathbf{i} + 2\mathbf{j} - \mathbf{k}$. The rate of change of f with respect to distance in this direction is

$$\frac{2\mathbf{i} + 2\mathbf{j} - \mathbf{k}}{\sqrt{4 + 4 + 1}} \bullet (2\mathbf{i} - 2\mathbf{j} + 4\mathbf{k}) = \frac{4 - 4 - 4}{3} = -\frac{4}{3},$$

that is, f decreases at rate $4/3$ of a unit per horizontal unit moved.

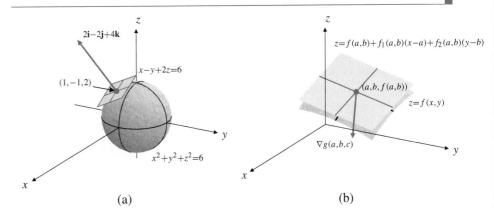

Figure 12.28

(a) The tangent plane to
$x^2 + y^2 + z^2 = 6$ at $(1, -1, 2)$

(b) The gradient of $f(x, y) - z$ at
$(a, b, f(a, b))$ is normal to the
tangent plane to $z = f(x, y)$ at that
point

! **BEWARE**

Make sure you understand the
difference between the graph of a
function and a level curve or level
surface of that function. (See the
discussion following this example.)
Here, the surface $z = f(x, y)$ is the
graph of the function f, but it is
also a *level surface* of a *different*
function g.

Example 7   The graph of a *function* $f(x, y)$ of two variables is the graph of the
equation $z = f(x, y)$ in 3-space. This surface is also the level surface $g(x, y, z) = 0$
of the 3-variable function

$$g(x, y, z) = f(x, y) - z.$$

If f is differentiable at (a, b) and $c = f(a, b)$, then g is differentiable at (a, b, c), and

$$\nabla g(a, b, c) = f_1(a, b)\mathbf{i} + f_2(a, b)\mathbf{j} - \mathbf{k}$$

is normal to $g(x, y, z) = 0$ at (a, b, c). (Note that $\nabla g(a, b, c) \neq \mathbf{0}$, since its z
component is -1.) It follows that the graph of f has nonvertical tangent plane at (a, b)
given by

$$f_1(a, b)(x - a) + f_2(a, b)(y - b) - (z - c) = 0,$$

or

$$z = f(a, b) + f_1(a, b)(x - a) + f_2(a, b)(y - b).$$

(See Figure 12.28(b).) This result was obtained by a different argument in Section 12.3.

Students sometimes confuse graphs of functions with level curves or surfaces of those
functions. In the above example we are talking about a *level surface* of the function
$g(x, y, z)$ that happens to coincide with the *graph* of a different function, $f(x, y)$. Do
not confuse that surface with the graph of g, which is a three-dimensional *hypersurface*
in 4-space having equation $w = g(x, y, z)$. Similarly, do not confuse the tangent *plane*
to the graph of $f(x, y)$ (i.e., the plane obtained in the above example) with the tangent
line to the level curve of $f(x, y)$ passing through (a, b) and lying in the xy-plane. This
line has an equation involving only x and y: $f_1(a, b)(x - a) + f_2(a, b)(y - b) = 0$.

Example 8   Find a vector tangent to the curve of intersection of the two surfaces

$$z = x^2 - y^2 \qquad \text{and} \qquad xyz + 30 = 0$$

at the point $(-3, 2, 5)$.

Solution   The coordinates of the given point satisfy the equations of both surfaces so
the point lies on the curve of intersection of the two surfaces. A vector tangent to this
curve at that point will be perpendicular to the normals to both surfaces, that is, to the
vectors

$$\mathbf{n}_1 = \nabla(x^2 - y^2 - z)\Big|_{(-3,2,5)} = 2x\mathbf{i} - 2y\mathbf{j} - \mathbf{k}\Big|_{(-3,2,5)} = -6\mathbf{i} - 4\mathbf{j} - \mathbf{k},$$

$$\mathbf{n}_2 = \nabla(xyz + 30)\Big|_{(-3,2,5)} = (yz\mathbf{i} + xz\mathbf{j} + xy\mathbf{k})\Big|_{(-3,2,5)} = 10\mathbf{i} - 15\mathbf{j} - 6\mathbf{k}.$$

For the tangent vector **T** we can therefore use the cross product of these normals:

$$\mathbf{T} = \mathbf{n}_1 \times \mathbf{n}_2 = \begin{vmatrix} \mathbf{i} & \mathbf{j} & \mathbf{k} \\ -6 & -4 & -1 \\ 10 & -15 & -6 \end{vmatrix} = 9\mathbf{i} - 46\mathbf{j} + 130\mathbf{k}.$$

Remark Maple's **VectorCalculus** package defines a function **Gradient** that takes a pair of arguments—an expression and a list of variables—and produces the gradient of the expression with respect to those variables:

```
> with(VectorCalculus):
> f := x^2+y^3+z^4; G := Gradient(f, [x,y,z]);
```

$$f := x^2 + y^3 + z^4$$

$$G := 2 x \, \bar{e}_x + 3 \, y^2 \, \bar{e}_y + 4 \, z^3 \, \bar{e}_z$$

Although the result for G looks like a vector, it is actually something different, namely a **vector field** which is a vector-valued function of a vector variable. This fact is conveyed by the bars that appear over the basis vectors in the output. We will deal extensively with vector fields in Chapters 15 and 16 and will say little about them here except to note that evaluating the Gradient at a particular point requires the evalVF function, which takes two arguments: a vector field and a vector at which to evaluate it.

```
> evalVF(G,<2,3,-1>);
```

$$4 \, e_x + 27 \, e_y - 4 \, e_z$$

Observe that the output is a vector, not a vector field; there are no bars on the basis vectors.

If you want to define a gradient function (let us call it grad) such that you would get the above value by using the input grad(f)(2,3,-1), you could use

```
> grad := g -> ((u,v,w) ->
>    evalVF(Gradient(g,[x,y,z]),<u,v,w>));
```

▌▌▌▌▌▌ Exercises 12.7 ▌▌▌▌▌▌

In Exercises 1–6, find:

(a) the gradient of the given function at the point indicated,

(b) an equation of the plane tangent to the graph of the given function at the point whose x and y coordinates are given, and

(c) an equation of the straight line tangent, at the given point, to the level curve of the given function passing through that point.

1. $f(x, y) = x^2 - y^2$ at $(2, -1)$

2. $f(x, y) = \dfrac{x - y}{x + y}$ at $(1, 1)$

3. $f(x, y) = \dfrac{x}{x^2 + y^2}$ at $(1, 2)$

4. $f(x, y) = e^{xy}$ at $(2, 0)$

5. $f(x, y) = \ln(x^2 + y^2)$ at $(1, -2)$

6. $f(x, y) = \sqrt{1 + xy^2}$ at $(2, -2)$

In Exercises 7–9, find an equation of the tangent plane to the level surface of the given function that passes through the given point.

7. $f(x, y, z) = x^2y + y^2z + z^2x$ at $(1, -1, 1)$

8. $f(x, y, z) = \cos(x + 2y + 3z)$ at $\left(\dfrac{\pi}{2}, \pi, \pi\right)$

9. $f(x, y, z) = y \, e^{-x^2} \sin z$ at $(0, 1, \pi/3)$

In Exercises 10–13, find the rate of change of the given function at the given point in the specified direction.

10. $f(x, y) = 3x - 4y$ at $(0, 2)$ in the direction of the vector $-2\mathbf{i}$

11. $f(x, y) = x^2y$ at $(-1, -1)$ in the direction of the vector $\mathbf{i} + 2\mathbf{j}$

12. $f(x, y) = \dfrac{x}{1 + y}$ at $(0, 0)$ in the direction of the vector $\mathbf{i} - \mathbf{j}$

13. $f(x, y) = x^2 + y^2$ at $(1, -2)$ in the direction making a (positive) angle of $60°$ with the positive x-axis

14. Let $f(x, y) = \ln |\mathbf{r}|$, where $\mathbf{r} = x\mathbf{i} + y\mathbf{j}$. Show that
$$\nabla f = \dfrac{\mathbf{r}}{|\mathbf{r}|^2}.$$

15. Let $f(x, y, z) = |\mathbf{r}|^{-n}$, where $\mathbf{r} = x\mathbf{i} + y\mathbf{j} + z\mathbf{k}$. Show that

$$\nabla f = \frac{-n\mathbf{r}}{|\mathbf{r}|^{n+2}}.$$

16. Show that, in terms of polar coordinates (r, θ) (where $x = r\cos\theta$ and $y = r\sin\theta$), the gradient of a function $f(r, \theta)$ is given by

$$\nabla f = \frac{\partial f}{\partial r}\hat{\mathbf{r}} + \frac{1}{r}\frac{\partial f}{\partial \theta}\hat{\boldsymbol{\theta}},$$

where $\hat{\mathbf{r}}$ is a unit vector in the direction of the position vector $\mathbf{r} = x\,\mathbf{i} + y\,\mathbf{j}$, and $\hat{\boldsymbol{\theta}}$ is a unit vector at right angles to $\hat{\mathbf{r}}$ in the direction of increasing θ.

17. In what directions at the point $(2, 0)$ does the function $f(x, y) = xy$ have rate of change -1? Are there directions in which the rate is -3? How about -2?

18. In what directions at the point (a, b, c) does the function $f(x, y, z) = x^2 + y^2 - z^2$ increase at half of its maximal rate at that point?

19. Find $\nabla f(a, b)$ for the differentiable function $f(x, y)$ given the directional derivatives

$$D_{(\mathbf{i}+\mathbf{j})/\sqrt{2}}f(a, b) = 3\sqrt{2} \text{ and } D_{(3\mathbf{i}-4\mathbf{j})/5}f(a, b) = 5.$$

20. If $f(x, y)$ is differentiable at (a, b), what condition should angles ϕ_1 and ϕ_2 satisfy in order that the gradient $\nabla f(a, b)$ can be determined from the values of the directional derivatives $D_{\phi_1}f(a, b)$ and $D_{\phi_2}f(a, b)$?

21. The temperature $T(x, y)$ at points of the xy-plane is given by $T(x, y) = x^2 - 2y^2$.

(a) Draw a contour diagram for T showing some isotherms (curves of constant temperature).

(b) In what direction should an ant at position $(2, -1)$ move if it wishes to cool off as quickly as possible?

(c) If the ant moves in that direction at speed k (units distance per unit time), at what rate does it experience the decrease of temperature?

(d) At what rate would the ant experience the decrease of temperature if it moved from $(2, -1)$ at speed k in the direction of the vector $-\mathbf{i} - 2\mathbf{j}$?

(e) Along what curve through $(2, -1)$ should the ant move in order to continue to experience maximum rate of cooling?

22. Find an equation of the curve in the xy-plane that passes through the point $(1, 1)$ and intersects all level curves of the function $f(x, y) = x^4 + y^2$ at right angles.

23. Find an equation of the curve in the xy-plane that passes through the point $(2, -1)$ and that intersects every curve with equation of the form $x^2y^3 = K$ at right angles.

24. Find the second directional derivative of $e^{-x^2-y^2}$ at the point $(a, b) \neq (0, 0)$ in the direction directly away from the origin.

25. Find the second directional derivative of $f(x, y, z) = xyz$ at $(2, 3, 1)$ in the direction of the vector $\mathbf{i} - \mathbf{j} - \mathbf{k}$.

26. Find a vector tangent to the curve of intersection of the two cylinders $x^2 + y^2 = 2$ and $y^2 + z^2 = 2$ at the point $(1, -1, 1)$.

27. Repeat Exercise 26 for the surfaces $x + y + z = 6$ and $x^2 + y^2 + z^2 = 14$ and the point $(1, 2, 3)$.

28. The temperature in 3-space is given by

$$T(x, y, z) = x^2 - y^2 + z^2 + xz^2.$$

At time $t = 0$ a fly passes through the point $(1, 1, 2)$, flying along the curve of intersection of the surfaces $z = 3x^2 - y^2$ and $2x^2 + 2y^2 - z^2 = 0$. If the fly's speed is 7, what rate of temperature change does it experience at $t = 0$?

29. State and prove a version of Theorem 6 for a function of three variables.

30. What is the level surface of $f(x, y, z) = \cos(x + 2y + 3z)$ that passes through (π, π, π)? What is the tangent plane to that level surface at that point? (Compare this exercise with Exercise 8 above.)

31. If $\nabla f(x, y) = 0$ throughout the disk $x^2 + y^2 < r^2$, prove that $f(x, y)$ is constant throughout the disk.

32. Theorem 6 implies that the level curve of $f(x, y)$ passing through (a, b) is smooth (has a tangent line) at (a, b) provided f is differentiable at (a, b) and satisfies $\nabla f(a, b) \neq \mathbf{0}$. Show that the level curve need not be smooth at (a, b) if $\nabla f(a, b) = \mathbf{0}$. (*Hint:* Consider $f(x, y) = y^3 - x^2$ at $(0, 0)$.)

33. If $\mathbf{v}$ is a nonzero vector, express $D_{\mathbf{v}}(D_{\mathbf{v}}f)$ in terms of the components of $\mathbf{v}$ and the second partials of f. What is the interpretation of this quantity for a moving observer?

∗34. An observer moves so that his position, velocity, and acceleration at time t are given by $\mathbf{r}(t) = x(t)\mathbf{i} + y(t)\mathbf{j} + z(t)\mathbf{k}$, $\mathbf{v}(t) = d\mathbf{r}/dt$, and $\mathbf{a}(t) = d\mathbf{v}/dt$. If the temperature in the vicinity of the observer depends only on position, $T = T(x, y, z)$, express the second time derivative of temperature as measured by the observer in terms of $D_{\mathbf{v}}$ and $D_{\mathbf{a}}$.

∗35. Repeat Exercise 34 but with T depending explicitly on time as well as position: $T = T(x, y, z, t)$.

36. Let $f(x, y) = \begin{cases} \dfrac{\sin(xy)}{\sqrt{x^2 + y^2}}, & \text{if } (x, y) \neq (0, 0) \\ 0, & \text{if } (x, y) = (0, 0). \end{cases}$

(a) Calculate $\nabla f(0, 0)$.

(b) Use the definition of directional derivative to calculate $D_{\mathbf{u}}f(0, 0)$, where $\mathbf{u} = (\mathbf{i} + \mathbf{j})/\sqrt{2}$.

(b) Is $f(x, y)$ differentiable at $(0, 0)$? Why?

37. Let $f(x, y) = \begin{cases} 2x^2y/(x^4 + y^2), & \text{if } (x, y) \neq (0, 0) \\ 0, & \text{if } (x, y) = (0, 0). \end{cases}$
Use the definition of directional derivative as a limit (Definition 7) to show that $D_{\mathbf{u}}f(0, 0)$ exists for every unit vector $\mathbf{u} = u\mathbf{i} + v\mathbf{j}$ in the plane. Specifically, show that $D_{\mathbf{u}}f(0, 0) = 0$ if $v = 0$, and $D_{\mathbf{u}}f(0, 0) = 2u^2/v$ if $v \neq 0$. However, as was shown in Example 4 in Section 12.2, $f(x, y)$ has no limit as $(x, y) \to (0, 0)$, so it is not continuous there. Even if a function has directional derivatives in all directions at a point, it may not be continuous at that point.

12.8 Implicit Functions

When we study the calculus of functions of one variable, we encounter examples of functions that are defined implicitly as solutions of equations in two variables. Suppose, for example, that $F(x, y) = 0$ is such an equation. Suppose that the point (a, b) satisfies the equation and that F has continuous first partial derivatives (and so is differentiable) at all points near (a, b). Can the equation be solved for y as a function of x near (a, b)? That is, does there exist a function $y(x)$ defined in some interval $I = (a - h, a + h)$ (where $h > 0$) satisfying $y(a) = b$ and such that

$$F\left(x, y(x)\right) = 0$$

holds for all x in the interval I? If there is such a function $y(x)$, we can try to find its derivative at $x = a$ by differentiating the equation $F(x, y) = 0$ implicitly with respect to x, and evaluating the result at (a, b):

$$F_1(x, y) + F_2(x, y)\,\frac{dy}{dx} = 0,$$

so that

$$\left.\frac{dy}{dx}\right|_{x=a} = -\,\frac{F_1(a, b)}{F_2(a, b)}, \qquad \text{provided} \qquad F_2(a, b) \neq 0.$$

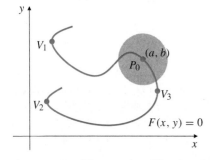

Figure 12.29   The equation $F(x, y) = 0$ can be solved for y as a function of x near P_0 or near any other point except the three points where the curve has a vertical tangent

Observe, however, that the condition $F_2(a, b) \neq 0$ required for the calculation of $y'(a)$ will itself guarantee that the solution $y(x)$ exists. This condition, together with the differentiability of $F(x, y)$ near (a, b), implies that the level curve $F(x, y) = F(a, b)$ has *nonvertical* tangent lines near (a, b), so some part of the level curve near (a, b) must be the graph of a function of x. (See Figure 12.29; the part of the curve $F(x, y) = 0$ in the shaded disk centred at $P_0 = (a, b)$ is the graph of a function $y(x)$ because vertical lines meet that part of the curve only once. The only points on the curve where a disk with that property cannot be drawn are the three points V_1, V_2, and V_3, where the curve has a vertical tangent, that is, where $F_2(x, y) = 0$.) This is a special case of the Implicit Function Theorem, which we will state more generally later in this section.

A similar situation holds for equations involving several variables. We can, for example, ask whether the equation

$$F(x, y, z) = 0$$

defines z as a function of x and y (say, $z = z(x, y)$) near some point P_0 with coordinates (x_0, y_0, z_0) satisfying the equation. If so, and if F has continuous first partials near P_0, then the partial derivatives of z can be found at (x_0, y_0) by implicit differentiation of the equation $F(x, y, z) = 0$ with respect to x and y:

$$F_1(x, y, z) + F_3(x, y, z)\,\frac{\partial z}{\partial x} = 0 \quad \text{and} \quad F_2(x, y, z) + F_3(x, y, z)\,\frac{\partial z}{\partial y} = 0,$$

so that

$$\left.\frac{\partial z}{\partial x}\right|_{(x_0, y_0)} = -\,\frac{F_1(x_0, y_0, z_0)}{F_3(x_0, y_0, z_0)} \quad \text{and} \quad \left.\frac{\partial z}{\partial y}\right|_{(x_0, y_0)} = -\,\frac{F_2(x_0, y_0, z_0)}{F_3(x_0, y_0, z_0)},$$

provided $F_3(x_0, y_0, z_0) \neq 0$. Since F_3 is the z component of the gradient of F, this condition implies that the level surface of F through P_0 does not have a horizontal normal vector, so it is not vertical (i.e., it is not parallel to the z-axis). Therefore, part of the surface near P_0 must indeed be the graph of a function $z = z(x, y)$. Similarly, $F(x, y, z) = 0$ can be solved for x as a function of y and z near points where $F_1 \neq 0$ and for $y = y(x, z)$ near points where $F_2 \neq 0$.

Example 1   Near what points on the sphere $x^2 + y^2 + z^2 = 1$ can the equation of the sphere be solved for z as a function of x and y? Find $\partial z / \partial x$ and $\partial z / \partial y$ at such points.

Solution   The sphere is the level surface $F(x, y, z) = 0$ of the function

$$F(x, y, z) = x^2 + y^2 + z^2 - 1.$$

The above equation can be solved for $z = z(x, y)$ near $P_0 = (x_0, y_0, z_0)$, provided that P_0 is not on the *equator* of the sphere, that is, the circle $x^2 + y^2 = 1$, $z = 0$. The equator consists of those points that satisfy $F_3(x, y, z) = 0$. If P_0 is not on the equator, then it is on either the upper or the lower hemisphere. The upper hemisphere has equation $z = z(x, y) = \sqrt{1 - x^2 - y^2}$, and the lower hemisphere has equation $z = z(x, y) = -\sqrt{1 - x^2 - y^2}$.

If $z \neq 0$, we can calculate the partial derivatives of the solution $z = z(x, y)$ by implicitly differentiating the equation of the sphere: $x^2 + y^2 + z^2 = 1$:

$$2x + 2z \frac{\partial z}{\partial x} = 0, \qquad \text{so} \qquad \frac{\partial z}{\partial x} = -\frac{x}{z},$$

$$2y + 2z \frac{\partial z}{\partial y} = 0, \qquad \text{so} \qquad \frac{\partial z}{\partial y} = -\frac{y}{z}.$$

Systems of Equations

Experience with linear equations shows us that systems of such equations can generally be solved for as many variables as there are equations in the system. We would expect, therefore, that a pair of equations in several variables might determine two of those variables as functions of the remaining ones. For instance, we might expect the equations

$$\begin{cases} F(x, y, z, w) = 0 \\ G(x, y, z, w) = 0 \end{cases}$$

to possess, near some point that satisfies them, solutions of one or more of the forms

$$\begin{cases} x = x(z, w) \\ y = y(z, w), \end{cases} \qquad \begin{cases} x = x(y, w) \\ z = z(y, w), \end{cases} \qquad \begin{cases} x = x(y, z) \\ w = w(y, z), \end{cases}$$

$$\begin{cases} y = y(x, w) \\ z = z(x, w), \end{cases} \qquad \begin{cases} y = y(x, z) \\ w = w(x, z), \end{cases} \qquad \begin{cases} z = z(x, y) \\ w = w(x, y). \end{cases}$$

Where such solutions exist, we should be able to differentiate the given system of equations implicitly to find partial derivatives of the solutions.

If you are given a single equation $F(x, y, z) = 0$ and asked to find $\partial x / \partial z$, you would understand that x is intended to be a function of the remaining variables y and z, so there would be no chance of misinterpreting which variable is to be held constant in calculating the partial derivative. Suppose, however, that you are asked to calculate $\partial x / \partial z$ given the system $F(x, y, z, w) = 0$, $G(x, y, z, w) = 0$. The question implies that x is one of the dependent variables and z is one of the independent variables, but does not imply which of y and w is the other dependent variable and which is the other independent variable. In short, which of the situations

$$\begin{cases} x = x(z, w) \\ y = y(z, w) \end{cases} \qquad \text{and} \qquad \begin{cases} x = x(y, z) \\ w = w(y, z) \end{cases}$$

are we dealing with? As it stands, the question is ambiguous. To avoid this ambiguity, we can specify *in the notation for the partial derivative* which variable is to be regarded as the other independent variable and therefore *held fixed* during the differentiation. Thus,

$$\left(\frac{\partial x}{\partial z}\right)_w \quad \text{implies the interpretation} \quad \begin{cases} x = x(z, w) \\ y = y(z, w), \end{cases}$$

$$\left(\frac{\partial x}{\partial z}\right)_y \quad \text{implies the interpretation} \quad \begin{cases} x = x(y, z) \\ w = w(y, z). \end{cases}$$

Example 2   Given the equations $F(x, y, z, w) = 0$ and $G(x, y, z, w) = 0$, where F and G have continuous first partial derivatives, calculate $(\partial x / \partial z)_w$.

Solution   We differentiate the two equations with respect to z, regarding x and y as functions of z and w, and holding w fixed:

$$F_1 \frac{\partial x}{\partial z} + F_2 \frac{\partial y}{\partial z} + F_3 = 0$$

$$G_1 \frac{\partial x}{\partial z} + G_2 \frac{\partial y}{\partial z} + G_3 = 0$$

(Note that the terms $F_4(\partial w / \partial z)$ and $G_4(\partial w / \partial z)$ are not present because w and z are independent variables, and w is being held fixed during the differentiation.) The pair of equations above is linear in $\partial x / \partial z$ and $\partial y / \partial z$. Eliminating $\partial y / \partial z$ (or using Cramer's Rule, Theorem 6 of Section 10.6), we obtain

$$\left(\frac{\partial x}{\partial z}\right)_w = -\frac{F_3 G_2 - F_2 G_3}{F_1 G_2 - F_2 G_1}.$$

In the light of the examples considered above, you should not be too surprised to learn that the nonvanishing of the denominator $F_1 G_2 - F_2 G_1$ at some point $P_0 = (x_0, y_0, z_0, w_0)$ satisfying the system $F = 0$, $G = 0$ is sufficient to guarantee that the system does indeed have a solution of the form $x = x(z, w)$, $y = y(z, w)$ near P_0. We will not, however, attempt to prove this fact here.

Example 3   Let x, y, u, and v be related by the equations

$$\begin{cases} u = x^2 + xy - y^2 \\ v = 2xy + y^2 \end{cases}$$

Find (a) $(\partial x / \partial u)_v$ and (b) $(\partial x / \partial u)_y$ at the point where $x = 2$ and $y = -1$.

Solution

(a) To calculate $(\partial x / \partial u)_v$ we regard x and y as functions of u and v and differentiate the given equations with respect to u, holding v constant:

$$1 = \frac{\partial u}{\partial u} = (2x + y)\frac{\partial x}{\partial u} + (x - 2y)\frac{\partial y}{\partial u}$$

$$0 = \frac{\partial v}{\partial u} = 2y\frac{\partial x}{\partial u} + (2x + 2y)\frac{\partial y}{\partial u}$$

At $x = 2$, $y = -1$ we have

$$1 = 3\frac{\partial x}{\partial u} + 4\frac{\partial y}{\partial u}$$

$$0 = -2\frac{\partial x}{\partial u} + 2\frac{\partial y}{\partial u}.$$

Eliminating $\partial y / \partial u$ leads to the result $(\partial x / \partial u)_v = 1/7$.

(b) To calculate $(\partial x / \partial u)_y$ we regard x and v as functions of y and u and differentiate the given equations with respect to u, holding y constant:

$$1 = \frac{\partial u}{\partial u} = (2x + y)\frac{\partial x}{\partial u}, \qquad \frac{\partial v}{\partial u} = 2y\frac{\partial x}{\partial u}.$$

At $x = 2$, $y = -1$ the first equation immediately gives $(\partial x / \partial u)_y = 1/3$.

It often happens that the independent variables in a problem are either clear from the context or can be chosen at the outset. In either case, ambiguity is not likely to occur, and we can safely omit subscripts showing which variables are being held constant. The following example is taken from the thermodynamics of an ideal gas.

Example 4   (**Reversible changes in an ideal gas**) The state of a closed system containing n moles of an ideal gas is characterized by three state variables—pressure, volume, and temperature (P, V, and T, respectively)—which satisfy the equation of state for an ideal gas:

$$PV = nRT,$$

where R is the universal gas constant. This equation can be regarded as defining one of the three variables as a function of the other two. The internal energy, E, and the entropy, S, of the system are thermodynamic quantities that depend on the state variables and hence may be expressed as functions of any two of P, V, and T. Let us choose V and T for the independent variables and so write $E = E(V, T)$ and $S = S(V, T)$. For reversible processes in the system, the first and second laws of thermodynamics imply that infinitesimal changes in these quantities satisfy the differential equation

$$T\,dS = dE + P\,dV.$$

Deduce that, for such processes, E is independent of V and depends only on the temperature T.

Solution   We calculate the differentials dS and dE and substitute them into the differential equation to obtain

$$T\left(\frac{\partial S}{\partial V}\,dV + \frac{\partial S}{\partial T}\,dT\right) = \frac{\partial E}{\partial V}\,dV + \frac{\partial E}{\partial T}\,dT + P\,dV.$$

Divide by T, substitute nR/V for P/T (from the equation of state), and collect coefficients of dV and dT on opposite sides of the equation to get

$$\left(\frac{\partial S}{\partial V} - \frac{1}{T}\frac{\partial E}{\partial V} - \frac{nR}{V}\right)dV = \left(\frac{1}{T}\frac{\partial E}{\partial T} - \frac{\partial S}{\partial T}\right)dT.$$

Since dV and dT are independent variables, both coefficients must vanish. Hence

$$\frac{\partial S}{\partial V} = \frac{1}{T}\frac{\partial E}{\partial V} + \frac{nR}{V}$$

$$\frac{\partial S}{\partial T} = \frac{1}{T}\frac{\partial E}{\partial T}.$$

Now differentiate the first of these equations with respect to T and the second with respect to V. Using equality of mixed partials for both S and E, we obtain the desired result:

$$\frac{\partial}{\partial T}\left(\frac{1}{T}\frac{\partial E}{\partial V} + \frac{nR}{V}\right) = \frac{\partial^2 S}{\partial T\,\partial V} = \frac{\partial^2 S}{\partial V\,\partial T} = \frac{\partial}{\partial V}\left(\frac{1}{T}\frac{\partial E}{\partial T}\right)$$

$$\frac{-1}{T^2}\frac{\partial E}{\partial V} + \frac{1}{T}\frac{\partial^2 E}{\partial T\,\partial V} = \frac{1}{T}\frac{\partial^2 E}{\partial V\,\partial T}$$

$$\frac{-1}{T^2}\frac{\partial E}{\partial V} = 0.$$

It follows that $\partial E / \partial V = 0$, so E is independent of V.

Jacobian Determinants

Partial derivatives obtained by implicit differentiation of systems of equations are fractions, the numerators and denominators of which are conveniently expressed in terms of certain determinants called *Jacobians*.

DEFINITION 8

The **Jacobian determinant** (or simply **the Jacobian**) of the two functions, $u = u(x, y)$ and $v = v(x, y)$, with respect to two variables, x and y, is the determinant

$$\frac{\partial(u, v)}{\partial(x, y)} = \begin{vmatrix} \dfrac{\partial u}{\partial x} & \dfrac{\partial u}{\partial y} \\ \dfrac{\partial v}{\partial x} & \dfrac{\partial v}{\partial y} \end{vmatrix}.$$

Similarly, the Jacobian of two functions, $F(x, y, \ldots)$ and $G(x, y, \ldots)$, with respect to the variables, x and y, is the determinant

$$\frac{\partial(F, G)}{\partial(x, y)} = \begin{vmatrix} \dfrac{\partial F}{\partial x} & \dfrac{\partial F}{\partial y} \\ \dfrac{\partial G}{\partial x} & \dfrac{\partial G}{\partial y} \end{vmatrix} = \begin{vmatrix} F_1 & F_2 \\ G_1 & G_2 \end{vmatrix}.$$

The definition above can be extended in the obvious way to give the Jacobian of n functions (or variables) with respect to n variables. For example, the Jacobian of three functions, F, G, and H, with respect to three variables, x, y, and z, is the determinant

$$\frac{\partial(F, G, H)}{\partial(x, y, z)} = \begin{vmatrix} F_1 & F_2 & F_3 \\ G_1 & G_2 & G_3 \\ H_1 & H_2 & H_3 \end{vmatrix}.$$

Jacobians are the determinants of the square Jacobian matrices corresponding to transformations of $\mathbb{R}^n$ to $\mathbb{R}^n$ as discussed briefly in Section 12.6.

Example 5   In terms of Jacobians, the value of $(\partial x/\partial z)_w$, obtained from the system of equations

$$F(x, y, z, w) = 0, \qquad G(x, y, z, w) = 0$$

in Example 2, can be expressed in the form

$$\left(\frac{\partial x}{\partial z}\right)_w = -\frac{\dfrac{\partial(F, G)}{\partial(z, y)}}{\dfrac{\partial(F, G)}{\partial(x, y)}}.$$

Observe the pattern here. The denominator is the Jacobian of F and G with respect to the two *dependent* variables, x and y. The numerator is the same Jacobian except that the dependent variable x is replaced by the independent variable z.

The pattern observed above is general. We state it formally in the Implicit Function Theorem below.

The Implicit Function Theorem

The Implicit Function Theorem guarantees that systems of equations can be solved for certain variables as functions of other variables under certain circumstances, and it provides formulas for the partial derivatives of the solution functions. Before stating it, we consider a simple illustrative example.

Example 6   Consider the system of linear equations

$$F(x, y, s, t) = a_1 x + b_1 y + c_1 s + d_1 t + e_1 = 0$$
$$G(x, y, s, t) = a_2 x + b_2 y + c_2 s + d_2 t + e_2 = 0.$$

This system can be written in matrix form:

$$\mathcal{A} \begin{pmatrix} x \\ y \end{pmatrix} + \mathcal{C} \begin{pmatrix} s \\ t \end{pmatrix} + E = \begin{pmatrix} 0 \\ 0 \end{pmatrix},$$

where

$$\mathcal{A} = \begin{pmatrix} a_1 & b_1 \\ a_2 & b_2 \end{pmatrix}, \qquad \mathcal{C} = \begin{pmatrix} c_1 & d_1 \\ c_2 & d_2 \end{pmatrix}, \quad \text{and} \quad E = \begin{pmatrix} e_1 \\ e_2 \end{pmatrix}.$$

The equations can be solved for x and y as functions of s and t provided $\det(\mathcal{A}) \neq 0$; this implies the existence of the inverse matrix $\mathcal{A}^{-1}$ (Theorem 4 of Section 10.6), so

$$\begin{pmatrix} x \\ y \end{pmatrix} = -\mathcal{A}^{-1} \left(\mathcal{C} \begin{pmatrix} s \\ t \end{pmatrix} + E \right).$$

Observe that $\det(\mathcal{A}) = \partial(F, G)/\partial(x, y)$, so the nonvanishing of this Jacobian guarantees that the equations can be solved for x and y.

THEOREM 8

The Implicit Function Theorem

Consider a system of n equations in $n + m$ variables,

$$\begin{cases} F_{(1)}(x_1, x_2, \ldots, x_m, y_1, y_2, \ldots, y_n) = 0 \\ F_{(2)}(x_1, x_2, \ldots, x_m, y_1, y_2, \ldots, y_n) = 0 \\ \quad \vdots \\ F_{(n)}(x_1, x_2, \ldots, x_m, y_1, y_2, \ldots, y_n) = 0, \end{cases}$$

and a point $P_0 = (a_1, a_2, \ldots, a_m, b_1, b_2, \ldots, b_n)$ that satisfies the system. Suppose each of the functions $F_{(i)}$ has continuous first partial derivatives with respect to each of the variables x_j and y_k, $(i = 1, \ldots, n, \ j = 1, \ldots, m, \ k = 1, \ldots, n)$, near P_0. Finally, suppose that

$$\left. \frac{\partial(F_{(1)}, F_{(2)}, \ldots, F_{(n)})}{\partial(y_1, y_2, \ldots, y_n)} \right|_{P_0} \neq 0.$$

Then the system can be solved for $y_1, y_2, \ldots, y_n$ as functions of $x_1, x_2, \ldots, x_m$ near P_0. That is, there exist functions

$$\phi_1(x_1, \ldots, x_m), \ldots, \phi_n(x_1, \ldots, x_m)$$

such that

$$\phi_j(a_1, \ldots, a_m) = b_j, \qquad (j = 1, \ldots, n),$$

and such that the equations

$$F_{(1)}\Big(x_1, \ldots, x_m, \phi_1(x_1, \ldots, x_m), \ldots, \phi_n(x_1, \ldots, x_m)\Big) = 0,$$

$$F_{(2)}\Big(x_1, \ldots, x_m, \phi_1(x_1, \ldots, x_m), \ldots, \phi_n(x_1, \ldots, x_m)\Big) = 0,$$

$$\vdots$$

$$F_{(n)}\Big(x_1, \ldots, x_m, \phi_1(x_1, \ldots, x_m), \ldots, \phi_n(x_1, \ldots, x_m)\Big) = 0,$$

hold for all $(x_1, \ldots, x_m)$ sufficiently near $(a_1, \ldots, a_m)$.

   Moreover,

$$\frac{\partial \phi_i}{\partial x_j} = \left(\frac{\partial y_i}{\partial x_j} \right)_{x_1, \ldots, x_{j-1}, x_{j+1}, \ldots, x_m} = -\frac{\dfrac{\partial(F_{(1)}, F_{(2)}, \ldots, F_{(n)})}{\partial(y_1, \ldots, x_j, \ldots, y_n)}}{\dfrac{\partial(F_{(1)}, F_{(2)}, \ldots, F_{(n)})}{\partial(y_1, \ldots, y_i, \ldots, y_n)}}.$$

Remark   The formula for the partial derivatives is a consequence of Cramer's Rule (Theorem 6 of Section 10.6) applied to the n linear equations in the n unknowns $\partial y_1/\partial x_j, \ldots, \partial y_n/\partial x_j$ obtained by differentiating each of the equations in the given system with respect to x_j.

Example 7   Show that the system

$$\begin{cases} xy^2 + xzu + yv^2 = 3 \\ x^3yz + 2xv - u^2v^2 = 2 \end{cases}$$

can be solved for (u, v) as a (vector) function of (x, y, z) near the point P_0 where $(x, y, z, u, v) = (1, 1, 1, 1, 1)$, and find the value of $\partial v/\partial y$ for the solution at $(x, y, z) = (1, 1, 1)$.

Solution   Let $\begin{cases} F(x, y, z, u, v) = xy^2 + xzu + yv^2 - 3 \\ G(x, y, z, u, v) = x^3yz + 2xv - u^2v^2 - 2 \end{cases}$. Then

$$\frac{\partial(F, G)}{\partial(u, v)}\bigg|_{P_0} = \begin{vmatrix} xz & 2yv \\ -2uv^2 & 2x - 2u^2v \end{vmatrix}\bigg|_{P_0} = \begin{vmatrix} 1 & 2 \\ -2 & 0 \end{vmatrix} = 4.$$

Since this Jacobian is not zero, the Implicit Function Theorem assures us that the given equations can be solved for u and v as functions of x, y, and z, that is, for $(u, v) = \mathbf{f}(x, y, z)$. Since

$$\frac{\partial(F, G)}{\partial(u, y)}\bigg|_{P_0} = \begin{vmatrix} xz & 2xy + v^2 \\ -2uv^2 & x^3z \end{vmatrix}\bigg|_{P_0} = \begin{vmatrix} 1 & 3 \\ -2 & 1 \end{vmatrix} = 7,$$

we have

$$\left(\frac{\partial v}{\partial y}\right)_{x,z} = -\frac{\dfrac{\partial(F, G)}{\partial(u, y)}}{\dfrac{\partial(F, G)}{\partial(u, v)}}\bigg|_{P_0} = -\frac{7}{4}.$$

∎

Remark   If all we wanted in this example was to calculate $\partial v/\partial y$, it would have been easier to use the technique of Example 3 and differentiate the given equations directly with respect to y, holding x and z fixed.

Example 8   If the equations $x = u^2 + v^2$ and $y = uv$ are solved for u and v in terms of x and y, find, where possible,

$$\frac{\partial u}{\partial x}, \quad \frac{\partial u}{\partial y}, \quad \frac{\partial v}{\partial x}, \quad \text{and} \quad \frac{\partial v}{\partial y}.$$

Hence, show that $\dfrac{\partial(u, v)}{\partial(x, y)} = 1 \bigg/ \dfrac{\partial(x, y)}{\partial(u, v)}$, provided the denominator is not zero.

Solution   The given equations can be rewritten in the form

$$F(u, v, x, y) = u^2 + v^2 - x = 0$$
$$G(u, v, x, y) = uv - y = 0.$$

Let

$$J = \frac{\partial(F, G)}{\partial(u, v)} = \begin{vmatrix} 2u & 2v \\ v & u \end{vmatrix} = 2(u^2 - v^2) = \frac{\partial(x, y)}{\partial(u, v)}.$$

If $u^2 \neq v^2$, then $J \neq 0$ and we can calculate the required partial derivatives:

$$\frac{\partial u}{\partial x} = -\frac{1}{J}\frac{\partial(F,G)}{\partial(x,v)} = -\frac{1}{J}\begin{vmatrix} -1 & 2v \\ 0 & u \end{vmatrix} = \frac{u}{2(u^2 - v^2)}$$

$$\frac{\partial u}{\partial y} = -\frac{1}{J}\frac{\partial(F,G)}{\partial(y,v)} = -\frac{1}{J}\begin{vmatrix} 0 & 2v \\ -1 & u \end{vmatrix} = \frac{-2v}{2(u^2 - v^2)}$$

$$\frac{\partial v}{\partial x} = -\frac{1}{J}\frac{\partial(F,G)}{\partial(u,x)} = -\frac{1}{J}\begin{vmatrix} 2u & -1 \\ v & 0 \end{vmatrix} = \frac{-v}{2(u^2 - v^2)}$$

$$\frac{\partial v}{\partial y} = -\frac{1}{J}\frac{\partial(F,G)}{\partial(u,y)} = -\frac{1}{J}\begin{vmatrix} 2u & 0 \\ v & -1 \end{vmatrix} = \frac{2u}{2(u^2 - v^2)}.$$

Thus,

$$\frac{\partial(u,v)}{\partial(x,y)} = \frac{1}{J^2}\begin{vmatrix} u & -2v \\ -v & 2u \end{vmatrix} = \frac{J}{J^2} = \frac{1}{J} = \frac{1}{\dfrac{\partial(x,y)}{\partial(u,v)}}.$$

Remark   Note in the above example that $\partial u/\partial x \neq 1/(\partial x/\partial u)$. This should be contrasted with the single-variable situation where, if $y = f(x)$ and $dy/dx \neq 0$, then $x = f^{-1}(y)$ and $dx/dy = 1/(dy/dx)$. This is another reason for distinguishing between ∂ and d. It is the Jacobian rather than any single partial derivative that takes the place of the ordinary derivative in such situations.

Remark   Let us look briefly at the general case of invertible transformations from $\mathbb{R}^n$ to $\mathbb{R}^n$. Suppose that $\mathbf{y} = \mathbf{f}(\mathbf{x})$ and $\mathbf{z} = \mathbf{g}(\mathbf{y})$ are both functions from $\mathbb{R}^n$ to $\mathbb{R}^n$ whose components have continuous first partial derivatives. As shown in Section 12.6, the Chain Rule implies that

$$\begin{pmatrix} \dfrac{\partial z_1}{\partial x_1} & \cdots & \dfrac{\partial z_1}{\partial x_n} \\ \vdots & \ddots & \vdots \\ \dfrac{\partial z_n}{\partial x_1} & \cdots & \dfrac{\partial z_n}{\partial x_n} \end{pmatrix} = \begin{pmatrix} \dfrac{\partial z_1}{\partial y_1} & \cdots & \dfrac{\partial z_1}{\partial y_n} \\ \vdots & \ddots & \vdots \\ \dfrac{\partial z_n}{\partial y_1} & \cdots & \dfrac{\partial z_n}{\partial y_n} \end{pmatrix} \begin{pmatrix} \dfrac{\partial y_1}{\partial x_1} & \cdots & \dfrac{\partial y_1}{\partial x_n} \\ \vdots & \ddots & \vdots \\ \dfrac{\partial y_n}{\partial x_1} & \cdots & \dfrac{\partial y_n}{\partial x_n} \end{pmatrix}.$$

This is just the Chain Rule for the composition $\mathbf{z} = \mathbf{g}\big(\mathbf{f}(\mathbf{x})\big)$. It follows from Theorem 3(b) of Section 10.6 that the determinants of these matrices satisfy a similar equation:

$$\frac{\partial(z_1 \cdots z_n)}{\partial(x_1 \cdots x_n)} = \frac{\partial(z_1 \cdots z_n)}{\partial(y_1 \cdots y_n)}\frac{\partial(y_1 \cdots y_n)}{\partial(x_1 \cdots x_n)}.$$

If $\mathbf{f}$ is one-to-one and $\mathbf{g}$ is the inverse of $\mathbf{f}$, then $\mathbf{z} = \mathbf{g}\big(\mathbf{f}(\mathbf{x})\big) = \mathbf{x}$, and $\partial(z_1 \cdots z_n)/\partial(x_1 \cdots x_n) = 1$, the determinant of the identity matrix. Thus

$$\frac{\partial(x_1 \cdots x_n)}{\partial(y_1 \cdots y_n)} = \frac{1}{\dfrac{\partial(y_1 \cdots y_n)}{\partial(x_1 \cdots x_n)}}.$$

In fact, the nonvanishing of either of these determinants is sufficient to guarantee that $\mathbf{f}$ is one-to-one and has an inverse. This is a special case of the Implicit Function Theorem.

We will encounter Jacobians again when we study transformations of coordinates in multiple integrals in Chapter 14.

Exercises 12.8

In Exercises 1–12, calculate the indicated derivative from the given equation(s). What condition on the variables will guarantee the existence of a solution that has the indicated derivative? Assume that any general functions F, G, and H have continuous first partial derivatives.

1. $\dfrac{dx}{dy}$ if $xy^3 + x^4y = 2$

2. $\dfrac{\partial x}{\partial y}$ if $xy^3 = y - z$

3. $\dfrac{\partial z}{\partial y}$ if $z^2 + xy^3 = \dfrac{xz}{y}$

4. $\dfrac{\partial y}{\partial z}$ if $e^{yz} - x^2z \ln y = \pi$

5. $\dfrac{\partial x}{\partial w}$ if $x^2y^2 + y^2z^2 + z^2t^2 + t^2w^2 - xw = 0$

6. $\dfrac{dy}{dx}$ if $F(x, y, x^2 - y^2) = 0$

7. $\dfrac{\partial u}{\partial x}$ if $G(x, y, z, u, v) = 0$

8. $\dfrac{\partial z}{\partial x}$ if $F(x^2 - z^2, y^2 + xz) = 0$

9. $\dfrac{\partial w}{\partial t}$ if $H(u^2w, v^2t, wt) = 0$

10. $\left(\dfrac{\partial y}{\partial x}\right)_u$ if $xyuv = 1$ and $x + y + u + v = 0$

11. $\left(\dfrac{\partial x}{\partial y}\right)_z$ if $x^2 + y^2 + z^2 + w^2 = 1$, and $x + 2y + 3z + 4w = 2$

12. $\dfrac{du}{dx}$ if $x^2y + y^2u - u^3 = 0$ and $x^2 + yu = 1$

13. If $x = u^3 + v^3$ and $y = uv - v^2$ are solved for u and v in terms of x and y, evaluate

$$\frac{\partial u}{\partial x}, \quad \frac{\partial u}{\partial y}, \quad \frac{\partial v}{\partial x}, \quad \frac{\partial v}{\partial y}, \quad \text{and} \quad \frac{\partial(u, v)}{\partial(x, y)}$$

at the point where $u = 1$ and $v = 1$.

14. Near what points (r, s) can the transformation

$$x = r^2 + 2s, \qquad y = s^2 - 2r$$

be solved for r and s as functions of x and y? Calculate the values of the first partial derivatives of the solution at the origin.

15. Evaluate the Jacobian $\partial(x, y)/\partial(r, \theta)$ for the transformation to polar coordinates: $x = r \cos \theta$, $y = r \sin \theta$. Near what points (r, θ) is the transformation one-to-one and therefore invertible to give r and θ as functions of x and y?

16. Evaluate the Jacobian $\partial(x, y, z)/\partial(\rho, \phi, \theta)$, where

$$x = \rho \sin \phi \cos \theta, \quad y = \rho \sin \phi \sin \theta, \quad \text{and } z = \rho \cos \phi.$$

This is the transformation from Cartesian to spherical polar coordinates in 3-space that we will consider in Section 14.6. Near what points is the transformation one-to-one and hence invertible to give ρ, ϕ, and θ as functions of x, y, and z?

17. Show that the equations

$$\begin{cases} xy^2 + zu + v^2 = 3 \\ x^3z + 2y - uv = 2 \\ xu + yv - xyz = 1 \end{cases}$$

can be solved for x, y, and z as functions of u and v near the point P_0 where $(x, y, z, u, v) = (1, 1, 1, 1, 1)$, and find $(\partial y/\partial u)_v$ at $(u, v) = (1, 1)$.

18. Show that the equations $\begin{cases} xe^y + uz - \cos v = 2 \\ u \cos y + x^2v - yz^2 = 1 \end{cases}$ can be solved for u and v as functions of x, y, and z near the point P_0 where $(x, y, z) = (2, 0, 1)$ and $(u, v) = (1, 0)$, and find $(\partial u/\partial z)_{x,y}$ at $(x, y, z) = (2, 0, 1)$.

19. Find dx/dy from the system

$$F(x, y, z, w) = 0, \quad G(x, y, z, w) = 0, \quad H(x, y, z, w) = 0.$$

20. Given the system

$$F(x, y, z, u, v) = 0$$
$$G(x, y, z, u, v) = 0$$
$$H(x, y, z, u, v) = 0,$$

how many possible interpretations are there for $\partial x/\partial y$? Evaluate them.

21. Given the system

$$F(x_1, x_2, \dots, x_8) = 0$$
$$G(x_1, x_2, \dots, x_8) = 0$$
$$H(x_1, x_2, \dots, x_8) = 0,$$

how many possible interpretations are there for the partial $\dfrac{\partial x_1}{\partial x_2}$? Evaluate $\left(\dfrac{\partial x_1}{\partial x_2}\right)_{x_4, x_6, x_7, x_8}$.

22. If $F(x, y, z) = 0$ determines z as a function of x and y, calculate $\partial^2 z/\partial x^2$, $\partial^2 z/\partial x \partial y$, and $\partial^2 z/\partial y^2$ in terms of the partial derivatives of F.

23. If $x = u + v$, $y = uv$, and $z = u^2 + v^2$ define z as a function of x and y, find $\partial z/\partial x$, $\partial z/\partial y$, and $\partial^2 z/\partial x \partial y$.

24. A certain gas satisfies the law $pV = T - \dfrac{4p}{T^2}$, where p = pressure, V = volume, and T = temperature.

(a) Calculate $\partial T/\partial p$ and $\partial T/\partial V$ at the point where $p = V = 1$ and $T = 2$.

(b) If measurements of p and V yield the values $p = 1 \pm 0.001$ and $V = 1 \pm 0.002$, find the approximate maximum error in the calculated value $T = 2$.

25. If $F(x, y, z) = 0$, show that $\left(\dfrac{\partial x}{\partial y}\right)_z \left(\dfrac{\partial y}{\partial z}\right)_x \left(\dfrac{\partial z}{\partial x}\right)_y = -1$.

Derive analogous results for $F(x, y, z, u) = 0$ and for $F(x, y, z, u, v) = 0$. What is the general case?

*26. If the equations $F(x, y, u, v) = 0$ and $G(x, y, u, v) = 0$ are solved for x and y as functions of u and v, show that

$$\frac{\partial(x, y)}{\partial(u, v)} = \frac{\partial(F, G)}{\partial(u, v)} \bigg/ \frac{\partial(F, G)}{\partial(x, y)}.$$

*27. If the equations $x = f(u, v)$, $y = g(u, v)$ can be solved for u and v in terms of x and y, show that

$$\frac{\partial(u, v)}{\partial(x, y)} = 1 \bigg/ \frac{\partial(x, y)}{\partial(u, v)}.$$

Hint: Use the result of Exercise 26.

*28. If $x = f(u, v)$, $y = g(u, v)$, $u = h(r, s)$, and $v = k(r, s)$, then x and y can be expressed as functions of r and s. Verify by direct calculation that

$$\frac{\partial(x, y)}{\partial(r, s)} = \frac{\partial(x, y)}{\partial(u, v)} \frac{\partial(u, v)}{\partial(r, s)}.$$

This is a special case of the Chain Rule for Jacobians.

*29. Two functions, $f(x, y)$ and $g(x, y)$, are said to be *functionally dependent* if one is a function of the other; that is, if there exists a single-variable function $k(t)$ such that $f(x, y) = k\left(g(x, y)\right)$ for all x and y. Show that in this case $\partial(f, g)/\partial(x, y)$ vanishes identically. Assume that all necessary derivatives exist.

*30. Prove the converse of Exercise 29 as follows: Let $u = f(x, y)$ and $v = g(x, y)$, and suppose that $\partial(u, v)/\partial(x, y) = \partial(f, g)/\partial(x, y)$ is identically zero for all x and y. Show that $(\partial u/\partial x)_v$ is identically zero. Hence u, considered as a function of x and v, is independent of x; that is, $u = k(v)$ for some function k of one variable. Why does this imply that f and g are functionally dependent?

12.9 Taylor Series and Approximations

As is the case for functions of one variable, power series representations and their partial sums (Taylor polynomials) can provide an efficient method for determining the behaviour of a smooth function of several variables near a point in its domain. In this section we will look briefly at the extension of Taylor series to such functions. As usual, we will develop the machinery for functions of two variables, but the extension to more variables should be clear.

As a starting point, recall Taylor's Formula for a function $F(x)$ with continuous derivatives of order up to $n + 1$ on the interval $[a, a + h]$:

$$F(a + h) = F(a) + F'(a)h + \frac{F''(a)}{2!} h^2 + \cdots + \frac{F^{(n)}(a)}{n!} h^n + \frac{F^{(n+1)}(X)}{(n + 1)!} h^{n+1},$$

where X is some number between a and $a + h$. (The last term in the formula is the *Lagrange* form of the remainder.) In the special case where $a = 0$ and $h = 1$, this formula becomes

$$F(1) = F(0) + F'(0) + \frac{F''(0)}{2!} + \cdots + \frac{F^{(n)}(0)}{n!} + \frac{F^{(n+1)}(\theta)}{(n + 1)!}$$

for some θ between 0 and 1.

Now suppose that $f(x, y)$ has continuous partial derivatives up to order $n + 1$ at all points in an open set containing the line segment joining the points (a, b) and $(a + h, b + k)$ in its domain. We can find the Taylor Formula for $f(a + h, b + k)$ in powers of h and k by applying the one-variable formula above to the function

$$F(t) = f(a + th, b + tk), \qquad 0 \le t \le 1.$$

Clearly, $F(0) = f(a, b)$ and $F(1) = f(a + h, b + k)$. Let us calculate some derivatives of F:

$$F'(t) = hf_1(a + th, b + tk) + kf_2(a + th, b + tk),$$

$$F''(t) = h^2 f_{11}(a + th, b + tk) + 2hk f_{12}(a + th, b + tk) + k^2 f_{22}(a + th, b + tk),$$

$$F'''(t) = \left(h^3 f_{111} + 3h^2 k f_{112} + 3hk^2 f_{122} + k^3 f_{222}\right)\bigg|_{(a+th, b+tk)}.$$

The pattern of binomial coefficients is pretty obvious here, but the notation, involving subscripts to denote partial derivatives of f, becomes more and more unwieldy as the order of the derivatives increases. The notation can be simplified greatly by using $D_1 f$ and $D_2 f$ to denote the first partials of f with respect to its first and second variables. Since h and k are constant and mixed partials commute ($D_1 D_2 f = D_2 D_1 f$), we have

$$h^2 D_1^2 f + 2hk D_1 D_2 f + k^2 D_2^2 f = (hD_1 + kD_2)^2 f,$$

and so on. Therefore,

$$F'(t) = (hD_1 + kD_2) f(a + th, b + tk),$$
$$F''(t) = (hD_1 + kD_2)^2 f(a + th, b + tk),$$
$$F'''(t) = (hD_1 + kD_2)^3 f(a + th, b + tk),$$
$$\vdots$$
$$F^{(m)}(t) = (hD_1 + kD_2)^m f(a + th, b + tk).$$

In particular, $F^{(m)}(0) = (hD_1 + kD_2)^m f(a, b)$. Hence, the Taylor Formula for $f(a + h, b + k)$ is

Note that
$$(hD_1 + kD_2)^m f(a, b) \text{ means}$$
$$\left. \left((hD_1 + kD_2)^m f \right) \right|_{(a,b)}, \text{ not}$$
$$(hD_1 + kD_2)^m \left(f(a, b) \right),$$
which would be zero.

$$f(a + h, b + k) = \sum_{m=0}^{n} \frac{1}{m!} (hD_1 + kD_2)^m f(a, b) + R_n(h, k)$$

$$= \sum_{m=0}^{n} \sum_{j=0}^{m} C_{mj} D_1^j D_2^{m-j} f(a, b) h^j k^{m-j} + R_n(h, k),$$

where, using the binomial expansion, we have

$$C_{mj} = \frac{1}{m!} \binom{m}{j} = \frac{1}{m!} \frac{m!}{j!(m-j)!} = \frac{1}{j!(m-j)!},$$

and where the remainder term is given by

$$R_n(h, k) = \frac{1}{(n+1)!} (hD_1 + kD_2)^{n+1} f(a + \theta h, b + \theta k)$$

for some θ between 0 and 1. If f has partial derivatives of all orders and

$$\lim_{n \to \infty} R_n(h, k) = 0,$$

then $f(a + h, b + k)$ can be expressed as a Taylor series in powers of h and k:

$$f(a + h, b + k) = \sum_{m=0}^{\infty} \sum_{j=0}^{m} \frac{1}{j!(m-j)!} D_1^j D_2^{m-j} f(a, b) h^j k^{m-j}.$$

As for functions of one variable, the Taylor polynomial of degree n,

$$P_n(x, y) = \sum_{m=0}^{n} \sum_{j=0}^{m} \frac{1}{j!(m-j)!} D_1^j D_2^{m-j} f(a, b) (x - a)^j (y - b)^{m-j},$$

provides the "best" nth-degree polynomial approximation to $f(x, y)$ near (a, b). For $n = 1$ this approximation reduces to the tangent plane approximation

$$f(x, y) \approx f(a, b) + f_1(a, b)(x - a) + f_2(a, b)(y - b).$$

Example 1   Find a second-degree polynomial approximation to the function $f(x, y) = \sqrt{x^2 + y^3}$ near the point $(1, 2)$, and use it to estimate $\sqrt{(1.02)^2 + (1.97)^3}$.

Solution   For the second-degree approximation we need the values of the partial derivatives of f up to second order at $(1, 2)$. We have

$$f(x, y) = \sqrt{x^2 + y^3} \qquad\qquad f(1, 2) = 3$$

$$f_1(x, y) = \frac{x}{\sqrt{x^2 + y^3}} \qquad\qquad f_1(1, 2) = \frac{1}{3}$$

$$f_2(x, y) = \frac{3y^2}{2\sqrt{x^2 + y^3}} \qquad\qquad f_2(1, 2) = 2$$

$$f_{11}(x, y) = \frac{y^3}{(x^2 + y^3)^{3/2}} \qquad\qquad f_{11}(1, 2) = \frac{8}{27}$$

$$f_{12}(x, y) = \frac{-3xy^2}{2(x^2 + y^3)^{3/2}} \qquad\qquad f_{12}(1, 2) = -\frac{2}{9}$$

$$f_{22}(x, y) = \frac{12x^2 y + 3y^4}{4(x^2 + y^3)^{3/2}} \qquad\qquad f_{22}(1, 2) = \frac{2}{3}.$$

Thus,

$$f(1 + h, 2 + k) \approx 3 + \frac{1}{3}h + 2k + \frac{1}{2!}\left(\frac{8}{27}h^2 + 2\left(-\frac{2}{9}\right)hk + \frac{2}{3}k^2\right)$$

or, setting $x = 1 + h$ and $y = 2 + k$,

$$f(x, y) = 3 + \frac{1}{3}(x - 1) + 2(y - 2) + \frac{4}{27}(x - 1)^2 - \frac{2}{9}(x - 1)(y - 2) + \frac{1}{3}(y - 2)^2.$$

This is the required second-degree Taylor polynomial for f near $(1, 2)$. Therefore,

$$\sqrt{(1.02)^2 + (1.97)^3} = f(1 + 0.02, 2 - 0.03)$$

$$\approx 3 + \frac{1}{3}(0.02) + 2(-0.03) + \frac{4}{27}(0.02)^2$$

$$- \frac{2}{9}(0.02)(-0.03) + \frac{1}{3}(-0.03)^2$$

$$\approx 2.947\,159\,3\,.$$

(For comparison purposes: the true value is $2.947\,163\,6\ldots$ The approximation is accurate to 6 significant figures.)

As observed for functions of one variable, it is not usually necessary to calculate derivatives in order to determine the coefficients in a Taylor series or Taylor polynomial. It is often much easier to perform algebraic manipulations on known series. For instance, the above example could have been done by writing f in the form

$$f(1 + h, 2 + k) = \sqrt{(1 + h)^2 + (2 + k)^3}$$

$$= \sqrt{9 + 2h + h^2 + 12k + 6k^2 + k^3}$$

$$= 3\sqrt{1 + \frac{2h + h^2 + 12k + 6k^2 + k^3}{9}}$$

and then applying the binomial expansion

$$\sqrt{1 + t} = 1 + \frac{1}{2}t + \frac{1}{2!}\left(\frac{1}{2}\right)\left(-\frac{1}{2}\right)t^2 + \cdots$$

with $t = \dfrac{2h + h^2 + 12k + 6k^2 + k^3}{9}$ to obtain the terms up to second degree in the variables h and k.

Example 2   Find the Taylor polynomial of degree 3 in powers of x and y for the function $f(x, y) = e^{x-2y}$.

Solution   The required Taylor polynomial will be the Taylor polynomial of degree 3 for e^t evaluated at $t = x - 2y$:

$$P_3(x, y) = 1 + (x - 2y) + \frac{1}{2!}(x - 2y)^2 + \frac{1}{3!}(x - 2y)^3$$

$$= 1 + x - 2y + \frac{1}{2}x^2 - 2xy + 2y^2 + \frac{1}{6}x^3 - x^2y + 2xy^2 - \frac{4}{3}y^3.$$

Remark   Maple can, of course, be used to compute multivariate Taylor polynomials with its function **mtaylor**, which, depending on the Maple version, may have to be read in from the Maple library before it can be used if it is not part of the Maple kernel.

```
>  readlib(mtaylor):
```

Arguments fed to `mtaylor` are as follows:

(a) an expression involving the expansion variables

(b) a list whose elements are either variable names or equations of the form `variable=value` giving the coordinates of the point about which the expansion is calculated. (Just naming a variable is equivalent to using the equation `variable=0`.)

(c) (optionally) a positive integer m forcing the order of the computed Taylor polynomial to be less than m. If m is not specified, the value of Maple's global variable "Order" is used. The default value is 6.

A few examples should suffice.

```
>  mtaylor(cos(x+y^2),[x,y]);
```

$$1 - \frac{1}{2}x^2 - y^2x + \frac{1}{24}x^4 - \frac{1}{2}y^4 + \frac{1}{6}y^2x^3$$

```
>  mtaylor(cos(x+y^2),[x=Pi,y],5);
```

$$-1 + \frac{1}{2}(x - \pi)^2 + y^2(x - \pi) - \frac{1}{24}(x - \pi)^4 + \frac{1}{2}y^4$$

```
>  mtaylor(g(x,y),[x=a,y=b],3);
```

$$g(a, b) + D_1(g)(a, b)(x - a) + D_2(g)(a, b)(y - b) + \frac{1}{2}D_{1,1}(g)(a, b)(x - a)^2$$

$$+ (x - a)D_{1,2}(g)(a, b)(y - b) + \frac{1}{2}D_{2,2}(g)(a, b)(y - b)^2$$

The function `mtaylor` can be a bit quirky. It has a tendency to expand linear terms; for example, in an expansion about $x = 1$ and $y = -2$, it may rewrite terms $2 + (x - 1) + 2(y + 2)$ in the form $5 + x + 2y$.

Approximating Implicit Functions

In the previous section we saw how to determine whether an equation in several variables could be solved for one of those variables as a function of the others. Even when such a solution is known to exist, it is not usually possible to find an exact formula for it. However, if the equation involves only smooth functions, then the solution will have a Taylor series. We can determine at least the first several coefficients in that series and thus obtain a useful approximation to the solution. The following example shows the technique.

Example 3 Show that the equation

$$\sin(x + y) = xy + 2x$$

has a solution of the form $y = f(x)$ near $x = 0$ satisfying $f(0) = 0$, and find the terms up to fourth degree for the Taylor series for $f(x)$ in powers of x.

Solution The given equation can be written in the form $F(x, y) = 0$, where

$$F(x, y) = \sin(x + y) - xy - 2x.$$

Since $F(0, 0) = 0$ and $F_2(0, 0) = \cos(0) = 1 \neq 0$, the equation has a solution $y = f(x)$ near $x = 0$ satisfying $f(0) = 0$ by the Implicit Function Theorem. It is not possible to calculate $f(x)$ exactly, but it will have a Maclaurin series of the form

$$y = f(x) = a_1 x + a_2 x^2 + a_3 x^3 + a_4 x^4 + \cdots.$$

(There is no constant term because $f(0) = 0$.) We can substitute this series into the given equation and keep track of terms up to degree 4 in order to calculate the coefficients a_1, a_2, a_3, and a_4. For the left side we use the Maclaurin series for sin to obtain

$$\sin(x + y) = \sin\left((1 + a_1)x + a_2 x^2 + a_3 x^3 + a_4 x^4 + \cdots\right)$$

$$= (1 + a_1)x + a_2 x^2 + a_3 x^3 + a_4 x^4 + \cdots$$

$$- \frac{1}{3!}\left((1 + a_1)x + a_2 x^2 + \cdots\right)^3 + \cdots$$

$$= (1 + a_1)x + a_2 x^2 + \left(a_3 - \frac{1}{6}(1 + a_1)^3\right)x^3$$

$$+ \left(a_4 - \frac{3}{6}(1 + a_1)^2 a_2\right)x^4 + \cdots.$$

The right side is

$$xy + 2x = 2x + a_1 x^2 + a_2 x^3 + a_3 x^4 + \cdots.$$

Equating coefficients of like powers of x, we obtain

$$1 + a_1 = 2 \qquad a_1 = 1$$

$$a_2 = a_1 \qquad a_2 = 1$$

$$a_3 - \frac{1}{6}(1 + a_1)^3 = a_2 \qquad a_3 = \frac{7}{3}$$

$$a_4 - \frac{1}{2}(1 + a_1)^2 a_2 = a_3 \qquad a_4 = \frac{13}{3}.$$

Thus

$$y = f(x) = x + x^2 + \frac{7}{3}x^3 + \frac{13}{3}x^4 + \cdots.$$

(We could have obtained more terms in the series by keeping track of higher powers of x in the substitution process.)

Remark From the series for $f(x)$ obtained above, we can determine the values of the first four derivatives of f at $x = 0$. Remember that

$$a_k = \frac{f^{(k)}(0)}{k!}.$$

We have, therefore,

$$f'(0) = a_1 = 1 \qquad f''(0) = 2!a_2 = 2$$

$$f'''(0) = 3!a_3 = 14 \qquad f^{(4)}(0) = 4!a_4 = 104.$$

We could have done the example by first calculating these derivatives by implicit differentiation of the given equation and then determining the series coefficients from them. This would have been a much more difficult way to do it. (Try it and see.)

Exercises 12.9

In Exercises 1–6, find the Taylor series for the given function about the indicated point.

1. $f(x, y) = \dfrac{1}{2 + xy^2}$, $(0, 0)$

2. $f(x, y) = \ln(1 + x + y + xy)$, $(0, 0)$

3. $f(x, y) = \tan^{-1}(x + xy)$, $(0, -1)$

4. $f(x, y) = x^2 + xy + y^3$, $(1, -1)$

5. $f(x, y) = e^{x^2 + y^2}$, $(0, 0)$

6. $f(x, y) = \sin(2x + 3y)$, $(0, 0)$

In Exercises 7–12, find Taylor polynomials of the indicated degree for the given functions near the given point. After calculating them by hand, try to get the same results using Maple's `mtaylor` function.

7. $f(x, y) = \dfrac{1}{2 + x - 2y}$, degree 3, near $(2, 1)$

8. $f(x, y) = \ln(x^2 + y^2)$, degree 3, near $(1, 0)$

9. $f(x, y) = \displaystyle\int_0^{x+y^2} e^{-t^2}\, dt$, degree 3, near $(0, 0)$

10. $f(x, y) = \cos(x + \sin y)$, degree 4, near $(0, 0)$

11. $f(x, y) = \dfrac{\sin x}{y}$, degree 2, near $\left(\frac{\pi}{2}, 1\right)$

12. $f(x, y) = \dfrac{1 + x}{1 + x^2 + y^4}$, degree 2, near $(0, 0)$

In Exercises 13–14, show that, for x near the indicated point $x = a$, the given equation has a solution of the form $y = f(x)$ taking on the indicated value at that point. Find the first three nonzero terms of the Taylor series for $f(x)$ in powers of $x - a$.

∗13. $x \sin y = y + \sin x$, near $x = 0$, with $f(0) = 0$

∗14. $\sqrt{1 + xy} = 1 + x + \ln(1 + y)$, near $x = 0$, with $f(0) = 0$

∗15. Show that the equation $x + 2y + z + e^{2z} = 1$ has a solution of the form $z = f(x, y)$ near $x = 0$, $y = 0$, where $f(0, 0) = 0$. Find the Taylor polynomial of degree 2 for $f(x, y)$ in powers of x and y.

∗16. Use series methods to find the value of the partial derivative $f_{112}(0, 0)$ given that $f(x, y) = \arctan(x + y)$.

∗17. Use series methods to evaluate

$$\left.\frac{\partial^{4n}}{\partial x^{2n} \partial y^{2n}}\, \frac{1}{1 + x^2 + y^2}\right|_{(0,0)}.$$

Chapter Review

Key Ideas

- **What do the following sentences and phrases mean?**
  - ◇ $\mathcal{S}$ is the graph of $f(x, y)$.
  - ◇ $\mathcal{C}$ is a level curve of $f(x, y)$.
  - ◇ $\lim_{(x,y)\to(a,b)} f(x, y) = L$.
  - ◇ $f(x, y)$ is continuous at (a, b).
  - ◇ the partial derivative $(\partial/\partial x)f(x, y)$
  - ◇ the tangent plane to $z = f(x, y)$ at (a, b)
  - ◇ pure second partials   ◇ mixed second partials
  - ◇ $f(x, y)$ is a harmonic function.
  - ◇ $L(x, y)$ is the linearization of $f(x, y)$ at (a, b).
  - ◇ the differential of $z = f(x, y)$
  - ◇ $f(x, y)$ is differentiable at (a, b).
  - ◇ the gradient of $f(x, y)$ at (a, b)
  - ◇ the directional derivative of $f(x, y)$ at (a, b) in direction $\mathbf{v}$
  - ◇ the Jacobian determinant $\partial(x, y)/\partial(u, v)$

- **Under what conditions are two mixed partial derivatives equal?**

- **State the Chain Rule for $z = f(x, y)$, where $x = g(u, v)$, and $y = h(u, v)$.**

- **Describe the process of calculating partial derivatives of implicitly defined functions.**

- **What is the Taylor series of $f(x, y)$ about (a, b)?**

Review Exercises

1. Sketch some level curves of the function $x + \dfrac{4y^2}{x}$.

2. Sketch some isotherms (curves of constant temperature) for the temperature function

$$T = \frac{140 + 30x^2 - 60x + 120y^2}{8 + x^2 - 2x + 4y^2} \quad (\,^\circ\text{C}).$$

What is the coolest location?

3. Sketch some level curves of the polynomial function $f(x, y) = x^3 - 3xy^2$. Why do you think the graph of this function is called a *monkey saddle*?

4. Let $f(x, y) = \begin{cases} \dfrac{x^3}{x^2 + y^2}, & \text{if } (x, y) \neq (0, 0) \\ 0, & \text{if } (x, y) = (0, 0). \end{cases}$

Calculate each of the following partial derivatives or explain why it does not exist: $f_1(0, 0)$, $f_2(0, 0)$, $f_{21}(0, 0)$, $f_{12}(0, 0)$.

5. Let $f(x, y) = \dfrac{x^3 - y^3}{x^2 - y^2}$. Where is $f(x, y)$ continuous? To what additional set of points does $f(x, y)$ have a continuous extension? In particular, can f be extended to be continuous at the origin? Can f be defined at the origin in such a way that its first partial derivatives exist there?

6. The surface $\mathcal{S}$ is the graph of the function $z = f(x, y)$, where $f(x, y) = e^{x^2 - 2x - 4y^2 + 5}$.

(a) Find an equation of the tangent plane to $\mathcal{S}$ at the point $(1, -1, 1)$.

(b) Sketch a representative sample of the level curves of the function $f(x, y)$.

7. Consider the surface $\mathcal{S}$ with equation $x^2 + y^2 + 4z^2 = 16$.

(a) Find an equation for the tangent plane to $\mathcal{S}$ at the point (a, b, c) on $\mathcal{S}$.

(b) For which points (a, b, c) on $\mathcal{S}$ does the tangent plane to $\mathcal{S}$ at (a, b, c) pass through the point $(0, 0, 4)$? Describe this set of points geometrically.

(c) For which points (a, b, c) on $\mathcal{S}$ is the tangent plane to $\mathcal{S}$ at (a, b, c) parallel to the plane $x + y + 2\sqrt{2}z = 97$?

8. Two variable resistors R_1 and R_2 are connected in parallel so that their combined resistance R is given by

$$\frac{1}{R} = \frac{1}{R_1} + \frac{1}{R_2}.$$

If $R_1 = 100$ ohms $\pm 5\%$ and $R_2 = 25$ ohms $\pm 2\%$, by approximately what percentage can the calculated value of their combined resistance $R = 20$ ohms be in error?

9. You have measured two sides of a triangular field and the angle between them. The side measurements are 150 m and 200 m, each accurate to within ± 1 m. The angle measurement is $30°$, accurate to within $\pm 2°$. What area do you calculate for the field, and what is your estimate of the maximum percentage error in this area?

10. Suppose that $T(x, y, z) = x^3 y + y^3 z + z^3 x$ gives the temperature at the point (x, y, z) in 3-space.

(a) Calculate the directional derivative of T at $(2, -1, 0)$ in the direction toward the point $(1, 1, 2)$.

(b) A fly is moving through space with constant speed 5. At time $t = 0$ the fly crosses the surface $2x^2 + 3y^2 + z^2 = 11$ at right angles at the point $(2, -1, 0)$, moving in the direction of increasing temperature. Find dT/dt at $t = 0$ as experienced by the fly.

11. Consider the function $f(x, y, z) = x^2 y + yz + z^2$.

(a) Find the directional derivative of f at $(1, -1, 1)$ in the direction of the vector $\mathbf{i} + \mathbf{k}$.

(b) An ant is crawling on the plane $x + y + z = 1$ through $(1, -1, 1)$. Suppose it crawls so as to keep f constant. In what direction is it going as it passes through $(1, -1, 1)$?

(c) Another ant crawls on the plane $x + y + z = 1$, moving in the direction of the greatest rate of increase of f. Find its direction as it goes through $(1, -1, 1)$.

12. Let $f(x, y, z) = (x^2 + z^2) \sin \dfrac{\pi x y}{2} + y z^2$. Let P_0 be the point $(1, 1, -1)$.

(a) Find the gradient of f at P_0.

(b) Find the linearization $L(x, y, z)$ of f at P_0.

(c) Find an equation for the tangent plane at P_0 to the level surface of f through P_0.

(d) If a bird flies through P_0 with speed 5, heading directly toward the point $(2, -1, 1)$, what is the rate of change of f as seen by the bird as it passes through P_0?

(e) In what direction from P_0 should the bird fly at speed 5 to experience the greatest rate of increase of f?

13. Verify that for any constant k the function
$$u(x, y) = k\Big(\ln \cos(x/k) - \ln \cos(y/k)\Big) \text{ satisfies the } minimal$$

surface equation

$$(1 + u_x^2)u_{yy} - u u_x u_y u_{xy} + (1 + u_y^2)u_{xx} = 0.$$

14. The equations $F(x, y, z) = 0$ and $G(x, y, z) = 0$ can define any two of the variables x, y, and z as functions of the remaining variable. Show that

$$\frac{dx}{dy}\frac{dy}{dz}\frac{dz}{dx} = 1.$$

15. The equations $\begin{cases} x = u^3 - uv \\ y = 3uv + 2v^2 \end{cases}$ define u and v as functions of x and y near the point P where $(u, v, x, y) = (-1, 2, 1, 2)$.

(a) Find $\dfrac{\partial u}{\partial x}$ and $\dfrac{\partial u}{\partial y}$ at P.

(b) Find the approximate value of u when $x = 1.02$ and $y = 1.97$.

16. The equations $\begin{cases} u = x^2 + y^2 \\ v = x^2 - 2xy^2 \end{cases}$ define x and y implicitly as functions of u and v for values of (x, y) near $(1, 2)$ and values of (u, v) near $(5, -7)$.

(a) Find $\dfrac{\partial x}{\partial u}$ and $\dfrac{\partial y}{\partial u}$ at $(u, v) = (5, -7)$.

(b) If $z = \ln(y^2 - x^2)$, find $\dfrac{\partial z}{\partial u}$ at $(u, v) = (5, -7)$.

Challenging Problems

1. (a) If the graph of a function $f(x, y)$ that is differentiable at (a, b) contains part of a straight line through (a, b), show that the line lies in the tangent plane to $z = f(x, y)$ at (a, b).

(b) If $g(t)$ is a differentiable function of t, describe the surface $z = yg(x/y)$ and show that all its tangent planes pass through the origin.

2. A particle moves in 3-space in such a way that its direction of motion at any point is perpendicular to the level surface of

$$f(x, y, z) = 4 - x^2 - 2y^2 + 3z^2$$

through that point. If the path of the particle passes through the point $(1, 1, 8)$, show that it also passes through $(2, 4, 1)$. Does it pass through $(3, 7, 0)$?

3. **(The Laplace operator in spherical coordinates)** If $u(x, y, z)$ has continuous second partial derivatives and

$$v(\rho, \phi, \theta) = u(\rho \sin \phi \cos \theta, \rho \sin \phi \sin \theta, \rho \cos \phi),$$

show that

$$\frac{\partial^2 v}{\partial \rho^2} + \frac{2}{\rho}\frac{\partial v}{\partial \rho} + \frac{\cot \phi}{\rho^2}\frac{\partial v}{\partial \phi} + \frac{1}{\rho^2}\frac{\partial^2 v}{\partial \phi^2} + \frac{1}{\rho^2 \sin^2 \phi}\frac{\partial^2 v}{\partial \theta^2}$$

$$= \frac{\partial^2 u}{\partial x^2} + \frac{\partial^2 u}{\partial y^2} + \frac{\partial^2 u}{\partial z^2}.$$

You can do this by hand, but it is a lot easier using computer algebra.

4. (Spherically expanding waves) If f is a twice differentiable function of one variable and $\rho = \sqrt{x^2 + y^2 + z^2}$, show that $u(x, y, z, t) = \dfrac{f(\rho - ct)}{\rho}$ satisfies the three-dimensional wave equation

$$\frac{\partial^2 u}{\partial t^2} = c^2 \left(\frac{\partial^2 u}{\partial x^2} + \frac{\partial^2 u}{\partial y^2} + \frac{\partial^2 u}{\partial z^2} \right).$$

What is the geometric significance of this solution as a function of increasing time t? *Hint:* You may want to use the result of Exercise 3. In this case $v(\rho, \phi, \theta)$ is independent of ϕ and θ.

CHAPTER 13

Applications of Partial Derivatives

I don't know what I may seem to the world, but as to myself, I seem to have been only like a boy playing on the sea-shore and diverting myself in now and then finding a smoother pebble or a prettier shell than ordinary, whilst the great ocean of truth lay all undiscovered before me.

Isaac Newton   1642–1727

Introduction In this chapter we will discuss some of the ways partial derivatives contribute to the understanding and solution of problems in applied mathematics. Many such problems can be put in the context of determining maximum or minimum values for functions of several variables, and the first four sections of this chapter deal with that subject. The remaining sections discuss some miscellaneous problems involving the differentiation of functions with respect to parameters, and also Newton's Method for approximating solutions of systems of nonlinear equations. Much of the material in this chapter may be considered *optional*. Only Sections 13.1–13.3 contain *core material*, and even parts of those sections can be omitted (e.g., the discussion of linear programming in Section 13.2).

13.1 Extreme Values

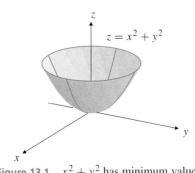

Figure 13.1 $x^2 + y^2$ has minimum value 0 at the origin

The function $f(x, y) = x^2 + y^2$, part of whose graph is shown in Figure 13.1, has a minimum value of 0; this value occurs at the origin $(0, 0)$ where the graph has a horizontal tangent plane. Similarly, the function $g(x, y) = 1 - x^2 - y^2$, part of whose graph appears in Figure 13.2, has a maximum value of 1 at $(0, 0)$. What techniques could be used to discover these facts if they were not evident from a diagram? Finding maximum and minimum values of functions of several variables is, like its single-variable counterpart, the crux of many applications of advanced calculus to problems that arise in other disciplines. Unfortunately, this problem is often much more complicated than in the single-variable case. Our discussion will begin by developing the techniques for functions of two variables. Some of the techniques extend to functions of more variables in obvious ways. The extension of those that do not will be discussed later in this section.

Let us begin by reviewing what we know about the single-variable case. Recall that a function $f(x)$ has a *local maximum value* (or a *local minimum value*) at a point

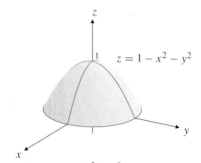

Figure 13.2 $1 - x^2 - y^2$ has maximum value 1 at the origin

a in its domain if $f(x) \le f(a)$ (or $f(x) \ge f(a)$) for all x in the domain of f that are *sufficiently close* to a. If the appropriate inequality holds *for all x* in the domain of f, then we say that f has an *absolute maximum* (or *absolute minimum*) value at a. Moreover, such local or absolute extreme values can occur only at points of one of the following three types:

(a) critical points, where $f'(x) = 0$,

(b) singular points, where $f'(x)$ does not exist, or

(c) endpoints of the domain of f.

A similar situation exists for functions of several variables. We say that a function of two variables has a **local maximum** or **relative maximum** value at the point (a, b) in its domain if $f(x, y) \le f(a, b)$ for all points (x, y) in the domain of f that are *sufficiently close* to the point (a, b). If the inequality holds *for all (x, y)* in the domain of f, then we say that f has a **global maximum** or **absolute maximum** value at (a, b). Similar definitions obtain for local (relative) and absolute (global) minimum values. In practice, the word *absolute* or *global* is usually omitted, and we refer simply to *the* maximum or *the* minimum value of f.

The following theorem shows that there are three possibilities for points where extreme values can occur, analogous to those for the single-variable case.

THEOREM 1

Necessary conditions for extreme values

A function $f(x, y)$ can have a local or absolute extreme value at a point (a, b) in its domain only if (a, b) is one of the following:

(a) a **critical point** of f, that is, a point satisfying $\nabla f(a, b) = \mathbf{0}$,

(b) a **singular point** of f, that is, a point where $\nabla f(a, b)$ does not exist, or

(c) a **boundary point** of the domain of f.

PROOF   Suppose that (a, b) belongs to the domain of f. If (a, b) is not on the boundary of the domain of f, then it must belong to the interior of that domain, and if (a, b) is not a singular point of f, then $\nabla f(a, b)$ exists. Finally, if (a, b) is not a critical point of f, then $\nabla f(a, b) \ne \mathbf{0}$, so f has a positive directional derivative in the direction of $\nabla f(a, b)$ and a negative directional derivative in the direction of $-\nabla f(a, b)$; that is, f is increasing as we move from (a, b) in one direction and decreasing as we move in the opposite direction. Hence, f cannot have either a maximum or a minimum value at (a, b). Therefore, any point where an extreme value occurs must be either a critical point or a singular point of f, or a boundary point of the domain of f.

Note that Theorem 1 remains valid with unchanged proof for functions of any number of variables. Of course, Theorem 1 does not guarantee that a given function will have any extreme values. It only tells us where to look to find any that may exist. Theorem 2, below, provides conditions that guarantee the existence of absolute maximum and minimum values for a continuous function. It is analogous to the Max-Min Theorem for functions of one variable. The proof is beyond the scope of this book; an interested student should consult an elementary text on mathematical analysis.

A set in $\mathbb{R}^n$ is **bounded** if it is contained inside some *ball* $x_1^2 + x_2^2 + \cdots + x_n^2 \le R^2$ of finite radius R. A set on the real line is bounded if it is contained in an interval of finite length.

THEOREM 2

Sufficient conditions for extreme values

If f is a *continuous* function of n variables whose domain is a *closed* and *bounded* set in $\mathbb{R}^n$, then the range of f is a bounded set of real numbers, and there are points in its domain where f takes on absolute maximum and minimum values.

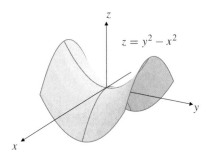

Figure 13.3 $y^2 - x^2$ has a saddle point at $(0, 0)$

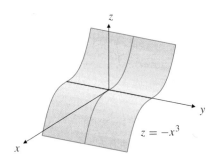

Figure 13.4 A line of saddle points

Example 1 The function $f(x, y) = x^2 + y^2$ (see Figure 13.1) has a critical point at $(0, 0)$, since $\nabla f = 2x\mathbf{i} + 2y\mathbf{j}$ and both components of ∇f vanish at $(0, 0)$. Since

$$f(x, y) > 0 = f(0, 0) \quad \text{if} \quad (x, y) \neq (0, 0),$$

f must have (absolute) minimum value 0 at that point. If the domain of f is not restricted, f has no maximum value. Similarly, $g(x, y) = 1 - x^2 - y^2$ has (absolute) maximum value 1 at its critical point $(0, 0)$. (See Figure 13.2.)

Example 2 The function $h(x, y) = y^2 - x^2$ also has a critical point at $(0, 0)$ but has neither a local maximum nor a local minimum value at that point. Observe that $h(0, 0) = 0$ but $h(x, 0) < 0$ and $h(0, y) > 0$ for all nonzero values of x and y. (See Figure 13.3.) The graph of h is a hyperbolic paraboloid. In view of the shape of this surface, we call the critical point $(0, 0)$ a *saddle point* of h.

In general, we will somewhat loosely call any *interior critical point* of the domain of a function f of several variables a **saddle point** if f does not have a local maximum or minimum value there. Even for functions of two variables, the graph will not always look like a saddle near a saddle point. For instance, the function $f(x, y) = -x^3$ has a whole line of saddle points along the y-axis (see Figure 13.4), although its graph does not resemble a saddle anywhere. These points resemble inflection points of a function of one variable. Saddle points are higher-dimensional analogues of such horizontal inflection points.

Example 3 The function $f(x, y) = \sqrt{x^2 + y^2}$ has no critical points but does have a singular point at $(0, 0)$ where it has a local (and absolute) minimum value, zero. The graph of f is (one nappe of) a circular cone. (See Figure 13.5(a).)

Example 4 The function $f(x, y) = 1 - x$ is defined everywhere in the xy-plane and has no critical or singular points. ($\nabla f(x, y) = -\mathbf{i}$ at every point (x, y).) Therefore f has no extreme values. However, if we restrict the domain of f to the points in the disk $x^2 + y^2 \leq 1$ (a closed bounded set in the xy-plane), then f does have absolute maximum and minimum values, as it must by Theorem 2. The maximum value is 2 at the boundary point $(-1, 0)$ and the minimum value is 0 at $(1, 0)$. (See Figure 13.5(b).)

Figure 13.5

(a) $\sqrt{x^2 + y^2}$ has a minimum value at the singular point $(0, 0)$

(b) When restricted to the disk $x^2 + y^2 \leq 1$, the function $1 - x$ has maximum and minimum values at boundary points

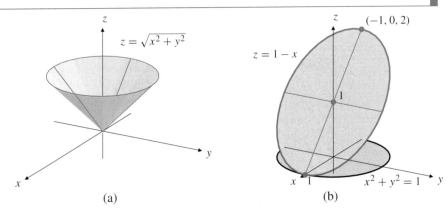

(a)   (b)

Classifying Critical Points

The above examples were very simple ones; it was immediately obvious in each case whether the function had a local maximum, local minimum, or a saddle point at the critical or singular point. For more complicated functions, it may be harder to classify the interior critical points. In theory, such a classification can always be made by considering the difference

$$\Delta f = f(a+h, b+k) - f(a, b)$$

for small values of h and k, where (a, b) is the critical point in question. If the difference is always nonnegative (or nonpositive) for small h and k, then f must have a local minimum (or maximum) at (a, b); if the difference is negative for some points (h, k) arbitrarily near $(0, 0)$ and positive for others, then f must have a saddle point at (a, b).

Example 5 Find and classify the critical points of $f(x, y) = 2x^3 - 6xy + 3y^2$.

Solution The critical points must satisfy the system of equations:

$$0 = f_1(x, y) = 6x^2 - 6y \quad \Longleftrightarrow \quad x^2 = y$$
$$0 = f_2(x, y) = -6x + 6y \quad \Longleftrightarrow \quad x = y.$$

Together, these equations imply that $x^2 = x$ so that $x = 0$ or $x = 1$. Therefore, the critical points are $(0, 0)$ and $(1, 1)$.

Consider $(0, 0)$. Here Δf is given by

$$\Delta f = f(h, k) - f(0, 0) = 2h^3 - 6hk + 3k^2.$$

Since $f(h, 0) - f(0, 0) = 2h^3$ is positive for small positive h and negative for small negative h, f cannot have a maximum or minimum value at $(0, 0)$. Therefore $(0, 0)$ is a saddle point.

Now consider $(1, 1)$. Here Δf is given by

$$\begin{aligned}
\Delta f &= f(1+h, 1+k) - f(1, 1) \\
&= 2(1+h)^3 - 6(1+h)(1+k) + 3(1+k)^2 - (-1) \\
&= 2 + 6h + 6h^2 + 2h^3 - 6 - 6h - 6k - 6hk + 3 + 6k + 3k^2 + 1 \\
&= 6h^2 - 6hk + 3k^2 + 2h^3 \\
&= 3(h-k)^2 + h^2(3 + 2h).
\end{aligned}$$

Both terms in the latter expression are nonnegative if $|h| < 3/2$, and they are not both zero unless $h = k = 0$. Hence, $\Delta f > 0$ for small h and k, and f has a local minimum value -1 at $(1, 1)$.

The method used to classify critical points in the above example takes on a "brute force" aspect if the function involved is more complicated. However, there is a *second derivative test* similar to that for functions of one variable. The n-variable version is the subject of the following theorem, the proof of which is based on properties of quadratic forms presented in Section 10.6.

THEOREM 3

A second derivative test

Suppose that $\mathbf{a} = (a_1, a_2, \ldots, a_n)$ is a critical point of $f(\mathbf{x}) = f(x_1, x_2, \ldots, x_n)$ and is interior to the domain of f. Also, suppose that all the second partial derivatives of f are continuous throughout a neighbourhood of $\mathbf{a}$, so that the **Hessian matrix**

$$\mathcal{H}(\mathbf{x}) = \begin{pmatrix} f_{11}(\mathbf{x}) & f_{12}(\mathbf{x}) & \cdots & f_{1n}(\mathbf{x}) \\ f_{21}(\mathbf{x}) & f_{22}(\mathbf{x}) & \cdots & f_{2n}(\mathbf{x}) \\ \vdots & \vdots & \ddots & \vdots \\ f_{n1}(\mathbf{x}) & f_{n2}(\mathbf{x}) & \cdots & f_{nn}(\mathbf{x}) \end{pmatrix}$$

is also continuous in that neighbourhood. Note that the continuity of the partials guarantees that $\mathcal{H}$ is a symmetric matrix.

(a) If $\mathcal{H}(\mathbf{a})$ is positive definite, then f has a local minimum at $\mathbf{a}$.

(b) If $\mathcal{H}(\mathbf{a})$ is negative definite, then f has a local maximum at $\mathbf{a}$.

(c) If $\mathcal{H}(\mathbf{a})$ is indefinite, then f has a saddle point at $\mathbf{a}$.

(d) If $\mathcal{H}(\mathbf{a})$ is neither positive nor negative definite nor indefinite, this test gives no information.

PROOF   Let $g(t) = f(\mathbf{a} + t\mathbf{h})$ for $0 \le t \le 1$, where $\mathbf{h}$ is an n-vector. Then

$$g'(t) = \sum_{i=1}^{n} f_i(\mathbf{a} + t\mathbf{h}) \, h_i$$

$$g''(t) = \sum_{i=1}^{n} \sum_{j=1}^{n} f_{ij}(\mathbf{a} + t\mathbf{h}) \, h_i \, h_j = \mathbf{h}^T \mathcal{H}(\mathbf{a} + t\mathbf{h})\mathbf{h}.$$

(In the latter expression, $\mathbf{h}$ is being treated as a column vector.) We apply Taylor's Formula with Lagrange remainder to g to write

$$g(1) = g(0) + g'(0) + \frac{1}{2} g''(\theta)$$

for some θ between 0 and 1. Thus,

$$f(\mathbf{a} + \mathbf{h}) = f(\mathbf{a}) + \sum_{i=1}^{n} f_i(\mathbf{a}) \, h_i + \frac{1}{2} \mathbf{h}^T \mathcal{H}(\mathbf{a} + \theta\mathbf{h})\mathbf{h}.$$

Since $\mathbf{a}$ is a critical point of f, $f_i(\mathbf{a}) = 0$ for $1 \le i \le n$, so

$$f(\mathbf{a} + \mathbf{h}) - f(\mathbf{a}) = \frac{1}{2} \mathbf{h}^T \mathcal{H}(\mathbf{a} + \theta\mathbf{h})\mathbf{h}.$$

If $\mathcal{H}(\mathbf{a})$ is positive definite, then, by the continuity of $\mathcal{H}$, so is $\mathcal{H}(\mathbf{a} + \theta\mathbf{h})$ for $|\mathbf{h}|$ sufficiently small. Therefore, $f(\mathbf{a} + \mathbf{h}) - f(\mathbf{a}) > 0$ for nonzero $\mathbf{h}$, proving (a).

Parts (b) and (c) are proved similarly. The functions $f(x, y) = x^4 + y^4$, $g(x, y) = -x^4 - y^4$, and $h(x, y) = x^4 - y^4$ all fall under part (d) and show that in this case a function can have a minimum, a maximum, or a saddle point.

Example 6   Find and classify the critical points of the function
$f(x, y, z) = x^2 y + y^2 z + z^2 - 2x$.

Solution   The equations that determine the critical points are

$$0 = f_1(x, y, z) = 2xy - 2,$$
$$0 = f_2(x, y, z) = x^2 + 2yz,$$
$$0 = f_3(x, y, z) = y^2 + 2z.$$

The third equation implies $z = -y^2/2$, and the second then implies $y^3 = x^2$. From the first equation we get $y^{5/2} = 1$. Thus $y = 1$ and $z = -\frac{1}{2}$. Since $xy = 1$, we must have $x = 1$. The only critical point is $P = (1, 1, -\frac{1}{2})$. Evaluating the second partial derivatives of f at this point, we obtain the Hessian matrix

$$\mathcal{H} = \begin{pmatrix} 2 & 2 & 0 \\ 2 & -1 & 2 \\ 0 & 2 & 2 \end{pmatrix}.$$

Since

$$2 > 0, \qquad \begin{vmatrix} 2 & 2 \\ 2 & -1 \end{vmatrix} = -6 < 0, \qquad \begin{vmatrix} 2 & 2 & 0 \\ 2 & -1 & 2 \\ 0 & 2 & 2 \end{vmatrix} = -20 < 0,$$

$\mathcal{H}$ is indefinite by Theorem 8 of Section 10.6, so P is a saddle point of f.

Remark   Applying the test for positive or negative definiteness or indefiniteness given in Theorem 8 of Section 10.6, we can paraphrase the second derivative test for a function of two variables as follows:

Suppose that (a, b) is a critical point of the function $f(x, y)$ that is interior to the domain of f. Suppose also that the second partial derivatives of f are continuous in a neighbourhood of (a, b) and have at that point the values

$$A = f_{11}(a, b), \quad B = f_{12}(a, b) = f_{21}(a, b), \quad \text{and} \quad C = f_{22}(a, b).$$

(a) If $B^2 - AC < 0$ and $A > 0$, then f has a local minimum value at (a, b).

(b) If $B^2 - AC < 0$ and $A < 0$, then f has a local maximum value at (a, b).

(c) If $B^2 - AC > 0$, then f has a saddle point at (a, b).

(d) If $B^2 - AC = 0$, this test provides no information; f may have a local maximum or a local minimum value or a saddle point at (a, b).

Example 7   Reconsider Example 5 and use the second derivative test to classify the two critical points $(0, 0)$ and $(1, 1)$ of $f(x, y) = 2x^3 - 6xy + 3y^2$.

Solution   We have

$$f_{11}(x, y) = 12x, \quad f_{12}(x, y) = -6, \quad \text{and} \quad f_{22}(x, y) = 6.$$

At $(0, 0)$ we therefore have

$$A = 0, \quad B = -6, \quad C = 6, \quad \text{and} \quad B^2 - AC = 36 > 0,$$

so $(0, 0)$ is a saddle point. At $(1, 1)$ we have

$$A = 12 > 0, \quad B = -6, \quad C = 6, \quad \text{and} \quad B^2 - AC = -36 < 0,$$

so f must have a local minimum at $(1, 1)$.

Example 8   Find and classify the critical points of

$$f(x, y) = xy\, e^{-(x^2+y^2)/2}.$$

Does f have absolute maximum and minimum values? Why?

Solution   We begin by calculating the first- and second-order partial derivatives of the function f:

$$f_1(x, y) = y(1 - x^2)\, e^{-(x^2+y^2)/2},$$
$$f_2(x, y) = x(1 - y^2)\, e^{-(x^2+y^2)/2},$$
$$f_{11}(x, y) = xy(x^2 - 3)\, e^{-(x^2+y^2)/2},$$
$$f_{12}(x, y) = (1 - x^2)(1 - y^2)\, e^{-(x^2+y^2)/2},$$
$$f_{22}(x, y) = xy(y^2 - 3)\, e^{-(x^2+y^2)/2}.$$

At any critical point $f_1 = 0$ and $f_2 = 0$, so the critical points are the solutions of the system of equations

$$y(1 - x^2) = 0$$
$$x(1 - y^2) = 0.$$

The first of these equations says that $y = 0$ or $x = \pm 1$. The second equation says that $x = 0$ or $y = \pm 1$. There are five points satisfying both conditions: $(0, 0)$, $(1, 1)$, $(1, -1)$, $(-1, 1)$, and $(-1, -1)$. We classify them using the second derivative test.

At $(0, 0)$ we have $A = C = 0$, $B = 1$, so that $B^2 - AC = 1 > 0$. Thus, f has a saddle point at $(0, 0)$.

At $(1, 1)$ and $(-1, -1)$ we have $A = C = -2/e < 0$, $B = 0$. It follows that $B^2 - AC = -4/e^2 < 0$. Thus, f has local maximum values at these points. The value of f is $1/e$ at each point.

At $(1, -1)$ and $(-1, 1)$ we have $A = C = 2/e > 0$, $B = 0$. If follows that $B^2 - AC = -4/e^2 < 0$. Thus, f has local minimum values at these points. The value of f at each of them is $-1/e$.

Indeed, f has absolute maximum and minimum values, namely, the values obtained above as local extrema. To see why, observe that $f(x, y)$ approaches 0 as the point (x, y) recedes to infinity in any direction because the negative exponential dominates the power factor xy for large $x^2 + y^2$. Pick a number between 0 and the local maximum value $1/e$ found above, say, the number $1/(2e)$. For some R, we must have $|f(x, y)| \leq 1/(2e)$ whenever $x^2 + y^2 \geq R^2$. On the closed disk $x^2 + y^2 \leq R^2$, f must have absolute maximum and minimum values by Theorem 2. These cannot occur on the boundary circle $x^2 + y^2 = R^2$ because $|f|$ is smaller there ($\leq 1/(2e)$) than it is at the critical points considered above. Since f has no singular points, the absolute maximum and minimum values for the disk, and therefore for the whole plane, must occur at those critical points.

Example 9   Find the shape of a rectangular box with no top having given volume V and the least possible total surface area of its five faces.

Solution   If the horizontal dimensions of the box are x, y, and its height is z (see Figure 13.6), then we want to minimize

$$S = xy + 2yz + 2xz$$

subject to the restriction that $xyz = V$, the required volume. We can use this restriction to reduce the number of variables on which S depends, for instance, by substituting

$$z = \frac{V}{xy}.$$

Then S becomes a function of the two variables x and y:

$$S = S(x, y) = xy + \frac{2V}{x} + \frac{2V}{y}.$$

A real box has positive dimensions, so the domain of S should consist of only those points (x, y) that satisfy $x > 0$ and $y > 0$. If either x or y approaches 0 or ∞, then $S \to \infty$, so the minimum value of S must occur at a critical point. (S has no singular points.) For critical points we solve the equations

$$0 = \frac{\partial S}{\partial x} = y - \frac{2V}{x^2} \qquad \Longleftrightarrow \qquad x^2 y = 2V,$$

$$0 = \frac{\partial S}{\partial y} = x - \frac{2V}{y^2} \qquad \Longleftrightarrow \qquad xy^2 = 2V.$$

Thus $x^2 y - xy^2 = 0$, or $xy(x - y) = 0$. Since $x > 0$ and $y > 0$, this implies that $x = y$. Therefore, $x^3 = 2V$, $x = y = (2V)^{1/3}$, and $z = V/(xy) = 2^{-2/3}V^{1/3} = x/2$. Since there is only one critical point it must minimize S. (Why?) The box having minimal surface area has a square base but is only half as high as its horizontal dimensions.

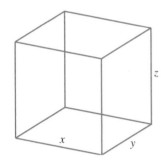

Figure 13.6

Remark   The preceding problem is a *constrained* extreme value problem in three variables; the equation $xyz = V$ is a *constraint* limiting the freedom of x, y, and z. We used the constraint to eliminate one variable, z, and so to reduce the problem to a *free* (i.e., *unconstrained*) problem in two variables. In Section 13.3 we will develop a more powerful method for solving constrained extreme value problems.

Exercises 13.1

In Exercises 1–17, find and classify the critical points of the given functions.

1. $f(x, y) = x^2 + 2y^2 - 4x + 4y$

2. $f(x, y) = xy - x + y$

3. $f(x, y) = x^3 + y^3 - 3xy$

4. $f(x, y) = x^4 + y^4 - 4xy$

5. $f(x, y) = \dfrac{x}{y} + \dfrac{8}{x} - y$

6. $f(x, y) = \cos(x + y)$

7. $f(x, y) = x \sin y$

8. $f(x, y) = \cos x + \cos y$

9. $f(x, y) = x^2 y e^{-(x^2 + y^2)}$

10. $f(x, y) = \dfrac{xy}{2 + x^4 + y^4}$

11. $f(x, y) = x e^{-x^3 + y^3}$

12. $f(x, y) = \dfrac{x^2}{x^2 + y^2}$

13. $f(x, y) = \dfrac{xy}{x^2 + y^2}$

14. $f(x, y) = \dfrac{1}{1 - x + y + x^2 + y^2}$

15. $f(x, y) = \left(1 + \dfrac{1}{x}\right)\left(1 + \dfrac{1}{y}\right)\left(\dfrac{1}{x} + \dfrac{1}{y}\right)$

*16. $f(x, y, z) = xyz - x^2 - y^2 - z^2$

*17. $f(x, y, z) = xy + x^2 z - x^2 - y - z^2$

*18. Show that $f(x, y, z) = 4xyz - x^4 - y^4 - z^4$ has a local maximum value at the point $(1, 1, 1)$.

19. Find the maximum and minimum values of $f(x, y) = xy\, e^{-x^2 - y^4}$.

20. Find the maximum and minimum values of $f(x, y) = \dfrac{x}{(1 + x^2 + y^2)}$.

*21. Find the maximum and minimum values of $f(x, y, z) = xyz\, e^{-x^2 - y^2 - z^2}$. How do you know that such extreme values exist?

22. Find the minimum value of $f(x, y) = x + 8y + \dfrac{1}{xy}$ in the first quadrant $x > 0$, $y > 0$. How do you know that a minimum exists?

23. Postal regulations require that the sum of the height and girth (horizontal perimeter) of a package should not exceed L units. Find the largest volume of a rectangular box that can satisfy this requirement.

24. The material used to make the bottom of a rectangular box is twice as expensive per unit area as the material used to make the top or side walls. Find the dimensions of the box of given volume V for which the cost of materials is minimum.

25. Find the volume of the largest rectangular box (with faces parallel to the coordinate planes) that can be inscribed inside the ellipsoid

$$\frac{x^2}{a^2} + \frac{y^2}{b^2} + \frac{z^2}{c^2} = 1.$$

26. Find the three positive numbers a, b, and c, whose sum is 30 and for which the expression $ab^2 c^3$ is maximum.

27. Find the critical points of the function $z = g(x, y)$ that satisfies the equation $e^{2zx - x^2} - 3e^{2zy + y^2} = 2$.

*28. Classify the critical points of the function g in the previous exercise.

*29. Let $f(x, y) = (y - x^2)(y - 3x^2)$. Show that the origin is a critical point of f and that the restriction of f to every straight line through the origin has a local minimum value at the origin. (That is, show that $f(x, kx)$ has a local minimum value at $x = 0$ for every k and that $f(0, y)$ has a local minimum value at $y = 0$.) Does $f(x, y)$ have a local minimum value at the origin? What happens to f on the curve $y = 2x^2$? What does the second derivative test say about this situation?

30. Verify by completing the square (i.e., without appealing to Theorem 8 of Section 10.6) that the quadratic form

$$Q(u, v) = (x, y) \begin{pmatrix} A & B \\ B & C \end{pmatrix} \begin{pmatrix} x \\ y \end{pmatrix} = Au^2 + 2Buv + Cv^2$$

is positive definite if $A > 0$ and $\begin{vmatrix} A & B \\ B & C \end{vmatrix} > 0$, negative definite if $A < 0$ and $\begin{vmatrix} A & B \\ B & C \end{vmatrix} > 0$, and indefinite if $\begin{vmatrix} A & B \\ B & C \end{vmatrix} < 0$. This gives independent confirmation of the assertion in the remark preceding Example 7.

*31. State and prove (using square completion arguments rather than appealing to Theorem 8 of Section 10.6) a result analogous to that of Exercise 30 for a quadratic form $Q(u, v, w)$ involving three variables. What are the implications of this for a critical point (a, b, c) of a function $f(x, y, z)$ all of whose second partial derivatives are known at (a, b, c)?

13.2 Extreme Values of Functions Defined on Restricted Domains

Much of the previous section was concerned with techniques for determining whether a critical point of a function provides a local maximum or minimum value or is a saddle point. In this section we address the problem of determining absolute maximum and minimum values for functions that have them—usually functions whose domains are restricted to subsets of $\mathbb{R}^2$ (or $\mathbb{R}^n$) having nonempty interiors. In Example 8 of Section 13.1 we had to *prove* that the given function had absolute extreme values. If, however, we are dealing with a continuous function on a domain that is closed and bounded, then we can rely on Theorem 2 to guarantee the existence of such extreme values, but we will always have to check boundary points as well as any interior critical or singular points to find them. The following examples illustrate the technique.

Example 1   Find the maximum and minimum values of $f(x, y) = 2xy$ on the closed disk $x^2 + y^2 \leq 4$. (See Figure 13.7.)

Solution   Since f is continuous and the disk is closed, f must have absolute maximum and minimum values at some points of the disk. The first partial derivatives of f are

$$f_1(x, y) = 2y \qquad \text{and} \qquad f_2(x, y) = 2x,$$

so there are no singular points, and the only critical point is $(0, 0)$, where f has the value 0.

We must still consider values of f on the boundary circle $x^2 + y^2 = 4$. We can express f as a function of a single variable on this circle by using a convenient parametrization of the circle, say,

$$x = 2\cos t, \qquad y = 2\sin t, \qquad (-\pi \leq t \leq \pi).$$

We have

$$f\left(2\cos t, 2\sin t\right) = 8\cos t \sin t = g(t).$$

We must find any extreme values of $g(t)$. We can do this in either of two ways. If we rewrite $g(t) = 4\sin 2t$, it is clear that $g(t)$ has maximum value 4 (at $t = \frac{\pi}{4}$ and $-\frac{3\pi}{4}$) and minimum value -4 (at $t = -\frac{\pi}{4}$ and $\frac{3\pi}{4}$). Alternatively, we can differentiate g to find its critical points:

$$0 = g'(t) = -8\sin^2 t + 8\cos^2 t \quad \Longleftrightarrow \quad \tan^2 t = 1$$
$$\Longleftrightarrow \quad t = \pm\frac{\pi}{4} \text{ or } \pm\frac{3\pi}{4},$$

which again yield the maximum value 4 and the minimum value -4. (It is not necessary to check the endpoints $t = -\pi$ and $t = \pi$; since g is everywhere differentiable and is periodic with period π, any absolute maximum or minimum will occur at a critical point.)

In any event, f has maximum value 4 at the boundary points $(\sqrt{2}, \sqrt{2})$ and $(-\sqrt{2}, -\sqrt{2})$ and minimum value -4 at the boundary points $(\sqrt{2}, -\sqrt{2})$ and $(-\sqrt{2}, \sqrt{2})$. It is easily shown by the Second Derivative Test (or otherwise) that the interior critical point $(0, 0)$ is a saddle point. (See Figure 13.7.)

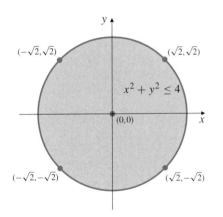

Figure 13.7   Points that are candidates for extreme values in Example 1

Example 2   Find the extreme values of the function $f(x, y) = x^2 y e^{-(x+y)}$ on the triangular region T given by $x \geq 0$, $y \geq 0$, and $x + y \leq 4$.

Solution   First, we look for critical points:

$$0 = f_1(x, y) = xy(2 - x)e^{-(x+y)} \qquad \Longleftrightarrow \quad x = 0, \ y = 0, \text{ or } x = 2,$$
$$0 = f_2(x, y) = x^2(1 - y)e^{-(x+y)} \qquad \Longleftrightarrow \quad x = 0 \text{ or } y = 1.$$

The critical points are $(0, y)$ for any y and $(2, 1)$. Only $(2, 1)$ is an interior point of T. (See Figure 13.8.) $f(2, 1) = 4/e^3 \approx 0.199$. The boundary of T consists of three straight line segments. On two of these, the coordinate axes, f is identically zero. The third segment is given by

$$y = 4 - x, \qquad 0 \le x \le 4,$$

so the values of f on this segment can be expressed as a function of x alone:

$$g(x) = f(x, 4 - x) = x^2(4 - x)e^{-4}, \qquad 0 \le x \le 4.$$

Note that $g(0) = g(4) = 0$ and $g(x) > 0$ if $0 < x < 4$. The critical points of g are given by $0 = g'(x) = (8x - 3x^2)e^{-4}$, so they are $x = 0$ and $x = 8/3$. We have

$$g\left(\frac{8}{3}\right) = f\left(\frac{8}{3}, \frac{4}{3}\right) = \frac{256}{27} e^{-4} \approx 0.174 < f(2, 1).$$

We conclude that the maximum value of f over the region T is $4/e^3$ and that it occurs at the interior critical point $(2, 1)$. The minimum value of f is zero and occurs at all points of the two perpendicular boundary segments. Note that f has neither a local maximum nor a local minimum at the boundary point $(8/3, 4/3)$, although g has a local maximum there. Of course, that point is not a saddle point of f either; it is not a critical point of f.

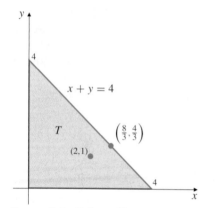

Figure 13.8   Points of interest in Example 2

Example 3   Among all triangles with vertices on a given circle, find those that have the largest area.

Solution   Intuition tells us that the equilateral triangles must have the largest area. However, proving this can be quite difficult unless a good choice of variables in which to set up the problem analytically is made. With a suitable choice of units and axes we can assume the circle is $x^2 + y^2 = 1$ and that one vertex of the triangle is the point P with coordinates $(1, 0)$. Let the other two vertices, Q and R, be as shown in Figure 13.9. There is no harm in assuming that Q lies on the upper semicircle and R on the lower, and that the origin O is inside triangle PQR. Let PQ and PR make angles θ and ϕ, respectively, with the negative direction of the x-axis. Clearly $0 \le \theta \le \pi/2$ and $0 \le \phi \le \pi/2$. The lines from O to Q and R make equal angles ψ with the line QR, where $2\theta + 2\phi + 2\psi = \pi$. Dropping perpendiculars from O to the three sides of the triangle PQR, we can write the area A of the triangle as the sum of the areas of six small, right-angled triangles:

$$A = 2 \times \frac{1}{2} \sin\theta \cos\theta + 2 \times \frac{1}{2} \sin\phi \cos\phi + 2 \times \frac{1}{2} \sin\psi \cos\psi$$
$$= \frac{1}{2}(\sin 2\theta + \sin 2\phi + \sin 2\psi).$$

Since $2\psi = \pi - 2(\theta + \phi)$, we express A as a function of the two variables θ and ϕ:

$$A = A(\theta, \phi) = \frac{1}{2}(\sin 2\theta + \sin 2\phi + \sin 2(\theta + \phi)).$$

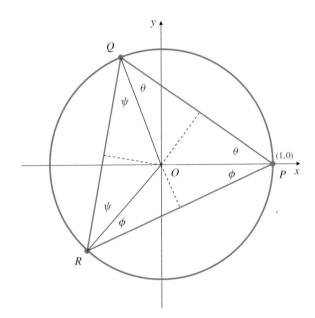

Figure 13.9 Where should Q and R be to ensure that triangle PQR has maximum area?

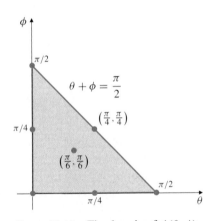

Figure 13.10 The domain of $A(\theta, \phi)$

The domain of A is the triangle $\theta \geq 0$, $\phi \geq 0$, $\theta + \phi \leq \pi/2$. $A = 0$ at the vertices of the triangle and is positive elsewhere. (See Figure 13.10.) We show that the maximum value of $A(\theta, \phi)$ on any edge of the triangle is 1 and occurs at the midpoint of that edge. On the edge $\theta = 0$ we have

$$A(0, \phi) = \frac{1}{2}\left(\sin 2\phi + \sin 2\phi\right) = \sin 2\phi \leq 1 = A(0, \pi/4).$$

Similarly, on $\phi = 0$, $A(\theta, 0) \leq 1 = A(\pi/4, 0)$. On the edge $\theta + \phi = \pi/2$ we have

$$A\left(\theta, \frac{\pi}{2} - \theta\right) = \frac{1}{2}\left(\sin 2\theta + \sin(\pi - 2\theta)\right)$$
$$= \sin 2\theta \leq 1 = A\left(\frac{\pi}{4}, \frac{\pi}{4}\right).$$

We must now check for any interior critical points of $A(\theta, \phi)$. (There are no singular points.) For critical points we have

$$0 = \frac{\partial A}{\partial \theta} = \cos 2\theta + \cos(2\theta + 2\phi),$$

$$0 = \frac{\partial A}{\partial \phi} = \cos 2\phi + \cos(2\theta + 2\phi),$$

so the critical points satisfy $\cos 2\theta = \cos 2\phi$ and, hence, $\theta = \phi$. We now substitute this equation into either of the above equations to determine θ:

$$\cos 2\theta + \cos 4\theta = 0$$
$$2\cos^2 2\theta + \cos 2\theta - 1 = 0$$
$$(2\cos 2\theta - 1)(\cos 2\theta + 1) = 0$$
$$\cos 2\theta = \frac{1}{2} \quad \text{or} \quad \cos 2\theta = -1.$$

The only solution leading to an interior point of the domain of A is $\theta = \phi = \pi/6$. Note that

$$A\left(\frac{\pi}{6}, \frac{\pi}{6}\right) = \frac{1}{2}\left(\frac{\sqrt{3}}{2} + \frac{\sqrt{3}}{2} + \frac{\sqrt{3}}{2}\right) = \frac{3\sqrt{3}}{4} > 1;$$

this interior critical point maximizes the area of the inscribed triangle. Finally, observe that for $\theta = \phi = \pi/6$, we also have $\psi = \pi/6$, so the largest triangle is indeed equilateral.

Remark   Since the area A of the inscribed triangle must have a maximum value (A is continuous and its domain is closed and bounded), a strictly geometric argument can be used to show that the largest triangle is equilateral. If an inscribed triangle has two unequal sides, its area can be made larger by moving the common vertex of these two sides along the circle to increase its perpendicular distance from the opposite side of the triangle.

Linear Programming

Linear programming is a branch of linear algebra that develops systematic techniques for finding maximum or minimum values of a *linear function* subject to several *linear inequality constraints*. Such problems arise frequently in management science and operations research. Because of their linear nature they do not usually involve calculus in their solution; linear programming is frequently presented in courses on *finite mathematics*. We will not attempt any formal study of linear programming here, but we will make a few observations for comparison with the more general nonlinear extreme-value problems considered above that involve calculus in their solution.

The inequality $ax + by \le c$ is an example of a linear inequality in two variables. The *solution set* of this inequality consists of a half-plane lying on one side of the straight line $ax + by = c$. The solution set of a system of several two-variable linear inequalities is an intersection of such half-planes, so it is a *convex* region of the plane bounded by a *polygonal line*. If it is a bounded set, then it is a convex polygon together with its interior. (A set is called **convex** if it contains the entire line segment between any two of its points. On the real line the convex sets are intervals.)

Let us examine a simple concrete example that involves only two variables and a few constraints.

Example 4   Find the maximum value of $F(x, y) = 2x + 7y$ subject to the constraints $x + 2y \le 6$, $2x + y \le 6$, $x \ge 0$,   and $y \ge 0$.

Solution   The solution set $\mathcal{S}$ of the system of four constraint inequalities is shown in Figure 13.11. It is the quadrilateral region with vertices $(0, 0)$, $(3, 0)$, $(2, 2)$, and $(0, 3)$. Several level curves of the linear function F are also shown in the figure. They are parallel straight lines with slope $-\frac{2}{7}$. We want the line that gives F the greatest value and that still intersects $\mathcal{S}$. Evidently this is the line $F = 21$ that passes through the vertex $(0, 3)$ of $\mathcal{S}$. The maximum value of F subject to the constraints is 21.

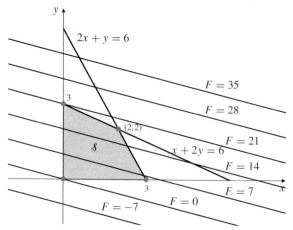

Figure 13.11   The shaded region is the solution set for the constraint inequalities in Example 4

As this simple example illustrates, a *linear* function with domain restricted by *linear inequalities* does not achieve maximum or minimum values at points in the interior of its domain (if that domain has an interior). Any such extreme value occurs at a boundary point of the domain or a set of such boundary points. Where an extreme value occurs at a set of boundary points, that set will *always* contain at least one vertex.

This phenomenon holds in general for extreme-value problems for linear functions in any number of variables with domains restricted by any number of linear inequalities. For problems involving three variables the domain will be a convex region of $\mathbb{R}^3$ bounded by planes. For a problem involving n variables the domain will be a convex region in $\mathbb{R}^n$ bounded by $(n-1)$-dimensional hyperplanes. Such *polyhedral* regions still have vertices (where n hyperplanes intersect), and maximum or minimum values of linear functions subject to the constraints will still occur at subsets of the boundary containing such vertices. These problems can therefore be solved by evaluating the linear function to be extremized (it is called the **objective function**) at all the vertices and selecting the greatest or least value.

In practice, linear programming problems can involve hundreds or even thousands of variables and even more constraints. Such problems need to be solved with computers, but even then it is extremely inefficient, if not impossible, to calculate all the vertices of the constraint solution set and the values of the objective function at them. Much of the study of linear programming therefore centres on devising techniques for getting to (or at least near) the optimizing vertex in as few steps as possible. Usually this involves criteria whereby large numbers of vertices can be rejected on geometric grounds. We will not delve into such techniques here but will content ourselves with one more example to illustrate, in a very simple case, how the underlying geometry of a problem can be used to reduce the number of vertices that must be considered.

Example 5   A tailor has 230 m of a certain fabric and has orders for up to 20 suits, up to 30 jackets, and up to 40 pairs of slacks to be made from the fabric. Each suit requires 6 m, each jacket 3 m, and each pair of slacks 2 m of the fabric. If the tailor's profit is $20 per suit, $14 per jacket, and $12 per pair of slacks, how many of each should he make to realize the maximum profit from his supply of the fabric?

Solution   Suppose he makes x suits, y jackets, and z pairs of slacks. Then his profit will be

$$P = 20x + 14y + 12z.$$

The constraints posed in the problem are

$$x \geq 0, \qquad x \leq 20,$$
$$y \geq 0, \qquad y \leq 30,$$
$$z \geq 0, \qquad z \leq 40,$$

$$6x + 3y + 2z \leq 230.$$

The last inequality is due to the limited supply of fabric. The solution set is shown in Figure 13.12. It has 10 vertices, $A, B, \ldots, J$. Since P increases in the direction of the vector $\nabla P = 20\mathbf{i} + 14\mathbf{j} + 12\mathbf{k}$, which points into the first octant, its maximum value cannot occur at any of the vertices $A, B, \ldots, G$. (Think about why.) Thus we need look only at the vertices H, I, and J.

$$H = (20, 10, 40), \qquad P = 1{,}020 \text{ at } H.$$
$$I = (10, 30, 40), \qquad P = 1{,}100 \text{ at } I.$$
$$J = (20, 30, 10), \qquad P = 940 \text{ at } J.$$

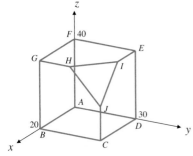

Figure 13.12   The convex set of points satisfying the constraints in Example 5

Thus, the tailor should make 10 suits, 30 jackets, and 40 pairs of slacks to realize the maximum profit, $1,100, from the fabric.

▌▌▌▌▌▌ Exercises 13.2 ▌▌▌▌▌▌▌▌

1. Find the maximum and minimum values of
$f(x, y) = x - x^2 + y^2$ on the rectangle $0 \le x \le 2$,
$0 \le y \le 1$.

2. Find the maximum and minimum values of
$f(x, y) = xy - 2x$ on the rectangle $-1 \le x \le 1, 0 \le y \le 1$.

3. Find the maximum and minimum values of
$f(x, y) = xy - y^2$ on the disk $x^2 + y^2 \le 1$.

4. Find the maximum and minimum values of
$f(x, y) = x + 2y$ on the disk $x^2 + y^2 \le 1$.

5. Find the maximum value of $f(x, y) = xy - x^3 y^2$ over the
square $0 \le x \le 1, 0 \le y \le 1$.

6. Find the maximum and minimum values of
$f(x, y) = xy(1 - x - y)$ over the triangle with vertices
$(0, 0)$, $(1, 0)$, and $(0, 1)$.

7. Find the maximum and minimum values of
$f(x, y) = \sin x \cos y$ on the closed triangular region
bounded by the coordinate axes and the line $x + y = 2\pi$.

8. Find the maximum value of

$$f(x, y) = \sin x \sin y \sin(x + y)$$

over the triangle bounded by the coordinate axes and the line
$x + y = \pi$.

9. The temperature at all points in the disk $x^2 + y^2 \le 1$ is given
by

$$T = (x + y) e^{-x^2 - y^2}.$$

Find the maximum and minimum temperatures at points of
the disk.

10. Find the maximum and minimum values of

$$f(x, y) = \frac{x - y}{1 + x^2 + y^2}$$

on the upper half-plane $y \ge 0$.

11. Find the maximum and minimum values of $xy^2 + yz^2$ over
the ball $x^2 + y^2 + z^2 \le 1$.

12. Find the maximum and minimum values of $xz + yz$ over the
ball $x^2 + y^2 + z^2 \le 1$.

13. Consider the function $f(x, y) = xy e^{-xy}$ with domain the
first quadrant: $x \ge 0$, $y \ge 0$. Show that
$\lim_{x \to \infty} f(x, kx) = 0$. Does f have a limit as (x, y) recedes
arbitrarily far from the origin in the first quadrant? Does f
have a maximum value in the first quadrant?

14. Repeat Exercise 13 for the function $f(x, y) = xy^2 e^{-xy}$.

15. In a certain community there are two breweries in
competition, so that sales of each negatively affect the profits
of the other. If brewery A produces x litres of beer per month
and brewery B produces y litres per month, then brewery A's
monthly profit $\$P$ and brewery B's monthly profit $\$Q$ are
assumed to be

$$P = 2x - \frac{2x^2 + y^2}{10^6},$$

$$Q = 2y - \frac{4y^2 + x^2}{2 \times 10^6}.$$

Find the sum of the profits of the two breweries if each
brewery independently sets its own production level to
maximize its own profit and assumes its competitor does
likewise. Find the sum of the profits if the two breweries
cooperate to determine their respective productions to
maximize that sum.

16. Equal angle bends are made at equal distances from the two
ends of a 100 m long straight length of fence so the resulting
three-segment fence can be placed along an existing wall to
make an enclosure of trapezoidal shape. What is the largest
possible area for such an enclosure?

17. Maximize $Q(x, y) = 2x + 3y$ subject to the constraints
$x \ge 0$, $y \ge 0$, $y \le 5$, $x + 2y \le 12$, and $4x + y \le 12$.

18. Minimize $F(x, y, z) = 2x + 3y + 4z$ subject to the
constraints $x \ge 0$, $y \ge 0$, $z \ge 0$, $x + y \ge 2$, $y + z \ge 2$, and
$x + z \ge 2$.

19. A textile manufacturer produces two grades of
wool-cotton-polyester fabric. The deluxe grade has
composition (by weight) 20% wool, 50% cotton, and 30%
polyester, and it sells for \$3 per kilogram. The standard
grade has composition 10% wool, 40% cotton, and 50%
polyester, and sells for \$2 per kilogram. If he has in stock
2,000 kg of wool and 6,000 kg each of cotton and polyester,
how many kilograms of fabric of each grade should he
manufacture to maximize his revenue?

20. A 10 hectare parcel of land is zoned for building densities of
6 detached houses per hectare, 8 duplex units per hectare, or
12 apartments per hectare. The developer who owns the land
can make a profit of \$40,000 per house, \$20,000 per duplex
unit, and \$16,000 per apartment that he builds. Municipal
bylaws require him to build at least as many apartments as
the total of houses and duplex units. How many of each type
of dwelling should he build to maximize his profit?

13.3 Lagrange Multipliers

A constrained extreme-value problem is one in which the variables of the function to
be maximized or minimized are not completely independent of one another, but must
satisfy one or more constraint equations or inequalities. For instance, the problems

$$\textbf{maximize} \quad f(x, y) \quad \textbf{subject to} \quad g(x, y) = C$$

and

$$\text{minimize} \quad f(x, y, z, w) \quad \textbf{subject to} \quad g(x, y, z, w) = C_1,$$
$$\textbf{and} \quad h(x, y, z, w) = C_2$$

have, respectively, one and two constraint equations, while the problem

$$\text{maximize} \quad f(x, y, z) \quad \textbf{subject to} \quad g(x, y, z) \le C$$

has a single constraint inequality.

Generally, inequality constraints can be regarded as restricting the domain of the function to be extremized to a smaller set that still has interior points. Section 13.2 was devoted to such problems. In each of the first three examples of that section we looked for *free* (i.e., *unconstrained*) extreme values in the interior of the domain, and we also examined the boundary of the domain, which was specified by one or more *constraint equations*. In Example 1 we parametrized the boundary and expressed the function to be extremized as a function of the parameter, thus reducing the boundary case to a free problem in one variable instead of a constrained problem in two variables. In Example 2 the boundary consisted of three line segments, on two of which the function was obviously zero. We solved the equation for the third boundary segment for y in terms of x, again in order to express the values of $f(x, y)$ on that segment as a function of one free variable. A similar approach was used in Example 3 to deal with the triangular boundary of the domain of the area function $A(\theta, \phi)$.

The reduction of extremization problems with equation constraints to free problems with fewer independent variables is only feasible when the constraint equations can be solved either explicitly for some variables in terms of others or parametrically for all variables in terms of some parameters. It is often very difficult or impossible to solve the constraint equations, so we need another technique.

The Method of Lagrange Multipliers

A technique for finding extreme values of $f(x, y)$ subject to the equality constraint $g(x, y) = 0$ is based on the following theorem:

THEOREM 4

Suppose that f and g have continuous first partial derivatives near the point $P_0 = (x_0, y_0)$ on the curve $\mathcal{C}$ with equation $g(x, y) = 0$. Suppose also that, when restricted to points on $\mathcal{C}$, the function $f(x, y)$ has a local maximum or minimum value at P_0. Finally, suppose that

    (i) P_0 is not an endpoint of $\mathcal{C}$, and

    (ii) $\nabla g(P_0) \ne \mathbf{0}$.

Then there exists a number λ_0 such that (x_0, y_0, λ_0) is a critical point of the **Lagrangian function**

$$L(x, y, \lambda) = f(x, y) + \lambda g(x, y).$$

PROOF Together, (i) and (ii) imply that $\mathcal{C}$ is smooth enough to have a tangent line at P_0 and that $\nabla g(P_0)$ is normal to that tangent line. If $\nabla f(P_0)$ is not parallel to $\nabla g(P_0)$, then $\nabla f(P_0)$ has a nonzero vector projection $\mathbf{v}$ along the tangent line to $\mathcal{C}$ at P_0. (See Figure 13.13.) Therefore, f has a positive directional derivative at P_0 in the direction of $\mathbf{v}$ and a negative directional derivative in the opposite direction. Thus, $f(x, y)$ increases or decreases as we move away from P_0 along $\mathcal{C}$ in the direction of $\mathbf{v}$ or $-\mathbf{v}$, and f cannot have a maximum or minimum value at P_0. Since we are assuming that f *does* have an extreme value at P_0, it must be that $\nabla f(P_0)$ is parallel to $\nabla g(P_0)$. Since $\nabla g(P_0) \ne \mathbf{0}$, there must exist a real number λ_0 such that $\nabla f(P_0) = -\lambda_0 \nabla g(P_0)$, or

$$\nabla(f + \lambda_0 g)(P_0) = \mathbf{0}.$$

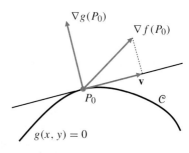

Figure 13.13 If $\nabla f(P_0)$ is not a multiple of $\nabla g(P_0)$, then $\nabla f(P_0)$ has a nonzero projection $\mathbf{v}$ tangent to the level curve of g through P_0

The two components of the above vector equation assert that $\partial L/\partial x = 0$ and $\partial L/\partial y = 0$ at (x_0, y_0, λ_0). The third equation that must be satisfied by a critical point of L is $\partial L/\partial \lambda = g(x, y) = 0$. This is satisfied at (x_0, y_0, λ_0) because P_0 lies on $\mathcal{C}$. Thus (x_0, y_0, λ_0) is a critical point of $L(x, y, \lambda)$.

Theorem 4 suggests that to find candidates for points on the curve $g(x, y) = 0$ at which $f(x, y)$ is maximum or minimum, we should look for critical points of the Lagrangian function

$$L(x, y, \lambda) = f(x, y) + \lambda g(x, y).$$

At any critical point of L we must have

$$0 = \frac{\partial L}{\partial x} = f_1(x, y) + \lambda g_1(x, y),$$
$$0 = \frac{\partial L}{\partial y} = f_2(x, y) + \lambda g_2(x, y),$$

that is, ∇f is parallel to ∇g,

and $\quad 0 = \dfrac{\partial L}{\partial \lambda} = g(x, y) \qquad$ the constraint equation.

Note, however, that it is *assumed* that the constrained problem *has a solution*. Theorem 4 does not guarantee that a solution exists; it only provides a means for finding a solution already known to exist. It is usually necessary to satisfy yourself that the problem you are trying to solve has a solution before using this method to find the solution.

Let us put the method to a concrete test:

Example 1   Find the shortest distance from the origin to the curve $x^2 y = 16$.

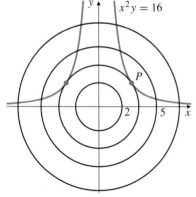

Figure 13.14   The level curve of the function representing the square of distance from the origin is tangent to the curve $x^2 y = 16$ at the two points on that curve that are closest to the origin

Solution   The graph of $x^2 y = 16$ is shown in Figure 13.14. There appear to be two points on the curve that are closest to the origin and no points that are farthest from the origin. (The curve is unbounded.) To find the closest points it is sufficient to minimize the *square* of the distance from the point (x, y) on the curve to the origin. (It is easier to work with the square of the distance rather than the distance itself, which involves a square root and so is harder to differentiate.) Thus we want to solve the problem

minimize $\quad f(x, y) = x^2 + y^2 \quad$ **subject to** $\quad g(x, y) = x^2 y - 16 = 0$.

Let $L(x, y, \lambda) = x^2 + y^2 + \lambda(x^2 y - 16)$. For critical points of L we want

$$0 = \frac{\partial L}{\partial x} = 2x + 2\lambda xy = 2x(1 + \lambda y) \tag{A}$$

$$0 = \frac{\partial L}{\partial y} = 2y + \lambda x^2 \tag{B}$$

$$0 = \frac{\partial L}{\partial \lambda} = x^2 y - 16. \tag{C}$$

Equation (A) requires that either $x = 0$ or $\lambda y = -1$. However, $x = 0$ is inconsistent with equation (C). Therefore $\lambda y = -1$. From equation (B) we now have

$$0 = 2y^2 + \lambda y x^2 = 2y^2 - x^2.$$

Thus, $x = \pm\sqrt{2}\, y$, and (C) now gives $2y^3 = 16$, so $y = 2$. There are, therefore, two candidates for points on $x^2 y = 16$ closest to the origin, $(\pm 2\sqrt{2}, 2)$. Both of these points are at distance $\sqrt{8 + 4} = 2\sqrt{3}$ units from the origin, so this must be the minimum distance from the origin to the curve. Some level curves of $x^2 + y^2$ are shown, along with the constraint curve $x^2 y = 16$, in Figure 13.14. Observe how the constraint curve is tangent to the level curve passing through the minimizing points $(\pm 2\sqrt{2}, 2)$, reflecting the fact that the two curves have parallel normals there.

Remark In the above example we could, of course, have solved the constraint equation for $y = 16/x^2$, substituted into f, and thus reduced the problem to one of finding the (unconstrained) minimum value of

$$F(x) = f\left(x, \frac{16}{x^2}\right) = x^2 + \frac{256}{x^4}.$$

The reader is invited to verify that this gives the same result.

The number λ that occurs in the Lagrangian function is called a **Lagrange multiplier**. The technique for solving an extreme-value problem with equation constraints by looking for critical points of an unconstrained problem in more variables (the original variables plus a Lagrange multiplier corresponding to each constraint equation) is called the **method of Lagrange multipliers**. It can be expected to give results as long as the function to be maximized or minimized (called the **objective function**) and the constraint equations have *smooth* graphs in a neighbourhood of the points where the extreme values occur, and these points are not on *edges* of those graphs. See Example 3 and Exercise 26 below.

Example 2 Find the points on the curve $17x^2 + 12xy + 8y^2 = 100$ that are closest to and farthest away from the origin.

Solution The quadratic form on the left side of the equation above is positive definite, as can be seen by completing a square. Hence the curve is bounded and must have points closest to and farthest from the origin. (In fact, the curve is an ellipse with centre at the origin and oblique principal axes. The problem asks us to find the ends of the major and minor axes.)

Again, we want to extremize $x^2 + y^2$ subject to an equation constraint. The Lagrangian in this case is

$$L(x, y, \lambda) = x^2 + y^2 + \lambda(17x^2 + 12xy + 8y^2 - 100),$$

and its critical points are given by

$$0 = \frac{\partial L}{\partial x} = 2x + \lambda(34x + 12y) \tag{A}$$

$$0 = \frac{\partial L}{\partial y} = 2y + \lambda(12x + 16y) \tag{B}$$

$$0 = \frac{\partial L}{\partial \lambda} = 17x^2 + 12xy + 8y^2 - 100. \tag{C}$$

Solving each of equations (A) and (B) for λ and equating the two expressions for λ obtained, we get

$$\frac{-2x}{34x + 12y} = \frac{-2y}{12x + 16y} \qquad \text{or} \qquad 12x^2 + 16xy = 34xy + 12y^2.$$

This equation simplifies to

$$2x^2 - 3xy - 2y^2 = 0. \tag{D}$$

We multiply equation (D) by 4 and add the result to equation (C) to get $25x^2 = 100$, so that $x = \pm 2$. Finally, we substitute each of these values of x into (D) and obtain (for each) two values of y from the resulting quadratics:

For $x = 2$: $\quad y^2 + 3y - 4 = 0,$ $\qquad\qquad$ For $x = -2$: $\quad y^2 - 3y - 4 = 0,$
$\qquad\qquad (y - 1)(y + 4) = 0.$ $\qquad\qquad\qquad\qquad (y + 1)(y - 4) = 0.$

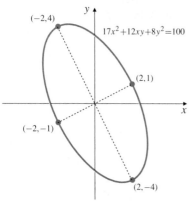

Figure 13.15   The points on the ellipse that are closest to and farthest from the origin

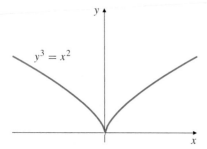

Figure 13.16   The minimum of y occurs at a point on the curve where the curve has no tangent line

We therefore obtain four candidate points: $(2, 1)$, $(-2, -1)$, $(2, -4)$, and $(-2, 4)$. The first two points are closest to the origin (they are the ends of the minor axis of the ellipse); the other two are farthest from the origin (the ends of the major axis). (See Figure 13.15.)

Considering the geometric underpinnings of the method of Lagrange multipliers, we would not expect the method to work if the level curves of the functions involved are not smooth or if the maximum or minimum occurs at an endpoint of the constraint curve. One of the pitfalls of the method is that the level curves of functions may not be smooth, even though the functions themselves have partial derivatives. Problems can occur where a gradient vanishes, as the following example shows.

Example 3   Find the minimum value of $f(x, y) = y$ subject to the constraint equation $g(x, y) = y^3 - x^2 = 0$.

Solution   The semicubical parabola $y^3 = x^2$ has a cusp at the origin. (See Figure 13.16.) Clearly, $f(x, y) = y$ has minimum value 0 at that point. Suppose, however, that we try to solve the problem using the method of Lagrange multipliers. The Lagrangian here is

$$L(x, y, \lambda) = y + \lambda(y^3 - x^2),$$

which has critical points given by

$$-2\lambda x = 0,$$
$$1 + 3\lambda y^2 = 0,$$
$$y^3 - x^2 = 0.$$

Observe that $y = 0$ cannot satisfy the second equation, and, in fact, the three equations have *no solution* (x, y, λ). (The first equation implies either $\lambda = 0$ or $x = 0$, but neither of these is consistent with the other two equations.)

Remark   The method of Lagrange multipliers breaks down in the above example because $\nabla g = \mathbf{0}$ at the solution point, and therefore the curve $g(x, y) = 0$ need not be smooth there. (In this case, it isn't smooth!) The geometric condition that ∇f should be parallel to ∇g at the solution point is meaningless in this case. When applying the method of Lagrange multipliers, be aware that an extreme value may occur at

　　(i) a critical point of the Lagrangian,

　　(ii) a point where $\nabla g = \mathbf{0}$,

　　(iii) a point where ∇f or ∇g does not exist, or

　　(iv) an "endpoint" of the constraint set.

This situation is similar to that for extreme values of a function f of one variable, which can occur at a critical point of f, a singular point of f, or an endpoint of the domain of f.

Problems with More than One Constraint

Next consider a three-dimensional problem requiring us to find a maximum or minimum value of a function of three variables subject to two equation constraints:

extremize $f(x, y, z)$ **subject to** $g(x, y, z) = 0$ and $h(x, y, z) = 0$.

Again, we assume that the problem has a solution, say, at the point $P_0 = (x_0, y_0, z_0)$, and that the functions f, g, and h have continuous first partial derivatives near P_0. Also, we assume that $\mathbf{T} = \nabla g(P_0) \times \nabla h(P_0) \neq \mathbf{0}$. These conditions imply that the surfaces $g(x, y, z) = 0$ and $h(x, y, z) = 0$ are smooth near P_0 and are not tangent to each other there, so they must intersect in a curve $\mathcal{C}$ that is smooth near P_0. The curve $\mathcal{C}$ has tangent vector $\mathbf{T}$ at P_0. The same geometric argument used in the proof of Theorem 4 again shows that $\nabla f(P_0)$ must be perpendicular to $\mathbf{T}$. (If not, then it would have a nonzero vector projection along $\mathbf{T}$, and f would have nonzero directional derivatives in the directions $\pm \mathbf{T}$ and would therefore increase and decrease as we moved away from P_0 along $\mathcal{C}$ in opposite directions.)

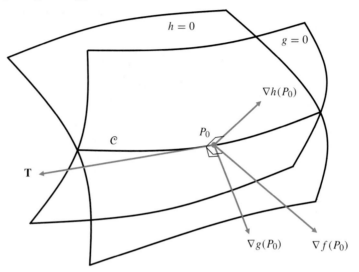

Figure 13.17   At P_0, ∇f, ∇g, and ∇h are all perpendicular to $\mathbf{T}$. Thus, ∇f is in the plane spanned by ∇g and ∇h

Since $\nabla g(P_0)$ and $\nabla h(P_0)$ are nonzero and both are perpendicular to $\mathbf{T}$ (see Figure 13.17), $\nabla f(P_0)$ must lie in the plane spanned by these two vectors and hence must be a linear combination of them:

$$\nabla f(x_0, y_0, z_0) = -\lambda_0 \nabla g(x_0, y_0, z_0) - \mu_0 \nabla h(x_0, y_0, z_0)$$

for some constants λ_0 and μ_0. It follows that $(x_0, y_0, z_0, \lambda_0, \mu_0)$ is a critical point of the Lagrangian function

$$L(x, y, z, \lambda, \mu) = f(x, y, z) + \lambda g(x, y, z) + \mu h(x, y, z).$$

We look for triples (x, y, z) that extremize $f(x, y, z)$ subject to the two constraints $g(x, y, z) = 0$ and $h(x, y, z) = 0$ among the points (x, y, z, λ, μ) that are critical points of the above Lagrangian function, and we therefore solve the system of equations

$$f_1(x, y, z) + \lambda g_1(x, y, z) + \mu h_1(x, y, z) = 0,$$
$$f_2(x, y, z) + \lambda g_2(x, y, z) + \mu h_2(x, y, z) = 0,$$
$$f_3(x, y, z) + \lambda g_3(x, y, z) + \mu h_3(x, y, z) = 0,$$
$$g(x, y, z) = 0,$$
$$h(x, y, z) = 0.$$

Solving such a system can be very difficult. It should be noted that, in using the method of Lagrange multipliers instead of solving the constraint equations, we have traded the problem of having to solve two equations for two variables as *functions* of a third one for a problem of having to solve five equations for *numerical* values of five unknowns.

Example 4   Find the maximum and minimum values of $f(x, y, z) = xy + 2z$ on the circle that is the intersection of the plane $x + y + z = 0$ and the sphere $x^2 + y^2 + z^2 = 24$.

Solution The function f is continuous, and the circle is a closed bounded set in 3-space. Therefore, maximum and minimum values must exist. We look for critical points of the Lagrangian

$$L = xy + 2z + \lambda(x + y + z) + \mu(x^2 + y^2 + z^2 - 24).$$

Setting the first partial derivatives of L equal to zero, we obtain

$$y + \lambda + 2\mu x = 0, \tag{A}$$
$$x + \lambda + 2\mu y = 0, \tag{B}$$
$$2 + \lambda + 2\mu z = 0, \tag{C}$$
$$x + y + z = 0, \tag{D}$$
$$x^2 + y^2 + z^2 - 24 = 0. \tag{E}$$

When none of the equations factors, try to combine two or more of them to produce an equation that does factor.

Subtracting (A) from (B) we get $(x - y)(1 - 2\mu) = 0$. Therefore, either $\mu = \frac{1}{2}$ or $x = y$. We analyze both possibilities.

CASE I If $\mu = \frac{1}{2}$, we obtain from (B) and (C)

$$x + \lambda + y = 0 \quad \text{and} \quad 2 + \lambda + z = 0.$$

Thus $x + y = 2 + z$. Combining this with (D), we get $z = -1$ and $x + y = 1$. Now, by (E), $x^2 + y^2 = 24 - z^2 = 23$. Since $x^2 + y^2 + 2xy = (x + y)^2 = 1$, we have $2xy = 1 - 23 = -22$ and $xy = -11$. Now $(x - y)^2 = x^2 + y^2 - 2xy = 23 + 22 = 45$, so $x - y = \pm 3\sqrt{5}$. Combining this with $x + y = 1$, we obtain two critical points arising from $\mu = \frac{1}{2}$, namely, $\left((1 + 3\sqrt{5})/2, (1 - 3\sqrt{5})/2, -1\right)$ and $\left((1 - 3\sqrt{5})/2, (1 + 3\sqrt{5})/2, -1\right)$. At both of these points we find that $f(x, y, z) = xy + 2z = -11 - 2 = -13$.

CASE II If $x = y$, then (D) implies that $z = -2x$, and (E) then gives $6x^2 = 24$, so $x = \pm 2$. Therefore, points $(2, 2, -4)$ and $(-2, -2, 4)$ must be considered. We have $f(2, 2, -4) = 4 - 8 = -4$ and $f(-2, -2, 4) = 4 + 8 = 12$.

We conclude that the maximum value of f on the circle is 12, and the minimum value is -13.

The method of Lagrange multipliers can be applied to find extreme values of a function of n variables, that is, of a vector variable $\mathbf{x} = (x_1, x_2, \ldots, x_n)$ subject to $m \le n - 1$ constraints:

extremize $f(\mathbf{x})$ **subject to** $g_{(1)}(\mathbf{x}) = 0, \quad \ldots \quad g_{(m)}(\mathbf{x}) = 0.$

Assuming that the problem has a solution at the point P_0, that f and all of the functions $g_{(j)}$ have continuous first partial derivatives in a neighbourhood of P_0, and that the intersection of the constraint (hyper)surfaces is smooth near P_0, then we should look for P_0 among the critical points of the $(n + m)$-variable Lagrangian function

$$L(\mathbf{x}, \lambda_1, \ldots, \lambda_m) = f(\mathbf{x}) + \sum_{j=1}^{m} \lambda_j g_{(j)}(\mathbf{x}).$$

We will not attempt to prove this general assertion. (A proof can be based on the Implicit Function Theorem.) Any critical points must satisfy the $n + m$ equations

$$\frac{\partial L}{\partial x_i} = 0, \quad (1 \le i \le n), \qquad \frac{\partial L}{\partial \lambda_j} = g_{(j)}(\mathbf{x}) = 0, \quad (1 \le j \le m).$$

Nonlinear Programming

When we looked for extreme values of functions f on restricted domains R in the previous section, we had to look separately for critical points of f in the interior of R and then for critical points of the restriction of f to the boundary of R. The interior of R is typically specified by one or more inequality constraints of the form $g < 0$, while the boundary corresponds to equation constraints of the form $g = 0$ (for which Lagrange multipliers can be used).

It is possible to unify these approaches into a single method for finding extreme values of functions defined on regions specified by inequalities of the form $g \le 0$.

Consider, for example, the problem of finding extreme values of $f(x, y)$ over the region R specified by $g(x, y) \le 0$. We can proceed by trying to find critical points of the four-variable function

$$L(x, y, \lambda, u) = f(x, y) + \lambda\big(g(x, y) + u^2\big).$$

Such critical points must satisfy the four equations

$$0 = \frac{\partial L}{\partial x} = f_1(x, y) + \lambda g_1(x, y), \tag{A}$$

$$0 = \frac{\partial L}{\partial y} = f_2(x, y) + \lambda g_2(x, y), \tag{B}$$

$$0 = \frac{\partial L}{\partial \lambda} = g(x, y) + u^2, \tag{C}$$

$$0 = \frac{\partial L}{\partial u} = 2\lambda u. \tag{D}$$

Suppose that (x, y, λ, u) satisfies these equations. We consider two cases:

CASE I $u \ne 0$. Then (D) implies that $\lambda = 0$, (C) implies that $g(x, y) = -u^2 < 0$, and (A) and (B) imply that $f_1(x, y) = 0$ and $f_2(x, y) = 0$. Thus, (x, y) is an interior critical point of f.

CASE II $u = 0$. Then (C) implies that $g(x, y) = 0$, and (A) and (B) imply that $\nabla f(x, y) = -\lambda \nabla g(x, y)$, so that (x, y) is a boundary point candidate for the location of the extreme value.

This technique can be extended to the problem of finding extreme values of a function of n variables, $\mathbf{x} = (x_1, x_2, \ldots, x_n)$, over the intersection R of m regions R_j defined by inequality constraints of the form $g_{(j)}(\mathbf{x}) \le 0$.

> **extremize** $f(\mathbf{x})$ **subject to** $g_{(1)}(\mathbf{x}) \le 0$, $\ldots$ $g_{(m)}(\mathbf{x}) \le 0$.

In this case we look for critical points of the $(n + 2m)$-variable Lagrangian

$$L(\mathbf{x}, \lambda_1, \ldots, \lambda_m, u_1, \ldots, u_m) = f(\mathbf{x}) + \sum_{j=1}^{m} \lambda_j\big(g_{(j)}(\mathbf{x}) + u_j^2\big).$$

The critical points will satisfy $n + 2m$ equations

$$\nabla f(\mathbf{x}) = -\sum_{j=1}^{m} \lambda_j \nabla g_{(j)}(\mathbf{x}), \qquad (n \text{ equations})$$

$$g_{(j)}(\mathbf{x}) = -u_j^2, \qquad (1 \le j \le m), \qquad (m \text{ equations})$$
$$2\lambda_j u_j = 0, \qquad (1 \le j \le m). \qquad (m \text{ equations})$$

The last m equations show that $\lambda_j = 0$ for any j for which $u_j \ne 0$. If all $u_j \ne 0$ then $\mathbf{x}$ is a critical point of f interior to R. Otherwise, some of the u_j will be zero, say, those corresponding to j in a subset J of $\{1, 2, \ldots, m\}$. In this case $\mathbf{x}$ will lie on the part of the boundary of R consisting of points lying on the boundaries of each of the regions R_j for which $j \in J$, and ∇f will be a linear combination of the corresponding gradients $\nabla g_{(j)}$:

$$\nabla f(\mathbf{x}) = -\sum_{j \in J} \lambda_j \nabla g_{(j)}(\mathbf{x}).$$

These are known as **Kuhn-Tucker conditions**, and this technique for solving extreme-value problems on restricted domains is called **nonlinear programming**.

▌▌▌▌▌▌▌ Exercises 13.3 ▌▌▌▌▌▌▌

1. Use the method of Lagrange multipliers to maximize $x^3 y^5$ subject to the constraint $x + y = 8$.

2. Find the shortest distance from the point $(3, 0)$ to the parabola $y = x^2$,

    (a) by reducing to an unconstrained problem in one variable, and

    (b) by using the method of Lagrange multipliers.

3. Find the distance from the origin to the plane $x + 2y + 2z = 3$,

    (a) using a geometric argument (no calculus),

    (b) by reducing the problem to an unconstrained problem in two variables, and

    (c) using the method of Lagrange multipliers.

4. Find the maximum and minimum values of the function $f(x, y, z) = x + y - z$ over the sphere $x^2 + y^2 + z^2 = 1$.

5. Use the Lagrange multiplier method to find the greatest and least distances from the point $(2, 1, -2)$ to the sphere with equation $x^2 + y^2 + z^2 = 1$. (Of course, the answer could be obtained more easily using a simple geometric argument.)

6. Find the shortest distance from the origin to the surface $xyz^2 = 2$.

7. Find a, b, and c so that the volume $V = 4\pi abc / 3$ of an ellipsoid $\dfrac{x^2}{a^2} + \dfrac{y^2}{b^2} + \dfrac{z^2}{c^2} = 1$ passing through the point $(1, 2, 1)$ is as small as possible.

8. Find the ends of the major and minor axes of the ellipse $3x^2 + 2xy + 3y^2 = 16$.

9. Find the maximum and minimum values of $f(x, y, z) = xyz$ on the sphere $x^2 + y^2 + z^2 = 12$.

10. Find the maximum and minimum values of $x + 2y - 3z$ over the ellipsoid $x^2 + 4y^2 + 9z^2 \le 108$.

11. Find the maximum and minimum values of the function $f(x, y, z) = x$ over the curve of intersection of the plane $z = x + y$ and the ellipsoid $x^2 + 2y^2 + 2z^2 = 8$.

12. Find the maximum and minimum values of $f(x, y, z) = x^2 + y^2 + z^2$ on the ellipse formed by the intersection of the cone $z^2 = x^2 + y^2$ and the plane $x - 2z = 3$.

13. Find the maximum and minimum values of $f(x, y, z) = 4 - z$ on the ellipse formed by the intersection of the cylinder $x^2 + y^2 = 8$ and the plane $x + y + z = 1$.

14. Find the maximum and minimum values of $f(x, y, z) = x + y^2 z$ subject to the constraints $y^2 + z^2 = 2$ and $z = x$.

*15. Use the method of Lagrange multipliers to find the shortest distance between the straight lines $x = y = z$ and $x = -y$, $z = 2$. (There are, of course, much easier ways to get the answer. This is an object lesson in the folly of shooting sparrows with cannons.)

16. Find the maximum and minimum values of the n-variable function $x_1 + x_2 + \cdots + x_n$ subject to the constraint $x_1^2 + x_2^2 + \cdots + x_n^2 = 1$.

17. Repeat Exercise 16 for the function $x_1 + 2x_2 + 3x_3 + \cdots + nx_n$ with the same constraint.

18. Find the most economical shape of a rectangular box with no top.

19. Find the maximum volume of a rectangular box with faces parallel to the coordinate planes if one corner is at the origin and the diagonally opposite corner lies on the plane $4x + 2y + z = 2$.

20. Find the maximum volume of a rectangular box with faces parallel to the coordinate planes if one corner is at the origin and the diagonally opposite corner is on the first octant part of the surface $xy + 2yz + 3xz = 18$.

21. A rectangular box having no top and having a prescribed volume $V \, \mathrm{m}^3$ is to be constructed using two different materials. The material used for the bottom and front of the box is five times as costly (per square metre) as the material used for the back and the other two sides. What should be the dimensions of the box to minimize the cost of materials?

*22. Find the maximum and minimum values of $xy + z^2$ on the ball $x^2 + y^2 + z^2 \le 1$. Use Lagrange multipliers to treat the boundary case.

*23. Repeat Exercise 22 but handle the boundary case by parametrizing the sphere $x^2 + y^2 + z^2 = 1$ using

$$x = \sin\phi \cos\theta, \quad y = \sin\phi \sin\theta, \quad z = \cos\phi,$$

where $0 \le \phi \le \pi$ and $0 \le \theta \le 2\pi$.

*24. If α, β, and γ are the angles of a triangle, show that

$$\sin\frac{\alpha}{2} \sin\frac{\beta}{2} \sin\frac{\gamma}{2} \le \frac{1}{8}.$$

For what triangles does equality occur?

*25. Suppose that f and g have continuous first partial derivatives throughout the xy-plane, and suppose that $g_2(a, b) \ne 0$. This implies that the equation $g(x, y) = g(a, b)$ defines y implicitly as a function of x near the point (a, b). Use the Chain Rule to show that if $f(x, y)$ has a local extreme value at (a, b) subject to the constraint $g(x, y) = g(a, b)$, then for some number λ the point (a, b, λ) is a critical point of the function

$$L(x, y, \lambda) = f(x, y) + \lambda g(x, y).$$

This constitutes a more formal justification of the method of Lagrange multipliers in this case.

26. What is the shortest distance from the point $(0, -1)$ to the curve $y = \sqrt{1 - x^2}$? Can this problem be solved by the Lagrange multiplier method? Why?

27. Example 3 showed that the method of Lagrange multipliers might fail to find a point that extremizes $f(x, y)$ subject to the constraint $g(x, y) = 0$ if $\nabla g = \mathbf{0}$ at the extremizing point. Can the method also fail if $\nabla f = \mathbf{0}$ at the extremizing point? Why?

13.4 The Method of Least Squares

Important optimization problems arise in the statistical analysis of experimental data. Frequently, experiments are designed to measure the values of one or more quantities supposed to be constant, or to demonstrate a supposed functional relationship between variable quantities. Experimental error is usually present in the measurements, and experiments need to be repeated several times in order to arrive at *mean* or *average* values of the quantities being measured.

Consider a very simple example. An experiment to measure a certain physical constant c is repeated n times, yielding the values $c_1, c_2, \ldots, c_n$. If none of the measurements is suspected of being faulty, intuition tells us that we should use the mean value $\bar{c} = (c_1 + c_2 + \cdots + c_n)/n$ as the value of c determined by the experiments. Let us see how this intuition can be justified.

Various methods for determining c from the data values are possible. We could, for instance, choose c to minimize the sum T of its distances from the data points:

$$T = |c - c_1| + |c - c_2| + \cdots + |c - c_n|.$$

This is unsatisfactory for a number of reasons. Since absolute values have singular points, it is difficult to determine the minimizing value of c. More importantly, c may not be determined uniquely. If $n = 2$, any point in the interval between c_1 and c_2 will give the same minimum value to T. (See Exercise 24 below for a generalization of this phenomenon.)

A more promising approach is to minimize the sum S of *squares* of the distances from c to the data points:

$$S = (c - c_1)^2 + (c - c_2)^2 + \cdots + (c - c_n)^2 = \sum_{i=1}^{n}(c - c_i)^2.$$

This function of c is smooth, and its (unconstrained) minimum value will occur at a critical point $\bar{c}$ given by

$$0 = \left.\frac{dS}{dc}\right|_{c=\bar{c}} = \sum_{i=1}^{n} 2(\bar{c} - c_i) = 2n\bar{c} - 2\sum_{i=1}^{n} c_i.$$

Thus $\bar{c}$ is the *mean* of the data values:

$$\bar{c} = \frac{1}{n}\sum_{i=1}^{n} c_i = \frac{c_1 + c_2 + \cdots + c_n}{n}.$$

The technique used to obtain $\bar{c}$ above is an example of what is called the **method of least squares**. It has the following geometric interpretation. If the data values $c_1, c_2, \ldots, c_n$ are regarded as components of a vector $\mathbf{c}$ in $\mathbb{R}^n$, and $\mathbf{w}$ is the vector with components $1, 1, \ldots, 1$, then the vector projection of $\mathbf{c}$ in the direction of $\mathbf{w}$,

$$\mathbf{c_w} = \frac{\mathbf{c} \bullet \mathbf{w}}{|\mathbf{w}|^2}\mathbf{w} = \frac{c_1 + c_2 + \cdots + c_n}{n}\mathbf{w},$$

has all its components equal to the average of the data values. Thus, determining c from the data by the method of least squares corresponds to finding the vector projection of the data vector onto the one-dimensional subspace of $\mathbb{R}^n$ spanned by $\mathbf{w}$. Had there been no error in the measurements c_i, then $\mathbf{c}$ would have been equal to $c\mathbf{w}$.

Linear Regression

In scientific investigations it is often believed that the response of a system is a certain kind of function of one or more input variables. An investigator can set up an experiment to measure the response of the system for various values of those variables in order to determine the parameters of the function.

For example, suppose that the response y of a system is suspected to depend on the input x according to the linear relationship

$$y = ax + b,$$

where the values of a and b are unknown. An experiment set up to measure values of y corresponding to several values of x yields n data points, (x_i, y_i), $i = 1, 2, \ldots, n$. If the supposed linear relationship is valid, these data points should lie *approximately* along a straight line, but not exactly on one because of experimental error. Suppose the points are as shown in Figure 13.18. The linear relationship seems reasonable in this case. We want to find values of a and b so that the straight line $y = ax + b$ "best" fits the data.

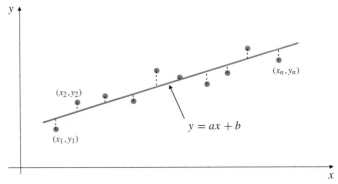

Figure 13.18   Fitting a straight line through experimental data

In this situation the method of least squares requires that a and b be chosen to minimize the sum S of the squares of the vertical displacements of the data points from the line:

$$S = \sum_{i=1}^{n} (y_i - ax_i - b)^2.$$

This is an unconstrained minimum problem in two variables, a and b. The minimum will occur at a critical point of S that satisfies

$$0 = \frac{\partial S}{\partial a} = -2 \sum_{i=1}^{n} x_i (y_i - ax_i - b),$$

$$0 = \frac{\partial S}{\partial b} = -2 \sum_{i=1}^{n} (y_i - ax_i - b).$$

These equations can be rewritten

$$\left(\sum_{i=1}^{n} x_i^2 \right) a + \left(\sum_{i=1}^{n} x_i \right) b = \sum_{i=1}^{n} x_i y_i,$$

$$\left(\sum_{i=1}^{n} x_i \right) a + n b = \sum_{i=1}^{n} y_i.$$

Solving this pair of linear equations, we obtain the desired parameters:

$$a = \frac{n\left(\sum\limits_{i=1}^{n} x_i y_i\right) - \left(\sum\limits_{i=1}^{n} x_i\right)\left(\sum\limits_{i=1}^{n} y_i\right)}{n\left(\sum\limits_{i=1}^{n} x_i^2\right) - \left(\sum\limits_{i=1}^{n} x_i\right)^2} = \frac{\overline{xy} - \bar{x}\bar{y}}{\overline{x^2} - (\bar{x})^2},$$

$$b = \frac{\left(\sum\limits_{i=1}^{n} x_i^2\right)\left(\sum\limits_{i=1}^{n} y_i\right) - \left(\sum\limits_{i=1}^{n} x_i\right)\left(\sum\limits_{i=1}^{n} x_i y_i\right)}{n\left(\sum\limits_{i=1}^{n} x_i^2\right) - \left(\sum\limits_{i=1}^{n} x_i\right)^2} = \frac{\overline{x^2}\bar{y} - \bar{x}\,\overline{xy}}{\overline{x^2} - (\bar{x})^2}.$$

In these formulas we have used a bar to indicate the mean value of a quantity; thus $\overline{xy} = (1/n)\sum_{i=1}^{n} x_i y_i$, and so on.

This procedure for fitting the "best" straight line through data points by the method of least squares is called **linear regression**, and the line $y = ax + b$ obtained in this way is called the **empirical regression line** corresponding to the data. Some scientific calculators with statistical features provide for linear regression by accumulating the sums of x_i, y_i, x_i^2, and $x_i y_i$ in various registers and keeping track of the number n of data points entered in another register. At any time it has available the information necessary to calculate a and b and the value of y corresponding to any given x.

Example 1   Find the empirical regression line for the data $(x, y) = (0, 2.10)$, $(1, 1.92)$, $(2, 1.84)$, and $(3, 1.71)$, $(4, 1.64)$. What is the predicted value of y at $x = 5$?

Solution   We have

$$\bar{x} = \frac{0 + 1 + 2 + 3 + 4}{5} = 2,$$

$$\bar{y} = \frac{2.10 + 1.92 + 1.84 + 1.71 + 1.64}{5} = 1.842,$$

$$\overline{xy} = \frac{(0)(2.10) + (1)(1.92) + (2)(1.84) + (3)(1.71) + (4)(1.64)}{5} = 3.458,$$

$$\overline{x^2} = \frac{0^2 + 1^2 + 2^2 + 3^2 + 4^2}{5} = 6.$$

Therefore,

$$a = \frac{3.458 - (2)(1.842)}{6 - 2^2} = -0.113,$$

$$b = \frac{(6)(1.842) - (2)(3.458)}{6 - 2^2} = 2.068,$$

and the empirical regression line is

$$y = 2.068 - 0.113x.$$

The predicted value of y at $x = 5$ is $2.068 - 0.113 \times 5 = 1.503$.

Figure 13.19 $\mathbf{p} = a\mathbf{x} + b\mathbf{w}$ is the projection of $\mathbf{y}$ onto the plane spanned by $\mathbf{x}$ and $\mathbf{w}$

Remark Linear regression can also be interpreted in terms of vector projection. The data points define two vectors $\mathbf{x}$ and $\mathbf{y}$ in $\mathbb{R}^n$ with components $x_1, x_2, \ldots, x_n$ and $y_1, y_2, \ldots, y_n$, respectively. Let $\mathbf{w}$ be the vector with components $1, 1, \ldots, 1$. Finding the coefficients a and b for the regression line corresponds to finding the orthogonal projection of $\mathbf{y}$ onto the two-dimensional subspace (plane) in $\mathbb{R}^n$ spanned by $\mathbf{x}$ and $\mathbf{w}$. (See Figure 13.19.) This projection is $\mathbf{p} = a\mathbf{x} + b\mathbf{w}$. In fact, the two equations obtained above by setting the partial derivatives of S equal to zero are just the two conditions

$$(\mathbf{y} - \mathbf{p}) \bullet \mathbf{x} = 0,$$
$$(\mathbf{y} - \mathbf{p}) \bullet \mathbf{w} = 0,$$

stating that $\mathbf{y}$ minus its projection onto the subspace is perpendicular to the subspace. The angle between $\mathbf{y}$ and this $\mathbf{p}$ provides a measure of how well the empirical regression line fits the data; the smaller the angle, the better the fit.

Linear regression can be used to find specific functional relationships of types other than linear if suitable transformations are applied to the data.

Example 2 Find the values of constants K and s for which the curve

$$y = Kx^s$$

best fits the experimental data points (x_i, y_i), $i = 1, 2, \ldots, n$. (Assume all data values are positive.)

Solution Observe that the required functional form corresponds to a linear relationship between $\ln y$ and $\ln x$:

$$\ln y = \ln K + s \ln x.$$

If we determine the parameters a and b of the empirical regression line $\eta = a\xi + b$ corresponding to the transformed data $(\xi_i, \eta_i) = (\ln x_i, \ln y_i)$, then $s = a$ and $K = e^b$ are the required values.

$\blacksquare$

Remark It should be stressed that the constants K and s obtained by the method used in the solution above are not the same as those that would be obtained by direct application of the least squares method to the untransformed problem, that is, by minimizing $\sum_{i=1}^{n}(y_i - Kx_i^s)^2$. This latter problem cannot readily be solved. (Try it!)

Generally, the method of least squares is applied to fit an equation in which the response is expressed as a sum of constants times functions of one or more input variables. The constants are determined as critical points of the sum of squared deviations of the actual response values from the values predicted by the equation.

Applications of the Least Squares Method to Integrals

The method of least squares can be used to find approximations to reasonably well-behaved (say, piecewise continuous) functions as sums of constants times specified functions. The idea is to choose the constants to minimize the *integral* of the square of the difference.

For example, suppose we want to approximate the continuous function $f(x)$ over the interval $[0, 1]$ by a linear function $g(x) = px + q$. The method of least squares would require that p and q be chosen to minimize the integral

$$I(p, q) = \int_0^1 \Big(f(x) - px - q \Big)^2 \, dx.$$

Assuming that we can "differentiate through the integral" (we will investigate this issue in Section 13.5), the critical point of $I(p, q)$ can be found from

$$0 = \frac{\partial I}{\partial p} = -2 \int_0^1 x\left(f(x) - px - q \right) dx,$$

$$0 = \frac{\partial I}{\partial q} = -2 \int_0^1 \left(f(x) - px - q \right) dx.$$

Thus,

$$\frac{p}{3} + \frac{q}{2} = \int_0^1 xf(x)\, dx,$$

$$\frac{p}{2} + q = \int_0^1 f(x)\, dx,$$

and solving this linear system for p and q we get

$$p = \int_0^1 (12x - 6) f(x)\, dx,$$

$$q = \int_0^1 (4 - 6x) f(x)\, dx.$$

The following example concerns the approximation of a function by a **trigonometric polynomial**. Such approximations form the basis for the study of **Fourier series**, which are of fundamental importance in the solution of boundary-value problems for the Laplace, heat, and wave equations and other partial differential equations that arise in applied mathematics. (See Section 9.9.)

Example 3   Use a least squares integral to approximate $f(x)$ by the sum

$$\sum_{k=1}^{n} b_k \sin kx$$

on the interval $0 \le x \le \pi$.

Solution   We want to choose the constants to minimize

$$I = \int_0^\pi \left(f(x) - \sum_{k=1}^{n} b_k \sin kx \right)^2 dx.$$

For each $1 \le j \le n$ we have

$$0 = \frac{\partial I}{\partial b_j} = -2 \int_0^\pi \left(f(x) - \sum_{k=1}^{n} b_k \sin kx \right) \sin jx\, dx.$$

Thus,

$$\sum_{k=1}^{n} b_k \int_0^\pi \sin kx \sin jx\, dx = \int_0^\pi f(x) \sin jx\, dx.$$

However, if $j \neq k$, then $\sin kx \sin jx$ is an even function, so that

$$\int_0^\pi \sin kx \sin jx\, dx = \frac{1}{2} \int_{-\pi}^\pi \sin kx \sin jx\, dx$$

$$= \frac{1}{4} \int_{-\pi}^\pi \left(\cos(k - j)x - \cos(k + j)x \right) dx = 0.$$

If $j = k$, then we have

$$\int_0^\pi \sin^2 jx\, dx = \frac{1}{2} \int_0^\pi (1 - \cos 2jx)\, dx = \frac{\pi}{2},$$

so that

$$b_j = \frac{2}{\pi} \int_0^\pi f(x) \sin jx\, dx.$$

Remark   The series

$$\sum_{k=1}^{\infty} b_k \sin kx, \quad \text{where} \quad b_k = \frac{2}{\pi} \int_0^{\pi} f(x) \sin kx \, dx, \quad k = 1, 2, \ldots,$$

is called the **Fourier sine series** representation of $f(x)$ on the interval $(0, \pi)$. If f is continuous on $[0, \pi]$, it can be shown that

$$\lim_{n \to \infty} \int_0^{\pi} \left(f(x) - \sum_{k=1}^{n} b_k \sin kx \right)^2 dx = 0,$$

but more than just continuity is required of f to ensure that this Fourier sine series converges to $f(x)$ at each point of $(0, \pi)$. Such questions are studied in *harmonic analysis*. Similarly, the series

$$\frac{a_0}{2} + \sum_{k=1}^{\infty} a_k \cos kx, \quad \text{where} \quad a_k = \frac{2}{\pi} \int_0^{\pi} f(x) \cos kx \, dx, \quad k = 0, 1, 2, \ldots,$$

is called the **Fourier cosine series** representation of $f(x)$ on the interval $(0, \pi)$.

Remark   Representing a function as the sum of a Fourier series is analogous to representing a vector as a linear combination of basis vectors. If we think of continuous functions on the interval $[0, \pi]$ as "vectors" with addition and scalar multiplication defined pointwise:

$$(f + g)(x) = f(x) + g(x), \qquad (cf)(x) = cf(x),$$

and with the "dot product" defined as

$$f \bullet g = \int_0^{\pi} f(x)g(x) \, dx,$$

then the functions $e_k(x) = \sqrt{2/\pi} \sin kx$ form a "basis." As shown in the example above, $e_j \bullet e_j = 1$, and if $k \neq j$, then $e_k \bullet e_j = 0$. Thus, these "basis vectors" are "mutually perpendicular unit vectors." The Fourier sine coefficients b_j of a function f are the components of f with respect to that basis.

Exercises 13.4

1. A generator is to be installed in a factory to supply power to n machines located at positions (x_i, y_i), $i = 1, 2, \ldots, n$. Where should the generator be located to minimize the sum of the squares of its distances from the machines?

2. The relationship $y = ax^2$ is known to hold between certain variables. Given the experimental data (x_i, y_i), $i = 1, 2, \ldots, n$, determine a value for a by the method of least squares.

3. Repeat Exercise 2 but with the relationship $y = ae^x$.

4. Use the method of least squares to find the plane $z = ax + by + c$ that best fits the data (x_i, y_i, z_i), $i = 1, 2, \ldots, n$.

5. Repeat Exercise 4 using a vector projection argument instead of the method of least squares.

In Exercises 6–11, show how to adapt linear regression to determine the two parameters p and q so that the given relationship fits the experimental data (x_i, y_i), $i = 1, 2, \ldots, n$. In which of these situations are the values of p and q obtained identical to those obtained by direct application of the method of least squares with no change of variable?

6. $y = p + qx^2$

7. $y = pe^{qx}$

8. $y = \ln(p + qx)$

9. $y = px + qx^2$

10. $y = \sqrt{px + q}$

11. $y = pe^x + qe^{-x}$

12. Find the parabola of the form $y = p + qx^2$ that best fits the data $(x, y) = (1, 0.11)$, $(2, 1.62)$, $(3, 4.07)$, $(4, 7.55)$, $(6, 17.63)$, and $(7, 24.20)$. No value of y was measured at $x = 5$. What value would you predict at this point?

13. Use the method of least squares to find constants a, b, and c so that the relationship $y = ax^2 + bx + c$ best describes the experimental data (x_i, y_i), $i = 1, 2, \ldots, n$, $(n \geq 3)$. How is this situation interpreted in terms of vector projection?

14. How can the result of Exercise 13 be used to fit a curve of the form $y = pe^x + q + re^{-x}$ through the same data points?

15. Find the value of the constant a for which the function $f(x) = ax^2$ best approximates the function $g(x) = x^3$ on the interval $[0, 1]$, in the sense that the integral

$$I = \int_0^1 \Big(f(x) - g(x)\Big)^2 dx$$

is minimized. What is the minimum value of I?

16. Find a to minimize $I = \int_0^\pi \Big(ax(\pi - x) - \sin x\Big)^2 dx$. What is the minimum value of the integral?

17. Repeat Exercise 15 with the function $f(x) = ax^2 + b$ and the same g. Find a and b.

∗18. Find a, b, and c to minimize $\int_0^1 (x^3 - ax^2 - bx - c)^2 dx$. What is the minimum value of the integral?

∗19. Find a and b to minimize $\int_0^\pi (\sin x - ax^2 - bx)^2 dx$.

20. Find a, b, and c to minimize the integral

$$J = \int_{-1}^1 \Big(x - a\sin \pi x - b\sin 2\pi x - c\sin 3\pi x\Big)^2 dx.$$

∗21. Find constants a_j, $j = 0, 1, \ldots, n$, to minimize

$$\int_0^\pi \left(f(x) - \frac{a_0}{2} - \sum_{k=1}^n a_k \cos kx\right)^2 dx.$$

22. Find the Fourier sine series for the function $f(x) = x$ on $0 < x < \pi$. Assuming the series does converge to x on the interval $(0, \pi)$, to what function would you expect the series to converge on $(-\pi, 0)$?

23. Repeat Exercise 22 but obtaining instead a Fourier cosine series.

24. Suppose $x_1, x_2, \ldots, x_n$ satisfy $x_i \leq x_j$ whenever $i < j$. Find x that minimizes $\sum_{i=1}^n |x - x_i|$. Treat the cases n odd and n even separately. For what values of n is x unique? *Hint:* Use no calculus in this problem.

13.5 Parametric Problems

In this section we will briefly examine three unrelated situations in which we want to differentiate a function with respect to a parameter rather than one of the basic variables of the function. Such situations arise frequently in mathematics and its applications.

Differentiating Integrals with Parameters

The Fundamental Theorem of Calculus shows how to differentiate a definite integral with respect to the upper limit of integration:

$$\frac{d}{dx} \int_a^x f(t)\, dt = f(x).$$

We are going to look at a different problem about differentiating integrals. If the integrand of a definite integral also depends on variables other than the variable of integration, then the integral will be a function of those other variables. How are we to find the derivative of such a function? For instance, consider the function $F(x)$ defined by

$$F(x) = \int_a^b f(x, t)\, dt.$$

We would like to be able to calculate $F'(x)$ by taking the derivative inside the integral:

$$F'(x) = \frac{d}{dx} \int_a^b f(x, t)\, dt = \int_a^b \frac{\partial}{\partial x} f(x, t)\, dt.$$

Observe that we use d/dx outside the integral and $\partial/\partial x$ inside; this is because the integral is a function of x only, but the integrand f is a function of both x and t. If the integrand depends on more than one parameter, then partial derivatives would be needed inside and outside the integral:

$$\frac{\partial}{\partial x} \int_a^b f(x, y, t)\, dt = \int_a^b \frac{\partial}{\partial x} f(x, y, t)\, dt.$$

The operation of taking a derivative with respect to a parameter inside the integral, or *differentiating through the integral*, as it is usually called, seems plausible. We differentiate sums term by term, and integrals are the limits of sums. However, both the differentiation and integration operations involve the taking of limits (limits of Newton quotients for derivatives, limits of Riemann sums for integrals). Differentiating through the integral requires changing the order in which the two limits are taken and, therefore, requires justification.

We have already seen another example of change of order of limits. When we assert that two mixed partial derivatives with respect to the same variables are equal,

$$\frac{\partial^2 f}{\partial x \partial y} = \frac{\partial^2 f}{\partial y \partial x},$$

we are, in fact, saying that limits corresponding to differentiation with respect to x and y can be taken in either order with the same result. This is not true in general; we proved it under the assumption that both of the mixed partials were *continuous*. (See Theorem 1 and Exercise 16 of Section 12.4.) In general, some assumptions are required to justify the interchange of limits. The following theorem gives one set of conditions that justify the interchange of limits involved in differentiating through the integral.

THEOREM 5

Differentiating through an integral

Suppose that for every x satisfying $c < x < d$ the following conditions hold:

(i) the integrals

$$\int_a^b f(x, t)\, dt \qquad \text{and} \qquad \int_a^b f_1(x, t)\, dt$$

both exist (either as proper or convergent improper integrals).

(ii) $f_{11}(x, t)$ exists and satisfies

$$|f_{11}(x, t)| \le g(t), \qquad a < t < b,$$

where

$$\int_a^b g(t)\, dt = K < \infty.$$

Then for each x satisfying $c < x < d$ we have

$$\frac{d}{dx} \int_a^b f(x, t)\, dt = \int_a^b \frac{\partial}{\partial x} f(x, t)\, dt.$$

PROOF Let

$$F(x) = \int_a^b f(x, t)\, dt.$$

If $c < x < d$, $h \ne 0$, and $|h|$ is sufficiently small that $c < x + h < d$, then, by Taylor's Formula,

$$f(x + h, t) = f(x, t) + h f_1(x, t) + \frac{h^2}{2} f_{11}(x + \theta h, t)$$

for some θ between 0 and 1. Therefore,

$$\left| \frac{F(x + h) - F(x)}{h} - \int_a^b f_1(x, t)\, dt \right|$$

$$= \left| \int_a^b \frac{f(x + h, t) - f(x, t)}{h}\, dt - \int_a^b f_1(x, t)\, dt \right|$$

$$\le \int_a^b \left| \frac{f(x + h, t) - f(x, t)}{h} - f_1(x, t) \right| dt$$

$$= \int_a^b \left| \frac{h}{2} f_{11}(x + \theta h, t) \right| dt$$

$$\le \frac{h}{2} \int_a^b g(t)\, dt = \frac{Kh}{2} \to 0 \text{ as } h \to 0.$$

Therefore,

$$F'(x) = \lim_{h \to 0} \frac{F(x+h) - F(x)}{h} = \int_a^b f_1(x,t)\,dt,$$

which is the desired result.

Remark   It can be shown that the conclusion of Theorem 5 also holds under the sole assumption that $f_1(x,t)$ is continuous on the *closed, bounded* rectangle $c \le x \le d$, $a \le t \le b$. We cannot prove this here; the proof depends on a subtle property called *uniform continuity* possessed by continuous functions on closed bounded sets in $\mathbb{R}^n$. (See Appendix IV for the case $n = 1$.) In any event, Theorem 5 is more useful for our purposes because it allows for improper integrals.

Example 1   Evaluate $\int_0^\infty t^n e^{-t}\,dt$.

Solution   Starting with the convergent improper integral

$$\int_0^\infty e^{-s}\,ds = \lim_{R \to \infty} \frac{e^{-s}}{-1}\Big|_0^R = \lim_{R \to \infty}(1 - e^{-R}) = 1,$$

we introduce a parameter by substituting $s = xt$, $ds = x\,dt$ (where $x > 0$) and get

$$\int_0^\infty e^{-xt}\,dt = \frac{1}{x}.$$

Now differentiate n times (each resulting integral converges):

$$\int_0^\infty -t\,e^{-xt}\,dt = -\frac{1}{x^2},$$
$$\int_0^\infty (-t)^2 e^{-xt}\,dt = (-1)^2\frac{2}{x^3},$$
$$\vdots$$
$$\int_0^\infty (-t)^n e^{-xt}\,dt = (-1)^n\frac{n!}{x^{n+1}}.$$

Putting $x = 1$ we get

$$\int_0^\infty t^n e^{-t}\,dt = n!.$$

Note that this result could be obtained by integration by parts (n times) or a reduction formula. This method is a little easier.

Remark   The reader should check that the function $f(x,t) = t^k e^{-xt}$ satisfies the conditions of Theorem 5 for $x > 0$ and $k \ge 0$. We will normally not make a point of this.

Example 2   Evaluate $F(x,y) = \int_0^\infty \frac{e^{-xt} - e^{-yt}}{t}\,dt$ for $x > 0$, $y > 0$.

Solution   We have

$$\frac{\partial F}{\partial x} = -\int_0^\infty e^{-xt}\, dt = -\frac{1}{x} \quad \text{and} \quad \frac{\partial F}{\partial y} = \int_0^\infty e^{-yt}\, dt = \frac{1}{y}.$$

It follows that

$$F(x, y) = -\ln x + C_1(y) \quad \text{and} \quad F(x, y) = \ln y + C_2(x).$$

Comparing these two formulas for F, we are forced to conclude that $C_1(y) = \ln y + C$ for some constant C. Therefore,

$$F(x, y) = \ln y - \ln x + C = \ln \frac{y}{x} + C.$$

Since $F(1, 1) = 0$, we must have $C = 0$ and $F(x, y) = \ln(y/x)$.

Remark   We can combine Theorem 5 and the Fundamental Theorem of Calculus to differentiate an integral with respect to a parameter that appears in the limits of integration as well as in the integrand. If

$$F(x, b, a) = \int_a^b f(x, t)\, dt,$$

then, by the Chain Rule,

$$\frac{d}{dx} F\big(x, b(x), a(x)\big) = \frac{\partial F}{\partial x} + \frac{\partial F}{\partial b}\frac{db}{dx} + \frac{\partial F}{\partial a}\frac{da}{dx}.$$

Accordingly, we have

$$\frac{d}{dx} \int_{a(x)}^{b(x)} f(x, t)\, dt$$

$$= \int_{a(x)}^{b(x)} \frac{\partial}{\partial x} f(x, t)\, dt + f\big(x, b(x)\big)b'(x) - f\big(x, a(x)\big)a'(x).$$

We require that $a(x)$ and $b(x)$ be differentiable at x, and for the application of Theorem 5, that $a \le a(x) \le b$ and $a \le b(x) \le b$ for all x satisfying $c < x < d$.

Example 3   Solve the *integral equation*

$$f(x) = a - \int_b^x (x - t) f(t)\, dt.$$

Solution   Assume, for the moment, that the equation has a sufficiently well-behaved solution to allow for differentiation through the integral. Differentiating twice, we get

$$f'(x) = -(x - x)f(x) - \int_b^x f(t)\, dt = -\int_b^x f(t)\, dt,$$

$$f''(x) = -f(x).$$

The latter equation is the differential equation of simple harmonic motion. Observe that the given equation for f and that for f' imply the initial conditions

$$f(b) = a \quad \text{and} \quad f'(b) = 0.$$

Accordingly, we write the general solution of $f''(x) = -f(x)$ in the form

$$f(x) = A \cos(x - b) + B \sin(x - b).$$

The initial conditions then imply $A = a$ and $B = 0$, so the required solution is $f(x) = a \cos(x - b)$. Finally, we note that this function is indeed smooth enough to allow the differentiations through the integral and is, therefore, the solution of the given integral equation. (If you wish, verify it in the integral equation.)

Envelopes

An equation $f(x, y, c) = 0$ that involves a parameter c as well as the variables x and y represents a family of curves in the xy-plane. Consider, for instance, the family

$$f(x, y, c) = \frac{x}{c} + cy - 2 = 0.$$

This family consists of straight lines with intercepts $(2c, 2/c)$ on the coordinate axes. Several of these lines are sketched in Figure 13.20. It appears that there is a curve to which all these lines are tangent. This curve is called the *envelope* of the family of lines.

In general, a curve $\mathcal{C}$ is called the **envelope** of the family of curves with equations $f(x, y, c) = 0$ if, for each value of c, the curve $f(x, y, c) = 0$ is tangent to $\mathcal{C}$ at some point depending on c.

For the family of lines in Figure 13.20 it appears that the envelope may be the rectangular hyperbola $xy = 1$. We will verify this after developing a method for determining the equation of the envelope of a family of curves. We assume that the function $f(x, y, c)$ has continuous first partials and that the envelope is a smooth curve.

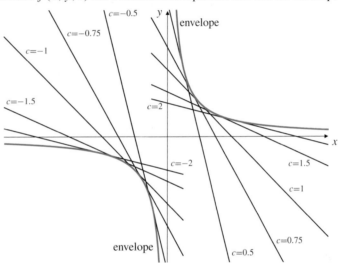

Figure 13.20 A family of straight lines and their envelope

For each c, the curve $f(x, y, c) = 0$ is tangent to the envelope at a point (x, y) that depends on c. Let us express this dependence in the explicit form $x = g(c)$, $y = h(c)$; these equations are parametric equations of the envelope. Since (x, y) lies on the curve $f(x, y, c) = 0$, we have

$$f\big(g(c), h(c), c\big) = 0.$$

Differentiating this equation with respect to c, we obtain

$$f_1 g'(c) + f_2 h'(c) + f_3 = 0, \qquad\qquad\qquad (*)$$

where the partials of f are evaluated at $\big(g(c), h(c), c\big)$.

The slope of the curve $f(x, y, c) = 0$ at $\big(g(c), h(c), c\big)$ can be obtained by differentiating its equation implicitly with respect to x:

$$f_1 + f_2 \frac{dy}{dx} = 0.$$

On the other hand, the slope of the envelope $x = g(c)$, $y = h(c)$ at that point is $dy/dx = h'(c)/g'(c)$. Since the curve and the envelope are tangent at $f\big(g(c), h(c), c\big)$, these slopes must be equal. Therefore,

$$f_1 + f_2 \frac{h'(c)}{g'(c)} = 0, \qquad \text{so} \qquad f_1 g'(c) + f_2 h'(c) = 0.$$

Combining this with equation $(*)$ we get $f_3(x, y, c) = 0$ at all points of the envelope.

> The equation of the envelope can be found by eliminating c between the two equations
>
> $$f(x, y, c) = 0 \qquad \text{and} \qquad \frac{\partial}{\partial c} f(x, y, c) = 0.$$

Example 4   Find the envelope of the family of straight lines

$$f(x, y, c) = \frac{x}{c} + cy - 2 = 0.$$

Solution   We eliminate c between the equations

$$f(x, y, c) = \frac{x}{c} + cy - 2 = 0 \qquad \text{and} \qquad f_3(x, y, c) = -\frac{x}{c^2} + y = 0.$$

These equations can be easily solved and give $x = c$ and $y = 1/c$. Hence, they imply that the envelope is $xy = 1$, as we conjectured earlier.

Example 5   Find the envelope of the family of circles

$$(x - c)^2 + y^2 = c.$$

Solution   Here, $f(x, y, c) = (x - c)^2 + y^2 - c$. The equation of the envelope is obtained by eliminating c from the pair of equations

$$f(x, y, c) = (x - c)^2 + y^2 - c = 0,$$

$$\frac{\partial}{\partial c} f(x, y, c) = -2(x - c) - 1 = 0.$$

From the second equation, $x = c - \frac{1}{2}$, and then from the first, $y^2 = c - \frac{1}{4}$. Hence, the envelope is the parabola

$$x = y^2 - \frac{1}{4}.$$

This envelope and some of the circles in the family are sketched in Figure 13.21.

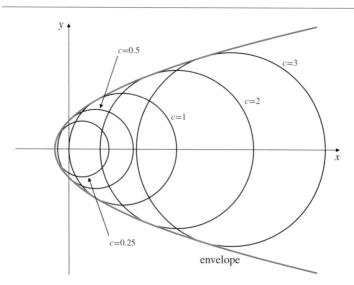

Figure 13.21   Circles $(x - c)^2 + y^2 = c$ and their envelope

A similar technique can be used to find the envelope of a family of surfaces. This will be a surface tangent to each member of the family.

Example 6   (**The Mach cone**) Suppose that sound travels at speed c in still air and that a supersonic aircraft is travelling at speed $v > c$ along the x-axis, so that its position at time t is $(vt, 0, 0)$. Find the envelope at time t of the sound waves created by the aircraft at previous times. See Figure 13.22.

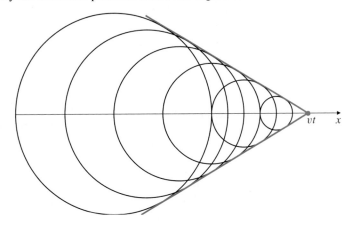

Figure 13.22   The Mach cone

Solution   The sound created by the aircraft at time $\tau < t$ spreads out as a spherical wave front at speed c. The centre of this wave front is $(v\tau, 0, 0)$, the position of the aircraft at time τ. At time t the radius of this wave front is $c(t - \tau)$, so its equation is

$$f(x, y, z, \tau) = (x - v\tau)^2 + y^2 + z^2 - c^2(t - \tau)^2 = 0. \qquad (*)$$

At time t the envelope of all these wave fronts created at earlier times τ is obtained by eliminating the parameter τ from the above equation and the equation

$$\frac{\partial}{\partial \tau} f(x, y, z, \tau) = -2v(x - v\tau) + 2c^2(t - \tau) = 0.$$

Solving this latter equation for τ, we get $\tau = \dfrac{vx - c^2 t}{v^2 - c^2}$. Thus,

$$x - v\tau = x - \frac{v^2 x - vc^2 t}{v^2 - c^2} = \frac{c^2}{v^2 - c^2}(vt - x)$$

$$t - \tau = t - \frac{vx - c^2 t}{v^2 - c^2} = \frac{v}{v^2 - c^2}(vt - x).$$

We substitute these two expressions into equation $(*)$ to eliminate τ:

$$\frac{c^4}{(v^2 - c^2)^2}(vt - x)^2 + y^2 + z^2 - \frac{c^2 v^2}{(v^2 - c^2)^2}(vt - x)^2 = 0$$

$$y^2 + z^2 = \frac{c^2}{(v^2 - c^2)^2}(v^2 - c^2)(vt - x)^2 = \frac{c^2}{v^2 - c^2}(vt - x)^2.$$

The envelope is the cone

$$x = vt - \frac{\sqrt{v^2 - c^2}}{c}\sqrt{y^2 + z^2},$$

which extends backward in the x direction from its vertex at $(vt, 0, 0)$, the position of the aircraft at time t. This is called the **Mach cone**. The sound of the aircraft cannot be heard at any point until the cone reaches that point.

Equations with Perturbations

In applied mathematics one frequently encounters intractable equations for which at least approximate solutions are desired. Sometimes such equations result from adding an extra term to what would otherwise be a simple and easily solved equation. This extra term is called a **perturbation** of the simpler equation. Often the perturbation has a coefficient smaller than the other terms in the equation, that is, it is a **small perturbation**. When this is the case you can find approximate solutions to the perturbed equation by replacing the small coefficient by a parameter and calculating Maclaurin polynomials in that parameter. One example should serve to clarify the method.

Example 7 Find an approximate solution of the equation

$$y + \frac{1}{50} \ln(1 + y) = x^2.$$

Solution Without the logarithm term the equation would clearly have the solution $y = x^2$. Let us replace the coefficient $1/50$ with the parameter ϵ and look for a solution $y = y(x, \epsilon)$ to the equation

$$y + \epsilon \ln(1 + y) = x^2 \tag{*}$$

in the form

$$y = y(x, \epsilon) = y(x, 0) + \epsilon y_\epsilon(x, 0) + \frac{\epsilon^2}{2!} y_{\epsilon\epsilon}(x, 0) + \cdots,$$

where the subscripts ϵ denote derivatives with respect to ϵ. We shall calculate the terms up to second order in ϵ. Evidently $y(x, 0) = x^2$. Differentiating equation (*) twice with respect to ϵ and evaluating the results at $\epsilon = 0$, we obtain

$$\frac{\partial y}{\partial \epsilon} + \ln(1 + y) + \frac{\epsilon}{1 + y}\frac{\partial y}{\partial \epsilon} = 0,$$

$$\frac{\partial^2 y}{\partial \epsilon^2} + \frac{2}{1 + y}\frac{\partial y}{\partial \epsilon} + \epsilon\frac{\partial}{\partial \epsilon}\left(\frac{1}{1 + y}\frac{\partial y}{\partial \epsilon}\right) = 0,$$

$$y_\epsilon(x, 0) = -\ln(1 + x^2),$$

$$y_{\epsilon\epsilon}(x, 0) = \frac{2}{1 + x^2}\ln(1 + x^2).$$

Hence,

$$y(x, \epsilon) = x^2 - \epsilon \ln(1 + x^2) + \frac{\epsilon^2}{1 + x^2}\ln(1 + x^2) + \cdots,$$

and the given equation has the approximate solution

$$y \approx x^2 - \frac{\ln(1 + x^2)}{50} + \frac{\ln(1 + x^2)}{2{,}500(1 + x^2)}.$$

Similar perturbation techniques can be used for systems of equations and for differential equations.

▌▌▌▌▌▌▌▌▌ **Exercises 13.5** ▌▌▌▌▌▌▌▌

1. Let $F(x) = \int_0^1 t^x \, dt = \dfrac{1}{x+1}$ for $x > -1$. By repeated differentiation of F evaluate the integral

$$\int_0^1 t^x (\ln t)^n \, dt.$$

2. By replacing t with xt in the well-known integral

$$\int_{-\infty}^\infty e^{-t^2} \, dt = \sqrt{\pi},$$

and differentiating with respect to x, evaluate

$$\int_{-\infty}^\infty t^2 e^{-t^2} \, dt \quad \text{and} \quad \int_{-\infty}^\infty t^4 e^{-t^2} \, dt.$$

3. Evaluate $\displaystyle\int_{-\infty}^\infty \frac{e^{-xt^2} - e^{-yt^2}}{t^2} \, dt$ for $x > 0, \ y > 0$.

4. Evaluate $\displaystyle\int_0^1 \frac{t^x - t^y}{\ln t} \, dt$ for $x > -1, \ y > -1$.

5. Given that $\displaystyle\int_0^\infty e^{-xt} \sin t \, dt = \frac{1}{1+x^2}$ for $x > 0$ (which can be shown by integration by parts), evaluate

$$\int_0^\infty t e^{-xt} \sin t \, dt \quad \text{and} \quad \int_0^\infty t^2 e^{-xt} \sin t \, dt.$$

* **6.** Referring to Exercise 5, for $x > 0$ evaluate

$$F(x) = \int_0^\infty e^{-xt} \frac{\sin t}{t} \, dt.$$

Show that $\lim_{x \to \infty} F(x) = 0$ and hence evaluate the integral

$$\int_0^\infty \frac{\sin t}{t} \, dt = \lim_{x \to 0} F(x).$$

7. Evaluate $\displaystyle\int_0^\infty \frac{dt}{x^2 + t^2}$ and use the result to help you evaluate

$$\int_0^\infty \frac{dt}{(x^2 + t^2)^2} \quad \text{and} \quad \int_0^\infty \frac{dt}{(x^2 + t^2)^3}.$$

* **8.** Evaluate $\displaystyle\int_0^x \frac{dt}{x^2 + t^2}$ and use the result to help you evaluate

$$\int_0^x \frac{dt}{(x^2 + t^2)^2} \quad \text{and} \quad \int_0^x \frac{dt}{(x^2 + t^2)^3}.$$

9. Find $f^{(n+1)}(a)$ if $f(x) = 1 + \displaystyle\int_a^x (x-t)^n f(t) \, dt$.

Solve the integral equations in Exercises 10–12.

10. $f(x) = Cx + D + \displaystyle\int_0^x (x-t) f(t) \, dt$

11. $f(x) = x + \displaystyle\int_0^x (x - 2t) f(t) \, dt$

12. $f(x) = 1 + \displaystyle\int_0^1 (x+t) f(t) \, dt$

Find the envelopes of the families of curves in Exercises 13–18.

13. $y = 2cx - c^2$

14. $y - (x-c) \cos c = \sin c$

15. $x \cos c + y \sin c = 1$

16. $\dfrac{x}{\cos c} + \dfrac{y}{\sin c} = 1$

17. $y = c + (x-c)^2$

18. $(x-c)^2 + (y-c)^2 = 1$

19. Does every one-parameter family of curves in the plane have an envelope? Try to find the envelope of $y = x^2 + c$.

20. For what values of k does the family of curves $x^2 + (y-c)^2 = kc^2$ have an envelope?

21. Try to find the envelope of the family $y^3 = (x+c)^2$. Are the curves of the family tangent to the envelope? What have you actually found in this case? Compare with Example 3 of Section 13.3.

*\ **22.** Show that if a two-parameter family of surfaces $f(x, y, z, \lambda, \mu) = 0$ has an envelope, then the equation of that envelope can be obtained by eliminating λ and μ from the three equations

$$f(x, y, z, \lambda, \mu) = 0,$$
$$\frac{\partial}{\partial \lambda} f(x, y, z, \lambda, \mu) = 0,$$
$$\frac{\partial}{\partial \mu} f(x, y, z, \lambda, \mu) = 0.$$

23. Find the envelope of the two-parameter family of planes

$$x \sin \lambda \cos \mu + y \sin \lambda \sin \mu + z \cos \lambda = 1.$$

24. Find the envelope of the two-parameter family of spheres

$$(x - \lambda)^2 + (y - \mu)^2 + z^2 = \frac{\lambda^2 + \mu^2}{2}.$$

In Exercises 25–27, find the terms up to second power in ϵ in the solution y of the given equation.

25. $y + \epsilon \sin \pi y = x$

26. $y^2 + \epsilon e^{-y^2} = 1 + x^2$

27. $2y + \dfrac{\epsilon x}{1 + y^2} = 1$

28. Use perturbation methods to evaluate y with error less than 10^{-8} given that $y + (y^5/100) = 1/2$.

*\ **29.** Use perturbation methods to find approximate values for x and y from the system $x + 2y + \dfrac{1}{100} e^{-x} = 3$,

$x - y + \dfrac{1}{100} e^{-y} = 0$. Calculate all terms up to second order in $\epsilon = 1/100$.

13.6 Newton's Method

A frequently encountered problem in applied mathematics is to determine, to some desired degree of accuracy, a root (i.e., a solution r) of an equation of the form

$$f(r) = 0.$$

Such a root is called a **zero** of the function f. In Section 4.6 we introduced Newton's Method, a simple but powerful method for determining roots of functions that are sufficiently smooth. The method involves *guessing* an approximate value x_0 for a root r of the function f, and then calculating successive approximations $x_1, x_2, \ldots$, using the formula

$$x_{n+1} = x_n - \frac{f(x_n)}{f'(x_n)}, \qquad n = 0, 1, 2, \cdots.$$

If the initial guess x_0 is not too far from r, and if $|f'(x)|$ is *not too small* and $|f''(x)|$ is *not too large* near r, then the successive approximations $x_1, x_2, \ldots$ will converge very rapidly to r. Recall that each new approximation x_{n+1} is obtained as the x-intercept of the tangent line drawn to the graph of f at the previous approximation, x_n. The tangent line to the graph $y = f(x)$ at $x = x_n$ has equation

$$y - f(x_n) = f'(x_n)(x - x_n).$$

(See Figure 13.23.) The x-intercept, x_{n+1}, of this line is determined by setting $y = 0$, $x = x_{n+1}$ in this equation, so is given by the formula in the shaded box above.

Newton's Method can be extended to finding solutions of systems of m equations in m variables. We will show here how to adapt the method to find approximations to a solution (x, y) of the pair of equations

$$\begin{cases} f(x, y) = 0 \\ g(x, y) = 0 \end{cases}$$

starting from an initial guess (x_0, y_0). Under auspicious circumstances, we will observe the same rapid convergence of approximations to the root that typifies the single-variable case.

The idea is as follows. The two surfaces $z = f(x, y)$ and $z = g(x, y)$ intersect in a curve which itself intersects the xy-plane at the point whose coordinates are the desired solution. If (x_0, y_0) is near that point, then the tangent planes to the two surfaces at (x_0, y_0) will intersect in a straight line. This line meets the xy-plane at a point (x_1, y_1) that should be even closer to the solution point than was (x_0, y_0). We can easily determine (x_1, y_1). The tangent planes to $z = f(x, y)$ and $z = g(x, y)$ at (x_0, y_0) have equations

$$z = f(x_0, y_0) + f_1(x_0, y_0)(x - x_0) + f_2(x_0, y_0)(y - y_0),$$
$$z = g(x_0, y_0) + g_1(x_0, y_0)(x - x_0) + g_2(x_0, y_0)(y - y_0).$$

The line of intersection of these two planes meets the xy-plane at the point (x_1, y_1) satisfying

$$f_1(x_0, y_0)(x_1 - x_0) + f_2(x_0, y_0)(y_1 - y_0) + f(x_0, y_0) = 0,$$
$$g_1(x_0, y_0)(x_1 - x_0) + g_2(x_0, y_0)(y_1 - y_0) + g(x_0, y_0) = 0.$$

Solving these two equations for x_1 and y_1, we obtain

$$x_1 = x_0 - \frac{f g_2 - f_2 g}{f_1 g_2 - f_2 g_1}\Bigg|_{(x_0, y_0)} = x_0 - \frac{\begin{vmatrix} f & f_2 \\ g & g_2 \end{vmatrix}}{\begin{vmatrix} f_1 & f_2 \\ g_1 & g_2 \end{vmatrix}}\Bigg|_{(x_0, y_0)},$$

$$y_1 = y_0 - \frac{f_1 g - f g_1}{f_1 g_2 - f_2 g_1}\Bigg|_{(x_0, y_0)} = y_0 - \frac{\begin{vmatrix} f_1 & f \\ g_1 & g \end{vmatrix}}{\begin{vmatrix} f_1 & f_2 \\ g_1 & g_2 \end{vmatrix}}\Bigg|_{(x_0, y_0)}.$$

Figure 13.23 x_{n+1} is the x-intercept of the tangent at x_n

Observe that the denominator in each of these expressions is the Jacobian determinant $\partial(f, g)/\partial(x, y)\big|_{(x_0, y_0)}$. This is another instance where the Jacobian is the appropriate multivariable analogue of the derivative of a function of one variable.

Continuing in this way we generate successive approximations (x_n, y_n) according to the formulas

$$x_{n+1} = x_n - \frac{\begin{vmatrix} f & f_2 \\ g & g_2 \end{vmatrix}}{\begin{vmatrix} f_1 & f_2 \\ g_1 & g_2 \end{vmatrix}}\Bigg|_{(x_n, y_n)},$$

$$y_{n+1} = y_n - \frac{\begin{vmatrix} f_1 & f \\ g_1 & g \end{vmatrix}}{\begin{vmatrix} f_1 & f_2 \\ g_1 & g_2 \end{vmatrix}}\Bigg|_{(x_n, y_n)}.$$

We stop when the desired accuracy has been achieved.

Example 1 Find the root of the system

$$x(1 + y^2) - 1 = 0, \qquad y(1 + x^2) - 2 = 0$$

with sufficient accuracy to ensure that the left sides of the equations vanish to the sixth decimal place.

Solution A sketch of the graphs of the two equations (see Figure 13.24) in the xy-plane indicates that the system has only one root near the point $(0.2, 1.8)$. Application of Newton's Method requires successive computations of the quantities

$$f(x, y) = x(1 + y^2) - 1, \quad f_1(x, y) = 1 + y^2, \quad f_2(x, y) = 2xy,$$
$$g(x, y) = y(1 + x^2) - 2, \quad g_1(x, y) = 2xy, \quad g_2(x, y) = 1 + x^2.$$

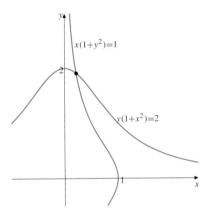

Figure 13.24 The two graphs intersect near $(0.2, 1.8)$

Using a calculator or computer, we can calculate successive values of (x_n, y_n) starting from $x_0 = 0.2$, $y_0 = 1.8$:

Table 1. Root near $(0.2, 1.8)$

n	x_n	y_n	$f(x_n, y_n)$	$g(x_n, y_n)$
0	0.200 000	1.800 000	−0.152 000	−0.128 000
1	0.216 941	1.911 349	0.009 481	0.001 303
2	0.214 827	1.911 779	−0.000 003	0.000 008
3	0.214 829	1.911 769	0.000 000	0.000 000

The values in this table were calculated sequentially in a spreadsheet by the method suggested below. They were rounded for inclusion in the table but the unrounded values were used in subsequent calculations. If you actually use the (rounded) values of x_n and y_n given in the table to calculate $f(x_n, y_n)$ and $g(x_n, y_n)$, your results may vary slightly.

The desired approximations to the root are the x_n and y_n values in the last line of the above table. Note the rapidity of convergence. However, many function evaluations are needed for each iteration of the method. For large systems Newton's Method is computationally too inefficient to be practical. Other methods requiring more iterations but many fewer calculations per iteration are used in practice.

Implementing Newton's Method Using a Spreadsheet

A computer spreadsheet is an ideal environment in which to calculate Newton's Method approximations. For a pair of equations in two unknowns such as the system in Example 1, you can proceed as follows:

(i) In the first nine cells of the first row (A1–I1) put the labels n, x, y, f, g, f1, f2, g1, and g2.

(ii) In cells A2–A9 put the numbers 0, 1, 2, . . . , 7.

(iii) In cells B2 and C2 put the starting values x_0 and y_0.

(iv) In cells D2–I2 put formulas for calculating $f(x, y)$, $g(x, y)$, . . . , $g_2(x, y)$ in terms of values of x and y assumed to be in B2 and C2.

(v) In cells B3 and C3 store the Newton's Method formulas for calculating x_1 and y_1 in terms of the values x_0 and y_0, using values calculated in the second row. For instance, cell B3 should contain the formula

```
+B2-(D2*I2-G2*E2)/(F2*I2-G2*H2).
```

(vi) Replicate the formulas in cells D2–I2 to cells D3–I3.

(vii) Replicate the formulas in cells B3–I3 to the cells B4–I9.

You can now inspect the successive approximations x_n and y_n in columns B and C. To use different starting values, just replace the numbers in cells B2 and C2. To solve a different system of (two) equations, replace the contents of cells D2–I2. You may wish to save this spreadsheet for reuse with the exercises below or other systems you may want to solve later.

Remark While a detailed analysis of the convergence of Newton's Method approximations is beyond the scope of this book, a few observations can be made. At each step in the approximation process we must divide by J, the Jacobian determinant of f and g with respect to x and y evaluated at the most recently obtained approximation. Assuming that the functions and partial derivatives involved in the formulas are continuous, the larger the value of J at the actual solution, the more likely are the approximations to converge to the solution, and to do so rapidly. If J vanishes (or is very small) at the solution, the successive approximations may not converge, even if the initial guess is quite close to the solution. Even if the first partials of f and g are large at the solution, their Jacobian may be small if their gradients are nearly parallel there. Thus, we cannot expect convergence to be rapid when the curves $f(x, y) = 0$ and $g(x, y) = 0$ intersect at a very small angle.

Newton's Method can be applied to systems of m equations in m variables; the formulas are the obvious generalizations of those for two functions given above.

Exercises 13.6

Find the solutions of the systems in Exercises 1–6, so that the left-hand sides of the equations vanish up to 6 decimal places. These can be done with the aid of a scientific calculator, but that approach will be very time consuming. It is much easier to program the Newton's Method formulas on a computer to generate the required approximations. In each case try to determine reasonable *initial guesses* by sketching graphs of the equations.

1. $y - e^x = 0$, $\quad x - \sin y = 0$

2. $x^2 + y^2 - 1 = 0$, $\quad y - e^x = 0$ (two solutions)

3. $x^4 + y^2 - 16 = 0$, $\quad xy - 1 = 0$ (four solutions)

4. $x^2 - xy + 2y^2 = 10$, $\quad x^3 y^2 = 2$ (four solutions)

5. $y - \sin x = 0$, $\quad x^2 + (y + 1)^2 - 2 = 0$ (two solutions)

6. $\sin x + \sin y - 1 = 0$, $\quad y^2 - x^3 = 0$ (two solutions)

7. Write formulas for obtaining successive Newton's Method approximations to a solution of the system

$$f(x, y, z) = 0, \quad g(x, y, z) = 0, \quad h(x, y, z) = 0,$$

starting from an initial guess (x_0, y_0, z_0).

8. Use the formulas from Exercise 7 to find the first octant intersection point of the surfaces $y^2 + z^2 = 3$, $x^2 + z^2 = 2$, and $x^2 - z = 0$.

9. The equations $y - x^2 = 0$ and $y - x^3 = 0$ evidently have the solutions $x = y = 0$ and $x = y = 1$. Try to obtain these

solutions using the two-variable form of Newton's Method with starting values:

(a) $x_0 = y_0 = 0.1$, and   (b) $x_0 = y_0 = 0.9$.

How many iterations are required to obtain 6-decimal-place accuracy for the appropriate solution in each case? How do you account for the difference in the behaviour of Newton's Method for these equations near $(0, 0)$ and $(1, 1)$?

13.7 Calculations with Maple

The calculations involved in solving systems of equations involving several variables can be very lengthy, even if the number of variables is small. In particular, locating critical points of a function of n variables involves solving a system of n (usually nonlinear) equations in n unknowns. In such situations the effective use of a computer algebra system like Maple can be very helpful. In this optional (and brief) section we present examples of how to use Maple's "fsolve" routine to solve systems of nonlinear equations and to find and classify critical points and thereby solve extreme-value problems.

Solving Systems of Equations

Maple has a procedure called **fsolve** built into its kernel (no package needs to be loaded to access it) that attempts to find floating-point real solutions to systems n equations in n variables. (For a single polynomial equation in one variable it will try to find all the real roots, but it may miss some.) For our purposes, an equation consists of either a single expression f in the variables (in which case the equation is taken to be $f = 0$) or else two expressions joined by an equal sign as in $f = g$. The procedure takes two or three arguments. The first is a set of n equations separated by commas. The set is enclosed in braces. The second argument is a set (also enclosed in braces) listing the n variables for which the equations are to be solved. (The number of variables in the equations must equal the number of equations.) The elements of the second set may consist of equations of the form "variable = initial guess," where the initial guess is a number we have reason to believe is *close* to the actual solution. It may not always be possible to make a good initial guess at the values of the variables, so, if we like, we can include a third argument specifing intervals of values of the variables in which to search for a solution. For example, to find a solution to the system $x^2 + y^3 = 3$, $x \sin(y) - y \cos(x) = 0$ near $(1, 2)$ we could try

```
>   Digits := 6:
>   fsolve({x^2+y^2=3, x*sin(y)-y*cos(x)}, {x=1, y=2});
```

$$\{x = 0.909510, \ y = 1.47404\}$$

If we had been unable to specify an initial guess, but instead had looked for a solution with x and y in $[0, 2]$, we would have got the same answer:

```
>   fsolve({x^2+y^2=3, x*sin(y)-y*cos(x)},
    {x,y}, {x=0..2, y=0..2});
```

$$\{x = 0.909510, \ y = 1.47404\}$$

In fact, not specifying an initial guess or even search intervals would have led to the same outcome

```
>   fsolve({x^2+y^2=3, x*sin(y)-y*cos(x)}, {x,y});
```

$$\{y = 1.47404, \ x = 0.909510\}$$

although, for its own private reasons, Maple chose to report the values of x and y in the opposite order this time. Had we specified a different search interval, we might have gotten a different result:

```
>   fsolve({x^2+y^2=3, x*sin(y)-y*cos(x)},
    {x,y}, {x=0..2, y=0..1});
```

$$\{y = 0., x = 1.73205\}$$

or even no solution at all, if there is in fact no solution in the given intervals.

```
>  fsolve({x^2+y^2=3, x*sin(y)-y*cos(x)},
   {x,y}, {x=0..1, y=0..1});
```

$$fsolve(\{x^2 + y^2 = 3, s\sin(y) - y\sin(x)\}, \{x, y\}, \{x = (0..1), y = (0..1)\})$$

Using fsolve efficiently usually requires us to have some idea where solutions can be found. If the number of variables is 2 or 3, Maple's graphical routines can often be used to help us find approximate locations of solutions.

Example 1   Solve the system

$$\begin{cases} x^2 + y^4 = 1 \\ z = x^3 y \\ e^x = 2y - z. \end{cases}$$

Solution   We begin by defining the set of equations.

```
>  eqns := {x^2+y^4=1, z=x^3*y, exp(x)=2*y-z};
```

$$eqns := \{x^2 + y^4 = 1, z = x^3\,y, \mathbf{e}^x = 2y - z\}$$

What are we to use for initial guesses? The first equation cannot be satisfied by any points outside the square $-1 \le x \le 1$, $-1 \le y \le 1$, so we need only consider starting values for x and y inside this square. The second equation then forces z to lie between -1 and 1 also. We could just try many initial guesses that satisfy these conditions and see what we get using fsolve. Alternatively, we can make several implicit plots of the three equations for fixed values of z between -1 and 1, looking for cases where the three curves come close to having a common intersection point:

```
>  with(plots):
   for z from -1 by .2 to 1 do print("z =", z);
   implicitplot({x^2+y^4-1, z-x^3*y, exp(x)-2*y+z},
   x=-1.5 .. 1.5, y=-1.5 .. 1.5) od;
```

These commands produce 11 graphs of the three equations, considered as depending on x and y for z values ranging from -1 to 1 in steps of 0.2. Two of them are shown in Figure 13.25 and Figure 13.26. They correspond to $z = -0.2$ and $z = 0.2$ and indicate that the three equations likely have solutions near $(-1, 0.2, -0.2)$ and $(0.5, 0.9, 0.2)$. We run fsolve with these starting values and then substitute the resulting output into the three equations to check that the equations are satisfied. We limit Maple's output to 6 significant figures rather than the default 10:

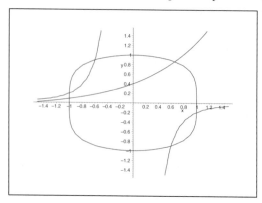

Figure 13.25 $z = -0.2$

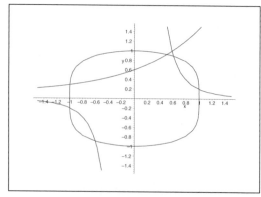

Figure 13.26 $z = 0.2$

```
> Digits := 6:
  vars := {x=-1, y=0.2, z=-0.2}:
  sols := fsolve(eqns,vars);
  evalf(subs(sols,eqns));
```

$$sols := \{x = -.999887, y = 0.122654, z = -.122613\}$$

$$\{-.122613 = -.122612, 1.00000 = 1., 0.367921 = 0.367921\}$$

```
> vars := {x=0.5, y=0.9, z=0.2}:
  sols := fsolve(eqns,vars);
  evalf(subs(sols,eqns));
```

$$sols := \{z = 0.138432, x = 0.531836, y = 0.920243\}$$

$$\{0.138432 = 0.138432, 1.00000 = 1., 1.70205 = 1.70206\}$$

We have found the two solutions to 6 significant digits.   ∎

Finding and Classifying Critical Points

Finding the critical points of a function of several variables amounts to solving the system of equations obtained by setting the first partial derivatives of the function to zero. The following example illustrates how this can be accomplished using Maple's fsolve routine. Since we also want to classify the critical points, we will find the eigenvalues of the Hessian matrix of the function at each critical point to determine whether that matrix is positive or negative definite, or indefinite

Because the VectorCalculus package contains a procedure Hessian for calculating the Hessian matrix and the LinearAlgebra contains a procedure Eigenvalues for determining the eigenvalues of a square matrix, we will either have to load both these packages or else call the procedures using VectorCalculus[Hessian] and LinearAlgebra[Eigenvalues], respectively. As we need nothing else from these packages here, we will do it the latter way. If you have a version earlier than Maple 8, be aware that the older linalg package has procedures hessian and eigenvals that will do the same job.

Example 2   Find and classify the critical points of

$$(x^2 + xy + 5y^2 + x - y)e^{-(x^2+y^2)}.$$

Solution   We begin by defining f to be the expression above, which involves only the two variables x and y. We don't need f to be a function, so just define it as an expression.

```
> f := (x^2+x*y+5*y^2+x-y)*exp(-(x^2+y^2));
```

$$f := (x^2 + xy + 5y^2 + x - y)\mathbf{e}^{-(x^2+y^2)}$$

Next, we define H to be the Hessian matrix for f with respect to the variables x and y. Since this produces several lines of output, we will suppress the output.

```
> H := VectorCalculus[Hessian](f,[x,y]):
```
The equations we want to solve to find the critical points of f are

```
> eqns := {diff(f,x)=0, diff(f,y)=0}:
```

Again we have surpressed output. We could have omitted the " $= 0$" from each equation; it would have been assumed.

Now comes the hard part: where do we look for solutions? Plotting some level curves of f can suggest likely locations for critical points.

```
> plots[contourplot](f,x=-3..3, y=-3..3, grid=[50,50],
  contours=16);
```

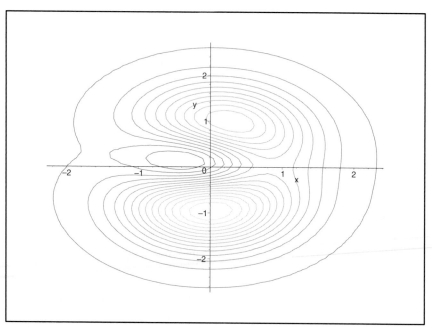

Figure 13.27 Contours of $f(x, y)$ in Example 2.

The contour plot (Figure 13.27) suggests that there are five critical points, three local extrema near $(0.3, 1)$, $(0, -1)$, and $(-0.6, 0.1)$ and two saddle points near $(1, 0)$ and $(-1.6, 0.2)$. We use each of these as initial guesses with fsolve. For each we first run fsolve to find the critical point. Then we find the value of f at that point. Finally, we calculate the eigenvalues of the Hessian of f to determine the nature of the critical point. We set Maple for 6 significant figures again.

```
>  Digits := 6:
```

(a) Near the point $(0.3, 1)$:

```
>  sols := fsolve(eqns,{x=0.3, y=1}); evalf(subs(sols,f));
```
$$sols := \{x = 0.275057, y = 1.00132\}$$
$$1.57773$$

```
>  LinearAlgebra[Eigenvalues](subs(sols,H));
```
$$\begin{bmatrix} -2.41894 \\ -6.61497 \end{bmatrix}$$

Since both eigenvalues are negative, f has a local maximum value 1.577 73 at the critical point $(0.275\,057, 1.001\,32)$.

(b) Near the point $(0, -1)$:

```
>  sols := fsolve(eqns,{x=0, y=-1}); evalf(subs(sols,f));
```
$$sols := \{y = -.955506, x = 0.00492113\}$$
$$2.21553$$

```
>  LinearAlgebra[Eigenvalues](subs(sols,H));
```
$$\begin{bmatrix} -3.58875 \\ -8.54885 \end{bmatrix}$$

Since both eigenvalues are negative, f has a local maximum value 2.215 533 at the critical point $(0.004\,921\,13, -0.955\,506)$.

(c) Near the point $(-0.6, 0.1)$:

```
>  sols := fsolve(eqns,{x=-0.6, y=0.1});
   evalf(subs(sols,f));
```
$$sols := \{y = 0.132977, x = -.421365\}$$
$$-.283329$$

```
>  LinearAlgebra[Eigenvalues](subs(sols,H));
```

$$\begin{bmatrix} 8.90194 \\ 2.32438 \end{bmatrix}$$

Since both eigenvalues are positive, f has a local minimum value $-0.283\,329$ at the critical point $(-0.421\,365, 0.132\,977)$.

(d) Near the point $(1, 0)$:

```
>   sols := fsolve(eqns,{x=1, y=0}); evalf(subs(sols,f));
```
$$sols := \{y = 0.0207852, x = 0.858435\}$$
$$0.762810$$

```
>   LinearAlgebra[Eigenvalues](subs(sols,H));
```
$$\begin{bmatrix} 3.28636 \\ -2.84680 \end{bmatrix}$$

Since the Hessian has both positive and negative eigenvalues, f has a saddle point at $(0.858\,435, 0.020\,785\,2)$. Its value there is $0.762\,810$.

(e) Near the point $(-1.6, 0.2)$:

```
>   sols := fsolve(eqns,{x=-1.6, y=0.2});
    evalf(subs(sols,f));
```
$$sols := \{y = 0.292686, x = -1.58082\}$$
$$0.0445843$$

```
>   LinearAlgebra[Eigenvalues](subs(sols,H));
```
$$\begin{bmatrix} 0.673365 \\ -.407579 \end{bmatrix}$$

Since the Hessian has both positive and negative eigenvalues, f has a saddle point at $(-1.580\,82, 0.292\,686)$. Its value there is $0.044\,584\,3$.

The negative exponential in the definition of f ensures that $f \to 0$ as $x^2 + y^2 \to \infty$. Assuming that we have found all the critical points of f, the value at the critical point in (b) must be an absolute maximum and that in (c) must be an absolute minimum. ∎

Remark   The most difficult part of using `fsolve` for large systems is determining suitable starting values for the roots or critical points. Graphical means are really only suitable for small systems (one, two, or three equations), and even then it is important to analyze the equations or functions involved for clues on where the roots or critical points may be. Here are some possibilities to consider:

1. Sometimes some of the equations will be simple enough that they can be solved for some variables and thus used to reduce the size of the system. We could have used the second equation in Example 1 to eliminate z from the first and third equations and, hence, reduced the system to two equations in two unknowns.

2. The system might result from adding a small extra term to a simpler system, the location of whose roots is known. In this case you can use those known roots as initial guesses.

3. Always be alert for equations limiting the possible values of some variables. For instance, in Example 1 the equation $x^2 + y^4 = 1$ limited x and y to the interval $[-1, 1]$.

▌▌▌▌▌▌▌ Exercises 13.7 ▌▌▌▌▌▌▌

In Exercises 1–2, solve the given systems of equations by using Maple's `fsolve` routine. Quote the solutions to 5 significant figures. Be alert for simple substitutions that can reduce the number of equations that must be fed to `fsolve`.

1. $\begin{cases} x^2 + y^2 + z^2 = 1 \\ z = xy \\ 6xz = 1 \end{cases}$

2. $\begin{cases} x^4 + y^2 + z^2 = 1 \\ y = \sin z \\ z + z^3 + z^4 = x + y \end{cases}$

In Exercises 3–6, use `fsolve` to calculate the requested results. In each case quote the results to 5 significant digits.

3. Find the maximum and minimum values and their locations for $f(x, y) = (xy - x - 2y)/((1 + x^2 + y^2)^2)$. Use a contour plot to help you determine suitable starting points.

4. Evidently $f = 1 - 10x^4 - 8y^4 - 7z^4$ has maximum value 1 at $(0, 0, 0)$. Find the absolute maximum value of $h = f + g$, where $g = yz - xyz - x - 2y + z$ by starting at various points near $(0, 0, 0)$.

5. Find the minimum value of
$$f = x^2 + y^2 + z^2 + 0.2xy - 0.3xz + 4x - y.$$

6. Find the maximum and minimum values of
$$f(x, y, z) = \frac{x + 1.1y - 0.9z + 1}{1 + x^2 + y^2 + z^2}.$$

Chapter Review

Key Ideas

- **What is meant by the following terms?**
  - ◇ a critical point of $f(x, y)$
  - ◇ a singular point of $f(x, y)$
  - ◇ an absolute maximum value of $f(x, y)$
  - ◇ a local minimum value of $f(x, y)$
  - ◇ a saddle point of $f(x, y)$
  - ◇ a quadratic form
  - ◇ a constraint
  - ◇ linear programming
  - ◇ an envelope of a family of curves
- **State the second derivative test for a critical point of $f(x, y)$.**
- **Describe the method of Lagrange multipliers.**
- **Describe the method of least squares.**
- **Describe Newton's Method for two equations.**

Review Exercises

In Exercises 1–4, find and classify all the critical points of the given functions.

1. $xy\,e^{-x+y}$

2. $x^2y - 2xy^2 + 2xy$

3. $\dfrac{1}{x} + \dfrac{4}{y} + \dfrac{9}{4 - x - y}$

4. $x^2y(2 - x - y)$

5. Let $f(x, y, z) = x^2 + y^2 + z^2 + \dfrac{1}{x^2 + y^2 + z^2}$. Does f have a minimum value? If so, what is it and where is it assumed?

6. Show that $x^2 + y^2 + z^2 - xy - xz - yz$ has a local minimum value at $(0, 0, 0)$. Is the minimum value assumed anywhere else?

7. Find the absolute maximum and minimum values of $f(x, y) = xye^{-x^2 - 4y^2}$. Justify your answer.

8. Let $f(x, y) = (4x^2 - y^2)e^{-x^2 + y^2}$.

  (a) Find the maximum and minimum values of $f(x, y)$ on the xy-plane.

  (b) Find the maximum and minimum values of $f(x, y)$ on the wedge-shaped region $0 \le y \le 3x$.

9. A wire of length L cm is cut into at most three pieces, and each piece is bent into a square. What is the (a) minimum and (b) maximum of the sum of the areas of the squares?

10. A delivery service will accept parcels in the shape of rectangular boxes the sum of whose girth and height is at most 120 in. (The girth is the perimeter of a horizontal cross-section.) What is the largest possible volume of such a box?

11. Find the area of the smallest ellipse $\dfrac{x^2}{a^2} + \dfrac{y^2}{b^2} = 1$ that contains the rectangle $-1 \le x \le 1, -2 \le y \le 2$.

12. Find the volume of the smallest ellipsoid
$$\frac{x^2}{a^2} + \frac{y^2}{b^2} + \frac{z^2}{c^2} = 1$$
that contains the rectangular box $-1 \le x \le 1, -2 \le y \le 2, -3 \le z \le 3$.

13. Find the volume of the smallest region of the form
$$0 \le z \le a\left(1 - \frac{x^2}{b^2} - \frac{y^2}{c^2}\right)$$
that contains the box $-1 \le x \le 1, -2 \le y \le 2, 0 \le z \le 2$.

14. A window has the shape of a rectangle surmounted by an isosceles triangle. What are the dimensions x, y, and z of the window (see Figure 13.28) if its perimeter is L and its area is maximum?

Figure 13.28

15. A widget manufacturer determines that if she manufactures x thousands of widgets per month and sells the widgets for y dollars each, then her monthly profit (in thousands of dollars) will be $P = xy - \frac{1}{27}x^2y^3 - x$. If her factory is capable of producing at most 3,000 widgets per month, and government regulations prevent her from charging more than $2 per widget, how many should she manufacture, and how much should she charge for each, to maximize her monthly profit?

16. Find the envelope of the family of curves $y = (x - c)^3 + 3c$.

17. Find an approximate solution $y(x, \epsilon)$ of the equation $y + \epsilon x e^y = -2x$ having terms up to second degree in ϵ.

18. (a) Calculate $G'(y)$ if $G(y) = \displaystyle\int_0^\infty \frac{\tan^{-1}(xy)}{x}\, dx$.

(b) Evaluate $\displaystyle\int_0^\infty \frac{\tan^{-1}(\pi x) - \tan^{-1} x}{x}\, dx$. *Hint:* This integral is $G(\pi) - G(1)$.

Challenging Problems

1. (Fourier series)
Show that the constants a_k, $(k = 0, 1, 2, \ldots, n)$, and b_k, $(k = 1, 2, \ldots, n)$, which minimize the integral

$$I_n = \int_{-\pi}^{\pi} \left[f(x) - \frac{a_0}{2} - \sum_{k=0}^{n}\left(a_k \cos kx + b_k \sin kx\right) \right]^2 dx,$$

are given by

$$a_k = \frac{1}{\pi}\int_{\pi}^{\pi} f(x) \cos kx\, dx, \quad b_k = \frac{1}{\pi}\int_{\pi}^{\pi} f(x) \sin kx\, dx.$$

Note that these numbers, called the **Fourier coefficients** of f on $[-\pi, \pi]$, do not depend on n. If they can be calculated for all positive integers k, then the series

$$\frac{a_0}{2} + \sum_{k=0}^{\infty}\left(a_k \cos kx + b_k \sin kx\right)$$

is called the **(full-range) Fourier series** of f on $[-\pi, \pi]$. (See Section 9.9.)

2. This is a continuation of Problem 1. Find the (full range) Fourier coefficients a_k and b_k of

$$f(x) = \begin{cases} 0 & \text{if } -\pi \le x < 0 \\ x & \text{if } 0 \le x \le \pi. \end{cases}$$

What is the minimum value of I_n in this case? How does it behave as $n \to \infty$?

3. Evaluate $\displaystyle\int_0^x \frac{\ln(tx + 1)}{1 + t^2}\, dt$.

4. (Steiner's problem) The problem of finding a point in the plane (or a higher-dimensional space) that minimizes the sum of its distances from n given points is very difficult. The case $n = 3$ is known as Steiner's problem. If $P_1 P_2 P_3$ is a triangle whose largest angle is less than $120°$, there is a point Q inside the triangle so that the lines QP_1, QP_2, and QP_3 make equal $120°$ angles with one another. Show that the sum of the distances from the vertices of the triangle to a point P is minimum when $P = Q$. *Hint:* First show that if $P = (x, y)$ and $P_i = (x_i, y_i)$, then

$$\frac{d|P P_i|}{dx} = \cos\theta_i \quad \text{and} \quad \frac{d|P P_i|}{dy} = \sin\theta_i,$$

where θ_i is the angle between $\overrightarrow{P_i P}$ and the positive direction of the x-axis. Hence, show that the minimal point P satisfies two trigonometric equations involving θ_1, θ_2, and θ_3. Then try to show that any two of those angles differ by $\pm 2\pi/3$. Where should P be taken if the triangle has an angle of $120°$ or greater?

CHAPTER 14

Multiple Integration

"Do you know what a mathematician is?" Lord Kelvin asked a class. He then stepped to the board and wrote

$$\int_{-\infty}^{\infty} e^{-x^2}\, dx = \sqrt{\pi}.$$

Putting his finger on what he had written, he turned to the class. "A mathematician is one to whom *that* is as obvious as that 'twice two makes four' is to you."

William Thomson Kelvin 1824–1907
anecdote from *Men of Mathematics* by E. Bell

Introduction   In this chapter we extend the concept of the definite integral to functions of several variables. Defined as limits of Riemann sums, like the one-dimensional definite integral, such multiple integrals can be evaluated using successive single definite integrals. They are used to represent and calculate quantities specified in terms of densities in regions of the plane or spaces of higher dimension. In the simplest instance, the volume of a three-dimensional region is given by a *double integral* of its height over the two-dimensional plane region that is its base.

14.1 Double Integrals

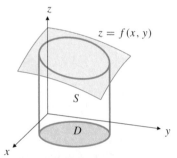

Figure 14.1   A *solid* region S lying above domain D in the xy-plane and below the surface $z = f(x, y)$

The definition of the definite integral, $\int_a^b f(x)\, dx$, is motivated by the *standard area problem*, namely, the problem of finding the area of the plane region bounded by the curve $y = f(x)$, the x-axis, and the lines $x = a$ and $x = b$. Similarly, we can motivate the double integral of a function of two variables over a domain D in the plane by means of the *standard volume problem* of finding the volume of the three-dimensional region S bounded by the surface $z = f(x, y)$, the xy-plane, and the cylinder parallel to the z-axis passing through the boundary of D. (See Figure 14.1. D is called the **domain of integration**.) We will call such a three-dimensional region S a "solid," although we are not implying that it is filled with any particular substance. We will define the double integral of $f(x, y)$ over the domain D,

$$\iint_D f(x, y)\, dA,$$

in such a way that its value will give the volume of the solid S whenever D is a "reasonable" domain and f is a "reasonable" function with positive values.

Let us start with the case where D is a closed rectangle with sides parallel to the coordinate axes in the xy-plane, and f is a bounded function on D. If D consists of the points (x, y) such that $a \le x \le b$ and $c \le y \le d$, we can form a **partition** P of D into small rectangles by partitioning each of the intervals $[a, b]$ and $[c, d]$, say by points

$$a = x_0 < x_1 < x_2 < \cdots < x_{m-1} < x_m = b,$$
$$c = y_0 < y_1 < y_2 < \cdots < y_{n-1} < y_n = d.$$

The partition P of D then consists of the mn rectangles R_{ij} ($1 \le i \le m$, $1 \le j \le n$), consisting of points (x, y) for which $x_{i-1} \le x \le x_i$ and $y_{j-1} \le y \le y_j$. (See Figure 14.2.)

Figure 14.2 A partition of D (the large shaded rectangle) into smaller rectangles R_{ij} ($1 \le i \le m$, $1 \le j \le n$)

The rectangle R_{ij} has area

$$\Delta A_{ij} = \Delta x_i \Delta y_j = (x_i - x_{i-1})(y_j - y_{j-1})$$

and *diameter* (i.e., diagonal length)

$$\text{diam}(R_{ij}) = \sqrt{(\Delta x_i)^2 + (\Delta y_j)^2} = \sqrt{(x_i - x_{i-1})^2 + (y_j - y_{j-1})^2}.$$

The **norm** of the partition P is the largest of these subrectangle diameters:

$$\|P\| = \max_{\substack{1 \le i \le m \\ 1 \le j \le n}} \text{diam}(R_{ij}).$$

Now we pick an arbitrary point (x_{ij}^*, y_{ij}^*) in each of the rectangles R_{ij} and form the **Riemann sum**

$$R(f, P) = \sum_{i=1}^{m} \sum_{j=1}^{n} f(x_{ij}^*, y_{ij}^*) \, \Delta A_{ij},$$

which is the sum of mn terms, one for each rectangle in the partition. (Here, the *double summation* indicates the sum as i goes from 1 to m of terms, each of which is itself a sum as j goes from 1 to n.) The term corresponding to rectangle R_{ij} is, if $f(x_{ij}^*, y_{ij}^*) \ge 0$, the volume of the rectangular box whose base is R_{ij} and whose height is the value of f at (x_{ij}^*, y_{ij}^*). (See Figure 14.3.) Therefore, for positive functions f, the Riemann sum $R(f, P)$ approximates the volume above D and under the graph of f. The double integral of f over D is defined to be the limit of such Riemann sums, provided the limit exists as $\|P\| \to 0$ independently of how the points (x_{ij}^*, y_{ij}^*) are chosen. We make this precise in the following definition.

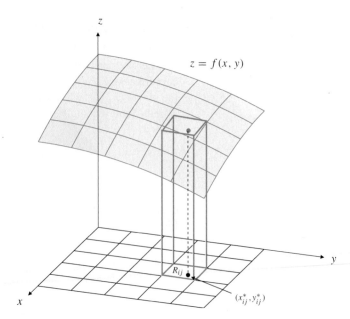

Figure 14.3 A rectangular box above rectangle R_{ij}. The Riemann sum is a sum of volumes of such boxes

DEFINITION	1

The double integral over a rectangle

We say that f is **integrable** over the rectangle D and has **double integral**

$$I = \iint_D f(x, y) \, dA,$$

if for every positive number ϵ there exists a number δ depending on ϵ, such that

$$|R(f, P) - I| < \epsilon$$

holds for every partition P of D satisfying $\|P\| < \delta$ and for all choices of the points (x_{ij}^*, y_{ij}^*) in the subrectangles of P.

The dA that appears in the expression for the double integral is an *area element*. It represents the limit of the $\Delta A = \Delta x \, \Delta y$ in the Riemann sum and can also be written $dx \, dy$ or $dy \, dx$, the order being unimportant. When we evaluate double integrals by *iteration* in the next section, dA will be replaced with a product of differentials dx and dy, and the order will be important.

As is true for functions of one variable, functions that are continuous on D are integrable on D. Of course, many bounded but discontinuous functions are also integrable, but an exact description of the class of integrable functions is beyond the scope of this text.

Example 1   Let D be the square $0 \le x \le 1, 0 \le y \le 1$. Use a Riemann sum corresponding to the partition of D into four smaller squares with points selected at the centre of each to find an approximate value for

$$\iint_D (x^2 + y) \, dA.$$

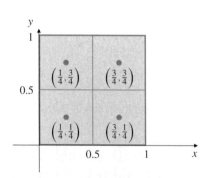

Figure 14.4 The partitioned square of Example 1

Solution   The required partition P is formed by the lines $x = 1/2$ and $y = 1/2$, which divide D into four squares, each of area $\Delta A = 1/4$. The centres of these squares are the points $\left(\frac{1}{4}, \frac{1}{4}\right)$, $\left(\frac{1}{4}, \frac{3}{4}\right)$, $\left(\frac{3}{4}, \frac{1}{4}\right)$, and $\left(\frac{3}{4}, \frac{3}{4}\right)$. (See Figure 14.4.) Therefore, the required approximation is

$$\iint_D (x^2 + y)\, dA \approx R(x^2 + y, P) = \left(\frac{1}{16} + \frac{1}{4}\right)\frac{1}{4} + \left(\frac{1}{16} + \frac{3}{4}\right)\frac{1}{4}$$

$$+ \left(\frac{9}{16} + \frac{1}{4}\right)\frac{1}{4} + \left(\frac{9}{16} + \frac{3}{4}\right)\frac{1}{4} = \frac{13}{16} = 0.8125.$$

Double Integrals over More General Domains

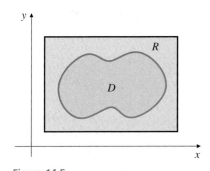

Figure 14.5

It is often necessary to use double integrals of bounded functions $f(x, y)$ over domains that are not rectangles. If the domain D is *bounded*, we can choose a rectangle R with sides parallel to the coordinate axes such that D is contained inside R. (See Figure 14.5.) If $f(x, y)$ is defined on D, we can extend its domain to be R by defining $f(x, y) = 0$ for points in R that are outside of D. The integral of f over D can then be defined to be the integral of the extended function over the rectangle R.

DEFINITION 2

If $f(x, y)$ is defined and bounded on domain D, let $\hat{f}$ be the extension of f that is zero everywhere outside D:

$$\hat{f}(x, y) = \begin{cases} f(x, y), & \text{if } (x, y) \text{ belongs to } D \\ 0, & \text{if } (x, y) \text{ does not belong to } D. \end{cases}$$

If D is a *bounded* domain, then it is contained in some rectangle R with sides parallel to the coordinate axes. If $\hat{f}$ is integrable over R, we say that f is **integrable** over D and define the **double integral** of f over D to be

$$\iint_D f(x, y)\, dA = \iint_R \hat{f}(x, y)\, dA.$$

This definition makes sense because the values of $\hat{f}$ in the part of R outside of D are all zero, so do not contribute anything to the value of the integral. However, even if f is continuous on D, $\hat{f}$ will not be continuous on R unless $f(x, y) \to 0$ as (x, y) approaches the boundary of D. Nevertheless, if f and D are "well-behaved," the integral will exist. We cannot delve too deeply into what constitutes *well-behaved*, but assert, without proof, the following theorem that will assure us that most of the double integrals we encounter do, in fact, exist.

THEOREM 1

If f is continuous on a *closed, bounded* domain D whose boundary consists of finitely many curves of finite length, then f is integrable on D.

According to Theorem 2 of Section 13.1, a continuous function is bounded if its domain is closed and bounded. Generally, however, it is not necessary to restrict our domains to be closed. If D is a bounded domain and int(D) is its interior (an open set), and if f is integrable on D, then

$$\iint_D f(x, y)\, dA = \iint_{\text{int}(D)} f(x, y)\, dA.$$

We will discuss *improper double integrals* of unbounded functions or over unbounded domains in Section 14.3.

Properties of the Double Integral

Some properties of double integrals are analogous to properties of the one-dimensional definite integral and require little comment: if f and g are integrable over D, and if L and M are constants, then

(a) $\displaystyle\iint_D f(x, y)\, dA = 0$ if D has zero area.

(b) **Area of a domain:** $\displaystyle\iint_D 1\, dA = $ area of D (because it is the volume of a cylinder with base D and height 1).

(c) **Integrals representing volumes:**
If $f(x, y) \geq 0$ on D, then $\displaystyle\iint_D f(x, y)\, dA = V \geq 0$, where V is the volume of the solid lying vertically above D and below the surface $z = f(x, y)$.

(d) If $f(x, y) \leq 0$ on D, then $\displaystyle\iint_D f(x, y)\, dA = -V \leq 0$, where V is the volume of the solid lying vertically below D and above the surface $z = f(x, y)$.

(e) **Linear dependence on the integrand:**
$$\iint_D \Big(Lf(x, y)+Mg(x, y)\Big)\, dA = L\iint_D f(x, y)\, dA + M\iint_D g(x, y)\, dA.$$

(f) **Inequalities are preserved:**
If $f(x, y) \leq g(x, y)$ on D, then $\displaystyle\iint_D f(x, y)\, dA \leq \iint_D g(x, y)\, dA.$

(g) **The triangle inequality:** $\displaystyle\left|\iint_D f(x, y)\, dA\right| \leq \iint_D |f(x, y)|\, dA.$

(h) **Additivity of domains:** If $D_1, D_2, \ldots, D_k$ are nonoverlapping domains on each of which f is integrable, then f is integrable over the union $D = D_1 \cup D_2 \cup \cdots \cup D_k$ and

$$\iint_D f(x, y)\, dA = \sum_{j=1}^{k} \iint_{D_j} f(x, y)\, dA.$$

Nonoverlapping domains can share boundary points but have no interior points in common.

Double Integrals by Inspection

As yet, we have not said anything about how to *evaluate* a double integral. The main technique for doing this, called *iteration*, will be developed in the next section, but it is worth pointing out that double integrals can sometimes be evaluated using symmetry arguments or by interpreting them as volumes that we already know.

Example 2 If R is the rectangle $a \leq x \leq b, c \leq y \leq d$, then

$$\iint_R 3\, dA = 3 \times \text{area of } R = 3(b - a)(d - c).$$

Here, the integrand is $f(x, y) = 3$, and the integral is equal to the volume of the solid box of height 3 whose base is the rectangle R. (See Figure 14.6.)

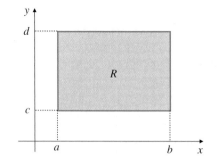

Figure 14.6

Example 3 Evaluate $\displaystyle I = \iint_{x^2+y^2\leq 1} (\sin x + y^3 + 4)\, dA.$

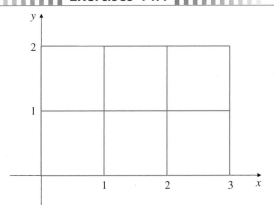

$x^2 + y^2 = 1$

D

Figure 14.7

Solution   The integral can be expressed as the sum of three integrals by property (e) of double integrals:

$$I = \iint_{x^2+y^2\leq 1} \sin x \, dA + \iint_{x^2+y^2\leq 1} y^3 \, dA + \iint_{x^2+y^2\leq 1} 4 \, dA$$
$$= I_1 + I_2 + I_3.$$

The domain of integration (Figure 14.7) is a circular disk of radius 1 centred at the origin. Since $f(x, y) = \sin x$ is an *odd* function of x, its graph bounds as much volume below the xy-plane in the region $x < 0$ as it does above the xy-plane in the region $x > 0$. These two contributions to the double integral cancel, so $I_1 = 0$. Note that symmetry of *both* the domain *and* the integrand is necessary for this argument.

Similarly, $I_2 = 0$ because y^3 is an odd function and D is symmetric about the x-axis.

Finally,

$$I_3 = \iint_D 4 \, dA = 4 \times \text{area of } D = 4\pi.$$

Thus $I = 0 + 0 + 4\pi = 4\pi$.

Example 4   If D is the disk of Example 3, the integral

$$\iint_D \sqrt{1 - x^2 - y^2} \, dA$$

represents the volume of a hemisphere of radius 1 and so has the value $2\pi/3$.

When evaluating double integrals, always be alert for situations such as those in the above examples. You can save much time by not trying to calculate an integral whose value should be obvious without calculation.

Exercises 14.1

Figure 14.8

Exercises 1–6 refer to the double integral

$$I = \iint_D (5 - x - y) \, dA,$$

where D is the rectangle $0 \leq x \leq 3, 0 \leq y \leq 2$. P is the partition of D into six squares of side 1 as shown in Figure 14.8.

In Exercises 1–5, calculate the Riemann sums for I corresponding to the given choices of points (x^*_{ij}, y^*_{ij}).

1. (x^*_{ij}, y^*_{ij}) is the upper-left corner of each square.

2. (x^*_{ij}, y^*_{ij}) is the upper-right corner of each square.

3. (x^*_{ij}, y^*_{ij}) is the lower-left corner of each square.

4. (x^*_{ij}, y^*_{ij}) is the lower-right corner of each square.

5. (x^*_{ij}, y^*_{ij}) is the centre of each square.

6. Evaluate I by interpreting it as a volume.

In Exercises 7–10, D is the disk $x^2 + y^2 \leq 25$, and P is the partition of the square $-5 \leq x \leq 5, -5 \leq y \leq 5$ into one hundred 1×1 squares, as shown in Figure 14.9. Approximate the double integral

$$J = \iint_D f(x, y) \, dA,$$

where $f(x, y) = 1$ by calculating the Riemann sums $R(f, P)$ corresponding to the indicated choice of points in the small squares. *Hint:* Using symmetry will make the job easier.

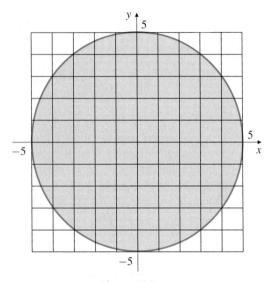

Figure 14.9

7. (x_{ij}^*, y_{ij}^*) is the corner of each square closest to the origin.

8. (x_{ij}^*, y_{ij}^*) is the corner of each square farthest from the origin.

9. (x_{ij}^*, y_{ij}^*) is the centre of each square.

10. Evaluate J.

11. Repeat Exercise 5 using the integrand e^x instead of $5 - x - y$.

12. Repeat Exercise 9 using $f(x, y) = x^2 + y^2$ instead of $f(x, y) = 1$.

In Exercises 13–22, evaluate the given double integral by inspection.

13. $\iint_R dA$, where R is the rectangle $-1 \le x \le 3$, $-4 \le y \le 1$

14. $\iint_D (x + 3) \, dA$, where D is the half-disk $0 \le y \le \sqrt{4 - x^2}$

15. $\iint_T (x + y) \, dA$, where T is the parallelogram having the points $(2, 2)$, $(1, -1)$, $(-2, -2)$, and $(-1, 1)$ as vertices

16. $\iint_{|x|+|y|\le 1} \left(x^3 \cos(y^2) + 3 \sin y - \pi\right) dA$

17. $\iint_{x^2+y^2\le 1} (4x^2 y^3 - x + 5) \, dA$

18. $\iint_{x^2+y^2\le a^2} \sqrt{a^2 - x^2 - y^2} \, dA$

19. $\iint_{x^2+y^2\le a^2} (a - \sqrt{x^2 + y^2}) \, dA$

20. $\iint_S (x + y) \, dA$, where S is the square $0 \le x \le a, 0 \le y \le a$

21. $\iint_T (1 - x - y) \, dA$, where T is the triangle with vertices $(0, 0)$, $(1, 0)$, and $(0, 1)$

22. $\iint_R \sqrt{b^2 - y^2} \, dA$, where R is the rectangle $0 \le x \le a, 0 \le y \le b$

14.2 Iteration of Double Integrals in Cartesian Coordinates

The existence of the double integral $\iint_D f(x, y) \, dA$ depends on f and the domain D. As we shall see, evaluation of double integrals is easiest when the domain of integration is of *simple* type.

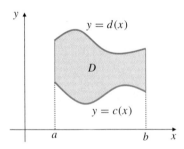

Figure 14.10 A *y*-simple domain

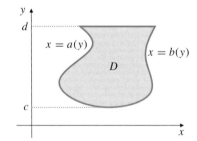

Figure 14.11 An *x*-simple domain

We say that the domain D in the xy-plane is *y*-**simple** if it is bounded by two vertical lines $x = a$ and $x = b$, and two continuous graphs $y = c(x)$ and $y = d(x)$ between these lines. (See Figure 14.10.) Lines parallel to the y-axis intersect a y-simple domain in an *interval* (possibly a single point) if at all. Similarly, D is *x*-**simple** if it is bounded by horizontal lines $y = c$ and $y = d$, and two continuous graphs $x = a(y)$ and $x = b(y)$ between these lines. (See Figure 14.11.) Many of the domains over which we will take integrals are y-simple, x-simple, or both. For example, rectangles, triangles, and disks are both x-simple and y-simple. Those domains that are neither one nor the other will usually be unions of finitely many nonoverlapping subdomains that are both x-simple and y-simple. We will call such domains **regular**. The shaded

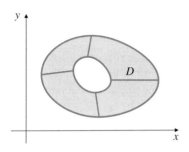

Figure 14.12   A regular domain

region in Figure 14.12 is divided into four subregions, each of which is both x-simple and y-simple.

It can be shown that a bounded, continuous function $f(x, y)$ is integrable over a bounded x-simple or y-simple domain and, therefore, over any regular domain.

Unlike the examples in the previous section, most double integrals cannot be evaluated by inspection. We need a technique for evaluating double integrals similar to the technique for evaluating single definite integrals in terms of antiderivatives. Since the double integral represents a volume, we can evaluate it for simple domains by a slicing technique.

Suppose, for instance, that D is y-simple and is bounded by $x = a$, $x = b$, $y = c(x)$, and $y = d(x)$, as shown in Figure 14.13(a). Then $\iint_D f(x, y)\, dA$ represents (at least for positive f) the volume of the solid region inside the vertical cylinder through the boundary of D and between the xy-plane and the surface $z = f(x, y)$. Consider the cross-section of this solid in the vertical plane perpendicular to the x-axis at position x. Note that x is constant in that plane. If we use the projections of the y- and z-axes onto the plane as coordinate axes there, the cross-section is a plane region bounded by vertical lines $y = c(x)$ and $y = d(x)$, by the horizontal line $z = 0$, and by the curve $z = f(x, y)$. The area of the cross-section is therefore given by

$$A(x) = \int_{c(x)}^{d(x)} f(x, y)\, dy.$$

The double integral $\iint_D f(x, y)\, dA$ is obtained by summing the volumes of "thin" slices of area $A(x)$ and thickness dx between $x = a$ and $x = b$ and is therefore given by

$$\iint_D f(x, y)\, dA = \int_a^b A(x)\, dx = \int_a^b \left(\int_{c(x)}^{d(x)} f(x, y)\, dy \right) dx.$$

Notationally, it is common to omit the large parentheses and write

$$\iint_D f(x, y)\, dA = \int_a^b \int_{c(x)}^{d(x)} f(x, y)\, dy\, dx,$$

or

$$\iint_D f(x, y)\, dA = \int_a^b dx \int_{c(x)}^{d(x)} f(x, y)\, dy.$$

The latter form shows more clearly which variable corresponds to which limits of integration.

Figure 14.13

(a) In integrals over y-simple domains, slices should be perpendicular to the x-axis

(b) In integrals over x-simple domains, slices should be perpendicular to the y-axis

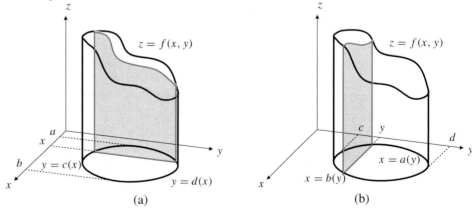

(a)   (b)

The expressions on the right-hand sides of the above formulas are called **iterated** integrals. **Iteration** is the process of reducing the problem of evaluating a double (or multiple) integral to one of evaluating two (or more) successive single definite integrals. In the above iteration, the integral

$$\int_{c(x)}^{d(x)} f(x, y)\, dy$$

is called the **inner** integral since it must be evaluated first. It is evaluated using standard techniques, treating x as a constant. The result of this evaluation is a function of x alone (note that both the integrand and the limits of the inner integral can depend on x) and is the integrand of the **outer** integral in which x is the variable of integration.

For double integrals over x-simple domains we can slice perpendicularly to the y-axis and obtain an iterated integral with the outer integral in the y direction. (See Figure 14.13(b).) We summarize the above discussion in the following theorem whose formal proof we will, however, not give.

THEOREM 2

Iteration of double integrals

If $f(x, y)$ is continuous on the bounded y-simple domain D given by $a \leq x \leq b$ and $c(x) \leq y \leq d(x)$, then

$$\iint_D f(x, y)\, dA = \int_a^b dx \int_{c(x)}^{d(x)} f(x, y)\, dy.$$

Similarly, if f is continuous on the x-simple domain D given by $c \leq y \leq d$ and $a(y) \leq x \leq b(y)$, then

$$\iint_D f(x, y)\, dA = \int_c^d dy \int_{a(y)}^{b(y)} f(x, y)\, dx.$$

Remark The symbol dA in the double integral is replaced in the iterated integrals by the dx and the dy. Accordingly, dA is frequently written $dx\, dy$ or $dy\, dx$ even in the double integral. The three expressions

$$\iint_D f(x, y)\, dx\, dy, \qquad \iint_D f(x, y)\, dy\, dx, \quad \text{and} \quad \iint_D f(x, y)\, dA$$

all stand for the double integral of f over D. Only when the double integral is iterated does the order of dx and dy become important. Later in this chapter we will iterate double integrals in polar coordinates, and dA will take the form $r\, dr\, d\theta$.

It is not always necessary to make a three-dimensional sketch of the solid volume represented by a double integral. In order to iterate the integral properly (in one direction or the other) it is usually sufficient to make a sketch of *the domain D* over which the integral is taken. The direction of iteration can be shown by a line along which the inner integral is taken. The following examples illustrate this.

Example 1 Find the volume of the solid lying above the square Q defined by $0 \leq x \leq 1$ and $1 \leq y \leq 2$ and below the plane $z = 4 - x - y$.

Solution The square Q is both x-simple and y-simple, so the double integral giving the volume can be iterated in either direction. We will do it both ways just for practice. The horizontal line at height y in Figure 14.14 suggests that we first integrate with respect to x along this line (from 0 to 1) and then integrate the result with respect to y from 1 to 2. Iterating the double integral in this direction, we calculate

Figure 14.14 The horizontal line through Q indicates iteration with the inner integral in the x direction

$$\text{Volume above } Q = \iint_Q (4 - x - y)\, dA$$

$$= \int_1^2 dy \int_0^1 (4 - x - y)\, dx$$

$$= \int_1^2 dy \left(4x - \frac{x^2}{2} - xy \right) \Big|_{x=0}^{x=1}$$

$$= \int_1^2 \left(\frac{7}{2} - y \right) dy$$

$$= \left(\frac{7y}{2} - \frac{y^2}{2} \right) \Big|_1^2 = 2 \text{ cubic units.}$$

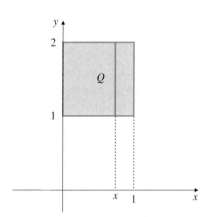

Figure 14.15 The vertical line through Q indicates iteration with the inner integral in the y direction

Using the opposite iteration, as illustrated in Figure 14.15, we calculate

$$\text{Volume above } Q = \iint_Q (4 - x - y)\, dA$$

$$= \int_0^1 dx \int_1^2 (4 - x - y)\, dy$$

$$= \int_0^1 dx \left(4y - xy - \frac{y^2}{2} \right) \Big|_{y=1}^{y=2}$$

$$= \int_0^1 \left(\frac{5}{2} - x \right) dx$$

$$= \left(\frac{5x}{2} - \frac{x^2}{2} \right) \Big|_0^1 = 2 \text{ cubic units.}$$

It is comforting to get the same answer both ways! Note that because Q is a rectangle with sides parallel to the coordinate axes, the limits of the inner integrals do not depend on the variables of the outer integrals in either iteration. This cannot be expected to happen with more general domains. ∎

Example 2 Evaluate $\displaystyle\iint_T xy\, dA$ over the triangle T with vertices $(0, 0)$, $(1, 0)$, and $(1, 1)$.

Solution The triangle T is shown in Figure 14.16. It is both x-simple and y-simple. Using the iteration corresponding to slicing in the direction shown in the figure, we obtain:

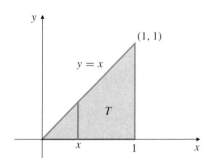

Figure 14.16 The triangular domain T with vertical line indicating iteration with inner integral in the y direction

$$\iint_T xy\, dA = \int_0^1 dx \int_0^x xy\, dy$$

$$= \int_0^1 dx \left(\frac{xy^2}{2} \right) \Big|_{y=0}^{y=x}$$

$$= \int_0^1 \frac{x^3}{2}\, dx = \frac{x^4}{8} \Big|_0^1 = \frac{1}{8}.$$

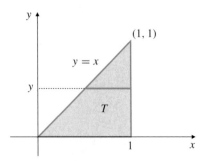

Figure 14.17   The triangular domain T with horizontal line indicating iteration with inner integral in the x direction

Iteration in the other direction (Figure 14.17) leads to the same value:

$$
\begin{aligned}
\iint_T xy\, dA &= \int_0^1 dy \int_y^1 xy\, dx \\
&= \int_0^1 dy \left(\frac{yx^2}{2}\right)\Bigg|_{x=y}^{x=1} \\
&= \int_0^1 \frac{y}{2}(1 - y^2)\, dy \\
&= \left(\frac{y^2}{4} - \frac{y^4}{8}\right)\Bigg|_0^1 = \frac{1}{8}.
\end{aligned}
$$

In both of the examples above, the double integral could be evaluated easily using either possible iteration. (We did them both ways just to illustrate that fact.) It often occurs, however, that a double integral is easily evaluated if iterated in one direction and very difficult, or impossible, if iterated in the other direction. Sometimes you will even encounter iterated integrals whose evaluation requires that they be expressed as double integrals and then reiterated in the opposite direction.

Example 3   Evaluate the iterated integral $I = \int_0^1 dx \int_{\sqrt{x}}^1 e^{y^3}\, dy$.

Solution   We cannot antidifferentiate e^{y^3} to evaluate the inner integral in this iteration, so we express I as a double integral and identify the region over which it is taken:

$$
I = \iint_D e^{y^3}\, dA,
$$

where D is the region shown in Figure 14.18. Reiterating with the x integration on the inside we get

$$
\begin{aligned}
I &= \int_0^1 dy \int_0^{y^2} e^{y^3}\, dx \\
&= \int_0^1 e^{y^3}\, dy \int_0^{y^2} dx \\
&= \int_0^1 y^2 e^{y^3}\, dy = \frac{e^{y^3}}{3}\Bigg|_0^1 = \frac{e - 1}{3}.
\end{aligned}
$$

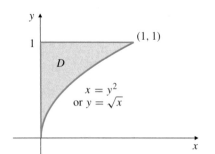

Figure 14.18   The region corresponding to the iterated integral in Example 3

The following is an example of the calculation of the volume of a somewhat awkward solid. Even though it is not always necessary to sketch solids to find their volumes, you are encouraged to sketch them whenever possible. When we encounter triple integrals over three-dimensional regions later in this chapter it will usually be necessary to sketch the regions. Get as much practice as you can.

Example 4   Sketch and find the volume of the solid bounded by the planes $y = 0$, $z = 0$, and $z = a - x + y$ and the parabolic cylinder $y = a - (x^2/a)$, where a is a positive constant.

Solution   The solid is shown in Figure 14.19. Its base is the parabolic segment D in the xy-plane bounded by $y = 0$ and $y = a - (x^2/a)$, so the volume of the solid is given by

$$
V = \iint_D (a - x + y)\, dA = \iint_D (a + y)\, dA.
$$

(Note how we used symmetry to drop the x term from the integrand. This term is an odd function of x, and D is symmetric about the y-axis.) Iterating the double integral in the direction suggested by the slice shown in the figure, we obtain

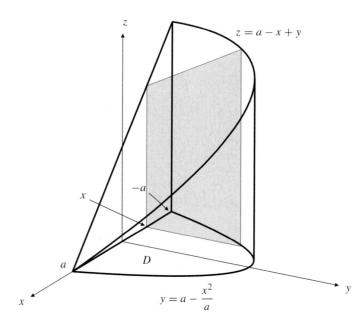

Figure 14.19   The solid in Example 4, sliced perpendicularly to the x-axis

$$V = \int_{-a}^{a} dx \int_{0}^{a-(x^2/a)} (a+y)\,dy$$

$$= \int_{-a}^{a} \left(ay + \frac{y^2}{2}\right)\Big|_{y=0}^{y=a-(x^2/a)} dx$$

$$= \int_{-a}^{a} \left[a^2 - x^2 + \frac{1}{2}\left(a^2 - 2x^2 + \frac{x^4}{a^2}\right)\right] dx$$

$$= 2 \int_{0}^{a} \left[\frac{3}{2}a^2 - 2x^2 + \frac{x^4}{2a^2}\right] dx$$

$$= \left(3a^2 x - \frac{4x^3}{3} + \frac{x^5}{5a^2}\right)\Big|_{0}^{a}$$

$$= 3a^3 - \frac{4}{3}a^3 + \frac{1}{5}a^3 = \frac{28}{15}a^3 \text{ cubic units.}$$

Remark   Maple's **int** routine can be nested to evaluate iterated double (or multiple) integrals symbolically. For instance, the iterated integral for the volume V calculated in Example 4 above can be calculated via the Maple command

```
>  V = int(int(a+y, y=0..a - x^2/a), x=-a..a);
```

$$V = \frac{28}{15}a^3$$

Recall that "int" has an *inert* form "Int," which prints the integral without attempting to evaluate it symbolically. For instance, we can print an equation for the reiterated integral in the solution of Example 3 using the command

```
>  Int(Int(exp(y^3),x=0..y^2),y=0..1)
   =int(int(exp(y^3),x=0..y^2),y=0..1);
```

$$\int_{0}^{1}\int_{0}^{y^2} e^{(y^3)}\,dx\,dy = \frac{1}{3}e - \frac{1}{3}$$

If you want Maple to approximate an iterated integral without first trying to evaluate it symbolically, just ask it to "evalf" the inert form.

```
>  evalf(Int(Int(exp(y^3),x=0..y^2),y=0..1));
```

$$.5727606095$$

Of course, Maple can't evaluate all integrals in symbolic form. If we replace $\exp(y^3)$ in the iterated integral above with $\exp(x^3)$, Maple struggles valiantly with the inner integral (of $\exp(x^3)$ with respect to x and manages to express that in terms of the gamma function and a related function of two variables called the incomplete gamma function, but it fails to evaluate the outer (y) integral.

```
>   Int(Int(exp(x^3),x=0..y^2),y=0..1)
    =int(int(exp(x^3),x=0..y^2),y=0..1);
```

$$\int_0^1 \int_0^{y^2} e^{(x^3)} \, dx \, dy = \int_0^1 \frac{1}{9} \frac{y^2 \left(-2\pi \sqrt{3} + 3\Gamma\left(\frac{1}{3}, -y^6\right) \Gamma\left(\frac{2}{3}\right) \right)}{\Gamma\left(\frac{2}{3}\right) (y^6)^{\left(\frac{1}{3}\right)}} \, dy$$

Again, we can force numerical approximation by using "evalf" on the inert form.

```
>   Int(Int(exp(x^3),x=0..y^2),y=0..1)
    =evalf(Int(Int(exp(x^3),x=0..y^2),y=0..1));
```

$$\int_0^1 \int_0^{y^2} e^{(x^3)} \, dx \, dy = .3668032540$$

Always use the inert form when you want numerical approximation. In some versions of Maple, trying to use "evalf" with "int(int(..." can produce unexpected results such as complex values for integrals of functions that are real:

```
>   evalf(int(int(exp(x^3),x=0..y^2),y=0..1));
```

$$-.1834016270 - .3176609362 \, I$$

Exercises 14.2

In Exercises 1–4, calculate the given iterated integrals.

1. $\displaystyle\int_0^1 dx \int_0^x (xy + y^2) \, dy$ **2.** $\displaystyle\int_0^1 \int_0^y (xy + y^2) \, dx \, dy$

3. $\displaystyle\int_0^\pi \int_{-x}^x \cos y \, dy \, dx$ **4.** $\displaystyle\int_0^2 dy \int_0^y y^2 e^{xy} \, dx$

In Exercises 5–14, evaluate the double integrals by iteration.

5. $\displaystyle\iint_R (x^2 + y^2) \, dA$, where R is the rectangle $0 \le x \le a$, $0 \le y \le b$

6. $\displaystyle\iint_R x^2 y^2 \, dA$, where R is the rectangle of Exercise 5

7. $\displaystyle\iint_S (\sin x + \cos y) \, dA$, where S is the square $0 \le x \le \pi/2, 0 \le y \le \pi/2$

8. $\displaystyle\iint_T (x - 3y) \, dA$, where T is the triangle with vertices $(0, 0)$, $(a, 0)$, and $(0, b)$

9. $\displaystyle\iint_R xy^2 \, dA$, where R is the finite region in the first quadrant bounded by the curves $y = x^2$ and $x = y^2$

10. $\displaystyle\iint_D x \cos y \, dA$, where D is the finite region in the first quadrant bounded by the coordinate axes and the curve $y = 1 - x^2$

11. $\displaystyle\iint_D \ln x \, dA$, where D is the finite region in the first quadrant bounded by the line $2x + 2y = 5$ and the hyperbola $xy = 1$

12. $\displaystyle\iint_T \sqrt{a^2 - y^2} \, dA$, where T is the triangle with vertices $(0, 0)$, $(a, 0)$, and (a, a)

13. $\displaystyle\iint_R \frac{x}{y} e^y \, dA$, where R is the region $0 \le x \le 1, x^2 \le y \le x$

14. $\displaystyle\iint_T \frac{xy}{1 + x^4} \, dA$, where T is the triangle with vertices $(0, 0)$, $(1, 0)$, and $(1, 1)$

In Exercises 15–18, sketch the domain of integration and evaluate the given iterated integrals.

15. $\displaystyle\int_0^1 dy \int_y^1 e^{-x^2} \, dx$ **16.** $\displaystyle\int_0^{\pi/2} dy \int_y^{\pi/2} \frac{\sin x}{x} \, dx$

17. $\displaystyle\int_0^1 dx \int_x^1 \frac{y^\lambda}{x^2 + y^2} \, dy \quad (\lambda > 0)$

18. $\displaystyle\int_0^1 dx \int_x^{x^{1/3}} \sqrt{1 - y^4} \, dy$

In Exercises 19–28, find the volumes of the indicated solids.

19. Under $z = 1 - x^2$ and above the region $0 \le x \le 1$, $0 \le y \le x$

20. Under $z = 1 - x^2$ and above the region $0 \le y \le 1$, $0 \le x \le y$

21. Under $z = 1 - x^2 - y^2$ and above the region $x \ge 0$, $y \ge 0$, $x + y \le 1$

22. Under $z = 1 - y^2$ and above $z = x^2$

23. Under the surface $z = 1/(x + y)$ and above the region in the xy-plane bounded by $x = 1$, $x = 2$, $y = 0$, and $y = x$

24. Under the surface $z = x^2 \sin(y^4)$ and above the triangle in the xy-plane with vertices $(0, 0)$, $(0, \pi^{1/4})$, and $(\pi^{1/4}, \pi^{1/4})$

25. Above the xy-plane and under the surface $z = 1 - x^2 - 2y^2$

26. Above the triangle with vertices $(0, 0)$, $(a, 0)$, and $(0, b)$, and under the plane $z = 2 - (x/a) - (y/b)$

27. Inside the two cylinders $x^2 + y^2 = a^2$ and $y^2 + z^2 = a^2$

28. Inside the cylinder $x^2 + 2y^2 = 8$, above the plane $z = y - 4$, and below the plane $z = 8 - x$

∗29. Suppose that $f(x, t)$ and $f_1(x, t)$ are continuous on the rectangle $a \le x \le b$ and $c \le t \le d$. Let

$$g(x) = \int_c^d f(x, t)\, dt \quad \text{and} \quad G(x) = \int_c^d f_1(x, t)\, dt.$$

Show that $g'(x) = G(x)$ for $a < x < b$. *Hint:* Evaluate

$\int_a^x G(u)\, du$ by reversing the order of iteration. Then differentiate the result. This is a different version of Theorem 5 of Section 13.5.

∗30. Let $F'(x) = f(x)$ and $G'(x) = g(x)$ on the interval $a \le x \le b$. Let T be the triangle with vertices (a, a), (b, a), and (b, b). By iterating $\iint_T f(x)g(y)\, dA$ in both directions, show that

$$\int_a^b f(x)G(x)\, dx$$

$$= F(b)G(b) - F(a)G(a) - \int_a^b g(y)F(y)\, dy.$$

(This is an alternative derivation of the formula for integration by parts.)

31. Use Maple's **int** routine or similar routines in other computer algebra systems to evaluate the iterated integrals in Exercises 1–4 or the iterated integrals you constructed in the remaining exercises above.

14.3 Improper Integrals and a Mean-Value Theorem

To simplify matters, the definition of the double integral given in Section 14.1 required that the domain D be bounded and that the integrand f be bounded on D. As in the single-variable case, **improper double integrals** can arise if either the domain of integration is unbounded or the integrand is unbounded near any point of the domain or its boundary.

Improper Integrals of Positive Functions

An improper integral of a function f satisfying $f(x, y) \ge 0$ on the domain D must either exist (i.e., converge to a finite value) or be infinite (diverge to infinity). Convergence or divergence of improper double integrals of such *nonnegative* functions can be determined by iterating them and determining the convergence or divergence of any single improper integrals that result.

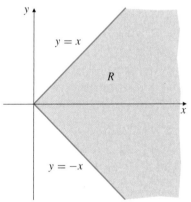

Figure 14.20   An unbounded sector of the plane

Example 1   Evaluate $I = \iint_R e^{-x^2}\, dA$. Here, R is the region where $x \ge 0$ and $-x \le y \le x$. (See Figure 14.20.)

Solution   We iterate with the outer integral in the x direction:

$$I = \int_0^\infty dx \int_{-x}^x e^{-x^2}\, dy$$

$$= \int_0^\infty e^{-x^2}\, dx \int_{-x}^x dy$$

$$= 2 \int_0^\infty x e^{-x^2}\, dx.$$

This is an improper integral that can be expressed as a limit:

$$I = 2 \lim_{r \to \infty} \int_0^r x e^{-x^2}\, dx$$

$$= 2 \lim_{r \to \infty} \left(-\frac{1}{2} e^{-x^2} \right) \Big|_0^r$$

$$= \lim_{r \to \infty} (1 - e^{-r^2}) = 1.$$

The given integral converges; its value is 1.

Example 2   If D is the region lying above the x-axis, under the curve $y = 1/x$, and to the right of the line $x = 1$, determine whether the double integral

$$\iint_D \frac{dA}{x + y}$$

converges or diverges.

Solution   The region D is sketched in Figure 14.21. We have

$$
\begin{aligned}
\iint_D \frac{dA}{x + y} &= \int_1^\infty dx \int_0^{1/x} \frac{dy}{x + y} \\
&= \int_1^\infty \ln(x + y)\Big|_{y=0}^{y=1/x} dx \\
&= \int_1^\infty \left(\ln\left(x + \frac{1}{x}\right) - \ln x \right) dx \\
&= \int_1^\infty \ln\left(\frac{x + \frac{1}{x}}{x}\right) dx = \int_1^\infty \ln\left(1 + \frac{1}{x^2}\right) dx.
\end{aligned}
$$

It happens that this integral can be evaluated exactly (see Exercise 28 below), but we are only asked to determine whether it converges, and that is more easily accomplished by estimating it. Since $0 < \ln(1 + u) < u$ if $u > 0$, we have

$$0 < \iint_D \frac{dA}{x + y} < \int_1^\infty \frac{1}{x^2} dx = 1.$$

Therefore, the given integral converges, and its value lies between 0 and 1.

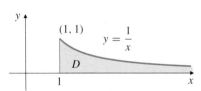

Figure 14.21   The domain of the integrand in Example 2

Example 3   Evaluate $\displaystyle\iint_D \frac{1}{(x + y)^2}\, dA$, where D is the region $0 \le x \le 1$, $0 \le y \le x^2$.

Solution   The integral is improper because the integrand is unbounded as (x, y) approaches $(0, 0)$, a boundary point of D. (See Figure 14.22.) Nevertheless, iteration leads to a proper integral:

$$
\begin{aligned}
\iint_D \frac{1}{(x + y)^2}\, dA &= \lim_{c \to 0+} \int_c^1 dx \int_0^{x^2} \frac{1}{(x + y)^2}\, dy \\
&= \lim_{c \to 0+} \int_c^1 dx \left(-\frac{1}{x + y} \right)\Big|_{y=0}^{y=x^2} \\
&= \lim_{c \to 0+} \int_c^1 \left(\frac{1}{x} - \frac{1}{x^2 + x} \right) dx \\
&= \int_0^1 \frac{1}{x + 1}\, dx = \ln(x + 1)\Big|_0^1 = \ln 2.
\end{aligned}
$$

Figure 14.22   The function $\dfrac{1}{(x + y)^2}$ is unbounded on D

Example 4   Determine the convergence or divergence of $I = \displaystyle\iint_D \frac{dA}{xy}$, where D is the bounded region in the first quadrant lying between the line $y = x$ and the parabola $y = x^2$.

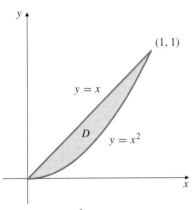

Figure 14.23 $\dfrac{1}{xy}$ is unbounded
on the domain D

Solution The domain D is shown in Figure 14.23. Again, the integral is improper because the integrand $1/(xy)$ is unbounded as (x, y) approaches the boundary point $(0, 0)$. We have

$$
I = \iint_D \frac{dA}{xy} = \int_0^1 \frac{dx}{x} \int_{x^2}^x \frac{dy}{y}
$$

$$
= \int_0^1 \frac{1}{x} (\ln x - \ln x^2)\, dx = -\int_0^1 \frac{\ln x}{x}\, dx.
$$

If we substitute $x = e^{-t}$ in this integral, we obtain

$$
I = -\int_\infty^0 \frac{-t}{e^{-t}} (-e^{-t})\, dt = \int_0^\infty t\, dt,
$$

which diverges to infinity. ∎

Remark In each of the examples above, the integrand was nonnegative on the domain of integration. Nonpositive integrands could have been handled similarly, but we cannot deal here with the convergence of general improper double integrals with integrands $f(x, y)$ that take both positive and negative values on the domain D of the integral. We remark, however, that such an integral cannot converge unless

$$
\iint_E f(x, y)\, dA
$$

is finite for every bounded, regular subdomain E of D. We cannot, in general, determine the convergence of the given integral by looking at the convergence of iterations. The double integral may diverge even if its iterations converge. (See Exercise 21 below.) In fact, opposite iterations may even give different values. This happens because of cancellation of infinite volumes of opposite sign. (Similar behaviour in one dimension is exemplified by the integral $\int_{-1}^1 dx/x$, which does not exist, although it represents the difference between "equal" but infinite areas.) It can be shown (for a large class of functions containing, for example, continuous functions) that an improper double integral of $f(x, y)$ over D converges if the integral of $|f(x, y)|$ over D converges:

$$
\iint_D |f(x, y)|\, dA \text{ converges} \quad \Rightarrow \quad \iint_D f(x, y)\, dA \text{ converges}.
$$

In this case any iterations will converge to the same value. Such double integrals are called **absolutely convergent** by analogy with absolutely convergent infinite series.

A Mean-Value Theorem for Double Integrals

Let D be a set in the xy-plane that is closed and bounded and has positive area $A = \iint_D dA$. Suppose that $f(x, y)$ is continuous on D. Then there exist points (x_1, y_1) and (x_2, y_2) in D where f assumes minimum and maximum values (see Theorem 2 of Section 13.1); that is,

$$
f(x_1, y_1) \le f(x, y) \le f(x_2, y_2)
$$

for all points (x, y) in D. If we integrate this inequality over D, we obtain

$$
f(x_1, y_1)A = \iint_D f(x_1, y_1)\, dA
$$

$$
\le \iint_D f(x, y)\, dA \le \iint_D f(x_2, y_2)\, dA = f(x_2, y_2)A.
$$

Therefore, dividing by A, we find that the *number*

$$\bar{f} = \frac{1}{A} \iint_D f(x, y) \, dA$$

lies between the minimum and maximum values of f on D:

$$f(x_1, y_1) \le \bar{f} \le f(x_2, y_2).$$

A set D in the plane is said to be **connected** if any two points in it can be joined by a continuous parametric curve $x = x(t)$, $y = y(t)$, $(0 \le t \le 1)$, lying in D. Suppose this curve joins (x_1, y_1) (where $t = 0$) and (x_2, y_2) (where $t = 1$). Let $g(t)$ satisfy

$$g(t) = f(x(t), y(t)), \quad 0 \le t \le 1.$$

Then g is continuous and takes the values $f(x_1, y_1)$ at $t = 0$ and $f(x_2, y_2)$ at $t = 1$. By the Intermediate-Value Theorem there exists a number t_0 between 0 and 1 such that $\bar{f} = g(t_0) = f(x_0, y_0)$, where $x_0 = x(t_0)$ and $y_0 = y(t_0)$. Thus, we have found a point (x_0, y_0) in D such that

$$\frac{1}{\text{area of } D} \iint_D f(x, y) \, dA = f(x_0, y_0).$$

We have therefore proved the following version of the Mean-Value Theorem.

THEOREM 3

A Mean-Value Theorem for double integrals

If the function $f(x, y)$ is continuous on a closed, bounded, connected set D in the xy-plane, then there exists a point (x_0, y_0) in D such that

$$\iint_D f(x, y) \, dA = f(x_0, y_0) \times (\text{area of } D).$$

By analogy with the definition of average value for one-variable functions, we make the following definition:

DEFINITION 3

The **average value** or **mean value** of an integrable function $f(x, y)$ over the set D is the number

$$\bar{f} = \frac{1}{\text{area of } D} \iint_D f(x, y) \, dA.$$

If $f(x, y) \ge 0$ on D, then the cylinder with base D and constant height $\bar{f}$ has volume equal to that of the solid region lying above D and below the surface $z = f(x, y)$. It is often very useful to interpret a double integral in terms of the average value of the function which is its integrand.

Example 5   The average value of x over a domain D having area A is

$$\bar{x} = \frac{1}{A} \iint_D x \, dA.$$

Of course, $\bar{x}$ is just the x-coordinate of the centroid of the region D.

Example 6   A large number of points (x, y) are chosen at random in the triangle T with vertices $(0, 0)$, $(1, 0)$, and $(1, 1)$. What is the approximate average value of $x^2 + y^2$ for these points?

Solution   The approximate average value of $x^2 + y^2$ for the randomly chosen points will be the average value of that function over the triangle, namely,

$$\frac{1}{1/2} \iint_T (x^2 + y^2)\, dA = 2 \int_0^1 dx \int_0^x (x^2 + y^2)\, dy$$

$$= 2 \int_0^1 \left(x^2 y + \frac{1}{3} y^3 \right) \Big|_{y=0}^{y=x} dx = \frac{8}{3} \int_0^1 x^3\, dx = \frac{2}{3}.$$

Example 7   Let (a, b) be an interior point of a domain D on which $f(x, y)$ is continuous. For sufficiently small positive r, the closed circular disk D_r with centre at (a, b) and radius r is contained in D. Show that

$$\lim_{r \to 0} \frac{1}{\pi r^2} \iint_{D_r} f(x, y)\, dA = f(a, b).$$

Solution   If D_r is contained in D, then by Theorem 3

$$\frac{1}{\pi r^2} \iint_{D_r} f(x, y)\, dA = f(x_0, y_0)$$

for some point (x_0, y_0) in D_r. As $r \to 0$, the point (x_0, y_0) approaches (a, b). Since f is continuous at (a, b), we have $f(x_0, y_0) \to f(a, b)$. Thus,

$$\lim_{r \to 0} \frac{1}{\pi r^2} \iint_{D_r} f(x, y)\, dA = f(a, b).$$

Exercises 14.3

In Exercises 1–12, determine whether the given integral converges or diverges. Try to evaluate those that converge.

1. $\iint_Q e^{-x-y}\, dA$, where Q is the first quadrant of the xy-plane

2. $\iint_Q \dfrac{dA}{(1 + x^2)(1 + y^2)}$, where Q is the first quadrant of the xy-plane

3. $\iint_S \dfrac{y}{1 + x^2}\, dA$, where S is the strip $0 < y < 1$ in the xy-plane

4. $\iint_T \dfrac{1}{x\sqrt{y}}\, dA$ over the triangle T with vertices $(0, 0)$, $(1, 1)$, and $(1, 2)$

5. $\iint_Q \dfrac{x^2 + y^2}{(1 + x^2)(1 + y^2)}\, dA$, where Q is the first quadrant of the xy-plane

6. $\iint_H \dfrac{1}{1 + x + y}\, dA$, where H is the half-strip $0 \le x < \infty$, $0 < y < 1$

7. $\iint_{\mathbb{R}^2} e^{-(|x| + |y|)}\, dA$

8. $\iint_{\mathbb{R}^2} e^{-|x+y|}\, dA$

9. $\iint_T \dfrac{1}{x^3} e^{-y/x}\, dA$, where T is the region satisfying $x \ge 1$ and $0 \le y \le x$

10. $\iint_T \dfrac{dA}{x^2 + y^2}$, where T is the region in Exercise 9

11. $\iint_Q e^{-xy}\, dA$, where Q is the first quadrant of the xy-plane

12. $\iint_R \dfrac{1}{x} \sin \dfrac{1}{x}\, dA$, where R is the region $2/\pi \le x < \infty$, $0 \le y \le 1/x$

13. Evaluate

$$I = \iint_S \frac{dA}{x + y},$$

where S is the square $0 \le x \le 1$, $0 \le y \le 1$,

(a) by direct iteration of the double integral,

(b) by using the symmetry of the integrand and the domain to write

$$I = 2 \iint_T \frac{dA}{x+y},$$

where T is the triangle with vertices $(0, 0)$, $(1, 0)$, and $(1, 1)$.

14. Find the volume of the solid lying above the square S of Exercise 13 and under the surface $z = 2xy/(x^2 + y^2)$.

In Exercises 15–20, a and b are given real numbers, D_k is the region $0 \le x \le 1$, $0 \le y \le x^k$, and R_k is the region $1 \le x < \infty$, $0 \le y \le x^k$. Find all real values of k for which the given integral converges.

15. $\displaystyle\iint_{D_k} \frac{dA}{x^a}$

16. $\displaystyle\iint_{D_k} y^b \, dA$

17. $\displaystyle\iint_{R_k} x^a \, dA$

18. $\displaystyle\iint_{R_k} \frac{dA}{y^b}$

19. $\displaystyle\iint_{D_k} x^a y^b \, dA$

20. $\displaystyle\iint_{R_k} x^a y^b \, dA$

∗21. Evaluate both iterations of the improper integral

$$\iint_S \frac{x-y}{(x+y)^3} \, dA,$$

where S is the square $0 < x < 1$, $0 < y < 1$. Show that the above improper double integral does not exist, by considering

$$\iint_T \frac{x-y}{(x+y)^3} \, dA,$$

where T is that part of the square S lying under the line $x = y$.

In Exercises 22–24, find the average value of the given function over the given region.

22. x^2 over the rectangle $a \le x \le b$, $c \le y \le d$

23. $x^2 + y^2$ over the triangle $0 \le x \le a$, $0 \le y \le a - x$

24. $1/x$ over the region $0 \le x \le 1$, $x^2 \le y \le \sqrt{x}$

∗25. Find the average distance from points in the quarter-disk $x^2 + y^2 \le a^2$, $x \ge 0$, $y \ge 0$, to the line $x + y = 0$.

26. Does $f(x, y) = x$ have an average value over the region $0 \le x < \infty$, $0 \le y \le \dfrac{1}{1 + x^2}$? If so, what is it?

27. Does $f(x, y) = xy$ have an average value over the region $0 \le x < \infty$, $0 \le y \le \dfrac{1}{1 + x^2}$? If so, what is it?

∗28. Find the exact value of the integral in Example 2. *Hint:* Integrate by parts in $\int_1^\infty \ln\left(1 + (1/x^2)\right) dx$.

∗29. Let (a, b) be an interior point of a domain D on which the function $f(x, y)$ is continuous. For small enough $h^2 + k^2$ the rectangle R_{hk} with vertices (a, b), $(a + h, b)$, $(a, b + k)$, and $(a + h, b + k)$ is contained in D. Show that

$$\lim_{(h,k)\to(0,0)} \frac{1}{hk} \iint_{R_{hk}} f(x, y) = f(a, b).$$

Hint: See Example 7.

∗30. (**Another proof of equality of mixed partials**) Suppose that $f_{12}(x, y)$ and $f_{21}(x, y)$ are continuous in a neighbourhood of the point (a, b). Without assuming the equality of these mixed partial derivatives, show that

$$\iint_R f_{12}(x, y) \, dA = \iint_R f_{21}(x, y) \, dA,$$

where R is the rectangle with vertices (a, b), $(a + h, b)$, $(a, b + k)$, and $(a + h, b + k)$ and $h^2 + k^2$ is sufficiently small. Now use the result of Exercise 29 to show that $f_{12}(a, b) = f_{21}(a, b)$. (This reproves Theorem 1 of Section 12.4. However, in that theorem we only assumed continuity of the mixed partials *at* (a, b). Here, we assume the continuity *at all points sufficiently near* (a, b).)

14.4   Double Integrals in Polar Coordinates

For many applications of double integrals, either the domain of integration, the integrand function, or both may be more easily expressed in terms of polar coordinates than in terms of Cartesian coordinates. Recall that a point P with Cartesian coordinates (x, y) can also be located by its polar coordinates $[r, \theta]$, where r is the distance from P to the origin O, and θ is the angle OP makes with the positive direction of the x-axis. (Positive angles θ are measured counterclockwise.) The polar and Cartesian coordinates of P are related by the transformations (see Figure 14.24)

$$x = r \cos \theta, \qquad r^2 = x^2 + y^2,$$
$$y = r \sin \theta, \qquad \tan \theta = y/x.$$

Figure 14.24   Polar–Cartesian conversions

Consider the problem of finding the volume V of the solid region lying above the xy-plane and beneath the paraboloid $z = 1 - x^2 - y^2$. Since the paraboloid intersects the xy-plane in the circle $x^2 + y^2 = 1$, the volume is given in Cartesian coordinates by

$$V = \iint_{x^2+y^2\leq 1} (1 - x^2 - y^2)\, dA = \int_{-1}^{1} dx \int_{-\sqrt{1-x^2}}^{\sqrt{1-x^2}} (1 - x^2 - y^2)\, dy.$$

Evaluating this iterated integral would require considerable effort. However, we can express the same volume in terms of polar coordinates as

$$V = \iint_{r\leq 1} (1 - r^2)\, dA.$$

In order to iterate this integral, we have to know the form that the *area element dA* takes in polar coordinates.

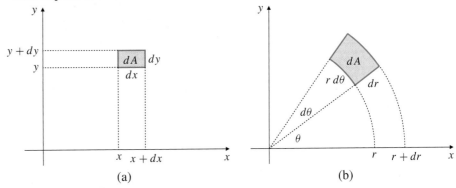

Figure 14.25

(a) $dA = dx\, dy$ in Cartesian coordinates

(b) $dA = r\, dr\, d\theta$ in polar coordinates

In the Cartesian formula for V, the area element $dA = dx\, dy$ represents the area of the "infinitesimal" region bounded by the coordinate lines at x, $x+dx$, y, and $y+dy$. (See Figure 14.25(a).) In the polar formula, the area element dA should represent the area of the "infinitesimal" region bounded by the coordinate circles with radii r and $r + dr$, and coordinate rays from the origin at angles θ and $\theta + d\theta$. (See Figure 14.25(b).) Observe that dA is approximately the area of a rectangle with dimensions dr and $r\, d\theta$. The error in this approximation becomes negligible *compared with the size of dA* as dr and $d\theta$ approach zero. Thus, in transforming a double integral between Cartesian and polar coordinates, the area element transforms according to the formula

$$dx\, dy = dA = r\, dr\, d\theta.$$

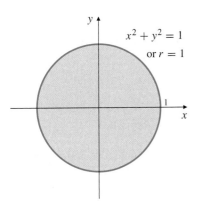

Figure 14.26   The domain in the xy-plane

Figure 14.27   The domain in the $r\theta$-plane

In order to iterate the polar form of the double integral for V considered above, we can regard the domain of integration as a set in a plane having *Cartesian coordinates r* and θ. In the xy Cartesian plane the domain is a disk $r \leq 1$ (see Figure 14.26), but in the $r\theta$ Cartesian plane (with perpendicular r- and θ-axes) the domain is the rectangle R specified by $0 \leq r \leq 1$ and $0 \leq \theta \leq 2\pi$. (See Figure 14.27.) The area element in the $r\theta$-plane is $dA^* = dr\, d\theta$, so area is not preserved under the transformation to polar coordinates ($dA = r\, dA^*$). Thus, the polar integral for V is really a Cartesian integral in the $r\theta$-plane, with integrand modified by the inclusion of an extra factor r to compensate for the change of area. It can be evaluated by standard iteration methods:

$$V = \iint_R (1 - r^2)r\, dA^* = \int_0^{2\pi} d\theta \int_0^1 (1 - r^2)r\, dr$$

$$= \int_0^{2\pi} \left(\frac{r^2}{2} - \frac{r^4}{4}\right)\Big|_0^1 d\theta = \frac{\pi}{2} \text{ units}^3.$$

Remark   It is not necessary to sketch the region R in the $r\theta$-plane. We are used to thinking of polar coordinates in terms of distances and angles in the xy-plane and can easily understand from looking at the disk in Figure 14.26 that the iteration of the integral in polar coordinates corresponds to $0 \leq \theta \leq 2\pi$ and $0 \leq r \leq 1$. That is, we should be able to write the iteration

$$V = \int_0^{2\pi} d\theta \int_0^1 (1 - r^2)r\, dr$$

directly from consideration of the domain of integration in the xy-plane.

Example 1   If R is that part of the annulus $0 < a^2 \le x^2 + y^2 \le b^2$ lying in the first quadrant and below the line $y = x$, evaluate

$$I = \iint_R \frac{y^2}{x^2}\, dA.$$

Solution   Figure 14.28 shows the region R. It is specified in polar coordinates by $0 \le \theta \le \pi/4$ and $a \le r \le b$. Since

$$\frac{y^2}{x^2} = \frac{r^2 \sin^2 \theta}{r^2 \cos^2 \theta} = \tan^2 \theta,$$

we have

$$I = \int_0^{\pi/4} \tan^2\theta\, d\theta \int_a^b r\, dr$$

$$= \frac{1}{2}(b^2 - a^2) \int_0^{\pi/4} \left(\sec^2\theta - 1 \right) d\theta$$

$$= \frac{1}{2}(b^2 - a^2)(\tan\theta - \theta)\Big|_0^{\pi/4}$$

$$= \frac{1}{2}(b^2 - a^2)\left(1 - \frac{\pi}{4}\right) = \frac{4 - \pi}{8}(b^2 - a^2).$$

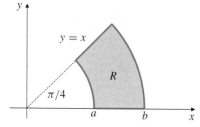

Figure 14.28

Example 2   (**Area of a polar region**) Derive the formula for the area of the polar region R bounded by the curve $r = f(\theta)$ and the rays $\theta = \alpha$ and $\theta = \beta$. (See Figure 14.29.)

Solution   The area A of R is numerically equal to the volume of a cylinder of height 1 above the region R:

$$A = \iint_R dx\, dy = \iint_R r\, dr\, d\theta$$

$$= \int_\alpha^\beta d\theta \int_0^{f(\theta)} r\, dr = \frac{1}{2} \int_\alpha^\beta \left(f(\theta) \right)^2 d\theta.$$

Observe that the inner integral in the iteration involves integrating r along the ray specified by θ from 0 to $f(\theta)$.

Figure 14.29   A standard area problem for polar coordinates

There is no firm rule as to whether one should or should not convert a double integral from Cartesian to polar coordinates. In Example 1 above, the conversion was strongly suggested by the shape of the domain but was also indicated by the fact that the integrand, y^2/x^2, becomes a function of θ alone when converted to polar coordinates. It is usually wise to switch to polar coordinates if the switch simplifies the iteration (i.e., if the *domain* is "simpler" when expressed in terms of polar coordinates), even if the form of the integrand is made more complicated.

Example 3   Find the volume of the solid lying in the first octant, inside the cylinder $x^2 + y^2 = a^2$, and under the plane $z = y$.

Solution   The solid is shown in Figure 14.30. The base is a quarter disk, which is expressed in polar coordinates by the inequalities $0 \le \theta \le \pi/2$ and $0 \le r \le a$. The height is given by $z = y = r \sin\theta$. The solid has volume

$$V = \int_0^{\pi/2} d\theta \int_0^a (r \sin\theta) r\, dr = \int_0^{\pi/2} \sin\theta\, d\theta \int_0^a r^2\, dr = \frac{1}{3} a^3 \text{ units}^3.$$

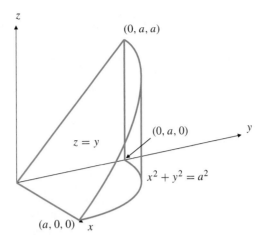

Figure 14.30 This volume is easily calculated using iteration in polar coordinates

The following example establishes the value of a definite integral that plays a very important role in probability theory and statistics. It is interesting that this single-variable integral cannot be evaluated by the techniques of single-variable calculus.

Example 4   (**A very important integral**) Show that

$$\int_{-\infty}^{\infty} e^{-x^2}\, dx = \sqrt{\pi}.$$

Solution   The improper integral (call it I) converges, and its value does not depend on what symbol we use for the variable of integration. Therefore, we can express the square of the integral as a product of two identical integrals but with their variables of integration named differently. We then interpret this product as an improper double integral and reiterate it in polar coordinates:

$$
\begin{aligned}
I^2 &= \left(\int_{-\infty}^{\infty} e^{-x^2}\, dx\right)^2 = \int_{-\infty}^{\infty} e^{-x^2}\, dx \int_{-\infty}^{\infty} e^{-y^2}\, dy \\
&= \iint_{\mathbb{R}^2} e^{-(x^2+y^2)}\, dA \\
&= \int_0^{2\pi} d\theta \int_0^{\infty} e^{-r^2} r\, dr \\
&= 2\pi \lim_{R\to\infty} \left(-\frac{1}{2} e^{-r^2}\right)\Bigg|_0^R = \pi.
\end{aligned}
$$

Thus $I = \sqrt{\pi}$ as asserted.

Note that the r integral in the iteration above is a convergent improper integral; it was evaluated with the aid of the substitution $u = r^2$.

As our final example of iteration in polar coordinates, let us try something a little more demanding.

Example 5   Find the volume of the solid region lying inside both the sphere $x^2 + y^2 + z^2 = 4a^2$ and the cylinder $x^2 + y^2 = 2ay$, where $a > 0$.

Solution   The sphere is centred at the origin and has radius $2a$. The equation of the cylinder becomes

$$x^2 + (y - a)^2 = a^2$$

if we complete the **square** in the y terms. Thus, it is a vertical circular cylinder of radius a having its axis along the vertical line through $(0, a, 0)$. The z-axis lies on the cylinder. One-quarter of the required volume lies in the first octant. This part is shown in Figure 14.31.

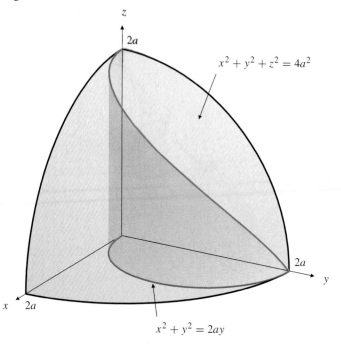

Figure 14.31 The first octant part of the intersection of the cylinder $x^2 + y^2 = 2ay$ and the sphere $x^2 + y^2 + z^2 = 4a^2$

If we use polar coordinates in the xy-plane, then the sphere has equation $r^2 + z^2 = 4a^2$ and the cylinder has equation $r^2 = 2ar \sin \theta$ or, more simply, $r = 2a \sin \theta$. The first octant portion of the volume lies above the region specified by the inequalities $0 \le \theta \le \pi/2$ and $0 \le r \le 2a \sin \theta$. Therefore, the total volume is

$$V = 4 \int_0^{\pi/2} d\theta \int_0^{2a \sin \theta} \sqrt{4a^2 - r^2}\, r\, dr \qquad \text{Let } u = 4a^2 - r^2$$

$$= 2 \int_0^{\pi/2} d\theta \int_{4a^2 \cos^2 \theta}^{4a^2} \sqrt{u}\, du$$

$$= \frac{4}{3} \int_0^{\pi/2} (8a^3 - 8a^3 \cos^3 \theta)\, d\theta \qquad \text{Let } v = \sin \theta$$

$$= \frac{16}{3}\pi a^3 - \frac{32}{3}a^3 \int_0^1 (1 - v^2)\, dv$$

$$= \frac{16}{3}\pi a^3 - \frac{64}{9}a^3 = \frac{16}{9}(3\pi - 4)a^3 \text{ cubic units.} \qquad \blacksquare$$

Change of Variables in Double Integrals

The transformation of a double integral to polar coordinates is just a special case of a general change of variables formula for double integrals. Suppose that x and y are expressed as functions of two other variables u and v by the equations

$$x = x(u, v)$$
$$y = y(u, v).$$

We regard these equations as defining a **transformation** (or mapping) from points (u, v) in a uv-Cartesian plane to points (x, y) in the xy-plane. (See Figure 14.32.) We

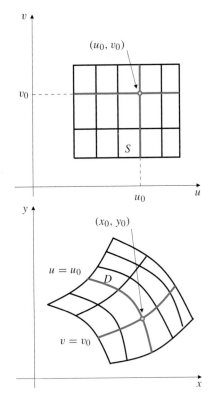

Figure 14.32 Under the transformation
$\begin{cases} x = x(u, v) \\ y = y(u, v) \end{cases}$ the lines $u = u_0$ and $v = v_0$ in the uv-plane get mapped to the curves $\begin{cases} x = x(u_0, v) \\ y = y(u_0, v) \end{cases}$ and $\begin{cases} x = x(u, v_0) \\ y = y(u, v_0) \end{cases}$ in the xy-plane, which we still label as $u = u_0$ and $v = v_0$. The point (u_0, v_0) is mapped to the point (x_0, y_0)

say that the transformation is *one-to-one* from the set S in the uv-plane *onto* the set D in the xy-plane provided:

(i) every point in S gets mapped to a point in D,

(ii) every point in D is the image of a point in S, and

(iii) different points in S get mapped to different points in D.

If the transformation is one-to-one, the defining equations can be solved for u and v as functions of x and y, and the resulting **inverse transformation**,

$$u = u(x, y)$$
$$v = v(x, y),$$

is one-to-one from D onto S.

Let us assume that the functions $x(u, v)$ and $y(u, v)$ have continuous first partial derivatives and that the Jacobian determinant

$$\frac{\partial(x, y)}{\partial(u, v)} \neq 0 \qquad \text{at} \quad (u, v).$$

As noted in Section 12.8, the Implicit Function Theorem implies that the transformation is one-to-one near (u, v) and the inverse transformation also has continuous first partial derivatives and nonzero Jacobian satisfying

$$\frac{\partial(u, v)}{\partial(x, y)} = \frac{1}{\dfrac{\partial(x, y)}{\partial(u, v)}} \qquad \text{on } D.$$

Example 6   The transformation $x = r\cos\theta$, $y = r\sin\theta$ to polar coordinates has Jacobian

$$\frac{\partial(x, y)}{\partial(r, \theta)} = \begin{vmatrix} \cos\theta & -r\sin\theta \\ \sin\theta & r\cos\theta \end{vmatrix} = r.$$

Near any point except the origin (where $r = 0$) the transformation is one-to-one. (In fact, it is one-to-one from any set in the $r\theta$-plane that does not contain more than one point where $r = 0$ and lies in, say, the strip $0 \leq \theta < 2\pi$.) ∎

A one-to-one transformation can be used to transform the double integral

$$\iint_D f(x, y)\, dA$$

to a double integral over the corresponding set S in the uv-plane. Under the transformation, the integrand $f(x, y)$ becomes $g(u, v) = f\big(x(u, v), y(u, v)\big)$. We must discover how to express the area element $dA = dx\, dy$ in terms of the area element $du\, dv$ in the uv-plane.

For any fixed value of u (say $u = c$), the equations

$$x = x(u, v) \quad \text{and} \quad y = y(u, v)$$

define a parametric curve (with v as parameter) in the xy-plane. This curve is called a u-curve corresponding to the value $u = c$. Similarly, for fixed v the equations define a parametric curve (with parameter u) called a v-curve. Consider the differential area element bounded by the u-curves corresponding to nearby values u and $u + du$ and the v-curves corresponding to nearby values v and $v + dv$. Since these curves are smooth, for small values of du and dv the area element is approximately a parallelogram, and its area is approximately

$$dA = |\vec{PQ} \times \vec{PR}|,$$

where P, Q, and R are the points shown in Figure 14.33. The error in this approximation becomes negligible compared with dA as du and dv approach zero.

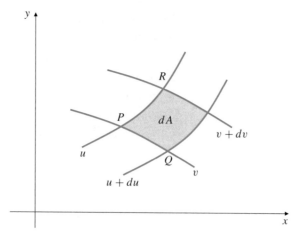

Figure 14.33 The image in the xy-plane of the area element $du\,dv$ in the uv-plane

Now $\overrightarrow{PQ} = dx\,\mathbf{i} + dy\,\mathbf{j}$, where

$$dx = \frac{\partial x}{\partial u}du + \frac{\partial x}{\partial v}dv \quad\text{and}\quad dy = \frac{\partial y}{\partial u}du + \frac{\partial y}{\partial v}dv.$$

However, $dv = 0$ along the v-curve PQ, so

$$\overrightarrow{PQ} = \frac{\partial x}{\partial u}du\,\mathbf{i} + \frac{\partial y}{\partial u}du\,\mathbf{j}.$$

Similarly,

$$\overrightarrow{PR} = \frac{\partial x}{\partial v}dv\,\mathbf{i} + \frac{\partial y}{\partial v}dv\,\mathbf{j}.$$

Hence,

$$dA = \left\| \begin{matrix} \mathbf{i} & \mathbf{j} & \mathbf{k} \\ \dfrac{\partial x}{\partial u}du & \dfrac{\partial y}{\partial u}du & 0 \\ \dfrac{\partial x}{\partial v}dv & \dfrac{\partial y}{\partial v}dv & 0 \end{matrix} \right\| = \left| \frac{\partial(x,y)}{\partial(u,v)} \right| du\,dv;$$

that is, the absolute value of the Jacobian $\partial(x,y)/\partial(u,v)$ is the ratio between corresponding area elements in the xy-plane and the uv-plane:

$$dA = dx\,dy = \left| \frac{\partial(x,y)}{\partial(u,v)} \right| du\,dv.$$

The following theorem summarizes the change of variables procedure for a double integral.

THEOREM 4

Change of variables formula for double integrals

Let $x = x(u, v)$, $y = y(u, v)$ be a one-to-one transformation from a domain S in the uv-plane onto a domain D in the xy-plane. Suppose that the functions x and y, and their first partial derivatives with respect to u and v, are continuous in S. If $f(x, y)$ is integrable on D, and if $g(u, v) = f(x(u, v), y(u, v))$, then g is integrable on S and

$$\iint_D f(x, y)\,dx\,dy = \iint_S g(u, v)\left| \frac{\partial(x,y)}{\partial(u,v)} \right| du\,dv.$$

Remark It is not necessary that S or D be closed or that the transformation be one-to-one on the boundary of S. The transformation to polar coordinates maps the rectangle $0 < r < 1, 0 \le \theta < 2\pi$ one-to-one onto the punctured disk $0 < x^2 + y^2 < 1$ and, as in the first example in this section, we can transform an integral over the closed disk $x^2 + y^2 \le 1$ to one over the closed rectangle $0 \le r \le 1, 0 \le \theta \le 2\pi$.

Example 7   Use an appropriate change of variables to find the area of the elliptic disk E given by

$$\frac{x^2}{a^2} + \frac{y^2}{b^2} \le 1.$$

Solution   Under the transformation $x = au$, $y = bv$, the elliptic disk E is the one-to-one image of the circular disk D given by $u^2 + v^2 \le 1$. Assuming $a > 0$ and $b > 0$, we have

$$dx\, dy = \left|\frac{\partial(x,\, y)}{\partial(u,\, v)}\right| du\, dv = \left\| \begin{matrix} a & 0 \\ 0 & b \end{matrix} \right\| du\, dv = ab\, du\, dv.$$

Therefore, the area of E is given by

$$\iint_E 1\, dx\, dy = \iint_D ab\, du\, dv = ab \times (\text{area of } D) = \pi ab \text{ square units.} \qquad \blacksquare$$

It is often tempting to try to use the change of variable formula to transform the domain of a double integral into a rectangle so that iteration will be easy. As the following example shows, this usually involves defining the inverse transformation (u and v in terms of x and y). Remember that inverse transformations have reciprocal Jacobians.

Example 8   Find the area of the finite plane region bounded by the four parabolas $y = x^2$, $y = 2x^2$, $x = y^2$, and $x = 3y^2$.

Solution   The region, call it D, is sketched in Figure 14.34. Let

$$u = \frac{y}{x^2} \qquad \text{and} \qquad v = \frac{x}{y^2}.$$

Then the region D corresponds to the rectangle R in the uv-plane given by $1 \le u \le 2$ and $1 \le v \le 3$. Since

$$\frac{\partial(u,\, v)}{\partial(x,\, y)} = \left| \begin{matrix} -2y/x^3 & 1/x^2 \\ 1/y^2 & -2x/y^3 \end{matrix} \right| = \frac{3}{x^2 y^2} = 3u^2 v^2,$$

it follows that

$$\left|\frac{\partial(x,\, y)}{\partial(u,\, v)}\right| = \frac{1}{3u^2 v^2}$$

and so the area of D is given by

$$\iint_D dx\, dy = \iint_R \frac{1}{3u^2 v^2}\, du\, dv$$
$$= \frac{1}{3} \int_1^2 \frac{du}{u^2} \int_1^3 \frac{dv}{v^2} = \frac{1}{3} \times \frac{1}{2} \times \frac{2}{3} = \frac{1}{9} \text{ square units.} \qquad \blacksquare$$

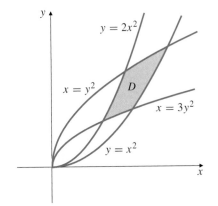

Figure 14.34

The following example shows what can happen if a transformation of the domain of a double integral is not one-to-one.

Example 9   Let D be the square $0 \le x \le 1, 0 \le y \le 1$ in the xy-plane, and let S be the square $0 \le u \le 1, 0 \le v \le 1$ in the uv-plane. Show that the transformation

$$x = 4u - 4u^2, \qquad y = v$$

maps S onto D, and use it to transform the integral $I = \iint_D dx\, dy$. Compare the value of I with that of the transformed integral.

Solution Since $x = 4u - 4u^2 = 1 - (1 - 2u)^2$, the minimum value of x on the interval $0 \leq u \leq 1$ is 0 (at $u = 0$ and $u = 1$), and the maximum value is 1 (at $u = \frac{1}{2}$). Therefore, $x = 4u - 4u^2$ maps the interval $0 \leq u \leq 1$ onto the interval $0 \leq x \leq 1$. Since $y = v$ clearly maps $0 \leq v \leq 1$ onto $0 \leq y \leq 1$, the given transformation maps S onto D. Since

$$dx \, dy = \left| \frac{\partial(x, y)}{\partial(u, v)} \right| du \, dv = \left\| \begin{matrix} 4 - 8u & 0 \\ 0 & 1 \end{matrix} \right\| du \, dv = |4 - 8u| \, du \, dv,$$

transforming I leads to the integral

$$J = \iint_S |4 - 8u| \, du \, dv = 4 \int_0^1 dv \int_0^1 |1 - 2u| \, du = 8 \int_0^{1/2} (1 - 2u) \, du = 2.$$

However, $I = \iint_D dx \, dy = $ area of $D = 1$. The reason that $J \neq I$ is that the transformation is not one-to-one from S onto D; it actually maps S onto D twice. The rectangle R defined by $0 \leq u \leq \frac{1}{2}$ and $0 \leq v \leq 1$ is mapped one-to-one onto D by the transformation, so the appropriate transformed integral is $\iint_R |4 - 8u| \, du \, dv$, which is equal to I.

Exercises 14.4

In Exercises 1–6, evaluate the given double integral over the disk D given by $x^2 + y^2 \leq a^2$, where $a > 0$.

1. $\displaystyle\iint_D (x^2 + y^2) \, dA$

2. $\displaystyle\iint_D \sqrt{x^2 + y^2} \, dA$

3. $\displaystyle\iint_D \frac{1}{\sqrt{x^2 + y^2}} \, dA$

4. $\displaystyle\iint_D |x| \, dA$

5. $\displaystyle\iint_D x^2 \, dA$

6. $\displaystyle\iint_D x^2 y^2 \, dA$

In Exercises 7–10, evaluate the given double integral over the quarter-disk Q given by $x \geq 0$, $y \geq 0$, and $x^2 + y^2 \leq a^2$, where $a > 0$.

7. $\displaystyle\iint_Q y \, dA$

8. $\displaystyle\iint_Q (x + y) \, dA$

9. $\displaystyle\iint_Q e^{x^2 + y^2} \, dA$

10. $\displaystyle\iint_Q \frac{2xy}{x^2 + y^2} \, dA$

11. Evaluate $\displaystyle\iint_S (x + y) \, dA$, where S is the region in the first quadrant lying inside the disk $x^2 + y^2 \leq a^2$ and under the line $y = \sqrt{3}x$.

12. Find $\displaystyle\iint_S x \, dA$, where S is the disk segment $x^2 + y^2 \leq 2$, $x \geq 1$.

13. Evaluate $\displaystyle\iint_T (x^2 + y^2) \, dA$, where T is the triangle with vertices $(0, 0)$, $(1, 0)$, and $(1, 1)$.

14. Evaluate $\displaystyle\iint_{x^2 + y^2 \leq 1} \ln(x^2 + y^2) \, dA$.

15. Find the average distance from the origin to points in the disk $x^2 + y^2 \leq a^2$.

16. Find the average value of $e^{-(x^2 + y^2)}$ over the annular region $0 < a \leq \sqrt{x^2 + y^2} \leq b$.

17. For what values of k, and to what value, does the integral
$$\iint_{x^2 + y^2 \leq 1} \frac{dA}{(x^2 + y^2)^k}$$ converge?

18. For what values of k, and to what value, does the integral
$$\iint_{\mathbb{R}^2} \frac{dA}{(1 + x^2 + y^2)^k}$$ converge?

19. Evaluate $\displaystyle\iint_D xy \, dA$, where D is the plane region satisfying $x \geq 0$, $0 \leq y \leq x$, and $x^2 + y^2 \leq a^2$.

20. Evaluate $\displaystyle\iint_C y \, dA$, where C is the upper half of the cardioid disk $r \leq 1 + \cos\theta$.

21. Find the volume lying between the paraboloids $z = x^2 + y^2$ and $3z = 4 - x^2 - y^2$.

22. Find the volume lying inside both the sphere $x^2 + y^2 + z^2 = a^2$ and the cylinder $x^2 + y^2 = ax$.

23. Find the volume lying inside both the sphere $x^2 + y^2 + z^2 = 2a^2$ and the cylinder $x^2 + y^2 = a^2$.

24. Find the volume of the region lying above the xy-plane, inside the cylinder $x^2 + y^2 = 4$ and below the plane $z = x + y + 4$.

*25. Find the volume of the region lying inside all three of the circular cylinders $x^2 + y^2 = a^2$, $x^2 + z^2 = a^2$, and $y^2 + z^2 = a^2$. *Hint:* Make a good sketch of the first octant part of the region, and use symmetry whenever possible.

26. Find the volume of the region lying inside the circular cylinder $x^2 + y^2 = 2y$ and inside the parabolic cylinder $z^2 = y$.

*27. Many points are chosen at random in the disk $x^2 + y^2 \leq 1$. Find the approximate average value of the distance from these points to the nearest side of the smallest square that contains the disk.

28. Find the average value of x over the segment of the disk $x^2 + y^2 \leq 4$ lying to the right of $x = 1$. What is the centroid of the segment?

29. Find the volume enclosed by the ellipsoid

$$\frac{x^2}{a^2} + \frac{y^2}{b^2} + \frac{z^2}{c^2} = 1.$$

30. Find the volume of the region in the first octant below the paraboloid

$$z = 1 - \frac{x^2}{a^2} - \frac{y^2}{b^2}.$$

Hint: Use the change of variables $x = au$, $y = bv$.

31. Evaluate $\displaystyle\iint_{|x|+|y|\leq a} e^{x+y}\, dA$.

32. Find $\displaystyle\iint_P (x^2 + y^2)\, dA$, where P is the parallelogram bounded by the lines $x + y = 1$, $x + y = 2$, $3x + 4y = 5$, and $3x + 4y = 6$.

33. Find the area of the region in the first quadrant bounded by the curves $xy = 1$, $xy = 4$, $y = x$, and $y = 2x$.

34. Evaluate $\displaystyle\iint_R (x^2 + y^2)\, dA$, where R is the region in the first quadrant bounded by $y = 0$, $y = x$, $xy = 1$, and $x^2 - y^2 = 1$.

35. Let T be the triangle with vertices $(0, 0)$, $(1, 0)$, and $(0, 1)$. Evaluate the integral $\displaystyle\iint_T e^{(y-x)/(y+x)}\, dA$,

    (a) by transforming to polar coordinates, and

    (b) by using the transformation $u = y - x$, $v = y + x$.

36. Use the method of Example 7 to find the area of the region inside the ellipse $4x^2 + 9y^2 = 36$ and above the line $2x + 3y = 6$.

37. (**The error function**) The error function, $\mathrm{Erf}(x)$, is defined for $x \geq 0$ by

$$\mathrm{Erf}(x) = \frac{2}{\sqrt{\pi}} \int_0^x e^{-t^2}\, dt.$$

Show that $\left(\mathrm{Erf}(x)\right)^2 = \dfrac{4}{\pi} \displaystyle\int_0^{\pi/4} \left(1 - e^{-x^2/\cos^2\theta}\right) d\theta$.

Hence deduce that $\mathrm{Erf}(x) \geq \sqrt{1 - e^{-x^2}}$.

38. (**The gamma and beta functions**) The gamma function $\Gamma(x)$ and the beta function $B(x, y)$ are defined by

$$\Gamma(x) = \int_0^\infty t^{x-1} e^{-t}\, dt, \quad (x > 0),$$

$$B(x, y) = \int_0^1 t^{x-1}(1 - t)^{y-1}\, dt, \quad (x > 0,\ y > 0).$$

The gamma function satisfies

$$\Gamma(x + 1) = x\Gamma(x) \qquad \text{and}$$
$$\Gamma(n + 1) = n!, \qquad (n = 0, 1, 2, \ldots).$$

Deduce the following further properties of these functions:

    (a) $\Gamma(x) = 2 \displaystyle\int_0^\infty s^{2x-1} e^{-s^2}\, ds, \qquad (x > 0)$,

    (b) $\Gamma\left(\dfrac{1}{2}\right) = \sqrt{\pi}, \qquad \Gamma\left(\dfrac{3}{2}\right) = \dfrac{1}{2}\sqrt{\pi}$,

    (c) If $x > 0$ and $y > 0$, then

$$B(x, y) = 2 \int_0^{\pi/2} \cos^{2x-1}\theta \sin^{2y-1}\theta\, d\theta,$$

    (d) $B(x, y) = \dfrac{\Gamma(x)\Gamma(y)}{\Gamma(x + y)}$.

14.5 Triple Integrals

Now that we have seen how to extend definite integration to two-dimensional domains, the extension to three (or more) dimensions is straightforward. For a bounded function $f(x, y, z)$ defined on a rectangular box B ($x_0 \leq x \leq x_1$, $y_0 \leq y \leq y_1$, $z_0 \leq z \leq z_1$), the **triple integral** of f over B,

$$\iiint_B f(x, y, z)\, dV \qquad \text{or} \qquad \iiint_B f(x, y, z)\, dx\, dy\, dz,$$

can be defined as a suitable limit of Riemann sums corresponding to partitions of B into subboxes by planes parallel to each of the coordinate planes. We omit the details. Triple integrals over more general domains are defined by extending the function to be zero outside the domain and integrating over a rectangular box containing the domain.

All the properties of double integrals mentioned in Section 14.1 have analogues for triple integrals. In particular, a continuous function is integrable over a closed, bounded domain. If $f(x, y, z) = 1$ on the domain D, then the triple integral gives the volume of D:

$$\text{Volume of } D = \iiint_D dV.$$

The triple integral of a positive function $f(x, y, z)$ can be interpreted as the "hypervolume" (i.e., the four-dimensional volume) of a region in 4-space having the set D as its three-dimensional "base" and having its top on the hypersurface $w = f(x, y, z)$. This is not a particularly useful interpretation; many more useful ones arise in applications. For instance, if $\delta(x, y, z)$ represents the density (mass per unit volume) at position (x, y, z) in a substance occupying the domain D in 3-space, then the mass m of the solid is the "sum" of mass elements $dm = \delta(x, y, z)\, dV$ occupying volume elements dV:

$$\text{mass} = \iiint_D \delta(x, y, z)\, dV.$$

Some triple integrals can be evaluated by inspection, using symmetry and known volumes.

Example 1   Evaluate

$$\iiint_{x^2+y^2+z^2 \leq a^2} (2 + x - \sin z)\, dV.$$

Solution   The domain of integration is the ball of radius a centred at the origin. The integral of 2 over this ball is twice the ball's volume, that is, $8\pi a^3/3$. The integrals of x and $\sin z$ over the ball are both zero, since both functions are odd in one of the variables and the domain is symmetric about each coordinate plane. (For instance, for every volume element dV in the half of the ball where $x > 0$, there is a corresponding element in the other half where x has the same size but the opposite sign. The contributions from these two elements cancel one another.) Thus,

$$\iiint_{x^2+y^2+z^2 \leq a^2} (2 + x - \sin z)\, dV = \frac{8}{3}\pi a^3 + 0 + 0 = \frac{8}{3}\pi a^3.$$

Most triple integrals are evaluated by an iteration procedure similar to that used for double integrals. We slice the domain D with a plane parallel to one of the coordinate planes, double integrate the function with respect to two variables over that slice, and then integrate the result with respect to the remaining variable. Some examples should clarify the procedure.

Example 2   Let B be the rectangular box $0 \leq x \leq a$, $0 \leq y \leq b$, $0 \leq z \leq c$. Evaluate

$$I = \iiint_B (xy^2 + z^3)\, dV.$$

Solution   As indicated in Figure 14.35(a), we will slice with planes perpendicular to the z-axis, so the z integral will be outermost in the iteration. The slices are rectangles, so the double integrals over them can be immediately iterated also. We do it with the y integral outer and the x integral inner, as suggested by the line shown in the slice.

$$I = \int_0^c dz \int_0^b dy \int_0^a (xy^2 + z^3)\, dx$$

$$= \int_0^c dz \int_0^b dy \left(\frac{x^2 y^2}{2} + xz^3\right)\Big|_{x=0}^{x=a}$$

$$= \int_0^c dz \int_0^b \left(\frac{a^2 y^2}{2} + az^3\right) dy$$

$$= \int_0^c dz \left(\frac{a^2 y^3}{6} + ayz^3\right)\Big|_{y=0}^{y=b}$$

$$= \int_0^c \left(\frac{a^2 b^3}{6} + abz^3 \right) dz$$

$$= \left(\frac{a^2 b^3 z}{6} + \frac{abz^4}{4} \right) \Bigg|_{z=0}^{z=c} = \frac{a^2 b^3 c}{6} + \frac{abc^4}{4}.$$

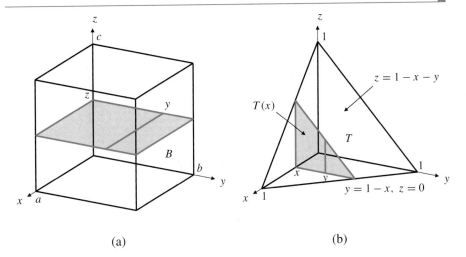

Figure 14.35

(a) The iteration in Example 2

(b) The iteration in Example 3

(a)   (b)

Example 3   If T is the tetrahedron with vertices $(0, 0, 0)$, $(1, 0, 0)$, $(0, 1, 0)$, and $(0, 0, 1)$, evaluate

$$I = \iiint_T y \, dV.$$

Solution   The tetrahedron is shown in Figure 14.35(b). The plane slice in the plane normal to the x-axis at position x is the triangle $T(x)$ shown in that figure; x is constant and y and z are variables in the slice. The double integral of y over $T(x)$ is a function of x. We evaluate it by integrating first in the z direction and then in the y direction as suggested by the vertical line shown in the slice:

$$\iint_{T(x)} y \, dA = \int_0^{1-x} dy \int_0^{1-x-y} y \, dz$$

$$= \int_0^{1-x} y(1 - x - y) \, dy$$

$$= \left((1 - x) \frac{y^2}{2} - \frac{y^3}{3} \right) \Bigg|_0^{1-x} = \frac{1}{6} (1 - x)^3.$$

The value of the triple integral I is the integral of this expression with respect to the remaining variable x, to sum the contributions from all such slices between $x = 0$ and $x = 1$:

$$I = \int_0^1 \frac{1}{6} (1 - x)^3 \, dx = -\frac{1}{24} (1 - x)^4 \Bigg|_0^1 = \frac{1}{24}.$$

In the above solution we carried out the iteration in two steps in order to show the procedure clearly. In practice, triple integrals are iterated in one step, with no explicit mention made of the double integral over the slice. Thus, using the iteration suggested by Figure 14.35(b), we would immediately write

$$I = \int_0^1 dx \int_0^{1-x} dy \int_0^{1-x-y} y \, dz.$$

The evaluation proceeds as above, starting with the right (i.e., inner) integral, followed by the middle integral and then the left (outer) integral. The triple integral represents the "sum" of elements $y\,dV$ over the three-dimensional region T. The above iteration corresponds to "summing" (i.e., integrating) first along a vertical line (the z integral), then summing these one-dimensional sums in the y direction to get the double sum of all elements in the plane slice, and finally summing these double sums in the x direction to add up the contributions from all the slices. The iteration can be carried out in other directions; there are six possible iterations corresponding to different orders of doing the x, y, and z integrals. The other five are

$$I = \int_0^1 dx \int_0^{1-x} dz \int_0^{1-x-z} y\,dy,$$

$$I = \int_0^1 dy \int_0^{1-y} dx \int_0^{1-x-y} y\,dz,$$

$$I = \int_0^1 dy \int_0^{1-y} dz \int_0^{1-y-z} y\,dx,$$

$$I = \int_0^1 dz \int_0^{1-z} dx \int_0^{1-x-z} y\,dy,$$

$$I = \int_0^1 dz \int_0^{1-z} dy \int_0^{1-y-z} y\,dx.$$

You should verify these by drawing diagrams analogous to Figure 14.35(b). Of course, all six iterations give the same result.

It is sometimes difficult to visualize the region of 3-space over which a given triple integral is taken. In such situations try to determine the *projection* of that region on one or other of the coordinate planes. For instance, if a region R is bounded by two surfaces with given equations, combining these equations to eliminate one variable will yield the equation of a cylinder (not necessarily circular) with axis parallel to the axis of the eliminated variable. This cylinder will then determine the projection of R onto the coordinate plane perpendicular to that axis. The following example illustrates the use of this technique to find a volume bounded by two surfaces. The volume is expressed as a triple integral with unit integrand.

Example 4   Find the volume of the region R lying below the plane $z = 3 - 2y$ and above the paraboloid $z = x^2 + y^2$.

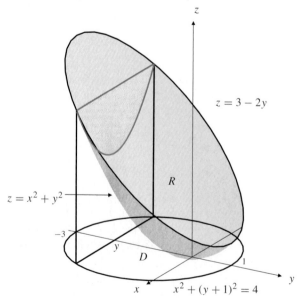

Figure 14.36 The volume above a paraboloid and under a slanting plane

Solution   The region R is shown in Figure 14.36. The two surfaces bounding R intersect on the vertical cylinder $x^2 + y^2 = 3 - 2y$, or $x^2 + (y+1)^2 = 4$. If D is the circular disk in which this cylinder intersects the xy-plane, then partial iteration gives

$$V = \iiint_R dV = \iint_D dx\,dy \int_{x^2+y^2}^{3-2y} dz.$$

Figure 14.36 shows a slice of R corresponding to a further iteration of the double integral over D:

$$V = \int_{-3}^{1} dy \int_{-\sqrt{3-2y-y^2}}^{\sqrt{3-2y-y^2}} dx \int_{x^2+y^2}^{3-2y} dz,$$

but there is an easier way to iterate the double integral. Since D is a circular disk of radius 2 and centre $(0, -1)$, we can use polar coordinates with centre at that point (i.e., $x = r\cos\theta$, $y = -1 + r\sin\theta$). Thus,

$$\begin{aligned}
V &= \iint_D (3 - 2y - x^2 - y^2)\,dx\,dy \\
&= \iint_D \left(4 - x^2 - (y+1)^2\right) dx\,dy \\
&= \int_0^{2\pi} d\theta \int_0^2 (4 - r^2) r\,dr = 2\pi \left(2r^2 - \frac{r^4}{4} \right)\Big|_0^2 = 8\pi \text{ cubic units.}
\end{aligned}$$

As was the case for double integrals, it is sometimes necessary to reiterate a given iterated integral so that the integrations are performed in a different order. This task is most easily accomplished if we can translate the given iteration into a sketch of the region of integration. The ability to deduce the shape of the region from the limits in the iterated integral is a skill that you can acquire with a little practice. You should first determine the projection of the region on a coordinate plane, namely, the plane of the two variables in the outer integrals of the given iteration.

It is also possible to reiterate an iterated integral in a different order by manipulating the limits of integration algebraically. We will illustrate both approaches (graphical and algebraic) in the following examples.

Example 5   Express the iterated integral

$$I = \int_0^1 dy \int_y^1 dz \int_0^z f(x, y, z)\,dx$$

as a triple integral, and sketch the region over which it is taken. Reiterate the integral in such a way that the integrations are performed in the order: first y, then z, then x (i.e., the opposite order to the given iteration).

Solution   We express I as an uniterated triple integral:

$$I = \iiint_R f(x, y, z)\,dV.$$

The outer integral in the given iteration shows that the region R lies between the planes $y = 0$ and $y = 1$. For each such value of y, z must lie between y and 1. Therefore, R lies below the plane $z = 1$ and above the plane $z = y$, and the projection of R onto the yz-plane is the triangle with vertices $(0, 0, 0)$, $(0, 0, 1)$, and $(0, 1, 1)$. Through any point $(0, y, z)$ in this triangle, a line parallel to the x-axis intersects R between $x = 0$ and $x = z$. Thus, the solid is bounded by the five planes $x = 0$, $y = 0$, $z = 1$, $y = z$, and $z = x$. It is sketched in Figure 14.37(a), with slice and line corresponding to the given iteration.

Figure 14.37

(a) The solid region for the triple integral in Example 5 sliced corresponding to the given iteration

(b) The same solid sliced to conform to the desired iteration

(a)   (b)

The required iteration corresponds to the slice and line shown in Figure 14.37(b). Therefore, it is

$$I = \int_0^1 dx \int_x^1 dz \int_0^z f(x, y, z)\, dy.$$

Example 6   Use algebra to write an iteration of the integral

$$I = \int_0^1 dx \int_x^1 dy \int_x^y f(x, y, z)\, dz$$

with the order of integrations reversed.

Solution   From the given iteration we can write three sets of inequalities satisfied by the outer variable x, the middle variable y, and the inner variable z. We write these in order as follows:

$$0 \leq x \leq 1 \qquad \text{inequalities for } x$$
$$x \leq y \leq 1 \qquad \text{inequalities for } y$$
$$x \leq z \leq y \qquad \text{inequalities for } z.$$

Note that the limits for each variable can be constant or can depend only on variables whose inequalities are on lines above the line for that variable. (In this case, the limits for x must both be constant, those for y can depend on x, and those for z can depend on both x and y.) This is a requirement for iterated integrals; outer integrals cannot depend on the variables of integration of the inner integrals.

We want to construct an equivalent set of inequalities with those for z on the top line, then those for y, then those for x on the bottom line. The limits for z must be constants. From the inequalities above we determine that $0 \leq x \leq z$ and $z \leq y \leq 1$. Thus z must satisfy $0 \leq z \leq 1$. The inequalities for y can depend on z. Since $z \leq y$ and $y \leq 1$, we have $z \leq y \leq 1$. Finally, the limits for x can depend on both y and z. We have $0 \leq x$, $x \leq y$, and $x \leq z$. Since we have already determined that $z \leq y$, we must have $0 \leq x \leq z$. Thus, the revised inequalities are

$$0 \leq z \leq 1 \qquad \text{inequalities for } z$$
$$z \leq y \leq 1 \qquad \text{inequalities for } y$$
$$0 \leq x \leq z \qquad \text{inequalities for } x$$

and the required iteration is

$$I = \int_0^1 dz \int_z^1 dy \int_0^z f(x, y, z)\, dx.$$

▮▮▮▮▮▮ Exercises 14.5 ▮▮▮▮▮▮▮▮

In Exercises 1–12, evaluate the triple integrals over the indicated region. Be alert for simplifications and auspicious orders of iteration.

1. $\iiint_R (1 + 2x - 3y)\,dV$, over the box $-a \le x \le a$, $-b \le y \le b$, $-c \le z \le c$

2. $\iiint_B xyz\,dV$, over the box B given by $0 \le x \le 1$, $-2 \le y \le 0$, $1 \le z \le 4$

3. $\iiint_D (3 + 2xy)\,dV$, over the solid hemispherical dome D given by $x^2 + y^2 + z^2 \le 4$ and $z \ge 0$

4. $\iiint_R x\,dV$, over the tetrahedron bounded by the coordinate planes and the plane $\dfrac{x}{a} + \dfrac{y}{b} + \dfrac{z}{c} = 1$

5. $\iiint_R (x^2 + y^2)\,dV$, over the cube $0 \le x, y, z \le 1$

6. $\iiint_R (x^2 + y^2 + z^2)\,dV$, over the cube of Exercise 5

7. $\iiint_R (xy + z^2)\,dV$, over the set $0 \le z \le 1 - |x| - |y|$

8. $\iiint_R yz^2 e^{-xyz}\,dV$, over the cube $0 \le x, y, z \le 1$

9. $\iiint_R \sin(\pi y^3)\,dV$, over the pyramid with vertices $(0, 0, 0)$, $(0, 1, 0)$, $(1, 1, 0)$, $(1, 1, 1)$, and $(0, 1, 1)$

10. $\iiint_R y\,dV$, over that part of the cube $0 \le x, y, z \le 1$ lying above the plane $y + z = 1$ and below the plane $x + y + z = 2$

11. $\iiint_R \dfrac{1}{(x + y + z)^3}\,dV$, over the region bounded by the six planes $z = 1$, $z = 2$, $y = 0$, $y = z$, $x = 0$, and $x = y + z$

12. $\iiint_R \cos x \cos y \cos z\,dV$, over the tetrahedron defined by $x \ge 0$, $y \ge 0$, $z \ge 0$, and $x + y + z \le \pi$

13. Evaluate $\iiint_{\mathbb{R}^3} e^{-x^2 - 2y^2 - 3z^2}\,dV$. *Hint:* Use the result of Example 4 of Section 14.4.

14. Find the volume of the region lying inside the cylinder $x^2 + 4y^2 = 4$, above the xy-plane, and below the plane $z = 2 + x$.

15. Find $\iiint_T x\,dV$, where T is the tetrahedron bounded by the

planes $x = 1$, $y = 1$, $z = 1$, and $x + y + z = 2$.

16. Sketch the region R in the first octant of 3-space that has finite volume and is bounded by the surfaces $x = 0$, $z = 0$, $x + y = 1$, and $z = y^2$. Write six different iterations of the triple integral of $f(x, y, z)$ over R.

In Exercises 17–20, express the given iterated integral as a triple integral and sketch the region over which it is taken. Reiterate the integral, so that the outermost integral is with respect to x and the innermost is with respect to z.

∗17. $\displaystyle\int_0^1 dz \int_0^{1-z} dy \int_0^1 f(x, y, z)\,dx$

∗18. $\displaystyle\int_0^1 dz \int_z^1 dy \int_0^y f(x, y, z)\,dx$

∗19. $\displaystyle\int_0^1 dz \int_z^1 dx \int_0^{x-z} f(x, y, z)\,dy$

∗20. $\displaystyle\int_0^1 dy \int_0^{\sqrt{1-y^2}} dz \int_{y^2+z^2}^1 f(x, y, z)\,dx$

21. Repeat Exercise 17 using the method of Example 6.

22. Repeat Exercise 18 using the method of Example 6.

23. Repeat Exercise 19 using the method of Example 6.

24. Repeat Exercise 20 using the method of Example 6.

25. Rework Example 5 using the method of Example 6.

26. Rework Example 6 using the method of Example 5.

In Exercises 27–28, evaluate the given iterated integral by reiterating it in a different order. (You will need to make a good sketch of the region.)

∗27. $\displaystyle\int_0^1 dz \int_z^1 dx \int_0^x e^{x^3}\,dy$

∗28. $\displaystyle\int_0^1 dx \int_0^{1-x} dy \int_y^1 \dfrac{\sin(\pi z)}{z(2 - z)}\,dz$

29. Define the average value of an integrable function $f(x, y, z)$ over a region R of 3-space. Find the average value of $x^2 + y^2 + z^2$ over the cube $0 \le x \le 1$, $0 \le y \le 1$, $0 \le z \le 1$.

∗30. State a Mean-Value Theorem for triple integrals analogous to Theorem 3 of Section 14.3. Use it to prove that if $f(x, y, z)$ is continuous near the point (a, b, c) and if $B_\epsilon(a, b, c)$ is the ball of radius ϵ centred at (a, b, c), then

$$\lim_{\epsilon \to 0} \frac{3}{4\pi\epsilon^3} \iiint_{B_\epsilon(a,b,c)} f(x, y, z)\,dV = f(a, b, c).$$

14.6 Change of Variables in Triple Integrals

The change of variables formula for a double integral extends to triple (and higher-order) integrals. Consider the transformation

$$x = x(u, v, w),$$
$$y = y(u, v, w),$$
$$z = z(u, v, w),$$

where x, y, and z have continuous first partial derivatives with respect to u, v, and w. Near any point where the Jacobian $\partial(x, y, z)/\partial(u, v, w)$ is nonzero, the transformation scales volume elements according to the formula

$$dV = dx\,dy\,dz = \left|\frac{\partial(x, y, z)}{\partial(u, v, w)}\right| du\,dv\,dw.$$

Thus, if the transformation is one-to-one from a domain S in uvw-space onto a domain D in xyz-space, and if

$$g(u, v, w) = f\big(x(u, v, w), y(u, v, w), z(u, v, w)\big),$$

then

$$\iiint_D f(x, y, z)\,dx\,dy\,dz = \iiint_S g(u, v, w) \left|\frac{\partial(x, y, z)}{\partial(u, v, w)}\right| du\,dv\,dw.$$

The proof is similar to that of the two-dimensional case given in Section 14.4. See Exercise 35 below.

Example 1 Under the change of variables $x = au$, $y = bv$, $z = cw$, where $a, b, c > 0$, the solid ellipsoid E given by

$$\frac{x^2}{a^2} + \frac{y^2}{b^2} + \frac{z^2}{c^2} \leq 1$$

becomes the ball B given by $u^2 + v^2 + w^2 \leq 1$. The Jacobian of this transformation is

$$\frac{\partial(x, y, z)}{\partial(u, v, w)} = \begin{vmatrix} a & 0 & 0 \\ 0 & b & 0 \\ 0 & 0 & c \end{vmatrix} = abc,$$

so the volume of the ellipsoid is given by

$$\begin{aligned} \text{Volume of } E &= \iiint_E dx\,dy\,dz \\ &= \iiint_B abc\,du\,dv\,dw = abc \times (\text{Volume of } B) \\ &= \frac{4}{3}\pi abc \text{ cubic units.} \end{aligned}$$

Cylindrical Coordinates

Among the most useful alternatives to Cartesian coordinates in 3-space are two coordinate systems that generalize plane polar coordinates. The simpler of these is the system of **cylindrical coordinates**. It uses ordinary plane polar coordinates in the xy-plane while retaining the Cartesian z coordinate for measuring vertical distances. Thus, each point in 3-space has cylindrical coordinates $[r, \theta, z]$ related to its Cartesian coordinates (x, y, z) by the transformation

$$x = r\cos\theta, \quad y = r\sin\theta, \quad z = z.$$

Figure 14.38 shows how a point P is located by its cylindrical coordinates $[r, \theta, z]$ as well as by its Cartesian coordinates (x, y, z). Note that the distance from the origin to P is

$$d = \sqrt{r^2 + z^2} = \sqrt{x^2 + y^2 + z^2}.$$

Example 2 The point with Cartesian coordinates $(1, 1, 1)$ has cylindrical coordinates $[\sqrt{2}, \pi/4, 1]$. The point with Cartesian coordinates $(0, 2, -3)$ has cylindrical coordinates $[2, \pi/2, -3]$. The point with cylindrical coordinates $[4, -\pi/3, 5]$ has Cartesian coordinates $(2, -2\sqrt{3}, 5)$.

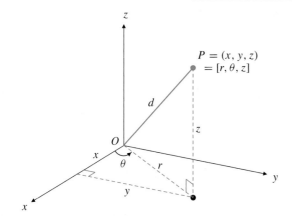

Figure 14.38 The cylindrical coordinates of a point

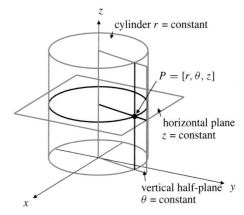

Figure 14.39 The coordinate surfaces for cylindrical coordinates

The coordinate surfaces in cylindrical coordinates are the r-surfaces (vertical circular cylinders centred on the z-axis), the θ-surfaces (vertical half-planes with edge along the z-axis), and the z-surfaces (horizontal planes). (See Figure 14.39.) Cylindrical coordinates lend themselves to representing domains that are bounded by such surfaces and, in general, to problems with axial symmetry (around the z-axis).

The **volume element in cylindrical coordinates** is

$$dV = r\, dr\, d\theta\, dz,$$

which is easily seen by examining the infinitesimal "box" bounded by the coordinate surfaces corresponding to values r, $r + dr$, θ, $\theta + d\theta$, z, and $z + dz$ (see Figure 14.40) or by calculating the Jacobian

$$\frac{\partial(x, y, z)}{\partial(r, \theta, z)} = \begin{vmatrix} \cos\theta & -r\sin\theta & 0 \\ \sin\theta & r\cos\theta & 0 \\ 0 & 0 & 1 \end{vmatrix} = r.$$

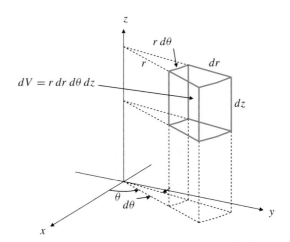

Figure 14.40 The volume element in cylindrical coordinates

Example 3 Evaluate $\iiint_D (x^2 + y^2)\, dV$ over the first octant region bounded by the cylinders $x^2 + y^2 = 1$ and $x^2 + y^2 = 4$ and the planes $z = 0$, $z = 1$, $x = 0$, and $x = y$.

Figure 14.41

Solution   In terms of cylindrical coordinates the region is bounded by $r = 1, r = 2$, $\theta = \pi/4, \theta = \pi/2, z = 0$, and $z = 1$. (See Figure 14.41. It is a rectangular coordinate box in $r\theta z$-space.) Since the integrand is $x^2 + y^2 = r^2$, the integral is

$$\iiint_D (x^2 + y^2)\, dV = \int_0^1 dz \int_{\pi/4}^{\pi/2} d\theta \int_1^2 r^2\, r\, dr$$

$$= (1 - 0)\left(\frac{\pi}{2} - \frac{\pi}{4}\right)\left(\frac{2^4}{4} - \frac{1^4}{4}\right) = \frac{15}{16}\pi.$$

This integral would have been much more difficult to evaluate using Cartesian coordinates.

Example 4   Use a triple integral to find the volume of the solid region inside the sphere $x^2 + y^2 + z^2 = 6$ and above the paraboloid $z = x^2 + y^2$.

Solution   One-quarter of the required volume lies in the first octant. (See region R in Figure 14.42.) The two surfaces intersect on the vertical cylinder

$$6 - x^2 - y^2 = z^2 = (x^2 + y^2)^2,$$

or, in terms of cylindrical coordinates, $6 - r^2 = r^4$, that is,

$$r^4 + r^2 - 6 = 0$$
$$(r^2 + 3)(r^2 - 2) = 0.$$

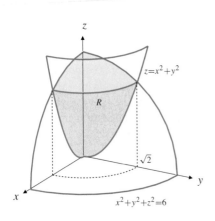

Figure 14.42

The only relevant solution to this equation is $r = \sqrt{2}$. Thus, the required volume lies above the disk of radius $\sqrt{2}$ centred at the origin in the xy-plane. The total volume V of the region is

$$V = \iiint_R dV = \int_0^{2\pi} d\theta \int_0^{\sqrt{2}} r\, dr \int_{r^2}^{\sqrt{6-r^2}} dz$$

$$= 2\pi \int_0^{\sqrt{2}} \left(r\sqrt{6 - r^2} - r^3\right) dr$$

$$= 2\pi \left[-\frac{1}{3}(6 - r^2)^{3/2} - \frac{r^4}{4}\right]\Bigg|_0^{\sqrt{2}}$$

$$= 2\pi \left[\frac{6\sqrt{6}}{3} - \frac{8}{3} - 1\right] = \frac{2\pi}{3}(6\sqrt{6} - 11) \text{ cubic units.}$$

Spherical Coordinates

In the system of **spherical coordinates** a point P in 3-space is represented by the ordered triple $[\rho, \phi, \theta]$, where ρ (Greek "rho") is the distance from P to the origin O, ϕ (Greek "phi") is the angle the radial line OP makes with the positive direction of the z-axis, and θ is the angle between the plane containing P and the z-axis and the xz-plane. (See Figure 14.43.) It is conventional to consider spherical coordinates restricted in such a way that $\rho \geq 0, 0 \leq \phi \leq \pi$, and $0 \leq \theta < 2\pi$ (or $-\pi < \theta \leq \pi$). Every point not on the z-axis then has exactly one spherical coordinate representation, and the transformation from Cartesian coordinates (x, y, z) to spherical coordinates $[\rho, \phi, \theta]$ is one-to-one off the z-axis. Using the right-angled triangles in the figure, we can see that this transformation is given by:

$$x = \rho \sin \phi \cos \theta$$
$$y = \rho \sin \phi \sin \theta$$
$$z = \rho \cos \phi.$$

Observe that

$$\rho^2 = x^2 + y^2 + z^2 = r^2 + z^2$$

and that the r coordinate in cylindrical coordinates is related to ρ and ϕ by

$$r = \sqrt{x^2 + y^2} = \rho \sin \phi.$$

Thus, also

$$\tan \phi = \frac{r}{z} = \frac{\sqrt{x^2 + y^2}}{z} \qquad \text{and} \qquad \tan \theta = \frac{y}{x}.$$

If $\phi = 0$ or $\phi = \pi$, then $r = 0$, so the θ coordinate is irrelevant at points on the z-axis.

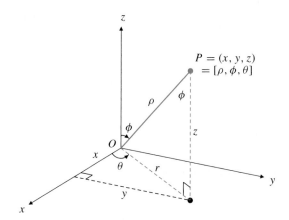

Figure 14.43   The spherical coordinates of a point

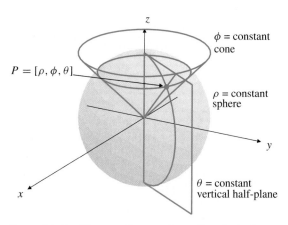

Figure 14.44   The coordinate surfaces for spherical coordinates

Some coordinate surfaces for spherical coordinates are shown in Figure 14.44. The ρ-surfaces ($\rho = $ constant) are spheres centred at the origin; the ϕ-surfaces ($\phi = $ constant) are nappes of circular cones with the z-axis as axis; the θ-surfaces ($\theta = $ constant) are vertical half-planes with edge along the z-axis. If we take a coordinate system with origin at the centre of the earth, z-axis through the north pole, and x-axis through the intersection of the Greenwich meridian and the equator, then the intersections of the surface of the earth with the ϕ-surfaces are the *parallels of latitude*, and the intersections with the θ-surfaces are the *meridians of longitude*. Since latitude is measured from $90°$ at the north pole to $-90°$ at the south pole, while ϕ is measured from 0 at the north pole to π ($= 180°$) at the south pole, the coordinate ϕ is frequently referred to as the **colatitude** coordinate; θ is the **longitude** coordinate. Observe that θ has the same significance in spherical coordinates as it does in cylindrical coordinates.

Example 5   Find:

(a) the Cartesian coordinates of the point P with spherical coordinates $[2, \pi/3, \pi/2]$ and

(b) the spherical coordinates of the point Q with Cartesian coordinates $(1, 1, \sqrt{2})$.

Solution

(a) If $\rho = 2$, $\phi = \pi/3$, and $\theta = \pi/2$, then

$$x = 2\sin(\pi/3)\cos(\pi/2) = 0$$
$$y = 2\sin(\pi/3)\sin(\pi/2) = \sqrt{3}$$
$$z = 2\cos(\pi/3) = 1.$$

The Cartesian coordinates of P are $(0, \sqrt{3}, 1)$.

(b) Given that

$$\rho\sin\phi\cos\theta = x = 1$$
$$\rho\sin\phi\sin\theta = y = 1$$
$$\rho\cos\phi = z = \sqrt{2},$$

we calculate that $\rho^2 = 1 + 1 + 2 = 4$, so $\rho = 2$. Also $r^2 = 1 + 1 = 2$, so $r = \sqrt{2}$. Thus $\tan\phi = r/z = 1$, so $\phi = \pi/4$. Also, $\tan\theta = y/x = 1$, so $\theta = \pi/4$ or $5\pi/4$. Since $x > 0$, we must have $\theta = \pi/4$. The spherical coordinates of Q are $[2, \pi/4, \pi/4]$. ∎

Remark You may wonder why we write spherical coordinates in the order ρ, ϕ, θ rather than ρ, θ, ϕ. The reason, which will not become apparent until Chapter 16, is so that the triad of unit vectors at any point P pointing in the directions of increasing ρ, increasing ϕ, and increasing θ form a *right-handed basis* rather than a left-handed one.

The **volume element in spherical coordinates** is

$$dV = \rho^2 \sin\phi\, d\rho\, d\phi\, d\theta.$$

To see this, observe that the infinitesimal coordinate box bounded by the coordinate surfaces corresponding to values ρ, $\rho + d\rho$, ϕ, $\phi + d\phi$, θ, and $\theta + d\theta$ has dimensions $d\rho$, $\rho\, d\phi$, and $\rho\sin\phi\, d\theta$. (See Figure 14.45.) Alternatively, the Jacobian of the transformation can be calculated:

$$\frac{\partial(x, y, z)}{\partial(\rho, \phi, \theta)} = \begin{vmatrix} \sin\phi\cos\theta & \rho\cos\phi\cos\theta & -\rho\sin\phi\sin\theta \\ \sin\phi\sin\theta & \rho\cos\phi\sin\theta & \rho\sin\phi\cos\theta \\ \cos\phi & -\rho\sin\phi & 0 \end{vmatrix}$$

$$= \cos\phi \begin{vmatrix} \rho\cos\phi\cos\theta & -\rho\sin\phi\sin\theta \\ \rho\cos\phi\sin\theta & \rho\sin\phi\cos\theta \end{vmatrix}$$

$$+ \rho\sin\phi \begin{vmatrix} \sin\phi\cos\theta & -\rho\sin\phi\sin\theta \\ \sin\phi\sin\theta & \rho\sin\phi\cos\theta \end{vmatrix}$$

$$= \cos\phi(\rho^2\sin\phi\cos\phi) + \rho\sin\phi(\rho\sin^2\phi)$$

$$= \rho^2\sin\phi.$$

Spherical coordinates are suited to problems involving spherical symmetry and, in particular, to regions bounded by spheres centred at the origin, circular cones with axes along the z-axis, and vertical planes containing the z-axis.

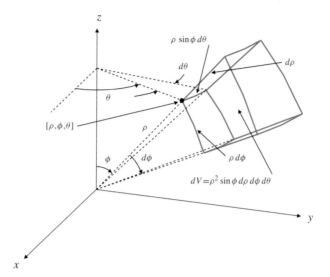

Figure 14.45   The volume element in spherical coordinates

Example 6   A solid half-ball H of radius a has density depending on the distance ρ from the centre of the base disk. The density is given by $k(2a - \rho)$, where k is a constant. Find the mass of the half-ball.

Solution   Choosing coordinates with origin at the centre of the base, and so that the half-ball lies above the xy-plane, we calculate the mass m as follows:

$$
\begin{aligned}
m &= \iiint_H k(2a - \rho)\, dV = \iiint_H k(2a - \rho)\, \rho^2 \sin\phi\, d\rho\, d\phi\, d\theta \\
&= k \int_0^{2\pi} d\theta \int_0^{\pi/2} \sin\phi\, d\phi \int_0^a (2a - \rho)\, \rho^2\, d\rho \\
&= 2k\pi \times 1 \times \left(\frac{2a}{3}\rho^3 - \frac{1}{4}\rho^4 \right)\bigg|_0^a = \frac{5}{6}\pi ka^4 \text{ units.}
\end{aligned}
$$

∎

Remark   In the above example both the integrand and the region of integration exhibited spherical symmetry, so the choice of spherical coordinates to carry out the integration was most appropriate. The mass could have been evaluated in cylindrical coordinates. The iteration in that system is

$$
m = \int_0^{2\pi} d\theta \int_0^a r\, dr \int_0^{\sqrt{a^2 - r^2}} k\left(2a - \sqrt{r^2 + z^2} \right) dz
$$

and is difficult to evaluate. It is even more difficult in Cartesian coordinates:

$$
m = 4 \int_0^a dx \int_0^{\sqrt{a^2 - x^2}} dy \int_0^{\sqrt{a^2 - x^2 - y^2}} k\left(2a - \sqrt{x^2 + y^2 + z^2} \right) dz.
$$

The choice of coordinate system can greatly affect the difficulty of computation of a multiple integral.

Many problems will have elements of spherical and axial symmetry. In such cases it may not be clear whether it would be better to use spherical or cylindrical coordinates. In such doubtful cases the integrand is usually the best guide. Use cylindrical or spherical coordinates according to whether the integrand involves $x^2 + y^2$ or $x^2 + y^2 + z^2$.

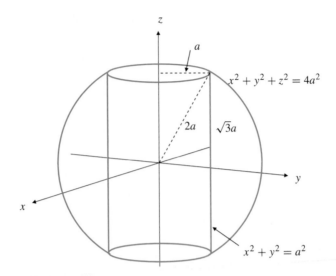

Figure 14.46 A solid ball with a cylindrical hole through it

Example 7 The moment of inertia about the z-axis of a solid of density δ occupying the region R is given by the integral

$$I = \iiint_R (x^2 + y^2)\delta \, dV.$$

(See Section 14.7.) Calculate that moment of inertia for a solid of unit density occupying the region inside the sphere $x^2 + y^2 + z^2 = 4a^2$ and outside the cylinder $x^2 + y^2 = a^2$.

Solution See Figure 14.46. In terms of spherical coordinates the required moment of inertia is

$$I = 2 \int_0^{2\pi} d\theta \int_{\pi/6}^{\pi/2} \sin\phi \, d\phi \int_{a/\sin\phi}^{2a} \rho^2 \sin^2\phi \, \rho^2 \, d\rho.$$

In terms of cylindrical coordinates it is

$$I = 2 \int_0^{2\pi} d\theta \int_a^{2a} r \, dr \int_0^{\sqrt{4a^2-r^2}} r^2 \, dz.$$

The latter formula looks somewhat easier to evaluate. We continue with it. Evaluating the θ and z integrals, we get

$$I = 4\pi \int_a^{2a} r^3 \sqrt{4a^2 - r^2} \, dr.$$

Making the substitution $u = 4a^2 - r^2$, $du = -2r \, dr$, we obtain

$$I = 2\pi \int_0^{3a^2} (4a^2 - u)\sqrt{u} \, du = 2\pi \left(4a^2 \frac{u^{3/2}}{3/2} - \frac{u^{5/2}}{5/2} \right)\Big|_0^{3a^2} = \frac{44}{5}\sqrt{3}\pi a^5.$$

░░░░░░░░░░ **Exercises 14.6** ░░░░░░░░░░

1. Convert the spherical coordinates $[4, \pi/3, 2\pi/3]$ to Cartesian coordinates and to cylindrical coordinates.

2. Convert the Cartesian coordinates $(2, -2, 1)$ to cylindrical coordinates and to spherical coordinates.

3. Convert the cylindrical coordinates $[2, \pi/6, -2]$ to Cartesian coordinates and to spherical coordinates.

4. A point P has spherical coordinates $[1, \phi, \theta]$ and cylindrical coordinates $[r, \pi/4, r]$. Find the Cartesian coordinates of the point.

Describe the sets of points in 3-space that satisfy the equations in Exercises 5–14. Here, r, θ, ρ, and ϕ denote the appropriate cylindrical or spherical coordinates.

5. $\theta = \pi/2$ **6.** $\phi = 2\pi/3$

7. $\phi = \pi/2$ **8.** $\rho = 4$

9. $r = 4$ **10.** $\rho = z$

11. $\rho = r$ **12.** $\rho = 2x$

13. $\rho = 2\cos\phi$ **14.** $r = 2\cos\theta$

In Exercises 15–23, find the volumes of the indicated regions.

15. Inside the cone $z = \sqrt{x^2 + y^2}$ and inside the sphere $x^2 + y^2 + z^2 = a^2$

16. Above the surface $z = (x^2 + y^2)^{1/4}$ and inside the sphere $x^2 + y^2 + z^2 = 2$

17. Between the paraboloids $z = 10 - x^2 - y^2$ and $z = 2(x^2 + y^2 - 1)$

18. Inside the paraboloid $z = x^2 + y^2$ and inside the sphere $x^2 + y^2 + z^2 = 12$

19. Above the xy-plane, inside the cone $z = 2a - \sqrt{x^2 + y^2}$, and inside the cylinder $x^2 + y^2 = 2ay$

20. Above the xy-plane, under the paraboloid $z = 1 - x^2 - y^2$, and in the wedge $-x \le y \le \sqrt{3}x$

21. In the first octant, between the planes $y = 0$ and $y = x$, and inside the ellipsoid $\dfrac{x^2}{a^2} + \dfrac{y^2}{b^2} + \dfrac{z^2}{c^2} = 1$. *Hint:* Use the change of variables suggested in Example 1.

$*$**22.** Bounded by the hyperboloid $\dfrac{x^2}{a^2} + \dfrac{y^2}{b^2} - \dfrac{z^2}{c^2} = 1$ and the planes $z = -c$ and $z = c$

23. Above the xy-plane and below the paraboloid

$$z = 1 - \frac{x^2}{a^2} - \frac{y^2}{b^2}$$

24. Evaluate $\displaystyle\iiint_R (x^2 + y^2 + z^2)\,dV$, where R is the cylinder $0 \le x^2 + y^2 \le a^2$, $0 \le z \le h$.

25. Find $\displaystyle\iiint_B (x^2 + y^2)\,dV$, where B is the ball given by $x^2 + y^2 + z^2 \le a^2$.

26. Find $\displaystyle\iiint_B (x^2 + y^2 + z^2)\,dV$, where B is the ball of Exercise 25.

27. Find $\displaystyle\iiint_R (x^2 + y^2 + z^2)\,dV$, where R is the region that lies above the cone $z = c\sqrt{x^2 + y^2}$ and inside the sphere $x^2 + y^2 + z^2 = a^2$.

28. Evaluate $\displaystyle\iiint_R (x^2 + y^2)\,dV$ over the region R of Exercise 27.

29. Find $\displaystyle\iiint_R z\,dV$, over the region R satisfying $x^2 + y^2 \le z \le \sqrt{2 - x^2 - y^2}$.

30. Find $\displaystyle\iiint_R x\,dV$ and $\displaystyle\iiint_R z\,dV$, over that part of the hemisphere $0 \le z \le \sqrt{a^2 - x^2 - y^2}$ that lies in the first octant.

$*$**31.** Find $\displaystyle\iiint_R x\,dV$ and $\displaystyle\iiint_R z\,dV$ over that part of the cone

$$0 \le z \le h\left(1 - \frac{\sqrt{x^2 + y^2}}{a}\right)$$

that lies in the first octant.

$*$**32.** Find the volume of the region inside the ellipsoid $\dfrac{x^2}{a^2} + \dfrac{y^2}{b^2} + \dfrac{z^2}{c^2} = 1$ and above the plane $z = b - y$.

$*$**33.** Show that for cylindrical coordinates the Laplace equation

$$\frac{\partial^2 u}{\partial x^2} + \frac{\partial^2 u}{\partial y^2} + \frac{\partial^2 u}{\partial z^2} = 0$$

is given by

$$\frac{\partial^2 u}{\partial r^2} + \frac{1}{r}\frac{\partial u}{\partial r} + \frac{1}{r^2}\frac{\partial^2 u}{\partial \theta^2} + \frac{\partial^2 u}{\partial z^2} = 0.$$

$*$**34.** Show that in spherical coordinates the Laplace equation is given by

$$\frac{\partial^2 u}{\partial \rho^2} + \frac{2}{\rho}\frac{\partial u}{\partial \rho} + \frac{\cot\phi}{\rho^2}\frac{\partial u}{\partial \phi} + \frac{1}{\rho^2}\frac{\partial^2 u}{\partial \phi^2} + \frac{1}{\rho^2 \sin^2\phi}\frac{\partial^2 u}{\partial \theta^2} = 0.$$

$*$**35.** If x, y, and z are functions of u, v, and w with continuous first partial derivatives and nonvanishing Jacobian at (u, v, w), show that they map an infinitesimal volume element in uvw-space bounded by the coordinate planes u, $u + du$, v, $v + dv$, w, and $w + dw$ into an infinitesimal "parallelepiped" in xyz-space having volume

$$dx\,dy\,dz = \left|\frac{\partial(x, y, z)}{\partial(u, v, w)}\right|\,du\,dv\,dw.$$

Hint: Adapt the two-dimensional argument given in Section 14.4. What three vectors from the point $P = (x(u, v, w), y(u, v, w), z(u, v, w))$ span the parallelepiped?

14.7 Applications of Multiple Integrals

When we express the volume V of a region R in 3-space as an integral,

$$V = \iiint_R dV,$$

we are regarding V as a "sum" of infinitely many *infinitesimal elements of volume,* that is, as the limit of the sum of volumes of smaller and smaller nonoverlapping subregions into which we subdivide R. This idea of representing sums of infinitesimal elements of quantities by integrals has many applications.

For example, if a rigid body of constant density δ g/cm^3 occupies a volume V cm^3, then its mass is $m = \delta V$ g. If the density is not constant but varies continuously over the region R of 3-space occupied by the rigid body, say $\delta = \delta(x, y, z)$, we can still regard the density as being constant on an infinitesimal element of R having volume dV. The mass of this element is therefore $dm = \delta(x, y, z)\,dV$, and the mass of the whole body is calculated by integrating these mass elements over R:

$$m = \iiint_R \delta(x, y, z)\,dV.$$

Similar formulas apply when the rigid body is one- or two-dimensional, and its density is given in units of mass per unit length or per unit area. In such cases single or double integrals are needed to sum the individual elements of mass. All this works because mass is "additive," that is, the mass of a composite object is the sum of the masses of the parts that comprise the object. The surface areas, gravitational forces, moments, and energies we consider in this section all have this additivity property.

The Surface Area of a Graph

We can use a double integral over a domain D in the xy-plane to add up surface area elements and thereby calculate the total area of the surface $\mathcal{S}$ with equation $z = f(x, y)$ defined for (x, y) in D. We assume that f has continuous first partial derivatives in D, so that $\mathcal{S}$ is smooth and has a nonvertical tangent plane at $P = \big(x, y, f(x, y)\big)$ for any (x, y) in D. The vector

$$\mathbf{n} = -f_1(x, y)\mathbf{i} - f_2(x, y)\mathbf{j} + \mathbf{k}$$

is an upward normal to $\mathcal{S}$ at P. An area element dA at position (x, y) in the xy-plane has a *vertical projection* onto $\mathcal{S}$ whose area dS is $\sec\gamma$ times the area dA, where γ is the angle between $\mathbf{n}$ and $\mathbf{k}$. (See Figure 14.47.)

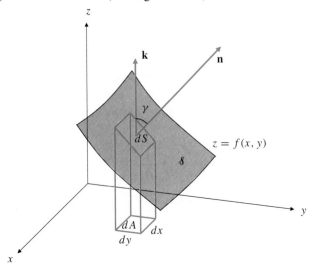

Figure 14.47   The surface area element dS on the surface $z = f(x, y)$ is $\sec\gamma$ times as large as its vertical projection dA onto the xy-plane

Since

$$\cos \gamma = \frac{\mathbf{n} \bullet \mathbf{k}}{|\mathbf{n}||\mathbf{k}|} = \frac{1}{\sqrt{1 + \big(f_1(x, y)\big)^2 + \big(f_2(x, y)\big)^2}},$$

we have

$$dS = \sqrt{1 + \left(\frac{\partial z}{\partial x}\right)^2 + \left(\frac{\partial z}{\partial y}\right)^2} \, dA.$$

Therefore, the area of $\mathcal{S}$ is

$$S = \iint_D \sqrt{1 + \left(\frac{\partial z}{\partial x}\right)^2 + \left(\frac{\partial z}{\partial y}\right)^2} \, dA.$$

Example 1   Find the area of that part of the hyperbolic paraboloid $z = x^2 - y^2$ that lies inside the cylinder $x^2 + y^2 = a^2$.

Solution   Since $\partial z/\partial x = 2x$ and $\partial z/\partial y = -2y$, the surface area element is

$$dS = \sqrt{1 + 4x^2 + 4y^2} \, dA = \sqrt{1 + 4r^2} \, r \, dr \, d\theta.$$

The required surface area is the integral of dS over the disk $r \le a$:

$$S = \int_0^{2\pi} d\theta \int_0^a \sqrt{1 + 4r^2} \, r \, dr \qquad \text{Let } u = 1 + 4r^2$$

$$= (2\pi) \frac{1}{8} \int_1^{1+4a^2} \sqrt{u} \, du$$

$$= \frac{\pi}{4} \left(\frac{2}{3}\right) u^{3/2} \Big|_1^{1+4a^2} = \frac{\pi}{6} \left((1 + 4a^2)^{3/2} - 1\right) \text{ square units.} \qquad \blacksquare$$

The Gravitational Attraction of a Disk

Newton's universal law of gravitation asserts that two point masses m_1 and m_2, separated by a distance s, attract one another with a force

$$F = \frac{km_1 m_2}{s^2},$$

k being a universal constant. The force on each mass is directed toward the other, along the line joining the two masses. Suppose that a flat disk D of radius a, occupying the region $x^2 + y^2 \le a^2$ of the xy-plane, has constant *areal density* σ (units of mass per unit area). Let us calculate the total force of attraction that this disk exerts upon a mass m located at the point $(0, 0, b)$ on the positive z-axis. The total force is a vector quantity. Although the various mass elements on the disk are in different directions from the mass m, symmetry indicates that the net force will be in the direction toward the centre of the disk, that is, toward the origin. Thus, the total force will be $-F\mathbf{k}$, where F is the magnitude of the force.

We will calculate F by integrating the vertical component dF of the force of attraction on m due to the mass $\sigma \, dA$ in an area element dA on the disk. If the area element is at the point with polar coordinates $[r, \theta]$, and if the line from this point to $(0, 0, b)$ makes angle ψ with the z-axis as shown in Figure 14.48, then the vertical component of the force of attraction of the mass element $\sigma \, dA$ on m is

$$dF = \frac{km\sigma \, dA}{r^2 + b^2} \cos \psi = km\sigma b \frac{dA}{(r^2 + b^2)^{3/2}}.$$

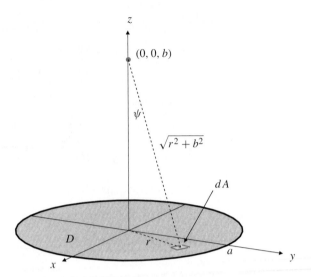

Figure 14.48 Each mass element $\sigma \, dA$ attracts m along a different line

Accordingly, the total vertical force of attraction of the disk on m is

$$
\begin{aligned}
F &= km\sigma b \iint_D \frac{dA}{(r^2 + b^2)^{3/2}} \\
&= km\sigma b \int_0^{2\pi} d\theta \int_0^a \frac{r \, dr}{(r^2 + b^2)^{3/2}} \qquad \text{Let } u = r^2 + b^2 \\
&= \pi km\sigma b \int_{b^2}^{a^2 + b^2} u^{-3/2} \, du \\
&= \pi km\sigma b \left(\frac{-2}{\sqrt{u}} \right) \Big|_{b^2}^{a^2 + b^2} = 2\pi km\sigma \left(1 - \frac{b}{\sqrt{a^2 + b^2}} \right).
\end{aligned}
$$

Remark If we let a approach infinity in the above formula, we obtain the formula $F = 2\pi km\sigma$ for the force of attraction of a plane of areal density σ on a mass m located at distance b from the plane. Observe that F does not depend on b. Try to reason on physical grounds why this should be so.

Remark The force of attraction on a point mass due to suitably symmetric solid objects (such as balls, cylinders, and cones) having constant density δ (units of mass per unit volume) can be found by integrating elements of force contributed by thin, disk-shaped slices of the solid. See Exercises 14–17 below.

Moments and Centres of Mass

The centre of mass of a rigid body is that point (fixed in the body) at which the body can be supported so that in the presence of a constant gravitational field it will not experience any unbalanced torques that will cause it to rotate. The torques experienced by a mass element dm in the body can be expressed in terms of the **moments** of dm about the three coordinate planes. If the body occupies a region R in 3-space and has continuous volume density $\delta(x, y, z)$, then the mass element $dm = \delta(x, y, z) \, dV$ that occupies the volume element dV is said to have **moments** $(x - x_0) \, dm$, $(y - y_0) \, dm$, and $(z - z_0) \, dm$ about the planes $x = x_0$, $y = y_0$, and $z = z_0$, respectively. Thus, the total moments of the body about these three planes are

$$
\begin{aligned}
M_{x=x_0} &= \iiint_R (x - x_0)\delta(x, y, z) \, dV = M_{x=0} - x_0 m \\
M_{y=y_0} &= \iiint_R (y - y_0)\delta(x, y, z) \, dV = M_{y=0} - y_0 m \\
M_{z=z_0} &= \iiint_R (z - z_0)\delta(x, y, z) \, dV = M_{z=0} - z_0 m,
\end{aligned}
$$

where $m = \iiint_R \delta\, dV$ is the mass of the body and $M_{x=0}$, $M_{y=0}$, and $M_{z=0}$ are the moments about the coordinate planes $x = 0$, $y = 0$, and $z = 0$, respectively. The **centre of mass** $\bar{P} = (\bar{x}, \bar{y}, \bar{z})$ of the body is that point for which $M_{x=\bar{x}}$, $M_{y=\bar{y}}$, and $M_{z=\bar{z}}$ are all equal to zero. Thus,

> **Centre of mass**
>
> The centre of mass of a solid occupying region R of 3-space and having continuous density $\delta(x, y, z)$ (units of mass per unit volume) is the point $(\bar{x}, \bar{y}, \bar{z})$ with coordinates given by
>
> $$\bar{x} = \frac{M_{x=0}}{m} = \frac{\iiint_R x\delta\, dV}{\iiint_R \delta\, dV}, \qquad \bar{y} = \frac{M_{y=0}}{m} = \frac{\iiint_R y\delta\, dV}{\iiint_R \delta\, dV},$$
>
> $$\bar{z} = \frac{M_{z=0}}{m} = \frac{\iiint_R z\delta\, dV}{\iiint_R \delta\, dV}.$$

These formulas can be combined into a single vector formula for the position vector $\bar{\mathbf{r}} = \bar{x}\mathbf{i} + \bar{y}\mathbf{j} + \bar{z}\mathbf{k}$ of the centre of mass in terms of the position vector $\mathbf{r} = x\mathbf{i} + y\mathbf{j} + z\mathbf{k}$ of an arbitrary point in R,

> $$\bar{\mathbf{r}} = \frac{M_{x=0}\mathbf{i} + M_{y=0}\mathbf{j} + M_{z=0}\mathbf{k}}{m} = \frac{\iiint_R \delta\, \mathbf{r}\, dV}{\iiint_R \delta\, dV},$$

where the integral of the vector function $\delta\,\mathbf{r}$ is understood to mean the vector whose components are the integrals of the components of $\delta\,\mathbf{r}$.

Remark   Similar expressions hold for distributions of mass over regions in the plane or over intervals on a line. We use the appropriate areal or line densities and double or single definite integrals.

Remark   If the density is constant, it cancels out of the expressions for the centre of mass. In this case the centre of mass is a *geometric* property of the region R and is called the **centroid** or **centre of gravity** of that region.

Example 2   Find the centroid of the tetrahedron T bounded by the coordinate planes and the plane

$$\frac{x}{a} + \frac{y}{b} + \frac{z}{c} = 1.$$

Solution   The density is assumed to be constant, so we may take it to be unity. The mass of T is thus equal to its volume: $m = V = abc/6$. The moment of T about the yz-plane is (see Figure 14.49):

$$M_{x=0} = \iiint_T x\, dV$$

$$= \int_0^a x\, dx \int_0^{b(1-\frac{x}{a})} dy \int_0^{c(1-\frac{x}{a}-\frac{y}{b})} dz$$

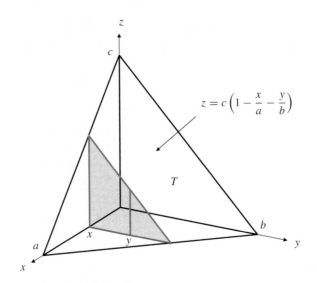

$$z = c\left(1 - \frac{x}{a} - \frac{y}{b}\right)$$

Figure 14.49   Iteration diagram for a triple integral over the tetrahedron of Example 2

$$= c \int_0^a x \, dx \int_0^{b(1-\frac{x}{a})} \left(1 - \frac{x}{a} - \frac{y}{b}\right) dy$$

$$= c \int_0^a x \left[\left(1 - \frac{x}{a}\right)y - \frac{y^2}{2b}\right]\Bigg|_{y=0}^{y=b(1-\frac{x}{a})} dx$$

$$= \frac{bc}{2} \int_0^a x \left(1 - \frac{x}{a}\right)^2 dx$$

$$= \frac{bc}{2} \left[\frac{x^2}{2} - \frac{2}{3}\frac{x^3}{a} + \frac{x^4}{4a^2}\right]\Bigg|_0^a = \frac{a^2bc}{24}.$$

Thus, $\bar{x} = M_{x=0}/m = a/4$. By symmetry, the centroid of T is $\left(\frac{a}{4}, \frac{b}{4}, \frac{c}{4}\right)$.

Example 3   Find the centre of mass of a solid occupying the region S that satisfies $x \geq 0$, $y \geq 0$, $z \geq 0$, and $x^2 + y^2 + z^2 \leq a^2$, if the density at distance ρ from the origin is $k\rho$.

Solution   The mass of the solid is distributed symmetrically in the first octant part of the ball $\rho \leq a$ so that the centre of mass, $(\bar{x}, \bar{y}, \bar{z})$, must satisfy $\bar{x} = \bar{y} = \bar{z}$. The mass of the solid is

$$m = \iiint_S k\rho \, dV = \int_0^{\pi/2} d\theta \int_0^{\pi/2} \sin\phi \, d\phi \int_0^a (k\rho)\rho^2 \, d\rho = \frac{\pi k a^4}{8}.$$

The moment about the xy-plane is

$$M_{z=0} = \iiint_S zk\rho \, dV = \iiint_S (k\rho)\rho\cos\phi\,\rho^2 \sin\phi \, d\rho \, d\phi \, d\theta$$

$$= \frac{k}{2} \int_0^{\pi/2} d\theta \int_0^{\pi/2} \sin(2\phi) \, d\phi \int_0^a \rho^4 \, d\rho = \frac{k\pi a^5}{20}.$$

Hence, $\bar{z} = \dfrac{k\pi a^5}{20} \Big/ \dfrac{k\pi a^4}{8} = \dfrac{2a}{5}$, and the centre of mass is $\left(\dfrac{2a}{5}, \dfrac{2a}{5}, \dfrac{2a}{5}\right)$.

Moment of Inertia

The **kinetic energy** of a particle of mass m moving with speed v is

$$KE = \frac{1}{2} mv^2.$$

The mass of the particle measures its *inertia*, which is twice the energy it has when its speed is one unit.

If the particle is moving in a circle of radius D, its motion can be described in terms of its **angular speed**, Ω, measured in radians per unit time. In one revolution the particle travels a distance $2\pi D$ in time $2\pi/\Omega$. Thus, its (translational) speed v is related to its angular speed by

$$v = \Omega D.$$

Suppose that a rigid body is rotating with angular speed Ω about an axis L. If (at some instant) the body occupies a region R and has density $\delta = \delta(x, y, z)$, then each mass element $dm = \delta\, dV$ in the body has kinetic energy

$$d\mathrm{KE} = \frac{1}{2} v^2\, dm = \frac{1}{2} \delta \Omega^2 D^2\, dV,$$

where $D = D(x, y, z)$ is the perpendicular distance from the volume element dV to the axis of rotation L. The total kinetic energy of the rotating body is therefore

$$\mathrm{KE} = \frac{1}{2}\Omega^2 \iiint_R D^2 \delta\, dV = \frac{1}{2} I\Omega^2,$$

where

$$I = \iiint_R D^2 \delta\, dV.$$

I is called the **moment of inertia** of the rotating body about the axis L. The moment of inertia plays the same role in the expression for kinetic energy of rotation (in terms of angular speed) that the mass does in the expression for kinetic energy of translation (in terms of linear speed). The moment of inertia is twice the kinetic energy of the body when it is rotating with unit angular speed.

If the entire mass of the rotating body were concentrated at a distance D_0 from the axis of rotation, then its kinetic energy would be $\frac{1}{2} m D_0^2 \Omega^2$. The **radius of gyration** $\bar{D}$ is the value of D_0 for which this energy is equal to the actual kinetic energy $\frac{1}{2} I\Omega^2$ of the rotating body. Thus, $m\bar{D}^2 = I$, and the radius of gyration is

$$\bar{D} = \sqrt{I/m} = \left(\frac{\iiint_R D^2 \delta\, dV}{\iiint_R \delta\, dV} \right)^{1/2}.$$

Example 4 (The acceleration of a rolling ball)

(a) Find the moment of inertia and radius of gyration of a solid ball of radius a and constant density δ about a diameter of that ball.

(b) With what linear acceleration will the ball roll (without slipping) down a plane inclined at angle α to the horizontal?

Solution

(a) We take the z-axis as the diameter and integrate in cylindrical coordinates over the ball B of radius a centred at the origin. Since the density δ is constant, we have

$$I = \delta \iiint_B r^2 \, dV$$

$$= \delta \int_0^{2\pi} d\theta \int_0^a r^3 \, dr \int_{-\sqrt{a^2-r^2}}^{\sqrt{a^2-r^2}} dz$$

$$= 4\pi\delta \int_0^a r^3 \sqrt{a^2 - r^2} \, dr \qquad \text{Let } u = a^2 - r^2$$

$$= 2\pi\delta \int_0^{a^2} (a^2 - u)\sqrt{u} \, du$$

$$= 2\pi\delta \left(\frac{2}{3}a^2 u^{3/2} - \frac{2}{5}u^{5/2} \right)\Big|_0^{a^2} = \frac{8}{15}\pi\delta a^5.$$

Since the mass of the ball is $m = \frac{4}{3}\pi\delta a^3$, the radius of gyration is

$$\bar{D} = \sqrt{\frac{I}{m}} = \sqrt{\frac{2}{5}}\,a.$$

(b) We can determine the acceleration of the ball by using conservation of total (kinetic plus potential) energy. When the ball is rolling down the plane with speed v, its centre is moving with speed v and losing height at a rate $v\sin\alpha$. (See Figure 14.50.) Since the ball is not slipping, it is rotating about a horizontal axis through its centre with angular speed $\Omega = v/a$. Hence, its kinetic energy (due to translation and rotation) is

$$\text{KE} = \frac{1}{2}mv^2 + \frac{1}{2}I\Omega^2$$

$$= \frac{1}{2}mv^2 + \frac{1}{2}\frac{2}{5}ma^2 \frac{v^2}{a^2} = \frac{7}{10}mv^2.$$

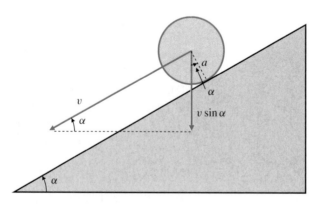

Figure 14.50   The actual velocity and the vertical velocity of a ball rolling down an incline as in Example 4

When the centre of the ball is at height h (above some reference height), the ball has (gravitational) potential energy

$$\text{PE} = mgh.$$

(This is the work that must be done against a constant gravitational force $F = mg$ to raise it to height h.) Since total energy is conserved,

$$\frac{7}{10}mv^2 + mgh = \text{constant}.$$

Differentiating with respect to time t, we obtain

$$0 = \frac{7}{10} m \, 2v\frac{dv}{dt} + mg\frac{dh}{dt} = \frac{7}{5} mv\frac{dv}{dt} - mgv \sin \alpha.$$

Thus, the ball rolls down the incline with acceleration $\dfrac{dv}{dt} = \dfrac{5}{7} g \sin \alpha$.

Exercises 14.7

Surface area problems

Use double integrals to calculate the areas of the surfaces in Exercises 1–9.

1. The part of the plane $z = 2x + 2y$ inside the cylinder $x^2 + y^2 = 1$

2. The part of the plane $5z = 3x - 4y$ inside the elliptic cylinder $x^2 + 4y^2 = 4$

3. The hemisphere $z = \sqrt{a^2 - x^2 - y^2}$

4. The half-ellipsoidal surface $z = 2\sqrt{1 - x^2 - y^2}$

5. The conical surface $3z^2 = x^2 + y^2$, $0 \le z \le 2$

6. The paraboloid $z = 1 - x^2 - y^2$ in the first octant

7. The part of the surface $z = y^2$ above the triangle with vertices $(0, 0)$, $(0, 1)$, and $(1, 1)$

8. The part of the surface $z = \sqrt{x}$ above the region $0 \le x \le 1$, $0 \le y \le \sqrt{x}$

9. The part of the cylindrical surface $x^2 + z^2 = 4$ that lies above the region $0 \le x \le 2$, $0 \le y \le x$

10. Show that the parts of the surfaces $z = 2xy$ and $z = x^2 + y^2$ that lie in the same vertical cylinder have the same area.

11. Show that the area S of the part of the paraboloid $z = \frac{1}{2}(x^2 + y^2)$ lying above the square $-1 \le x \le 1$, $-1 \le y \le 1$ is given by

$$S = \frac{8}{3} \int_0^{\pi/4} (1 + \sec^2 \theta)^{3/2} \, d\theta - \frac{2\pi}{3},$$

and use numerical methods to evaluate the area to 3 decimal places.

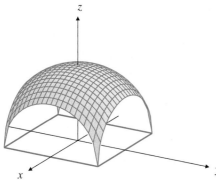

Figure 14.51

*12. The canopy shown in Figure 14.51 is the part of the hemisphere of radius $\sqrt{2}$ centred at the origin that lies above the square $-1 \le x \le 1$, $-1 \le y \le 1$. Find its area. *Hint:* It is possible to get an exact solution by first finding the area of

the part of the sphere $x^2 + y^2 + z^2 = 2$ that lies above the plane $z = 1$. If you do the problem directly by integrating the surface area element over the square, you may encounter an integral that you can't evaluate exactly, and you will have to use numerical methods.

Mass and gravitational attraction

13. Find the mass of a spherical planet of radius a whose density at distance R from the centre is $\delta = A/(B + R^2)$.

In Exercises 14–17, find the gravitational attraction that the given object exerts on a mass m located at $(0, 0, b)$. Assume the object has constant density δ. In each case you can obtain the answer by integrating the contributions made by disks of thickness dz, making use of the formula for the attraction exerted by the disk obtained in the text.

14. The ball $x^2 + y^2 + z^2 \le a^2$, where $a < b$.

15. The cylinder $x^2 + y^2 \le a^2$, $0 \le z \le h$, where $h < b$.

16. The cone $0 \le z \le b - (\sqrt{x^2 + y^2})/a$

17. The half-ball $0 \le z \le \sqrt{a^2 - x^2 - y^2}$, where $a < b$.

Centres of mass and centroids

18. Find the centre of mass of an object occupying the cube $0 \le x, y, z \le a$ with density given by $\delta = x^2 + y^2 + z^2$.

Find the centroids of the regions in Exercises 19–22.

19. The prism $x \ge 0$, $y \ge 0$, $x + y \le 1$, $0 \le z \le 1$

20. The unbounded region $0 \le z \le e^{-(x^2+y^2)}$

21. The first octant part of the ball $x^2 + y^2 + z^2 \le a^2$

22. The region inside the cube $0 \le x, y, z \le 1$ and under the plane $x + y + z = 2$

Moments of inertia

23. Explain *in physical terms* why the acceleration of the ball rolling down the incline in Example 4 does not approach g (the acceleration due to gravity) as the angle of incline, α, approaches $90°$.

Find the moments of inertia and radii of gyration of the solid objects in Exercises 24–32. Assume constant density in all cases.

24. A circular cylinder of base radius a and height h about the axis of the cylinder

25. A circular cylinder of base radius a and height h about a diameter of the base of the cylinder

26. A right circular cone of base radius a and height h about the axis of the cone

27. A right circular cone of base radius a and height h about a diameter of the base of the cone

28. A cube of edge length a about an edge of the cube

29. A cube of edge length a about a diagonal of a face of the cube

30. A cube of edge length a about a diagonal of the cube

31. The rectangular box $-a \le x \le a$, $-b \le y \le b$, $-c \le z \le c$ about the z-axis

32. The region between the two concentric cylinders $x^2 + y^2 = a^2$ and $x^2 + y^2 = b^2$ (where $0 < a < b$) and between $z = 0$ and $z = c$ about the z-axis

33. A ball of radius a has constant density δ. A cylindrical hole of radius $b < a$ is drilled through the centre of the ball. Find the mass of the remaining part of the ball and its moment of inertia about the axis of the hole.

34. With what acceleration will a solid cylinder of base radius a, height h, and constant density δ roll (without slipping) down a plane inclined at angle α to the horizontal?

35. Repeat Exercise 34 for the ball with the cylindrical hole in Exercise 33. Assume that the axis of the hole remains horizontal while the ball rolls.

∗36. A rigid pendulum of mass m swings about point A on a horizontal axis. Its moment of inertia about that axis is I. The centre of mass C of the pendulum is at distance a from A. When the pendulum hangs at rest, C is directly under A. (Why?) Suppose the pendulum is swinging. Let $\theta = \theta(t)$ measure the angular displacement of the line AC from the vertical at time t. ($\theta = 0$ when the pendulum is in its rest position.) Use a conservation of energy argument similar to that in Example 4 to show that

$$\frac{1}{2} I \left(\frac{d\theta}{dt}\right)^2 - mga \cos\theta = \text{constant}$$

and, hence, differentiating with respect to t, that

$$\frac{d^2\theta}{dt^2} + \frac{mga}{I}\sin\theta = 0.$$

This is a nonlinear differential equation, and it is not easily solved. However, for small oscillations ($|\theta|$ small) we can use the approximation $\sin\theta \approx \theta$. In this case the differential equation is that of simple harmonic motion. What is the period?

∗37. Let L_0 be a straight line passing through the centre of mass of a rigid body B of mass m. Let L_k be a straight line parallel to and k units distant from L_0. If I_0 and I_k are the moments of inertia of B about L_0 and L_k, respectively, show that $I_k = I_0 + k^2 m$. Hence, a body always has smallest moment of inertia about an axis through its centre of mass. *Hint:* Assume that the z-axis coincides with L_0 and that L_k passes through the point $(k, 0, 0)$.

∗38. Reestablish the expression for the total kinetic energy of the rolling ball in Example 4 by regarding the ball at any instant as rotating about a horizontal line through its point of contact with the inclined plane. Use the result of Exercise 37.

∗39. (**Products of inertia**) A rigid body with density δ is placed with its centre of mass at the origin and occupies a region R of 3-space. Suppose the six second moments P_{xx}, P_{yy}, P_{zz}, P_{xy}, P_{xz}, and P_{yz} are all known, where

$$P_{xx} = \iiint_R x^2 \delta\, dV, \quad P_{xy} = \iiint_R xy\delta\, dV, \quad \cdots .$$

(There exist tables giving these six moments for bodies of many standard shapes. They are called products of inertia.) Show how to express the moment of inertia of the body about any axis through the origin in terms of these six second moments. (If this result is combined with that of Exercise 37, the moment of inertia about *any* axis can be found.)

Chapter Review

Key Ideas

- **What do the following terms and phrases mean?**
  - ◇ a Riemann sum for $f(x, y)$ on $a \le x \le b, c \le y \le d$
  - ◇ $f(x, y)$ is integrable on $a \le x \le b, c \le y \le d$
  - ◇ the double integral of $f(x, y)$ over $a \le x \le b, c \le y \le d$
  - ◇ iteration of a double integral
  - ◇ the average value of $f(x, y)$ over region R
  - ◇ the area element in polar coordinates
  - ◇ a triple integral
  - ◇ the volume element in cylindrical coordinates
  - ◇ the volume element in spherical coordinates
  - ◇ the surface area of the graph of $z = f(x, y)$
  - ◇ the moment of inertia of a solid about an axis

- **Describe how to change variables in a double integral.**

- **How do you calculate the centroid of a solid region?**

- **How do you calculate the moment of inertia of a solid about an axis?**

Review Exercises

1. Evaluate $\iint_R (x + y)\, dA$, over the first-quadrant region lying under $x = y^2$ and above $y = x^2$.

2. Evaluate $\iint_P (x^2 + y^2)\, dA$, where P is the parallelogram with vertices $(0, 0)$, $(2, 0)$, $(3, 1)$, and $(1, 1)$.

3. Find $\iint_S (y/x)\, dA$, where S is the part of the disk $x^2 + y^2 \le 4$ in the first quadrant and under the line $y = x$.

4. Consider the iterated integral

$$I = \int_0^{\sqrt{3}} dy \int_{y/\sqrt{3}}^{\sqrt{4-y^2}} e^{-x^2-y^2}\, dx.$$

(a) Write I as a double integral $\iint_R e^{-x^2-y^2}\, dA$, and sketch the region R over which the double integral is taken.

(b) Write I as an iterated integral with the order of integrations reversed from that of the given iteration.

(c) Write I as an iterated integral in polar coordinates.

(d) Evaluate I.

5. Find the constant $k > 0$ such that the volume of the region lying inside the sphere $x^2 + y^2 + z^2 = a^2$ and above the cone $z = k\sqrt{x^2 + y^2}$ is one-quarter of the volume contained by the whole sphere.

6. Reiterate the integral

$$I = \int_0^2 dy \int_0^y f(x, y)\, dx + \int_2^6 dy \int_0^{\sqrt{6-y}} f(x, y)\, dx$$

with the y integral on the inside.

7. Let $J = \int_0^1 dz \int_0^z dy \int_0^y f(x, y, z)\, dx$. Express J as an iterated integral where the integrations are to be performed in the following order: first z, then y, then x.

8. An object in the shape of a right-circular cone has height 10 m and base radius 5 m. Its density is proportional to the square of the distance from the base and equals 3,000 kg/m^3 at the vertex.

(a) Find the mass of the object.

(b) Express the moment of inertia of the object about its central axis as an iterated integral.

9. Find the average value of $f(t) = \int_t^a e^{-x^2}\, dx$ over the interval $0 \le t \le a$.

10. Find the average value of the function $f(x, y) = \lfloor x + y \rfloor$ over the quarter-disk $x \ge 0$, $y \ge 0$, $x^2 + y^2 \le 4$. (Recall that $\lfloor x \rfloor$ denotes the greatest integer less than or equal to x.)

11. Let D be the smaller of the two solid regions bounded by the surfaces

$$z = \frac{x^2 + y^2}{a} \quad \text{and} \quad x^2 + y^2 + z^2 = 6a^2,$$

where a is a positive constant. Find $\iiint_D (x^2 + y^2)\, dV$.

12. Find the moment of inertia about the z-axis of a solid V of density 1 if V is specified by the inequalities $0 \le z \le \sqrt{x^2 + y^2}$ and $x^2 + y^2 \le 2ay$, where $a > 0$.

13. The rectangular solid $0 \le x \le 1$, $0 \le y \le 2$, $0 \le z \le 1$ is cut into two pieces by the plane $2x + y + z = 2$. Let D be the piece that includes the origin. Find the volume of D and $\bar{z}$, the z-coordinate of the centroid of D.

14. A solid S consists of those points (x, y, z) that lie in the first octant and satisfy $x + y + 2z \le 2$ and $y + z \le 1$. Find the volume of S and the x-coordinate of its centroid.

15. Find $\iiint_S z\, dV$, where S is the portion of the first octant that is above the plane $x + y - z = 1$ and below the plane $z = 1$.

16. Find the area of that part of the plane $z = 2x$ that lies inside the paraboloid $z = x^2 + y^2$.

17. Find the area of that part of the paraboloid $z = x^2 + y^2$ that lies below the plane $z = 2x$. Express the answer as a single integral, and evaluate it to 3 decimal places.

∗18. Find the volume of the smaller of the two regions into which the plane $x + y + z = 1$ divides the interior of the ellipsoid $x^2 + 4y^2 + 9z^2 = 36$. *Hint:* First change variables so that the ellipsoid becomes a ball. Then replace the plane by a plane with a simpler equation passing the same distance from the origin.

Challenging Problems

1. The plane $(x/a) + (y/b) + (z/c) = 1$ (where $a > 0$, $b > 0$, and $c > 0$) divides the solid ellipsoid

$$\frac{x^2}{a^2} + \frac{y^2}{b^2} + \frac{z^2}{c^2} \le 1$$

into two unequal pieces. Find the volume of the smaller piece.

2. Find the area of the part of the plane $(x/a) + (y/b) + (z/c) = 1$ (where $a > 0$, $b > 0$, and $c > 0$) that lies inside the ellipsoid

$$\frac{x^2}{a^2} + \frac{y^2}{b^2} + \frac{z^2}{c^2} \le 1.$$

3. (a) Expand $1/(1 - xy)$ as a geometric series, and hence show that

$$\int_0^1 \int_0^1 \frac{1}{1 - xy}\, dx\, dy = \sum_{n=1}^{\infty} \frac{1}{n^2}.$$

(b) Similarly, express the following integrals as sums of series:

(i) $\displaystyle \int_0^1 \int_0^1 \frac{1}{1 + xy}\, dx\, dy$,

(ii) $\displaystyle \int_0^1 \int_0^1 \int_0^1 \frac{1}{1 - xyz}\, dx\, dy\, dz$,

(iii) $\displaystyle \int_0^1 \int_0^1 \int_0^1 \frac{1}{1 + xyz}\, dx\, dy\, dz$.

∗ 4. Let P be the parallelepiped bounded by the three pairs of parallel planes $\mathbf{a} \bullet \mathbf{r} = 0$, $\mathbf{a} \bullet \mathbf{r} = d_1 > 0$, $\mathbf{b} \bullet \mathbf{r} = 0$, $\mathbf{b} \bullet \mathbf{r} = d_2 > 0$, $\mathbf{c} \bullet \mathbf{r} = 0$, and $\mathbf{c} \bullet \mathbf{r} = d_3 > 0$, where $\mathbf{a}$, $\mathbf{b}$, and $\mathbf{c}$ are constant vectors, and $\mathbf{r} = x\mathbf{i} + y\mathbf{j} + z\mathbf{k}$. Show that

$$\iiint_P (\mathbf{a} \bullet \mathbf{r})(\mathbf{b} \bullet \mathbf{r})(\mathbf{c} \bullet \mathbf{r})\, dx\, dy\, dz = \frac{(d_1 d_2 d_3)^2}{8|\mathbf{a} \bullet (\mathbf{b} \times \mathbf{c})|}.$$

Hint: Make the change of variables $u = \mathbf{a} \bullet \mathbf{r}$, $v = \mathbf{b} \bullet \mathbf{r}$, $w = \mathbf{c} \bullet \mathbf{r}$.

5. A hole whose cross-section is a square of side 2 is punched through the middle of a ball of radius 2. Find the volume of the remaining part of the ball.

∗ 6. Find the volume bounded by the surface with equation $x^{2/3} + y^{2/3} + z^{2/3} = a^{2/3}$.

∗ 7. Find the volume bounded by the surface $|x|^{1/3} + |y|^{1/3} + |z|^{1/3} = |a|^{1/3}$.

CHAPTER 15

Vector Fields

"Take some more tea," the March Hare said to Alice, very earnestly.

"I've had nothing yet," Alice replied, in an offended tone, "so I can't take more."

"You mean you can't take less," said the Hatter: "it's very easy to take more than nothing."

Lewis Carroll (Charles Lutwidge Dodgson) 1832–1898
from *Alice's Adventures in Wonderland*

Introduction This chapter and the next are concerned mainly with vector-valued functions of a vector variable, typically functions whose domains and ranges lie in the plane or in 3-space. Such functions are frequently called *vector fields*. Applications of vector fields often involve integrals taken, not along axes or over regions in the plane or 3-space, but rather over curves and surfaces. We will introduce such line and surface integrals in this chapter. The next chapter will be devoted to developing versions of the Fundamental Theorem of Calculus for integrals of vector fields.

15.1 Vector and Scalar Fields

A function whose domain and range are subsets of Euclidean 3-space, $\mathbb{R}^3$, is called a **vector field**. Thus, a vector field $\mathbf{F}$ associates a vector $\mathbf{F}(x, y, z)$ with each point (x, y, z) in its domain. The three components of $\mathbf{F}$ are scalar-valued (real-valued) functions $F_1(x, y, z)$, $F_2(x, y, z)$, and $F_3(x, y, z)$, and $\mathbf{F}(x, y, z)$ can be expressed in terms of the standard basis in $\mathbb{R}^3$ as

$$\mathbf{F}(x, y, z) = F_1(x, y, z)\mathbf{i} + F_2(x, y, z)\mathbf{j} + F_3(x, y, z)\mathbf{k}.$$

(Note that the subscripts here represent *components* of a vector, *not* partial derivatives.) If $F_3(x, y, z) = 0$ and F_1 and F_2 are independent of z, then $\mathbf{F}$ reduces to

$$\mathbf{F}(x, y) = F_1(x, y)\mathbf{i} + F_2(x, y)\mathbf{j}$$

and so is called a **plane vector field**, or a vector field in the xy-plane. We will frequently make use of position vectors in the arguments of vector fields. The position vector of (x, y, z) is $\mathbf{r} = x\mathbf{i} + y\mathbf{j} + z\mathbf{k}$, and we can write $\mathbf{F}(\mathbf{r})$ as a shorthand for $\mathbf{F}(x, y, z)$. In the context of discussion of vector fields, a scalar-valued function of a vector variable (i.e., a function of several real variables as considered in the context of Chapters 12–14) is frequently called a **scalar field**. Thus, the components of a vector field are scalar fields.

Many of the results we prove about vector fields require that the field be smooth in some sense. We will call a vector field **smooth** wherever its component scalar fields have continuous partial derivatives of all orders. (For most purposes, however, second order would be sufficient.)

Vector fields arise in many situations in applied mathematics. Let us list some:

(a) The gravitational field $\mathbf{F}(x, y, z)$ due to some object is the force of attraction that the object exerts on a unit mass located at position (x, y, z).

(b) The electrostatic force field $\mathbf{E}(x, y, z)$ due to an electrically charged object is the electrical force that the object exerts on a unit charge at position (x, y, z). (The force may be either an attraction or a repulsion.)

(c) The velocity field $\mathbf{v}(x, y, z)$ in a moving fluid (or solid) is the velocity of motion of the particle at position (x, y, z). If the motion is not "steady state," then the velocity field will also depend on time: $\mathbf{v} = \mathbf{v}(x, y, z, t)$.

(d) The gradient $\nabla f(x, y, z)$ of any scalar field f gives the direction and magnitude of the greatest rate of increase of f at (x, y, z). In particular, a *temperature gradient*, $\nabla T(x, y, z)$, is a vector field giving the direction and magnitude of the greatest rate of increase of temperature T at the point (x, y, z) in a heat-conducting medium. *Pressure gradients* provide similar information about the variation of pressure in a fluid such as an air mass or an ocean.

(e) The unit radial and unit transverse vectors $\hat{\mathbf{r}}$ and $\hat{\boldsymbol{\theta}}$ are examples of vector fields in the xy-plane. Both are defined at all points of the plane except the origin.

Example 1   **(The gravitational field of a point mass)** The gravitational force field due to a point mass m located at point P_0 having position vector $\mathbf{r}_0$ is

$$\mathbf{F}(x, y, z) = \mathbf{F}(\mathbf{r}) = \frac{-km}{|\mathbf{r} - \mathbf{r}_0|^3} (\mathbf{r} - \mathbf{r}_0)$$
$$= -km \frac{(x - x_0)\mathbf{i} + (y - y_0)\mathbf{j} + (z - z_0)\mathbf{k}}{\left((x - x_0)^2 + (y - y_0)^2 + (z - z_0)^2\right)^{3/2}},$$

where $k > 0$ is a constant. $\mathbf{F}$ points toward the point $\mathbf{r}_0$ and has magnitude

$$|\mathbf{F}| = km/|\mathbf{r} - \mathbf{r}_0|^2.$$

Some vectors in a plane section of the field are shown graphically in Figure 15.1. Each represents the value of the field at the position of its tail. The lengths of the vectors indicate that the strength of the force increases the closer you get to P_0.

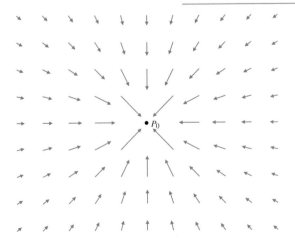

Figure 15.1   The gravitational field of a point mass located at P_0

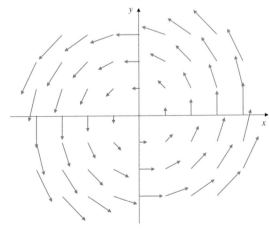

Figure 15.2   The velocity field of a rigid body rotating about the z-axis

Remark   The electrostatic field **F** due to a point charge q at P_0 is given by the same formula as the gravitational field above, except with $-m$ replaced by q. The reason for the opposite sign is that like charges repel each other whereas masses attract each other.

Example 2   The velocity field of a solid rotating about the z-axis with angular velocity $\mathbf{\Omega} = \Omega\mathbf{k}$ is

$$\mathbf{v}(x, y, z) = \mathbf{v}(\mathbf{r}) = \mathbf{\Omega} \times \mathbf{r} = -\Omega y\mathbf{i} + \Omega x\mathbf{j}.$$

Being the same in all planes normal to the z-axis, **v** can be regarded as a plane vector field. Some vectors of the field are shown in Figure 15.2. ∎

Field Lines (Integral Curves)

The graphical representations of vector fields such as those shown in Figures 15.1 and 15.2 and the wind velocity field over a hill shown in Figure 15.3 suggest a pattern of motion through space or in the plane. Whether or not the field is a velocity field, we can interpret it as such and ask what path will be followed by a particle, initially at some point, whose velocity is given by the field. The path will be a curve to which the field is tangent at every point. Such curves are called **field lines** or **integral curves** for the given vector field. In the specific case where the vector field gives the velocity in a fluid flow, the field lines are called **streamlines** or **flow lines** of the flow; some of these are shown for the air flow in Figure 15.3. For a force field, the field lines are called **lines of force**.

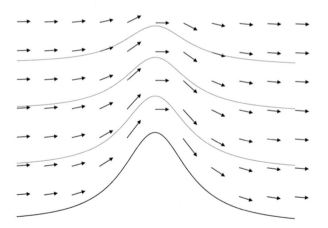

Figure 15.3   The velocity field and some streamlines of wind blowing over a hill

The field lines of **F** do not depend on the magnitude of **F** at any point but only on the direction of the field. If the field line through some point has parametric equation $\mathbf{r} = \mathbf{r}(t)$, then its tangent vector $d\mathbf{r}/dt$ must be parallel to $\mathbf{F}(\mathbf{r}(t))$ for all t. Thus

$$\frac{d\mathbf{r}}{dt} = \lambda(t)\mathbf{F}(\mathbf{r}(t)).$$

For *some* vector fields this differential equation can be integrated to find the field lines. If we break the equation into components,

$$\frac{dx}{dt} = \lambda(t)F_1(x, y, z), \quad \frac{dy}{dt} = \lambda(t)F_2(x, y, z), \quad \frac{dz}{dt} = \lambda(t)F_3(x, y, z),$$

we can obtain equivalent differential expressions for $\lambda(t)\,dt$ and hence write the differential equation for the field lines in the form

$$\frac{dx}{F_1(x, y, z)} = \frac{dy}{F_2(x, y, z)} = \frac{dz}{F_3(x, y, z)}.$$

If multiplication of these differential equations by some function puts them in the form

$$P(x)\,dx = Q(y)\,dy = R(z)\,dz,$$

then we can integrate all three expressions to find the field lines.

Example 3 Find the field lines of the gravitational force field of Example 1:

$$\mathbf{F}(x, y, z) = -km\,\frac{(x - x_0)\mathbf{i} + (y - y_0)\mathbf{j} + (z - z_0)\mathbf{k}}{\left((x - x_0)^2 + (y - y_0)^2 + (z - z_0)^2\right)^{3/2}}.$$

Solution The vector in the numerator of the fraction gives the direction of $\mathbf{F}$. Therefore, the field lines satisfy the system

$$\frac{dx}{x - x_0} = \frac{dy}{y - y_0} = \frac{dz}{z - z_0}.$$

Integrating all three expressions leads to

$$\ln|x - x_0| + \ln C_1 = \ln|y - y_0| + \ln C_2 = \ln|z - z_0| + \ln C_3,$$

or, on taking exponentials,

$$C_1(x - x_0) = C_2(y - y_0) = C_3(z - z_0).$$

This represents two families of planes all passing through $P_0 = (x_0, y_0, z_0)$. The field lines are the intersections of planes from each of the families, so they are straight lines through the point P_0. (This is a *two-parameter* family of lines; any one of the constants C_i that is nonzero can be divided out of the equations above.) The nature of the field lines should also be apparent from the plot of the vector field in Figure 15.1.

Example 4 Find the field lines of the velocity field $\mathbf{v} = \Omega(-y\mathbf{i}+x\mathbf{j})$ of Example 2.

Solution The field lines satisfy the differential equation

$$\frac{dx}{-y} = \frac{dy}{x}.$$

We can separate variables in this equation to get $x\,dx = -y\,dy$. Integration then gives $x^2/2 = -y^2/2 + C/2$, or $x^2 + y^2 = C$. Thus, the field lines are circles centred at the origin in the xy-plane, as is also apparent from the vector field plot in Figure 15.2. If we regard $\mathbf{v}$ as a vector field in 3-space, we find that the field lines are horizontal circles centred on the z-axis:

$$x^2 + y^2 = C_1, \qquad z = C_2.$$

Our ability to find field lines depends on our ability to solve differential equations and, in 3-space, systems of differential equations.

Example 5 Find the field lines of $\mathbf{F} = xz\mathbf{i} + 2x^2z\mathbf{j} + x^2\mathbf{k}$.

Solution The field lines satisfy $\dfrac{dx}{xz} = \dfrac{dy}{2x^2z} = \dfrac{dz}{x^2}$, or, equivalently

$$dy = 2x\,dx \qquad \text{and} \qquad dy = 2z\,dz.$$

The field lines are the curves of intersection of the two families $y = x^2 + C_1$ and $y = z^2 + C_2$ of parabolic cylinders.

Vector Fields in Polar Coordinates

A vector field in the plane can be expressed in terms of polar coordinates in the form

$$\mathbf{F} = \mathbf{F}(r, \theta) = F_r(r, \theta)\hat{\mathbf{r}} + F_\theta(r, \theta)\hat{\boldsymbol{\theta}},$$

where $\hat{\mathbf{r}}$ and $\hat{\boldsymbol{\theta}}$, defined everywhere except at the origin by

$$\hat{\mathbf{r}} = \quad \cos\theta\,\mathbf{i} + \sin\theta\,\mathbf{j}$$
$$\hat{\boldsymbol{\theta}} = -\sin\theta\,\mathbf{i} + \cos\theta\,\mathbf{j},$$

are unit vectors in the direction of increasing r and θ at $[r, \theta]$. Note that $d\hat{\mathbf{r}}/d\theta = \hat{\boldsymbol{\theta}}$ and that $\hat{\boldsymbol{\theta}}$ is just $\hat{\mathbf{r}}$ rotated 90° counterclockwise. Also note that we are using F_r and F_θ to denote the components of $\mathbf{F}$ with respect to the basis $\{\hat{\mathbf{r}}, \hat{\boldsymbol{\theta}}\}$; the subscripts do not indicate partial derivatives. Here, $F_r(r, \theta)$ is called the *radial* component of $\mathbf{F}$, and $F_\theta(r, \theta)$ is called the *transverse* component.

A curve with polar equation $r = r(\theta)$ can be expressed in vector parametric form,

$$\mathbf{r} = r\hat{\mathbf{r}},$$

as we did in Section 11.6. This curve is a field line of $\mathbf{F}$ if its differential tangent vector

$$d\mathbf{r} = dr\,\hat{\mathbf{r}} + r\,\frac{d\hat{\mathbf{r}}}{d\theta}\,d\theta = dr\,\hat{\mathbf{r}} + r\,d\theta\,\hat{\boldsymbol{\theta}}$$

is parallel to the field vector $\mathbf{F}(r, \theta)$ at any point except the origin, that is, if $r = f(\theta)$ satisfies the differential equation

$$\frac{dr}{F_r(r, \theta)} = \frac{r\,d\theta}{F_\theta(r, \theta)}.$$

In specific cases we can find the field lines by solving this equation.

Example 6   Sketch the vector field $\mathbf{F}(r, \theta) = \hat{\mathbf{r}} + \hat{\boldsymbol{\theta}}$, and find its field lines.

Solution   At each point $[r, \theta]$, the field vector bisects the angle between $\hat{\mathbf{r}}$ and $\hat{\boldsymbol{\theta}}$, making a counterclockwise angle of 45° with $\hat{\mathbf{r}}$. All of the vectors in the field have the same length, $\sqrt{2}$. Some of the field vectors are shown in Figure 15.4(a). They suggest that the field lines will spiral outward from the origin. Since $F_r(r, \theta) = F_\theta(r, \theta) = 1$ for this field, the field lines satisfy $dr = r\,d\theta$, or, dividing by $d\theta$, $dr/d\theta = r$. This is the differential equation of exponential growth and has solution $r = Ke^\theta$, or, equivalently, $r = e^{\theta + \alpha}$, where $\alpha = \ln K$ is a constant. Several such curves are shown in Figure 15.4(b).

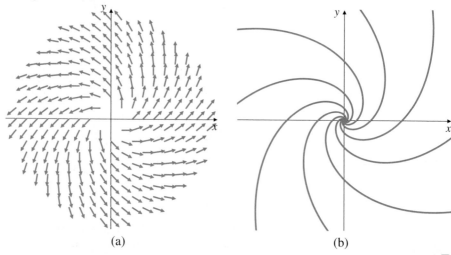

Figure 15.4

(a) The vector field $\mathbf{F} = \hat{\mathbf{r}} + \hat{\boldsymbol{\theta}}$

(b) Field lines of $\mathbf{F} = \hat{\mathbf{r}} + \hat{\boldsymbol{\theta}}$

(a)   (b)

Exercises 15.1

In Exercises 1–8, sketch the given plane vector field and determine its field lines.

1. $\mathbf{F}(x, y) = x\mathbf{i} + x\mathbf{j}$

2. $\mathbf{F}(x, y) = x\mathbf{i} + y\mathbf{j}$

3. $\mathbf{F}(x, y) = y\mathbf{i} + x\mathbf{j}$

4. $\mathbf{F}(x, y) = \mathbf{i} + \sin x\,\mathbf{j}$

5. $\mathbf{F}(x, y) = e^x\mathbf{i} + e^{-x}\mathbf{j}$

6. $\mathbf{F}(x, y) = \nabla(x^2 - y)$

7. $\mathbf{F}(x, y) = \nabla \ln(x^2 + y^2)$

8. $\mathbf{F}(x, y) = \cos y\,\mathbf{i} - \cos x\,\mathbf{j}$

In Exercises 9–16, describe the streamlines of the given velocity fields.

9. $\mathbf{v}(x, y, z) = y\mathbf{i} - y\mathbf{j} - y\mathbf{k}$

10. $\mathbf{v}(x, y, z) = x\mathbf{i} + y\mathbf{j} - x\mathbf{k}$

11. $\mathbf{v}(x, y, z) = y\mathbf{i} - x\mathbf{j} + \mathbf{k}$

12. $\mathbf{v}(x, y, z) = \dfrac{x\mathbf{i} + y\mathbf{j}}{(1 + z^2)(x^2 + y^2)}$

13. $\mathbf{v}(x, y, z) = xz\mathbf{i} + yz\mathbf{j} + x\mathbf{k}$

14. $\mathbf{v}(x, y, z) = e^{xyz}(x\mathbf{i} + y^2\mathbf{j} + z\mathbf{k})$

15. $\mathbf{v}(x, y) = x^2\mathbf{i} - y\mathbf{j}$

∗16. $\mathbf{v}(x, y) = x\mathbf{i} + (x + y)\mathbf{j}$ *Hint:* Let $y = xv(x)$.

In Exercises 17–20, determine the field lines of the given polar vector fields.

17. $\mathbf{F} = \hat{\mathbf{r}} + r\hat{\boldsymbol{\theta}}$

18. $\mathbf{F} = \hat{\mathbf{r}} + \theta\hat{\boldsymbol{\theta}}$

19. $\mathbf{F} = 2\hat{\mathbf{r}} + \theta\hat{\boldsymbol{\theta}}$

20. $\mathbf{F} = r\hat{\mathbf{r}} - \hat{\boldsymbol{\theta}}$

15.2 Conservative Fields

Since the gradient of a scalar field is a vector field, it is natural to ask whether every vector field is the gradient of a scalar field. Given a vector field $\mathbf{F}(x, y, z)$, does there exist a scalar field $\phi(x, y, z)$ such that

$$\mathbf{F}(x, y, z) = \nabla\phi(x, y, z) = \frac{\partial\phi}{\partial x}\mathbf{i} + \frac{\partial\phi}{\partial y}\mathbf{j} + \frac{\partial\phi}{\partial z}\mathbf{k} ?$$

The answer in general is "no." Only special vector fields can be written in this way.

DEFINITION 1

> If $\mathbf{F}(x, y, z) = \nabla\phi(x, y, z)$ in a domain D, then we say that $\mathbf{F}$ is a **conservative** vector field in D, and we call the function ϕ a **(scalar) potential** for $\mathbf{F}$ on D. Similar definitions hold in the plane or in n-space.

Like antiderivatives, potentials are not determined uniquely; arbitrary constants can be added to them. Note that $\mathbf{F}$ is **conservative in a domain** D if and only if $\mathbf{F} = \nabla\phi$ at *every* point of D; the potential ϕ cannot have any singular points in D.

The equation $F_1(x, y, z)\,dx + F_2(x, y, z)\,dy + F_3(x, y, z)\,dz = 0$ is called an **exact** differential equation if the left side is the differential of a scalar function $\phi(x, y, z)$:

$$d\phi = F_1(x, y, z)\,dx + F_2(x, y, z)\,dy + F_3(x, y, z)\,dz.$$

In this case the differential equation has solutions given by $\phi(x, y, z) = C$ (constant). (See Section 17.3 for a discussion of exact equations in the plane.) Observe that the differential equation is exact if and only if the vector field $\mathbf{F} = F_1\mathbf{i} + F_2\mathbf{j} + F_3\mathbf{k}$ is conservative and that ϕ is the potential of $\mathbf{F}$.

Being scalar fields rather than vector fields, potentials for conservative vector fields are easier to manipulate algebraically than the vector fields themselves. For instance, a sum of potential functions is the potential function for the sum of the corresponding vector fields. A vector field can always be computed from its potential function by taking the gradient.

Example 1   (**The gravitational field of a point mass is conservative**) Show that the gravitational field $\mathbf{F}(\mathbf{r}) = -km(\mathbf{r} - \mathbf{r}_0)/|\mathbf{r} - \mathbf{r}_0|^3$ of Example 1 in Section 15.1 is conservative wherever it is defined (i.e., everywhere in $\mathbb{R}^3$ except at $\mathbf{r}_0$), by showing that

$$\phi(x, y, z) = \frac{km}{|\mathbf{r} - \mathbf{r}_0|} = \frac{km}{\sqrt{(x - x_0)^2 + (y - y_0)^2 + (z - z_0)^2}}$$

is a potential function for $\mathbf{F}$.

Solution Observe that

$$\frac{\partial \phi}{\partial x} = \frac{-km(x - x_0)}{\left((x - x_0)^2 + (y - y_0)^2 + (z - z_0)^2\right)^{3/2}} = \frac{-km(x - x_0)}{|\mathbf{r} - \mathbf{r}_0|^3} = F_1(\mathbf{r}),$$

and similar formulas hold for the other partial derivatives of ϕ. It follows that $\nabla\phi(x, y, z) = \mathbf{F}(x, y, z)$ for $(x, y, z) \neq (x_0, y_0, z_0)$, and $\mathbf{F}$ is conservative except at $\mathbf{r}_0$.

Remark It is not necessary to write the expression $km/|\mathbf{r} - \mathbf{r}_0|$ in terms of the components of $\mathbf{r} - \mathbf{r}_0$ as we did in Example 1 in order to calculate its partial derivatives. Here is a useful formula for the derivative of the length of a vector function $\mathbf{F}$ with respect to a variable x:

$$\frac{\partial}{\partial x}|\mathbf{F}| = \frac{\mathbf{F} \bullet \left(\dfrac{\partial}{\partial x}\mathbf{F}\right)}{|\mathbf{F}|}.$$

To see why this is true, express $|\mathbf{F}| = \sqrt{\mathbf{F} \bullet \mathbf{F}}$, and calculate its derivative using the Chain Rule and the Product Rule:

$$\frac{\partial}{\partial x}|\mathbf{F}| = \frac{\partial}{\partial x}\sqrt{\mathbf{F} \bullet \mathbf{F}} = \frac{1}{2\sqrt{\mathbf{F} \bullet \mathbf{F}}}2\mathbf{F} \bullet \left(\frac{\partial}{\partial x}\mathbf{F}\right) = \frac{\mathbf{F} \bullet \left(\dfrac{\partial}{\partial x}\mathbf{F}\right)}{|\mathbf{F}|}.$$

Compare this with the derivative of an absolute value of a function of one variable:

$$\frac{d}{dx}|f(x)| = \text{sgn}(f(x))\,f'(x) = \frac{f(x)}{|f(x)|}\,f'(x).$$

In the context of Example 1 we have

$$\frac{\partial}{\partial x}\frac{km}{|\mathbf{r} - \mathbf{r}_0|} = \frac{-km}{|\mathbf{r} - \mathbf{r}_0|^2}\frac{\partial}{\partial x}|\mathbf{r} - \mathbf{r}_0| = \frac{-km}{|\mathbf{r} - \mathbf{r}_0|^2}\frac{(\mathbf{r} - \mathbf{r}_0) \bullet \mathbf{i}}{|\mathbf{r} - \mathbf{r}_0|} = \frac{-km(x - x_0)}{|\mathbf{r} - \mathbf{r}_0|^3},$$

with similar expressions for the other partials of $km/|\mathbf{r} - \mathbf{r}_0|$.

Example 2 Show that the velocity field $\mathbf{v} = -\Omega y\mathbf{i} + \Omega x\mathbf{j}$ of rigid body rotation about the z-axis (see Example 2 of Section 15.1) is not conservative if $\Omega \neq 0$.

Solution There are two ways to show that no potential for $\mathbf{v}$ can exist. One way is to try to find a potential $\phi(x, y)$ for the vector field. We require

$$\frac{\partial \phi}{\partial x} = -\Omega y \qquad \text{and} \qquad \frac{\partial \phi}{\partial y} = \Omega x.$$

The first of these equations implies that $\phi(x, y) = -\Omega xy + C_1(y)$. (We have integrated with respect to x; the constant can still depend on y.) Similarly, the second equation implies that $\phi(x, y) = \Omega xy + C_2(x)$. Therefore, we must have $-\Omega xy + C_1(y) = \Omega xy + C_2(x)$, or $2\Omega xy = C_1(y) - C_2(x)$ for all (x, y). This is not possible for any choice of the single-variable functions $C_1(y)$ and $C_2(x)$ unless $\Omega = 0$.

Alternatively, if **v** has a potential ϕ, then we can form the mixed partial derivatives of ϕ from the two equations above and get

$$\frac{\partial^2 \phi}{\partial y \partial x} = -\Omega \quad \text{and} \quad \frac{\partial^2 \phi}{\partial x \partial y} = \Omega.$$

This is not possible if $\Omega \neq 0$ because the smoothness of **v** implies that its potential should be smooth, so the mixed partials should be equal. Thus, no such ϕ can exist; **v** is not conservative. ∎

Example 2 suggests a condition that must be satisfied by any conservative plane vector field.

Necessary condition for a conservative plane vector field

If $\mathbf{F}(x, y) = F_1(x, y)\mathbf{i} + F_2(x, y)\mathbf{j}$ is a conservative vector field in a domain D of the xy-plane, then the condition

$$\frac{\partial}{\partial y} F_1(x, y) = \frac{\partial}{\partial x} F_2(x, y)$$

must be satisfied at all points of D.

To see this, observe that

$$F_1\mathbf{i} + F_2\mathbf{j} = \mathbf{F} = \nabla\phi = \frac{\partial\phi}{\partial x}\mathbf{i} + \frac{\partial\phi}{\partial y}\mathbf{j}$$

implies the two scalar equations

$$F_1 = \frac{\partial\phi}{\partial x} \quad \text{and} \quad F_2 = \frac{\partial\phi}{\partial y},$$

and since the mixed partial derivatives of ϕ should be equal,

$$\frac{\partial F_1}{\partial y} = \frac{\partial^2 \phi}{\partial y \partial x} = \frac{\partial^2 \phi}{\partial x \partial y} = \frac{\partial F_2}{\partial x}.$$

A similar condition obtains for vector fields in 3-space.

Necessary conditions for a conservative vector field in 3-space

If $\mathbf{F}(x, y, z) = F_1(x, y, z)\mathbf{i} + F_2(x, y, z)\mathbf{j} + F_3(x, y, z)\mathbf{k}$ is a conservative vector field in a domain D in 3-space, then we must have, everywhere in D,

$$\frac{\partial F_1}{\partial y} = \frac{\partial F_2}{\partial x}, \quad \frac{\partial F_1}{\partial z} = \frac{\partial F_3}{\partial x}, \quad \frac{\partial F_2}{\partial z} = \frac{\partial F_3}{\partial y}.$$

Equipotential Surfaces and Curves

If $\phi(x, y, z)$ is a potential function for the conservative vector field **F**, then the *level surfaces* $\phi(x, y, z) = C$ of ϕ are called **equipotential surfaces** of **F**. Since $\mathbf{F} = \nabla\phi$ is normal to these surfaces (wherever it does not vanish), the field lines of **F** always intersect the equipotential surfaces at right angles. For instance, the equipotential surfaces of the gravitational force field of a point mass are spheres centred at the point; these spheres are normal to the field lines, which are straight lines passing through the point. Similarly, for a conservative plane vector field, the *level curves* of the potential function are called **equipotential curves** of the vector field. They are the **orthogonal trajectories** of the field lines; that is, they intersect the field lines at right angles.

Example 3   Show that the vector field $\mathbf{F}(x, y) = x\mathbf{i} - y\mathbf{j}$ is conservative and find a potential function for it. Describe the field lines and the equipotential curves.

Solution   Since $\partial F_1 / \partial y = 0 = \partial F_2 / \partial x$ everywhere in $\mathbb{R}^2$, we would expect $\mathbf{F}$ to be conservative. Any potential function ϕ must satisfy

$$\frac{\partial \phi}{\partial x} = F_1 = x \qquad \text{and} \qquad \frac{\partial \phi}{\partial y} = F_2 = -y.$$

The first of these equations gives

$$\phi(x, y) = \int x \, dx = \frac{1}{2} x^2 + C_1(y).$$

Observe that, since the integral is taken with respect to x, the "constant" of integration is allowed to depend on the other variable. Now we use the second equation to get

$$-y = \frac{\partial \phi}{\partial y} = C_1'(y) \quad \Rightarrow \quad C_1(y) = -\frac{1}{2} y^2 + C_2.$$

Thus, $\mathbf{F}$ is conservative and, for any constant C_2,

$$\phi(x, y) = \frac{x^2 - y^2}{2} + C_2$$

is a potential function for $\mathbf{F}$. The field lines of $\mathbf{F}$ satisfy

$$\frac{dx}{x} = -\frac{dy}{y} \quad \Rightarrow \quad \ln|x| = -\ln|y| + \ln C_3 \quad \Rightarrow \quad xy = C_3.$$

The field lines of $\mathbf{F}$ are thus rectangular hyperbolas with the coordinate axes as asymptotes. The equipotential curves constitute another family of rectangular hyperbolas, $x^2 - y^2 = C_4$, with the lines $x = \pm y$ as asymptotes. Curves of the two families intersect at right angles. (See Figure 15.5.) Note, however, that $\mathbf{F}$ does not specify a direction at the origin and the orthogonality breaks down there; in fact, neither family has a unique curve through that point. ∎

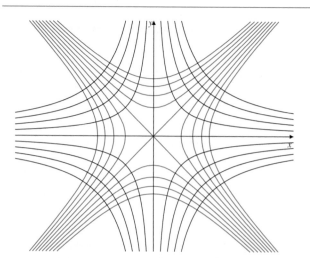

Figure 15.5   The field lines (black) and equipotential curves (colour) for the field $\mathbf{F} = x\mathbf{i} - y\mathbf{j}$

Remark   In the above example we constructed the potential ϕ by first integrating $\partial \phi / \partial x = F_1$. We could equally well have started by integrating $\partial \phi / \partial y = F_2$, in which case the constant of integration would have depended on x. In the end the same ϕ would have emerged.

Example 4   Decide whether the vector field

$$\mathbf{F} = \left(xy - \sin z\right)\mathbf{i} + \left(\frac{1}{2}x^2 - \frac{e^y}{z}\right)\mathbf{j} + \left(\frac{e^y}{z^2} - x\cos z\right)\mathbf{k}$$

is conservative in $D = \{(x, y, z) : z \neq 0\}$, and find a potential if it is.

Solution   Note that $\mathbf{F}$ is not defined when $z = 0$. However, since

$$\frac{\partial F_1}{\partial y} = x = \frac{\partial F_2}{\partial x}, \quad \frac{\partial F_1}{\partial z} = -\cos z = \frac{\partial F_3}{\partial x}, \quad \text{and} \quad \frac{\partial F_2}{\partial z} = \frac{e^y}{z^2} = \frac{\partial F_3}{\partial y},$$

$\mathbf{F}$ may still be conservative in domains not intersecting the xy-plane $z = 0$. If so, its potential ϕ should satisfy

$$\frac{\partial \phi}{\partial x} = xy - \sin z, \quad \frac{\partial \phi}{\partial y} = \frac{1}{2}x^2 - \frac{e^y}{z}, \quad \text{and} \quad \frac{\partial \phi}{\partial z} = \frac{e^y}{z^2} - x\cos z. \qquad (*)$$

From the first equation of $(*)$,

$$\phi(x, y, z) = \int (xy - \sin z)\, dx = \frac{1}{2}x^2 y - x\sin z + C_1(y, z).$$

(Again, note that the constant of integration can be a function of any parameters of the integrand; it is constant only with respect to the variable of integration.) Using the second equation of $(*)$, we obtain

$$\frac{1}{2}x^2 - \frac{e^y}{z} = \frac{\partial \phi}{\partial y} = \frac{1}{2}x^2 + \frac{\partial C_1(y, z)}{\partial y}.$$

Thus,

$$C_1(y, z) = -\int \frac{e^y}{z}\, dy = -\frac{e^y}{z} + C_2(z)$$

and

$$\phi(x, y, z) = \frac{1}{2}x^2 y - x\sin z - \frac{e^y}{z} + C_2(z).$$

Finally, using the third equation of $(*)$,

$$\frac{e^y}{z^2} - x\cos z = \frac{\partial \phi}{\partial z} = -x\cos z + \frac{e^y}{z^2} + C_2'(z).$$

Thus $C_2'(z) = 0$ and $C_2(z) = C$ (a constant). Indeed, $\mathbf{F}$ is conservative and, for any constant C,

$$\phi(x, y, z) = \frac{1}{2}x^2 y - x\sin z - \frac{e^y}{z} + C$$

is a potential function for $\mathbf{F}$ in the given domain D. C may have different values in the two regions $z > 0$ and $z < 0$ whose union constitutes D. ∎

Remark   If, in the above solution, the differential equation for $C_1(y, z)$ had involved x or if that for $C_2(z)$ had involved either x or y, we would not have been able to find ϕ. This did not happen because of the three conditions on the partials of F_1, F_2, and F_3 verified at the outset.

Remark The existence of a potential for a vector field depends on the *topology* of the domain of the field (i.e., whether the domain has *holes* in it and what kind of holes) as well as on the structure of the components of the field itself. (Even if the necessary conditions given above are satisfied, a vector field may not be conservative in a domain that has *holes*.) We will be probing further into the nature of conservative vector fields in Section 15.4 and in the next chapter; we will eventually show that the above *necessary conditions* are also *sufficient* to guarantee that $\mathbf{F}$ is conservative if the domain of $\mathbf{F}$ satisfies certain conditions. At this point, however, we give an example in which a plane vector field fails to be conservative on a domain where the necessary condition is, nevertheless, satisfied.

Example 5   For $(x, y) \neq (0, 0)$, define a vector field $\mathbf{F}(x, y)$ and a scalar field $\theta(x, y)$ as follows:

$$\mathbf{F}(x, y) = \left(\frac{-y}{x^2 + y^2} \right) \mathbf{i} + \left(\frac{x}{x^2 + y^2} \right) \mathbf{j}$$

$\theta(x, y) =$ the polar angle θ of (x, y) such that $0 \le \theta < 2\pi$.

Thus, $x = r \cos \theta(x, y)$ and $y = r \sin \theta(x, y)$, where $r^2 = x^2 + y^2$. Verify the following:

(a) $\dfrac{\partial}{\partial y} F_1(x, y) = \dfrac{\partial}{\partial x} F_2(x, y)$ for $(x, y) \neq (0, 0)$.

(b) $\nabla\theta(x, y) = \mathbf{F}(x, y)$ for all $(x, y) \neq (0, 0)$ such that $0 < \theta < 2\pi$.

(c) $\mathbf{F}$ is not conservative on the whole xy-plane excluding the origin.

Solution

(a) We have $F_1 = \dfrac{-y}{x^2 + y^2}$ and $F_2 = \dfrac{x}{x^2 + y^2}$. Thus

$$\frac{\partial}{\partial y} F_1(x, y) = \frac{\partial}{\partial y}\left(-\frac{y}{x^2 + y^2} \right) = \frac{y^2 - x^2}{(x^2 + y^2)^2} = \frac{\partial}{\partial x}\left(\frac{x}{x^2 + y^2} \right)$$

$$= \frac{\partial}{\partial x} F_2(x, y)$$

for all $(x, y) \neq (0, 0)$.

(b) Differentiate the equations $x = r \cos \theta$ and $y = r \sin \theta$ implicitly with respect to x to obtain

$$1 = \frac{\partial x}{\partial x} = \frac{\partial r}{\partial x} \cos \theta - r \sin \theta \frac{\partial \theta}{\partial x},$$

$$0 = \frac{\partial y}{\partial x} = \frac{\partial r}{\partial x} \sin \theta + r \cos \theta \frac{\partial \theta}{\partial x}.$$

Eliminating $\partial r/\partial x$ from this pair of equations and solving for $\partial \theta/\partial x$ leads to

$$\frac{\partial \theta}{\partial x} = -\frac{r \sin \theta}{r^2} = -\frac{y}{x^2 + y^2} = F_1.$$

Similarly, differentiation with respect to y produces

$$\frac{\partial \theta}{\partial y} = \frac{x}{x^2 + y^2} = F_2.$$

These formulas hold only if $0 < \theta < 2\pi$; θ is not even continuous on the positive x-axis; if $x > 0$, then

$$\lim_{y \to 0+} \theta(x, y) = 0 \qquad \text{but} \qquad \lim_{y \to 0-} \theta(x, y) = 2\pi.$$

Thus, $\nabla\theta = \mathbf{F}$ holds everywhere in the plane except at points $(x, 0)$ where $x \geq 0$.

(c) Suppose that $\mathbf{F}$ is conservative on the whole plane excluding the origin. Then $\mathbf{F} = \nabla\phi$ there, for some scalar function $\phi(x, y)$. Then $\nabla(\theta - \phi) = \mathbf{0}$ for $0 < \theta < 2\pi$, and $\theta - \phi = C$ (constant), or $\theta = \phi + C$. The left side of this equation is discontinuous along the positive x-axis but the right side is not. Therefore the two sides cannot be equal. This contradiction shows that $\mathbf{F}$ cannot be conservative on the whole plane excluding the origin.

∎

Remark Observe that the origin $(0, 0)$ is a *hole* in the domain of $\mathbf{F}$ in the above example. While $\mathbf{F}$ satisfies the necessary condition for being conservative everywhere except at this hole, you must remove from the domain of $\mathbf{F}$ a half-line (ray), or, more generally, a curve from the origin to infinity in order to get a potential function for $\mathbf{F}$. $\mathbf{F}$ is *not* conservative on any domain containing a curve that surrounds the origin. Exercises 22–24 of Section 15.4 will shed further light on this situation.

Sources, Sinks, and Dipoles

Imagine that 3-space is filled with an incompressible fluid emitted by a point source at the origin at a volume rate $dV/dt = 4\pi m$. (We say that the origin is a **source** of strength m.) By symmetry, the fluid flows outward on radial lines from the origin with equal speed at equal distances from the origin in all directions, and the fluid emitted at the origin at some instant $t = 0$ will at later time t be spread over a spherical surface of radius $r = r(t)$. All the fluid inside that sphere was emitted in the time interval $[0, t]$, so we have

$$\frac{4}{3}\pi r^3 = 4\pi m t.$$

Differentiating this equation with respect to t we obtain $r^2(dr/dt) = m$, and the outward speed of the fluid at distance r from the origin is $v(r) = m/r^2$. The velocity field of the moving fluid is therefore

$$\mathbf{v}(\mathbf{r}) = v(r)\frac{\mathbf{r}}{|\mathbf{r}|} = \frac{m}{r^3}\mathbf{r}.$$

This velocity field is conservative (except at the origin) and has potential

$$\phi(\mathbf{r}) = -\frac{m}{r}.$$

A **sink** is a negative source. A sink of strength m at the origin (which annihilates or sucks up fluid at a rate $dV/dt = 4\pi m$) has velocity field and potential given by

$$\mathbf{v}(\mathbf{r}) = -\frac{m}{r^3}\mathbf{r} \qquad \text{and} \qquad \phi(\mathbf{r}) = \frac{m}{r}.$$

The potentials or velocity fields of sources or sinks located at other points are obtained by translation of these formulas; for instance, the velocity field of a source of strength m at the point with position vector $\mathbf{r}_0$ is

$$\mathbf{v}(\mathbf{r}) = -\nabla\left(\frac{m}{|\mathbf{r} - \mathbf{r}_0|}\right) = \frac{m}{|\mathbf{r} - \mathbf{r}_0|^3}(\mathbf{r} - \mathbf{r}_0).$$

This should be compared with the gravitational force field due to a mass m at the origin. The two are the same except for sign and a constant related to units of measurement. For this reason we regard a point mass as a sink for its own gravitational field. Similarly, the electrostatic field due to a point charge q at $\mathbf{r}_0$ is the field of a source (or sink if $q < 0$) of strength proportional to q; if units of measurement are suitably chosen we have

$$\mathbf{E}(\mathbf{r}) = -\nabla\left(\frac{q}{|\mathbf{r} - \mathbf{r}_0|}\right) = \frac{q}{|\mathbf{r} - \mathbf{r}_0|^3}(\mathbf{r} - \mathbf{r}_0).$$

In general, the field lines of a vector field converge at a source or sink of that field.

A **dipole** is a system consisting of a source and a sink of equal strength m separated by a short distance ℓ. The product $\mu = m\ell$ is called the **dipole moment**, and the line containing the source and sink is called the **axis** of the dipole. Real physical dipoles, such as magnets, are frequently modelled by *ideal* dipoles that are the limits of such real dipoles as $m \to \infty$ and $\ell \to 0$ in such a way that the dipole moment μ remains constant.

Example 6   Calculate the velocity field, $\mathbf{v}(x, y, z)$, associated with a dipole of moment μ located at the origin and having axis along the z-axis.

Solution   We start with a source of strength m at position $(0, 0, \ell/2)$ and a sink of strength m at $(0, 0, -\ell/2)$. The potential of this system is

$$\phi(\mathbf{r}) = -m \left(\frac{1}{|\mathbf{r} - \frac{1}{2}\ell\mathbf{k}|} - \frac{1}{|\mathbf{r} + \frac{1}{2}\ell\mathbf{k}|} \right).$$

The potential of the ideal dipole is the limit of the potential of this system as $m \to \infty$ and $\ell \to 0$ in such a way that $m\ell = \mu$:

$$\phi(\mathbf{r}) = \lim_{\substack{\ell \to 0 \\ m\ell = \mu}} -m \left(\frac{|\mathbf{r} + \frac{1}{2}\ell\mathbf{k}| - |\mathbf{r} - \frac{1}{2}\ell\mathbf{k}|}{|\mathbf{r} + \frac{1}{2}\ell\mathbf{k}| \, |\mathbf{r} - \frac{1}{2}\ell\mathbf{k}|} \right)$$

$$= -\frac{\mu}{|\mathbf{r}|^2} \lim_{\ell \to 0} \frac{|\mathbf{r} + \frac{1}{2}\ell\mathbf{k}| - |\mathbf{r} - \frac{1}{2}\ell\mathbf{k}|}{\ell}$$

(now use l'Hôpital's Rule and the rule for differentiating lengths of vectors)

$$= -\frac{\mu}{|\mathbf{r}|^2} \lim_{\ell \to 0} \frac{\dfrac{\left(\mathbf{r} + \frac{1}{2}\ell\mathbf{k}\right) \bullet \frac{1}{2}\mathbf{k}}{|\mathbf{r} + \frac{1}{2}\ell\mathbf{k}|} - \dfrac{\left(\mathbf{r} - \frac{1}{2}\ell\mathbf{k}\right) \bullet \left(-\frac{1}{2}\mathbf{k}\right)}{|\mathbf{r} - \frac{1}{2}\ell\mathbf{k}|}}{1}$$

$$= -\frac{\mu}{|\mathbf{r}|^2} \lim_{\ell \to 0} \left(\frac{\frac{1}{2}z + \frac{1}{4}\ell}{|\mathbf{r} + \frac{1}{2}\ell\mathbf{k}|} + \frac{\frac{1}{2}z - \frac{1}{4}\ell}{|\mathbf{r} - \frac{1}{2}\ell\mathbf{k}|} \right)$$

$$= -\frac{\mu z}{|\mathbf{r}|^3}.$$

The required velocity field is the gradient of this potential. We have

$$\frac{\partial \phi}{\partial x} = \frac{3\mu z}{|\mathbf{r}|^4} \frac{\mathbf{r} \bullet \mathbf{i}}{|\mathbf{r}|} = \frac{3\mu x z}{|\mathbf{r}|^5}$$

$$\frac{\partial \phi}{\partial y} = \frac{3\mu y z}{|\mathbf{r}|^5}$$

$$\frac{\partial \phi}{\partial z} = -\frac{\mu}{|\mathbf{r}|^3} + \frac{3\mu z^2}{|\mathbf{r}|^5} = \frac{\mu(2z^2 - x^2 - y^2)}{|\mathbf{r}|^5}$$

$$\mathbf{v}(\mathbf{r}) = \nabla\phi(\mathbf{r}) = \frac{\mu}{|\mathbf{r}|^5} \left(3xz\mathbf{i} + 3yz\mathbf{j} + (2z^2 - x^2 - y^2)\mathbf{k} \right).$$

Some streamlines for a plane cross-section containing the z-axis are shown in Figure 15.6.

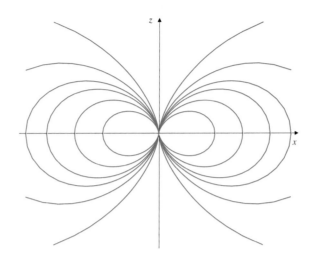

Figure 15.6   Streamlines of a dipole

Exercises 15.2

In Exercises 1–6, determine whether the given vector field is conservative, and find a potential if it is.

1. $\mathbf{F}(x, y, z) = x\mathbf{i} - 2y\mathbf{j} + 3z\mathbf{k}$

2. $\mathbf{F}(x, y, z) = y\mathbf{i} + x\mathbf{j} + z^2\mathbf{k}$

3. $\mathbf{F}(x, y) = \dfrac{x\mathbf{i} - y\mathbf{j}}{x^2 + y^2}$ **4.** $\mathbf{F}(x, y) = \dfrac{x\mathbf{i} + y\mathbf{j}}{x^2 + y^2}$

5. $\mathbf{F}(x, y, z) = (2xy - z^2)\mathbf{i} + (2yz + x^2)\mathbf{j} - (2zx - y^2)\mathbf{k}$

6. $\mathbf{F}(x, y, z) = e^{x^2+y^2+z^2}(xz\mathbf{i} + yz\mathbf{j} + xy\mathbf{k})$

7. Find the three-dimensional vector field with potential
$$\phi(\mathbf{r}) = \frac{1}{|\mathbf{r} - \mathbf{r}_0|^2}.$$

8. Calculate $\nabla \ln |\mathbf{r}|$, where $\mathbf{r} = x\mathbf{i} + y\mathbf{j} + z\mathbf{k}$.

9. Show that the vector field
$$\mathbf{F}(x, y, z) = \frac{2x}{z}\mathbf{i} + \frac{2y}{z}\mathbf{j} - \frac{x^2 + y^2}{z^2}\mathbf{k}$$
is conservative, and find its potential. Describe the equipotential surfaces. Find the field lines of $\mathbf{F}$.

10. Repeat Exercise 9 for the field
$$\mathbf{F}(x, y, z) = \frac{2x}{z}\mathbf{i} + \frac{2y}{z}\mathbf{j} + \left(1 - \frac{x^2 + y^2}{z^2}\right)\mathbf{k}.$$

11. Find the velocity field due to two sources of strength m, one located at $(0, 0, \ell)$ and the other at $(0, 0, -\ell)$. Where is the velocity zero? Find the velocity at any point $(x, y, 0)$ in the xy-plane. Where in the xy-plane is the speed greatest?

∗12. Find the velocity field for a system consisting of a source of strength 2 at the origin and a sink of strength 1 at $(0, 0, 1)$. Show that the velocity is vertical at all points of a certain sphere. Sketch the streamlines of the flow.

Exercises 13–18 provide an analysis of two-dimensional sources and dipoles similar to that developed for three dimensions in the text.

13. In 3-space filled with an incompressible fluid, we say that the z-axis is a **line source** of strength m if every interval Δz

along that axis emits fluid at volume rate $dV/dt = 2\pi m \Delta z$. The fluid then spreads out symmetrically in all directions perpendicular to the z-axis. Show that the velocity field of the flow is
$$\mathbf{v} = \frac{m}{x^2 + y^2}(x\mathbf{i} + y\mathbf{j}).$$

14. The flow in Exercise 13 is two-dimensional because $\mathbf{v}$ depends only on x and y and has no component in the z direction. Regarded as a *plane* vector field, it is the field of a two-dimensional point source of strength m located at the origin (i.e., fluid is emitted at the origin at the *areal rate* $dA/dt = 2\pi m$). Show that the vector field is conservative, and find a potential function $\phi(x, y)$ for it.

∗15. Find the potential, ϕ, and the field, $\mathbf{F} = \nabla\phi$, for a two-dimensional dipole at the origin, with axis in the y direction and dipole moment μ. Such a dipole is the limit of a system consisting of a source of strength m at $(0, \ell/2)$ and a sink of strength m at $(0, -\ell/2)$, as $\ell \to 0$ and $m \to \infty$ such that $m\ell = \mu$.

∗16. Show that the equipotential curves of the two-dimensional dipole in Exercise 15 are circles tangent to the x-axis at the origin.

∗17. Show that the streamlines (field lines) of the two-dimensional dipole in Exercises 15 and 16 are circles tangent to the y-axis at the origin. *Hint:* It is possible to do this geometrically. If you choose to do it by setting up a differential equation, you may find the change of dependent variable
$$y = vx, \qquad \frac{dy}{dx} = v + x\frac{dv}{dx}$$
useful for integrating the equation.

∗18. Show that the velocity field of a line source of strength $2m$ can be found by integrating the (three-dimensional) velocity field of a point source of strength $m\,dz$ at $(0, 0, z)$ over the whole z-axis. Why does the integral correspond to a line source of strength $2m$ rather than strength m? Can the potential of the line source be obtained by integrating the potentials of the point sources?

19. Show that the gradient of a function expressed in terms of polar coordinates in the plane is

$$\nabla\phi(r,\theta) = \frac{\partial\phi}{\partial r}\hat{\mathbf{r}} + \frac{1}{r}\frac{\partial\phi}{\partial\theta}\hat{\boldsymbol{\theta}}.$$

(This is a repeat of Exercise 16 in Section 12.7.)

20. Use the result of Exercise 19 to show that a necessary condition for the vector field

$$\mathbf{F}(r,\theta) = F_r(r,\theta)\hat{\mathbf{r}} + F_\theta(r,\theta)\hat{\boldsymbol{\theta}}$$

(expressed in terms of polar coordinates) to be conservative is that

$$\frac{\partial F_r}{\partial\theta} - r\frac{\partial F_\theta}{\partial r} = F_\theta.$$

21. Show that $\mathbf{F} = r\sin 2\theta\,\hat{\mathbf{r}} + r\cos 2\theta\,\hat{\boldsymbol{\theta}}$ is conservative, and find a potential for it.

22. For what values of the constants α and β is the vector field

$$\mathbf{F} = r^2\cos\theta\,\hat{\mathbf{r}} + \alpha r^\beta \sin\theta\,\hat{\boldsymbol{\theta}}$$

conservative? Find a potential for $\mathbf{F}$ if α and β have these values.

15.3 Line Integrals

The definite integral, $\int_a^b f(x)\,dx$, represents the *total amount* of a quantity distributed along the x-axis between a and b in terms of the *line density*, $f(x)$, of that quantity at point x. The amount of the quantity in an *infinitesimal* interval of length dx at x is $f(x)\,dx$, and the integral adds up these infinitesimal contributions (or *elements*) to give the total amount of the quantity. Similarly, the integrals $\iint_D f(x,y)\,dA$ and $\iiint_R f(x,y,z)\,dV$ represent the total amounts of quantities distributed over regions D in the plane and R in 3-space in terms of the *areal* or *volume* densities of these quantities.

It may happen that a quantity is distributed with specified line density along a *curve* in the plane or in 3-space, or with specified areal density over a *surface* in 3-space. In such cases we require *line integrals* or *surface integrals* to add up the contributing elements and calculate the total quantity. We examine line integrals in this section and the next and surface integrals in Sections 15.5 and 15.6.

Let $\mathcal{C}$ be a bounded, continuous parametric curve in $\mathbb{R}^3$. Recall (from Section 11.1) that $\mathcal{C}$ is a *smooth curve* if it has a parametrization of the form

$$\mathbf{r} = \mathbf{r}(t) = x(t)\mathbf{i} + y(t)\mathbf{j} + z(t)\mathbf{k}, \qquad t \text{ in interval } I,$$

with "velocity" vector $\mathbf{v} = d\mathbf{r}/dt$ continuous and nonzero. We will call $\mathcal{C}$ a **smooth arc** if it is a smooth curve with *finite* parameter interval $I = [a, b]$.

In Section 11.3 we saw how to calculate the length of $\mathcal{C}$ by subdividing it into short arcs using points corresponding to parameter values

$$a = t_0 < t_1 < t_2 < \cdots < t_{n-1} < t_n = b,$$

adding up the lengths $|\Delta\mathbf{r}_i| = |\mathbf{r}_i - \mathbf{r}_{i-1}|$ of line segments joining these points, and taking the limit as the maximum distance between adjacent points approached zero. The length was denoted

$$\int_\mathcal{C} ds$$

and is a special example of a line integral along $\mathcal{C}$ having integrand 1.

The line integral of a general function $f(x, y, z)$ can be defined similarly. We choose a point (x_i^*, y_i^*, z_i^*) on the ith subarc and form the Riemann sum

$$S_n = \sum_{i=1}^n f(x_i^*, y_i^*, z_i^*)\,|\Delta\mathbf{r}_i|.$$

If this sum has a limit as max $|\Delta \mathbf{r}_i| \to 0$, independent of the particular choices of the points (x_i^*, y_i^*, z_i^*), then we call this limit the **line integral** of f along $\mathcal{C}$ and denote it

$$\int_{\mathcal{C}} f(x, y, z)\, ds.$$

If $\mathcal{C}$ is a smooth arc and if f is continuous on $\mathcal{C}$, then the limit will certainly exist; its value is given by a definite integral of a continuous function, as shown in the next paragraph. It will also exist (for continuous f) if $\mathcal{C}$ is **piecewise smooth**, consisting of finitely many smooth arcs linked end to end; in this case the line integral of f along $\mathcal{C}$ is the sum of the line integrals of f along each of the smooth arcs. Improper line integrals can also be considered, where f has discontinuities or where the length of a curve is not finite.

Evaluating Line Integrals

The length of $\mathcal{C}$ was evaluated by expressing the arc length element $ds = |d\mathbf{r}/dt|\, dt$ in terms of a parametrization $\mathbf{r} = \mathbf{r}(t)$, $(a \le t \le b)$ of the curve, and integrating this from $t = a$ to $t = b$:

$$\text{length of } \mathcal{C} = \int_{\mathcal{C}} ds = \int_a^b \left|\frac{d\mathbf{r}}{dt}\right| dt.$$

More general line integrals are evaluated similarly:

$$\int_{\mathcal{C}} f(x, y, z)\, ds = \int_a^b f\big(\mathbf{r}(t)\big) \left|\frac{d\mathbf{r}}{dt}\right| dt.$$

Of course, all of the above discussion applies equally well to line integrals of functions $f(x, y)$ along curves $\mathcal{C}$ in the xy-plane.

Remark It should be noted that the value of the line integral of a function f along a curve $\mathcal{C}$ depends on f and $\mathcal{C}$ but not on the particular way $\mathcal{C}$ is parametrized. If $\mathbf{r} = \mathbf{r}^*(u)$, $\alpha \le u \le \beta$, is another parametrization of the same smooth curve $\mathcal{C}$, then any point $\mathbf{r}(t)$ on $\mathcal{C}$ can be expressed in terms of the new parametrization as $\mathbf{r}^*(u)$, where u depends on t: $u = u(t)$. If $\mathbf{r}^*(u)$ traces $\mathcal{C}$ in the same direction as $\mathbf{r}(t)$, then $u(a) = \alpha$, $u(b) = \beta$, and $du/dt \ge 0$; if $\mathbf{r}^*(u)$ traces $\mathcal{C}$ in the opposite direction, then $u(a) = \beta$, $u(b) = \alpha$, and $du/dt \le 0$. In either event,

$$\int_a^b f\big(\mathbf{r}(t)\big) \left|\frac{d\mathbf{r}}{dt}\right| dt = \int_a^b f\big(\mathbf{r}^*(u(t))\big) \left|\frac{d\mathbf{r}^*}{du}\frac{du}{dt}\right| dt = \int_\alpha^\beta f\big(\mathbf{r}^*(u)\big) \left|\frac{d\mathbf{r}^*}{du}\right| du.$$

Thus, the line integral is *independent of parametrization* of the curve $\mathcal{C}$. The following example illustrates this fact.

Example 1 A circle of radius $a > 0$ has centre at the origin in the xy-plane. Let $\mathcal{C}$ be the half of this circle lying in the half-plane $y \ge 0$. Use two different parametrizations of $\mathcal{C}$ to find the moment of $\mathcal{C}$ about $y = 0$.

Solution We are asked to calculate $\int_{\mathcal{C}} y\, ds$. $\mathcal{C}$ can be parametrized as $\mathbf{r} = a \cos t\, \mathbf{i} + a \sin t\, \mathbf{j}$, $(0 \le t \le \pi)$. Therefore,

$$\frac{d\mathbf{r}}{dt} = -a \sin t\, \mathbf{i} + a \cos t\, \mathbf{j} \qquad \text{and} \qquad \left|\frac{d\mathbf{r}}{dt}\right| = a,$$

and the moment of $\mathcal{C}$ about $y = 0$ is

$$\int_{\mathcal{C}} y\, ds = \int_0^\pi a \sin t\, a\, dt = -a^2 \cos t \Big|_0^\pi = 2a^2.$$

$\mathcal{C}$ can also be parametrized $\mathbf{r} = x\mathbf{i} + \sqrt{a^2 - x^2}\,\mathbf{j}$, $(-a \le x \le a)$, for which we have

$$\frac{d\mathbf{r}}{dx} = \mathbf{i} - \frac{x}{\sqrt{a^2 - x^2}}\mathbf{j},$$

$$\left|\frac{d\mathbf{r}}{dx}\right| = \sqrt{1 + \frac{x^2}{a^2 - x^2}} = \frac{a}{\sqrt{a^2 - x^2}}.$$

Thus, the moment of $\mathcal{C}$ about $y = 0$ is

$$\int_{\mathcal{C}} y\,ds = \int_{-a}^{a} \sqrt{a^2 - x^2}\,\frac{a}{\sqrt{a^2 - x^2}}\,dx = a\int_{-a}^{a} dx = 2a^2.$$

It is comforting to get the same answer using different parametrizations. Unlike the line integrals of vector fields considered in the next section, the line integrals of scalar fields considered here do not depend on the direction (orientation) of $\mathcal{C}$. The two parametrizations of the semicircle were in opposite directions but still gave the same result.

Line integrals frequently lead to definite integrals that are very difficult or impossible to evaluate without using numerical techniques. Only very simple curves and ones that have been contrived to lead to simple expressions for ds are amenable to exact calculation of line integrals.

Example 2   Find the centroid of the circular helix $\mathcal{C}$ given by

$$\mathbf{r} = a\,\cos t\,\mathbf{i} + a\,\sin t\,\mathbf{j} + bt\,\mathbf{k}, \qquad 0 \le t \le 2\pi.$$

Solution   As we observed in Example 5 of Section 11.3, for this helix $ds = \sqrt{a^2 + b^2}\,dt$. On the helix we have $z = bt$, so its moment about $z = 0$ is

$$M_{z=0} = \int_{\mathcal{C}} z\,ds = b\sqrt{a^2 + b^2}\int_0^{2\pi} t\,dt = 2\pi^2 b\sqrt{a^2 + b^2}.$$

Since the helix has length $L = 2\pi\sqrt{a^2 + b^2}$, the z-component of its centroid is $M_{z=0}/L = \pi b$. The moment of the helix about $x = 0$ is

$$M_{x=0} = \int_{\mathcal{C}} x\,ds = a\sqrt{a^2 + b^2}\int_0^{2\pi} \cos t\,dt = 0,$$

$$M_{y=0} = \int_{\mathcal{C}} y\,ds = a\sqrt{a^2 + b^2}\int_0^{2\pi} \sin t\,dt = 0.$$

Thus the centroid is $(0, 0, \pi b)$.

Sometimes a curve, along which a line integral is to be taken, is specified as the intersection of two surfaces with given equations. It is normally necessary to parametrize the curve in order to evaluate a line integral. Recall from Section 11.3 that if one of the surfaces is a cylinder parallel to one of the coordinate axes, it is usually easiest to begin by parametrizing that cylinder. (Otherwise, combine the equations to eliminate one variable and thus obtain such a cylinder on which the curve lies.)

Figure 15.7 The curve of intersection of $z = x^2$ and $z = 2 - x^2 - 2y^2$

Example 3 Find the mass of a wire lying along the first octant part $\mathcal{C}$ of the curve of intersection of the elliptic paraboloid $z = 2 - x^2 - 2y^2$ and the parabolic cylinder $z = x^2$ between $(0, 1, 0)$ and $(1, 0, 1)$ (see Figure 15.7) if the density of the wire at position (x, y, z) is $\delta(x, y, z) = xy$.

Solution We need a convenient parametrization of $\mathcal{C}$. Since the curve $\mathcal{C}$ lies on the cylinder $z = x^2$ and x goes from 0 to 1, we can let $x = t$ and $z = t^2$. Thus, $2y^2 = 2 - x^2 - z = 2 - 2t^2$, so $y^2 = 1 - t^2$. Since $\mathcal{C}$ lies in the first octant, it can be parametrized by

$$x = t, \qquad y = \sqrt{1 - t^2}, \qquad z = t^2, \qquad (0 \le t \le 1).$$

Then $dx/dt = 1$, $dy/dt = -t/\sqrt{1 - t^2}$, and $dz/dt = 2t$, so

$$ds = \sqrt{1 + \frac{t^2}{1 - t^2} + 4t^2}\, dt = \frac{\sqrt{1 + 4t^2 - 4t^4}}{\sqrt{1 - t^2}}\, dt.$$

Hence, the mass of the wire is

$$
\begin{aligned}
m &= \int_{\mathcal{C}} xy\, ds = \int_0^1 t\sqrt{1 - t^2}\, \frac{\sqrt{1 + 4t^2 - 4t^4}}{\sqrt{1 - t^2}}\, dt \\
&= \int_0^1 t\sqrt{1 + 4t^2 - 4t^4}\, dt && \text{Let } u = t^2 \\
&= \frac{1}{2} \int_0^1 \sqrt{1 + 4u - 4u^2}\, du \\
&= \frac{1}{2} \int_0^1 \sqrt{2 - (2u - 1)^2}\, du && \text{Let } v = 2u - 1 \\
&= \frac{1}{4} \int_{-1}^1 \sqrt{2 - v^2}\, dv = \frac{1}{2} \int_0^1 \sqrt{2 - v^2}\, dv \\
&= \frac{1}{2}\left(\frac{\pi}{4} + \frac{1}{2}\right) = \frac{\pi + 2}{8}.
\end{aligned}
$$

(The final integral above was evaluated by interpreting it as the area of part of a circle. You are invited to supply the details. It can also be done by the substitution $v = \sqrt{2} \sin w$.)

▌▌▌▌▌▌ Exercises 15.3 ▌▌▌▌▌▌▌▌

1. Show that the curve $\mathcal{C}$ given by

$$\mathbf{r} = a \cos t \sin t\, \mathbf{i} + a \sin^2 t\, \mathbf{j} + a \cos t\, \mathbf{k}, \quad (0 \le t \le \tfrac{\pi}{2}),$$

   lies on a sphere centred at the origin. Find $\displaystyle\int_{\mathcal{C}} z\, ds$.

2. Let $\mathcal{C}$ be the conical helix with parametric equations

   $x = t \cos t$, $y = t \sin t$, $z = t$, $(0 \le t \le 2\pi)$. Find $\displaystyle\int_{\mathcal{C}} z\, ds$.

3. Find the mass of a wire along the curve

$$\mathbf{r} = 3t\mathbf{i} + 3t^2\mathbf{j} + 2t^3\mathbf{k}, \quad (0 \le t \le 1),$$

   if the density at $\mathbf{r}(t)$ is $1 + t$ g/unit length.

4. Show that the curve $\mathcal{C}$ in Example 3 also has parametrization $x = \cos t$, $y = \sin t$, $z = \cos^2 t$, $(0 \le t \le \pi/2)$, and recalculate the mass of the wire in that example using this parametrization.

5. Find the moment of inertia about the z-axis (i.e., the value of $\delta \displaystyle\int_{\mathcal{C}} (x^2 + y^2)\, ds$), for a wire of constant density δ lying along the curve $\mathcal{C}$: $\mathbf{r} = e^t \cos t\mathbf{i} + e^t \sin t\mathbf{j} + t\mathbf{k}$, from $t = 0$ to $t = 2\pi$.

6. Evaluate $\displaystyle\int_{\mathcal{C}} e^z\, ds$, where $\mathcal{C}$ is the curve in Exercise 5.

7. Find $\displaystyle\int_{\mathcal{C}} x^2\, ds$ along the line of intersection of the two planes $x - y + z = 0$ and $x + y + 2z = 0$, from the origin to the point $(3, 1, -2)$.

8. Find $\displaystyle\int_{\mathcal{C}} \sqrt{1 + 4x^2z^2}\, ds$, where $\mathcal{C}$ is the curve of intersection of the surfaces $x^2 + z^2 = 1$ and $y = x^2$.

9. Find the mass and centre of mass of a wire bent in the shape of the circular helix $x = \cos t$, $y = \sin t$, $z = t$, $(0 \le t \le 2\pi)$, if the wire has line density given by $\delta(x, y, z) = z$.

10. Repeat Exercise 9 for the part of the wire corresponding to $0 \le t \le \pi$.

11. Find the moment of inertia about the y-axis, that is,

$$\int_{\mathcal{C}} (x^2 + z^2)\, ds,$$

   of the curve $x = e^t$, $y = \sqrt{2}\,t$, $z = e^{-t}$, $(0 \le t \le 1)$.

12. Find the centroid of the curve in Exercise 11.

*13. Find $\displaystyle\int_{\mathcal{C}} x\, ds$ along the first octant part of the curve of intersection of the cylinder $x^2 + y^2 = a^2$ and the plane $z = x$.

*14. Find $\displaystyle\int_{\mathcal{C}} z\, ds$ along the part of the curve $x^2 + y^2 + z^2 = 1$, $x + y = 1$, where $z \ge 0$.

*15. Find $\displaystyle\int_{\mathcal{C}} \frac{ds}{(2y^2 + 1)^{3/2}}$, where $\mathcal{C}$ is the parabola $z^2 = x^2 + y^2$, $x + z = 1$. *Hint:* Use $y = t$ as parameter.

16. Express as a definite integral, but do not try to evaluate, the value of $\displaystyle\int_{\mathcal{C}} xyz\, ds$, where $\mathcal{C}$ is the curve $y = x^2$, $z = y^2$ from $(0, 0, 0)$ to $(2, 4, 16)$.

*17. The function

$$E(k, \phi) = \int_0^\phi \sqrt{1 - k^2 \sin^2 t}\, dt$$

   is called the **elliptic integral function of the second kind**. The **complete elliptic integral** of the second kind is the function $E(k) = E(k, \pi/2)$. In terms of these functions, express the length of one complete revolution of the elliptic helix

$$x = a \cos t, \qquad y = b \sin t, \qquad z = ct,$$

   where $0 < a < b$. What is the length of that part of the helix lying between $t = 0$ and $t = T$, where $0 < T < \pi/2$?

*18. Evaluate $\displaystyle\int_L \frac{ds}{x^2 + y^2}$, where L is the entire straight line with equation $Ax + By = C$, $(C \ne 0)$. *Hint:* Use the symmetry of the integrand to replace the line with a line having a simpler equation but giving the same value to the integral.

15.4 Line Integrals of Vector Fields

In elementary physics the **work** done by a constant force of magnitude F in moving an object a distance d is defined to be the product of F and d: $W = Fd$. There is, however, a catch to this; it is understood that the force is exerted in the direction of motion of the object. If the object moves in a direction different from that of the force (because of some other forces acting on it), then the work done by the particular force is the product of the distance moved and the component of the force in the direction of motion. For instance, the work done by gravity in causing a 10 kg crate to slide 5 m down a ramp inclined at $45°$ to the horizontal is $W = 50g/\sqrt{2}$ N·m (where $g = 9.8$ m/s^2), since the scalar projection of the $10g$ N gravitational force on the crate in the direction of the ramp is $10g/\sqrt{2}$ N.

The work done by a *variable* force $\mathbf{F}(x, y, z) = \mathbf{F}(\mathbf{r})$, which depends continuously on position, in moving an object along a smooth curve $\mathcal{C}$ is the integral of *work elements* dW. The element dW corresponding to arc length element ds at position $\mathbf{r}$ on $\mathcal{C}$ is ds times the tangential component of the force $\mathbf{F}(\mathbf{r})$ along $\mathcal{C}$ in the direction of motion (see Figure 15.8); since $\hat{\mathbf{T}} = d\mathbf{r}/ds$ is the unit tangent to $\mathcal{C}$,

$$dW = \mathbf{F}(\mathbf{r}) \bullet \hat{\mathbf{T}} \, ds = \mathbf{F} \bullet d\mathbf{r}.$$

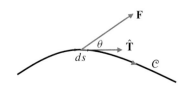

Figure 15.8 $dW = |\mathbf{F}| \cos \theta \, ds$
$\qquad\qquad = \mathbf{F} \bullet \hat{\mathbf{T}} \, ds$

Thus, the total work done by $\mathbf{F}$ in moving the object along $\mathcal{C}$ is

$$W = \int_{\mathcal{C}} \mathbf{F} \bullet \hat{\mathbf{T}} \, ds = \int_{\mathcal{C}} \mathbf{F} \bullet d\mathbf{r} = \int_{\mathcal{C}} F_1 \, dx + F_2 \, dy + F_3 \, dz.$$

In general, if $\mathbf{F} = F_1 \mathbf{i} + F_2 \mathbf{j} + F_3 \mathbf{k}$ is a continuous vector field, and $\mathcal{C}$ is an oriented smooth curve, then the **line integral of the tangential component of F** along $\mathcal{C}$ is

$$\int_{\mathcal{C}} \mathbf{F} \bullet d\mathbf{r} = \int_{\mathcal{C}} \mathbf{F} \bullet \hat{\mathbf{T}} \, ds$$

$$= \int_{\mathcal{C}} F_1(x, y, z) \, dx + F_2(x, y, z) \, dy + F_3(x, y, z) \, dz.$$

Such a line integral is sometimes called, somewhat improperly, the line integral of $\mathbf{F}$ along $\mathcal{C}$. (It is not the line integral of $\mathbf{F}$, which should have a vector value, but rather the line integral of the *tangential component* of $\mathbf{F}$, which has a scalar value.) Unlike the line integral considered in the previous section, this line integral depends on the direction of the orientation of $\mathcal{C}$; reversing the direction of $\mathcal{C}$ causes this line integral to change sign.

If $\mathcal{C}$ is a closed curve, the line integral of the tangential component of $\mathbf{F}$ around $\mathcal{C}$ is also called the **circulation** of $\mathbf{F}$ around $\mathcal{C}$. The fact that the curve is closed is often indicated by a small circle drawn on the integral sign;

$$\oint_{\mathcal{C}} \mathbf{F} \bullet d\mathbf{r} \quad \text{denotes the circulation of } \mathbf{F} \text{ around the closed curve } \mathcal{C}.$$

Like the line integrals studied in the previous section, a line integral of a continuous vector field is converted into an ordinary definite integral by using a parametrization of the path of integration. For a smooth arc $\mathbf{r} = \mathbf{r}(t) = x(t)\mathbf{i} + y(t)\mathbf{j} + z(t)\mathbf{k}$, $(a \le t \le b)$, we have

$$\int_{\mathcal{C}} \mathbf{F} \bullet d\mathbf{r} = \int_a^b \mathbf{F} \bullet \frac{d\mathbf{r}}{dt} \, dt$$

$$= \int_a^b \left[F_1\Big(x(t), y(t), z(t)\Big) \frac{dx}{dt} + F_2\Big(x(t), y(t), z(t)\Big) \frac{dy}{dt} \right.$$

$$\left. + F_3\Big(x(t), y(t), z(t)\Big) \frac{dz}{dt} \right] dt.$$

Although this type of line integral changes sign if the orientation of $\mathcal{C}$ is reversed, it is otherwise independent of the particular parametrization used for $\mathcal{C}$. Again, a line integral over a piecewise smooth path is the sum of the line integrals over the individual smooth arcs constituting that path.

Example 1   Let $\mathbf{F}(x, y) = y^2 \mathbf{i} + 2xy\mathbf{j}$. Evaluate the line integral

$$\int_{\mathcal{C}} \mathbf{F} \bullet d\mathbf{r}$$

from $(0, 0)$ to $(1, 1)$ along

(a) the straight line $y = x$,

(b) the curve $y = x^2$, and

(c) the piecewise smooth path consisting of the straight line segments from $(0, 0)$ to $(0, 1)$ and from $(0, 1)$ to $(1, 1)$.

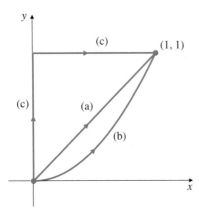

Figure 15.9 Three paths from $(0, 0)$ to $(1, 1)$

Solution The three paths are shown in Figure 15.9. The straight path (a) can be parametrized $\mathbf{r} = t\mathbf{i} + t\mathbf{j}, 0 \le t \le 1$. Thus $d\mathbf{r} = dt\mathbf{i} + dt\mathbf{j}$ and

$$\mathbf{F} \bullet d\mathbf{r} = (t^2\mathbf{i} + 2t^2\mathbf{j}) \bullet (\mathbf{i} + \mathbf{j})dt = 3t^2\,dt.$$

Therefore,

$$\int_{\mathcal{C}} \mathbf{F} \bullet d\mathbf{r} = \int_0^1 3t^2\,dt = t^3\Big|_0^1 = 1.$$

The parabolic path (b) can be parametrized $\mathbf{r} = t\mathbf{i} + t^2\mathbf{j}, 0 \le t \le 1$, so that $d\mathbf{r} = dt\mathbf{i} + 2t\,dt\mathbf{j}$. Thus,

$$\mathbf{F} \bullet d\mathbf{r} = (t^4\mathbf{i} + 2t^3\mathbf{j}) \bullet (\mathbf{i} + 2t\mathbf{j})\,dt = 5t^4\,dt,$$

and

$$\int_{\mathcal{C}} \mathbf{F} \bullet d\mathbf{r} = \int_0^1 5t^4\,dt = t^5\Big|_0^1 = 1.$$

The third path (c) is made up of two segments, and we parametrize each separately. Let us use y as the parameter on the vertical segment (where $x = 0$ and $dx = 0$) and x as the parameter on the horizontal segment (where $y = 1$ and $dy = 0$):

$$\int_{\mathcal{C}} \mathbf{F} \bullet d\mathbf{r} = \int_{\mathcal{C}} y^2\,dx + 2xy\,dy$$
$$= \int_0^1 (0)\,dy + \int_0^1 (1)\,dx = 1.$$

In view of these results, we might ask whether $\int_{\mathcal{C}} \mathbf{F} \bullet d\mathbf{r}$ is the same along *every* path from $(0, 0)$ to $(1, 1)$.

Example 2 Let $\mathbf{F} = y\mathbf{i} - x\mathbf{j}$. Find $\int_{\mathcal{C}} \mathbf{F} \bullet d\mathbf{r}$ from $(1, 0)$ to $(0, -1)$ along

(a) the straight line segment joining these points and

(b) three-quarters of the circle of unit radius centred at the origin and traversed counterclockwise.

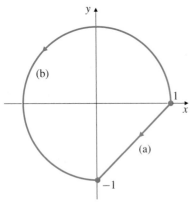

Figure 15.10 Two paths from $(1, 0)$ to $(0, -1)$

Solution Both paths are shown in Figure 15.10. The straight path (a) can be parametrized:

$$\mathbf{r} = (1 - t)\mathbf{i} - t\mathbf{j}, \qquad 0 \le t \le 1.$$

Thus, $d\mathbf{r} = -dt\mathbf{i} - dt\mathbf{j}$, and

$$\int_{\mathcal{C}} \mathbf{F} \bullet d\mathbf{r} = \int_0^1 \big((-t)(-dt) - (1 - t)(-dt)\big) = \int_0^1 dt = 1.$$

The circular path (b) can be parametrized:

$$\mathbf{r} = \cos t\,\mathbf{i} + \sin t\,\mathbf{j}, \qquad 0 \le t \le \frac{3\pi}{2},$$

so that $d\mathbf{r} = -\sin t\,dt\mathbf{i} + \cos t\,dt\mathbf{j}$. Therefore,

$$\mathbf{F} \bullet d\mathbf{r} = -\sin^2 t\,dt - \cos^2 t\,dt = -dt,$$

and we have

$$\int_{\mathcal{C}} \mathbf{F} \bullet d\mathbf{r} = -\int_{0}^{3\pi/2} dt = -\frac{3\pi}{2}.$$

In this case the line integral depends on the path from $(1, 0)$ to $(0, -1)$ along which the integral is taken.

Some readers may have noticed that in Example 1 above the vector field **F** is conservative, while in Example 2 it is not. Theorem 1 below confirms the link between *independence of path* for a line integral of the tangential component of a vector field and the existence of a scalar potential function for that field. This and subsequent theorems require specific assumptions on the nature of the domain of the vector field **F**, so we need to formulate some topological definitions.

Connected and Simply Connected Domains

Recall that a set S in the plane (or in 3-space) is open if every point in S is the centre of a disk (or a ball) having positive radius and contained in S. If S is open and B is a set (possibly empty) of boundary points of S, then the set $D = S \cup B$ is called a **domain**. A domain cannot contain isolated points. It may be closed, but it must have interior points near any of its boundary points. (See Section 10.1 for a discussion of open and closed sets and interior and boundary points.)

DEFINITION 2

> A domain D is said to be **connected**, if every pair of points P and Q in D can be joined by a piecewise smooth curve lying in D.

For instance, the set of points (x, y) in the plane satisfying $x > 0$, $y > 0$, and $x^2 + y^2 \leq 4$ is a connected domain, but the set of points satisfying $|x| > 1$ is not connected. (There is no path from $(-2, 0)$ to $(2, 0)$ lying entirely in $|x| > 1$.) The set of points (x, y, z) in 3-space satisfying $0 < z < 1$ is a connected domain, but the set satisfying $z \neq 0$ is not.

A closed curve is **simple** if it has no self-intersections other than beginning and ending at the same point. (For example, a circle is a simple closed curve.) Imagine an elastic band stretched in the shape of such a curve. If the elastic is infinitely shrinkable, it can contract down to a single point.

DEFINITION 3

> A **simply connected domain** D is a connected domain in which every *simple closed curve* can be continuously shrunk to a point in D without any part ever passing out of D.

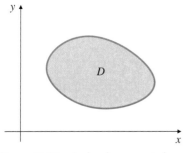

Figure 15.11   A simply connected domain

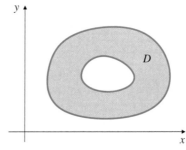

Figure 15.12   A connected domain that is not simply connected

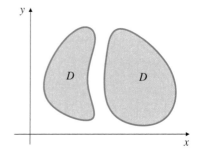

Figure 15.13   A domain that is not connected

Figure 15.11 shows a simply connected domain in the plane. Figure 15.12 shows a connected but not simply connected domain. (A closed curve surrounding the hole cannot be shrunk to a point without passing out of D.) The domain in Figure 15.13 is not even connected. It has two *components*; points in different components cannot be joined by a curve that lies in D.

In the plane, a simply connected domain D can have no holes, not even a hole consisting of a single point. The interior of every non-self-intersecting closed curve in such a domain D lies in D. For instance, the domain of the function $1/(x^2 + y^2)$ is not simply connected because the origin does not belong to it. (The origin is a "hole" in that domain.) In 3-space, a simply connected domain can have holes. The set of all points in $\mathbb{R}^3$ different from the origin is simply connected, as is the exterior of a ball. But the set of all points in $\mathbb{R}^3$ satisfying $x^2 + y^2 > 0$ is not simply connected. Neither is the interior of a doughnut (a *torus*). In general, each of the following conditions characterizes simply connected domains D:

(i) Any simple closed curve in D is the boundary of a "surface" lying in D.

(ii) If $\mathcal{C}_1$ and $\mathcal{C}_2$ are two curves in D having the same endpoints, then $\mathcal{C}_1$ can be continuously deformed into $\mathcal{C}_2$, remaining in D throughout the deformation process.

Independence of Path

THEOREM 1

Independence of path

Let D be an open, connected domain, and let $\mathbf{F}$ be a smooth vector field defined on D. Then the following three statements are *equivalent* in the sense that, if any one of them is true, so are the other two:

(a) $\mathbf{F}$ is conservative in D.

(b) $\oint_{\mathcal{C}} \mathbf{F} \bullet d\mathbf{r} = 0$ for every piecewise smooth, closed curve $\mathcal{C}$ in D.

(c) Given any two points P_0 and P_1 in D, $\int_{\mathcal{C}} \mathbf{F} \bullet d\mathbf{r}$ has the same value for all piecewise smooth curves in D starting at P_0 and ending at P_1.

PROOF We will show that (a) implies (b), that (b) implies (c), and that (c) implies (a). It then follows that any one implies the other two.

Suppose (a) is true. Then $\mathbf{F} = \nabla\phi$ for some scalar potential function ϕ defined in D. Therefore,

$$\mathbf{F} \bullet d\mathbf{r} = \left(\frac{\partial\phi}{\partial x}\mathbf{i} + \frac{\partial\phi}{\partial y}\mathbf{j} + \frac{\partial\phi}{\partial z}\mathbf{k}\right) \bullet \left(dx\,\mathbf{i} + dy\,\mathbf{j} + dz\,\mathbf{k}\right)$$

$$= \frac{\partial\phi}{\partial x}\,dx + \frac{\partial\phi}{\partial y}\,dy + \frac{\partial\phi}{\partial z}\,dz = d\phi.$$

If $\mathcal{C}$ is any piecewise smooth, closed curve, parametrized, say, by $\mathbf{r} = \mathbf{r}(t)$, ($a \le t \le b$), then $\mathbf{r}(a) = \mathbf{r}(b)$, and

$$\int_{\mathcal{C}} \mathbf{F} \bullet d\mathbf{r} = \int_a^b \frac{d\phi(\mathbf{r}(t))}{dt}\,dt = \phi(\mathbf{r}(b)) - \phi(\mathbf{r}(a)) = 0.$$

Thus (a) implies (b).

Now suppose (b) is true. Let P_0 and P_1 be two points in D, and let $\mathcal{C}_1$ and $\mathcal{C}_2$ be two piecewise smooth curves in D from P_0 to P_1. Let $\mathcal{C} = \mathcal{C}_1 - \mathcal{C}_2$ denote the closed curve going from P_0 to P_1 along $\mathcal{C}_1$ and then back to P_0 along $\mathcal{C}_2$ in the opposite direction. (See Figure 15.14.) Since we are assuming that (b) is true, we have

$$0 = \oint_{\mathcal{C}} \mathbf{F} \bullet d\mathbf{r} = \int_{\mathcal{C}_1} \mathbf{F} \bullet d\mathbf{r} - \int_{\mathcal{C}_2} \mathbf{F} \bullet d\mathbf{r}.$$

Therefore,

$$\int_{\mathcal{C}_1} \mathbf{F} \bullet d\mathbf{r} = \int_{\mathcal{C}_2} \mathbf{F} \bullet d\mathbf{r},$$

Figure 15.14 $\mathcal{C}_1 - \mathcal{C}_2 = \mathcal{C}_1 + (-\mathcal{C}_2)$ is a closed curve

and we have proved that (b) implies (c).

Finally, suppose that (c) is true. Let $P_0 = (x_0, y_0, z_0)$ be a fixed point in the domain D, and let $P = (x, y, z)$ be an arbitrary point in that domain. Define a function ϕ by

$$\phi(x, y, z) = \int_{\mathcal{C}} \mathbf{F} \bullet d\mathbf{r},$$

where $\mathcal{C}$ is some piecewise smooth curve in D from P_0 to P. (Under the hypotheses of the theorem such a curve exists, and, since we are assuming (c), the integral has the same value for all such curves. Therefore, ϕ is well defined in D.) We will show that $\nabla\phi = \mathbf{F}$ and thus establish that $\mathbf{F}$ is conservative and has potential ϕ.

It is sufficient to show that $\partial\phi/\partial x = F_1(x, y, z)$; the other two components are treated similarly. Since D is open, there is a ball of positive radius centred at P and contained in D. Pick a point (x_1, y, z) in this ball having $x_1 < x$. Note that the line from this point to P is parallel to the x-axis. Since we are free to choose the curve $\mathcal{C}$ in the integral defining ϕ, let us choose it to consist of two segments: $\mathcal{C}_1$, which is piecewise smooth and goes from (x_0, y_0, z_0) to (x_1, y, z), and $\mathcal{C}_2$, a straight line segment from (x_1, y, z) to (x, y, z). (See Figure 15.15.) Then

$$\phi(x, y, z) = \int_{\mathcal{C}_1} \mathbf{F} \bullet d\mathbf{r} + \int_{\mathcal{C}_2} \mathbf{F} \bullet d\mathbf{r}.$$

The first integral does not depend on x, so its derivative with respect to x is zero. The straight line path for the second integral is parametrized by $\mathbf{r} = t\mathbf{i} + y\mathbf{j} + z\mathbf{k}$, where $x_1 \leq t \leq x$ so $d\mathbf{r} = dt\mathbf{i}$. By the Fundamental Theorem of Calculus,

$$\frac{\partial\phi}{\partial x} = \frac{\partial}{\partial x} \int_{\mathcal{C}_2} \mathbf{F} \bullet d\mathbf{r} = \frac{\partial}{\partial x} \int_{x_1}^{x} F_1(t, y, z)\, dt = F_1(x, y, z),$$

which is what we wanted. Thus $\mathbf{F} = \nabla\phi$ is conservative, and (c) implies (a).

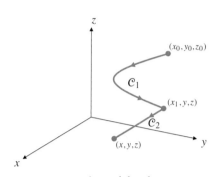

Figure 15.15   A special path from P_0 to P_1

Remark   It is very easy to evaluate the line integral of the tangential component of a *conservative* vector field along a curve $\mathcal{C}$, when you know a potential for $\mathbf{F}$. If $\mathbf{F} = \nabla\phi$, and $\mathcal{C}$ goes from P_0 to P_1, then

$$\int_{\mathcal{C}} \mathbf{F} \bullet d\mathbf{r} = \int_{\mathcal{C}} d\phi = \phi(P_1) - \phi(P_0).$$

As noted above, the value of the integral depends only on the endpoints of $\mathcal{C}$.

Remark   In the next chapter we will add another item to the list of three conditions shown to be equivalent in Theorem 1, provided that the domain D is *simply connected*. For such a domain each of the above three conditions in the theorem is equivalent to

$$\frac{\partial F_1}{\partial y} = \frac{\partial F_2}{\partial x}, \qquad \frac{\partial F_1}{\partial z} = \frac{\partial F_3}{\partial x}, \qquad \text{and} \qquad \frac{\partial F_2}{\partial z} = \frac{\partial F_3}{\partial y}.$$

We already know that these equations are satisfied on a domain where $\mathbf{F}$ is conservative. Theorem 4 of Section 16.2 states that if these three equations hold on a simply connected domain, then $\mathbf{F}$ is conservative on that domain.

Example 3   For what values of the constants A and B is the vector field

$$\mathbf{F} = Ax\sin(\pi y)\mathbf{i} + \left(x^2 \cos(\pi y) + Bye^{-z}\right)\mathbf{j} + y^2 e^{-z}\mathbf{k}$$

conservative? For this choice of A and B, evaluate $\displaystyle\int_{\mathcal{C}} \mathbf{F} \bullet d\mathbf{r}$, where $\mathcal{C}$ is

(a) the curve $\mathbf{r} = \cos t\mathbf{i} + \sin 2t\mathbf{j} + \sin^2 t\mathbf{k}$, $(0 \leq t \leq 2\pi)$, and

(b) the curve of intersection of the paraboloid $z = x^2 + 4y^2$ and the plane $z = 3x - 2y$ from $(0, 0, 0)$ to $(1, 1/2, 2)$.

Solution   **F** cannot be conservative unless

$$\frac{\partial F_1}{\partial y} = \frac{\partial F_2}{\partial x}, \qquad \frac{\partial F_1}{\partial z} = \frac{\partial F_3}{\partial x}, \qquad \text{and} \qquad \frac{\partial F_2}{\partial z} = \frac{\partial F_3}{\partial y},$$

that is, unless

$$A\pi x \cos(\pi y) = 2x \cos(\pi y), \qquad 0 = 0, \qquad \text{and} \quad -Bye^{-z} = 2ye^{-z}.$$

Thus, we require that $A = 2/\pi$ and $B = -2$. In this case, it is easily checked that

$$\mathbf{F} = \nabla\phi, \quad \text{where} \quad \phi(x, y, z) = \frac{x^2 \sin(\pi y)}{\pi} - y^2 e^{-z}.$$

For the curve (a) we have $\mathbf{r}(0) = \mathbf{i} = \mathbf{r}(2\pi)$, so this curve is a closed curve, and

$$\int_{\mathcal{C}} \mathbf{F} \bullet d\mathbf{r} = \oint_{\mathcal{C}} \nabla\phi \bullet d\mathbf{r} = 0.$$

Since the curve (b) starts at $(0, 0, 0)$ and ends at $(1, 1/2, 2)$, we have

$$\int_{\mathcal{C}} \mathbf{F} \bullet d\mathbf{r} = \left. \left(\frac{x^2 \sin(\pi y)}{\pi} - y^2 e^{-z} \right) \right|_{(0,0,0)}^{(1,1/2,2)} = \frac{1}{\pi} - \frac{1}{4e^2}.$$

The following example shows how to exploit the fact that

$$\int_{\mathcal{C}} \mathbf{F} \bullet d\mathbf{r}$$

is easily evaluated for conservative **F** even if the **F** we want to integrate isn't quite conservative.

Example 4   Evaluate $I = \oint_{\mathcal{C}} (e^x \sin y + 3y)dx + (e^x \cos y + 2x - 2y)dy$ counterclockwise around the ellipse $4x^2 + y^2 = 4$.

Solution $I = \oint_{\mathcal{C}} \mathbf{F} \bullet d\mathbf{r}$, where **F** is the vector field

$$\mathbf{F} = \left(e^x \sin y + 3y \right)\mathbf{i} + \left(e^x \cos y + 2x - 2y \right)\mathbf{j}.$$

This vector field is not conservative, but it would be if the $3y$ term in F_1 were $2y$ instead; specifically, if

$$\phi(x, y) = e^x \sin y + 2xy - y^2,$$

then $\mathbf{F} = \nabla\phi + y\mathbf{i}$, the sum of a conservative part and a nonconservative part. Therefore, we have

$$I = \oint_{\mathcal{C}} \nabla\phi \bullet d\mathbf{r} + \oint_{\mathcal{C}} y \, dx.$$

The first integral is zero since $\nabla\phi$ is conservative and $\mathcal{C}$ is closed. For the second integral we parametrize $\mathcal{C}$ by $x = \cos t$, $y = 2 \sin t$, $(0 \le t \le 2\pi)$, and obtain

$$I = \oint_{\mathcal{C}} y \, dx = -2 \int_0^{2\pi} \sin^2 t \, dt = -2 \int_0^{2\pi} \frac{1 - \cos(2t)}{2} \, dt = -2\pi.$$

Exercises 15.4

In Exercises 1–6, evaluate the line integral of the tangential component of the given vector field along the given curve.

1. $\mathbf{F}(x, y) = xy\mathbf{i} - x^2\mathbf{j}$ along $y = x^2$ from $(0, 0)$ to $(1, 1)$

2. $\mathbf{F}(x, y) = \cos x\,\mathbf{i} - y\mathbf{j}$ along $y = \sin x$ from $(0, 0)$ to $(\pi, 0)$

3. $\mathbf{F}(x, y, z) = y\mathbf{i} + z\mathbf{j} - x\mathbf{k}$ along the straight line from $(0, 0, 0)$ to $(1, 1, 1)$

4. $\mathbf{F}(x, y, z) = z\mathbf{i} - y\mathbf{j} + 2x\mathbf{k}$ along the curve $x = t$, $y = t^2$, $z = t^3$ from $(0, 0, 0)$ to $(1, 1, 1)$

5. $\mathbf{F}(x, y, z) = yz\mathbf{i} + xz\mathbf{j} + xy\mathbf{k}$ from $(-1, 0, 0)$ to $(1, 0, 0)$ along either direction of the curve of intersection of the cylinder $x^2 + y^2 = 1$ and the plane $z = y$

6. $\mathbf{F}(x, y, z) = (x - z)\mathbf{i} + (y - z)\mathbf{j} - (x + y)\mathbf{k}$ along the polygonal path from $(0, 0, 0)$ to $(1, 0, 0)$ to $(1, 1, 0)$ to $(1, 1, 1)$

7. Find the work done by the force field

$$\mathbf{F} = (x + y)\mathbf{i} + (x - z)\mathbf{j} + (z - y)\mathbf{k}$$

in moving an object from $(1, 0, -1)$ to $(0, -2, 3)$ along any smooth curve.

8. Evaluate $\oint_{\mathcal{C}} x^2 y^2\,dx + x^3 y\,dy$ counterclockwise around the square with vertices $(0, 0)$, $(1, 0)$, $(1, 1)$, and $(0, 1)$.

9. Evaluate

$$\int_{\mathcal{C}} e^{x+y}\sin(y + z)\,dx + e^{x+y}\left(\sin(y + z) + \cos(y + z)\right)dy$$
$$+ e^{x+y}\cos(y + z)\,dz$$

along the straight line segment from $(0,0,0)$ to $(1, \frac{\pi}{4}, \frac{\pi}{4})$.

10. The field $\mathbf{F} = (axy + z)\mathbf{i} + x^2\mathbf{j} + (bx + 2z)\mathbf{k}$ is conservative. Find a and b, and find a potential for $\mathbf{F}$. Also, evaluate $\int_{\mathcal{C}} \mathbf{F} \bullet d\mathbf{r}$, where $\mathcal{C}$ is the curve from $(1, 1, 0)$ to $(0, 0, 3)$ that lies on the intersection of the surfaces $2x + y + z = 3$ and $9x^2 + 9y^2 + 2z^2 = 18$ in the octant $x \geq 0$, $y \geq 0$, $z \geq 0$.

11. Determine the values of A and B for which the vector field

$$\mathbf{F} = Ax \ln z\,\mathbf{i} + By^2 z\,\mathbf{j} + \left(\frac{x^2}{z} + y^3\right)\mathbf{k}$$

is conservative. If $\mathcal{C}$ is the straight line from $(1, 1, 1)$ to $(2, 1, 2)$, find

$$\int_{\mathcal{C}} 2x \ln z\,dx + 2y^2 z\,dy + y^3\,dz.$$

12. Find the work done by the force field

$$\mathbf{F} = (y^2 \cos x + z^3)\mathbf{i} + (2y \sin x - 4)\mathbf{j} + (3xz^2 + 2)\mathbf{k}$$

in moving a particle along the curve $x = \sin^{-1} t$, $y = 1 - 2t$, $z = 3t - 1$, $(0 \leq t \leq 1)$.

13. If $\mathcal{C}$ is the intersection of $z = \ln(1 + x)$ and $y = x$ from $(0, 0, 0)$ to $(1, 1, \ln 2)$, evaluate

$$\int_{\mathcal{C}} \left(2x \sin(\pi y) - e^z\right)dx + \left(\pi x^2 \cos(\pi y) - 3e^z\right)dy - xe^z\,dz.$$

14. Is each of the following sets a domain? a connected domain? a simply connected domain?

(a) the set of points (x, y) in the plane such that $x > 0$ and $y \geq 0$

(b) the set of points (x, y) in the plane such that $x = 0$ and $y \geq 0$

(c) the set of points (x, y) in the plane such that $x \neq 0$ and $y > 0$

(d) the set of points (x, y, z) in 3-space such that $x^2 > 1$

(e) the set of points (x, y, z) in 3-space such that $x^2 + y^2 > 1$

(f) the set of points (x, y, z) in 3-space such that $x^2 + y^2 + z^2 > 1$

In Exercises 15–19, evaluate the closed line integrals

(a) $\oint_{\mathcal{C}} x\,dy$,   (b) $\oint_{\mathcal{C}} y\,dx$

around the given curves, all oriented counterclockwise.

15. The circle $x^2 + y^2 = a^2$

16. The ellipse $\dfrac{x^2}{a^2} + \dfrac{y^2}{b^2} = 1$

17. The boundary of the half-disk $x^2 + y^2 \leq a^2$, $y \geq 0$

18. The boundary of the square with vertices $(0, 0)$, $(1, 0)$, $(1, 1)$, and $(0, 1)$

19. The triangle with vertices $(0, 0)$, $(a, 0)$, and $(0, b)$

20. On the basis of your results for Exercises 15–19, guess the values of the closed line integrals

(a) $\oint_{\mathcal{C}} x\,dy$,   (b) $\oint_{\mathcal{C}} y\,dx$

for any non-self-intersecting closed curve in the xy-plane. Prove your guess in the case that $\mathcal{C}$ bounds a region of the plane that is both x-simple and y-simple. (See Section 14.2.)

21. If f and g are scalar fields with continuous first partial derivatives in a connected domain D, show that

$$\int_{\mathcal{C}} f\nabla g \bullet d\mathbf{r} + \int_{\mathcal{C}} g\nabla f \bullet d\mathbf{r} = f(Q)g(Q) - f(P)g(P)$$

for any piecewise smooth curve in D from P to Q.

22. Evaluate

$$\frac{1}{2\pi} \oint_{\mathcal{C}} \frac{-y\,dx + x\,dy}{x^2 + y^2}$$

(a) counterclockwise around the circle $x^2 + y^2 = a^2$,

(b) clockwise around the square with vertices $(-1, -1)$, $(-1, 1)$, $(1, 1)$, and $(1, -1)$,

(c) counterclockwise around the boundary of the region $1 \leq x^2 + y^2 \leq 4$, $y \geq 0$.

23. Review Example 5 in Section 15.2 in which it was shown that

$$\frac{\partial}{\partial y}\left(\frac{-y}{x^2+y^2}\right)=\frac{\partial}{\partial x}\left(\frac{x}{x^2+y^2}\right),$$

for all $(x, y) \neq (0, 0)$. Why does this result, together with that of Exercise 22, not contradict the final assertion in the remark following Theorem 1?

*24. **(Winding number)** Let $\mathcal{C}$ be a piecewise smooth curve in the xy-plane which does not pass through the origin. Let $\theta = \theta(x, y)$ be the polar angle coordinate of the point $P = (x, y)$ on $\mathcal{C}$, not restricted to an interval of length 2π, but varying continuously as P moves from one end of $\mathcal{C}$ to

the other. As in Example 5 of Section 15.2, it happens that

$$\nabla\theta = -\frac{y}{x^2+y^2}\,\mathbf{i}+\frac{x}{x^2+y^2}\,\mathbf{j}.$$

If, in addition, $\mathcal{C}$ is a closed curve, show that

$$w(\mathcal{C})=\frac{1}{2\pi}\oint_{\mathcal{C}}\frac{x\,dy-y\,dx}{x^2+y^2}$$

has an integer value. w is called the **winding number** of $\mathcal{C}$ about the origin.

15.5 Surfaces and Surface Integrals

This section and the next are devoted to integrals of functions defined over surfaces in 3-space. Before we can begin, it is necessary to make more precise just what is meant by the term "surface." Until now we have been treating surfaces in an intuitive way, either as the graphs of functions $f(x, y)$ or as the graphs of equations $f(x, y, z) = 0$.

A smooth curve is a *one-dimensional* object because points on it can be located by giving *one coordinate* (for instance, the distance from an endpoint). Therefore, the curve can be defined as the range of a vector-valued function of one real variable. A surface is a *two-dimensional* object; points on it can be located by using *two coordinates*, and it can be defined as the range of a vector-valued function of two real variables. We will call certain such functions parametric surfaces.

Parametric Surfaces

DEFINITION 4

A **parametric surface** in 3-space is a continuous function $\mathbf{r}$ defined on some rectangle R given by $a \leq u \leq b, c \leq v \leq d$ in the uv-plane and having values in 3-space:

$$\mathbf{r}(u, v) = x(u, v)\mathbf{i} + y(u, v)\mathbf{j} + z(u, v)\mathbf{k}, \qquad (u, v) \text{ in } R.$$

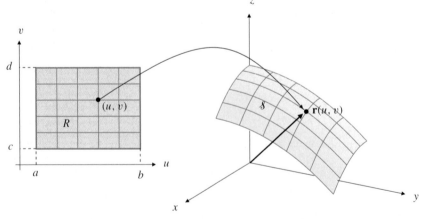

Figure 15.16 A parametric surface $\mathcal{S}$ defined on parameter region R. The *contour curves* on $\mathcal{S}$ correspond to the rulings of R

Actually, we think of the *range* of the function $\mathbf{r}(u, v)$ as being the parametric surface. It is a set $\mathcal{S}$ of points (x, y, z) in 3-space whose position vectors are the vectors $\mathbf{r}(u, v)$ for (u, v) in R. (See Figure 15.16.) If $\mathbf{r}$ is one-to-one, then the surface does not intersect itself. In this case $\mathbf{r}$ maps the boundary of the rectangle R (the four edges) onto a curve in 3-space, which we call the **boundary of the parametric surface**. The requirement that R be a rectangle is made only to simplify the discussion. Any connected, closed, bounded set in the uv-plane, having well-defined area and consisting of an open set together with its boundary points, would do as well. Thus, we will from time to time consider parametric surfaces over closed disks, triangles, or other such domains in the uv-plane. Being the range of a continuous function defined on a closed, bounded set, a parametric surface is always bounded in 3-space.

Example 1　The graph of $z = f(x, y)$, where f has the rectangle R as its domain, can be represented as the parametric surface

$$\mathbf{r} = \mathbf{r}(u, v) = u\mathbf{i} + v\mathbf{j} + f(u, v)\mathbf{k}$$

for (u, v) in R. Its scalar parametric equations are

$$x = u, \qquad y = v, \qquad z = f(u, v), \qquad (u, v) \text{ in } R.$$

For such graphs it is sometimes convenient to identify the uv-plane with the xy-plane and write the equation of the surface in the form

$$\mathbf{r} = x\mathbf{i} + y\mathbf{j} + f(x, y)\mathbf{k}, \qquad (x, y) \text{ in } R.$$

Example 2　Describe the surface

$$\mathbf{r} = a \cos u \sin v\, \mathbf{i} + a \sin u \sin v\, \mathbf{j} + a \cos v\, \mathbf{k}, \qquad (0 \le u \le 2\pi,\ 0 \le v \le \pi/2),$$

where $a > 0$. What is its boundary?

Solution　Observe that if $x = a \cos u \sin v$, $y = a \sin u \sin v$, and $z = a \cos v$, then $x^2 + y^2 + z^2 = a^2$. Thus, the given parametric surface lies on the sphere of radius a centred at the origin. (Observe that u and v are the spherical coordinates θ and ϕ on the sphere.) The restrictions on u and v allow (x, y) to be any point in the disk $x^2 + y^2 \le a^2$ but force $z \ge 0$. Thus, the surface is the *upper half* of the sphere. The given parametrization is one-to-one on the open rectangle $0 < u < 2\pi$, $0 < v < \pi/2$, but not on the closed rectangle, since the edges $u = 0$ and $u = 2\pi$ get mapped onto the same points, and the entire edge $v = 0$ collapses to a single point. The boundary of the surface is still the circle $x^2 + y^2 = a^2$, $z = 0$, and corresponds to the edge $v = \pi/2$ of the rectangle.

Remark　Surface parametrizations that are one-to-one only in the interior of the parameter domain R are still reasonable representations of the surface. However, as in Example 2, the boundary of the surface may be obtained from only part of the boundary of R, or there may be no boundary at all, in which case the surface is called a **closed surface**. For example, if the domain of $\mathbf{r}$ in Example 2 is extended to allow $0 \le v \le \pi$, then the surface becomes the entire sphere of radius a centred at the origin. The sphere is a closed surface, having no boundary curves.

Remark　Like parametrizations of curves, parametrizations of surfaces are not unique. The hemisphere in Example 2 can also be parametrized:

$$\mathbf{r}(u, v) = u\mathbf{i} + v\mathbf{j} + \sqrt{a^2 - u^2 - v^2}\,\mathbf{k} \qquad \text{for} \quad u^2 + v^2 \le a^2.$$

Here, the domain of $\mathbf{r}$ is a closed disk of radius a.

Figure 15.17 A tube in the shape of a trefoil knot

Example 3 (**A tube around a curve**) If $\mathbf{r} = \mathbf{F}(t)$, $a \leq t \leq b$, is a parametric curve $\mathcal{C}$ in 3-space having unit normal $\hat{\mathbf{N}}(t)$ and binormal $\hat{\mathbf{B}}(t)$, then the parametric surface

$$\mathbf{r} = \mathbf{F}(u) + s \cos v \, \hat{\mathbf{N}}(u) + s \sin v \, \hat{\mathbf{B}}(u), \quad a \leq u \leq b, \quad 0 \leq v \leq 2\pi,$$

is a tube-shaped surface of radius s centred along the curve $\mathcal{C}$. (Why?) Figure 15.17 shows such a tube, having radius $s = 0.25$, around the curve

$$\mathbf{r} = \big(1 + 0.3 \cos(3t)\big)\big(\cos(2t)\mathbf{i} + \sin(2t)\mathbf{j}\big) + 0.35 \sin(3t)\mathbf{k}, \quad 0 \leq t \leq 2\pi.$$

This closed curve is called a **trefoil knot**. ∎

Composite Surfaces

If two parametric surfaces are joined together along part or all of their boundary curves, the result is called a **composite surface**, or, thinking geometrically, just a **surface**. For example, a sphere can be obtained by joining two hemispheres along their boundary circles. In general, composite surfaces can be obtained by joining a finite number of parametric surfaces pairwise along edges. The surface of a cube consists of the six square faces joined in pairs along the edges of the cube. This surface is closed since there are no unjoined edges to comprise the boundary. If the top square face is removed, the remaining five form the surface of a cubical box with no top. The top edges of the four side faces now constitute the boundary of this composite surface. (See Figure 15.18.)

Figure 15.18 A composite surface obtained by joining five smooth parametric surfaces (squares) in pairs along edges. The four unpaired edges at the tops of the side faces make up the boundary of the composite surface

Surface Integrals

In order to define integrals of functions defined on a surface as limits of Riemann sums, we need to refer to the *areas* of regions on the surface. It is more difficult to define the area of a curved surface than it is to define the length of a curve. However, you will likely have a good idea of what area means for a region lying in a plane, and we examined briefly the problem of finding the area of the graph of a function $f(x, y)$ in Section 14.7. We will avoid difficulties by assuming that all the surfaces we will encounter are "smooth enough" that they can be subdivided into small pieces each of which is approximately planar. We can then approximate the surface area of each piece by a plane area and add up the approximations to get a Riemann sum approximation to the area of the whole surface. We will make more precise definitions of "smooth surface" and "surface area" later in this section. For the moment, we assume the reader has an intuitive feel for what they mean.

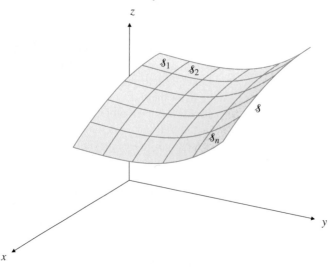

Figure 15.19 A partition of a parametric surface into many nonoverlapping pieces

Let δ be a smooth surface of finite area in $\mathbb{R}^3$, and let $f(x, y, z)$ be a bounded function defined at all points of δ. If we subdivide δ into small, nonoverlapping pieces, say δ_1, $\delta_2, \ldots, \delta_n$, where δ_i has area ΔS_i (see Figure 15.19), we can form a **Riemann sum** R_n for f on δ by choosing arbitrary points (x_i, y_i, z_i) in δ_i and letting

$$R_n = \sum_{i=1}^{n} f(x_i, y_i, z_i) \, \Delta S_i.$$

If such Riemann sums have a unique limit as the diameters of all the pieces δ_i approach zero, independently of how the points (x_i, y_i, z_i) are chosen, then we say that f is **integrable** on δ and call the limit the **surface integral** of f over δ, denoting it by

$$\iint_{\delta} f(x, y, z) \, dS.$$

Smooth Surfaces, Normals, and Area Elements

A surface is smooth if it has a unique tangent plane at any nonboundary point P. A nonzero vector $\mathbf{n}$ normal to that tangent plane at P is said to be normal to the surface at P. The following somewhat technical definition makes this precise.

DEFINITION 5

A set δ in 3-space is a **smooth surface** if any point P in δ has a neighbourhood N (an open ball of positive radius centred at P) that is the domain of a smooth function $g(x, y, z)$ satisfying:

(i) $N \cap S = \{Q \in N : g(Q) = 0\}$ and

(ii) $\nabla g(Q) \neq \mathbf{0}$, if Q is in $N \cap S$.

For example, the cone $x^2 + y^2 = z^2$, with the origin removed, is a smooth surface. Note that $\nabla(x^2 + y^2 - z^2) = \mathbf{0}$ at the origin, and the cone is not smooth there, since it does not have a unique tangent plane.

A parametric surface cannot satisfy the condition of the smoothness definition at its boundary points but will be called **smooth** if that condition is satisfied at all nonboundary points.

We can find the normal to a smooth parametric surface defined on parameter domain R as follows. If (u_0, v_0) is a point in the interior of R, then $\mathbf{r} = \mathbf{r}(u, v_0)$ and $\mathbf{r} = \mathbf{r}(u_0, v)$ are two curves on δ, intersecting at $\mathbf{r}_0 = \mathbf{r}(u_0, v_0)$ and having, at that point, tangent vectors (see Figure 15.20)

$$\left.\frac{\partial \mathbf{r}}{\partial u}\right|_{(u_0, v_0)} \qquad \text{and} \qquad \left.\frac{\partial \mathbf{r}}{\partial v}\right|_{(u_0, v_0)},$$

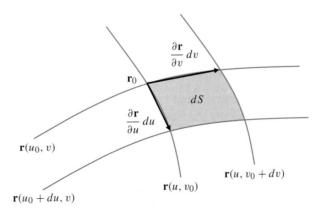

Figure 15.20   An area element dS on a parametric surface

respectively. Assuming these two tangent vectors are not parallel, their cross product **n**, which is not zero, is *normal* to $\mathcal{S}$ at $\mathbf{r}_0$. Furthermore, the *area element* on $\mathcal{S}$ bounded by the four curves $\mathbf{r} = \mathbf{r}(u_0, v)$, $\mathbf{r} = \mathbf{r}(u_0 + du, v)$, $\mathbf{r} = \mathbf{r}(u, v_0)$, and $\mathbf{r} = \mathbf{r}(u, v_0 + dv)$ is an infinitesimal parallelogram spanned by the vectors $(\partial \mathbf{r}/\partial u)\, du$ and $(\partial \mathbf{r}/\partial v)\, dv$ (at (u_0, v_0)), and hence has area

$$dS = \left| \frac{\partial \mathbf{r}}{\partial u} \times \frac{\partial \mathbf{r}}{\partial v} \right| du\, dv.$$

Let us express the normal vector **n** and the area element dS in terms of the components of **r**. Since

$$\frac{\partial \mathbf{r}}{\partial u} = \frac{\partial x}{\partial u}\mathbf{i} + \frac{\partial y}{\partial u}\mathbf{j} + \frac{\partial z}{\partial u}\mathbf{k} \qquad \text{and} \qquad \frac{\partial \mathbf{r}}{\partial v} = \frac{\partial x}{\partial v}\mathbf{i} + \frac{\partial y}{\partial v}\mathbf{j} + \frac{\partial z}{\partial v}\mathbf{k},$$

the **normal vector** to $\mathcal{S}$ at $\mathbf{r}(u, v)$ is

$$\mathbf{n} = \frac{\partial \mathbf{r}}{\partial u} \times \frac{\partial \mathbf{r}}{\partial v} = \begin{vmatrix} \mathbf{i} & \mathbf{j} & \mathbf{k} \\ \dfrac{\partial x}{\partial u} & \dfrac{\partial y}{\partial u} & \dfrac{\partial z}{\partial u} \\ \dfrac{\partial x}{\partial v} & \dfrac{\partial y}{\partial v} & \dfrac{\partial z}{\partial v} \end{vmatrix}$$

$$= \frac{\partial(y, z)}{\partial(u, v)}\mathbf{i} + \frac{\partial(z, x)}{\partial(u, v)}\mathbf{j} + \frac{\partial(x, y)}{\partial(u, v)}\mathbf{k}.$$

Also, the **area element** at a point $\mathbf{r}(u, v)$ on the surface is given by

$$dS = \left| \frac{\partial \mathbf{r}}{\partial u} \times \frac{\partial \mathbf{r}}{\partial v} \right| du\, dv$$

$$= \sqrt{\left(\frac{\partial(y, z)}{\partial(u, v)} \right)^2 + \left(\frac{\partial(z, x)}{\partial(u, v)} \right)^2 + \left(\frac{\partial(x, y)}{\partial(u, v)} \right)^2} \; du\, dv.$$

The area of the surface itself is the "sum" of these area elements:

$$\text{Area of } \mathcal{S} = \iint_{\mathcal{S}} dS.$$

Example 4   The graph $z = g(x, y)$ of a function g with continuous first partial derivatives in a domain D of the xy-plane can be regarded as a parametric surface $\mathcal{S}$ with parametrization

$$x = u, \qquad y = v, \qquad z = g(u, v), \qquad (u, v) \text{ in } D.$$

In this case

$$\frac{\partial(y, z)}{\partial(u, v)} = -g_1(u, v), \qquad \frac{\partial(z, x)}{\partial(u, v)} = -g_2(u, v), \qquad \text{and} \qquad \frac{\partial(x, y)}{\partial(u, v)} = 1,$$

and, since the parameter region coincides with the domain D of g, the surface integral of $f(x, y, z)$ over $\mathcal{S}$ can be expressed as a double integral over D:

$$\iint_{\mathcal{S}} f(x, y, z)\, dS$$

$$= \iint_{D} f\big(x, y, g(x, y)\big) \sqrt{1 + \big(g_1(x, y)\big)^2 + \big(g_2(x, y)\big)^2}\, dx\, dy.$$

As observed in Section 14.7, this formula can also be justified geometrically. The vector $\mathbf{n} = -g_1(x, y)\mathbf{i} - g_2(x, y)\mathbf{j} + \mathbf{k}$ is normal to $\mathcal{S}$ and makes angle γ with the positive z-axis, where

$$\cos \gamma = \frac{\mathbf{n} \bullet \mathbf{k}}{|\mathbf{n}|} = \frac{1}{\sqrt{1 + \big(g_1(x, y)\big)^2 + \big(g_2(x, y)\big)^2}}.$$

The surface area element dS must have area $1/\cos\gamma$ times the area $dx\,dy$ of its perpendicular projection onto the xy-plane. (See Figure 15.21.)

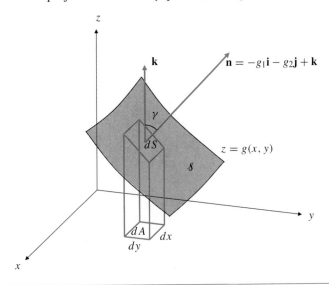

Figure 15.21 The surface area element dS and its projection onto the xy-plane

Evaluating Surface Integrals

We illustrate the use of the formulas given above for dS in calculating surface integrals.

Example 5 Evaluate $\iint_{\mathcal{S}} z\,dS$ over the conical surface $z = \sqrt{x^2 + y^2}$ between $z = 0$ and $z = 1$.

Solution Since $z^2 = x^2 + y^2$ on the surface $\mathcal{S}$, we have $\partial z/\partial x = x/z$ and $\partial z/\partial y = y/z$. Therefore,

$$dS = \sqrt{1 + \frac{x^2}{z^2} + \frac{y^2}{z^2}}\,dx\,dy = \sqrt{\frac{z^2 + z^2}{z^2}}\,dx\,dy = \sqrt{2}\,dx\,dy.$$

(Note that we could have anticipated this result, since the normal to the cone always makes an angle of $\gamma = 45°$ with the positive z-axis; see Figure 15.22. Therefore, $dS = dx\,dy/\cos 45° = \sqrt{2}\,dx\,dy$.) Since $z = \sqrt{x^2 + y^2} = r$ on the conical surface, it is easiest to carry out the integration in polar coordinates:

$$\iint_{\mathcal{S}} z\,dS = \sqrt{2} \iint_{x^2+y^2 \le 1} z\,dx\,dy$$

$$= \sqrt{2} \int_0^{2\pi} d\theta \int_0^1 r^2\,dr = \frac{2\sqrt{2}\pi}{3}.$$

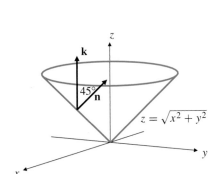

Figure 15.22 $dS = \sqrt{2}\,dx\,dy$ on this cone

Example 6 Find the moment of inertia about the z-axis of the parametric surface $x = 2uv$, $y = u^2 - v^2$, $z = u^2 + v^2$, where $u^2 + v^2 \le 1$.

Solution   We are asked to find $\iint_{\mathcal{S}} (x^2 + y^2)\, dS$. We have

$$\frac{\partial(x, y)}{\partial(u, v)} = \begin{vmatrix} 2v & 2u \\ 2u & -2v \end{vmatrix} = -4(u^2 + v^2),$$

$$\frac{\partial(z, x)}{\partial(u, v)} = \begin{vmatrix} 2u & 2v \\ 2v & 2u \end{vmatrix} = 4(u^2 - v^2),$$

$$\frac{\partial(y, z)}{\partial(u, v)} = \begin{vmatrix} 2u & -2v \\ 2u & 2v \end{vmatrix} = 8uv.$$

Therefore, the surface area element on $\mathcal{S}$ is given by

$$dS = 4\sqrt{(u^2 + v^2)^2 + (u^2 - v^2)^2 + 4u^2v^2}\, du\, dv$$
$$= 4\sqrt{2(u^4 + v^4 + 2u^2v^2)}\, du\, dv = 4\sqrt{2}(u^2 + v^2)\, du\, dv.$$

Now $x^2 + y^2 = 4u^2v^2 + (u^2 - v^2)^2 = (u^2 + v^2)^2$. Thus,

$$\iint_{\mathcal{S}} (x^2 + y^2)\, dS = \iint_{u^2 + v^2 \leq 1} (u^2 + v^2)^2\, 4\sqrt{2}(u^2 + v^2)\, du\, dv$$

$$= 4\sqrt{2} \int_0^{2\pi} d\theta \int_0^1 r^6\, r\, dr \qquad \text{(using polar coordinates)}$$

$$= \sqrt{2}\pi.$$

This is the required moment of inertia.

Even though most surfaces we encounter can be easily parametrized, it is usually possible to obtain the surface area element dS geometrically rather than relying on the parametric formula. As we have seen above, if a surface has a one-to-one projection onto a region in the xy-plane, then the area element dS on the surface can be expressed as

$$dS = \left| \frac{1}{\cos \gamma} \right| dx\, dy = \frac{|\mathbf{n}|}{|\mathbf{n} \bullet \mathbf{k}|}\, dx\, dy,$$

where γ is the angle between the normal vector $\mathbf{n}$ to $\mathcal{S}$ and the positive z-axis. This formula is useful no matter how we obtain $\mathbf{n}$.

Consider a surface $\mathcal{S}$ with equation of the form $G(x, y, z) = 0$. As we discovered in Section 12.7, if G has continuous first partial derivatives that do not all vanish at a point (x, y, z) on $\mathcal{S}$, then the nonzero vector

$$\mathbf{n} = \nabla G(x, y, z)$$

is normal to $\mathcal{S}$ at that point. Since $\mathbf{n} \bullet \mathbf{k} = G_3(x, y, z)$, if $\mathcal{S}$ has a one-to-one projection onto the domain D in the xy-plane, then

$$dS = \left| \frac{\nabla G(x, y, z)}{G_3(x, y, z)} \right| dx\, dy,$$

and the surface integral of $f(x, y, z)$ over $\mathcal{S}$ can be expressed as a double integral over the domain D:

$$\iint_{\mathcal{S}} f(x, y, z)\, dS = \iint_D f\big(x, y, g(x, y)\big) \left| \frac{\nabla G(x, y, z)}{G_3(x, y, z)} \right| dx\, dy.$$

Of course, there are analogous formulas for area elements of surfaces (and integrals over surfaces) with one-to-one projections onto the xz-plane or the yz-plane. (G_3 is replaced by G_2 and G_1, respectively.)

Example 7   Find the moment about $z = 0$, that is, $\iint_{\mathcal{S}} z \, dS$, where $\mathcal{S}$ is the hyperbolic bowl $z^2 = 1 + x^2 + y^2$ between the planes $z = 1$ and $z = \sqrt{5}$.

Solution $\mathcal{S}$ is given by $G(x, y, z) = 0$, where $G(x, y, z) = x^2 + y^2 - z^2 + 1$. It lies above the disk $x^2 + y^2 \le 4$ in the xy-plane. We have $\nabla G = 2x\mathbf{i} + 2y\mathbf{j} - 2z\mathbf{k}$, and $G_3 = -2z$. Hence, on $\mathcal{S}$, we have

$$z \, dS = z \, \frac{\sqrt{4x^2 + 4y^2 + 4z^2}}{2z} \, dx \, dy = \sqrt{1 + 2(x^2 + y^2)} \, dx \, dy,$$

and the required moment is

$$\iint_{\mathcal{S}} z \, dS = \iint_{x^2 + y^2 \le 4} \sqrt{1 + 2(x^2 + y^2)} \, dx \, dy$$

$$= \int_0^{2\pi} d\theta \int_0^2 \sqrt{1 + 2r^2} \, r \, dr = \frac{\pi}{3} (1 + 2r^2)^{3/2} \Big|_0^2 = \frac{26\pi}{3}.$$

The next example illustrates a technique that can often reduce the effort needed to integrate over a cylindrical surface.

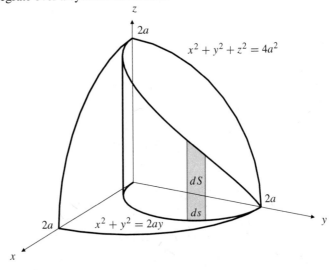

Figure 15.23   An area element on a cylinder. The z-coordinate has already been integrated

Example 8   Find the area of that part of the cylinder $x^2 + y^2 = 2ay$ that lies inside the sphere $x^2 + y^2 + z^2 = 4a^2$.

Solution   One quarter of the required area lies in the first octant. (See Figure 15.23.) Since the cylinder is generated by vertical lines, we can express an area element dS on it in terms of the length element ds along the curve $\mathcal{C}$ in the xy-plane having equation $x^2 + y^2 = 2ay$:

$$dS = z \, ds = \sqrt{4a^2 - x^2 - y^2} \, ds.$$

In expressing dS this way, we have already integrated dz, so only a single integral is needed to sum these area elements. Again, it is convenient to use polar coordinates in the xy-plane. In terms of polar coordinates, the curve $\mathcal{C}$ has equation $r = 2a \sin \theta$. Thus $dr/d\theta = 2a \cos \theta$ and $ds = \sqrt{r^2 + (dr/d\theta)^2} \, d\theta = 2a \, d\theta$. Therefore, the total surface area of that part of the cylinder that lies inside the sphere is given by

$$A = 4 \int_0^{\pi/2} \sqrt{4a^2 - r^2} \, 2a \, d\theta$$

$$= 8a \int_0^{\pi/2} \sqrt{4a^2 - 4a^2 \sin^2 \theta} \, d\theta$$

$$= 16a^2 \int_0^{\pi/2} \cos \theta \, d\theta = 16a^2 \text{ square units.}$$

Remark The area calculated in Example 8 can also be calculated by projecting the cylindrical surface in Figure 15.23 into the yz-plane. (This is the only coordinate plane you can use. Why?) See Exercise 6 below.

In spherical coordinates, ϕ and θ can be used as parameters on the spherical surface $\rho = a$. The area element on that surface can therefore be expressed in terms of these coordinates:

> Area element on the sphere $\rho = a$: $\quad dS = a^2 \sin \phi \, d\phi \, d\theta$.

(See Figure 14.43 in Section 14.6 and Exercise 2 below.)

Example 9 Find $\displaystyle\iint_S z^2 \, dS$ over the hemisphere $z = \sqrt{a^2 - x^2 - y^2}$.

Solution Since $z = a \cos \phi$ and the hemisphere corresponds to $0 \le \theta \le 2\pi$, and $0 \le \phi \le \dfrac{\pi}{2}$, we have

$$\iint_S z^2 \, dS = \int_0^{2\pi} d\theta \int_0^{\pi/2} a^2 \cos^2 \phi \, a^2 \sin \phi \, d\phi$$

$$= 2\pi a^4 \left(-\frac{1}{3} \cos^3 \phi \right) \Big|_0^{\pi/2} = \frac{2\pi a^4}{3}.$$

Finally, if a composite surface S is composed of *smooth parametric surfaces* joined pairwise along their edges, then we call S a **piecewise smooth surface**. The surface integral of a function f over a piecewise smooth surface S is the sum of the surface integrals of f over the individual smooth surfaces comprising S. We will encounter an example of this in the next section.

The Attraction of a Spherical Shell

In Section 14.7 we calculated the gravitational attraction of a disk in the xy-plane on a mass m located at position $(0, 0, b)$ on the z-axis. Here, we undertake a similar calculation of the attractive force exerted on m by a spherical shell of radius a and areal density σ (units of mass per unit area) centred at the origin. This calculation would be more difficult if we tried to do it by integrating the vertical component of the force on m as we did in Section 14.7. It is greatly simplified if, instead, we use an integral to find the total *gravitational potential* $\Phi(0, 0, z)$ due to the sphere at position $(0, 0, z)$ and then calculate the force on m as $\mathbf{F} = m\nabla\Phi(0, 0, b)$.

By the Cosine Law, the distance from the point with spherical coordinates $[a, \phi, \theta]$ to the point $(0, 0, z)$ on the positive z-axis (see Figure 15.24) is

$$D = \sqrt{a^2 + z^2 - 2az \cos \phi}.$$

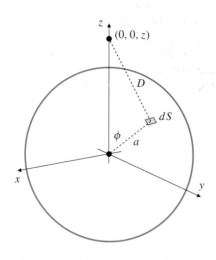

Figure 15.24 The attraction of a sphere

The area element $dS = a^2 \sin \phi \, d\phi \, d\theta$ at $[a, \phi, \theta]$ has mass $dm = \sigma \, dS$, and its gravitational potential at $(0, 0, z)$ (see Example 1 in Section 15.2) is

$$d\Phi(0, 0, z) = \frac{k \, dm}{D} = \frac{k\sigma a^2 \sin \phi \, d\phi \, d\theta}{\sqrt{a^2 + z^2 - 2az \cos \phi}}.$$

For the total potential at $(0, 0, z)$ due to the sphere, we integrate $d\Phi$ over the surface of the sphere. Making the change of variables $u = a^2 + z^2 - 2az \cos \phi$, $du = 2az \sin \phi \, d\phi$, we obtain

$$\Phi(0, 0, z) = k\sigma a^2 \int_0^{2\pi} d\theta \int_0^{\pi} \frac{\sin\phi \, d\phi}{\sqrt{a^2 + z^2 - 2az\cos\phi}}$$

$$= 2\pi k\sigma a^2 \int_{(z-a)^2}^{(z+a)^2} \frac{1}{\sqrt{u}} \frac{du}{2az}$$

$$= \frac{2\pi k\sigma a}{z} \sqrt{u} \Big|_{(z-a)^2}^{(z+a)^2}$$

$$= \frac{2\pi k\sigma a}{z} \left(z + a - |z - a|\right) = \begin{cases} 4\pi k\sigma a^2/z & \text{if } z > a \\ 4\pi k\sigma a & \text{if } z < a. \end{cases}$$

The potential is constant inside the sphere and decreases proportionally to $1/z$ outside. The force on a mass m located at $(0, 0, b)$ is, therefore,

$$\mathbf{F} = m\nabla\Phi(0, 0, b) = \begin{cases} -(4\pi km\sigma a^2/b^2)\mathbf{k} & \text{if } b > a \\ \mathbf{0} & \text{if } b < a. \end{cases}$$

We are led to the somewhat surprising result that, if the mass m is anywhere inside the sphere, the net force of attraction of the sphere on it is zero. This is to be expected at the centre of the sphere, but away from the centre it appears that the larger forces due to parts of the sphere close to m are exactly cancelled by smaller forces due to parts farther away; these farther parts have larger area and therefore larger total mass. If m is outside the sphere, the sphere attracts it with a force of magnitude

$$F = \frac{kmM}{b^2},$$

where $M = 4\pi\sigma a^2$ is the total mass of the sphere. This is the same force that would be exerted by a point mass with the same mass as the sphere and located at the centre of the sphere.

Remark   A solid ball of constant density, or density depending only on the distance from the centre (for instance, a planet), can be regarded as being made up of mass elements that are concentric spheres of constant density. Therefore, the attraction of such a ball on a mass m located outside the ball will also be the same as if the whole mass of the ball were concentrated at its centre. However, the attraction on a mass m located somewhere inside the ball will be that produced by only the part of the ball that is closer to the centre than m is. The maximum force of attraction will occur when m is right at the surface of the ball. If the density is constant, the magnitude of the force increases linearly with the distance from the centre (why?) up to the surface and then decreases with the square of the distance as m recedes from the ball. (See Figure 15.25.)

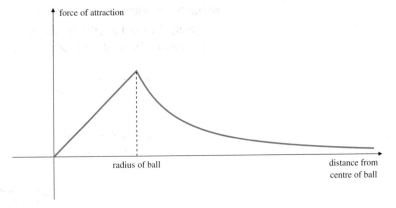

Figure 15.25   The force of attraction of a homogeneous solid ball on a particle located at varying distances from the centre of the ball

Remark   All of the above discussion also holds for the electrostatic attraction or repulsion of a point charge by a uniform charge density over a spherical shell, which is also governed by an inverse square law. In particular, there is no net electrostatic force on a charge located inside the shell.

▌▌▌▌▌▌▌▌ Exercises 15.5 ▌▌▌▌▌▌▌▌

1. Verify that on the curve with polar equation $r = g(\theta)$ the arc length element is given by

$$ds = \sqrt{(g(\theta))^2 + (g'(\theta))^2}\, d\theta.$$

What is the area element on the vertical cylinder given in terms of cylindrical coordinates by $r = g(\theta)$?

2. Verify that on the spherical surface $x^2 + y^2 + z^2 = a^2$ the area element is given in terms of spherical coordinates by $dS = a^2 \sin\phi\, d\phi\, d\theta$.

3. Find the area of the part of the plane $Ax + By + Cz = D$ lying inside the elliptic cylinder

$$\frac{x^2}{a^2} + \frac{y^2}{b^2} = 1.$$

4. Find the area of the part of the sphere $x^2 + y^2 + z^2 = 4a^2$ that lies inside the cylinder $x^2 + y^2 = 2ay$.

5. State formulas for the surface area element dS for the surface with equation $F(x, y, z) = 0$ valid for the case where the surface has a one-to-one projection on (a) the xz-plane and (b) the yz-plane.

6. Repeat the area calculation of Example 8 by projecting the part of the surface shown in Figure 15.23 onto the yz-plane and using the formula in Exercise 5(b).

7. Find $\iint_{\mathcal{S}} x\, dS$ over the part of the parabolic cylinder $z = x^2/2$ that lies inside the first octant part of the cylinder $x^2 + y^2 = 1$.

8. Find the area of the part of the cone $z^2 = x^2 + y^2$ that lies inside the cylinder $x^2 + y^2 = 2ay$.

9. Find the area of the part of the cylinder $x^2 + y^2 = 2ay$ that lies outside the cone $z^2 = x^2 + y^2$.

10. Find the area of the part of the cylinder $x^2 + z^2 = a^2$ that lies inside the cylinder $y^2 + z^2 = a^2$.

11. A circular cylinder of radius a is circumscribed about a sphere of radius a so that the cylinder is tangent to the sphere along the equator. Two planes, each perpendicular to the axis of the cylinder, intersect the sphere and the cylinder in circles. Show that the area of that part of the sphere between the two planes is equal to the area of the part of the cylinder between the two planes. Thus, the area of the part of a sphere between two parallel planes that intersect it depends only on the radius of the sphere and the distance between the planes, and not on the particular position of the planes.

12. Let $0 < a < b$. In terms of the elliptic integral functions defined in Exercise 17 of Section 15.3, find the area of that part of each of the cylinders $x^2 + z^2 = a^2$ and $y^2 + z^2 = b^2$ that lies inside the other cylinder.

13. Find $\iint_{\mathcal{S}} y\, dS$, where $\mathcal{S}$ is the part of the plane $z = 1 + y$ that lies inside the cone $z = \sqrt{2(x^2 + y^2)}$.

14. Find $\iint_{\mathcal{S}} y\, dS$, where $\mathcal{S}$ is the part of the cone $z = \sqrt{2(x^2 + y^2)}$ that lies below the plane $z = 1 + y$.

15. Find $\iint_{\mathcal{S}} xz\, dS$, where $\mathcal{S}$ is the part of the surface $z = x^2$ that lies in the first octant of 3-space and inside the paraboloid $z = 1 - 3x^2 - y^2$.

16. Find the mass of the part of the surface $z = \sqrt{2xy}$ that lies above the region $0 \le x \le 5, 0 \le y \le 2$, if the areal density of the surface is $\sigma(x, y, z) = kz$.

17. Find the total charge on the surface

$$\mathbf{r} = e^u \cos v\,\mathbf{i} + e^u \sin v\,\mathbf{j} + u\mathbf{k}, \quad (0 \le u \le 1,\ 0 \le v \le \pi),$$

if the charge density on the surface is $\delta = \sqrt{1 + e^{2u}}$.

Exercises 18–19 concern **spheroids**, which are ellipsoids with two of their three semi-axes equal, say $a = b$:

$$\frac{x^2}{a^2} + \frac{y^2}{a^2} + \frac{z^2}{c^2} = 1.$$

∗18. Find the surface area of a **prolate spheroid**, where $0 < a < c$. A prolate spheroid has its two shorter semi-axes equal, like an American "pro football."

∗19. Find the surface area of an **oblate spheroid**, where $0 < c < a$. An oblate spheroid has its two longer semi-axes equal, like the earth.

20. Describe the parametric surface

$$x = au \cos v, \quad y = au \sin v, \quad z = bv,$$

$(0 \le u \le 1,\ 0 \le v \le 2\pi)$, and find its area.

∗21. Evaluate $\iint_{\mathcal{P}} \dfrac{dS}{(x^2 + y^2 + z^2)^{3/2}}$, where $\mathcal{P}$ is the plane with equation $Ax + By + Cz = D$, $(D \ne 0)$.

22. A spherical shell of radius a is centred at the origin. Find the centroid of that part of the sphere that lies in the first octant.

23. Find the centre of mass of a right-circular conical shell of base radius a, height h, and constant areal density σ.

∗24. Find the gravitational attraction of a hemispherical shell of radius a and constant areal density σ on a mass m located at the centre of the base of the hemisphere.

∗25. Find the gravitational attraction of a circular cylindrical shell of radius a, height h, and constant areal density σ on a mass m located on the axis of the cylinder b units above the base.

In Exercises 26–28, find the moment of inertia and radius of gyration of the given object about the given axis. Assume constant areal density σ in each case.

26. A cylindrical shell of radius a and height h about the axis of the cylinder

27. A spherical shell of radius a about a diameter

28. A right-circular conical shell of base radius a and height h about the axis of the cone

29. With what acceleration will the spherical shell of Exercise 27 roll down a plane inclined at angle α to the horizontal? (Compare your result with that of Example 4(b) of Section 16.7.)

15.6 ## Oriented Surfaces and Flux Integrals

Surface integrals of normal components of vector fields play a very important role in vector calculus, similar to the role played by line integrals of tangential components of vector fields. Before we consider such surface integrals we need to define the *orientation* of a surface.

Oriented Surfaces

A smooth surface $\mathcal{S}$ in 3-space is said to be **orientable** if there exists a *unit vector field* $\hat{\mathbf{N}}(P)$ defined on $\mathcal{S}$ that varies continuously as P ranges over $\mathcal{S}$ and that is everywhere normal to $\mathcal{S}$. Any such vector field $\hat{\mathbf{N}}(P)$ determines an **orientation** of $\mathcal{S}$. The surface must have two sides since $\hat{\mathbf{N}}(P)$ can have only one value at each point P. The side out of which $\hat{\mathbf{N}}$ points is called the **positive side**; the other side is the **negative side**. An **oriented surface** is a smooth surface together with a particular choice of orienting unit normal vector field $\hat{\mathbf{N}}(P)$.

For example, if we define $\hat{\mathbf{N}}$ on the smooth surface $z = f(x, y)$ by

$$\hat{\mathbf{N}} = \frac{-f_1(x, y)\mathbf{i} - f_2(x, y)\mathbf{j} + \mathbf{k}}{\sqrt{1 + (f_1(x, y))^2 + (f_2(x, y))^2}},$$

then the top of the surface is the positive side. (See Figure 15.26.)

A smooth or piecewise smooth surface may be **closed** (i.e., it may have no boundary), or it may have one or more boundary curves. (The unit normal vector field $\hat{\mathbf{N}}(P)$ need not be defined at points of the boundary curves.)

An oriented surface $\mathcal{S}$ **induces an orientation** on any of its boundary curves $\mathcal{C}$; if we stand on the positive side of the surface $\mathcal{S}$ and walk around $\mathcal{C}$ in the direction of its orientation, then $\mathcal{S}$ will be on our left side. (See Figure 15.26(a) and (b).)

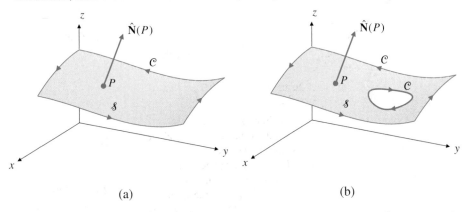

Figure 15.26 The boundary curves of an oriented surface are themselves oriented with the surface on the left

(a)   (b)

A *piecewise smooth* surface is **orientable** if, whenever two smooth component surfaces join along a common boundary curve $\mathcal{C}$, they induce *opposite* orientations along $\mathcal{C}$. This forces the normals $\hat{\mathbf{N}}$ to be on the same side of adjacent components. For instance, the surface of a cube is a piecewise smooth, closed surface, consisting of six smooth surfaces (the square faces) joined along edges. (See Figure 15.27.) If all of the faces are oriented so that their normals $\hat{\mathbf{N}}$ point out of the cube (or if they all point into the cube), then the surface of the cube itself is oriented.

Not every surface can be oriented, even if it appears smooth. An orientable surface must have two sides. For example, a **Möbius band**, consisting of a strip of paper with ends joined together to form a loop, but with one end given a half twist before the ends are joined, has only one side (make one and see), so it cannot be oriented. (See Figure 15.28.) If a nonzero vector is moved around the band, starting at point P, so that it is always normal to the surface, then it can return to its starting position pointing in the opposite direction.

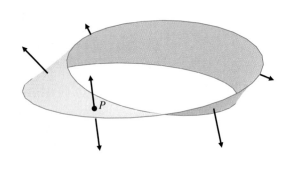

Figure 15.27 The surface of the cube is orientable; adjacent faces induce opposite orientations on their common edge

Figure 15.28 The Möbius band is not orientable; it has only one "side"

The Flux of a Vector Field Across a Surface

Suppose 3-space is filled with an incompressible fluid that flows with velocity field $\mathbf{v}$. Let $\mathcal{S}$ be an imaginary, smooth, oriented surface in 3-space. (We say $\mathcal{S}$ is *imaginary* because it does not impede the motion of the fluid; it is fixed in space and the fluid can move freely through it.) We calculate the rate at which fluid flows across $\mathcal{S}$. Let dS be a small area element at point P on the surface. The fluid crossing that element between time t and time $t + dt$ occupies a cylinder of base area dS and height $|\mathbf{v}(P)|\, dt\, \cos\theta$, where θ is the angle between $\mathbf{v}(P)$ and the normal $\hat{\mathbf{N}}(P)$. (See Figure 15.29.) This cylinder has (signed) volume $\mathbf{v}(P) \bullet \hat{\mathbf{N}}(P)\, dS\, dt$. The rate at which fluid crosses dS is $\mathbf{v}(P) \bullet \hat{\mathbf{N}}(P)\, dS$, and the total rate at which it crosses $\mathcal{S}$ is given by the surface integral

$$\iint_{\mathcal{S}} \mathbf{v} \bullet \hat{\mathbf{N}}\, dS \qquad \text{or} \qquad \iint_{\mathcal{S}} \mathbf{v} \bullet d\mathbf{S},$$

where we use $d\mathbf{S}$ to represent the vector surface area element $\hat{\mathbf{N}}\, dS$.

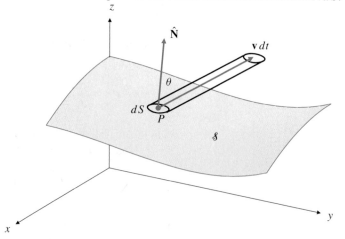

Figure 15.29 The fluid crossing dS in time dt fills the tube

DEFINITION 6

Flux of a vector field across an oriented surface

Given any continuous vector field $\mathbf{F}$, the **flux** of $\mathbf{F}$ across the orientable surface $\mathcal{S}$ is integral of the normal component of $\mathbf{F}$ over $\mathcal{S}$,

$$\iint_{\mathcal{S}} \mathbf{F} \bullet \hat{\mathbf{N}}\, dS \qquad \text{or} \qquad \iint_{\mathcal{S}} \mathbf{F} \bullet d\mathbf{S}.$$

When the surface is closed, the flux integral can be denoted by

$$\oiint_{\mathcal{S}} \mathbf{F} \cdot \hat{\mathbf{N}} \, dS \quad \text{or} \quad \oiint_{\mathcal{S}} \mathbf{F} \cdot d\mathbf{S}.$$

In this case we refer to the flux of $\mathbf{F}$ *out of* $\mathcal{S}$ if $\hat{\mathbf{N}}$ is the unit *exterior* normal, and the flux *into* $\mathcal{S}$ if $\hat{\mathbf{N}}$ is the unit *interior* normal.

Example 1   Find the flux of the vector field $\mathbf{F} = m\mathbf{r}/|\mathbf{r}|^3$ out of a sphere $\mathcal{S}$ of radius a centred at the origin. (Here $\mathbf{r} = x\mathbf{i} + y\mathbf{j} + z\mathbf{k}$.)

Solution   Since $\mathbf{F}$ is the field associated with a source of strength m at the origin (which produces $4\pi m$ units of fluid per unit time at the origin), the answer must be $4\pi m$. Let us calculate it anyway. We use spherical coordinates. At any point $\mathbf{r}$ on the sphere, with spherical coordinates $[a, \phi, \theta]$, the unit outward normal is $\hat{\mathbf{r}} = \mathbf{r}/|\mathbf{r}|$. Since the vector field is $\mathbf{F} = m\hat{\mathbf{r}}/a^2$ on the sphere, and since an area element is $dS = a^2 \sin\phi \, d\phi \, d\theta$, the flux of $\mathbf{F}$ out of the sphere is

$$\oiint_{\mathcal{S}} \left(\frac{m}{a^2}\hat{\mathbf{r}}\right) \cdot \hat{\mathbf{r}} \, a^2 \sin\phi \, d\phi \, d\theta = m \int_0^{2\pi} d\theta \int_0^{\pi} \sin\phi \, d\phi = 4\pi m. \quad \blacksquare$$

Example 2   Calculate the total flux of $\mathbf{F} = x\mathbf{i} + y\mathbf{j} + z\mathbf{k}$ outward through the surface of the solid cylinder $x^2 + y^2 \le a^2$, $-h \le z \le h$.

Solution   The cylinder is shown in Figure 15.30. Its surface consists of top and bottom disks and the cylindrical side wall. We calculate the flux of $\mathbf{F}$ out of each. Naturally, we use cylindrical coordinates. On the top disk we have $z = h$, $\hat{\mathbf{N}} = \mathbf{k}$, and $dS = r \, dr \, d\theta$. Therefore, $\mathbf{F} \cdot \hat{\mathbf{N}} \, dS = hr \, dr \, d\theta$ and

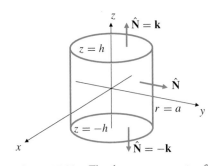

Figure 15.30   The three components of the surface of a solid cylinder with their outward normals

$$\iint_{\text{top}} \mathbf{F} \cdot \hat{\mathbf{N}} \, dS = h \int_0^{2\pi} d\theta \int_0^a r \, dr = \pi a^2 h.$$

On the bottom disk we have $z = -h$, $\hat{\mathbf{N}} = -\mathbf{k}$, and $dS = r \, dr \, d\theta$. Therefore, $\mathbf{F} \cdot \hat{\mathbf{N}} \, dS = hr \, dr \, d\theta$ and

$$\iint_{\text{bottom}} \mathbf{F} \cdot \hat{\mathbf{N}} \, dS = \iint_{\text{top}} \mathbf{F} \cdot \hat{\mathbf{N}} \, dS = \pi a^2 h.$$

On the cylindrical wall $\mathbf{F} = a\cos\theta \, \mathbf{i} + a\sin\theta \, \mathbf{j} + z\mathbf{k}$, $\hat{\mathbf{N}} = \cos\theta \, \mathbf{i} + \sin\theta \, \mathbf{j}$, and $dS = a \, d\theta \, dz$. Thus, $\mathbf{F} \cdot \hat{\mathbf{N}} \, dS = a^2 \, d\theta \, dz$ and

$$\iint_{\text{cylwall}} \mathbf{F} \cdot \hat{\mathbf{N}} \, dS = a^2 \int_0^{2\pi} d\theta \int_{-h}^h dz = 4\pi a^2 h.$$

The total flux of $\mathbf{F}$ out of the surface $\mathcal{S}$ of the cylinder is the sum of these three contributions:

$$\oiint_{\mathcal{S}} \mathbf{F} \cdot \hat{\mathbf{N}} \, dS = 6\pi a^2 h. \quad \blacksquare$$

Let $\mathcal{S}$ be a smooth, oriented surface with a one-to-one projection onto a domain D in the xy-plane, and with equation of the form $G(x, y, z) = 0$. In Section 15.5 we showed that the surface area element on $\mathcal{S}$ could be written in the form

$$dS = \left|\frac{\nabla G}{G_3}\right| \, dx \, dy,$$

and hence surface integrals over $\mathcal{S}$ could be reduced to double integrals over the domain D. Flux integrals can be treated likewise. Depending on the orientation of $\mathcal{S}$, the unit normal $\hat{\mathbf{N}}$ can be written as

$$\hat{\mathbf{N}} = \pm \frac{\nabla G}{|\nabla G|}.$$

Thus, the vector area element $d\mathbf{S}$ can be written

$$d\mathbf{S} = \hat{\mathbf{N}}\,dS = \pm \frac{\nabla G(x, y, z)}{G_3(x, y, z)}\,dx\,dy.$$

The sign must be chosen to give $\mathcal{S}$ the desired orientation. If $G_3 > 0$ and we want the positive side of $\mathcal{S}$ to face upward, we should use the $+$ sign. Of course, similar formulas apply for surfaces with one-to-one projections onto the other coordinate planes.

Example 3   Find the flux of $z\mathbf{i} + x^2\mathbf{k}$ upward through that part of the surface $z = x^2 + y^2$ lying above the square R defined by $-1 \le x \le 1$ and $-1 \le y \le 1$.

Solution   For $F(x, y, z) = z - x^2 - y^2$ we have $\nabla F = -2x\mathbf{i} - 2y\mathbf{j} + \mathbf{k}$ and $F_3 = 1$. Thus,

$$d\mathbf{S} = (-2x\mathbf{i} - 2y\mathbf{j} + \mathbf{k})\,dx\,dy,$$

and the required flux is

$$\iint_{\mathcal{S}} (z\mathbf{i} + x^2\mathbf{k}) \bullet d\mathbf{S} = \iint_R \left(-2x(x^2 + y^2) + x^2\right) dx\,dy$$

$$= \int_{-1}^{1} dx \int_{-1}^{1} (x^2 - 2x^3 - 2xy^2)\,dy$$

$$= \int_{-1}^{1} 2x^2\,dx = \frac{4}{3}.$$

(Two of the three terms in the double integral had zero integrals because of symmetry.) ∎

For a surface $\mathcal{S}$ with equation $z = f(x, y)$ we have

$$\hat{\mathbf{N}} = \pm \frac{-\dfrac{\partial f}{\partial x}\mathbf{i} - \dfrac{\partial f}{\partial y}\mathbf{j} + \mathbf{k}}{\sqrt{1 + \left(\dfrac{\partial f}{\partial x}\right)^2 + \left(\dfrac{\partial f}{\partial y}\right)^2}} \qquad \text{and}$$

$$dS = \sqrt{1 + \left(\frac{\partial f}{\partial x}\right)^2 + \left(\frac{\partial f}{\partial y}\right)^2}\,dx\,dy,$$

so that the vector area element on $\mathcal{S}$ is given by

$$d\mathbf{S} = \hat{\mathbf{N}}\,dS = \pm \left(-\frac{\partial f}{\partial x}\mathbf{i} - \frac{\partial f}{\partial y}\mathbf{j} + \mathbf{k}\right) dx\,dy.$$

Again, the $+$ sign corresponds to an upward normal.

For a general parametric surface $\mathbf{r} = \mathbf{r}(u, v)$, the unit normal $\hat{\mathbf{N}}$ and area element dS were calculated in Section 15.5:

$$\hat{\mathbf{N}} = \pm \frac{\dfrac{\partial \mathbf{r}}{\partial u} \times \dfrac{\partial \mathbf{r}}{\partial v}}{\left|\dfrac{\partial \mathbf{r}}{\partial u} \times \dfrac{\partial \mathbf{r}}{\partial v}\right|}, \qquad dS = \left|\frac{\partial \mathbf{r}}{\partial u} \times \frac{\partial \mathbf{r}}{\partial v}\right| du\,dv.$$

Thus, the vector area element is

$$dS = \hat{\mathbf{N}}\, dS = \pm \left(\frac{\partial \mathbf{r}}{\partial u} \times \frac{\partial \mathbf{r}}{\partial v} \right) du\, dv.$$

Example 4 Find the flux of $\mathbf{F} = y\mathbf{i} - x\mathbf{j} + 4\mathbf{k}$ upward through $\mathcal{S}$, where $\mathcal{S}$ is the part of the surface $z = 1 - x^2 - y^2$ lying in the first octant of 3-space.

Solution The vector area element corresponding to the upward normal on $\mathcal{S}$ is

$$d\mathbf{S} = \left(-\frac{\partial z}{\partial x}\mathbf{i} - \frac{\partial z}{\partial y}\mathbf{j} + \mathbf{k} \right) dx\, dy = (2x\mathbf{i} + 2y\mathbf{j} + \mathbf{k})\, dx\, dy.$$

The projection of $\mathcal{S}$ onto the xy-plane is the quarter-circular disk Q given by $x^2 + y^2 \le 1$, $x \ge 0$, and $y \ge 0$. Thus, the flux of $\mathbf{F}$ upward through $\mathcal{S}$ is

$$\iint_{\mathcal{S}} \mathbf{F} \bullet d\mathbf{S} = \iint_Q (2xy - 2xy + 4)\, dx\, dy$$
$$= 4 \times (\text{area of } Q) = \pi.$$

■

Example 5 Find the flux of $\mathbf{F} = \dfrac{2x\mathbf{i} + 2y\mathbf{j}}{x^2 + y^2} + \mathbf{k}$ downward through the surface $\mathcal{S}$ defined parametrically by

$$\mathbf{r} = u \cos v\,\mathbf{i} + u \sin v\,\mathbf{j} + u^2\mathbf{k}, \qquad (0 \le u \le 1,\ 0 \le v \le 2\pi).$$

Solution First we calculate $d\mathbf{S}$:

$$\frac{\partial \mathbf{r}}{\partial u} = \cos v\,\mathbf{i} + \sin v\,\mathbf{j} + 2u\mathbf{k}$$

$$\frac{\partial \mathbf{r}}{\partial v} = -u \sin v\,\mathbf{i} + u \cos v\,\mathbf{j}$$

$$\frac{\partial \mathbf{r}}{\partial u} \times \frac{\partial \mathbf{r}}{\partial v} = -2u^2 \cos v\,\mathbf{i} - 2u^2 \sin v\,\mathbf{j} + u\mathbf{k}.$$

Since $u \ge 0$ on $\mathcal{S}$, the latter expression is an upward normal. We want a downward normal, so we use

$$d\mathbf{S} = (2u^2 \cos v\,\mathbf{i} + 2u^2 \sin v\,\mathbf{j} - u\mathbf{k})\, du\, dv.$$

On $\mathcal{S}$ we have

$$\mathbf{F} = \frac{2x\mathbf{i} + 2y\mathbf{j}}{x^2 + y^2} + \mathbf{k} = \frac{2u \cos v\,\mathbf{i} + 2u \sin v\,\mathbf{j}}{u^2} + \mathbf{k},$$

so the downward flux of $\mathbf{F}$ through $\mathcal{S}$ is

$$\iint_{\mathcal{S}} \mathbf{F} \bullet d\mathbf{S} = \int_0^{2\pi} dv \int_0^1 (4u - u)\, du = 3\pi.$$

■

Exercises 15.6

1. Find the flux of $\mathbf{F} = x\mathbf{i} + z\mathbf{j}$ out of the tetrahedron bounded by the coordinate planes and the plane $x + 2y + 3z = 6$.

2. Find the flux of $\mathbf{F} = x\mathbf{i} + y\mathbf{j} + z\mathbf{k}$ outward across the sphere $x^2 + y^2 + z^2 = a^2$.

3. Find the flux of the vector field of Exercise 2 out of the surface of the box $0 \le x \le a, 0 \le y \le b, 0 \le z \le c$.

4. Find the flux of the vector field $\mathbf{F} = y\mathbf{i} + z\mathbf{k}$ out across the boundary of the solid cone $0 \le z \le 1 - \sqrt{x^2 + y^2}$.

5. Find the flux of $\mathbf{F} = x\mathbf{i} + y\mathbf{j} + z\mathbf{k}$ upward through the part of the surface $z = a - x^2 - y^2$ lying above plane $z = b < a$.

6. Find the flux of $\mathbf{F} = x\mathbf{i} + x\mathbf{j} + \mathbf{k}$ upward through the part of the surface $z = x^2 - y^2$ inside the cylinder $x^2 + y^2 = a^2$.

7. Find the flux of $\mathbf{F} = y^3\mathbf{i} + z^2\mathbf{j} + x\mathbf{k}$ downward through the part of the surface $z = 4 - x^2 - y^2$ that lies above the plane $z = 2x + 1$.

8. Find the flux of $\mathbf{F} = z^2\mathbf{k}$ upward through the part of the sphere $x^2 + y^2 + z^2 = a^2$ in the first octant of 3-space.

9. Find the flux of $\mathbf{F} = x\mathbf{i} + y\mathbf{j}$ upward through the part of the surface $z = 2 - x^2 - 2y^2$ that lies above the xy-plane.

10. Find the flux of $\mathbf{F} = 2x\mathbf{i} + y\mathbf{j} + z\mathbf{k}$ upward through the surface $\mathbf{r} = u^2 v\mathbf{i} + uv^2\mathbf{j} + v^3\mathbf{k}$, $(0 \le u \le 1, 0 \le v \le 1)$.

11. Find the flux of $\mathbf{F} = x\mathbf{i} + y\mathbf{j} + z^2\mathbf{k}$ upward through the surface $u \cos v\,\mathbf{i} + u \sin v\,\mathbf{j} + u\,\mathbf{k}$, $(0 \le u \le 2, 0 \le v \le \pi)$.

12. Find the flux of $\mathbf{F} = yz\mathbf{i} - xz\mathbf{j} + (x^2 + y^2)\mathbf{k}$ upward through the surface $\mathbf{r} = e^u \cos v\,\mathbf{i} + e^u \sin v\,\mathbf{j} + u\,\mathbf{k}$, where $0 \le u \le 1$ and $0 \le v \le \pi$.

13. Find the flux of $\mathbf{F} = m\mathbf{r}/|\mathbf{r}|^3$ out of the surface of the cube $-a \le x, y, z \le a$.

*14. Find the flux of the vector field of Exercise 13 out of the box $1 \le x, y, z \le 2$. *Note:* This problem can be solved very easily using the Divergence Theorem of Section 16.4; the required flux is, in fact, zero. However, the object here is to do it by direct calculation of the surface integrals involved, and as such it is quite difficult. By symmetry, it is sufficient to evaluate the net flux out of the cube through any one of the three pairs of opposite faces; that is, you must calculate the flux through only two faces, say $z = 1$ and $z = 2$. Be prepared to work very hard to evaluate these integrals! When they are done you may find the identities

$2 \arctan a = \arctan \left(2a/(1 - a^2) \right)$ and

$\arctan a + \arctan (1/a)\pi/2$ useful for showing that the net flux is zero.

15. Define the flux of a *plane* vector field across a piecewise smooth *curve*. Find the flux of $\mathbf{F} = x\mathbf{i} + y\mathbf{j}$ outward across

   (a) the circle $x^2 + y^2 = a^2$ and

   (b) the boundary of the square $-1 \le x, y \le 1$.

16. Find the flux of $\mathbf{F} = -(x\mathbf{i} + y\mathbf{j})/(x^2 + y^2)$ *inward* across each of the two curves in the previous exercise.

17. If $\mathcal{S}$ is a smooth, oriented surface in 3-space and $\hat{\mathbf{N}}$ is the unit vector field determining the orientation of $\mathcal{S}$, show that the flux of $\hat{\mathbf{N}}$ across $\mathcal{S}$ is the area of $\mathcal{S}$.

*18. The Divergence Theorem presented in Section 16.4 implies that the flux of a constant vector field across any oriented, piecewise smooth, closed surface is zero. Prove this now for (a) a rectangular box and (b) a sphere.

Chapter Review

Key Ideas

- **What do the following terms and phrases mean?**
  ◇ vector field
  ◇ scalar field
  ◇ field line
  ◇ conservative field
  ◇ scalar potential
  ◇ equipotential
  ◇ a source
  ◇ a dipole
  ◇ connected domain
  ◇ simply connected
  ◇ parametric surface
  ◇ orientable surface
  ◇ the line integral of f along curve $\mathcal{C}$
  ◇ the line integral of the tangential component of $\mathbf{F}$ along $\mathcal{C}$
  ◇ the flux of a vector field through a surface

- **How are the field lines of a conservative field related to its equipotential curves or surfaces?**

- **How is a line integral of a scalar field calculated?**

- **How is a line integral of the tangential component of a vector field calculated?**

- **When is a line integral between two points independent of the path joining those points?**

- **How is a surface integral of a scalar field calculated?**

- **How do you calculate the flux of a vector field through a surface?**

Review Exercises

1. Find $\int_{\mathcal{C}} \dfrac{1}{y}\,ds$, where $\mathcal{C}$ is the curve

   $$x = t, \quad y = 2e^t, \quad z = e^{2t}, \quad (-1 \le t \le 1).$$

2. Let $\mathcal{C}$ be the part of the curve of intersection of the surfaces $z = x + y^2$ and $y = 2x$ from the origin to the point $(2, 4, 18)$. Evaluate $\int_{\mathcal{C}} 2y\,dx + x\,dy + 2\,dz$.

3. Find $\displaystyle\iint_{\mathcal{S}} x\,dS$, where $\mathcal{S}$ is that part of the cone $z = \sqrt{x^2 + y^2}$ in the region $0 \le x \le 1 - y^2$.

4. Find $\displaystyle\iint_{\mathcal{S}} xyz\,dS$ over the part of the plane $x + y + z = 1$ lying in the first octant.

5. Find the flux of $x^2 y\mathbf{i} - 10xy^2\mathbf{j}$ upward through the surface $z = xy$, $0 \le x \le 1$, $0 \le y \le 1$.

6. Find the flux of $x\mathbf{i} + y\mathbf{j} + z\mathbf{k}$ downward through the part of the plane $x + 2y + 3z = 6$ lying in the first octant.

7. A bead of mass m slides down a wire in the shape of the curve $x = a\sin t$, $\quad y = a\cos t$, $\quad z = bt$, where $0 \le t \le 6\pi$.

(a) What is the work done by the gravitational force $\mathbf{F} = -mg\mathbf{k}$ on the bead during its descent?

(b) What is the work done against a resistance of constant magnitude R which directly opposes the motion of the bead during its descent?

8. For what values of the constants a, b, and c can you determine the value of the integral I of the tangential component of $\mathbf{F} = (axy + 3yz)\mathbf{i} + (x^2 + 3xz + by^2z)\mathbf{j} + (bxy + cy^3)\mathbf{k}$ along a curve from $(0, 1, -1)$ to $(2, 1, 1)$ without knowing exactly which curve? What is the value of the integral?

9. Let $\mathbf{F} = (x^2/y)\mathbf{i} + y\mathbf{j} + \mathbf{k}$.

(a) Find the field line of $\mathbf{F}$ that passes through $(1, 1, 0)$ and show that it also passes through $(e, e, 1)$.

(b) Find $\displaystyle\int_{\mathcal{C}} \mathbf{F} \cdot d\mathbf{r}$, where $\mathcal{C}$ is the part of the field line in (a) from $(1, 1, 0)$ to $(e, e, 1)$.

10. Consider the vector fields

$$\mathbf{F} = (1 + x)e^{x+y}\mathbf{i} + (xe^{x+y} + 2y)\mathbf{j} - 2z\mathbf{k},$$
$$\mathbf{G} = (1 + x)e^{x+y}\mathbf{i} + (xe^{x+y} + 2z)\mathbf{j} - 2y\mathbf{k}.$$

(a) Show that $\mathbf{F}$ is conservative by finding a potential for it.

(b) Evaluate $\displaystyle\int_{\mathcal{C}} \mathbf{G} \cdot d\mathbf{r}$, where $\mathcal{C}$ is given by

$$\mathbf{r} = (1 - t)e^t\mathbf{i} + t\mathbf{j} + 2t\mathbf{k}, \quad (0 \le t \le 1),$$

by taking advantage of the similarity between $\mathbf{F}$ and $\mathbf{G}$.

11. Find a plane vector field $\mathbf{F}(x, y)$ that satisfies the following conditions:

(i) The field lines of $\mathbf{F}$ are the curves $xy = C$.

(ii) $|\mathbf{F}(x, y)| = 1$ if $(x, y) \ne (0, 0)$.

(iii) $\mathbf{F}(1, 1) = (\mathbf{i} - \mathbf{j})/\sqrt{2}$.

(iv) $\mathbf{F}$ is continuous except at $(0, 0)$.

12. Let $\mathcal{S}$ be the part of the surface of the cylinder $y^2 + z^2 = 16$ that lies in the first octant and between the planes $x = 0$ and $x = 5$. Find the flux of $3z^2x\mathbf{i} - x\mathbf{j} - y\mathbf{k}$ away from the x-axis through $\mathcal{S}$.

Challenging Problems

$*$ **1.** Find the centroid of the surface

$$\mathbf{r} = (2 + \cos v)(\cos u\mathbf{i} + \sin u\mathbf{j}) + \sin v\mathbf{k},$$

where $0 \le u \le 2\pi$ and $0 \le v \le \pi$. Describe this surface.

$*$ **2.** A smooth surface $\mathcal{S}$ is given parametrically by

$$\mathbf{r} = (\cos 2u)(2 + v\cos u)\mathbf{i}$$
$$+ (\sin 2u)(2 + v\cos u)\mathbf{j} + v\sin u\mathbf{k},$$

where $0 \le u \le 2\pi$ and $-1 \le v \le 1$. Show that for *every* smooth vector field $\mathbf{F}$ on $\mathcal{S}$,

$$\iint_{\mathcal{S}} \mathbf{F} \cdot \hat{\mathbf{N}}\, dS = 0,$$

where $\hat{\mathbf{N}} = \hat{\mathbf{N}}(u, v)$ is a unit normal vector field on $\mathcal{S}$ that depends continuously on (u, v). How do you explain this? *Hint:* Try to describe what the surface $\mathcal{S}$ looks like.

$*$ **3.** Recalculate the gravitational force exerted by a sphere of radius a and areal density σ centred at the origin on a point mass located at $(0, 0, b)$ by directly integrating the vertical component of the force due to an area element dS, rather than by integrating the potential as we did in the last part of Section 15.5. You will have to be quite creative in dealing with the resulting integral.

CHAPTER 16

Vector Calculus

Mathematicians are like Frenchmen: whenever you say something to them, they translate it into their own language, and at once it is something entirely different.

Johann Wolfgang von Goethe   1749–1832
from *Maxims and Reflections, 1829*

Introduction   In this chapter we develop two- and three-dimensional analogues of the one-dimensional Fundamental Theorem of Calculus. These analogues—Green's Theorem, Gauss's Divergence Theorem, and Stokes's Theorem—are of great importance both theoretically and in applications. They are phrased in terms of certain differential operators, divergence and curl, which are related to the gradient operator encountered in Section 12.7. The operators are introduced and their properties are derived in Sections 16.1 and 16.2. The rest of the chapter deals with the generalizations of the Fundamental Theorem of Calculus and their applications.

16.1 Gradient, Divergence, and Curl

First-order information about the rate of change of a three-dimensional scalar field, $f(x, y, z)$, is contained in the three first partial derivatives $\partial f/\partial x$, $\partial f/\partial y$, and $\partial f/\partial z$. The gradient,

$$\mathbf{grad}\, f(x, y, z) = \nabla f(x, y, z) = \frac{\partial f}{\partial x}\mathbf{i} + \frac{\partial f}{\partial y}\mathbf{j} + \frac{\partial f}{\partial z}\mathbf{k},$$

collects this information into a single vector-valued "derivative" of f. We will develop similar ways of conveying information about the rate of change of vector fields.

First-order information about the rate of change of the vector field

$$\mathbf{F}(x, y, z) = F_1(x, y, z)\mathbf{i} + F_2(x, y, z)\mathbf{j} + F_3(x, y, z)\mathbf{k}$$

is contained in nine first partial derivatives, three for each of the three components of the vector field $\mathbf{F}$:

$$\frac{\partial F_1}{\partial x} \qquad \frac{\partial F_1}{\partial y} \qquad \frac{\partial F_1}{\partial z}$$

$$\frac{\partial F_2}{\partial x} \qquad \frac{\partial F_2}{\partial y} \qquad \frac{\partial F_2}{\partial z}$$

$$\frac{\partial F_3}{\partial x} \qquad \frac{\partial F_3}{\partial y} \qquad \frac{\partial F_3}{\partial z}.$$

(Again, we stress that F_1, F_2, and F_3 denote the components of $\mathbf{F}$, not partial derivatives.) Two special combinations of these derivatives organize this information in particularly useful ways, as the gradient does for scalar fields. These are the **divergence** of $\mathbf{F}$ (**div F**) and the **curl** of $\mathbf{F}$ (**curl F**), defined as follows:

Divergence and curl

$$\mathbf{div\,F} = \nabla \bullet \mathbf{F} = \frac{\partial F_1}{\partial x} + \frac{\partial F_2}{\partial y} + \frac{\partial F_3}{\partial z},$$

$$\mathbf{curl\,F} = \nabla \times \mathbf{F}$$

$$= \left(\frac{\partial F_3}{\partial y} - \frac{\partial F_2}{\partial z}\right)\mathbf{i} + \left(\frac{\partial F_1}{\partial z} - \frac{\partial F_3}{\partial x}\right)\mathbf{j} + \left(\frac{\partial F_2}{\partial x} - \frac{\partial F_1}{\partial y}\right)\mathbf{k}$$

$$= \begin{vmatrix} \mathbf{i} & \mathbf{j} & \mathbf{k} \\ \frac{\partial}{\partial x} & \frac{\partial}{\partial y} & \frac{\partial}{\partial z} \\ F_1 & F_2 & F_3 \end{vmatrix}.$$

Note that the divergence of a vector field is a scalar field, while the curl is another vector field. Also observe the notation $\nabla \bullet \mathbf{F}$ and $\nabla \times \mathbf{F}$, which we will sometimes use instead of **div F** and **curl F**. This makes use of the *vector differential operator*

$$\nabla = \mathbf{i}\,\frac{\partial}{\partial x} + \mathbf{j}\,\frac{\partial}{\partial y} + \mathbf{k}\,\frac{\partial}{\partial z},$$

frequently called *del* or *nabla*. Just as the gradient of the scalar field f can be regarded as *formal scalar multiplication* of ∇ and f, so also can the divergence and curl of $\mathbf{F}$ be regarded as *formal dot* and *cross products* of ∇ with $\mathbf{F}$. When using ∇ the order of "factors" is important; the quantities on which ∇ acts must appear to the right of ∇. For instance, $\nabla \bullet \mathbf{F}$ and $\mathbf{F} \bullet \nabla$ do not mean the same thing; the former is a scalar field and the latter is a scalar differential operator:

$$\mathbf{F} \bullet \nabla = F_1\,\frac{\partial}{\partial x} + F_2\,\frac{\partial}{\partial y} + F_3\,\frac{\partial}{\partial z}.$$

> ⚠️ **BEWARE**
>
> Do not confuse the scalar field $\nabla \bullet \mathbf{F}$ with the scalar differential operator $\mathbf{F} \bullet \nabla$. They are quite different objects.

Example 1 Find the divergence and curl of the vector field

$$\mathbf{F} = xy\mathbf{i} + (y^2 - z^2)\mathbf{j} + yz\mathbf{k}.$$

Solution We have

$$\mathbf{div\,F} = \nabla \bullet \mathbf{F} = \frac{\partial}{\partial x}(xy) + \frac{\partial}{\partial y}(y^2 - z^2) + \frac{\partial}{\partial z}(yz) = y + 2y + y = 4y,$$

$$\mathbf{curl\,F} = \nabla \times \mathbf{F} = \begin{vmatrix} \mathbf{i} & \mathbf{j} & \mathbf{k} \\ \frac{\partial}{\partial x} & \frac{\partial}{\partial y} & \frac{\partial}{\partial z} \\ xy & y^2 - z^2 & yz \end{vmatrix}$$

$$= \left[\frac{\partial}{\partial y}(yz) - \frac{\partial}{\partial z}(y^2 - z^2)\right]\mathbf{i} + \left[\frac{\partial}{\partial z}(xy) - \frac{\partial}{\partial x}(yz)\right]\mathbf{j}$$

$$+ \left[\frac{\partial}{\partial x}(y^2 - z^2) - \frac{\partial}{\partial y}(xy)\right]\mathbf{k} = 3z\mathbf{i} - x\mathbf{k}.$$

The divergence and curl of a two-dimensional vector field can also be defined: if $\mathbf{F}(x, y) = F_1(x, y)\mathbf{i} + F_2(x, y)\mathbf{j}$, then

$$\operatorname{div}\mathbf{F} = \frac{\partial F_1}{\partial x} + \frac{\partial F_2}{\partial y},$$

$$\operatorname{curl}\mathbf{F} = \left(\frac{\partial F_2}{\partial x} - \frac{\partial F_1}{\partial y}\right)\mathbf{k}.$$

Note that the curl of a two-dimensional vector field is still a 3-vector and is perpendicular to the plane of the field. Although **div** and **grad** are defined in all dimensions, **curl** is defined only in three dimensions and in the plane (provided we allow values in three dimensions).

Example 2   Find the divergence and curl of $\mathbf{F} = xe^y\mathbf{i} - ye^x\mathbf{j}$.

Solution   We have

$$\operatorname{div}\mathbf{F} = \nabla \bullet \mathbf{F} = \frac{\partial}{\partial x}(xe^y) + \frac{\partial}{\partial y}(-ye^x) = e^y - e^x,$$

$$\operatorname{curl}\mathbf{F} = \nabla \times \mathbf{F} = \left(\frac{\partial}{\partial x}(-ye^x) - \frac{\partial}{\partial y}(xe^y)\right)\mathbf{k}$$

$$= -(ye^x + xe^y)\mathbf{k}.$$

Interpretation of the Divergence

The value of the divergence of a vector field $\mathbf{F}$ at point P is, loosely speaking, a measure of the rate at which the field "diverges" or "spreads away" from P. This spreading away can be measured by the flux out of a small closed surface surrounding P. For instance, $\operatorname{div}\mathbf{F}(P)$ is the limit of the *flux per unit volume* out of smaller and smaller spheres centred at P.

THEOREM   1

The divergence as flux density

If $\hat{\mathbf{N}}$ is the unit outward normal on the sphere $\mathcal{S}_\epsilon$ of radius ϵ centred at point P, and if $\mathbf{F}$ is a smooth three-dimensional vector field, then

$$\operatorname{div}\mathbf{F}(P) = \lim_{\epsilon \to 0^+} \frac{3}{4\pi\epsilon^3} \oiint_{\mathcal{S}_\epsilon} \mathbf{F} \bullet \hat{\mathbf{N}}\, dS.$$

PROOF   Without loss of generality we assume that P is at the origin. We want to expand $\mathbf{F} = F_1\mathbf{i} + F_2\mathbf{j} + F_3\mathbf{k}$ in a Taylor series about the origin (a Maclaurin series). As shown in Section 12.9 for a function of two variables, the Maclaurin series for a scalar-valued function of three variables takes the form

$$f(x, y, z) = f(0, 0, 0) + \frac{\partial f}{\partial x}\bigg|_{(0,0,0)} x + \frac{\partial f}{\partial y}\bigg|_{(0,0,0)} y + \frac{\partial f}{\partial z}\bigg|_{(0,0,0)} z + \cdots,$$

where "$\cdots$" represents terms of second and higher degree in x, y, and z. If we apply this formula to the components of $\mathbf{F}$, we obtain

$$\mathbf{F}(x, y, z) = \mathbf{F}_0 + \mathbf{F}_{x0}\, x + \mathbf{F}_{y0}\, y + \mathbf{F}_{z0}\, z + \cdots,$$

So as not to confuse partial derivatives with components of vectors, we are using subscripts $x0$, $y0$, and $z0$ here to denote the values of the first partial derivatives of $\mathbf{F}$ at $(0, 0, 0)$.

where

$$\mathbf{F}_0 = \mathbf{F}(0, 0, 0)$$

$$\mathbf{F}_{x0} = \left.\frac{\partial \mathbf{F}}{\partial x}\right|_{(0,0,0)} = \left.\left(\frac{\partial F_1}{\partial x}\mathbf{i} + \frac{\partial F_2}{\partial x}\mathbf{j} + \frac{\partial F_3}{\partial x}\mathbf{k}\right)\right|_{(0,0,0)}$$

$$\mathbf{F}_{y0} = \left.\frac{\partial \mathbf{F}}{\partial y}\right|_{(0,0,0)} = \left.\left(\frac{\partial F_1}{\partial y}\mathbf{i} + \frac{\partial F_2}{\partial y}\mathbf{j} + \frac{\partial F_3}{\partial y}\mathbf{k}\right)\right|_{(0,0,0)}$$

$$\mathbf{F}_{z0} = \left.\frac{\partial \mathbf{F}}{\partial z}\right|_{(0,0,0)} = \left.\left(\frac{\partial F_1}{\partial z}\mathbf{i} + \frac{\partial F_2}{\partial z}\mathbf{j} + \frac{\partial F_3}{\partial z}\mathbf{k}\right)\right|_{(0,0,0)};$$

again, the "$\cdots$" represents the second- and higher-degree terms in x, y, and z. The unit normal on $\mathcal{S}_\epsilon$ is $\hat{\mathbf{N}} = (x\mathbf{i} + y\mathbf{j} + z\mathbf{k})/\epsilon$, so we have

$$\begin{aligned}
\mathbf{F} \bullet \hat{\mathbf{N}} = \frac{1}{\epsilon}\Big(& \mathbf{F}_0 \bullet \mathbf{i}\, x + \mathbf{F}_0 \bullet \mathbf{j}\, y + \mathbf{F}_0 \bullet \mathbf{k}\, z \\
& + \mathbf{F}_{x0} \bullet \mathbf{i}\, x^2 + \mathbf{F}_{x0} \bullet \mathbf{j}\, xy + \mathbf{F}_{x0} \bullet \mathbf{k}\, xz \\
& + \mathbf{F}_{y0} \bullet \mathbf{i}\, xy + \mathbf{F}_{y0} \bullet \mathbf{j}\, y^2 + \mathbf{F}_{y0} \bullet \mathbf{k}\, yz \\
& + \mathbf{F}_{z0} \bullet \mathbf{i}\, xz + \mathbf{F}_{z0} \bullet \mathbf{j}\, yz + \mathbf{F}_{z0} \bullet \mathbf{k}\, z^2 + \cdots \Big).
\end{aligned}$$

We integrate each term within the parentheses over $\mathcal{S}_\epsilon$. By symmetry,

$$\oiint_{\mathcal{S}_\epsilon} x\, dS = \oiint_{\mathcal{S}_\epsilon} y\, dS = \oiint_{\mathcal{S}_\epsilon} z\, dS = 0,$$

$$\oiint_{\mathcal{S}_\epsilon} xy\, dS = \oiint_{\mathcal{S}_\epsilon} xz\, dS = \oiint_{\mathcal{S}_\epsilon} yz\, dS = 0.$$

Also, by symmetry,

$$\begin{aligned}
\oiint_{\mathcal{S}_\epsilon} x^2\, dS &= \oiint_{\mathcal{S}_\epsilon} y^2\, dS = \oiint_{\mathcal{S}_\epsilon} z^2\, dS \\
&= \frac{1}{3}\oiint_{\mathcal{S}_\epsilon}(x^2 + y^2 + z^2)\, dS = \frac{1}{3}(\epsilon^2)(4\pi\epsilon^2) = \frac{4}{3}\pi\epsilon^4,
\end{aligned}$$

and the higher-degree terms have surface integrals involving ϵ^5 and higher powers. Thus,

$$\begin{aligned}
\frac{3}{4\pi\epsilon^3}\oiint_{\mathcal{S}_\epsilon} \mathbf{F} \bullet \hat{\mathbf{N}}\, dS &= \mathbf{F}_{x0} \bullet \mathbf{i} + \mathbf{F}_{y0} \bullet \mathbf{j} + \mathbf{F}_{z0} \bullet \mathbf{k} + \epsilon\,(\cdots) \\
&= \nabla \bullet \mathbf{F}(0, 0, 0) + \epsilon\,(\cdots) \\
&\to \nabla \bullet \mathbf{F}(0, 0, 0)
\end{aligned}$$

as $\epsilon \to 0^+$. This is what we wanted to show.

Remark The spheres $\mathcal{S}_\epsilon$ in the above theorem can be replaced by other contracting families of piecewise smooth surfaces. For instance, if B is the surface of a rectangular box with dimensions Δx, Δy, and Δz containing P, then

$$\mathbf{div}\, \mathbf{F}(P) = \lim_{\Delta x, \Delta y, \Delta z \to 0} \frac{1}{\Delta x\, \Delta y\, \Delta z}\oiint_B \mathbf{F} \bullet \hat{\mathbf{N}}\, dS.$$

See Exercise 12 below.

Remark   In two dimensions, the value **div F**(P) represents the limiting *flux per unit area* outward across small, non-self-intersecting closed curves that enclose P. See Exercise 13 at the end of this section.

Let us return again to the interpretation of a vector field as a velocity field of a moving incompressible fluid. If the total flux of the velocity field outward across the boundary surface of a domain is positive (or negative), then the fluid must be produced (or annihilated) within that domain.

The vector field $\mathbf{F} = x\mathbf{i} + y\mathbf{j} + z\mathbf{k}$ of Example 2 in Section 15.6 has constant divergence, $\nabla \bullet \mathbf{F} = 3$. In that example we showed that the flux of **F** out of a certain cylinder of base radius a and height $2h$ is $6\pi a^2 h$, which is three times the volume of the cylinder. Exercises 2 and 3 of Section 15.6 confirm similar results for the flux of **F** out of other domains. This leads to another interpretation for the divergence: **div F**(P) is the *source strength per unit volume* of **F** at P. With this interpretation, we would expect, even for a vector field **F** with nonconstant divergence, that the total flux of **F** out of the surface $\mathcal{S}$ of a domain D would be equal to the total source strength of **F** within D, that is,

$$\oiint_{\mathcal{S}} \mathbf{F} \bullet \hat{\mathbf{N}}\, dS = \iiint_D \nabla \bullet \mathbf{F}\, dV.$$

This is the **Divergence Theorem**, which we will prove in Section 16.4.

Example 3   Verify that the vector field $\mathbf{F} = m\mathbf{r}/|\mathbf{r}|^3$, due to a source of strength m at $(0, 0, 0)$, has zero divergence at all points in $\mathbb{R}^3$ except the origin. What would you expect to be the total flux of **F** outward across the boundary surface of a domain D if the origin lies outside D? if the origin is inside D?

Solution   Since

$$\mathbf{F}(x, y, z) = \frac{m}{r^3}\left(x\mathbf{i} + y\mathbf{j} + z\mathbf{k}\right), \quad \text{where} \quad r^2 = x^2 + y^2 + z^2,$$

and since $\partial r/\partial x = x/r$, we have

$$\frac{\partial F_1}{\partial x} = m\frac{\partial}{\partial x}\left(\frac{x}{r^3}\right) = m\frac{r^3 - 3xr^2\left(\frac{x}{r}\right)}{r^6} = m\frac{r^2 - 3x^2}{r^5}.$$

Similarly,

$$\frac{\partial F_2}{\partial y} = m\frac{r^2 - 3y^2}{r^5} \quad \text{and} \quad \frac{\partial F_3}{\partial z} = m\frac{r^2 - 3z^2}{r^5}.$$

Adding these up, we get $\nabla \bullet \mathbf{F}(x, y, z) = 0$ if $r > 0$.

If the origin lies outside the domain D, then the source density of **F** in D is zero, so we would expect the total flux of **F** out of D to be zero. If the origin lies inside D, then D contains a source of strength m (producing $4\pi m$ cubic units of fluid per unit time), so we would expect the flux out of D to be $4\pi m$. See Example 1 of Section 15.6 and also Exercises 9 and 10 of that section for specific examples.

Distributions and Delta Functions

If $\ell(x)$ represents the line density (mass per unit length) of mass distributed on the x-axis, then the total mass so distributed is

$$m = \int_{-\infty}^{\infty} \ell(x)\, dx.$$

Now suppose that the only mass on the axis is a "point mass" $m = 1$ located at the origin. Then at all other points $x \neq 0$, the density is $\ell(x) = 0$, but we must still have

$$\int_{-\infty}^{\infty} \ell(x)\, dx = m = 1,$$

so $\ell(0)$ must be infinite. This is an ideal situation, a mathematical model. No real *function* $\ell(x)$ can have such properties; if a function is zero everywhere except at a single point, then any integral of that function will be zero. (Why?) (Also, no real mass can occupy just a single point.) Nevertheless, it is very useful to model real, isolated masses as point masses and to model their densities using **generalized functions** (also called **distributions**).

We can think of the density of a point mass 1 at $x = 0$ as the limit of large densities concentrated on small intervals. For instance, if

$$d_n(x) = \begin{cases} n/2 & \text{if } |x| \leq 1/n \\ 0 & \text{if } |x| > 1/n \end{cases}$$

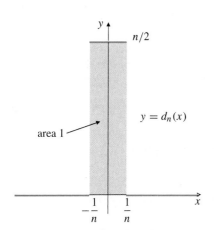

Figure 16.1 The functions $d_n(x)$ converge to $\delta(x)$ as $n \to \infty$

(see Figure 16.1), then for any smooth function $f(x)$ defined on $\mathbb{R}$ we have

$$\int_{-\infty}^{\infty} d_n(x)\, f(x)\, dx = \frac{n}{2} \int_{-1/n}^{1/n} f(x)\, dx.$$

Replace $f(x)$ in the integral on the right with its Maclaurin series:

$$f(x) = f(0) + \frac{f'(0)}{1!} x + \frac{f''(0)}{2!} x^2 + \cdots.$$

Since

$$\int_{-1/n}^{1/n} x^k\, dx = \begin{cases} 2/((k+1)n^{k+1}) & \text{if } k \text{ is even} \\ 0 & \text{if } k \text{ is odd,} \end{cases}$$

we can take the limit as $n \to \infty$ and obtain

$$\lim_{n \to \infty} \int_{-\infty}^{\infty} d_n(x)\, f(x)\, dx = f(0).$$

DEFINITION 1

The **Dirac distribution** $\delta(x)$ (also called the **Dirac delta function**, although it is really not a function) is the "limit" of the sequence $d_n(x)$ as $n \to \infty$. It is defined by the requirement that

$$\int_{-\infty}^{\infty} \delta(x) f(x)\, dx = f(0)$$

for every smooth function $f(x)$.

A formal change of variables shows that the delta function also satisfies

$$\int_{-\infty}^{\infty} \delta(x - t) f(t)\, dt = f(x).$$

Example 4   In view of the fact that $\mathbf{F}(\mathbf{r}) = m\mathbf{r}/|\mathbf{r}|^3$ satisfies $\mathbf{div}\, \mathbf{F}(x, y, z) = 0$ for $(x, y, z) \neq (0, 0, 0)$ but produces a flux of $4\pi m$ out of any sphere centred at the origin, we can regard $\mathbf{div}\, \mathbf{F}(x, y, z)$ as a distribution

$$\mathbf{div}\, \mathbf{F}(x, y, z) = 4\pi m \delta(x) \delta(y) \delta(z).$$

In particular, integrating this distribution against $f(x, y, z) = 1$ over $\mathbb{R}^3$, we have

$$\iiint_{\mathbb{R}^3} \mathbf{div\,F}(x, y, z)\, dV = 4\pi m \int_{-\infty}^{\infty} \delta(x)\, dx \int_{-\infty}^{\infty} \delta(y)\, dy \int_{-\infty}^{\infty} \delta(z)\, dz$$

$$= 4\pi m.$$

The integral can equally well be taken over *any domain* in $\mathbb{R}^3$ that contains the origin in its interior, and the result will be the same. If the origin is outside the domain, the result will be zero. We will reexamine this situation after establishing the Divergence Theorem in Section 16.4.

A formal study of distributions is beyond the scope of this book; refer to more advanced textbooks on differential equations and engineering mathematics.

Interpretation of the Curl

Roughly speaking, **curl F**(P) measures the extent to which the vector field **F** "swirls" around P.

Example 5   Consider the velocity field

$$\mathbf{v} = -\Omega y\mathbf{i} + \Omega x\mathbf{j}$$

of a solid rotating with angular speed Ω about the z-axis, that is, with angular velocity $\mathbf{\Omega} = \Omega\mathbf{k}$. (See Figure 15.2 in Section 15.1.) Calculate the circulation of this field around a circle $\mathcal{C}_\epsilon$ in the xy-plane centred at any point (x_0, y_0), having radius ϵ, and oriented counterclockwise. What is the relationship between this circulation and the curl of **v**?

Solution   The indicated circle has parametrization

$$\mathbf{r} = (x_0 + \epsilon \cos t)\mathbf{i} + (y_0 + \epsilon \sin t)\mathbf{j}, \qquad (0 \le t \le 2\pi),$$

and the circulation of **v** around it is given by

$$\oint_{\mathcal{C}_\epsilon} \mathbf{v} \bullet d\mathbf{r} = \int_0^{2\pi} \Big(-\Omega(y_0 + \epsilon \sin t)(-\epsilon \sin t) + \Omega(x_0 + \epsilon \cos t)(\epsilon \cos t)\Big)\, dt$$

$$= \int_0^{2\pi} \Big(\Omega\epsilon(y_0 \sin t + x_0 \cos t) + \Omega\epsilon^2\Big)\, dt$$

$$= 2\Omega\pi\epsilon^2.$$

Since

$$\mathbf{curl\,v} = \nabla \times \mathbf{v} = \left(\frac{\partial}{\partial x}(\Omega x) - \frac{\partial}{\partial y}(-\Omega y)\right)\mathbf{k} = 2\Omega\mathbf{k} = 2\mathbf{\Omega},$$

the circulation is the product of $(\mathbf{curl\,v}) \bullet \mathbf{k}$ and the area bounded by $\mathcal{C}_\epsilon$. Note that this circulation is constant for circles of any fixed radius; it does not depend on the position of the centre.

The calculations in the example above suggest that the curl of a vector field is a measure of the *circulation per unit area* in planes normal to the curl. A more precise version of this conjecture is stated in Theorem 2 below. We will not prove this theorem now because a proof at this stage would be quite complicated. (However, see Exercise 14 below for a special case.) A simple proof can be based on Stokes's Theorem; see Exercise 13 in Section 16.5.

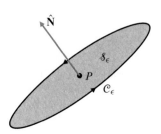

THEOREM **2**

The curl as circulation density

If **F** is a smooth vector field and $\mathcal{C}_\epsilon$ is a circle of radius ϵ centred at point P and bounding a disk $\mathcal{S}_\epsilon$ with unit normal $\hat{\mathbf{N}}$ (and orientation inherited from $\mathcal{C}_\epsilon$; see Figure 16.2), then

$$\lim_{\epsilon \to 0^+} \frac{1}{\pi \epsilon^2} \oint_{\mathcal{C}_\epsilon} \mathbf{F} \bullet d\mathbf{r} = \hat{\mathbf{N}} \bullet \mathbf{curl}\, \mathbf{F}(P).$$

Figure 16.2

Example 5 also suggests the following definition for the *local* angular velocity of a moving fluid:

The **local angular velocity** at point P in a fluid moving with velocity field $\mathbf{v}(P)$ is given by

$$\boldsymbol{\Omega}(P) = \frac{1}{2}\mathbf{curl}\, \mathbf{v}(P).$$

Theorem 2 states that the local angular velocity $\boldsymbol{\Omega}(P)$ is that vector whose component in the direction of any unit vector $\hat{\mathbf{N}}$ is one-half of the limiting circulation per unit area around the (oriented) boundary circles of small circular disks centred at P and having normal $\hat{\mathbf{N}}$.

Not all vector fields with nonzero curl *appear* to circulate. The velocity field for the rigid body rotation considered in Example 5 appears to circulate around the axis of rotation, but the circulation around a circle in a plane perpendicular to that axis turned out to be independent of the position of the circle; it depended only on its area. The circle need not even surround the axis of rotation. The following example investigates a fluid velocity field whose streamlines are *straight lines* but that still has nonzero, constant curl and, therefore, constant local angular velocity.

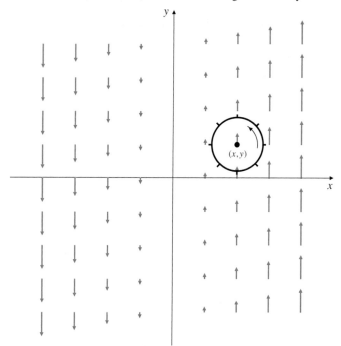

Figure 16.3   The paddle wheel is not only carried along but is set rotating by the flow

Example 6   Consider the velocity field $\mathbf{v} = x\mathbf{j}$ of a fluid moving in the xy-plane. Evidently, particles of fluid are moving along lines parallel to the y-axis. However, $\mathbf{curl}\, \mathbf{v}(x, y) = \mathbf{k}$, and $\boldsymbol{\Omega}(x, y) = \frac{1}{2}\mathbf{k}$. A small "paddle wheel" of radius ϵ (see Figure 16.3) placed with its centre at position (x, y) in the fluid will be carried along with the fluid at velocity $x\mathbf{j}$ but will also be set rotating with angular velocity $\mathbf{\Omega}(x, y) = \frac{1}{2}\mathbf{k}$, which is independent of its position. This angular velocity is due to the fact that the velocity of the fluid along the right side of the wheel exceeds that along the left side.

■

Exercises 16.1

In Exercises 1–11, calculate **div F** and **curl F** for the given vector fields.

1. $\mathbf{F} = x\mathbf{i} + y\mathbf{j}$

2. $\mathbf{F} = y\mathbf{i} + x\mathbf{j}$

3. $\mathbf{F} = y\mathbf{i} + z\mathbf{j} + x\mathbf{k}$

4. $\mathbf{F} = yz\mathbf{i} + xz\mathbf{j} + xy\mathbf{k}$

5. $\mathbf{F} = x\mathbf{i} + x\mathbf{k}$

6. $\mathbf{F} = xy^2\mathbf{i} - yz^2\mathbf{j} + zx^2\mathbf{k}$

7. $\mathbf{F} = f(x)\mathbf{i} + g(y)\mathbf{j} + h(z)\mathbf{k}$ **8.** $\mathbf{F} = f(z)\mathbf{i} - f(z)\mathbf{j}$

9. $\mathbf{F}(r, \theta) = r\mathbf{i} + \sin\theta\mathbf{j}$, where (r, θ) are polar coordinates in the plane

10. $\mathbf{F} = \hat{\mathbf{r}} = \cos\theta\mathbf{i} + \sin\theta\mathbf{j}$

11. $\mathbf{F} = \hat{\boldsymbol{\theta}} = -\sin\theta\mathbf{i} + \cos\theta\mathbf{j}$

*12. Let **F** be a smooth, three-dimensional vector field. If $B_{a,b,c}$ is the surface of the box $-a \le x \le a$, $-b \le y \le b$, $-c \le z \le c$, with outward normal $\hat{\mathbf{N}}$, show that

$$\lim_{a,b,c \to 0^+} \frac{1}{8abc} \oiint_{B_{a,b,c}} \mathbf{F} \bullet \hat{\mathbf{N}}\,dS = \nabla \bullet \mathbf{F}(0, 0, 0).$$

*13. Let **F** be a smooth two-dimensional vector field. If $\mathcal{C}_\epsilon$ is the circle of radius ϵ centred at the origin, and $\hat{\mathbf{N}}$ is the unit outward normal to $\mathcal{C}_\epsilon$, show that

$$\lim_{\epsilon \to 0^+} \frac{1}{\pi\epsilon^2} \oint_{\mathcal{C}_\epsilon} \mathbf{F} \bullet \hat{\mathbf{N}}\,ds = \mathbf{div}\,\mathbf{F}(0, 0).$$

*14. Prove Theorem 2 in the special case that $\mathcal{C}_\epsilon$ is the circle in the xy-plane with parametrization $x = \epsilon\cos\theta$, $y = \epsilon\sin\theta$, $(0 \le \theta \le 2\pi)$. In this case $\hat{\mathbf{N}} = \mathbf{k}$. *Hint:* Expand $\mathbf{F}(x, y, z)$ in a vector Taylor series about the origin as in the proof of Theorem 1, and calculate the circulation of individual terms around $\mathcal{C}_\epsilon$.

16.2 Some Identities Involving Grad, Div, and Curl

There are numerous identities involving the functions

$$\mathbf{grad}\,f(x, y, z) = \nabla f(x, y, z) = \frac{\partial f}{\partial x}\mathbf{i} + \frac{\partial f}{\partial y}\mathbf{j} + \frac{\partial f}{\partial z}\mathbf{k},$$

$$\mathbf{div}\,\mathbf{F}(x, y, z) = \nabla \bullet \mathbf{F}(x, y, z) = \frac{\partial F_1}{\partial x} + \frac{\partial F_2}{\partial y} + \frac{\partial F_3}{\partial z},$$

$$\mathbf{curl}\,\mathbf{F}(x, y, z) = \nabla \times \mathbf{F}(x, y, z) = \begin{vmatrix} \mathbf{i} & \mathbf{j} & \mathbf{k} \\ \frac{\partial}{\partial x} & \frac{\partial}{\partial y} & \frac{\partial}{\partial z} \\ F_1 & F_2 & F_3 \end{vmatrix},$$

and the **Laplacian operator**, $\nabla^2 = \nabla \bullet \nabla$, defined for a scalar field ϕ by

$$\nabla^2\phi = \nabla \bullet \nabla\phi = \mathbf{div}\,\mathbf{grad}\,\phi = \frac{\partial^2\phi}{\partial x^2} + \frac{\partial^2\phi}{\partial y^2} + \frac{\partial^2\phi}{\partial z^2},$$

and for a vector field $\mathbf{F} = F_1\mathbf{i} + F_2\mathbf{j} + F_3\mathbf{k}$ by

$$\nabla^2\mathbf{F} = (\nabla^2 F_1)\mathbf{i} + (\nabla^2 F_2)\mathbf{j} + (\nabla^2 F_3)\mathbf{k}.$$

(The Laplacian operator, $\nabla^2 = (\partial^2/\partial x^2) + (\partial^2/\partial y^2) + (\partial^2/\partial z^2)$, is denoted by Δ in some books.) Recall that a function ϕ is called **harmonic** in a domain D if $\nabla^2\phi = 0$ throughout D. (See Section 12.4.)

We collect the most important identities together in the following theorem. Most of them are forms of the Product Rule. We will prove a few of the identities to illustrate the techniques involved (mostly brute-force calculation) and leave the rest as exercises. Note that two of the identities involve quantities like $(\mathbf{G} \bullet \nabla)\mathbf{F}$; this represents the vector obtained by applying the scalar differential operator $\mathbf{G} \bullet \nabla$ to the vector field $\mathbf{F}$:

$$(\mathbf{G} \bullet \nabla)\mathbf{F} = G_1 \frac{\partial \mathbf{F}}{\partial x} + G_2 \frac{\partial \mathbf{F}}{\partial y} + G_3 \frac{\partial \mathbf{F}}{\partial z}.$$

THEOREM 3

Vector differential identities

Let ϕ and ψ be scalar fields and $\mathbf{F}$ and $\mathbf{G}$ be vector fields, all assumed to be sufficiently smooth that all the partial derivatives in the identities are continuous. Then the following identities hold:

(a) $\quad \nabla(\phi\psi) = \phi\nabla\psi + \psi\nabla\phi$

(b) $\quad \nabla \bullet (\phi\mathbf{F}) = (\nabla\phi) \bullet \mathbf{F} + \phi(\nabla \bullet \mathbf{F})$

(c) $\quad \nabla \times (\phi\mathbf{F}) = (\nabla\phi) \times \mathbf{F} + \phi(\nabla \times \mathbf{F})$

(d) $\quad \nabla \bullet (\mathbf{F} \times \mathbf{G}) = (\nabla \times \mathbf{F}) \bullet \mathbf{G} - \mathbf{F} \bullet (\nabla \times \mathbf{G})$

(e) $\quad \nabla \times (\mathbf{F} \times \mathbf{G}) = (\nabla \bullet \mathbf{G})\mathbf{F} + (\mathbf{G} \bullet \nabla)\mathbf{F} - (\nabla \bullet \mathbf{F})\mathbf{G} - (\mathbf{F} \bullet \nabla)\mathbf{G}$

(f) $\quad \nabla(\mathbf{F} \bullet \mathbf{G}) = \mathbf{F} \times (\nabla \times \mathbf{G}) + \mathbf{G} \times (\nabla \times \mathbf{F}) + (\mathbf{F} \bullet \nabla)\mathbf{G} + (\mathbf{G} \bullet \nabla)\mathbf{F}$

(g) $\quad \nabla \bullet (\nabla \times \mathbf{F}) = 0 \qquad$ (**div curl** $= 0$)

(h) $\quad \nabla \times (\nabla\phi) = \mathbf{0} \qquad$ (**curl grad** $= \mathbf{0}$)

(i) $\quad \nabla \times (\nabla \times \mathbf{F}) = \nabla(\nabla \bullet \mathbf{F}) - \nabla^2\mathbf{F}$

$\qquad\qquad\qquad$ (**curl curl** $=$ **grad div** $-$ Laplacian)

Identities (a)–(f) are versions of the Product Rule and are first-order identities involving only one application of ∇. Identities (g)–(i) are second-order identities. Identities (g) and (h) are equivalent to the equality of mixed partial derivatives and are especially important for the understanding of **div** and **curl**.

PROOF We will prove only identities (c), (e), and (g). The remaining proofs are similar to these.

(c) The first component (**i** component) of $\nabla \times (\phi\mathbf{F})$ is

$$\frac{\partial}{\partial y}(\phi F_3) - \frac{\partial}{\partial z}(\phi F_2) = \frac{\partial\phi}{\partial y}F_3 - \frac{\partial\phi}{\partial z}F_2 + \phi\frac{\partial F_3}{\partial y} - \phi\frac{\partial F_2}{\partial z}.$$

The first two terms on the right constitute the first component of $(\nabla\phi) \times \mathbf{F}$, and the last two terms constitute the first component of $\phi(\nabla \times \mathbf{F})$. Therefore, the first components of both sides of identity (c) are equal. The equality of the other components follows similarly.

(e) Again, it is sufficient to show that the first components of the vectors on both sides of the identity are equal. To calculate the first component of $\nabla \times (\mathbf{F} \times \mathbf{G})$ we need the second and third components of $\mathbf{F} \times \mathbf{G}$, which are

$$(\mathbf{F} \times \mathbf{G})_2 = F_3G_1 - F_1G_3 \quad \text{and} \quad (\mathbf{F} \times \mathbf{G})_3 = F_1G_2 - F_2G_1.$$

The first component of $\nabla \times (\mathbf{F} \times \mathbf{G})$ is therefore

$$\frac{\partial}{\partial y}(F_1G_2 - F_2G_1) - \frac{\partial}{\partial z}(F_3G_1 - F_1G_3)$$

$$= \frac{\partial F_1}{\partial y}G_2 + F_1\frac{\partial G_2}{\partial y} - \frac{\partial F_2}{\partial y}G_1 - F_2\frac{\partial G_1}{\partial y} - \frac{\partial F_3}{\partial z}G_1$$

$$- F_3\frac{\partial G_1}{\partial z} + \frac{\partial F_1}{\partial z}G_3 + F_1\frac{\partial G_3}{\partial z}.$$

The first components of the four terms on the right side of identity (e) are

$$((\nabla \bullet \mathbf{G})\mathbf{F})_1 = F_1 \frac{\partial G_1}{\partial x} + F_1 \frac{\partial G_2}{\partial y} + F_1 \frac{\partial G_3}{\partial z}$$

$$((\mathbf{G} \bullet \nabla)\mathbf{F})_1 = \frac{\partial F_1}{\partial x} G_1 + \frac{\partial F_1}{\partial y} G_2 + \frac{\partial F_1}{\partial z} G_3$$

$$-((\nabla \bullet \mathbf{F})\mathbf{G})_1 = -\frac{\partial F_1}{\partial x} G_1 - \frac{\partial F_2}{\partial y} G_1 - \frac{\partial F_3}{\partial z} G_1$$

$$-((\mathbf{F} \bullet \nabla)\mathbf{G})_1 = -F_1 \frac{\partial G_1}{\partial x} - F_2 \frac{\partial G_1}{\partial y} - F_3 \frac{\partial G_1}{\partial z}.$$

When we add up all the terms in these four expressions, some cancel out and we are left with the same terms as in the first component of $\nabla \times (\mathbf{F} \times \mathbf{G})$.

(g) This is a straightforward calculation involving the equality of mixed partial derivatives:

$$\nabla \bullet (\nabla \times \mathbf{F}) = \frac{\partial}{\partial x} \left(\frac{\partial F_3}{\partial y} - \frac{\partial F_2}{\partial z} \right) + \frac{\partial}{\partial y} \left(\frac{\partial F_1}{\partial z} - \frac{\partial F_3}{\partial x} \right)$$

$$+ \frac{\partial}{\partial z} \left(\frac{\partial F_2}{\partial x} - \frac{\partial F_1}{\partial y} \right)$$

$$= \frac{\partial^2 F_3}{\partial x \partial y} - \frac{\partial^2 F_2}{\partial x \partial z} + \frac{\partial^2 F_1}{\partial y \partial z} - \frac{\partial^2 F_3}{\partial y \partial x} + \frac{\partial^2 F_2}{\partial z \partial x} - \frac{\partial^2 F_1}{\partial z \partial y}$$

$$= 0.$$

Remark   Two *triple product* identities for vectors were previously presented in Exercises 18 and 23 of Section 10.3:

$$\mathbf{a} \bullet (\mathbf{b} \times \mathbf{c}) = \mathbf{b} \bullet (\mathbf{c} \times \mathbf{a}) = \mathbf{c} \bullet (\mathbf{a} \times \mathbf{b}),$$

$$\mathbf{a} \times (\mathbf{b} \times \mathbf{c}) = (\mathbf{a} \bullet \mathbf{c})\mathbf{b} - (\mathbf{a} \bullet \mathbf{b})\mathbf{c}.$$

While these are useful identities, they *cannot* be used to give simpler proofs of the identities in Theorem 3 by replacing one or other of the vectors with ∇. (Why?)

Scalar and Vector Potentials

Two special terms are used to describe vector fields for which either the divergence or the curl is identically zero.

DEFINITION 2

Solenoidal and irrotational vector fields

A vector field $\mathbf{F}$ is called **solenoidal** in a domain D if $\mathbf{div}\,\mathbf{F} = 0$ in D.
A vector field $\mathbf{F}$ is called **irrotational** in a domain D if $\mathbf{curl}\,\mathbf{F} = \mathbf{0}$ in D.

Part (h) of Theorem 3 says that $\mathbf{F} = \mathbf{grad}\,\phi \implies \mathbf{curl}\,\mathbf{F} = \mathbf{0}$. Thus,

Every conservative vector field is irrotational.

Part (g) of Theorem 3 says that $\mathbf{F} = \mathbf{curl}\,\mathbf{G} \implies \mathbf{div}\,\mathbf{F} = 0$. Thus,

The curl of any vector field is solenoidal.

The *converses* of these assertions hold if the domain of $\mathbf{F}$ satisfies certain conditions.

THEOREM 4

If $\mathbf{F}$ is a smooth, irrotational vector field on a simply connected domain D, then $\mathbf{F} = \nabla \phi$ for some scalar potential function defined on D, so $\mathbf{F}$ is conservative.

THEOREM 5

If $\mathbf{F}$ is a smooth, solenoidal vector field on a domain D with the property that every closed surface in D bounds a domain contained in D, then $\mathbf{F} = \mathbf{curl\,G}$ for some vector field $\mathbf{G}$ defined on D. Such a vector field $\mathbf{G}$ is called a **vector potential** of the vector field $\mathbf{F}$.

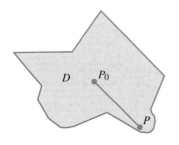

Figure 16.4 The line segment from P_0 to any point in D lies in D

We cannot prove these results in their full generality at this point. However, both theorems have simple proofs in the special case where the domain D is **star-like**. A star-like domain is one for which there exists a point P_0 such that the line segment from P_0 to any point P in D lies wholly in D. (See Figure 16.4.) Both proofs are *constructive* in that they tell you how to find a potential.

Proof of Theorem 4 for star-like domains. Without loss of generality, we can assume that P_0 is the origin. If $P = (x, y, z)$ is any point in D, then the straight line segment

$$\mathbf{r}(t) = tx\mathbf{i} + ty\mathbf{j} + tz\mathbf{k}, \qquad (0 \le t \le 1),$$

from P_0 to P lies in D. Define the function ϕ on D by

$$\phi(x, y, z) = \int_0^1 \mathbf{F}(\mathbf{r}(t)) \bullet \frac{d\mathbf{r}}{dt}\, dt$$

$$= \int_0^1 \Big(x F_1(\xi, \eta, \zeta) + y F_2(\xi, \eta, \zeta) + z F_3(\xi, \eta, \zeta) \Big)\, dt,$$

where $\xi = tx$, $\eta = ty$, and $\zeta = tz$. We calculate $\partial \phi / \partial x$, making use of the fact that $\mathbf{curl\,F} = \mathbf{0}$ to replace $(\partial/\partial\xi) F_2(\xi, \eta, \zeta)$ with $(\partial/\partial\eta) F_1(\xi, \eta, \zeta)$ and $(\partial/\partial\xi) F_3(\xi, \eta, \zeta)$ with $(\partial/\partial\zeta) F_1(\xi, \eta, \zeta)$:

$$\frac{\partial\phi}{\partial x} = \int_0^1 \left(F_1(\xi, \eta, \zeta) + tx\frac{\partial F_1}{\partial\xi} + ty\frac{\partial F_2}{\partial\xi} + tz\frac{\partial F_3}{\partial\xi} \right) dt$$

$$= \int_0^1 \left(F_1(\xi, \eta, \zeta) + tx\frac{\partial F_1}{\partial\xi} + ty\frac{\partial F_1}{\partial\eta} + tz\frac{\partial F_1}{\partial\zeta} \right) dt$$

$$= \int_0^1 \frac{d}{dt}\Big(t\, F_1(\xi, \eta, \zeta) \Big)\, dt$$

$$= \Big(t\, F_1(tx, ty, tz) \Big)\Big|_0^1 = F_1(x, y, z).$$

Similarly, $\partial\phi/\partial y = F_2$ and $\partial\phi/\partial z = F_3$. Thus $\nabla\phi = \mathbf{F}$.

The details of the proof of Theorem 5 for star-like domains are similar to those of Theorem 4, and we relegate the proof to Exercise 18 below.

Note that vector potentials, when they exist, are *very* nonunique. Since **curl grad** ϕ is identically zero (Theorem 3(h)), an arbitrary conservative field can be added to $\mathbf{G}$ without changing the value of **curl G**. The following example illustrates just how much freedom you have in making simplifying assumptions when trying to find a vector potential.

Example 1   Show that the vector field $\mathbf{F} = (x^2 + yz)\mathbf{i} - 2y(x + z)\mathbf{j} + (xy + z^2)\mathbf{k}$ is solenoidal in $\mathbb{R}^3$ and find a vector potential for it.

Solution   Since $\operatorname{\mathbf{div}}\mathbf{F} = 2x - 2(x + z) + 2z = 0$ in $\mathbb{R}^3$, $\mathbf{F}$ is solenoidal. A vector potential $\mathbf{G}$ for $\mathbf{F}$ must satisfy $\operatorname{\mathbf{curl}}\mathbf{G} = \mathbf{F}$, that is,

$$\frac{\partial G_3}{\partial y} - \frac{\partial G_2}{\partial z} = x^2 + yz,$$

$$\frac{\partial G_1}{\partial z} - \frac{\partial G_3}{\partial x} = -2xy - 2yz,$$

$$\frac{\partial G_2}{\partial x} - \frac{\partial G_1}{\partial y} = xy + z^2.$$

The three components of $\mathbf{G}$ have nine independent first partial derivatives, so there are nine "degrees of freedom" involved in their determination. The three equations above use up three of these nine degrees of freedom. That leaves six. Let us try to find a solution $\mathbf{G}$ with $G_2 = 0$ identically. This means that all three first partials of G_2 are zero, so we have used up three degrees of freedom in making this assumption. We have three left. The first equation now implies that

$$G_3 = \int (x^2 + yz)\,dy = x^2 y + \frac{1}{2}y^2 z + M(x, z).$$

(Since we were integrating with respect to y, the constant of integration can still depend on x and z.) We make a second simplifying assumption, that $M(x, z) = 0$. This uses up two more degrees of freedom, leaving one. From the second equation we have

$$\frac{\partial G_1}{\partial z} = \frac{\partial G_3}{\partial x} - 2xy - 2yz = 2xy - 2xy - 2yz = -2yz,$$

so

$$G_1 = -2 \int yz\,dz = -yz^2 + N(x, y).$$

We cannot assume that $N(x, y) = 0$ identically because that would require two degrees of freedom and we have only one. However, the third equation implies

$$xy + z^2 = -\frac{\partial G_1}{\partial y} = z^2 - \frac{\partial N}{\partial y}.$$

Thus, $(\partial/\partial y)N(x, y) = -xy$; observe that the terms involving z have cancelled out. This happened because $\operatorname{\mathbf{div}}\mathbf{F} = 0$. Had $\mathbf{F}$ not been solenoidal, we could not have determined N as a function of x and y only from the above equation. As it is, however, we have

$$N(x, y) = -\int xy\,dy = -\frac{1}{2}xy^2 + P(x).$$

We can use our last degree of freedom to choose $P(x)$ to be identically zero and hence obtain

$$\mathbf{G} = -\left(yz^2 + \frac{xy^2}{2}\right)\mathbf{i} + \left(x^2 y + \frac{y^2 z}{2}\right)\mathbf{k}$$

as the required vector potential for $\mathbf{F}$. You can check that $\operatorname{\mathbf{curl}}\mathbf{G} = \mathbf{F}$. Of course, other choices of simplifying assumptions would have led to very different functions $\mathbf{G}$, which would have been equally correct.

Maple Calculations

The Maple **VectorCalculus** package defines routines for creating a vector field as well as calculating the gradient of a scalar field and the divergences and curl of a vector field. It will also calculate the Laplacian of a scalar or vector field and allow the use of the "del" operator in dot and cross products. Some of these capabilities are restricted to 3-dimensional vector fields. Let us begin by loading the package and declaring the type of coordinate system we will use and the names of the coordinates:

```
>   with(VectorCalculus):
>   SetCoordinates('cartesian'[x,y,z]);
```
$$cartesian_{x,y,z}$$

Setting the coordinates at the outset means we don't have to do it every time we call one of the procedures for handling vector fields, such as the `Gradient` procedure, which we illustrated at the end of Section 12.7. To calculate the gradient of a scalar expression in the variables x, y, and z, we could simply enter

```
>   f := x^2 + x*y - z^3; G := Gradient(f);
```
$$f := x^2 + xy - z^3$$

$$G := (2x + y)\,\bar{e}_x + x\,\bar{e}_y - 3z^2\,\bar{e}_z$$

Maple shows that the result G is a *vector field* rather than just a *vector* by placing bars over the basis vectors. Maple treats vector fields and vectors as different kinds of objects; you can, for example, add two vector fields or two vectors, but you can't add a vector to a vector field. A vector field is a vector-valued function of a vector variable. To evaluate a vector field at a particular vector, you use the `evalVF` procedure:

```
>   evalVF(G,<1,1,1>);
```
$$3\,e_x + e_y - 3\,e_z$$

You can define a vector field **F** with the `VectorField` procedure:

```
>   F := VectorField(<x*y, 2*y*z, 3*x*z>);
```
$$F := x\,y\,\bar{e}_x + 2\,y\,z\,\bar{e}_y + 3\,x\,z\,\bar{e}_z$$

Then we can calculate the divergence or curl of **F** by using the `Divergence` or `Curl` procedures, or by dot or cross products with the `Del` operator:

```
>   Divergence(F); Del.F;
```
$$y + 2z + 3x$$

$$y + 2z + 3x$$

```
>   Curl(F); Del &x F;
```
$$-2\,y\,\bar{e}_x - 3\,z\,\bar{e}_y - x\,\bar{e}_z$$

$$-2\,y\,\bar{e}_x - 3\,z\,\bar{e}_y - x\,\bar{e}_z$$

We can verify the identities in Theorem 3 by using arbitrary scalar and vector fields:

```
>   H := VectorField(<u(x,y,z),v(x,y,z),w(x,y,z)>);
```
$$H := u(x, y, z)\,\bar{e}_x + v(x, y, z)\,\bar{e}_y + w(x, y, z)\,\bar{e}_z$$

```
>   Divergence(Curl(H)); Curl(Gradient(u(x,y,z)));
```
$$0$$

$$0\,\bar{e}_x$$

$0\,\bar{e}_x$ is VectorCalculus's way of denoting the zero vector field.

```
>  Curl(Curl(H)) - Gradient(Divergence(H)) + Laplacian(H);
```

$$0\,\bar{e}_x$$

VectorCalculus also has procedures for finding the scalar potential of an irrotational vector field and the vector potential of a solenoidal vector field:

```
>  ScalarPotential(VectorField(<x,y,z>));
```

$$\frac{1}{2}x^2 + \frac{1}{2}y^2 + \frac{1}{2}z^2$$

```
>  VectorPotential(VectorField(<x^2, -x*y, -x*z>));
```

$$-x\,y\,z\,\bar{e}_x - x^2\,z\,\bar{e}_y$$

Neither procedure gives any output if you fail to feed it a vector field satisfying the appropriate condition (irrotational or solenoidal).

Finally, let us note that VectorCalculus is quite happy to deal with coordinate systems other than `'cartesian'` [x,y,z]. For instance

```
>  SetCoordinates('cylindrical'[r,theta,z]);
```

$$cylindrical_{r,\theta,z}$$

```
>  Laplacian(u(r,theta,z));
```

$$\frac{\left(\frac{\partial}{\partial r}u(r,\theta,z)\right) + r\left(\frac{\partial^2}{\partial r^2}u(r,\theta,z)\right) + \dfrac{\frac{\partial^2}{\partial \theta^2}u(r,\theta,z)}{r} + r\left(\frac{\partial^2}{\partial z^2}u(r,\theta,z)\right)}{r}$$

which is not written as neatly as we would like, but is correct. Similarly, we can use coordinate systems `'spherical'` [rho,phi,theta] in 3-space and also `'polar'` [r,theta] in the plane.

⬛⬛⬛ Exercises 16.2 ⬛⬛⬛

1. Prove part (a) of Theorem 3.

2. Prove part (b) of Theorem 3.

3. Prove part (d) of Theorem 3.

4. Prove part (f) of Theorem 3.

5. Prove part (h) of Theorem 3.

6. Prove part (i) of Theorem 3.

✱ **7.** Given that the field lines of the vector field $\mathbf{F}(x, y, z)$ are parallel straight lines, can you conclude anything about **div F**? about **curl F**?

8. Let $\mathbf{r} = x\mathbf{i} + y\mathbf{j} + z\mathbf{k}$ and let $\mathbf{c}$ be a constant vector. Show that $\nabla \bullet (\mathbf{c} \times \mathbf{r}) = 0$, $\nabla \times (\mathbf{c} \times \mathbf{r}) = 2\mathbf{c}$, and $\nabla(\mathbf{c} \bullet \mathbf{r}) = \mathbf{c}$.

9. Let $\mathbf{r} = x\mathbf{i} + y\mathbf{j} + z\mathbf{k}$ and let $r = |\mathbf{r}|$. If f is a differentiable function of one variable, show that

$$\nabla \bullet (f(r)\mathbf{r}) = rf'(r) + 3f(r).$$

Find $f(r)$ if $f(r)\mathbf{r}$ is solenoidal for $r \neq 0$.

10. If the smooth vector field $\mathbf{F}$ is both irrotational and solenoidal on $\mathbb{R}^3$, show that the three components of $\mathbf{F}$ and the scalar potential for $\mathbf{F}$ are all harmonic functions in $\mathbb{R}^3$.

11. If $\mathbf{r} = x\mathbf{i} + y\mathbf{j} + z\mathbf{k}$ and $\mathbf{F}$ is smooth, show that

$$\nabla \times (\mathbf{F} \times \mathbf{r}) = \mathbf{F} - (\nabla \bullet \mathbf{F})\mathbf{r} + \nabla(\mathbf{F} \bullet \mathbf{r}) - \mathbf{r} \times (\nabla \times \mathbf{F}).$$

In particular, if $\nabla \bullet \mathbf{F} = 0$ and $\nabla \times \mathbf{F} = \mathbf{0}$, then

$$\nabla \times (\mathbf{F} \times \mathbf{r}) = \mathbf{F} + \nabla(\mathbf{F} \bullet \mathbf{r}).$$

12. If ϕ and ψ are harmonic functions, show that $\phi\nabla\psi - \psi\nabla\phi$ is solenoidal.

13. If ϕ and ψ are smooth scalar fields, show that

$$\nabla \times (\phi\nabla\psi) = -\nabla \times (\psi\nabla\phi) = \nabla\phi \times \nabla\psi.$$

14. Verify the identity

$$\nabla \bullet \left(f(\nabla g \times \nabla h) \right) = \nabla f \bullet (\nabla g \times \nabla h)$$

for smooth scalar fields f, g, and h.

15. If the vector fields $\mathbf{F}$ and $\mathbf{G}$ are smooth and conservative, show that $\mathbf{F} \times \mathbf{G}$ is solenoidal. Find a vector potential for $\mathbf{F} \times \mathbf{G}$.

16. Find a vector potential for $\mathbf{F} = -y\mathbf{i} + x\mathbf{j}$.

17. Show that $\mathbf{F} = xe^{2z}\mathbf{i} + ye^{2z}\mathbf{j} - e^{2z}\mathbf{k}$ is a solenoidal vector field, and find a vector potential for it.

*18. Suppose **div F** $= 0$ in a domain D any point P of which can by joined to the origin by a straight line segment in D. Let $\mathbf{r} = tx\mathbf{i} + ty\mathbf{j} + tz\mathbf{k}$, $(0 \le t \le 1)$, be a parametrization of the line segment from the origin to (x, y, z) in D. If

$$\mathbf{G}(x, y, z) = \int_0^1 t\mathbf{F}(\mathbf{r}(t)) \times \frac{d\mathbf{r}}{dt}\, dt,$$

show that **curl G** $=$ **F** throughout D. *Hint:* It is enough to check the first components of **curl G** and **F**. Proceed in a manner similar to the proof of Theorem 4.

🖳 **19.** Use the Maple VectorCalculus package to verify the identities (a)–(f) of Theorem 3. *Hint:* For expressions of the form $(\mathbf{F} \bullet \nabla)\mathbf{G}$ you will have to use

```
>   F[1]*diff(G,x)+F[2]*diff(G,y)
>     +F[3]*diff(G,z)
```

because `Del` cannot be applied to a vector field except via a dot or cross product.

16.3 Green's Theorem in the Plane

The Fundamental Theorem of Calculus,

$$\int_a^b \frac{d}{dx} f(x)\, dx = f(b) - f(a),$$

expresses the integral, taken over the interval $[a, b]$, of the derivative of a single-variable function, f, as a *sum* of values of that function at the *oriented boundary* of the interval $[a, b]$, that is, at the two endpoints a and b, the former providing a *negative* contribution and the latter a *positive* one. The line integral of a conservative vector field over a curve $\mathcal{C}$ from A to B,

$$\int_{\mathcal{C}} \nabla\phi \bullet d\mathbf{r} = \phi(B) - \phi(A),$$

has a similar interpretation; $\nabla\phi$ is a derivative, and the curve $\mathcal{C}$, although lying in a two- or three-dimensional space, is intrinsically a one-dimensional object, and the points A and B constitute its boundary.

Green's Theorem is a two-dimensional version of the Fundamental Theorem of Calculus that expresses the *double integral* of a certain kind of derivative of a two-dimensional vector field $\mathbf{F}(x, y)$, namely, the **k**-component of **curl F**, over a region R in the xy-plane as a line integral (i.e., a "sum") of the tangential component of **F** around the curve $\mathcal{C}$ which is the oriented boundary of R:

$$\iint_R \mathbf{curl\, F} \bullet \mathbf{k}\, dA = \oint_{\mathcal{C}} \mathbf{F} \bullet d\mathbf{r},$$

or, more explicitly,

$$\iint_R \left(\frac{\partial F_2}{\partial x} - \frac{\partial F_1}{\partial y} \right) dx\, dy = \oint_{\mathcal{C}} F_1(x, y)\, dx + F_2(x, y)\, dy.$$

For this formula to hold, $\mathcal{C}$ must be the oriented boundary of R considered as a surface with orientation provided by $\hat{\mathbf{N}} = \mathbf{k}$. Thus, $\mathcal{C}$ is oriented with R on the left as we move around $\mathcal{C}$ in the direction of its orientation. We will call such a curve positively oriented with respect to R. In particular, if $\mathcal{C}$ is a simple closed curve bounding R, then $\mathcal{C}$ is oriented counterclockwise. Of course, R may have holes, and the boundaries of the holes will be oriented clockwise. In any case, the unit tangent $\hat{\mathbf{T}}$ and unit exterior (pointing out of R) normal $\hat{\mathbf{N}}$ on $\mathcal{C}$ satisfy $\hat{\mathbf{N}} = \hat{\mathbf{T}} \times \mathbf{k}$. See Figure 16.5.

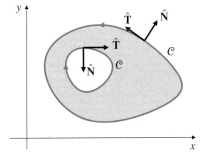

Figure 16.5   A plane domain with positively oriented boundary

THEOREM 6

Green's Theorem

Let R be a regular, closed region in the xy-plane whose boundary, $\mathcal{C}$, consists of one or more piecewise smooth, non-self-intersecting, closed curves that are positively oriented with respect to R. If $\mathbf{F} = F_1(x, y)\mathbf{i} + F_2(x, y)\mathbf{j}$ is a smooth vector field on R, then

$$\oint_{\mathcal{C}} F_1(x, y)\, dx + F_2(x, y)\, dy = \iint_R \left(\frac{\partial F_2}{\partial x} - \frac{\partial F_1}{\partial y} \right) dA.$$

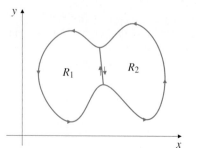

Figure 16.6 Green's Theorem holds for the union of R_1 and R_2 if it holds for each of those regions

PROOF Recall that a regular region can be divided into nonoverlapping subregions that are both x-simple and y-simple. (See Section 14.2.) When two such regions share a common boundary curve, they induce opposite orientations on that curve, so the sum of the line integrals over the boundaries of the subregions is just the line integral over the boundary of the whole region. (See Figure 16.6.) The double integrals over the subregions also add to give the double integral over the whole region. It therefore suffices to show that the formula holds for a region R that is both x-simple and y-simple.

Since R is y-simple, it is specified by inequalities of the form $a \le x \le b$, $f(x) \le y \le g(x)$, with the bottom boundary $y = f(x)$ oriented left to right and the upper boundary $y = g(x)$ oriented right to left. (See Figure 16.7.) Thus,

$$-\iint_R \frac{\partial F_1}{\partial y}\, dx\, dy = -\int_a^b dx \int_{f(x)}^{g(x)} \frac{\partial F_1}{\partial y}\, dy$$
$$= \int_a^b \left(-F_1(x, g(x)) + F_1(x, f(x))\right) dx.$$

On the other hand, since $dx = 0$ on the vertical sides of R, and the top boundary is traversed from b to a, we have

$$\oint_{\mathcal{C}} F_1(x, y)\, dx = \int_a^b \left(F_1(x, f(x)) - F_1(x, g(x))\right) dx = \iint_R -\frac{\partial F_1}{\partial y}\, dx\, dy.$$

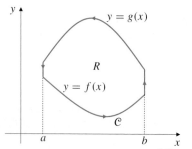

Figure 16.7 $\oint_{\mathcal{C}} F_1\, dx = -\iint_R \frac{\partial F_1}{\partial y}\, dA$ for this y-simple region R

Similarly, since R is x-simple, $\oint_{\mathcal{C}} F_2\, dy = \iint_R \frac{\partial F_2}{\partial x}\, dx\, dy$, so

$$\oint_{\mathcal{C}} F_1(x, y)\, dx + F_2(x, y)\, dy = \iint_R \left(\frac{\partial F_2}{\partial x} - \frac{\partial F_1}{\partial y}\right) dA.$$

Example 1 (**Area bounded by a simple closed curve**) For any of the three vector fields

$$\mathbf{F} = x\mathbf{j}, \qquad \mathbf{F} = -y\mathbf{i}, \qquad \text{and} \qquad \mathbf{F} = \frac{1}{2}(-y\mathbf{i} + x\mathbf{j}),$$

we have $(\partial F_2/\partial x) - (\partial F_1/\partial y) = 1$. If $\mathcal{C}$ is a positively oriented, piecewise smooth, simple closed curve bounding a region R in the plane, then, by Green's Theorem,

$$\oint_{\mathcal{C}} x\, dy = -\oint_{\mathcal{C}} y\, dx = \frac{1}{2} \oint_{\mathcal{C}} x\, dy - y\, dx = \iint_R 1\, dA = \text{area of } R.$$

Example 2 Evaluate $I = \oint_{\mathcal{C}} (x - y^3)\, dx + (y^3 + x^3)\, dy$, where $\mathcal{C}$ is the positively oriented boundary of the quarter-disk Q: $0 \le x^2 + y^2 \le a^2$, $x \ge 0$, $y \ge 0$.

Solution We use Green's Theorem to calculate I:

$$I = \iint_Q \left(\frac{\partial}{\partial x}(y^3 + x^3) - \frac{\partial}{\partial y}(x - y^3)\right) dA$$
$$= 3 \iint_Q (x^2 + y^2)\, dA = 3 \int_0^{\pi/2} d\theta \int_0^a r^3\, dr = \frac{3}{8}\pi a^4.$$

Example 3   Let $\mathcal{C}$ be a positively oriented, simple closed curve in the xy-plane, bounding a region R and not passing through the origin. Show that

$$\oint_{\mathcal{C}} \frac{-y\,dx + x\,dy}{x^2 + y^2} = \begin{cases} 0 & \text{if the origin is outside } R \\ 2\pi & \text{if the origin is inside } R. \end{cases}$$

Solution   First, if $(x, y) \neq (0, 0)$, then, by direct calculation,

$$\frac{\partial}{\partial x}\left(\frac{x}{x^2 + y^2}\right) - \frac{\partial}{\partial y}\left(\frac{-y}{x^2 + y^2}\right) = 0.$$

If the origin is not in R, then Green's Theorem implies that

$$\oint_{\mathcal{C}} \frac{-y\,dx + x\,dy}{x^2 + y^2} = \iint_R \left[\frac{\partial}{\partial x}\left(\frac{x}{x^2 + y^2}\right) - \frac{\partial}{\partial y}\left(\frac{-y}{x^2 + y^2}\right)\right] dx\,dy = 0.$$

Now suppose the origin is in R. Since it is assumed that the origin is not on $\mathcal{C}$, it must be an interior point of R. The interior of R is open, so there exists $\epsilon > 0$ such that the circle $\mathcal{C}_\epsilon$ of radius ϵ centred at the origin is in the interior of R. Let $\mathcal{C}_\epsilon$ be oriented negatively (clockwise). By direct calculation (see Exercise 22(a) of Section 15.4) it is easily shown that

$$\oint_{\mathcal{C}_\epsilon} \frac{-y\,dx + x\,dy}{x^2 + y^2} = -2\pi.$$

Together $\mathcal{C}$ and $\mathcal{C}_\epsilon$ form the positively oriented boundary of a region R_1 that excludes the origin. (See Figure 16.8.) So, by Green's Theorem,

$$\oint_{\mathcal{C}} \frac{-y\,dx + x\,dy}{x^2 + y^2} + \oint_{\mathcal{C}_\epsilon} \frac{-y\,dx + x\,dy}{x^2 + y^2} = 0.$$

The desired result now follows:

$$\oint_{\mathcal{C}} \frac{-y\,dx + x\,dy}{x^2 + y^2} = -\oint_{\mathcal{C}_\epsilon} \frac{-y\,dx + x\,dy}{x^2 + y^2} = -(-2\pi) = 2\pi.$$

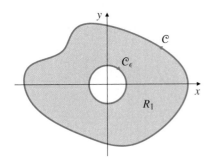

Figure 16.8

The Two-Dimensional Divergence Theorem

The following theorem is an alternative formulation of the two-dimensional Fundamental Theorem of Calculus. In this case we express the double integral of **div F** (a derivative of **F**) over R as a single integral of the outward normal component of **F** on the boundary $\mathcal{C}$ of R.

THEOREM 7

The Divergence Theorem in the Plane

Let R be a regular, closed region in the xy-plane whose boundary, $\mathcal{C}$, consists of one or more piecewise smooth, non-self-intersecting, closed curves. Let $\hat{\mathbf{N}}$ denote the unit outward (from R) normal field on $\mathcal{C}$. If $\mathbf{F} = F_1(x, y)\mathbf{i} + F_2(x, y)\mathbf{j}$ is a smooth vector field on R, then

$$\iint_R \mathbf{div\,F}\,dA = \oint_{\mathcal{C}} \mathbf{F} \bullet \hat{\mathbf{N}}\,ds.$$

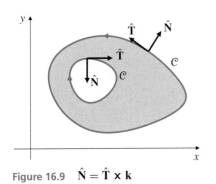

Figure 16.9 $\hat{\mathbf{N}} = \hat{\mathbf{T}} \times \mathbf{k}$

PROOF   As observed in the second paragraph of this section, $\hat{\mathbf{N}} = \hat{\mathbf{T}} \times \mathbf{k}$, where $\hat{\mathbf{T}}$ is the unit tangent field in the positive direction on $\mathcal{C}$. If $\hat{\mathbf{T}} = T_1\mathbf{i} + T_2\mathbf{j}$, then $\hat{\mathbf{N}} = T_2\mathbf{i} - T_1\mathbf{j}$. (See Figure 16.9.) Now let **G** be the vector field with components $G_1 = -F_2$ and $G_2 = F_1$. Then $\mathbf{G} \bullet \hat{\mathbf{T}} = \mathbf{F} \bullet \hat{\mathbf{N}}$ and, by Green's Theorem,

$$\iint_R \mathbf{div\,F}\,dA = \iint_R \left(\frac{\partial F_1}{\partial x} + \frac{\partial F_2}{\partial y}\right) dA$$

$$= \iint_R \left(\frac{\partial G_2}{\partial x} - \frac{\partial G_1}{\partial y}\right) dA$$

$$= \oint_{\mathcal{C}} \mathbf{G} \bullet d\mathbf{r} = \oint_{\mathcal{C}} \mathbf{G} \bullet \hat{\mathbf{T}}\,ds = \oint_{\mathcal{C}} \mathbf{F} \bullet \hat{\mathbf{N}}\,ds.$$

||||||||||| **Exercises 16.3** |||||||||||

1. Evaluate $\oint_{\mathcal{C}} (\sin x + 3y^2)\,dx + (2x - e^{-y^2})\,dy$, where $\mathcal{C}$ is the boundary of the half-disk $x^2 + y^2 \le a^2$, $y \ge 0$, oriented counterclockwise.

2. Evaluate $\oint_{\mathcal{C}} (x^2 - xy)\,dx + (xy - y^2)\,dy$ clockwise around the triangle with vertices $(0, 0)$, $(1, 1)$, and $(2, 0)$.

3. Evaluate $\oint_{\mathcal{C}} \left(x\sin(y^2) - y^2 \right) dx + \left(x^2 y \cos(y^2) + 3x \right) dy$, where $\mathcal{C}$ is the counterclockwise boundary of the trapezoid with vertices $(0, -2)$, $(1, -1)$, $(1, 1)$, and $(0, 2)$.

4. Evaluate $\oint_{\mathcal{C}} x^2 y\,dx - xy^2\,dy$, where $\mathcal{C}$ is the clockwise boundary of the region $0 \le y \le \sqrt{9 - x^2}$.

5. Use a line integral to find the plane area enclosed by the curve $\mathbf{r} = a\cos^3 t\,\mathbf{i} + b\sin^3 t\,\mathbf{j}$, $(0 \le t \le 2\pi)$.

6. We deduced the two-dimensional Divergence Theorem from Green's Theorem. Reverse the argument and use the two-dimensional Divergence Theorem to prove Green's Theorem.

7. Sketch the plane curve $\mathcal{C}$: $\mathbf{r} = \sin t\,\mathbf{i} + \sin 2t\,\mathbf{j}$, $(0 \le t \le 2\pi)$. Evaluate $\oint_{\mathcal{C}} \mathbf{F} \bullet d\mathbf{r}$, where $\mathbf{F} = ye^{x^2}\mathbf{i} + x^3 e^y\mathbf{j}$.

8. If $\mathcal{C}$ is the positively oriented boundary of a plane region R having area A and centroid $(\bar{x}, \bar{y})$, interpret geometrically the line integral $\oint_{\mathcal{C}} \mathbf{F} \bullet d\mathbf{r}$, where   (a) $\mathbf{F} = x^2\mathbf{j}$,   (b) $\mathbf{F} = xy\mathbf{i}$, and   (c) $\mathbf{F} = y^2\mathbf{i} + 3xy\mathbf{j}$.

* 9. **(Average values of harmonic functions)** If $u(x, y)$ is harmonic in a domain containing a disk of radius r with boundary C_r, then the average value of u around the circle is the value of u at the centre. Prove this by showing that the derivative of the average value with respect to r is zero (use the Divergence Theorem and harmonicity of u) and the fact that the limit of the average value as $r \to 0$ is the value of u at the centre.

16.4 The Divergence Theorem in 3-Space

The **Divergence Theorem** (also called **Gauss's Theorem**) is one of two important versions of the Fundamental Theorem of Calculus in $\mathbb{R}^3$. (The other is Stokes's Theorem, presented in the next section.)

In the Divergence Theorem, the integral of the *derivative* **div F** $= \nabla \bullet \mathbf{F}$ over a domain in 3-space is expressed as the flux of **F** out of the surface of that domain. It therefore closely resembles the two-dimensional version, Theorem 7, given in the previous section. The theorem holds for a general class of domains in $\mathbb{R}^3$ that are bounded by piecewise smooth closed surfaces. However, we will restrict our statement and proof of the theorem to domains of a special type. Extending the concept of an x-simple plane domain defined in Section 14.2, we say the three-dimensional domain D is x-**simple** if it is bounded by a piecewise smooth surface $\mathcal{S}$ and if every straight line parallel to the x-axis and passing through an interior point of D meets $\mathcal{S}$ at exactly two points. Similar definitions hold for y-simple and z-simple, and we call the domain D **regular** if it is a union of finitely many, nonoverlapping subdomains, each of which is x-simple, y-simple, and z-simple.

THEOREM 8

The Divergence Theorem (Gauss's Theorem)

Let D be a regular, three-dimensional domain whose boundary $\mathcal{S}$ is an oriented, closed surface with unit normal field $\hat{\mathbf{N}}$ pointing out of D. If $\mathbf{F}$ is a smooth vector field defined on D, then

$$\iiint_D \text{div}\,\mathbf{F}\,dV = \oiint_{\mathcal{S}} \mathbf{F} \bullet \hat{\mathbf{N}}\,dS.$$

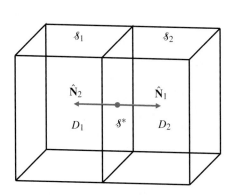

Figure 16.10 A union of abutting domains

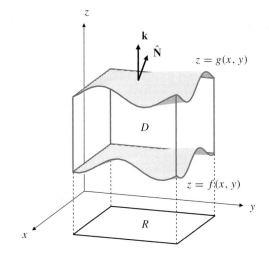

Figure 16.11 A z-simple domain

PROOF Since the domain D is a union of finitely many nonoverlapping domains that are x-simple, y-simple, and z-simple, it is sufficient to prove the theorem for a subdomain of D with this property. To see this, suppose, for instance, that D and $\mathscr{S}$ are each divided into two parts, D_1 and D_2, and $\mathscr{S}_1$ and $\mathscr{S}_2$, by a surface $\mathscr{S}^*$ slicing through D. (See Figure 16.10.) $\mathscr{S}^*$ is part of the boundary of both D_1 and D_2, but the exterior normals, $\hat{\mathbf{N}}_1$ and $\hat{\mathbf{N}}_2$, of the two subdomains point in opposite directions on either side of $\mathscr{S}^*$. If the formula in the theorem holds for both subdomains,

$$\iiint_{D_1} \operatorname{div}\mathbf{F}\, dV = \oiint_{\mathscr{S}_1 \cup \mathscr{S}^*} \mathbf{F} \bullet \hat{\mathbf{N}}_1\, dS$$

$$\iiint_{D_2} \operatorname{div}\mathbf{F}\, dV = \oiint_{\mathscr{S}_2 \cup \mathscr{S}^*} \mathbf{F} \bullet \hat{\mathbf{N}}_2\, dS,$$

then, adding these equations, we get

$$\iiint_{D} \operatorname{div}\mathbf{F}\, dV = \oiint_{\mathscr{S}_1 \cup \mathscr{S}_2} \mathbf{F} \bullet \hat{\mathbf{N}}\, dS = \oiint_{\mathscr{S}} \mathbf{F} \bullet \hat{\mathbf{N}}\, dS;$$

the contributions from $\mathscr{S}^*$ cancel out because on that surface $\hat{\mathbf{N}}_2 = -\hat{\mathbf{N}}_1$.

For the rest of this proof we assume, therefore, that D is x-, y-, and z-simple. Since D is z-simple, it lies between the graphs of two functions defined on a region R in the xy-plane; if (x, y, z) is in D, then (x, y) is in R and $f(x, y) \le z \le g(x, y)$. (See Figure 16.11.) We have

$$\iiint_{D} \frac{\partial F_3}{\partial z}\, dV = \iint_{R} dx\, dy \int_{f(x,y)}^{g(x,y)} \frac{\partial F_3}{\partial z}\, dz$$

$$= \iint_{R} \Big(F_3\big(x, y, g(x, y)\big) - F_3\big(x, y, f(x, y)\big) \Big)\, dx\, dy.$$

Now

$$\oiint_{\mathscr{S}} \mathbf{F} \bullet \hat{\mathbf{N}}\, dS = \oiint_{\mathscr{S}} \Big(F_1\, \mathbf{i} \bullet \hat{\mathbf{N}} + F_2\, \mathbf{j} \bullet \hat{\mathbf{N}} + F_3\, \mathbf{k} \bullet \hat{\mathbf{N}} \Big)\, dS.$$

Only the last term involves F_3, and it can be split into three integrals, over the top surface $z = g(x, y)$, the bottom surface $z = f(x, y)$, and vertical side wall lying above the boundary of R:

$$\oiint_{\mathscr{S}} F_3(x, y, z)\, \mathbf{k} \bullet \hat{\mathbf{N}}\, dS = \left(\iint_{\text{top}} + \iint_{\text{bottom}} + \iint_{\text{side}} \right) F_3(x, y, z)\, \mathbf{k} \bullet \hat{\mathbf{N}}\, dS.$$

On the side wall, $\mathbf{k} \bullet \hat{\mathbf{N}} = 0$, so that integral is zero. On the top surface, $z = g(x, y)$, and the vector area element is

$$\hat{\mathbf{N}} \, dS = \left(-\frac{\partial g}{\partial x} \mathbf{i} - \frac{\partial g}{\partial y} \mathbf{j} + \mathbf{k} \right) dx \, dy.$$

Accordingly,

$$\iint_{\text{top}} F_3(x, y, z) \, \mathbf{k} \bullet \hat{\mathbf{N}} \, dS = \iint_R F_3(x, y, g(x, y)) \, dx \, dy.$$

Similarly, we have

$$\iint_{\text{bottom}} F_3(x, y, z) \, \mathbf{k} \bullet \hat{\mathbf{N}} \, dS = -\iint_R F_3(x, y, f(x, y)) \, dx \, dy;$$

the negative sign occurs because $\hat{\mathbf{N}}$ points down rather than up on the bottom. Thus we have shown that

$$\iiint_D \frac{\partial F_3}{\partial z} \, dV = \oiint_{\mathcal{S}} F_3 \, \mathbf{k} \bullet \hat{\mathbf{N}} \, dS.$$

Similarly, because D is also x-simple and y-simple,

$$\iiint_D \frac{\partial F_1}{\partial x} \, dV = \oiint_{\mathcal{S}} F_1 \, \mathbf{i} \bullet \hat{\mathbf{N}} \, dS$$

$$\iiint_D \frac{\partial F_2}{\partial y} \, dV = \oiint_{\mathcal{S}} F_2 \, \mathbf{j} \bullet \hat{\mathbf{N}} \, dS.$$

Adding these three results we get

$$\iiint_D \operatorname{div} \mathbf{F} \, dV = \oiint_{\mathcal{S}} \mathbf{F} \bullet \hat{\mathbf{N}} \, dS.$$

The Divergence Theorem can be used in both directions to simplify explicit calculations of surface integrals or volumes. We give examples of each.

Example 1   Let $\mathbf{F} = bxy^2\mathbf{i} + bx^2 y\mathbf{j} + (x^2 + y^2)z^2\mathbf{k}$, and let $\mathcal{S}$ be the closed surface bounding the solid cylinder R defined by $x^2 + y^2 \le a^2$ and $0 \le z \le b$. Find $\oiint_{\mathcal{S}} \mathbf{F} \bullet d\mathbf{S}$.

Solution   By the Divergence Theorem,

$$\oiint_{\mathcal{S}} \mathbf{F} \bullet d\mathbf{S} = \iiint_R \operatorname{div} \mathbf{F} \, dV = \iiint_R (x^2 + y^2)(b + 2z) \, dV$$

$$= \int_0^b (b + 2z) \, dz \int_0^{2\pi} d\theta \int_0^a r^2 \, r \, dr$$

$$= (b^2 + b^2) 2\pi (a^4/4) = \pi a^4 b^2.$$

Example 2   Evaluate $\oiint_{\mathcal{S}} (x^2 + y^2) \, dS$, where $\mathcal{S}$ is the sphere $x^2 + y^2 + z^2 = a^2$. Use the Divergence Theorem.

Solution On $\mathcal{S}$ we have

$$\hat{\mathbf{N}} = \frac{\mathbf{r}}{a} = \frac{x\mathbf{i} + y\mathbf{j} + z\mathbf{k}}{a}.$$

We would like to choose $\mathbf{F}$ so that $\mathbf{F} \bullet \hat{\mathbf{N}} = x^2 + y^2$. Observe that $\mathbf{F} = a(x\mathbf{i} + y\mathbf{j})$ will do. If B is the ball bounded by $\mathcal{S}$, then

$$\oiint_{\mathcal{S}} (x^2 + y^2)\, dS = \oiint_{\mathcal{S}} \mathbf{F} \bullet \hat{\mathbf{N}}\, dS = \iiint_B \operatorname{div} \mathbf{F}\, dV$$
$$= \iiint_B 2a\, dV = (2a)\frac{4}{3}\pi a^3 = \frac{8}{3}\pi a^4.$$

Example 3 By using the Divergence Theorem with $\mathbf{F} = x\mathbf{i} + y\mathbf{j} + z\mathbf{k}$, calculate the volume of a cone having base area A and height h. The base can be any smoothly bounded plane region.

Solution Let the vertex of the cone be at the origin and the base in the plane $z = h$ as shown in Figure 16.12. The solid cone C has surface consisting of two parts: the conical wall $\mathcal{S}$ and the base region D that has area A. Since $\mathbf{F}(x, y, z)$ points directly away from the origin at any point $(x, y, z) \neq (0, 0, 0)$, we have $\mathbf{F} \bullet \hat{\mathbf{N}} = 0$ on $\mathcal{S}$. On D, we have $\hat{\mathbf{N}} = \mathbf{k}$ and $z = h$, so $\mathbf{F} \bullet \hat{\mathbf{N}} = z = h$ on the base of the cone. Since $\operatorname{div} \mathbf{F}(x, y, z) = 1 + 1 + 1 = 3$, we have, by the Divergence Theorem,

$$3V = \iiint_C \operatorname{div} \mathbf{F}\, dV = \iint_{\mathcal{S}} \mathbf{F} \bullet \hat{\mathbf{N}}\, dS + \iint_D \mathbf{F} \bullet \hat{\mathbf{N}}\, dS$$
$$= 0 + h \iint_D dS = Ah.$$

Thus $V = \frac{1}{3}Ah$, the well-known formula for the volume of a cone.

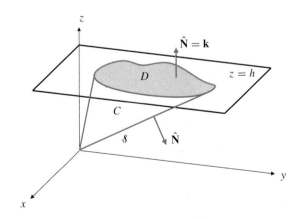

Figure 16.12 A cone with an arbitrarily shaped base

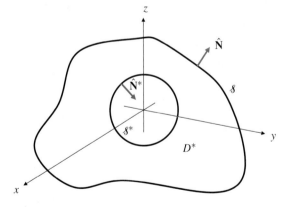

Figure 16.13 A solid domain with a spherical cavity

Example 4 Let $\mathcal{S}$ be the surface of an arbitrary regular domain D in 3-space that contains the origin in its interior. Find

$$\oiint_{\mathcal{S}} \mathbf{F} \bullet \hat{\mathbf{N}}\, dS,$$

where $\mathbf{F}(\mathbf{r}) = m\mathbf{r}/|\mathbf{r}|^3$ and $\hat{\mathbf{N}}$ is the unit outward normal on $\mathcal{S}$. (See Figure 16.13.)

Solution Since **F** and, therefore, **div F** are undefined at the origin, we cannot apply the Divergence Theorem directly. To overcome this problem we use a little trick. Let $\mathscr{S}^*$ be a small sphere centred at the origin bounding a ball contained wholly in D. (See Figure 16.13.) Let $\hat{\mathbf{N}}^*$ be the unit normal on $\mathscr{S}^*$ pointing *into* the sphere, and let D^* be that part of D that lies outside $\mathscr{S}^*$. As shown in Example 3 of Section 16.1, **div F** $= 0$ on D^*. Also,

$$\oiint_{\mathscr{S}^*} \mathbf{F} \bullet \hat{\mathbf{N}}^* \, dS = -4\pi m$$

is the flux of **F** *inward* through the sphere $\mathscr{S}^*$. (See Example 1 of Section 15.6.) Therefore,

$$0 = \iiint_{D^*} \mathbf{div}\,\mathbf{F} \, dV = \oiint_{\mathscr{S}} \mathbf{F} \bullet \hat{\mathbf{N}} \, dS + \oiint_{\mathscr{S}^*} \mathbf{F} \bullet \hat{\mathbf{N}}^* \, dS$$

$$= \oiint_{\mathscr{S}} \mathbf{F} \bullet \hat{\mathbf{N}} \, dS - 4\pi m,$$

so $\oiint_{\mathscr{S}} \mathbf{F} \bullet \hat{\mathbf{N}} \, dS = 4\pi m.$

Example 5 Find the flux of $\mathbf{F} = x\mathbf{i} + y^2\mathbf{j} + z\mathbf{k}$ upward through the first octant part $\mathscr{S}$ of the cylindrical surface $x^2 + z^2 = a^2, 0 \le y \le b$.

Solution $\mathscr{S}$ is one of five surfaces that form the boundary of the solid region D shown in Figure 16.14. The other four surfaces are planar: $\mathscr{S}_1$ lies in the plane $z = 0$, $\mathscr{S}_2$ lies in the plane $x = 0$, $\mathscr{S}_3$ lies in the plane $y = 0$, and $\mathscr{S}_4$ lies in the plane $y = b$. Orient all these surfaces with normal $\hat{\mathbf{N}}$ pointing out of D. On $\mathscr{S}_1$ we have $\hat{\mathbf{N}} = -\mathbf{k}$, so $\mathbf{F} \bullet \hat{\mathbf{N}} = -z = 0$ on $\mathscr{S}_1$. Similarly, $\mathbf{F} \bullet \hat{\mathbf{N}} = 0$ on $\mathscr{S}_2$ and $\mathscr{S}_3$. On $\mathscr{S}_4$, $y = b$ and $\hat{\mathbf{N}} = \mathbf{j}$, so $\mathbf{F} \bullet \hat{\mathbf{N}} = y^2 = b^2$ there. If $\mathscr{S}_{\text{tot}}$ denotes the whole boundary of D, then

$$\oiint_{\mathscr{S}_{\text{tot}}} \mathbf{F} \bullet \hat{\mathbf{N}} \, dS = \iint_{\mathscr{S}} \mathbf{F} \bullet \hat{\mathbf{N}} \, dS + 0 + 0 + 0 + \iint_{\mathscr{S}_4} \mathbf{F} \bullet \hat{\mathbf{N}} \, dS$$

$$= \iint_{\mathscr{S}} \mathbf{F} \bullet \hat{\mathbf{N}} \, dS + \frac{\pi a^2 b^2}{4}.$$

On the other hand, by the Divergence Theorem,

$$\oiint_{\mathscr{S}_{\text{tot}}} \mathbf{F} \bullet \hat{\mathbf{N}} \, dS = \iiint_D \mathbf{div}\,\mathbf{F} \, dV = \iiint_D (2 + 2y) \, dV = 2V + 2V\bar{y},$$

Figure 16.14 The boundary of domain D has five faces, one curved and four planar

where $V = \pi a^2 b/4$ is the volume of D, and $\bar{y} = b/2$ is the y-coordinate of the centroid of D. Combining these results, the flux of **F** upward through $\mathscr{S}$ is

$$\iint_{\mathscr{S}} \mathbf{F} \bullet \hat{\mathbf{N}} \, dS = \frac{2\pi a^2 b}{4}\left(1 + \frac{b}{2}\right) - \frac{\pi a^2 b^2}{4} = \frac{\pi a^2 b}{2}.$$

Among the examples above, Example 4 is the most significant and the one that best represents the way that the Divergence Theorem is used in practice. It is predominantly a theoretical tool, rather than a tool for calculation. We will look at some applications in Section 16.6.

Variants of the Divergence Theorem

Other versions of the Fundamental Theorem of Calculus can be derived from the Divergence Theorem. Two are given in the following theorem:

THEOREM **9**

If D satisfies the conditions of the Divergence Theorem and has surface $\mathcal{S}$, and if $\mathbf{F}$ is a smooth vector field and ϕ is a smooth scalar field, then

(a)
$$\iiint_D \mathbf{curl\, F}\, dV = -\oiint_{\mathcal{S}} \mathbf{F} \times \hat{\mathbf{N}}\, dS,$$

(b)
$$\iiint_D \mathbf{grad}\, \phi\, dV = \oiint_{\mathcal{S}} \phi \hat{\mathbf{N}}\, dS.$$

PROOF Observe that both of these formulas are equations of *vectors*. They are derived by applying the Divergence Theorem to $\mathbf{F} \times \mathbf{c}$ and $\phi \mathbf{c}$, respectively, where $\mathbf{c}$ is an arbitrary constant vector. We give the details for formula (a) and leave (b) as an exercise.

Using Theorem 3(d), we calculate

$$\nabla \bullet (\mathbf{F} \times \mathbf{c}) = (\nabla \times \mathbf{F}) \bullet \mathbf{c} - \mathbf{F} \bullet (\nabla \times \mathbf{c}) = (\nabla \times \mathbf{F}) \bullet \mathbf{c}.$$

Also, by the scalar triple product identity (see Exercise 18 of Section 10.3),

$$(\mathbf{F} \times \mathbf{c}) \bullet \hat{\mathbf{N}} = (\hat{\mathbf{N}} \times \mathbf{F}) \bullet \mathbf{c} = -(\mathbf{F} \times \hat{\mathbf{N}}) \bullet \mathbf{c}.$$

Therefore,

$$\left(\iiint_D \mathbf{curl\, F}\, dV + \oiint_{\mathcal{S}} \mathbf{F} \times \hat{\mathbf{N}}\, dS \right) \bullet \mathbf{c}$$
$$= \iiint_D (\nabla \times \mathbf{F}) \bullet \mathbf{c}\, dV - \oiint_{\mathcal{S}} (\mathbf{F} \times \mathbf{c}) \bullet \hat{\mathbf{N}}\, dS$$
$$= \iiint_D \mathbf{div}\, (\mathbf{F} \times \mathbf{c})\, dV - \oiint_{\mathcal{S}} (\mathbf{F} \times \mathbf{c}) \bullet \hat{\mathbf{N}}\, dS = 0.$$

Since $\mathbf{c}$ is arbitrary, the vector in the large parentheses must be the zero vector. (If $\mathbf{c} \bullet \mathbf{a} = 0$ for every vector $\mathbf{c}$, then $\mathbf{a} = \mathbf{0}$.) This establishes formula (a).

Exercises 16.4

In Exercises 1–4, use the Divergence Theorem to calculate the flux of the given vector field out of the sphere $\mathcal{S}$ with equation $x^2 + y^2 + z^2 = a^2$, where $a > 0$.

1. $\mathbf{F} = x\mathbf{i} - 2y\mathbf{j} + 4z\mathbf{k}$ **2.** $\mathbf{F} = ye^z\mathbf{i} + x^2 e^z\mathbf{j} + xy\mathbf{k}$

3. $\mathbf{F} = (x^2 + y^2)\mathbf{i} + (y^2 - z^2)\mathbf{j} + z\mathbf{k}$

4. $\mathbf{F} = x^3\mathbf{i} + 3yz^2\mathbf{j} + (3y^2z + x^2)\mathbf{k}$

In Exercises 5–8, evaluate the flux of $\mathbf{F} = x^2\mathbf{i} + y^2\mathbf{j} + z^2\mathbf{k}$ outward across the boundary of the given solid region.

5. The ball $(x - 2)^2 + y^2 + (z - 3)^2 \le 9$

6. The solid ellipsoid $x^2 + y^2 + 4(z - 1)^2 \le 4$

7. The tetrahedron $x + y + z \le 3, x \ge 0, y \ge 0, z \ge 0$

8. The cylinder $x^2 + y^2 \le 2y, 0 \le z \le 4$

9. Let A be the area of a region D forming part of the surface of a sphere of radius R centred at the origin, and let V be the volume of the solid cone C consisting of all points on line segments joining the centre of the sphere to points in D. Show that

$$V = \frac{1}{3} AR$$

by applying the Divergence Theorem to $\mathbf{F} = x\mathbf{i} + y\mathbf{j} + z\mathbf{k}$.

10. Let $\phi(x, y, z) = xy + z^2$. Find the flux of $\nabla \phi$ upward through the triangular planar surface $\mathcal{S}$ with vertices at $(a, 0, 0)$, $(0, b, 0)$, and $(0, 0, c)$.

11. A conical domain with vertex $(0, 0, b)$ and axis along the z-axis has as base a disk of radius a in the xy-plane. Find the flux of

$$\mathbf{F} = (x + y^2)\mathbf{i} + (3x^2y + y^3 - x^3)\mathbf{j} + (z + 1)\mathbf{k}$$

upward through the conical part of the surface of the domain.

12. Find the flux of $\mathbf{F} = (y + xz)\mathbf{i} + (y + yz)\mathbf{j} - (2x + z^2)\mathbf{k}$ upward through the first octant part of the sphere $x^2 + y^2 + z^2 = a^2$.

Therefore,

$$\iint_{\mathcal{S}} \mathbf{curl\,F} \bullet \hat{\mathbf{N}}\,dS = \iint_{R} \left[\left(\frac{\partial F_3}{\partial y} - \frac{\partial F_2}{\partial z} \right) \left(-\frac{\partial g}{\partial x} \right) + \left(\frac{\partial F_1}{\partial z} - \frac{\partial F_3}{\partial x} \right) \left(-\frac{\partial g}{\partial y} \right) \right. $$
$$\left. + \left(\frac{\partial F_2}{\partial x} - \frac{\partial F_1}{\partial y} \right) \right] dA.$$

Since $z = g(x, y)$ on $\mathcal{C}$, we have $dz = \dfrac{\partial g}{\partial x}\,dx + \dfrac{\partial g}{\partial y}\,dy$. Thus,

$$\oint_{\mathcal{C}} \mathbf{F} \bullet d\mathbf{r} = \oint_{\mathcal{C}^*} \left[F_1(x, y, z)\,dx + F_2(x, y, z)\,dy \right.$$
$$\left. + F_3(x, y, z) \left(\frac{\partial g}{\partial x}\,dx + \frac{\partial g}{\partial y}\,dy \right) \right]$$
$$= \oint_{\mathcal{C}^*} \left(\left[F_1(x, y, z) + F_3(x, y, z)\frac{\partial g}{\partial x} \right] dx \right.$$
$$\left. + \left[F_2(x, y, z) + F_3(x, y, z)\frac{\partial g}{\partial y} \right] dy \right).$$

We now apply Green's Theorem in the xy-plane to obtain

$$\oint_{\mathcal{C}} \mathbf{F} \bullet d\mathbf{r} = \iint_{R} \left(\frac{\partial}{\partial x} \left[F_2(x, y, z) + F_3(x, y, z)\frac{\partial g}{\partial y} \right] \right.$$
$$\left. - \frac{\partial}{\partial y} \left[F_1(x, y, z) + F_3(x, y, z)\frac{\partial g}{\partial x} \right] \right) dA$$
$$= \iint_{R} \left(\frac{\partial F_2}{\partial x} + \frac{\partial F_2}{\partial z}\frac{\partial g}{\partial x} + \frac{\partial F_3}{\partial x}\frac{\partial g}{\partial y} + \frac{\partial F_3}{\partial z}\frac{\partial g}{\partial x}\frac{\partial g}{\partial y} + F_3 \frac{\partial^2 g}{\partial x \partial y} \right.$$
$$\left. - \frac{\partial F_1}{\partial y} - \frac{\partial F_1}{\partial z}\frac{\partial g}{\partial y} - \frac{\partial F_3}{\partial y}\frac{\partial g}{\partial x} - \frac{\partial F_3}{\partial z}\frac{\partial g}{\partial y}\frac{\partial g}{\partial x} - F_3 \frac{\partial^2 g}{\partial y \partial x} \right) dA.$$

Observe that four terms in the final integrand cancel out, and the remaining terms are equal to the terms in the expression for $\iint_{\mathcal{S}} \mathbf{curl\,F} \bullet \hat{\mathbf{N}}\,dS$ calculated above. Therefore the proof is complete.

Remark   If $\mathbf{curl\,F} = \mathbf{0}$ on a domain D with the property that every piecewise smooth, non-self-intersecting, closed curve in D is the boundary of a piecewise smooth surface in D, then Stokes's Theorem assures us that $\oint_{\mathcal{C}} \mathbf{F} \bullet d\mathbf{r} = 0$ for *every* such curve $\mathcal{C}$; therefore $\mathbf{F}$ must be conservative. A simply connected domain D does have the property specified above. We will not attempt a formal proof of this topological fact here, but it should seem plausible if you recall the definition of simple connectedness. A closed curve $\mathcal{C}$ in a simply connected domain D must be able to shrink to a point in D without ever passing out of D. In so shrinking, it traces out a surface in D. This is why Theorem 4 of Section 16.2 is valid for simply connected domains.

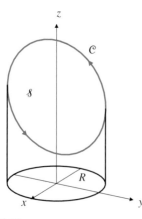

Figure 16.16

Example 1   Evaluate $\oint_{\mathcal{C}} \mathbf{F} \bullet d\mathbf{r}$, where $\mathbf{F} = -y^3\mathbf{i} + x^3\mathbf{j} - z^3\mathbf{k}$, and $\mathcal{C}$ is the curve of intersection of the cylinder $x^2 + y^2 = 1$ and the plane $2x + 2y + z = 3$ oriented so as to have a counterclockwise projection onto the xy-plane.

Solution $\mathcal{C}$ is the oriented boundary of an elliptic disk $\mathcal{S}$ that lies in the plane $2x + 2y + z = 3$ and has the circular disk R: $x^2 + y^2 \le 1$ as projection onto the xy-plane. (See Figure 16.16.) On $\mathcal{S}$ we have

$$\hat{\mathbf{N}}\,dS = (2\mathbf{i} + 2\mathbf{j} + \mathbf{k})\,dx\,dy.$$

Also,

$$\mathbf{curl\,F} = \begin{vmatrix} \mathbf{i} & \mathbf{j} & \mathbf{k} \\ \dfrac{\partial}{\partial x} & \dfrac{\partial}{\partial y} & \dfrac{\partial}{\partial z} \\ -y^3 & x^3 & -z^3 \end{vmatrix} = 3(x^2 + y^2)\mathbf{k}.$$

Thus, by Stokes's Theorem,

$$\oint_{\mathcal{C}} \mathbf{F} \bullet d\mathbf{r} = \iint_{\mathcal{S}} \mathbf{curl\,F} \bullet \hat{\mathbf{N}}\, dS$$

$$= \iint_{R} 3(x^2 + y^2)\, dx\, dy = 2\pi \int_0^1 3r^2\, r\, dr = \frac{3\pi}{2}.$$

As with the Divergence Theorem, the principal importance of Stokes's Theorem is as a theoretical tool. However, it can also simplify the calculation of circulation integrals such as the one in the previous example. It is not difficult to imagine integrals whose evaluation would be impossibly difficult without the use of Stokes's Theorem or the Divergence Theorem. In the following example we use Stokes's Theorem twice, but the result could be obtained just as easily by using the Divergence Theorem.

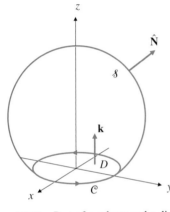

Figure 16.17   Part of a sphere and a disk with the same boundary

Example 2   Find $I = \iint_{\mathcal{S}} \mathbf{curl\,F} \bullet \hat{\mathbf{N}}\, dS$, where $\mathcal{S}$ is that part of the sphere $x^2 + y^2 + (z - 2)^2 = 8$ that lies above the xy-plane, $\hat{\mathbf{N}}$ is the unit outward normal field on $\mathcal{S}$, and

$$\mathbf{F} = y^2 \cos xz\,\mathbf{i} + x^3 e^{yz}\mathbf{j} - e^{-xyz}\mathbf{k}.$$

Solution   The boundary, $\mathcal{C}$, of $\mathcal{S}$ is the circle $x^2 + y^2 = 4$ in the xy-plane, oriented counterclockwise as seen from the positive z-axis. (See Figure 16.17.) This curve is also the oriented boundary of the plane disk D: $x^2 + y^2 \le 4$, $z = 0$, with normal field $\hat{\mathbf{N}} = \mathbf{k}$. Thus, two applications of Stokes's Theorem give

$$I = \iint_{\mathcal{S}} \mathbf{curl\,F} \bullet \hat{\mathbf{N}}\, dS = \oint_{\mathcal{C}} \mathbf{F} \bullet d\mathbf{r} = \iint_{D} \mathbf{curl\,F} \bullet \mathbf{k}\, dA.$$

On D we have

$$\mathbf{curl\,F} \bullet \mathbf{k} = \left(\frac{\partial}{\partial x}\left(x^3 e^{yz}\right) - \frac{\partial}{\partial y}\left(y^2 \cos xz\right) \right)\Bigg|_{z=0}$$

$$= 3x^2 - 2y.$$

By symmetry, $\iint_D y\, dA = 0$, so

$$I = 3 \iint_D x^2\, dA = 3 \int_0^{2\pi} \cos^2 \theta\, d\theta \int_0^2 r^3\, dr = 12\pi.$$

Remark   A surface $\mathcal{S}$ satisfying the conditions of Stokes's Theorem may no longer do so if a single point is removed from it. An isolated boundary point of a surface is not an orientable curve, and Stokes's Theorem may therefore break down for such a surface. Consider, for example, the vector field

$$\mathbf{F} = \frac{\hat{\boldsymbol{\theta}}}{r} = -\frac{y}{x^2 + y^2}\mathbf{i} + \frac{x}{x^2 + y^2}\mathbf{j},$$

which is defined on the *punctured disk D* satisfying $0 < x^2 + y^2 \le a^2$. (See Example 3 in Section 16.3.) If D is oriented with upward normal $\mathbf{k}$, then its boundary consists of the oriented, smooth, closed curve, $\mathcal{C}$, given by $x = a \cos \theta$, $y = a \sin \theta$, $(0 \le \theta \le 2\pi)$, and the isolated point $(0, 0)$. We have

$$\oint_{\mathcal{C}} \mathbf{F} \bullet d\mathbf{r} = \int_0^{2\pi} \left(\frac{-\sin \theta}{a} \mathbf{i} + \frac{\cos \theta}{a} \mathbf{j} \right) \bullet (-a \sin \theta \mathbf{i} + a \cos \theta \mathbf{j}) \, d\theta$$

$$= \int_0^{2\pi} (\sin^2 \theta + \cos^2 \theta) \, d\theta = 2\pi.$$

However,

$$\mathbf{curl}\, \mathbf{F} = \left[\frac{\partial}{\partial x} \left(\frac{x}{x^2 + y^2} \right) - \frac{\partial}{\partial y} \left(-\frac{y}{x^2 + y^2} \right) \right] \mathbf{k} = \mathbf{0}$$

identically on D. Thus,

$$\iint_D \mathbf{curl}\, \mathbf{F} \bullet \hat{\mathbf{N}} \, dS = 0,$$

and the conclusion of Stokes's Theorem fails in this case.

Exercises 16.5

1. Evaluate $\oint_{\mathcal{C}} xy \, dx + yz \, dy + zx \, dz$ around the triangle with vertices $(1, 0, 0)$, $(0, 1, 0)$, and $(0, 0, 1)$, oriented clockwise as seen from the point $(1, 1, 1)$.

2. Evaluate $\oint_{\mathcal{C}} y \, dx - x \, dy + z^2 \, dz$ around the curve $\mathcal{C}$ of intersection of the cylinders $z = y^2$ and $x^2 + y^2 = 4$, oriented counterclockwise as seen from a point high on the z-axis.

3. Evaluate $\iint_{\mathcal{S}} \mathbf{curl}\, \mathbf{F} \bullet \hat{\mathbf{N}} \, dS$, where $\mathcal{S}$ is the hemisphere $x^2 + y^2 + z^2 = a^2$, $z \ge 0$ with outward normal, and $\mathbf{F} = 3y\mathbf{i} - 2xz\mathbf{j} + (x^2 - y^2)\mathbf{k}$.

4. Evaluate $\iint_{\mathcal{S}} \mathbf{curl}\, \mathbf{F} \bullet \hat{\mathbf{N}} \, dS$, where $\mathcal{S}$ is the surface $x^2 + y^2 + 2(z - 1)^2 = 6$, $z \ge 0$, $\hat{\mathbf{N}}$ is the unit outward (away from the origin) normal on $\mathcal{S}$, and

$$\mathbf{F} = (xz - y^3 \cos z)\mathbf{i} + x^3 e^z \mathbf{j} + xyz \, e^{x^2 + y^2 + z^2} \mathbf{k}.$$

5. Use Stokes's Theorem to show that

$$\oint_{\mathcal{C}} y \, dx + z \, dy + x \, dz = \sqrt{3} \, \pi a^2,$$

where $\mathcal{C}$ is the suitably oriented intersection of the surfaces $x^2 + y^2 + z^2 = a^2$ and $x + y + z = 0$.

6. Evaluate $\oint_{\mathcal{C}} \mathbf{F} \bullet d\mathbf{r}$ around the curve

$$\mathbf{r} = \cos t \, \mathbf{i} + \sin t \, \mathbf{j} + \sin 2t \, \mathbf{k}, \quad (0 \le t \le 2\pi),$$

where

$$\mathbf{F} = (e^x - y^3)\mathbf{i} + (e^y + x^3)\mathbf{j} + e^z \mathbf{k}.$$

Hint: Show that $\mathcal{C}$ lies on the surface $z = 2xy$.

7. Find the circulation of $\mathbf{F} = -y\mathbf{i} + x^2 \mathbf{j} + z\mathbf{k}$ around the oriented boundary of the part of the paraboloid $z = 9 - x^2 - y^2$ lying above the xy-plane and having normal field pointing upward.

8. Evaluate $\oint_{\mathcal{C}} \mathbf{F} \bullet d\mathbf{r}$, where

$$\mathbf{F} = ye^x \mathbf{i} + (x^2 + e^x)\mathbf{j} + z^2 e^z \mathbf{k},$$

and $\mathcal{C}$ is the curve

$$\mathbf{r}(t) = (1 + \cos t)\mathbf{i} + (1 + \sin t)\mathbf{j} + (1 - \cos t - \sin t)\mathbf{k}$$

for $0 \le t \le 2\pi$. *Hint:* Use Stokes's Theorem, observing that $\mathcal{C}$ lies in a certain plane and has a circle as its projection onto the xy-plane. The integral can also be evaluated by using the techniques of Section 15.4.

9. Let $\mathcal{C}_1$ be the straight line joining $(-1, 0, 0)$ to $(1, 0, 0)$, and let $\mathcal{C}_2$ be the semicircle $x^2 + y^2 = 1$, $z = 0$, $y \ge 0$. Let $\mathcal{S}$ be a smooth surface joining $\mathcal{C}_1$ to $\mathcal{C}_2$ having upward normal, and let

$$\mathbf{F} = (\alpha x^2 - z)\mathbf{i} + (xy + y^3 + z)\mathbf{j} + \beta y^2 (z + 1)\mathbf{k}.$$

Find the values of α and β for which $I = \iint_{\mathcal{S}} \mathbf{F} \bullet d\mathbf{S}$ is independent of the choice of $\mathcal{S}$, and find the value of I for these values of α and β.

10. Let C be the curve $(x - 1)^2 + 4y^2 = 16$, $2x + y + z = 3$, oriented counterclockwise when viewed from high on the z-axis. Let

$$\mathbf{F} = (z^2 + y^2 + \sin x^2)\mathbf{i} + (2xy + z)\mathbf{j} + (xz + 2yz)\mathbf{k}.$$

Evaluate $\oint_C \mathbf{F} \bullet d\mathbf{r}$.

11. If C is the oriented boundary of surface $\mathcal{S}$, and ϕ and ψ are arbitrary smooth scalar fields, show that

$$\oint_C \phi \nabla \psi \bullet d\mathbf{r} = -\oint_C \psi \nabla \phi \bullet d\mathbf{r}$$
$$= \iint_{\mathcal{S}} (\nabla \phi \times \nabla \psi) \bullet \hat{\mathbf{N}} \, dS.$$

Is $\nabla \phi \times \nabla \psi$ solenoidal? Find a vector potential for it.

12. Let C be a closed, non-self-intersecting, piecewise smooth plane curve in $\mathbb{R}^3$, which lies in a plane with unit normal $\hat{\mathbf{N}} = a\mathbf{i} + b\mathbf{j} + c\mathbf{k}$ and has orientation inherited from that of the plane. Show that the plane area enclosed by C is

$$\frac{1}{2} \oint_C (bz - cy) \, dx + (cx - az) \, dy + (ay - bx) \, dz.$$

13. Use Stokes's Theorem to prove Theorem 2 of Section 16.1.

16.6 Some Physical Applications of Vector Calculus

In this section we will show how the theory developed in this chapter can be used to model concrete applied mathematical problems. We will look at two areas of application—fluid dynamics and electromagnetism—and will develop a few of the fundamental vector equations underlying these disciplines. Our purpose is to illustrate the techniques of vector calculus in applied contexts, rather than to provide any complete or even coherent introductions to the disciplines themselves.

Fluid Dynamics

Suppose that a region of 3-space is filled with a fluid (liquid or gas) in motion. Two approaches can be taken to describe the motion. We could attempt to determine the position, $\mathbf{r} = \mathbf{r}(a, b, c, t)$ at any time t, of a "particle" of fluid that was located at the point (a, b, c) at time $t = 0$. This is the Lagrange approach. Alternatively, we could attempt to determine the velocity, $\mathbf{v}(x, y, z, t)$, the density, $\delta(x, y, z, t)$, and other physical variables such as the pressure, $p(x, y, z, t)$, at any time t at any point (x, y, z) in the region occupied by the fluid. This is the Euler approach.

We will examine the latter method and describe how the Divergence Theorem can be used to translate some fundamental physical laws into equivalent mathematical equations. We assume throughout that the velocity, density, and pressure vary smoothly in all their variables and that the fluid is an *ideal fluid*, that is, nonviscous (it doesn't stick to itself), homogeneous, and isotropic (it has the same properties at all points and in all directions). Such properties are not always shared by real fluids, so we are dealing with a simplified mathematical model that does not always correspond exactly to the behaviour of real fluids.

Consider an imaginary closed surface $\mathcal{S}$ in the fluid, bounding a domain D. We call $\mathcal{S}$ "imaginary" because it is not a barrier that impedes the flow of the fluid in any way; it is just a means to concentrate our attention on a particular part of the fluid. It is fixed in space and does not move with the fluid. Let us assume that the fluid is being neither created nor destroyed anywhere (in particular, there are no sources or sinks), so the law of **conservation of mass** tells us that the rate of change of the mass of fluid in D equals the rate at which fluid enters D across $\mathcal{S}$.

The mass of fluid in volume element dV located at position (x, y, z) at time t is $\delta(x, y, z, t) \, dV$, so the mass in D at time t is $\iiint_D \delta \, dV$. This mass changes at rate

$$\frac{\partial}{\partial t} \iiint_D \delta \, dV = \iiint_D \frac{\partial \delta}{\partial t} \, dV.$$

As we noted in Section 15.6, the volume of fluid passing *out* of D through area element dS at position (x, y, z) in the interval from time t to $t + dt$ is given by $\mathbf{v}(x, y, z, t) \bullet \hat{\mathbf{N}}\, dS\, dt$, where $\hat{\mathbf{N}}$ is the unit normal at (x, y, z) on $\mathcal{S}$ pointing out of D. Hence, the mass crossing dS outward in that time interval is $\delta \mathbf{v} \bullet \hat{\mathbf{N}}\, dS\, dt$, and the *rate* at which mass is flowing out of D across $\mathcal{S}$ at time t is

$$\oiint_{\mathcal{S}} \delta \mathbf{v} \bullet \hat{\mathbf{N}}\, dS.$$

The rate at which mass is flowing *into D* is the negative of the above rate. Since mass is conserved, we must have

$$\iiint_D \frac{\partial \delta}{\partial t}\, dV = -\oiint_{\mathcal{S}} \delta \mathbf{v} \bullet \hat{\mathbf{N}}\, dS = -\iiint_D \mathbf{div}\,(\delta \mathbf{v})\, dV,$$

where we have used the Divergence Theorem to replace the surface integral with a volume integral. Thus,

$$\iiint_D \left(\frac{\partial \delta}{\partial t} + \mathbf{div}\,(\delta \mathbf{v}) \right) dV = 0.$$

This equation must hold for *any* domain D in the fluid.

If a continuous function f satisfies $\iiint_D f(P)\, dV = 0$ for every domain D, then $f(P) = 0$ at all points P, for if there were a point P_0 such that $f(P_0) \neq 0$ (say $f(P_0) > 0$), then, by continuity, f would be positive at all points in some sufficiently small ball B centred at P_0, and $\iiint_B f(P)\, dV$ would be greater than 0. Applying this principle, we must have

$$\frac{\partial \delta}{\partial t} + \mathbf{div}\,(\delta \mathbf{v}) = 0$$

throughout the fluid. This is called the **equation of continuity** for the fluid. It is equivalent to conservation of mass. Observe that if the fluid is **incompressible** then δ is a constant, independent of both time and spatial position. In this case $\partial \delta / \partial t = 0$, and $\mathbf{div}\,(\delta \mathbf{v}) = \delta\, \mathbf{div}\, \mathbf{v}$. Therefore, the equation of continuity for an incompressible fluid is simply

$$\mathbf{div}\, \mathbf{v} = 0.$$

The motion of the fluid is governed by Newton's Second Law, which asserts that the rate of change of momentum of any part of the fluid is equal to the sum of the forces applied to that part. Again, let us consider the part of the fluid in a domain D. At any time t its momentum is $\iiint_D \delta \mathbf{v}\, dV$ and is changing at the rate

$$\iiint_D \frac{\partial}{\partial t}\,(\delta \mathbf{v})\, dV.$$

This change is due partly to momentum crossing $\mathcal{S}$ into or out of D (the momentum of the fluid crossing $\mathcal{S}$), partly to the pressure exerted on the fluid in D by the fluid outside, and partly to any external *body forces* (such as gravity or electromagnetic forces) acting on the fluid. Let us examine each of these causes in turn.

Momentum is transferred across $\mathcal{S}$ into D at the rate

$$-\oiint_{\mathcal{S}} \mathbf{v}(\delta \mathbf{v} \bullet \hat{\mathbf{N}})\, dS.$$

The pressure on the fluid in D is exerted across $\mathcal{S}$ in the direction of the inward normal $-\hat{\mathbf{N}}$. Thus, this part of the force on the fluid in D is

$$-\oiint_{\mathcal{S}} p\hat{\mathbf{N}}\, dS.$$

The body forces are best expressed in terms of the *force density* (force per unit mass), **F**. The total body force on the fluid in D is therefore

$$\iiint_D \delta\mathbf{F}\,dV.$$

Newton's Second Law now implies that

$$\iiint_D \frac{\partial}{\partial t}(\delta\mathbf{v})\,dV = -\oiint_{\mathcal{S}} \mathbf{v}(\delta\mathbf{v}\bullet\hat{\mathbf{N}})\,dS - \oiint_{\mathcal{S}} p\hat{\mathbf{N}}\,dS + \iiint_D \delta\mathbf{F}\,dV.$$

Again, we would like to convert the surface integrals to triple integrals over D. If we use the results of Exercise 29 of Section 16.4 and Exercise 2 below, we get

$$\oiint_{\mathcal{S}} p\hat{\mathbf{N}}\,dS = \iiint_D \nabla p\,dV,$$

$$\oiint_{\mathcal{S}} \mathbf{v}(\delta\mathbf{v}\bullet\hat{\mathbf{N}})\,dS = \iiint_D \Big(\delta(\mathbf{v}\bullet\nabla)\mathbf{v} + \mathbf{v}\,\mathbf{div}\,(\delta\mathbf{v})\Big)\,dV.$$

Accordingly, we have

$$\iiint_D \left(\delta\frac{\partial\mathbf{v}}{\partial t} + \mathbf{v}\frac{\partial\delta}{\partial t} + \mathbf{v}\,\mathbf{div}\,(\delta\mathbf{v}) + \delta(\mathbf{v}\bullet\nabla)\mathbf{v} + \nabla p - \delta\mathbf{F}\right)\,dV = \mathbf{0}.$$

The second and third terms in the integrand cancel out by virtue of the continuity equation. Since D is arbitrary, we must therefore have

$$\delta\frac{\partial\mathbf{v}}{\partial t} + \delta(\mathbf{v}\bullet\nabla)\mathbf{v} = -\nabla p + \delta\mathbf{F}.$$

This is the **equation of motion** of the fluid. Observe that it is not a *linear* partial differential equation; the second term on the left is not linear in **v**.

Electromagnetism

In 3-space there are defined two vector fields that determine the electric and magnetic forces that would be experienced by a unit charge at a particular point if it is moving with unit speed. (These vector fields are determined by electric charges and currents present in the space.) A charge q_0 at position $\mathbf{r} = x\mathbf{i} + y\mathbf{j} + z\mathbf{k}$ moving with velocity $\mathbf{v}_0$ experiences an electric force $q_0\mathbf{E}(\mathbf{r})$, where **E** is the **electric field**, and a magnetic force $\mu_0 q_0 \mathbf{v}_0 \times \mathbf{H}(\mathbf{r})$, where **H** is the **magnetic field** and $\mu_0 \approx 1.26 \times 10^{-6}$ N/ampere^2 is a physical constant called the **permeability of free space**. We will look briefly at each of these fields but will initially restrict ourselves to considering *static* situations. Electric fields produced by static charge distributions and magnetic fields produced by static electric currents do not depend on time. Later we will consider the interaction between the two fields when they are time-dependent.

Electrostatics

Experimental evidence shows that the value of the electric field at any point **r** is the vector sum of the fields caused by any elements of charge located in 3-space. A "point charge" q at position $\mathbf{s} = \xi\mathbf{i} + \eta\mathbf{j} + \zeta\mathbf{k}$ generates the electric field

$$\mathbf{E}(\mathbf{r}) = \frac{q}{4\pi\epsilon_0}\frac{\mathbf{r}-\mathbf{s}}{|\mathbf{r}-\mathbf{s}|^3} \qquad \text{(Coulomb's Law)},$$

where $\epsilon_0 \approx 8.85 \times 10^{-12}$ coulombs^2/N·m^2 is a physical constant called the **permittivity of free space**. This is just the field due to a point source of strength $q/4\pi\epsilon_0$ at **s**. Except at $\mathbf{r} = \mathbf{s}$ the field is conservative, with potential

$$\phi(\mathbf{r}) = -\frac{q}{4\pi\epsilon_0}\frac{1}{|\mathbf{r}-\mathbf{s}|},$$

so for $\mathbf{r} \neq \mathbf{s}$ we have $\mathbf{curl\, E} = \mathbf{0}$. Also $\mathbf{div\, E} = 0$, except at $\mathbf{r} = \mathbf{s}$ where it is infinite; in terms of the Dirac distribution, $\mathbf{div\, E} = (q/\epsilon_0)\delta(x - \xi)\delta(y - \eta)\delta(z - \zeta)$. (See Section 16.1.) The flux of $\mathbf{E}$ outward across the surface $\mathcal{S}$ of any region R containing q is

$$\oiint_{\mathcal{S}} \mathbf{E} \bullet \hat{\mathbf{N}}\, dS = \frac{q}{\epsilon_0},$$

by analogy with Example 4 of Section 16.4.

Given a *charge distribution* of density $\rho(\xi, \eta, \zeta)$ in 3-space (so that the charge in volume element $dV = d\xi\, d\eta\, d\zeta$ at $\mathbf{s}$ is $dq = \rho\, dV$), the flux of $\mathbf{E}$ out of $\mathcal{S}$ due to the charge in R is

$$\oiint_{\mathcal{S}} \mathbf{E} \bullet \hat{\mathbf{N}}\, dS = \frac{1}{\epsilon_0} \iiint_R dq = \frac{1}{\epsilon_0} \iiint_R \rho\, dV.$$

If we apply the Divergence Theorem to the surface integral, we obtain

$$\iiint_R \left(\mathbf{div\, E} - \frac{\rho}{\epsilon_0} \right) dV = 0,$$

and since R is an arbitrary region,

$$\mathbf{div\, E} = \frac{\rho}{\epsilon_0}.$$

This is the differential form of Gauss's Law. See Exercise 3 below.

The potential due to a charge distribution of density $\rho(\mathbf{s})$ in the region R is

$$\begin{aligned}
\phi(\mathbf{r}) &= -\frac{1}{4\pi\epsilon_0} \iiint_R \frac{\rho(\mathbf{s})}{|\mathbf{r} - \mathbf{s}|}\, dV \\
&= -\frac{1}{4\pi\epsilon_0} \iiint_R \frac{\rho(\xi, \eta, \zeta)\, d\xi\, d\eta\, d\zeta}{\sqrt{(x - \xi)^2 + (y - \eta)^2 + (z - \zeta)^2}}.
\end{aligned}$$

If ρ is continuous and vanishes outside a bounded region, the triple integral is convergent everywhere (see Exercise 4 below), so $\mathbf{E} = \nabla\phi$ is conservative throughout 3-space. Thus, at all points,

$$\mathbf{curl\, E} = \mathbf{0}.$$

Since $\mathbf{div\, E} = \mathbf{div\,} \nabla\phi = \nabla^2\phi$, the potential ϕ satisfies **Poisson's equation**

$$\nabla^2\phi = \frac{\rho}{\epsilon_0}.$$

In particular, ϕ is harmonic in regions of space where no charge is distributed.

Magnetostatics

Magnetic fields are produced by moving charges, that is, by currents. Suppose that a constant electric current, I, is flowing in a filament along the curve $\mathcal{F}$. It has been determined experimentally that the magnetic fields produced at position $\mathbf{r} = x\mathbf{i} + y\mathbf{j} + z\mathbf{k}$ by the elements of current $dI = I\, ds$ along the filament add vectorially and that the element at position $\mathbf{s} = \xi\mathbf{i} + \eta\mathbf{j} + \zeta\mathbf{k}$ produces the field

$$d\mathbf{H}(\mathbf{r}) = \frac{I}{4\pi} \frac{d\mathbf{s} \times (\mathbf{r} - \mathbf{s})}{|\mathbf{r} - \mathbf{s}|^3} \qquad \text{(the Biot–Savart Law)},$$

where $d\mathbf{s} = \hat{\mathbf{T}} \, ds$, $\hat{\mathbf{T}}$ being the unit tangent to $\mathcal{F}$ in the direction of the current. Under the reasonable assumption that charge is not created or destroyed anywhere, the filament $\mathcal{F}$ must form a closed circuit, and the total magnetic field at $\mathbf{r}$ due to the current flowing in the circuit is

$$\mathbf{H} = \frac{I}{4\pi} \oint_{\mathcal{F}} \frac{d\mathbf{s} \times (\mathbf{r} - \mathbf{s})}{|\mathbf{r} - \mathbf{s}|^3}.$$

Let $\mathbf{A}$ be the vector field defined by

$$\mathbf{A}(\mathbf{r}) = \frac{I}{4\pi} \oint_{\mathcal{F}} \frac{d\mathbf{s}}{|\mathbf{r} - \mathbf{s}|},$$

for all $\mathbf{r}$ not on the filament $\mathcal{F}$. If we make use of the fact that

$$\nabla \left(\frac{1}{|\mathbf{r} - \mathbf{s}|} \right) = -\frac{\mathbf{r} - \mathbf{s}}{|\mathbf{r} - \mathbf{s}|^3},$$

and the vector identity $\nabla \times (\phi \mathbf{F}) = (\nabla \phi) \times \mathbf{F} + \phi (\nabla \times \mathbf{F})$ (with $\mathbf{F}$ the vector $d\mathbf{s}$, which does not depend on $\mathbf{r}$), we can calculate the curl of $\mathbf{A}$:

$$\nabla \times \mathbf{A} = \frac{I}{4\pi} \oint_{\mathcal{F}} \nabla \left(\frac{1}{|\mathbf{r} - \mathbf{s}|} \right) \times d\mathbf{s} = \frac{I}{4\pi} \oint_{\mathcal{F}} -\frac{\mathbf{r} - \mathbf{s}}{|\mathbf{r} - \mathbf{s}|^3} \times d\mathbf{s} = \mathbf{H}(\mathbf{r}).$$

Thus $\mathbf{A}$ is a vector potential for $\mathbf{H}$, and **div H** $= 0$ at points off the filament. We can also verify by calculation that **curl H** $= \mathbf{0}$ off the filament. (See Exercises 9–11 below.)

Imagine a circuit consisting of a straight filament along the z-axis with return at infinite distance. The field $\mathbf{H}$ at a finite point will then just be due to the current along the z-axis, where the current I is flowing in the direction of $\mathbf{k}$, say. The currents in all elements $d\mathbf{s}$ produce, at $\mathbf{r}$, fields in the same direction, normal to the plane containing $\mathbf{r}$ and the z-axis. (See Figure 16.18.)   Therefore, the field strength $H = |\mathbf{H}|$ at a distance a from the z-axis is obtained by integrating the elements

$$dH = \frac{I}{4\pi} \frac{\sin \theta \, d\zeta}{a^2 + (\zeta - z)^2} = \frac{I}{4\pi} \frac{a \, d\zeta}{\left(a^2 + (\zeta - z)^2 \right)^{3/2}}.$$

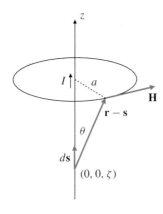

Figure 16.18

We have

$$
\begin{aligned}
H &= \frac{Ia}{4\pi} \int_{-\infty}^{\infty} \frac{d\zeta}{\left(a^2 + (\zeta - z)^2 \right)^{3/2}} \qquad (\text{Let } \zeta - z = a \tan \phi.) \\
&= \frac{I}{4\pi a} \int_{-\pi/2}^{\pi/2} \cos \phi \, d\phi = \frac{I}{2\pi a}.
\end{aligned}
$$

The field lines of $\mathbf{H}$ are evidently horizontal circles centred on the z-axis. If $\mathcal{C}_a$ is such a circle, having radius a, then the circulation of $\mathbf{H}$ around $\mathcal{C}_a$ is

$$\oint_{\mathcal{C}_a} \mathbf{H} \bullet d\mathbf{r} = \frac{I}{2\pi a} \, 2\pi a = I.$$

Observe that the circulation calculated above is independent of a. In fact, if $\mathcal{C}$ is any closed curve that encircles the z-axis once counterclockwise (as seen from above), then $\mathcal{C}$ and $-\mathcal{C}_a$ comprise the oriented boundary of a washer-like surface $\mathcal{S}$ with a hole in it through which the filament passes. Since **curl H** $= \mathbf{0}$ on $\mathcal{S}$, Stokes's Theorem guarantees that

$$\oint_{\mathcal{C}} \mathbf{H} \bullet d\mathbf{r} = \oint_{\mathcal{C}_a} \mathbf{H} \bullet d\mathbf{r} = I.$$

Furthermore, when $\mathcal{C}$ is very small (and therefore very close to the filament), most of the contribution to the circulation of **H** around it comes from the part of the filament that is very close to $\mathcal{C}$. It therefore does not matter whether the filament is straight or infinitely long. For any closed-loop filament carrying a current, the circulation of the magnetic field around the oriented boundary of a surface through which the filament passes is equal to the current flowing in the loop. This is **Ampère's Circuital Law**. The surface is oriented with normal on the side out of which the current is flowing.

Now let us replace the filament with a more general current specified by a vector density, **J**. This means that at any point **s** the current is flowing in the direction $\mathbf{J(s)}$ and that the current crossing an area element dS with unit normal $\hat{\mathbf{N}}$ is $\mathbf{J} \bullet \hat{\mathbf{N}} \, dS$. The circulation of **H** around the boundary $\mathcal{C}$ of surface $\mathcal{S}$ is equal to the total current flowing across $\mathcal{S}$, so

$$\oint_{\mathcal{C}} \mathbf{H} \bullet d\mathbf{r} = \iint_{\mathcal{S}} \mathbf{J} \bullet \hat{\mathbf{N}} \, dS.$$

By using Stokes's Theorem, we can replace the line integral with another surface integral and so obtain

$$\iint_{\mathcal{S}} (\operatorname{curl} \mathbf{H} - \mathbf{J}) \bullet \hat{\mathbf{N}} \, dS = 0.$$

Since $\mathcal{S}$ is arbitrary, we must have, at all points,

$$\operatorname{\mathbf{curl}} \mathbf{H} = \mathbf{J},$$

which is the pointwise version of Ampère's Circuital Law. It can be readily checked that, if

$$\mathbf{A(r)} = \frac{1}{4\pi} \iiint_{R} \frac{\mathbf{J(s)}}{|\mathbf{r} - \mathbf{s}|} \, dV,$$

then $\mathbf{H} = \operatorname{\mathbf{curl}} \mathbf{A}$ (so that **A** is a vector potential for the magnetic field **H**). Here, R is the region of 3-space where **J** is nonzero. If **J** is continuous and vanishes outside a bounded set, then the triple integral converges for all **r** (Exercise 4 below), and **H** is everywhere solenoidal:

$$\operatorname{\mathbf{div}} \mathbf{H} = 0.$$

Maxwell's Equations

The four equations obtained above for static electric and magnetic fields,

$$\operatorname{\mathbf{div}} \mathbf{E} = \rho/\epsilon_0 \qquad \operatorname{\mathbf{div}} \mathbf{H} = 0$$
$$\operatorname{\mathbf{curl}} \mathbf{E} = \mathbf{0} \qquad \operatorname{\mathbf{curl}} \mathbf{H} = \mathbf{J},$$

require some modification if the fields **E** and **H** depend on time. Gauss's Law $\operatorname{\mathbf{div}} \mathbf{E} = \rho/\epsilon_0$ remains valid, as does $\operatorname{\mathbf{div}} \mathbf{H} = 0$, which expresses the fact that there are no *known* magnetic *sources* or *sinks* (i.e., magnetic *monopoles*). The field lines of **H** must be closed curves.

It was observed by Michael Faraday that the circulation of an electric field around a simple closed curve $\mathcal{C}$ corresponds to a change in the magnetic flux

$$\Phi = \iint_{\mathcal{S}} \mathbf{H} \bullet \hat{\mathbf{N}} \, dS$$

through any oriented surface $\mathcal{S}$ having boundary $\mathcal{C}$, according to the formula

$$\frac{d\Phi}{dt} = -\frac{1}{\mu_0} \oint_{\mathcal{C}} \mathbf{E} \bullet d\mathbf{r}.$$

Applying Stokes's Theorem to the line integral, we obtain

$$\iint_{\mathcal{S}} \operatorname{curl} \mathbf{E} \bullet \hat{\mathbf{N}} \, dS = \oint_{\mathcal{C}} \mathbf{E} \bullet d\mathbf{r} = -\mu_0 \frac{d}{dt} \iint_{\mathcal{S}} \mathbf{H} \bullet \hat{\mathbf{N}} \, dS = -\mu_0 \iint_{\mathcal{S}} \frac{\partial \mathbf{H}}{\partial t} \bullet \hat{\mathbf{N}} \, dS.$$

Since $\mathcal{S}$ is arbitrary, we obtain the differential form of Faraday's Law:

$$\mathbf{curl}\,\mathbf{E} = -\mu_0 \frac{\partial \mathbf{H}}{\partial t}.$$

The electric field is irrotational only if the magnetic field is constant in time.

The differential form of Ampère's Law, $\mathbf{curl}\,\mathbf{H} = \mathbf{J}$, also requires modification. If the electric field depends on time, then so will the current density $\mathbf{J}$. Assuming conservation of charge (charges are not produced or destroyed), we can show, by an argument identical to that used to obtain the continuity equation for fluid motion earlier in this section, that the rate of change of charge density satisfies

$$\frac{\partial \rho}{\partial t} = -\mathbf{div}\,\mathbf{J}.$$

(See Exercise 5 below.) This is inconsistent with Ampère's Law because $\mathbf{div}\,\mathbf{curl}\,\mathbf{H} = 0$, while $\mathbf{div}\,\mathbf{J} \neq 0$ when ρ depends on time. Note, however, that $\rho = \epsilon_0 \mathbf{div}\,\mathbf{E}$ implies that

$$-\mathbf{div}\,\mathbf{J} = \frac{\partial \rho}{\partial t} = \epsilon_0 \mathbf{div}\,\frac{\partial \mathbf{E}}{\partial t},$$

so $\mathbf{div}\,(\mathbf{J} + \epsilon_0 \partial\mathbf{E}/\partial t) = 0$. This suggests that, for the nonstatic case, Ampère's Law becomes

$$\mathbf{curl}\,\mathbf{H} = \mathbf{J} + \epsilon_0 \frac{\partial \mathbf{E}}{\partial t},$$

which indicates (as was discovered by Maxwell) that magnetic fields are not just produced by currents, but also by changing electric fields.

Together, the four equations

$$\mathbf{div}\,\mathbf{E} = \rho/\epsilon_0 \qquad\qquad \mathbf{div}\,\mathbf{H} = 0$$

$$\mathbf{curl}\,\mathbf{E} = -\mu_0 \frac{\partial \mathbf{H}}{\partial t} \qquad \mathbf{curl}\,\mathbf{H} = \mathbf{J} + \epsilon_0 \frac{\partial \mathbf{E}}{\partial t}$$

are known as **Maxwell's equations**. They govern the way electric and magnetic fields are produced in 3-space by the presence of charges and currents.

Exercises 16.6

1. **(Archimedes' principle)** A solid occupying region R with surface $\mathcal{S}$ is immersed in a liquid of constant density δ. The pressure at depth h in the liquid is δgh, so the pressure satisfies $\nabla p = \delta \mathbf{g}$, where $\mathbf{g}$ is the (vector) constant acceleration of gravity. Over each surface element dS on $\mathcal{S}$ the pressure of the fluid exerts a force $-p\hat{\mathbf{N}}\,dS$ on the solid.

   (a) Show that the resultant "buoyancy force" on the solid is

   $$\mathbf{B} = -\iiint_R \delta\mathbf{g}\,dV.$$

   Thus, the buoyancy force has the same magnitude as, and opposite direction to, the weight of the liquid displaced by the solid. This is Archimedes' principle.

   (b) Extend the above result to the case where the solid is only partly submerged in the fluid.

2. By breaking the vector $\mathbf{F}(\mathbf{G} \bullet \hat{\mathbf{N}})$ into its separate components and applying the Divergence Theorem to each separately, show that

   $$\oiint_{\mathcal{S}} \mathbf{F}(\mathbf{G} \bullet \hat{\mathbf{N}})\,dS = \iiint_D \Big(\mathbf{F}\,\mathbf{div}\,\mathbf{G} + (\mathbf{G} \bullet \nabla)\mathbf{F}\Big)\,dV,$$

   where $\hat{\mathbf{N}}$ is the unit outward normal on the surface $\mathcal{S}$ of the domain D.

3. **(Gauss's Law)** Show that the flux of the electric field $\mathbf{E}$ outward through a closed surface $\mathcal{S}$ in 3-space is $1/\epsilon_0$ times the total charge enclosed by $\mathcal{S}$.

4. If $\mathbf{s} = \xi\mathbf{i} + \eta\mathbf{j} + \zeta\mathbf{k}$ and $f(\xi, \eta, \zeta)$ is continuous on $\mathbb{R}^3$ and vanishes outside a bounded region, show that, for any fixed $\mathbf{r}$,

   $$\iiint_{\mathbb{R}^3} \frac{|f(\xi, \eta, \zeta)|}{|\mathbf{r} - \mathbf{s}|}\,d\xi\,d\eta\,d\zeta \leq \text{constant}.$$

This shows that the potentials for the electric and magnetic fields corresponding to continuous charge and current densities that vanish outside bounded regions exist everywhere in $\mathbb{R}^3$. *Hint:* Without loss of generality you can assume $\mathbf{r} = \mathbf{0}$ and use spherical coordinates.

5. The electric charge density, ρ, in 3-space depends on time as well as position if charge is moving around. The motion is described by the current density, $\mathbf{J}$. Derive the **continuity equation**

$$\frac{\partial \rho}{\partial t} = -\mathbf{div}\,\mathbf{J}$$

from the fact that charge is conserved.

6. If $\mathbf{b}$ is a constant vector, show that

$$\nabla \left(\frac{1}{|\mathbf{r} - \mathbf{b}|} \right) = -\frac{\mathbf{r} - \mathbf{b}}{|\mathbf{r} - \mathbf{b}|^3}.$$

7. If $\mathbf{a}$ and $\mathbf{b}$ are constant vectors, show that for $\mathbf{r} \neq \mathbf{b}$,

$$\mathbf{div} \left(\mathbf{a} \times \frac{\mathbf{r} - \mathbf{b}}{|\mathbf{r} - \mathbf{b}|^3} \right) = 0.$$

Hint: Use identities (d) and (h) from Theorem 3 of Section 16.2.

8. Use the result of Exercise 7 to give an alternative proof that

$$\mathbf{div} \oint_{\mathcal{F}} \frac{d\mathbf{s} \times (\mathbf{r} - \mathbf{s})}{|\mathbf{r} - \mathbf{s}|^3} = 0.$$

Note that **div** refers to the $\mathbf{r}$ variable.

9. If $\mathbf{a}$ and $\mathbf{b}$ are constant vectors, show that for $\mathbf{r} \neq \mathbf{b}$,

$$\mathbf{curl} \left(\mathbf{a} \times \frac{\mathbf{r} - \mathbf{b}}{|\mathbf{r} - \mathbf{b}|^3} \right) = -(\mathbf{a} \bullet \nabla) \frac{\mathbf{r} - \mathbf{b}}{|\mathbf{r} - \mathbf{b}|^3}.$$

Hint: Use identity (e) from Theorem 3 of Section 16.2.

10. If $\mathbf{F}$ is any smooth vector field, show that

$$\oint_{\mathcal{F}} (d\mathbf{s} \bullet \nabla)\mathbf{F}(\mathbf{s}) = \mathbf{0}$$

around any closed loop $\mathcal{F}$. *Hint:* The gradients of the components of $\mathbf{F}$ are conservative.

11. Verify that if $\mathbf{r}$ does not lie on $\mathcal{F}$, then

$$\mathbf{curl} \oint_{\mathcal{F}} \frac{d\mathbf{s} \times (\mathbf{r} - \mathbf{s})}{|\mathbf{r} - \mathbf{s}|^3} = \mathbf{0}.$$

Here, **curl** is taken with respect to the $\mathbf{r}$ variable.

12. Verify the formula $\mathbf{curl}\,\mathbf{A} = \mathbf{H}$, where $\mathbf{A}$ is the magnetic vector potential defined in terms of the steady-state current density $\mathbf{J}$.

13. If $\mathbf{A}$ is the vector potential for the magnetic field produced by a steady current in a closed-loop filament, show that $\mathbf{div}\,\mathbf{A} = 0$ off the filament.

14. If $\mathbf{A}$ is the vector potential for the magnetic field produced by a steady, continuous current density, show that $\mathbf{div}\,\mathbf{A} = 0$ everywhere. Hence, show that $\mathbf{A}$ satisfies the vector Poisson equation $\nabla^2\mathbf{A} = -\mathbf{J}$.

15. Show that in a region of space containing no charges ($\rho = 0$) and no currents ($\mathbf{J} = \mathbf{0}$), both $\mathbf{U} = \mathbf{E}$ and $\mathbf{U} = \mathbf{H}$ satisfy the wave equation

$$\frac{\partial^2 \mathbf{U}}{\partial t^2} = c^2 \nabla^2 \mathbf{U},$$

where $c = \sqrt{1/(\epsilon_0 \mu_0)} \approx 3 \times 10^8$ m/s.

*16. **(Heat flow in 3-space)** The heat content of a volume element dV within a homogeneous solid is $\delta c T\, dV$, where δ and c are constants (the density and specific heat of the solid material), and $T = T(x, y, z, t)$ is the temperature at time t at position (x, y, z) in the solid. Heat always flows in the direction of the negative temperature gradient and at a rate proportional to the size of that gradient. Thus, the rate of flow of heat energy across a surface element dS with normal $\hat{\mathbf{N}}$ is $-k\nabla T \bullet \hat{\mathbf{N}}\, dS$, where k is also a constant depending on the material of the solid (the coefficient of thermal conductivity). Use "conservation of heat energy" to show that for any region R with surface $\mathcal{S}$ within the solid

$$\delta c \iiint_R \frac{\partial T}{\partial t}\, dV = k \oiint_{\mathcal{S}} \nabla T \bullet \hat{\mathbf{N}}\, dS,$$

where $\hat{\mathbf{N}}$ is the unit outward normal on $\mathcal{S}$. Hence, show that heat flow within the solid is governed by the partial differential equation

$$\frac{\partial T}{\partial t} = \frac{k}{\delta c} \nabla^2 T = \frac{k}{\delta c} \left(\frac{\partial^2 T}{\partial x^2} + \frac{\partial^2 T}{\partial y^2} + \frac{\partial^2 T}{\partial z^2} \right).$$

16.7 Orthogonal Curvilinear Coordinates

In this optional section we will derive formulas for the gradient of a scalar field and the divergence and curl of a vector field in terms of coordinate systems more general than the Cartesian coordinate system used in the earlier sections of this chapter. In particular, we will express these quantities in terms of the cylindrical and spherical coordinate systems introduced in Section 14.6.

We denote by xyz-space the usual system of Cartesian coordinates (x, y, z) in $\mathbb{R}^3$. A different system of coordinates $[u, v, w]$ in xyz-space can be defined by a continuous transformation of the form

$$x = x(u, v, w), \qquad y = y(u, v, w), \qquad z = z(u, v, w).$$

If the transformation is one-to-one from a region D in uvw-space onto a region R in xyz-space, then a point P in R can be represented by a triple $[u, v, w]$, the (Cartesian) coordinates of the unique point Q in uvw-space that the transformation maps to P. In this case we say that the transformation defines a **curvilinear coordinate system** in R and call $[u, v, w]$ the **curvilinear coordinates** of P with respect to that system. Note that $[u, v, w]$ are Cartesian coordinates in their own space (uvw-space); they are curvilinear coordinates in xyz-space.

Typically, we relax the requirement that the transformation defining a curvilinear coordinate system be one-to-one, that is, that every point P in R should have a unique set of curvilinear coordinates. It is reasonable to require the transformation to be only *locally one-to-one*. Thus, there may be more than one point Q that gets mapped to a point P by the transformation, but only one in any suitably small subregion of D. For example, in the plane polar coordinate system

$$x = r \cos\theta, \qquad y = r \sin\theta,$$

the transformation is locally one-to-one from D, the half of the $r\theta$-plane where $0 < r < \infty$, to the region R consisting of all points in the xy-plane except the origin. Although, say, $[1, 0]$ and $[1, 2\pi]$ are polar coordinates of the same point in the xy-plane, they are not close together in D. Observe, however, that there is still a problem with the origin, which can be represented by $[0, \theta]$ for *any* θ. Since the transformation is not even locally one-to-one at $r = 0$, we regard the origin of the xy-plane as a **singular point** for the polar coordinate system in the plane.

Example 1 The **cylindrical coordinate system** $[r, \theta, z]$ in $\mathbb{R}^3$ is defined by the transformation

$$x = r \cos\theta, \qquad y = r \sin\theta, \qquad z = z,$$

where $r \geq 0$. (See Section 14.6.) This transformation maps the half-space D given by $r > 0$ onto all of xyz-space excluding the z-axis, and it is locally one-to-one. We regard $[r, \theta, z]$ as cylindrical polar coordinates in all of xyz-space but call points on the z-axis singular points of the system since the points $[0, \theta, z]$ are identical for any θ.

Example 2 The **spherical coordinate system** $[\rho, \phi, \theta]$ is defined by the transformation

$$x = \rho \sin\phi \cos\theta, \qquad y = \rho \sin\phi \sin\theta, \qquad z = \rho \cos\phi,$$

where $\rho \geq 0$ and $0 \leq \phi \leq \pi$. (See Section 14.6.) The transformation maps the region D in $\rho\phi\theta$-space given by $\rho > 0, 0 < \phi < \pi$ in a locally one-to-one way onto xyz-space excluding the z-axis. The point with Cartesian coordinates $(0, 0, z)$ can be represented by the spherical coordinates $[0, \phi, \theta]$ for arbitrary ϕ and θ if $z = 0$, by $[z, 0, \theta]$ for arbitrary θ if $z > 0$, and by $[|z|, \pi, \theta]$ for arbitrary θ if $z < 0$. Thus, all points of the z-axis are singular for the spherical coordinate system.

Coordinate Surfaces and Coordinate Curves

Let $[u, v, w]$ be a curvilinear coordinate system in xyz-space, and let P_0 be a nonsingular point for the system. Thus, the transformation

$$x = x(u, v, w), \qquad y = y(u, v, w), \qquad z = z(u, v, w)$$

is locally one-to-one near P_0. Let P_0 have curvilinear coordinates $[u_0, v_0, w_0]$. The plane with equation $u = u_0$ in uvw-space gets mapped by the transformation to a surface in xyz-space passing through P_0. We call this surface a u-surface and still refer to it by the equation $u = u_0$; it has parametric equations

$$x = x(u_0, v, w), \qquad y = y(u_0, v, w), \qquad z = z(u_0, v, w)$$

with parameters v and w. Similarly, the v-surface $v = v_0$ and the w-surface $w = w_0$ pass through P_0; they are the images of the planes $v = v_0$ and $w = w_0$ in uvw-space.

Orthogonal curvilinear coordinates

We say that $[u, v, w]$ is an **orthogonal curvilinear coordinate system** in xyz-space if, for every nonsingular point P_0 in xyz-space, each of the three **coordinate surfaces** $u = u_0$, $v = v_0$, and $w = w_0$ intersects the other two at P_0 at right angles.

It is tacitly assumed that the coordinate surfaces are smooth at all nonsingular points, so we are really assuming that their normal vectors are mutually perpendicular. Figure 16.19 shows the coordinate surfaces through P_0 for a typical orthogonal curvilinear coordinate system.

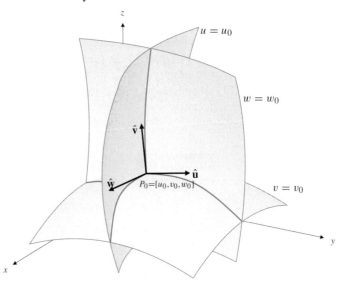

Figure 16.19 u-, v-, and w-coordinate surfaces

Pairs of coordinate surfaces through a point intersect along a **coordinate curve** through that point. For example, the coordinate surfaces $v = v_0$ and $w = w_0$ intersect along the u-**curve** with parametric equations

$$x = x(u, v_0, w_0), \quad y = y(u, v_0, w_0), \quad \text{and} \quad z = z(u, v_0, w_0),$$

where the parameter is u. A unit vector $\hat{\mathbf{u}}$ tangent to the u-curve through P_0 is normal to the coordinate surface $u = u_0$ there. Similar statements hold for unit vectors $\hat{\mathbf{v}}$ and $\hat{\mathbf{w}}$. For an orthogonal curvilinear coordinate system, the three vectors $\hat{\mathbf{u}}$, $\hat{\mathbf{v}}$, and $\hat{\mathbf{w}}$ form a basis of mutually perpendicular unit vectors at any nonsingular point P_0. (See Figure 16.19.) We call this basis the **local basis** at P_0.

Example 3   For the cylindrical coordinate system (see Figure 16.20), the coordinate surfaces are:

circular cylinders with axis along the z-axis	(r-surfaces),
vertical half-planes radiating from the z-axis	(θ-surfaces),
horizontal planes	(z-surfaces).

The coordinate curves are:

horizontal straight half-lines radiating from the z-axis	(r-curves),
horizontal circles with centres on the z-axis	(θ-curves),
vertical straight lines	(z-curves).

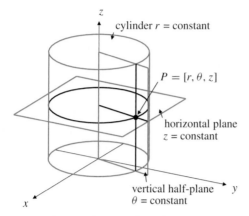

Figure 16.20 The coordinate surfaces for cylindrical coordinates

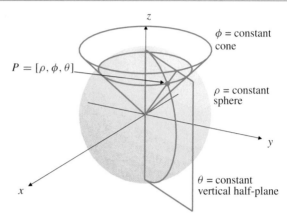

Figure 16.21 The coordinate surfaces for spherical coordinates

Example 4 For the spherical coordinate system (see Figure 16.21), the coordinate surfaces are:

spheres centred at the origin	(ρ-surfaces),
vertical circular cones with vertices at the origin	(ϕ-surfaces),
vertical half-planes radiating from the z-axis	(θ-surfaces).

The coordinate curves are:

half-lines radiating from the origin	(ρ-curves),
vertical semicircles with centres at the origin	(ϕ-curves),
horizontal circles with centres on the z-axis	(θ-curves).

Scale Factors and Differential Elements

For the rest of this section we assume that $[u, v, w]$ are **orthogonal** curvilinear coordinates in xyz-space defined via the transformation

$$x = x(u, v, w), \qquad y = y(u, v, w), \qquad z = z(u, v, w).$$

We also assume that the coordinate surfaces are smooth at any nonsingular point and that the local basis vectors $\hat{\mathbf{u}}$, $\hat{\mathbf{v}}$, and $\hat{\mathbf{w}}$ at any such point form a right-handed triad. This is the case for both cylindrical and spherical coordinates. For spherical coordinates, this is the reason we chose the order of the coordinates as $[\rho, \phi, \theta]$, rather than $[\rho, \theta, \phi]$.

The **position vector** of a point P in xyz-space can be expressed in terms of the curvilinear coordinates:

$$\mathbf{r} = x(u, v, w)\mathbf{i} + y(u, v, w)\mathbf{j} + z(u, v, w)\mathbf{k}.$$

If we hold $v = v_0$ and $w = w_0$ fixed and let u vary, then $\mathbf{r} = \mathbf{r}(u, v_0, w_0)$ defines a u-curve in xyz-space. At any point P on this curve, the vector

$$\frac{\partial \mathbf{r}}{\partial u} = \frac{\partial x}{\partial u}\mathbf{i} + \frac{\partial y}{\partial u}\mathbf{j} + \frac{\partial z}{\partial u}\mathbf{k}$$

is tangent to the u-curve at P. In general, the three vectors

$$\frac{\partial \mathbf{r}}{\partial u}, \qquad \frac{\partial \mathbf{r}}{\partial v}, \qquad \text{and} \qquad \frac{\partial \mathbf{r}}{\partial w}$$

are tangent, respectively, to the u-curve, the v-curve, and the w-curve through P. They are also normal, respectively, to the u-surface, the v-surface, and the w-surface through P, so they are mutually perpendicular. (See Figure 16.19.) The lengths of these tangent vectors are called the *scale factors* of the coordinate system.

> The **scale factors** of the orthogonal curvilinear coordinate system $[u, v, w]$ are the three functions
>
> $$h_u = \left| \frac{\partial \mathbf{r}}{\partial u} \right|, \qquad h_v = \left| \frac{\partial \mathbf{r}}{\partial v} \right|, \qquad h_w = \left| \frac{\partial \mathbf{r}}{\partial w} \right|.$$

The scale factors are nonzero at a nonsingular point P of the coordinate system, so the local basis at P can be obtained by dividing the tangent vectors to the coordinate curves by their lengths. As noted previously, we denote the local basis vectors by $\hat{\mathbf{u}}$, $\hat{\mathbf{v}}$, and $\hat{\mathbf{w}}$. Thus,

$$\frac{\partial \mathbf{r}}{\partial u} = h_u \hat{\mathbf{u}}, \qquad \frac{\partial \mathbf{r}}{\partial v} = h_v \hat{\mathbf{v}}, \qquad \text{and} \qquad \frac{\partial \mathbf{r}}{\partial w} = h_w \hat{\mathbf{w}}.$$

The basis vectors $\hat{\mathbf{u}}$, $\hat{\mathbf{v}}$, and $\hat{\mathbf{w}}$ will form a right-handed triad provided we have chosen a suitable order for the coordinates u, v, and w.

Example 5 For cylindrical coordinates we have $\mathbf{r} = r \cos\theta \, \mathbf{i} + r \sin\theta \, \mathbf{j} + z \mathbf{k}$, so

$$\frac{\partial \mathbf{r}}{\partial r} = \cos\theta \, \mathbf{i} + \sin\theta \, \mathbf{j}, \qquad \frac{\partial \mathbf{r}}{\partial \theta} = -r \sin\theta \, \mathbf{i} + r \cos\theta \, \mathbf{j}, \qquad \text{and} \qquad \frac{\partial \mathbf{r}}{\partial z} = \mathbf{k}.$$

Thus, the scale factors for the cylindrical coordinate system are given by

$$h_r = \left| \frac{\partial \mathbf{r}}{\partial r} \right| = 1, \qquad h_\theta = \left| \frac{\partial \mathbf{r}}{\partial \theta} \right| = r, \qquad \text{and} \qquad h_z = \left| \frac{\partial \mathbf{r}}{\partial z} \right| = 1,$$

and the local basis consists of the vectors

$$\hat{\mathbf{r}} = \cos\theta \, \mathbf{i} + \sin\theta \, \mathbf{j}, \qquad \hat{\boldsymbol{\theta}} = -\sin\theta \, \mathbf{i} + \cos\theta \, \mathbf{j}, \qquad \hat{\mathbf{z}} = \mathbf{k}.$$

See Figure 16.22. The local basis is right-handed. ∎

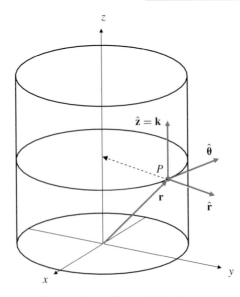

Figure 16.22 The local basis for cylindrical coordinates

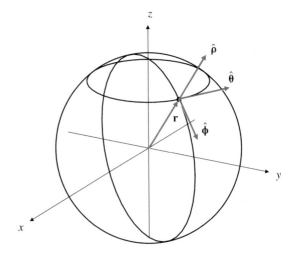

Figure 16.23 The local basis for spherical coordinates

Example 6   For spherical coordinates we have

$$\mathbf{r} = \rho \sin\phi \cos\theta \, \mathbf{i} + \rho \sin\phi \sin\theta \, \mathbf{j} + \rho \cos\phi \, \mathbf{k}.$$

Thus, the tangent vectors to the coordinate curves are

$$\frac{\partial \mathbf{r}}{\partial \rho} = \sin\phi \cos\theta \, \mathbf{i} + \sin\phi \sin\theta \, \mathbf{j} + \cos\phi \, \mathbf{k},$$

$$\frac{\partial \mathbf{r}}{\partial \phi} = \rho \cos\phi \cos\theta \, \mathbf{i} + \rho \cos\phi \sin\theta \, \mathbf{j} - \rho \sin\phi \, \mathbf{k},$$

$$\frac{\partial \mathbf{r}}{\partial \theta} = -\rho \sin\phi \sin\theta \, \mathbf{i} + \rho \sin\phi \cos\theta \, \mathbf{j},$$

and the scale factors are given by

$$h_\rho = \left| \frac{\partial \mathbf{r}}{\partial \rho} \right| = 1, \quad h_\phi = \left| \frac{\partial \mathbf{r}}{\partial \phi} \right| = \rho, \quad \text{and} \quad h_\theta = \left| \frac{\partial \mathbf{r}}{\partial \theta} \right| = \rho \sin\phi.$$

The local basis consists of the vectors

$$\hat{\boldsymbol{\rho}} = \sin\phi \cos\theta \, \mathbf{i} + \sin\phi \sin\theta \, \mathbf{j} + \cos\phi \, \mathbf{k}$$

$$\hat{\boldsymbol{\phi}} = \cos\phi \cos\theta \, \mathbf{i} + \cos\phi \sin\theta \, \mathbf{j} - \sin\phi \, \mathbf{k}$$

$$\hat{\boldsymbol{\theta}} = -\sin\theta \, \mathbf{i} + \cos\theta \, \mathbf{j}.$$

See Figure 16.23. The local basis is right-handed.

The volume element in an orthogonal curvilinear coordinate system is the volume of an infinitesimal *coordinate box* bounded by pairs of u-, v-, and w-surfaces corresponding to values u and $u+du$, v and $v+dv$, and w and $w+dw$, respectively. See Figure 16.24. Since these coordinate surfaces are assumed smooth, and since they intersect at right angles, the coordinate box is rectangular, and is spanned by the vectors

$$\frac{\partial \mathbf{r}}{\partial u} \, du = h_u \, du \, \hat{\mathbf{u}}, \quad \frac{\partial \mathbf{r}}{\partial v} \, dv = h_v \, dv \, \hat{\mathbf{v}}, \quad \text{and} \quad \frac{\partial \mathbf{r}}{\partial w} \, dw = h_w \, dw \, \hat{\mathbf{w}}.$$

Therefore, the volume element is given by

$$dV = h_u h_v h_w \, du \, dv \, dw.$$

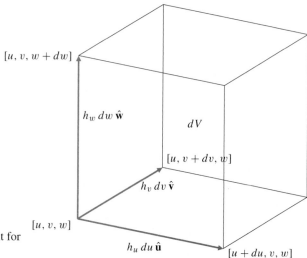

Figure 16.24   The volume element for orthogonal curvilinear coordinates

Furthermore, the surface area elements on the u-, v-, and w-surfaces are the areas of the appropriate faces of the coordinate box:

Area elements on coordinate surfaces

$$dS_u = h_v h_w \, dv \, dw, \qquad dS_v = h_u h_w \, du \, dw, \qquad dS_w = h_u h_v \, du \, dv.$$

The arc length elements along the u-, v-, and w-coordinate curves are the edges of the coordinate box:

Arc length elements on coordinate curves

$$ds_u = h_u \, du, \qquad ds_v = h_v \, dv, \qquad ds_w = h_w \, dw.$$

Example 7   For cylindrical coordinates, the volume element, as shown in Section 14.6, is

$$dV = h_r h_\theta h_z \, dr \, d\theta \, dz = r \, dr \, d\theta \, dz.$$

The surface area elements on the cylinder r = constant, the half-plane θ = constant, and the plane z = constant are, respectively,

$$dS_r = r \, d\theta \, dz, \qquad dS_\theta = dr \, dz, \qquad \text{and} \qquad dS_z = r \, dr \, d\theta.$$

■

Example 8   For spherical coordinates, the volume element, as developed in Section 14.6, is

$$dV = h_\rho h_\phi h_\theta \, d\rho \, d\phi \, d\theta = \rho^2 \sin \phi \, d\rho \, d\phi \, d\theta.$$

The area element on the sphere ρ = constant is

$$dS_\rho = h_\phi h_\theta \, d\phi \, d\theta = \rho^2 \sin \phi \, d\phi \, d\theta.$$

The area element on the cone ϕ = constant is

$$dS_\phi = h_\rho h_\theta \, d\rho \, d\theta = \rho \sin \phi \, d\rho \, d\theta.$$

The area element on the half-plane θ = constant is

$$dS_\theta = h_\rho h_\phi \, d\rho \, d\phi = \rho \, d\rho \, d\phi.$$

■

Grad, Div, and Curl in Orthogonal Curvilinear Coordinates

The gradient ∇f of a scalar field f can be expressed in terms of the local basis at any point P with curvilinear coordinates $[u, v, w]$ in the form

$$\nabla f = f_u \hat{\mathbf{u}} + f_v \hat{\mathbf{v}} + f_w \hat{\mathbf{w}}.$$

In order to determine the coefficients f_u, f_v, and f_w in this formula, we will compare two expressions for the directional derivative of f along an arbitrary curve in xyz-space.

If the curve $\mathcal{C}$ has parametrization $\mathbf{r} = \mathbf{r}(s)$ in terms of arc length, then the directional derivative of f along $\mathcal{C}$ is given by

$$\frac{df}{ds} = \frac{\partial f}{\partial u} \frac{du}{ds} + \frac{\partial f}{\partial v} \frac{dv}{ds} + \frac{\partial f}{\partial w} \frac{dw}{ds}.$$

On the other hand, this directional derivative is also given by $df/ds = \nabla f \bullet \hat{\mathbf{T}}$, where $\hat{\mathbf{T}}$ is the unit tangent vector to $\mathcal{C}$. We have

$$\hat{\mathbf{T}} = \frac{d\mathbf{r}}{ds} = \frac{\partial \mathbf{r}}{\partial u}\frac{du}{ds} + \frac{\partial \mathbf{r}}{\partial v}\frac{dv}{ds} + \frac{\partial \mathbf{r}}{\partial w}\frac{dw}{ds}$$
$$= h_u \frac{du}{ds}\,\hat{\mathbf{u}} + h_v \frac{dv}{ds}\,\hat{\mathbf{v}} + h_w \frac{dw}{ds}\,\hat{\mathbf{w}}.$$

Thus,

$$\frac{df}{ds} = \nabla f \bullet \hat{\mathbf{T}} = f_u h_u \frac{du}{ds} + f_v h_v \frac{dv}{ds} + f_w h_w \frac{dw}{ds}.$$

Comparing these two expressions for df/ds along $\mathcal{C}$, we see that

$$f_u h_u = \frac{\partial f}{\partial u}, \qquad f_v h_v = \frac{\partial f}{\partial v}, \qquad f_w h_w = \frac{\partial f}{\partial w}.$$

Therefore, we have shown that

The gradient in orthogonal curvilinear coordinates

$$\nabla f = \frac{1}{h_u}\frac{\partial f}{\partial u}\,\hat{\mathbf{u}} + \frac{1}{h_v}\frac{\partial f}{\partial v}\,\hat{\mathbf{v}} + \frac{1}{h_w}\frac{\partial f}{\partial w}\,\hat{\mathbf{w}}.$$

Example 9   In terms of cylindrical coordinates, the gradient of the scalar field $f(r, \theta, z)$ is

$$\nabla f(r, \theta, z) = \frac{\partial f}{\partial r}\,\hat{\mathbf{r}} + \frac{1}{r}\frac{\partial f}{\partial \theta}\,\hat{\boldsymbol{\theta}} + \frac{\partial f}{\partial z}\,\mathbf{k}.$$

Example 10   In terms of spherical coordinates, the gradient of the scalar field $f(\rho, \phi, \theta)$ is

$$\nabla f(\rho, \phi, \theta) = \frac{\partial f}{\partial \rho}\,\hat{\boldsymbol{\rho}} + \frac{1}{\rho}\frac{\partial f}{\partial \phi}\,\hat{\boldsymbol{\phi}} + \frac{1}{\rho \sin\phi}\frac{\partial f}{\partial \theta}\,\hat{\boldsymbol{\theta}}.$$

Now consider a vector field $\mathbf{F}$ expressed in terms of the curvilinear coordinates:

$$\mathbf{F}(u, v, w) = F_u(u, v, w)\hat{\mathbf{u}} + F_v(u, v, w)\hat{\mathbf{v}} + F_w(u, v, w)\hat{\mathbf{w}}.$$

The flux of $\mathbf{F}$ out of the infinitesimal coordinate box of Figure 16.24 is the sum of the fluxes of $\mathbf{F}$ out of the three pairs of opposite surfaces of the box. The flux out of the u-surfaces corresponding to u and $u + du$ is

$$\mathbf{F}(u + du, v, w) \bullet \hat{\mathbf{u}}\, dS_u - \mathbf{F}(u, v, w) \bullet \hat{\mathbf{u}}\, dS_u$$
$$= \big(F_u(u + du, v, w)h_v(u + du, v, w)h_w(u + du, v, w)$$
$$\quad - F_u(u, v, w)h_v(u, v, w)h_w(u, v, w) \big)\, dv\, dw$$
$$= \frac{\partial}{\partial u}\big(h_v h_w F_u \big)\, du\, dv\, dw.$$

Similar expressions hold for the fluxes out of the other pairs of coordinate surfaces.

The divergence at P of $\mathbf{F}$ is the flux *per unit volume* out of the infinitesimal coordinate box at P. Thus it is given by

The divergence in orthogonal curvilinear coordinates

$$\mathbf{div}\,\mathbf{F}(u, v, w) = \frac{1}{h_u h_v h_w}\left[\frac{\partial}{\partial u}\left(h_v h_w F_u(u, v, w)\right)\right.$$
$$\left. + \frac{\partial}{\partial v}\left(h_u h_w F_v(u, v, w)\right) + \frac{\partial}{\partial w}\left(h_u h_v F_w(u, v, w)\right)\right].$$

Example 11 For cylindrical coordinates, $h_r = h_z = 1$, and $h_\theta = r$. Thus, the divergence of $\mathbf{F} = F_r \hat{\mathbf{r}} + F_\theta \hat{\boldsymbol{\theta}} + F_z \mathbf{k}$ is

$$\mathbf{div}\,\mathbf{F} = \frac{1}{r}\left[\frac{\partial}{\partial r}\left(r F_r\right) + \frac{\partial}{\partial \theta} F_\theta + \frac{\partial}{\partial z}\left(r F_z\right)\right]$$
$$= \frac{\partial F_r}{\partial r} + \frac{1}{r} F_r + \frac{1}{r}\frac{\partial F_\theta}{\partial \theta} + \frac{\partial F_z}{\partial z}.$$

Example 12 For spherical coordinates, $h_\rho = 1$, $h_\phi = \rho$, and $h_\theta = \rho\sin\phi$. The divergence of the vector field $\mathbf{F} = F_\rho \hat{\boldsymbol{\rho}} + F_\phi \hat{\boldsymbol{\phi}} + F_\theta \hat{\boldsymbol{\theta}}$ is

$$\mathbf{div}\,\mathbf{F} = \frac{1}{\rho^2 \sin\phi}\left[\frac{\partial}{\partial \rho}\left(\rho^2 \sin\phi\, F_\rho\right) + \frac{\partial}{\partial \phi}\left(\rho\sin\phi\, F_\phi\right) + \frac{\partial}{\partial \theta}\left(\rho\, F_\theta\right)\right]$$
$$= \frac{1}{\rho^2}\frac{\partial}{\partial \rho}\left(\rho^2 F_\rho\right) + \frac{1}{\rho\sin\phi}\frac{\partial}{\partial \phi}\left(\sin\phi\, F_\phi\right) + \frac{1}{\rho\sin\phi}\frac{\partial F_\theta}{\partial \theta}$$
$$= \frac{\partial F_\rho}{\partial \rho} + \frac{2}{\rho} F_\rho + \frac{1}{\rho}\frac{\partial F_\phi}{\partial \phi} + \frac{\cot\phi}{\rho} F_\phi + \frac{1}{\rho\sin\phi}\frac{\partial F_\theta}{\partial \theta}.$$

To calculate the curl of a vector field expressed in terms of orthogonal curvilinear coordinates we can make use of some previously obtained vector identities. First, observe that the gradient of the scalar field $f(u, v, w) = u$ is $\hat{\mathbf{u}}/h_u$, so that $\hat{\mathbf{u}} = h_u \nabla u$. Similarly, $\hat{\mathbf{v}} = h_v \nabla v$ and $\hat{\mathbf{w}} = h_w \nabla w$. Therefore, the vector field

$$\mathbf{F} = F_u \hat{\mathbf{u}} + F_v \hat{\mathbf{v}} + F_w \hat{\mathbf{w}}$$

can be written in the form

$$\mathbf{F} = F_u h_u \nabla u + F_v h_v \nabla v + F_w h_w \nabla w.$$

Using the identity $\mathbf{curl}\,(f\nabla g) = \nabla f \times \nabla g$ (see Exercise 13 of Section 16.2), we can calculate the curl of each term in the expression above. We have

$$\mathbf{curl}\,\left(F_u h_u \nabla u\right) = \nabla(F_u h_u) \times \nabla u$$
$$= \left[\frac{1}{h_u}\frac{\partial}{\partial u}(F_u h_u)\hat{\mathbf{u}} + \frac{1}{h_v}\frac{\partial}{\partial v}(F_u h_u)\hat{\mathbf{v}} + \frac{1}{h_w}\frac{\partial}{\partial w}(F_u h_u)\hat{\mathbf{w}}\right] \times \frac{\hat{\mathbf{u}}}{h_u}$$
$$= \frac{1}{h_u h_w}\frac{\partial}{\partial w}(F_u h_u)\hat{\mathbf{v}} - \frac{1}{h_u h_v}\frac{\partial}{\partial v}(F_u h_u)\hat{\mathbf{w}}$$
$$= \frac{1}{h_u h_v h_w}\left[\frac{\partial}{\partial w}(F_u h_u)(h_v \hat{\mathbf{v}}) - \frac{\partial}{\partial v}(F_u h_u)(h_w \hat{\mathbf{w}})\right].$$

We have used the facts that $\hat{\mathbf{u}} \times \hat{\mathbf{u}} = \mathbf{0}$, $\hat{\mathbf{v}} \times \hat{\mathbf{u}} = -\hat{\mathbf{w}}$, and $\hat{\mathbf{w}} \times \hat{\mathbf{u}} = \hat{\mathbf{v}}$ to obtain the result above. This is why we assumed that the curvilinear coordinate system was right-handed.

Corresponding expressions can be calculated for the other two terms in the formula for **curl F**. Combining the three terms, we conclude that the curl of

$$\mathbf{F} = F_u \hat{\mathbf{u}} + F_v \hat{\mathbf{v}} + F_w \hat{\mathbf{w}}$$

is given by

The curl in orthogonal curvilinear coordinates

$$\mathbf{curl\,F}(u, v, w) = \frac{1}{h_u h_v h_w} \begin{vmatrix} h_u \hat{\mathbf{u}} & h_v \hat{\mathbf{v}} & h_w \hat{\mathbf{w}} \\ \dfrac{\partial}{\partial u} & \dfrac{\partial}{\partial v} & \dfrac{\partial}{\partial w} \\ F_u h_u & F_v h_v & F_w h_w \end{vmatrix}.$$

Example 13   For cylindrical coordinates, the curl of $\mathbf{F} = F_r \hat{\mathbf{r}} + F_\theta \hat{\boldsymbol{\theta}} + F_z \mathbf{k}$ is given by

$$\mathbf{curl\,F} = \frac{1}{r} \begin{vmatrix} \hat{\mathbf{r}} & r\hat{\boldsymbol{\theta}} & \mathbf{k} \\ \dfrac{\partial}{\partial r} & \dfrac{\partial}{\partial \theta} & \dfrac{\partial}{\partial z} \\ F_r & r F_\theta & F_z \end{vmatrix}$$

$$= \left(\frac{1}{r} \frac{\partial F_z}{\partial \theta} - \frac{\partial F_\theta}{\partial z} \right) \hat{\mathbf{r}} + \left(\frac{\partial F_r}{\partial z} - \frac{\partial F_z}{\partial r} \right) \hat{\boldsymbol{\theta}} + \left(\frac{\partial F_\theta}{\partial r} + \frac{F_\theta}{r} - \frac{1}{r} \frac{\partial F_r}{\partial \theta} \right) \mathbf{k}.$$

Example 14   For spherical coordinates, the curl of $\mathbf{F} = F_\rho \hat{\boldsymbol{\rho}} + F_\phi \hat{\boldsymbol{\phi}} + F_\theta \hat{\boldsymbol{\theta}}$ is given by

$$\mathbf{curl\,F} = \frac{1}{\rho^2 \sin \phi} \begin{vmatrix} \hat{\boldsymbol{\rho}} & \rho\hat{\boldsymbol{\phi}} & \rho \sin \phi \, \hat{\boldsymbol{\theta}} \\ \dfrac{\partial}{\partial \rho} & \dfrac{\partial}{\partial \phi} & \dfrac{\partial}{\partial \theta} \\ F_\rho & \rho F_\phi & \rho \sin \phi \, F_\theta \end{vmatrix}$$

$$= \frac{1}{\rho \sin \phi} \left[\frac{\partial}{\partial \phi} (\sin \phi \, F_\theta) - \frac{\partial F_\phi}{\partial \theta} \right] \hat{\boldsymbol{\rho}}$$

$$+ \frac{1}{\rho \sin \phi} \left[\frac{\partial F_\rho}{\partial \theta} - \sin \phi \frac{\partial}{\partial \rho} (\rho F_\theta) \right] \hat{\boldsymbol{\phi}}$$

$$+ \frac{1}{\rho} \left[\frac{\partial}{\partial \rho} (\rho F_\phi) - \frac{\partial F_\rho}{\partial \phi} \right] \hat{\boldsymbol{\theta}}$$

$$= \frac{1}{\rho \sin \phi} \left[(\cos \phi) F_\theta + (\sin \phi) \frac{\partial F_\theta}{\partial \phi} - \frac{\partial F_\phi}{\partial \theta} \right] \hat{\boldsymbol{\rho}}$$

$$+ \frac{1}{\rho \sin \phi} \left[\frac{\partial F_\rho}{\partial \theta} - (\sin \phi) F_\theta - (\rho \sin \phi) \frac{\partial F_\theta}{\partial \rho} \right] \hat{\boldsymbol{\phi}}$$

$$+ \frac{1}{\rho} \left[F_\phi + \rho \frac{\partial F_\phi}{\partial \rho} - \frac{\partial F_\rho}{\partial \phi} \right] \hat{\boldsymbol{\theta}}.$$

▮▮▮▮▮▮▮ Exercises 16.7 ▮▮▮▮▮▮▮▮

In Exercises 1–2, calculate the gradients of the given scalar fields expressed in terms of cylindrical or spherical coordinates.

1. $f(r, \theta, z) = r\theta z$ **2.** $f(\rho, \phi, \theta) = \rho\phi\theta$

In Exercises 3–8, calculate **div F** and **curl F** for the given vector fields expressed in terms of cylindrical coordinates or spherical coordinates.

3. $\mathbf{F}(r, \theta, z) = r\hat{\mathbf{r}}$ **4.** $\mathbf{F}(r, \theta, z) = r\hat{\boldsymbol{\theta}}$

5. $\mathbf{F}(\rho, \phi, \theta) = \sin\phi\,\hat{\boldsymbol{\rho}}$ **6.** $\mathbf{F}(\rho, \phi, \theta) = \rho\hat{\boldsymbol{\phi}}$

7. $\mathbf{F}(\rho, \phi, \theta) = \rho\hat{\boldsymbol{\theta}}$ **8.** $\mathbf{F}(\rho, \phi, \theta) = \rho^2\hat{\boldsymbol{\rho}}$

9. Let $x = x(u, v)$, $y = y(u, v)$ define orthogonal curvilinear coordinates (u, v) in the xy-plane. Find the scale factors, local basis vectors, and area element for the system of coordinates (u, v).

10. Continuing Exercise 9, express the gradient of a scalar field $f(u, v)$ and the divergence and curl of a vector field $\mathbf{F}(u, v)$ in terms of the curvilinear coordinates.

11. Express the gradient of the scalar field $f(r, \theta)$ and the divergence and curl of a vector field $\mathbf{F}(r, \theta)$ in terms of plane polar coordinates (r, θ).

12. The transformation

$$x = a\cosh u \cos v, \qquad y = a\sinh u \sin v$$

defines **elliptical coordinates** in the xy-plane. This coordinate system has singular points at $x = \pm a$, $y = 0$.

(a) Show that the v-curves, $u = $ constant, are ellipses with foci at the singular points.

(b) Show that the u-curves, $v = $ constant, are hyperbolas with foci at the singular points.

(c) Show that the u-curve and the v-curve through a nonsingular point intersect at right angles.

(d) Find the scale factors h_u and h_v and the area element dA for the elliptical coordinate system.

13. Describe the coordinate surfaces and coordinate curves of the system of elliptical cylindrical coordinates in xyz-space defined by

$$x = a\cosh u \cos v, \qquad y = a\sinh u \sin v, \qquad z = z.$$

14. The Laplacian $\nabla^2 f$ of a scalar field f can be calculated as **div** ∇f. Use this method to calculate the Laplacian of the function $f(r, \theta, z)$ expressed in terms of cylindrical coordinates. (This repeats Exercise 33 of Section 14.6.)

15. Calculate the Laplacian $\nabla^2 f = $ **div** ∇f for the function $f(\rho, \phi, \theta)$, expressed in terms of spherical coordinates. (This repeats Exercise 34 of Section 14.6 but is now much easier.)

16. Calculate the Laplacian $\nabla^2 f = $ **div** ∇f for a function $f(u, v, w)$ expressed in terms of arbitrary orthogonal curvilinear coordinates (u, v, w).

Chapter Review

Key Ideas

• **What do the following terms mean?**

◇ the divergence of a vector field **F**

◇ the curl of a vector field **F**

◇ **F** is solenoidal

◇ **F** is irrotational

◇ a scalar potential

◇ a vector potential

◇ orthogonal curvilinear coordinates

• **State the following theorems:**

◇ the Divergence Theorem

◇ Green's Theorem

◇ Stokes's Theorem

Review Exercises

1. If $\mathbf{F} = x^2 z\mathbf{i} + (y^2 z + 3y)\mathbf{j} + x^2\mathbf{k}$, find the flux of **F** across the part of the ellipsoid $x^2 + y^2 + 4z^2 = 16$, where $z \geq 0$, oriented with upward normal.

2. Let $\mathcal{S}$ be the part of the cylinder $x^2 + y^2 = 2ax$ between the horizontal planes $z = 0$ and $z = b$, where $b > 0$. Find the

flux of $\mathbf{F} = x\mathbf{i} + \cos(z^2)\mathbf{j} + e^z\mathbf{k}$ outward through $\mathcal{S}$.

3. Find $\oint_{\mathcal{C}} (3y^2 + 2xe^{y^2})dx + (2x^2 ye^{y^2})dy$ counterclockwise around the boundary of the parallelogram with vertices $(0, 0)$, $(2, 0)$, $(3, 1)$, and $(1, 1)$.

4. If $\mathbf{F} = -z\mathbf{i} + x\mathbf{j} + y\mathbf{k}$, what are the possible values of $\oint_{\mathcal{C}} \mathbf{F}\bullet d\mathbf{r}$ around circles of radius a in the plane $2x + y + 2z = 7$?

5. Let **F** be a smooth vector field in 3-space and suppose that, for every $a > 0$, the flux of **F** out of the sphere of radius a centred at the origin is $\pi(a^3 + 2a^4)$. Find the divergence of **F** at the origin.

6. Let $\mathbf{F} = -y\mathbf{i} + x\cos(1 - x^2 - y^2)\mathbf{j} + yz\mathbf{k}$. Find the flux of **curl F** upward through a surface whose boundary is the curve $x^2 + y^2 = 1$, $z = 2$.

7. Let $\mathbf{F}(\mathbf{r}) = r^\lambda \mathbf{r}$, where $\mathbf{r} = x\mathbf{i} + y\mathbf{j} + z\mathbf{k}$ and $r = |\mathbf{r}|$. For what value(s) of λ is **F** solenoidal on an open subset of 3-space? Is **F** solenoidal on all of 3-space for any value of λ?

8. Given that **F** satisfies **curl F** $= \mu\mathbf{F}$ on 3-space, where μ is a nonzero constant, show that $\nabla^2\mathbf{F} + \mu^2\mathbf{F} = \mathbf{0}$.

9. Let P be a polyhedron in 3-space having n planar faces, F_1, F_2, ..., F_n. Let $\mathbf{N}_i$ be normal to F_i in the direction outward from P, and let $\mathbf{N}_i$ have length equal to the area of face F_i. Show that

$$\sum_{i=1}^{n} \mathbf{N}_i = \mathbf{0}.$$

Also, state a version of this result for a plane polygon P.

10. Around what simple, closed curve C in the xy-plane does the vector field

$$\mathbf{F} = (2y^3 - 3y + xy^2)\mathbf{i} + (x - x^3 + x^2y)\mathbf{j}$$

have the greatest circulation?

11. Through what closed, oriented surface in $\mathbb{R}^3$ does the vector field

$$\mathbf{F} = (4x + 2x^3z)\mathbf{i} - y(x^2 + z^2)\mathbf{j} - (3x^2z^2 + 4y^2z)\mathbf{k}$$

have the greatest flux?

12. Find the maximum value of

$$\oint_C \mathbf{F} \bullet d\mathbf{r},$$

where $\mathbf{F} = xy^2\mathbf{i} + (3z - xy^2)\mathbf{j} + (4y - x^2y)\mathbf{k}$, and C is a simple closed curve in the plane $x + y + z = 1$ oriented counterclockwise as seen from high on the z-axis. What curve C gives this maximum?

Challenging Problems

1. **(The expanding universe)** Let $\mathbf{v}$ be the large-scale velocity field of matter in the universe. (*Large-scale* means on the scale of intergalactic distances; *small-scale* motion such as that of planetary systems about their suns, and even stars about galactic centres, has been averaged out.) Assume that $\mathbf{v}$ is a smooth vector field. According to present astronomical theory, the distance between any two points is increasing, and the rate of increase is proportional to the distance between the points. The constant of proportionality, C, is called *Hubble's constant*. In terms of $\mathbf{v}$, if $\mathbf{r}_1$ and $\mathbf{r}_2$ are two points, then

$$\left(\mathbf{v}(\mathbf{r}_2) - \mathbf{v}(\mathbf{r}_1)\right) \bullet (\mathbf{r}_2 - \mathbf{r}_1) = C|\mathbf{r}_2 - \mathbf{r}_1|^2.$$

Show that **div v** is constant, and find the value of the constant in terms of Hubble's constant. *Hint:* Find the flux of $\mathbf{v}(\mathbf{r})$ out of a sphere of radius ϵ centred at $\mathbf{r}_1$ and take the limit as ϵ approaches zero.

2. **(Solid angle)** Two rays from a point P determine an angle at P whose measure in radians is equal to the length of the arc of the circle of radius 1 with centre at P lying between the two rays. Similarly, an arbitrarily shaped half-cone K with vertex at P determines a **solid angle** at P whose measure in **steradians** (*stereo + radians*) is the area of that part of the sphere of radius 1 with centre at P lying within K. For example, the first octant of $\mathbb{R}^3$ is a half-cone with vertex at the origin. It determines a solid angle at the origin measuring

$$4\pi \times \frac{1}{8} = \frac{\pi}{2} \text{ steradians,}$$

since the area of the unit sphere is 4π. (See Figure 16.25.)

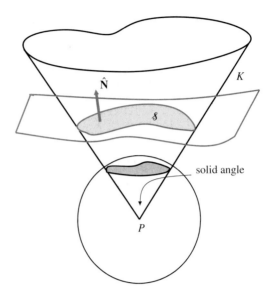

Figure 16.25

(a) Find the steradian measure of the solid angle at the vertex of a right-circular half-cone whose generators make angle α with its central axis.

(b) If a smooth, oriented surface intersects the general half-cone K but not at its vertex P, let $\mathcal{S}$ be the part of the surface lying within K. Orient $\mathcal{S}$ with normal pointing away from P. Show that the steradian measure of the solid angle at P determined by K is the flux of $\mathbf{r}/|\mathbf{r}|^3$ through $\mathcal{S}$, where $\mathbf{r}$ is the vector from P to the point (x, y, z).

Integrals over moving domains

By the Fundamental Theorem of Calculus, the derivative with respect to time t of an integral of $f(x, t)$ over a "moving interval" $[a(t), b(t)]$ is given by

$$\frac{d}{dt} \int_{a(t)}^{b(t)} f(x, t)\, dx = \int_{a(t)}^{b(t)} \frac{\partial}{\partial t} f(x, t)\, dx$$
$$+ f(b(t), t)\frac{db}{dt} - f(a(t), t)\frac{da}{dt}.$$

The next three problems, suggested by Luigi Quartapelle of the Politecnico di Milano, provide various extensions of this one-dimensional result to higher dimensions. The calculations are somewhat lengthy, so you may want to get some help from Maple or another computer algebra system.

* **3.** **(Rate of change of circulation along a moving curve)**

(a) Let $\mathbf{F}(\mathbf{r}, t)$ be a smooth vector field in $\mathbb{R}^3$ depending on a parameter t, and let

$$\mathbf{G}(s, t) = \mathbf{F}\left(\mathbf{r}(s, t), t\right) = \mathbf{F}\left(x(s, t), y(s, t), z(x, t), t\right),$$

where $\mathbf{r}(s, t) = x(s, t)\mathbf{i} + y(s, t)\mathbf{j} + z(s, t)\mathbf{k}$ has continuous partial derivatives of second order. Show that

$$\frac{\partial}{\partial t}\left(\mathbf{G} \bullet \frac{\partial \mathbf{r}}{\partial s}\right) - \frac{\partial}{\partial s}\left(\mathbf{G} \bullet \frac{\partial \mathbf{r}}{\partial t}\right)$$
$$= \frac{\partial \mathbf{F}}{\partial t} \bullet \frac{\partial \mathbf{r}}{\partial s} + \left((\nabla \times \mathbf{F}) \times \frac{\partial \mathbf{r}}{\partial t}\right) \bullet \frac{\partial \mathbf{r}}{\partial s}.$$

Here, the curl $\nabla \times \mathbf{F}$ is taken with respect to the position vector $\mathbf{r}$.

(b) For fixed t (which you can think of as time), $\mathbf{r} = \mathbf{r}(s, t)$, ($a \le s \le b$), represents parametrically a curve C_t in $\mathbb{R}^3$. The curve moves as t varies; the velocity of any point on C_t is $\mathbf{v}_C(s, t) = \partial\mathbf{r}/\partial t$. Show that

$$\frac{d}{dt}\int_{C_t} \mathbf{F} \bullet d\mathbf{r} = \int_{C_t} \frac{\partial\mathbf{F}}{\partial t} \bullet d\mathbf{r} + \int_{C_t}\left((\nabla \times \mathbf{F}) \times \mathbf{v}_C\right) \bullet d\mathbf{r}$$
$$+ \mathbf{F}\left(\mathbf{r}(b, t), t\right) \bullet \mathbf{v}_C(b, t) - \mathbf{F}\left(\mathbf{r}(a, t), t\right) \bullet \mathbf{v}_C(a, t).$$

Hint: Write

$$\frac{d}{dt}\int_{C_t} \mathbf{F} \bullet d\mathbf{r} = \int_a^b \frac{\partial}{\partial t}\left(\mathbf{G} \bullet \frac{\partial\mathbf{r}}{\partial s}\right) ds$$
$$= \int_a^b \left[\frac{\partial}{\partial s}\left(\mathbf{G} \bullet \frac{\partial\mathbf{r}}{\partial t}\right)\right.$$
$$\left. + \left(\frac{\partial}{\partial t}\left(\mathbf{G} \bullet \frac{\partial\mathbf{r}}{\partial s}\right) - \frac{\partial}{\partial s}\left(\mathbf{G} \bullet \frac{\partial\mathbf{r}}{\partial t}\right)\right)\right] ds.$$

Now use the result of (a).

* **4. (Rate of change of flux through a moving surface)** Let S_t be a moving surface in $\mathbb{R}^3$ smoothly parametrized (for each t) by

$$\mathbf{r} = \mathbf{r}(u, v, t) = x(u, v, t)\mathbf{i} + y(u, v, t)\mathbf{j} + z(u, v, t)\mathbf{k},$$

where (u, v) belongs to a parameter region R in the uv-plane. Let $\mathbf{F}(\mathbf{r}, t) = F_1\mathbf{i} + F_2\mathbf{j} + F_3\mathbf{k}$ be a smooth 3-vector function, and let $\mathbf{G}(u, v, t) = \mathbf{F}(\mathbf{r}(u, v, t), t)$.

(a) Show that

$$\frac{\partial}{\partial t}\left(\mathbf{G} \bullet \left[\frac{\partial\mathbf{r}}{\partial u} \times \frac{\partial\mathbf{r}}{\partial v}\right]\right) - \frac{\partial}{\partial u}\left(\mathbf{G} \bullet \left[\frac{\partial\mathbf{r}}{\partial t} \times \frac{\partial\mathbf{r}}{\partial v}\right]\right)$$
$$- \frac{\partial}{\partial v}\left(\mathbf{G} \bullet \left[\frac{\partial\mathbf{r}}{\partial u} \times \frac{\partial\mathbf{r}}{\partial t}\right]\right)$$
$$= \frac{\partial\mathbf{F}}{\partial t} \bullet \left[\frac{\partial\mathbf{r}}{\partial u} \times \frac{\partial\mathbf{r}}{\partial v}\right] + (\nabla \bullet \mathbf{F})\frac{\partial\mathbf{r}}{\partial t} \bullet \left[\frac{\partial\mathbf{r}}{\partial u} \times \frac{\partial\mathbf{r}}{\partial v}\right].$$

(b) If C_t is the boundary of S_t with orientation corresponding to that of S_t, use Green's Theorem to show that

$$\iint_R \left[\frac{\partial}{\partial u}\left(\mathbf{G} \bullet \left[\frac{\partial\mathbf{r}}{\partial t} \times \frac{\partial\mathbf{r}}{\partial v}\right]\right)\right.$$
$$\left. + \frac{\partial}{\partial v}\left(\mathbf{G} \bullet \left[\frac{\partial\mathbf{r}}{\partial u} \times \frac{\partial\mathbf{r}}{\partial t}\right]\right)\right] du\, dv$$
$$= \oint_{C_t} \left(\mathbf{F} \times \frac{\partial\mathbf{r}}{\partial t}\right) \bullet d\mathbf{r}.$$

(c) Combine the results of (a) and (b) to show that

$$\frac{d}{dt}\iint_{S_t} \mathbf{F} \bullet \hat{\mathbf{N}}\, dS$$
$$= \iint_{S_t} \frac{\partial\mathbf{F}}{\partial t} \bullet \hat{\mathbf{N}}\, dS + \iint_{S_t} (\nabla \bullet \mathbf{F})\mathbf{v}_S \bullet \hat{\mathbf{N}}\, dS$$
$$+ \oint_{C_t} (\mathbf{F} \times \mathbf{v}_C) \bullet d\mathbf{r},$$

where $\mathbf{v}_S = \partial\mathbf{r}/\partial t$ on S_t is the velocity of S_t, $\mathbf{v}_C = \partial\mathbf{r}/\partial t$ on C_t is the velocity of C_t, and $\hat{\mathbf{N}}$ is the unit normal field on S_t corresponding to its orientation.

* **5. (Rate of change of integrals over moving volumes)** Let S_t be the position at time t of a smooth, closed surface in $\mathbb{R}^3$ that varies smoothly with t and bounds at any time t a region D_t. If $\hat{\mathbf{N}}(\mathbf{r}, t)$ denotes the unit outward (from D_t) normal field on S_t, and $\mathbf{v}_S(\mathbf{r}, t)$ is the velocity of the point $\mathbf{r}$ on S_t at time t, show that

$$\frac{d}{dt}\iiint_{D_t} f\, dV = \iiint_{D_t} \frac{\partial f}{\partial t}\, dV + \oiint_{S_t} f\mathbf{v}_S \bullet \hat{\mathbf{N}}\, dS$$

holds for smooth functions $f(\mathbf{r}, t)$. *Hint:* Let ΔD_t consist of the points through which S_t passes as t increases to $t + \Delta t$. The volume element dV in ΔD_t can be expressed in terms of the area element dS on S_t by

$$dV = \mathbf{v} \bullet \hat{\mathbf{N}}\, dS\, \Delta t.$$

Show that

$$\frac{1}{\Delta t}\left[\iiint_{D_{t+\Delta t}} f(\mathbf{r}, t + \Delta t)\, dV - \iiint_{D_t} f(\mathbf{r}, t)\, dV\right]$$
$$= \iiint_{D_t} \frac{f(\mathbf{r}, t + \Delta t) - f(\mathbf{r}, t)}{\Delta t}\, dV$$
$$+ \frac{1}{\Delta t}\iiint_{\Delta D_t} f(\mathbf{r}, t)\, dV$$
$$+ \iiint_{\Delta D_t} \frac{f(\mathbf{r}, t + \Delta t) - f(\mathbf{r}, t)}{\Delta t}\, dV,$$

and show that the last integral $\to 0$ as $\Delta t \to 0$.

CHAPTER 17

Ordinary Differential Equations

> In order to solve this differential equation you look at it until the solution occurs to you.
>
> **George Polyá   1887–1985**
> from *How to Solve It*   Princeton, 1945

> Science is a differential equation. Religion is a boundary condition.
>
> **Alan Turing   1912–1954**
> quoted in *Theories of Everything* by J. D. Barrow

Introduction   A **differential equation** (or **DE**) is an equation that involves one or more derivatives of an unknown function. Solving the differential equation means finding a function (or every such function) that satisfies the differential equation.

Many physical laws and relationships between quantities studied in various scientific disciplines are expressed mathematically as differential equations. For example, Newton's second law of motion ($F = ma$) states that the position $x(t)$ at time t of an object of constant mass m subjected to a force $F(t)$ must satisfy the differential equation (equation of motion):

$$m\frac{d^2x}{dt^2} = F(t).$$

Similarly, the biomass $m(t)$ at time t of a bacterial culture growing in a uniformly supporting medium changes at a rate proportional to the biomass:

$$\frac{dm}{dt} = km(t),$$

which is the differential equation of exponential growth (or, if $k < 0$, exponential decay). Because differential equations arise so extensively in the abstract modelling of concrete phenomena, such equations and techniques for solving them are at the heart of applied mathematics. Indeed, most of the existing mathematical literature is either directly involved with differential equations or is motivated by problems arising in the study of such equations. Because of this, we have introduced various differential equations, terms for their description, and techniques for their solution at several places in the development of calculus throughout this book. This final chapter provides a more unified framework for a brief introduction to the study of ordinary differential equations. Some material from earlier sections (notably Sections 7.9 and 3.7) forms a natural part of this chapter; you will be referred back to these sections at the appropriate time. This chapter is, of necessity, relatively short. Students of mathematics and its applications usually take one or more full courses on differential equations, and even then hardly scratch the surface of the subject.

17.1 Classifying Differential Equations

Differential equations are classified in several ways. The most significant classification is based on the number of variables with respect to which derivatives appear in the equation. An **ordinary differential equation (ODE)** is one that involves derivatives with respect to only one variable. Both of the examples given above are ordinary differential equations. A **partial differential equation (PDE)** is one that involves partial derivatives of the unknown function with respect to more than one variable. For example, the **one-dimensional wave equation**

$$\frac{\partial^2 u}{\partial t^2} = c^2 \frac{\partial^2 u}{\partial x^2}$$

models the lateral displacement $u(x, t)$ at position x at time t of a stretched vibrating string. (See Section 12.4.) We will not discuss partial differential equations in this chapter.

Differential equations are also classified with respect to **order**. The order of a differential equation is the order of the highest-order derivative present in the equation. The one-dimensional wave equation is a second-order PDE. The following example records the order of two ODEs.

Example 1
$$\frac{d^2 y}{dx^2} + x^3 y = \sin x \qquad \text{has order 2,}$$

$$\frac{d^3 y}{dx^3} + 4x \left(\frac{dy}{dx}\right)^2 = y \frac{d^2 y}{dx^2} + e^y \qquad \text{has order 3.}$$

Like any equation, a differential equation can be written in the form $F = 0$, where F is a function. For an ODE, the function F can depend on the independent variable (usually called x or t), the unknown function (usually y), and any derivatives of the unknown function up to the order of the equation. For instance, an nth-order ODE can be written in the form

$$F(x, y, y', y'', \dots, y^{(n)}) = 0.$$

An important special class of differential equations consists of those that are **linear**. An nth-order linear ODE has the form

$$a_n(x)y^{(n)}(x) + a_{n-1}(x)y^{(n-1)}(x) + \cdots$$
$$+ a_2(x)y''(x) + a_1(x)y'(x) + a_0(x)y(x) = f(x).$$

Each term in the expression on the left side is the product of a *coefficient* that is a function of x, and a second factor that is either y or one of the derivatives of y. The term on the right does not depend on y; it is called the **nonhomogeneous term**. Observe that no term on the left side involves any power of y or its derivatives other than the first power, and y and its derivatives are never multiplied together.

A linear ODE is said to be **homogeneous** if all of its terms involve the unknown function y, that is, if $f(x) = 0$. If $f(x)$ is not identically zero, the equation is **nonhomogeneous**.

Example 2   In Example 1 the first DE,

$$\frac{d^2 y}{dx^2} + x^3 y = \sin x,$$

is linear. Here, the coefficients are $a_2(x) = 1$, $a_1(x) = 0$, $a_0(x) = x^3$, and the nonhomogeneous term is $f(x) = \sin x$. Although it can be written in the form

$$\frac{d^3 y}{dx^3} + 4x \left(\frac{dy}{dx}\right)^2 - y \frac{d^2 y}{dx^2} - e^y = 0,$$

the second equation is *not linear* (we say it is **nonlinear**) because the second term involves the square of a derivative of y, the third term involves the product of y and one of its derivatives, and the fourth term is not y times a function of x. The equation

$$(1 + x^2)\frac{d^3 y}{dx^3} + \sin x \frac{d^2 y}{dx^2} - 4\frac{dy}{dx} + y = 0$$

is a linear equation of order 3. The coefficients are $a_3(x) = 1 + x^2$, $a_2(x) = \sin x$, $a_1(x) = -4$, and $a_0(x) = 1$. Since $f(x) = 0$, this equation is *homogeneous*.

The following theorem states that any *linear combination* of solutions of a linear, homogeneous DE is also a solution. This is an extremely important fact about linear, homogeneous DEs.

THEOREM 1

If $y = y_1(x)$ and $y = y_2(x)$ are two solutions of the linear, homogeneous DE

$$a_n y^{(n)} + a_{n-1} y^{(n-1)} + \cdots + a_2 y'' + a_1 y' + a_0 y = 0,$$

then so is the linear combination

$$y = A y_1(x) + B y_2(x)$$

for any values of the constants A and B.

PROOF We are given that

$$a_n y_1^{(n)} + a_{n-1} y_1^{(n-1)} + \cdots + a_2 y_1'' + a_1 y_1' + a_0 y_1 = 0 \qquad \text{and}$$
$$a_n y_2^{(n)} + a_{n-1} y_2^{(n-1)} + \cdots + a_2 y_2'' + a_1 y_2' + a_0 y_2 = 0.$$

Multiplying the first equation by A and the second by B and adding the two gives

$$a_n(A y_1^{(n)} + B y_2^{(n)}) + a_{n-1}(A y_1^{(n-1)} + B y_2^{(n-1)})$$
$$+ \cdots + a_2(A y_1'' + B y_2'') + a_1(A y_1' + B y_2') + a_0(A y_1 + B y_2) = 0.$$

Thus, $y = A y_1(x) + B y_2(x)$ is also a solution of the equation.

The same kind of proof can be used to verify the following theorem.

THEOREM 2

If $y = y_1(x)$ is a solution of the linear, homogeneous equation

$$a_n y^{(n)} + a_{n-1} y^{(n-1)} + \cdots + a_2 y'' + a_1 y' + a_0 y = 0$$

and $y = y_2(x)$ is a solution of the linear, nonhomogeneous equation

$$a_n y^{(n)} + a_{n-1} y^{(n-1)} + \cdots + a_2 y'' + a_1 y' + a_0 y = f(x),$$

then $y = y_1(x) + y_2(x)$ is also a solution of the same linear, nonhomogeneous equation.

We will make extensive use of the two theorems above when we discuss second-order linear equations in Sections 17.6–17.8.

Example 3 Verify that $y = \sin 2x$ and $y = \cos 2x$ satisfy the DE $y'' + 4y = 0$. Find a solution $y(x)$ of that DE that satisfies $y(0) = 2$ and $y'(0) = -4$.

Solution   If $y = \sin 2x$, then $y'' = \dfrac{d}{dx}(2\cos 2x) = -4\sin 2x = -4y$. Thus $y'' + 4y = 0$. A similar calculation shows that $y = \cos 2x$ also satisfies the DE. Since the DE is linear and homogeneous, the function

$$y = A\sin 2x + B\cos 2x$$

is a solution for any values of the constants A and B. We want $y(0) = 2$, so we need $2 = A\sin 0 + B\cos 0 = B$. Thus $B = 2$. Also,

$$y' = 2A\cos 2x - 2B\sin 2x.$$

We want $y'(0) = -4$, so $-4 = 2A\cos 0 - 2B\sin 0 = 2A$. Thus, $A = -2$ and the required solution is $y = -2\sin 2x + 2\cos 2x$.   ∎

Remark   Let $P_n(r)$ be the nth-degree polynomial in the variable r given by

$$P_n(r) = a_n(x)r^n + a_{n-1}(x)r^{n-1} + \cdots + a_2(x)r^2 + a_1(x)r + a_0(x),$$

with coefficients depending on the variable x. We can write the nth-order linear ODE with coefficients $a_k(x)$, $(0 \le k \le n)$, and nonhomogeneous term $f(x)$ in the form

$$P_n(D)y(x) = f(x),$$

where D stands for the *differential operator* d/dx. The left side of the equation above denotes the application of the nth-order differential operator

$$P_n(D) = a_n(x)D^n + a_{n-1}(x)D^{n-1} + \cdots + a_2(x)D^2 + a_1(x)D + a_0(x)$$

to the function $y(x)$. For example,

$$a_k(x)D^k y(x) = a_k(x)\frac{d^k y}{dx^k}.$$

It is often useful to write linear DEs in terms of differential operators in this way.

Remark   Unfortunately, the term *homogeneous* is used in more than one way in the study of differential equations. Certain ODEs that are not necessarily linear are called homogeneous for a different reason than the one applying for linear equations above. We will encounter equations of this type in Section 17.2.

Exercises 17.1

In Exercises 1–10, state the order of the given DE, and whether it is linear or nonlinear. If it is linear, is it homogeneous or nonhomogeneous?

1. $\dfrac{dy}{dx} = 5y$

2. $\dfrac{d^2 y}{dx^2} + x = y$

3. $y\dfrac{dy}{dx} = x$

4. $y''' + xy' = x\sin x$

5. $y'' + x\sin x\, y' = y$

6. $y'' + 4y' - 3y = 2y^2$

7. $\dfrac{d^3 y}{dt^3} + t\dfrac{dy}{dt} + t^2 y = t^3$

8. $\cos x\dfrac{dx}{dt} + x\sin t = 0$

9. $y^{(4)} + e^x y'' = x^3 y'$

10. $x^2 y'' + e^x y' = \dfrac{1}{y}$

11. Verify that $y = \cos x$ and $y = \sin x$ are solutions of the DE $y'' + y = 0$. Are any of the following functions solutions? (a) $\sin x - \cos x$, (b) $\sin(x+3)$, (c) $\sin 2x$. Justify your answers.

12. Verify that $y = e^x$ and $y = e^{-x}$ are solutions of the DE $y'' - y = 0$. Are any of the following functions solutions? (a) $\cosh x = \frac{1}{2}(e^x + e^{-x})$, (b) $\cos x$, (c) x^e. Justify your answers.

13. $y_1 = \cos(kx)$ is a solution of $y'' + k^2 y = 0$. Guess and verify another solution y_2 that is not a multiple of y_1. Then find a solution that satisfies $y(\pi/k) = 3$ and $y'(\pi/k) = 3$.

14. $y_1 = e^{kx}$ is a solution of $y'' - k^2 y = 0$. Guess and verify another solution y_2 that is not a multiple of y_1. Then find a solution that satisfies $y(1) = 0$ and $y'(1) = 2$.

15. Find a solution of $y'' + y = 0$ that satisfies $y(\pi/2) = 2y(0)$ and $y(\pi/4) = 3$. *Hint:* See Exercise 11.

16. Find two values of r such that $y = e^{rx}$ is a solution of $y'' - y' - 2y = 0$. Then find a solution of the equation that

satisfies $y(0) = 1$, $y'(0) = 2$.

17. Verify that $y = x$ is a solution of $y'' + y = x$, and find a solution y of this DE that satisfies $y(\pi) = 1$ and $y'(\pi) = 0$. *Hint:* Use Exercise 11 and Theorem 2.

18. Verify that $y = -e$ is a solution of $y'' - y = e$, and find a solution y of this DE that satisfies $y(1) = 0$ and $y'(1) = 1$. *Hint:* Use Exercise 12 and Theorem 2.

17.2 Solving First-Order Equations

In this section we will develop techniques for solving several types of first-order ODEs, specifically

1. separable equations,

2. homogeneous equations,

3. exact equations, and

4. linear equations,

Most first-order equations are of the form

$$\frac{dy}{dx} = f(x, y).$$

Solving such differential equations typically involves integration; indeed, the process of solving a DE is called *integrating* the DE. Nevertheless, solving DEs is usually more complicated than just writing down an integral and evaluating it. The only kind of DE that can be solved that way is the simplest kind of first-order, linear DE that can be written in the form

$$\frac{dy}{dx} = f(x).$$

The solution is then just the antiderivative of f:

$$y = \int f(x)\, dx.$$

Separable Equations

Consider the logistic equation that models the growth of an animal population with a limited food supply:

$$\frac{dy}{dt} = ky\left(1 - \frac{y}{L}\right),$$

where $y(t)$ is the size of the population at time t, k is a positive constant related to the fertility of the population, and L is the steady-state population size that can be sustained by the available food supply. This equation is an example of a class of first-order differential equations called **separable equations** because when they are written in terms of differentials, they can be separated with only the dependent variable on one side of the equation and only the independent variable on the other. The logistic equation can be written in the form

$$\frac{L\, dy}{y(L - y)} = k\, dt,$$

and solved by integrating both sides. Expanding the left side in partial fractions and integrating, we get

$$\int \left(\frac{1}{y} + \frac{1}{L - y}\right) dy = kt + C.$$

Assuming that $0 < y < L$, we therefore obtain

$$\ln y - \ln(L - y) = kt + C,$$

$$\ln\left(\frac{y}{L - y}\right) = kt + C.$$

We can solve this equation for y by taking exponentials of both sides:

$$\frac{y}{L - y} = e^{kt+C} = C_1 e^{kt}$$

$$y = (L - y)C_1 e^{kt}$$

$$y = \frac{C_1 L e^{kt}}{1 + C_1 e^{kt}},$$

where $C_1 = e^C$.

Generally, separable equations are of the form

$$\frac{dy}{dx} = f(x)g(y).$$

We solve them by rewriting them in the form

$$\frac{dy}{g(y)} = f(x)\,dx$$

and integrating both sides.

Example 1   Solve the equation $\dfrac{dy}{dx} = \dfrac{x}{y}$.

Solution   We rewrite the equation in the form $y\,dy = x\,dx$ and integrate both sides to get

$$\frac{1}{2}y^2 = \frac{1}{2}x^2 + C,$$

or $y^2 - x^2 = C_1$, where $C_1 = 2C$ is an arbitrary constant. The solution curves are rectangular hyperbolas. Their asymptotes $y = x$ and $y = -x$ are also solutions corresponding to $C = 0$.

Example 2   Solve the initial-value problem

$$\begin{cases} \dfrac{dy}{dx} = x^2 y^3 \\ y(1) = 3. \end{cases}$$

Solution   Separating the differential equation gives $\dfrac{dy}{y^3} = x^2\,dx$. Thus,

$$\int \frac{dy}{y^3} = \int x^2\,dx, \qquad \text{so} \qquad \frac{-1}{2y^2} = \frac{x^3}{3} + C.$$

Since $y = 3$ when $x = 1$, we have $-\frac{1}{18} = \frac{1}{3} + C$ and $C = -\frac{7}{18}$. Substituting this value into the above solution and solving for y, we obtain

$$y(x) = \frac{3}{\sqrt{7 - 6x^3}}.$$

This solution is valid for $x < \left(\frac{7}{6}\right)^{1/3}$.

Example 3   Solve the **integral equation** $y(x) = 3 + 2 \int_1^x t y(t) \, dt$.

Solution   Differentiating the integral equation with respect to x gives

$$\frac{dy}{dx} = 2x \, y(x) \qquad \text{or} \qquad \frac{dy}{y} = 2x \, dx.$$

Thus $\ln |y(x)| = x^2 + C$, and solving for y, $y(x) = C_1 e^{x^2}$. Putting $x = 1$ in the integral equation provides an initial value: $y(1) = 3 + 0 = 3$, so $C_1 = 3/e$ and

$$y(x) = 3 e^{x^2 - 1}.$$

Example 4   (**A solution concentration problem**)   Initially a tank contains 1,000 L of brine with 50 kg of dissolved salt. Brine containing 10 g of salt per litre is flowing into the tank at a constant rate of 10 L/min. If the contents of the tank are kept thoroughly mixed at all times, and if the solution also flows out at 10 L/min, how much salt remains in the tank at the end of 40 min?

Solution   Let $x(t)$ be the number of kilograms of salt in solution in the tank after t min. Thus $x(0) = 50$. Salt is coming into the tank at a rate of 10 g/L $\times$ 10 L/min = 100 g/min = 1/10 kg/min. At all times the tank contains 1,000 L of liquid, so the concentration of salt in the tank at time t is $x/1{,}000$ kg/L. Since the contents flow out at 10 L/min, salt is being removed at a rate of $10x/1{,}000 = x/100$ kg/min. Therefore,

$$\frac{dx}{dt} = \text{rate in} - \text{rate out} = \frac{1}{10} - \frac{x}{100} = \frac{10 - x}{100}$$

or

$$\frac{dx}{10 - x} = \frac{dt}{100}.$$

Integrating both sides of this equation, we obtain

$$-\ln |10 - x| = \frac{t}{100} + C.$$

Observe that $x(t) \neq 10$ for any finite time t (since $\ln 0$ is not defined). Since $x(0) = 50 > 10$, it follows that $x(t) > 10$ for all $t > 0$. ($x(t)$ is necessarily continuous so it cannot take any value less than 10 without somewhere taking the value 10 by the Intermediate-Value Theorem.) Hence, we can drop the absolute value from the solution above and obtain

$$\ln(x - 10) = -\frac{t}{100} - C.$$

Since $x(0) = 50$, we have $-C = \ln 40$ and

$$x = x(t) = 10 + 40 e^{-t/100}.$$

After 40 min there will be $10 + 40 e^{-0.4} \approx 36.8$ kg of salt in the tank.

Example 5   (**A rate of reaction problem**)   In a chemical reaction that goes to completion in solution, one molecule of each of two reactants, A and B, combine to form each molecule of the product C. According to the law of mass action, the reaction proceeds at a rate proportional to the product of the concentrations of A and B in the solution. Thus, if there were initially present a molecules/cm^3 of A and b

molecules/cm^3 of B, then the number $x(t)$ of molecules/cm^3 of C present at time t thereafter is determined by the differential equation

$$\frac{dx}{dt} = k(a - x)(b - x).$$

We solve this equation by the technique of partial fraction decomposition under the assumption that $b \neq a$:

$$\int \frac{dx}{(a - x)(b - x)} = k \int dt = kt + C.$$

Since

$$\frac{1}{(a - x)(b - x)} = \frac{1}{b - a}\left(\frac{1}{a - x} - \frac{1}{b - x}\right),$$

and since necessarily $x \leq a$ and $x \leq b$, we have

$$\frac{1}{b - a}\left(-\ln(a - x) + \ln(b - x)\right) = kt + C,$$

or

$$\ln\left(\frac{b - x}{a - x}\right) = (b - a)kt + C_1, \quad \text{where } C_1 = (b - a)C.$$

By assumption, $x(0) = 0$, so $C_1 = \ln(b/a)$ and

$$\ln\frac{a(b - x)}{b(a - x)} = (b - a)kt.$$

This equation can be solved for x to yield $x = x(t) = \dfrac{ab(e^{(b-a)kt} - 1)}{be^{(b-a)kt} - a}.$ ∎

Example 6 Find a family of curves, each of which intersects every parabola with equation of the form $y = Cx^2$ at right angles.

Solution The family of parabolas $y = Cx^2$ satisfies the differential equation

$$\frac{d}{dx}\left(\frac{y}{x^2}\right) = \frac{d}{dx}C = 0,$$

that is,

$$x^2 \frac{dy}{dx} - 2xy = 0 \quad \text{or} \quad \frac{dy}{dx} = \frac{2y}{x}.$$

Any curve that meets the parabolas $y = Cx^2$ at right angles must, at any point (x, y) on it, have slope equal to the negative reciprocal of the slope of the particular parabola passing through that point. Thus such a curve must satisfy

$$\frac{dy}{dx} = -\frac{x}{2y}.$$

Separation of the variables leads to $2y\, dy = -x\, dx$, and integration of both sides then yields $y^2 = -\frac{1}{2}x^2 + C_1$ or $x^2 + 2y^2 = C$, where $C = 2C_1$. This equation represents a family of ellipses centred at the origin. Each ellipse meets each parabola at right angles, as shown in Figure 17.1. When the curves of one family intersect the curves of a second family at right angles, each family is called the family of **orthogonal trajectories** of the other family.

Figure 17.1 The parabolas $y = C_1 x^2$ and the ellipses $x^2 + 2y^2 = C_2$ intersect at right angles

First-Order Homogeneous Equations

A first-order DE of the form

$$\frac{dy}{dx} = f\left(\frac{y}{x}\right)$$

is said to be **homogeneous**. This is a *different* use of the term homogeneous from that in the previous section, which applied only to linear equations. Here homogeneous refers to the fact that y/x, and therefore $g(x, y) = f(y/x)$ is *homogeneous of degree 0* in the sense described after Example 7 in Section 12.5. Such a homogeneous equation can be transformed into a separable equation (and therefore solved) by means of a change of dependent variable. If we set

$$v = \frac{y}{x}, \qquad \text{or equivalently} \qquad y = xv(x),$$

then we have

$$\frac{dy}{dx} = v + x\frac{dv}{dx}$$

and the original differential equation transforms into

$$\frac{dv}{dx} = \frac{f(v) - v}{x},$$

which is separable.

Example 7   Solve the equation

$$\frac{dy}{dx} = \frac{x^2 + xy}{xy + y^2}.$$

Solution   The equation is homogeneous. (Divide the numerator and denominator of the right-hand side by x^2 to see this.) If $y = vx$ the equation becomes

$$v + x\frac{dv}{dx} = \frac{1 + v}{v + v^2} = \frac{1}{v},$$

or

$$x\frac{dv}{dx} = \frac{1-v^2}{v}.$$

Separating variables and integrating, we calculate

$$\int \frac{v\,dv}{1-v^2} = \int \frac{dx}{x} \qquad \text{let } u = 1 - v^2$$

$$-\frac{1}{2}\int \frac{du}{u} = \int \frac{dx}{x}$$

$$-\ln|u| = 2\ln|x| + C_1 = \ln C_2 x^2 \qquad (C_1 = \ln C_2)$$

$$\frac{1}{|u|} = C_2 x^2$$

$$|1 - v^2| = \frac{C_3}{x^2} \qquad (C_3 = 1/C_2)$$

$$\left|1 - \frac{y^2}{x^2}\right| = \frac{C_3}{x^2}.$$

The solution is best expressed in the form $x^2 - y^2 = C_4$. However, near points where $y \neq 0$, the equation can be solved for y as a function of x.

Exact Equations

A first-order differential equation expressed in differential form as

$$M(x, y)\,dx + N(x, y)\,dy = 0,$$

which is equivalent to $\dfrac{dy}{dx} = -\dfrac{M(x, y)}{N(x, y)}$, is said to be **exact** if the left-hand side is the differential of a function $\phi(x, y)$:

$$d\phi(x, y) = M(x, y)\,dx + N(x, y)\,dy.$$

The function ϕ is called an **integral function** of the differential equation. The level curves $\phi(x, y) = C$ of ϕ are the **solution curves** of the differential equation. For example, the differential equation

$$x\,dx + y\,dy = 0$$

has solution curves given by

$$x^2 + y^2 = C$$

since $d(x^2 + y^2) = 2(x\,dx + y\,dy) = 0$.

Remark The condition that the differential equation $M\,dx + N\,dy = 0$ should be exact is just the condition that the vector field

$$\mathbf{F} = M(x, y)\,\mathbf{i} + N(x, y)\,\mathbf{j}$$

should be *conservative*; the integral function of the differential equation is then the potential function of the vector field. (See Section 15.2.)

A **necessary condition** for the exactness of the DE $M\,dx + N\,dy = 0$ is that

$$\frac{\partial M}{\partial y} = \frac{\partial N}{\partial x};$$

this just says that the mixed partial derivatives $\dfrac{\partial^2 \phi}{\partial x \partial y}$ and $\dfrac{\partial^2 \phi}{\partial y \partial x}$ of the integral function ϕ must be equal.

Once you know that an equation is exact, you can often guess the integral function. In any event, ϕ can always be found by the same method used to find the potential of a conservative vector field in Section 15.2.

Example 8   Verify that the DE

$$(2x + \sin y - ye^{-x})\, dx + (x \cos y + \cos y + e^{-x})\, dy = 0$$

is exact and find its solution curves.

Solution   Here $M = 2x + \sin y - ye^{-x}$ and $N = x \cos y + \cos y + e^{-x}$. Since

$$\frac{\partial M}{\partial y} = \cos y - e^{-x} = \frac{\partial N}{\partial x},$$

the DE is exact. We want to find ϕ so that

$$\frac{\partial \phi}{\partial x} = M = 2x + \sin y - ye^{-x} \quad \text{and} \quad \frac{\partial \phi}{\partial y} = N = x \cos y + \cos y + e^{-x}.$$

Integrate the first equation with respect to x, being careful to allow the constant of integration to depend on y:

$$\phi(x, y) = \int (2x + \sin y - ye^{-x})\, dx = x^2 + x \sin y + ye^{-x} + C_1(y).$$

Now substitute this expression into the second equation:

$$x \cos y + \cos y + e^{-x} = \frac{\partial \phi}{\partial y} = x \cos y + e^{-x} + C_1'(y).$$

Thus $C_1'(y) = \cos y$, and $C_1(y) = \sin y + C_2$. (It is because the original DE was exact that the equation for $C_1'(y)$ turned out to be independent of x; this had to happen or we could not have found C_1 as a function of y only.) Choosing $C_2 = 0$, we find that $\phi(x, y) = x^2 + x \sin y + ye^{-x} + \sin y$ is an integral function for the given DE. The solution curves for the DE are the level curves

$$x^2 + x \sin y + ye^{-x} + \sin y = C.$$

■

Integrating Factors

Any ordinary differential equation of order 1 and degree 1 can be expressed in differential form: $M\, dx + N\, dy = 0$. However, this latter equation will usually not be exact. It *may* be possible to multiply the equation by an **integrating factor** $\mu(x, y)$ so that the resulting equation

$$\mu(x, y)\, M(x, y)\, dx + \mu(x, y)\, N(x, y)\, dy = 0$$

is exact. In general such integrating factors are difficult to find; they must satisfy the partial differential equation

$$M(x, y)\, \frac{\partial \mu}{\partial y} - N(x, y)\, \frac{\partial \mu}{\partial x} = \mu(x, y) \left(\frac{\partial N}{\partial x} - \frac{\partial M}{\partial y} \right),$$

which follows from the necessary condition for exactness stated above. We will not try to solve this equation here.

Sometimes it happens that a differential equation has an integrating factor depending on only one of the two variables. Suppose, for instance, that $\mu(x)$ is an integrating factor for $M\,dx + N\,dy = 0$. Then $\mu(x)$ must satisfy the ordinary differential equation

$$N(x, y)\frac{d\mu}{dx} = \mu(x)\left(\frac{\partial M}{\partial y} - \frac{\partial N}{\partial x}\right),$$

or,

$$\frac{1}{\mu(x)}\frac{d\mu}{dx} = \frac{\dfrac{\partial M}{\partial y} - \dfrac{\partial N}{\partial x}}{N(x, y)}.$$

This equation can be solved (by integration) for μ as a function of x alone *provided that the right-hand side is independent of y*.

Example 9 Show that $(x + y^2)\,dx + xy\,dy = 0$ has an integrating factor depending only on x, find it, and solve the equation.

Solution Here $M = x + y^2$ and $N = xy$. Since

$$\frac{\dfrac{\partial M}{\partial y} - \dfrac{\partial N}{\partial x}}{N(x, y)} = \frac{2y - y}{xy} = \frac{1}{x}$$

does not depend on y, the equation has an integrating factor depending only on x. This factor is given by $d\mu/\mu = dx/x$. Evidently $\mu = x$ is a suitable integrating factor; if we multiply the given differential equation by x, we obtain

$$0 = (x^2 + xy^2)\,dx + x^2y\,dy = d\left(\frac{x^3}{3} + \frac{x^2y^2}{2}\right).$$

The solution is therefore $2x^3 + 3x^2y^2 = C$.

Remark Of course, it may be possible to find an integrating factor depending on y instead of x. See Exercises 49–51 below. It is also possible to look for integrating factors that depend on specific combinations of x and y, for instance, xy. See Exercise 52.

First-Order Linear Equations

A first-order **linear** differential equation is one of the type

$$\frac{dy}{dx} + p(x)y = q(x),$$

where $p(x)$ and $q(x)$ are given functions, which we assume to be continuous. We can solve such equations (i.e., find y as a function of x) by the following procedure.

Let $\mu(x)$ be any antiderivative of $p(x)$:

$$\mu(x) = \int p(x)\,dx \qquad \text{and} \qquad \frac{d\mu}{dx} = p(x).$$

If $y = y(x)$ satisfies the given equation, then we calculate, using the Product Rule,

$$\frac{d}{dx}\left(e^{\mu(x)}y(x)\right) = e^{\mu(x)}\frac{dy}{dx} + e^{\mu(x)}\frac{d\mu}{dx}y(x)$$

$$= e^{\mu(x)}\left(\frac{dy}{dx} + p(x)y\right) = e^{\mu(x)}q(x).$$

Therefore, $e^{\mu(x)}y(x) = \int e^{\mu(x)}q(x)\,dx$, or

$$y(x) = e^{-\mu(x)}\int e^{\mu(x)}q(x)\,dx.$$

We reuse this method, rather than the final formula, in the examples below. $e^{\mu(x)}$ is an **integrating factor** of the given differential equation because, if we multiply the equation by $e^{\mu(x)}$, the left side becomes the derivative of $e^{\mu(x)}y(x)$.

Example 10 Solve $\dfrac{dy}{dx} + \dfrac{y}{x} = 1$ for $x > 0$.

Solution Here, $p(x) = 1/x$, so $\mu(x) = \int p(x)\,dx = \ln x$ (for $x > 0$) and $e^{\mu(x)} = x$. We calculate

$$\frac{d}{dx}(xy) = x\frac{dy}{dx} + y = x\left(\frac{dy}{dx} + \frac{y}{x}\right) = x$$

and

$$xy = \int x\,dx = \frac{1}{2}x^2 + C.$$

Finally,

$$y = \frac{1}{x}\left(\frac{1}{2}x^2 + C\right) = \frac{x}{2} + \frac{C}{x}.$$

This is a solution of the given equation for any value of the constant C.

Example 11 Solve $\dfrac{dy}{dx} + xy = x^3$.

Solution Here, $p(x) = x$, so $\mu(x) = x^2/2$ and $e^{\mu(x)} = e^{x^2/2}$. We calculate

$$\frac{d}{dx}\left(e^{x^2/2}y\right) = e^{x^2/2}\frac{dy}{dx} + e^{x^2/2}xy = e^{x^2/2}\left(\frac{dy}{dx} + xy\right) = x^3 e^{x^2/2}.$$

Thus,

$$e^{x^2/2}y = \int x^3 e^{x^2/2}\,dx \qquad \text{Let} \quad U = x^2, \qquad dV = x\,e^{x^2/2}\,dx.$$
$$\qquad\qquad\qquad\qquad\qquad \text{Then } dU = 2x\,dx, \qquad V = e^{x^2/2}.$$
$$= x^2 e^{x^2/2} - 2\int x\,e^{x^2/2}\,dx$$
$$= x^2 e^{x^2/2} - 2e^{x^2/2} + C,$$

and, finally, $y = x^2 - 2 + Ce^{-x^2/2}$.

Figure 17.2   An inductance-resistance circuit

Example 12   (**An inductance-resistance circuit**) An electric circuit (Figure 17.2) contains a constant DC voltage source of V volts, a switch, a resistor of size R ohms, and an inductor of size L henrys. The circuit has no capacitance. The switch, initially open so that no current is flowing, is closed at time $t = 0$ so that current begins to flow at that time. If the inductance L were zero, the current would suddenly jump from 0 amperes when $t < 0$ to $I = V/R$ amperes when $t > 0$. However, if $L > 0$ the current cannot change instantaneously; it will depend on time t. Let the current t seconds after the switch is closed be $I(t)$ amperes. It is known that $I(t)$ satisfies the initial-value problem

$$\begin{cases} L\dfrac{dI}{dt} + RI = V \\ \quad I(0) = 0. \end{cases}$$

Find $I(t)$. What is $\lim_{t \to \infty} I(t)$? How long does it take after the switch is closed for the current to rise to 90% of its limiting value?

Solution   The DE can be written in the form $\dfrac{dI}{dt} + \dfrac{R}{L}I = \dfrac{V}{L}$. It is linear and has integrating factor $e^{\mu(t)}$, where

$$\mu(t) = \int \frac{R}{L}\, dt = \frac{Rt}{L}.$$

Therefore,

$$\frac{d}{dt}\left(e^{Rt/L}I\right) = e^{Rt/L}\left(\frac{dI}{dt} + \frac{R}{L}I\right) = e^{Rt/L}\frac{V}{L}$$

$$e^{Rt/L}I = \frac{V}{L}\int e^{Rt/L}\, dt = \frac{V}{R}e^{Rt/L} + C$$

$$I(t) = \frac{V}{R} + Ce^{-Rt/L}.$$

Since $I(0) = 0$, we have $0 = (V/R) + C$, so $C = -V/R$. Thus the current flowing at any time $t > 0$ is

$$I(t) = \frac{V}{R}\left(1 - e^{-Rt/L}\right).$$

It is clear from this solution that $\lim_{t \to \infty} I(t) = V/R$; the *steady state* current is the current that would flow if the inductance were zero.

$I(t)$ will be 90% of this limiting value when

$$\frac{V}{R}\left(1 - e^{-Rt/L}\right) = \frac{90}{100}\frac{V}{R}.$$

This equation implies that $e^{-Rt/L} = 1/10$, or $t = (L \ln 10)/R$. The current will grow to 90% of its limiting value in $(L \ln 10)/R$ seconds.

Our final example reviews a typical *stream of payments* problem .

Example 13   A savings account is opened with a deposit of A dollars. At any time t years thereafter, money is being continually deposited into the account at a rate of $(C + Dt)$ dollars per year. If interest is also being paid into the account at a nominal rate of $100R$ percent per year, compounded continuously, find the balance $B(t)$ dollars in the account after t years. Illustrate the solution for the data $A = 5,000$, $C = 1,000$, $D = 200$, $R = 0.13$, and $t = 5$.

Solution Continuous compounding of interest causes the value of an account to grow exponentially; if the nominal rate of of interest is $100R$ percent, then $1.00 will grow to e^{Rt} dollars in t years. Without subsequent deposits, the balance in the account would grow according to the differential equation of exponential growth:

$$\frac{dB}{dt} = RB.$$

Allowing for additional growth due to the continual deposits, we observe that B must satisfy the differential equation

$$\frac{dB}{dt} = RB + (C + Dt)$$

or, equivalently, $dB/dt - RB = C + Dt$. This is a linear equation for B having $p(t) = -R$. Hence, we may take $\mu(t) = -Rt$ and $e^{\mu(t)} = e^{-Rt}$. We now calculate

$$\frac{d}{dt}\left(e^{-Rt} B(t)\right) = e^{-Rt} \frac{dB}{dt} - Re^{-Rt} B(t) = (C + Dt) e^{-Rt}$$

and

$$e^{-Rt} B(t) = \int (C + Dt)e^{-Rt} dt \qquad \text{Let} \quad U = C + Dt, \quad dV = e^{-Rt} dt.$$
$$\text{Then } dU = D\,dt, \qquad V = -e^{-Rt}/R.$$
$$= -\frac{C + Dt}{R} e^{-Rt} + \frac{D}{R} \int e^{-Rt} dt$$
$$= -\frac{C + Dt}{R} e^{-Rt} - \frac{D}{R^2} e^{-Rt} + K, \qquad (K = \text{constant}).$$

Hence

$$B(t) = -\frac{C + Dt}{R} - \frac{D}{R^2} + Ke^{Rt}.$$

Since $A = B(0) = -\dfrac{C}{R} - \dfrac{D}{R^2} + K$, we have $K = A + \dfrac{C}{R} + \dfrac{D}{R^2}$ and

$$B(t) = \left(A + \frac{C}{R} + \frac{D}{R^2}\right) e^{Rt} - \frac{C + Dt}{R} - \frac{D}{R^2}.$$

For the illustration $A = 5,000$, $C = 1,000$, $D = 200$, $R = 0.13$, and $t = 5$, we obtain, using a calculator, $B(5) = 19,762.82$. The account will contain $19,762.82, after 5 years, under these circumstances.

Exercises 17.2

Solve the separable equations in Exercises 1–10.

1. $\dfrac{dy}{dx} = \dfrac{y}{2x}$

2. $\dfrac{dy}{dx} = \dfrac{3y - 1}{x}$

3. $\dfrac{dy}{dx} = \dfrac{x^2}{y^2}$

4. $\dfrac{dy}{dx} = x^2 y^2$

5. $\dfrac{dY}{dt} = tY$

6. $\dfrac{dx}{dt} = e^x \sin t$

7. $\dfrac{dy}{dx} = 1 - y^2$

8. $\dfrac{dy}{dx} = 1 + y^2$

9. $\dfrac{dy}{dt} = 2 + e^y$

10. $\dfrac{dy}{dx} = y^2(1 - y)$

Solve the homogeneous differential equations in Exercises 11–16.

11. $\dfrac{dy}{dx} = \dfrac{x + y}{x - y}$

12. $\dfrac{dy}{dx} = \dfrac{xy}{x^2 + 2y^2}$

13. $\dfrac{dy}{dx} = \dfrac{x^2 + xy + y^2}{x^2}$

14. $\dfrac{dy}{dx} = \dfrac{x^3 + 3xy^2}{3x^2y + y^3}$

15. $x\dfrac{dy}{dx} = y + x\cos^2\left(\dfrac{y}{x}\right)$

16. $\dfrac{dy}{dx} = \dfrac{y}{x} - e^{-y/x}$

Solve the linear equations in Exercises 17–22.

17. $\dfrac{dy}{dx} - \dfrac{2y}{x} = x^2$

18. $\dfrac{dy}{dx} + \dfrac{2y}{x} = \dfrac{1}{x^2}$

19. $\dfrac{dy}{dx} + 2y = 3$ **20.** $\dfrac{dy}{dx} + y = e^x$

21. $\dfrac{dy}{dx} + y = x$ **22.** $\dfrac{dy}{dx} + 2e^x y = e^x$

Solve the initial-value problems in Exercises 23–26.

23. $\begin{cases} \dfrac{dy}{dt} + 10y = 1 \\ y(1/10) = 2/10 \end{cases}$ **24.** $\begin{cases} \dfrac{dy}{dx} + 3x^2 y = x^2 \\ y(0) = 1 \end{cases}$

25. $\begin{cases} x^2 y' + y = x^2 e^{1/x} \\ y(1) = 3e \end{cases}$ **26.** $\begin{cases} y' + (\cos x)y = 2xe^{-\sin x} \\ y(\pi) = 0 \end{cases}$

Solve the integral equations in Exercises 27–30.

27. $y(x) = 2 + \displaystyle\int_0^x \dfrac{t}{y(t)}\,dt$ **28.** $y(x) = 1 + \displaystyle\int_0^x \dfrac{\big(y(t)\big)^2}{1+t^2}\,dt$

29. $y(x) = 1 + \displaystyle\int_1^x \dfrac{y(t)\,dt}{t(t+1)}$ **30.** $y(x) = 3 + \displaystyle\int_0^x e^{-y(t)}\,dt$

31. Find an equation of the curve in the xy-plane which passes through the point $(2, 3)$, and has, at every point (x, y) on it, slope equal to $2x/(1 + y^2)$.

32. Repeat the previous problem for the point $(1, 3)$ and slope $1 + (2y/x)$.

33. Show that the change of variables $\xi = x - x_0$, $\eta = y - y_0$ transforms the equation

$$\frac{dy}{dx} = \frac{ax + by + c}{ex + fy + g}$$

into the homogeneous equation

$$\frac{d\eta}{d\xi} = \frac{a\xi + b\eta}{e\xi + f\eta}$$

provided (x_0, y_0) is the solution of the system

$$ax + by + c = 0$$
$$ex + fy + g = 0.$$

34. Use the technique of the previous exercise to solve the equation $\dfrac{dy}{dx} = \dfrac{x + 2y - 4}{2x - y - 3}$.

Show that the DEs in Exercises 35–38 are exact, and solve them.

35. $(xy^2 + y)\,dx + (x^2 y + x)\,dy = 0$

36. $(e^x \sin y + 2x)\,dx + (e^x \cos y + 2y)\,dy = 0$

37. $e^{xy}(1 + xy)\,dx + x^2 e^{xy}\,dy = 0$

38. $\left(2x + 1 - \dfrac{y^2}{x^2}\right) dx + \dfrac{2y}{x}\,dy = 0$

Show that the DEs in Exercises 39–40 admit integrating factors that are functions of x alone. Then solve the equations.

39. $(x^2 + 2y)\,dx - x\,dy = 0$

40. $(xe^x + x \ln y + y)\,dx + \left(\dfrac{x^2}{y} + x \ln x + x \sin y\right) dy = 0$

41. If $a > b > 0$ in Example 5, find $\lim_{t \to \infty} x(t)$.

42. If $b > a > 0$ in Example 5, find $\lim_{t \to \infty} x(t)$.

43. Why is the solution given in Example 5 not valid for $a = b$? Find the solution for the case $a = b$.

44. An object of mass m falling near the surface of the earth is retarded by air resistance proportional to its velocity so that, according to Newton's Second Law of Motion,

$$m\frac{dv}{dt} = mg - kv,$$

where $v = v(t)$ is the velocity of the object at time t, and g is the acceleration of gravity near the surface of the earth. Assuming that the object falls from rest at time $t = 0$, that is, $v(0) = 0$, find the velocity $v(t)$ for any $t > 0$ (up until the object strikes the ground). Show $v(t)$ approaches a limit as $t \to \infty$. Do you need the explicit formula for $v(t)$ to determine this limiting velocity?

45. Repeat Exercise 44 except assuming that the air resistance is proportional to the square of the velocity so that the equation of motion is

$$m\frac{dv}{dt} = mg - kv^2.$$

46. Find the amount in a savings account after one year if the initial balance in the account was $1,000, if the interest is paid continuously into the account at a nominal rate of 10% per annum, compounded continuously, and if the account is being continuously depleted (by taxes, say) at a rate of $y^2/1,000,000$ dollars per year, where $y = y(t)$ is the balance in the account after t years. How large can the account grow? How long will it take the account to grow to half this balance?

47. Find the family of curves each of which intersects all of the hyperbolas $xy = C$ at right angles.

48. Repeat the solution concentration problem in Example 4, changing the rate of inflow of brine into the tank to 12 L/min but leaving all the other data as they were in that example. Note that the volume of liquid in the tank is no longer constant as time increases.

49. What condition must the coefficients $M(x, y)$ and $N(x, y)$ satisfy if the equation $M\,dx + N\,dy = 0$ is to have an integrating factor of the form $\mu(y)$, and what DE must the integrating factor satisfy?

50. Find an integrating factor of the form $\mu(y)$ for the equation

$$2y^2(x + y^2)\,dx + xy(x + 6y^2)\,dy = 0,$$

and hence solve the equation. *Hint:* see Exercise 49.

51. Find an integrating factor of the form $\mu(y)$ for the equation $y\,dx - (2x + y^3 e^y)\,dy = 0$, and hence solve the equation. *Hint:* see Exercise 49.

52. What condition must the coefficients $M(x, y)$ and $N(x, y)$ satisfy if the equation $M\,dx + N\,dy = 0$ is to have an integrating factor of the form $\mu(xy)$, and what DE must the integrating factor satisfy?

53. Find an integrating factor of the form $\mu(xy)$ for the equation

$$\left(x \cos x + \dfrac{y^2}{x}\right) dx - \left(\dfrac{x \sin x}{y} + y\right) dy = 0,$$

and hence solve the equation. *Hint:* see Exercise 52.

17.3 Existence, Uniqueness, and Numerical Methods

A general first-order differential equation of the form

$$\frac{dy}{dx} = f(x, y)$$

specifies a slope $f(x, y)$ at every point (x, y) in the domain of f, and therefore represents a **slope field**. Such a slope field can be represented graphically by drawing short line segments of the indicated slope at many points in the xy-plane. Slope fields resemble vector fields, but the segments are usually drawn having the same length and without arrowheads. Figure 17.3 portrays the slope field for the differential equation

$$\frac{dy}{dx} = x - y.$$

Solving a typical initial-value problem

$$\begin{cases} \dfrac{dy}{dx} = f(x, y) \\ y(x_0) = y_0 \end{cases}$$

involves finding a function $y = \phi(x)$ such that

$$\phi'(x) = f\big(x, \phi(x)\big) \qquad \text{and} \qquad \phi(x_0) = y_0.$$

The graph of the equation $y = \phi(x)$ is a curve passing through (x_0, y_0) that is tangent to the slope-field at each point. Such curves are called **solution curves** of the differential equation. Figure 17.3 shows four solution curves for $y' = x - y$ corresponding to the initial conditions $y(0) = C$, where $C = -2, -1, 0$, and 1.

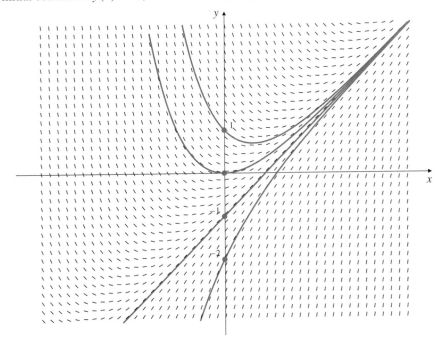

Figure 17.3   The slope field for the DE $y' = x - y$ and four solution curves for this DE

The DE $y' = x - y$ is linear and can be solved explicitly by the method of Section 17.2. Indeed, the solution satisfying $y(0) = C$ is $y = x - 1 + (C + 1)e^{-x}$. Most differential equations of the form $y' = f(x, y)$ cannot be solved for y as an explicit function of x, so we must use numerical approximation methods to find the value of a solution function $\phi(x)$ at particular points.

Existence and Uniqueness of Solutions

Even if we cannot calculate an explicit solution of an initial-value problem, it is important to know when the problem has a solution and whether that solution is unique.

THEOREM 3

An existence and uniqueness theorem for first-order initial-value problems

Suppose that $f(x, y)$ and $f_2(x, y) = (\partial/\partial y) f(x, y)$ are continuous on a rectangle R of the form $a \leq x \leq b, c \leq y \leq d$, containing the point (x_0, y_0) in its interior. Then there exists a number $\delta > 0$ and a *unique* function $\phi(x)$ defined and having a continuous derivative on the interval $(x_0 - \delta, x_0 + \delta)$ such that $\phi(x_0) = y_0$ and $\phi'(x) = f(x, \phi(x))$ for $x_0 - \delta < x < x_0 + \delta$. In other words, the initial-value problem

$$\begin{cases} \dfrac{dy}{dx} = f(x, y) \\ y(x_0) = y_0 \end{cases} \tag{$*$}$$

has a unique solution on $(x_0 - \delta, x_0 + \delta)$.

We give only an outline of the proof here. Any solution $y = \phi(x)$ of the initial-value problem $(*)$ must also satisfy the **integral equation**

$$\phi(x) = y_0 + \int_{x_0}^{x} f(t, \phi(t)) \, dt, \tag{$**$}$$

and, conversely, any solution of the integral equation $(**)$ must also satisfy the initial-value problem $(*)$. A sequence of approximations $\phi_n(x)$ to a solution of $(**)$ can be constructed as follows:

$$\phi_0(x) = y_0$$

$$\phi_{n+1}(x) = y_0 + \int_{x_0}^{x} f(t, \phi_n(t)) \, dt \qquad \text{for} \quad n = 0, 1, 2, \dots$$

(These are called **Picard iterations.**) The proof of Theorem 3 involves showing that

$$\lim_{n \to \infty} \phi_n(x) = \phi(x)$$

exists on an interval $(x_0 - \delta, x_0 + \delta)$ and that the resulting limit $\phi(x)$ satisfies the integral equation $(**)$. The details can be found in more advanced texts on differential equations and analysis.

Remark   Some initial-value problems can have nonunique solutions. For example, the functions $y_1(x) = x^3$ and $y_2(x) = 0$ both satisfy the initial-value problem

$$\begin{cases} \dfrac{dy}{dx} = 3y^{2/3} \\ y(0) = 0. \end{cases}$$

In this case $f(x, y) = 3y^{2/3}$ is continuous on the whole xy-plane. However, $\partial f/\partial y = 2y^{-1/3}$ is not continuous on the x-axis and is therefore not continuous on any rectangle containing $(0, 0)$ in its interior. The conditions of Theorem 3 are not satisfied, and the initial-value problem has a solution but not a unique one.

Remark   The unique solution $y = \phi(x)$ to the initial-value problem $(*)$ guaranteed by Theorem 3 may not be defined on the whole interval $[a, b]$ because it can "escape" from the rectangle R through the top or bottom edges. Even if $f(x, y)$ and $(\partial/\partial y) f(x, y)$ are continuous on the whole xy-plane, the solution may not be defined on the whole real line. For example,

$$y = \frac{1}{1 - x} \qquad \text{satisfies the initial-value problem} \qquad \begin{cases} \dfrac{dy}{dx} = y^2 \\ y(0) = 1 \end{cases}$$

but only for $x < 1$. Starting from $(0, 1)$, we can follow the solution curve as far as we want to the left of $x = 0$, but to the right of $x = 0$ the curve recedes to ∞ as $x \to 1-$. (See Figure 17.4.) It makes no sense to regard the part of the curve to the right of $x = 1$ as part of the solution curve to the initial-value problem.

Numerical Methods

Suppose that the conditions of Theorem 3 are satisfied, so we know that the initial-value problem

$$\begin{cases} \dfrac{dy}{dx} = f(x, y) \\ y(x_0) = y_0 \end{cases}$$

has a unique solution $y = \phi(x)$ on some interval containing x_0. Even if we cannot solve the differential equation and find $\phi(x)$ explicitly, we can still try to find approximate values y_n for $\phi(x_n)$ at a sequence of points

$$x_0, \quad x_1 = x_0 + h, \quad x_2 = x_0 + 2h, \quad x_3 = x_0 + 3h, \quad \dots$$

starting at x_0. Here, $h > 0$ (or $h < 0$) is called the **step size** of the approximation scheme. In the remainder of this section we will describe three methods for constructing the approximations $\{y_n\}$:

1. the Euler method,
2. the improved Euler method, and
3. the fourth-order Runge–Kutta method.

Each of these methods starts with the given value of y_0 and provides a formula for constructing y_{n+1} when you know y_n. The three methods are listed above in increasing order of the complexity of their formulas, but the more complicated formulas produce much better approximations for any given step size h.

The **Euler method** involves approximating the solution curve $y = \phi(x)$ by a polygonal line (a sequence of straight line segments joined end to end), where each segment has horizontal length h and slope determined by the value of $f(x, y)$ at the end of the previous segment. Thus, if $x_n = x_0 + nh$, then

$$y_1 = y_0 + f(x_0, y_0)h$$
$$y_2 = y_1 + f(x_1, y_1)h$$
$$y_3 = y_2 + f(x_2, y_2)h$$

and, in general,

> **Iteration formulas for Euler's method**
>
> $$x_{n+1} = x_n + h, \qquad y_{n+1} = y_n + hf(x_n, y_n).$$

Example 1   Use Euler's method to find approximate values for the solution of the initial-value problem

$$\begin{cases} \dfrac{dy}{dx} = x - y \\ y(0) = 1 \end{cases}$$

on the interval $[0, 1]$ using

(a) 5 steps of size $h = 0.2$, and

(b) 10 steps of size $h = 0.1$.

Calculate the error at each step, given that the problem (which involves a linear equation and so can be solved explicitly) has solution $y = \phi(x) = x - 1 + 2e^{-x}$.

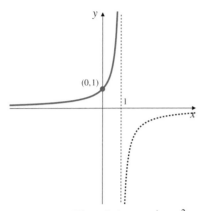

Figure 17.4 The solution to $y' = y^2$, $y(0) = 1$ is the part of the curve $y = 1/(1 - x)$ to the left of the vertical asymptote at $x = 1$

Solution

(a) Here we have $f(x, y) = x - y$, $x_0 = 0$, $y_0 = 1$, and $h = 0.2$, so that

$$x_n = \frac{n}{5}, \qquad y_{n+1} = y_n + 0.2(x_n - y_n),$$

and the error is $e_n = \phi(x_n) - y_n$ for $n = 0, 1, 2, 3, 4$, and 5. The results of the calculation, which was done easily using a computer spreadsheet program, are presented in Table 1.

Table 1.   Euler approximations with $h = 0.2$

n	x_n	y_n	$f(x_n, y_n)$	y_{n+1}	$e_n = \phi(x_n) - y_n$
0	0.0	1.000 000	−1.000 000	0.800 000	0.000 000
1	0.2	0.800 000	−0.600 000	0.680 000	0.037 462
2	0.4	0.680 000	−0.280 000	0.624 000	0.060 640
3	0.6	0.624 000	−0.024 000	0.619 200	0.073 623
4	0.8	0.619 200	0.180 800	0.655 360	0.079 458
5	1.0	0.655 360	0.344 640		0.080 399

The exact solution $y = \phi(x)$ and the polygonal line representing the Euler approximation are shown in Figure 17.5. The approximation lies below the solution curve, as is reflected in the positive values in the last column of Table 1, representing the error at each step.

(b) Here we have $h = 0.1$, so that

$$x_n = \frac{n}{10}, \qquad y_{n+1} = y_n + 0.1(x_n - y_n)$$

for $n = 0, 1, \ldots, 10$. Again we present the results in tabular form:

Table 2.   Euler approximations with $h = 0.1$

n	x_n	y_n	$f(x_n, y_n)$	y_{n+1}	$e_n = \phi(x_n) - y_n$
0	0.0	1.000 000	−1.000 000	0.900 000	0.000 000
1	0.1	0.900 000	−0.800 000	0.820 000	0.009 675
2	0.2	0.820 000	−0.620 000	0.758 000	0.017 462
3	0.3	0.758 000	−0.458 000	0.712 200	0.023 636
4	0.4	0.712 200	−0.312 200	0.680 980	0.028 440
5	0.5	0.680 980	−0.180 980	0.662 882	0.032 081
6	0.6	0.662 882	−0.062 882	0.656 594	0.034 741
7	0.7	0.656 594	0.043 406	0.660 934	0.036 577
8	0.8	0.660 934	0.139 066	0.674 841	0.037 724
9	0.9	0.674 841	0.225 159	0.697 357	0.038 298
10	1.0	0.697 357	0.302 643		0.038 402

Observe that the error at the end of the first step is about one-quarter of the error at the end of the first step in part (a), but the final error at $x = 1$ is only about half as large as in part (a). This behaviour is characteristic of Euler's method.

If we decrease the step size h, it takes more steps ($n = |x - x_0|/h$) to get from the starting point x_0 to a particular value x where we want to know the value of the solution. For Euler's method it can be shown that the error at each step decreases on average proportionally to h^2, but the errors can accumulate from step to step, so the error at x can be expected to decrease proportionally to $nh^2 = |x - x_0|h$. This is consistent with the results of Example 1. Decreasing h and so increasing n is costly in terms of computing resources, so we would like to find ways of reducing the error without decreasing the step size. This is similar to developing better techniques than the Trapezoid Rule for evaluating definite integrals numerically.

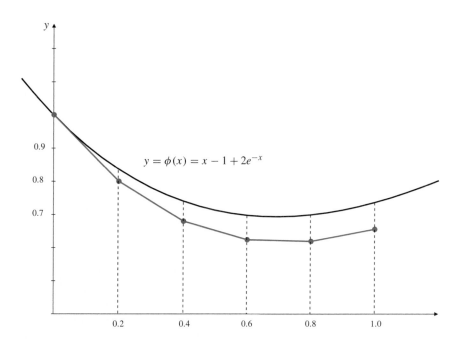

Figure 17.5   The solution $y = \phi(x)$ to $y' = x - y$, $y(0) = 1$ and an Euler approximation to it on $[0, 1]$ with step size $h = 0.2$

The **improved Euler method** is a step in this direction. The accuracy of the Euler method is hampered by the fact that the slope of each segment in the approximating polygonal line is determined by the value of $f(x, y)$ at one endpoint of the segment. Since f varies along the segment, we would expect to do better by using, say, the average value of $f(x, y)$ at the two ends of the segment, that is, by calculating y_{n+1} from the formula

$$y_{n+1} = y_n + h \frac{f(x_n, y_n) + f(x_{n+1}, y_{n+1})}{2}.$$

Unfortunately, y_{n+1} appears on both sides of this equation, and we can't usually solve the equation for y_{n+1}. We can get around this difficulty by replacing y_{n+1} on the right side by its Euler approximation $y_n + hf(x_n, y_n)$. The resulting formula is the basis for the improved Euler method.

Iteration formulas for the improved Euler method

$$x_{n+1} = x_n + h$$
$$u_{n+1} = y_n + h f(x_n, y_n)$$
$$y_{n+1} = y_n + h \frac{f(x_n, y_n) + f(x_{n+1}, u_{n+1})}{2}.$$

Example 2   Use the improved Euler method with $h = 0.2$ to find approximate values for the solution to the initial-value problem of Example 1 on $[0, 1]$. Compare the errors with those obtained by the Euler method.

Solution   Table 3 summarizes the calculation of five steps of the improved Euler method for $f(x, y) = x - y$, $x_0 = 0$, and $y_0 = 1$.

Table 3.   Improved Euler approximations with $h = 0.2$

n	x_n	y_n	u_{n+1}	y_{n+1}	$e_n = \phi(x_n) - y_n$
0	0.0	1.000 000	0.800 000	0.840 000	0.000 000
1	0.2	0.840 000	0.712 000	0.744 800	−0.002 538
2	0.4	0.744 800	0.675 840	0.702 736	−0.004 160
3	0.6	0.702 736	0.682 189	0.704 244	−0.005 113
4	0.8	0.704 244	0.723 395	0.741 480	−0.005 586
5	1.0	0.741 480	0.793 184		−0.005 721

Observe that the errors are considerably less than one-tenth those obtained in Example 1(a). Of course, more calculations are necessary at each step, but the number of evaluations of $f(x, y)$ required is only twice the number required for Example 1(a). As for numerical integration, if f is complicated, it is these function evaluations that constitute most of the computational "cost" of computing numerical solutions.

Remark   It can be shown for well-behaved functions f that the error at each step in the improved Euler method is bounded by a multiple of h^3 rather than h^2 as for the (unimproved) Euler method. Thus the cumulative error at x can be bounded by a constant times $|x - x_0|h^2$. If Example 2 is repeated with 10 steps of size $h = 0.1$, the error at $n = 10$ i.e., at $x = 1$) is $-0.001\,323$, which is about one-fourth the size of the error at $x = 1$ with $h = 0.2$.

The fourth-order Runge–Kutta method further improves upon the improved Euler method, but at the expense of requiring more complicated calculations at each step. It requires four evaluations of $f(x, y)$ at each step, but the error at each step is less than a constant times h^5, so the cumulative error decreases like h^4 as h decreases. Like the improved Euler method, this method involves calculating a certain kind of average slope for each segment in the polygonal approximation to the solution to the initial-value problem. We present the appropriate formulas below but cannot derive them here.

Iteration formulas for the Runge–Kutta method

$$x_{n+1} = x_n + h$$
$$p_n = f(x_n, y_n)$$
$$q_n = f\left(x_n + \frac{h}{2}, y_n + \frac{h}{2} p_n\right)$$
$$r_n = f\left(x_n + \frac{h}{2}, y_n + \frac{h}{2} q_n\right)$$
$$s_n = f(x_n + h, y_n + h r_n)$$
$$y_{n+1} = y_n + h\,\frac{p_n + 2q_n + 2r_n + s_n}{6}.$$

Example 3   Use the fourth-order Runge-Kutta method with $h = 0.2$ to find approximate values for the solution to the initial-value problem of Example 1 on $[0, 1]$. Compare the errors with those obtained by the Euler and improved Euler methods.

Solution   Table 4 summarizes the calculation of five steps of the Runge–Kutta method for $f(x, y) = x - y$, $x_0 = 0$, and $y_0 = 1$ according to the formulas above. The table does not show the values of the intermediate quantities p_n, q_n, r_n, and s_n, but columns for these quantities were included in the spreadsheet in which the calculations were made.

Table 4.   Fourth-order Runge–Kutta approximations with $h = 0.2$

n	x_n	y_n	$e_n = \phi(x_n) - y_n$
0	0.0	1.000 000	0.000 000 0
1	0.2	0.837 467	$-0.000\,005\,2$
2	0.4	0.740 649	$-0.000\,008\,5$
3	0.6	0.697 634	$-0.000\,010\,4$
4	0.8	0.698 669	$-0.000\,011\,3$
5	1.0	0.735 770	$-0.000\,011\,6$

The errors here are about 1/500 of the size of the errors obtained with the improved Euler method and about 1/7,000 of the size of the errors obtained with the Euler method. This great improvement was achieved at the expense of doubling the number of function evaluations required in the improved Euler method and quadrupling the number required in the Euler method. If we use 10 steps of size $h = 0.1$ in the Runge–Kutta method, the error at $x = 1$ is reduced to $-6.664\,82 \times 10^{-7}$, which is less than $1/16$ of its value when $h = 0.2$.

Our final example shows what can happen with numerical approximations to a solution that is unbounded.

Example 4 Obtain solutions at $x = 0.4$, $x = 0.8$, and $x = 1.0$ for solutions to the initial-value problem

$$\begin{cases} y' = y^2 \\ y(0) = 1 \end{cases}$$

using all three methods described above, and using step sizes $h = 0.2$, $h = 0.1$, and $h = 0.05$ for each method. What do the results suggest about the values of the solution at these points? Compare the results with the actual solution $y = 1/(1 - x)$.

Solution The various approximations are calculated using the various formulas described above for $f(x, y) = y^2$, $x_0 = 0$, and $y_0 = 1$. The results are presented in Table 5.

Table 5. Comparing methods and step sizes for $y' = y^2$, $y(0) = 1$

	$h = 0.2$	$h = 0.1$	$h = 0.05$
Euler			
$x = 0.4$	1.488 000	1.557 797	1.605 224
$x = 0.8$	2.676 449	3.239 652	3.793 197
$x = 1.0$	4.109 124	6.128 898	9.552 668
Improved Euler			
$x = 0.4$	1.640 092	1.658 736	1.664 515
$x = 0.8$	4.190 396	4.677 726	4.897 519
$x = 1.0$	11.878 846	22.290 765	43.114 668
Runge–Kutta			
$x = 0.4$	1.666 473	1.666 653	1.666 666
$x = 0.8$	4.965 008	4.996 663	4.999 751
$x = 1.0$	41.016 258	81.996 399	163.983 395

Little useful information can be read from the Euler results. The improved Euler results suggest that the solution exists at $x = 0.4$ and $x = 0.8$, but likely not at $x = 1$. The Runge–Kutta results confirm this and suggest that $y(0.4) = 5/3$ and $y(0.8) = 5$, which are the correct values provided by the actual solution $y = 1/(1 - x)$. They also suggest very strongly that the solution "blows up" at (or near) $x = 1$.

Exercises 17.3

A computer is almost essential for doing most of these exercises. The calculations are easily done with a spreadsheet program in which formulas for calculating the various quantities involved can be replicated down columns to automate the iteration process.

1. Use the Euler method with step sizes (a) $h = 0.2$, (b) $h = 0.1$, and (c) $h = 0.05$ to approximate $y(2)$ given that $y' = x + y$ and $y(1) = 0$.

2. Repeat Exercise 1 using the improved Euler method.

3. Repeat Exercise 1 using the Runge–Kutta method.

4. Use the Euler method with step sizes (a) $h = 0.2$ and (b) $h = 0.1$ to approximate $y(2)$ given that $y' = xe^{-y}$ and $y(0) = 0$.

5. Repeat Exercise 4 using the improved Euler method.

6. Repeat Exercise 4 using the Runge–Kutta method.

7. Use the Euler method with (a) $h = 0.2$, (b) $h = 0.1$, and (c) $h = 0.05$ to approximate $y(1)$ given that $y' = \cos y$ and $y(0) = 0$.

8. Repeat Exercise 7 using the improved Euler method.

9. Repeat Exercise 7 using the Runge–Kutta method.

10. Use the Euler method with (a) $h = 0.2$, (b) $h = 0.1$, and (c) $h = 0.05$ to approximate $y(1)$ given that $y' = \cos(x^2)$ and $y(0) = 0$.

11. Repeat Exercise 10 using the improved Euler method.

12. Repeat Exercise 10 using the Runge–Kutta method.

Solve the integral equations in Exercises 13–14 by rephrasing them as initial-value problems.

13. $y(x) = 2 + \int_1^x \big(y(t)\big)^2 dt$. *Hint:* Find $\dfrac{dy}{dx}$ and $y(1)$.

14. $u(x) = 1 + 3 \int_2^x t^2 u(t) \, dt$. *Hint:* Find $\dfrac{du}{dx}$ and $u(2)$.

15. The methods of this section can be used to approximate definite integrals numerically. For example,

$$I = \int_a^b f(x) \, dx$$

is given by $I = y(b)$, where

$$y' = f(x), \qquad \text{and} \qquad y(a) = 0.$$

Show that one step of the Runge–Kutta method with $h = b - a$ gives the same result for I as Simpson's Rule (Section 6.7) with two subintervals of length $h/2$.

16. If $\phi(0) = A \geq 0$ and $\phi'(x) \geq k\phi(x)$ on $[0, X]$, where $k > 0$ and $X > 0$ are constants, show that $\phi(x) \geq Ae^{kx}$ on $[0, X]$. *Hint:* Calculate $(d/dx)(\phi(x)/e^{kx})$.

*17. Consider the three initial-value problems

$$\text{(A)} \qquad u' = u^2 \qquad\qquad u(0) = 1$$

$$\text{(B)} \qquad y' = x + y^2 \qquad\qquad y(0) = 1$$

$$\text{(C)} \qquad v' = 1 + v^2 \qquad\qquad v(0) = 1$$

(a) Show that the solution of (B) remains between the solutions of (A) and (C) on any interval $[0, X]$ where solutions of all three problems exist. *Hint:* We must have $u(x) \geq 1$, $y(x) \geq 1$, and $v(x) \geq 1$ on $[0, X]$. (Why?) Apply the result of Exercise 16 to $\phi = y - u$ and to $\phi = v - y$.

(b) Find explicit solutions for problems (A) and (C). What can you conclude about the solution to problem (B)?

(c) Use the Runge–Kutta method with $h = 0.05$, $h = 0.02$, and $h = 0.01$ to approximate the solution to (B) on $[0, 1]$. What can you conclude now?

17.4 Differential Equations of Second Order

The general second-order ordinary differential equation is of the form

$$F\left(\frac{d^2y}{dx^2}, \frac{dy}{dx}, y, x\right) = 0$$

for some function F of four variables. When such an equation can be solved explicitly for y as a function of x, the solution typically involves two integrations and therefore two arbitrary constants. A unique solution usually results from prescribing the values of the solution y and its first derivative $y' = dy/dx$ at a particular point. Such a prescription constitutes an **initial-value problem** for the second-order equation.

Equations Reducible to First Order

A second-order equation of the form

$$F\left(\frac{d^2y}{dx^2}, \frac{dy}{dx}, x\right) = 0$$

that does not involve the unknown function y explicitly (except through its derivatives) can be reduced to a first-order equation by a change of dependent variable; if $v = dy/dx$, then the equation can be written

$$F\left(\frac{dv}{dx}, v, x\right) = 0.$$

This first-order equation in v may be amenable to the techniques described in earlier sections. If an explicit solution $v = v(x)$ can be found and integrated, then the function

$$y = \int v(x) \, dx$$

is an explicit solution of the given equation.

Example 1   Solve the initial-value problem

$$\frac{d^2y}{dx^2} = x\left(\frac{dy}{dx}\right)^2, \qquad y(0) = 1, \qquad y'(0) = -2.$$

Solution   If we let $v = dy/dx$, the given differential equation becomes

$$\frac{dv}{dx} = xv^2,$$

which is a separable first-order equation. Thus

$$\frac{dv}{v^2} = x\,dx$$

$$-\frac{1}{v} = \frac{x^2}{2} + \frac{C_1}{2}$$

$$v = -\frac{2}{x^2 + C_1}.$$

The initial condition $y'(0) = -2$ implies that $v(0) = -2$ and so $C_1 = 1$. Therefore,

$$y = -2\int \frac{dx}{x^2 + 1} = -2\tan^{-1}x + C_2.$$

The initial condition $y(0) = 1$ implies that $C_2 = 1$, so the solution of the given initial-value problem is $y = 1 - 2\tan^{-1}x$.

A second-order equation of the form

$$F\left(\frac{d^2y}{dx^2}, \frac{dy}{dx}, y\right) = 0$$

that does not explicitly involve the independent variable x can be reduced to a first-order equation by a change of both dependent and independent variables. Again let $v = dy/dx$, but regard v as a function of y rather than x; $v = v(y)$. Then

$$\frac{d^2y}{dx^2} = \frac{dv}{dx} = \frac{dv}{dy}\frac{dy}{dx} = v\frac{dv}{dy}$$

by the Chain Rule. Hence, the given differential equation becomes

$$F\left(v\frac{dv}{dy}, v, y\right) = 0,$$

which is a first-order equation for v as a function of y. If this equation can be solved for $v = v(y)$, there still remains the problem of solving the separable equation $(dy/dx) = v(y)$ for y as a function of x.

Example 2   Solve the equation $y\dfrac{d^2y}{dx^2} = \left(\dfrac{dy}{dx}\right)^2$.

Solution   The change of variable $dy/dx = v(y)$ leads to the equation

$$yv\frac{dv}{dy} = v^2,$$

which is separable, $dv/v = dy/y$, and has solution $v = C_1y$. The equation

$$\frac{dy}{dx} = C_1y$$

is again separable and leads to

$$\frac{dy}{y} = C_1\,dx$$

$$\ln|y| = C_1x + C_2$$

$$y = \pm e^{C_1x + C_2} = C_3e^{C_1x}.$$

Second-Order Linear Equations

The most frequently encountered ordinary differential equations arising in applications are second-order linear equations. The general second-order linear equation is of the form

$$a_2(x)\frac{d^2y}{dx^2} + a_1(x)\frac{dy}{dx} + a_0(x)y = f(x).$$

As remarked in Section 17.1, if $f(x) = 0$ identically, then we say that the equation is **homogeneous**. If the coefficients $a_2(x)$, $a_1(x)$, and $a_0(x)$ are continuous on an interval and $a_2(x) \neq 0$ there, then the homogeneous equation

$$a_2(x)\frac{d^2y}{dx^2} + a_1(x)\frac{dy}{dx} + a_0(x)y = 0$$

has a general solution of the form

$$y_h = C_1y_1(x) + C_2y_2(x),$$

where $y_1(x)$ and $y_2(x)$ are two **independent** solutions, that is, two solutions with the property that $C_1y_1(x) + C_2y_2(x) = 0$ for all x in the interval only if $C_1 = C_2 = 0$. (We will not prove this here.)

Whenever one solution, $y_1(x)$, of a homogeneous linear second-order equation is known, another independent solution (and therefore the general solution) can be found by substituting $y = v(x)y_1(x)$ into the differential equation. This leads to a first-order, linear, separable equation for v'.

Example 3   Show that $y_1 = e^{-2x}$ is a solution of $y'' + 4y' + 4y = 0$, and find the general solution of this equation.

Solution   Since $y_1' = -2e^{-2x}$ and $y_1'' = 4e^{-2x}$, we have

$$y_1'' + 4y_1' + 4y_1 = e^{-2x}(4 - 8 + 4) = 0,$$

so y_1 is indeed a solution of the given differential equation. To find the general solution, try $y = y_1v = e^{-2x}v(x)$. We have

$$y' = -2e^{-2x}v + e^{-2x}v'$$
$$y'' = 4e^{-2x}v - 4e^{-2x}v' + e^{-2x}v''.$$

Substituting these expressions into the given DE we obtain

$$0 = y'' + 4y' + 4y$$
$$= e^{-2x}(4v - 4v' + v'' - 8v + 4v' + 4v) = e^{-2x}v''.$$

Thus, $y = y_1v$ is a solution provided $v''(x) = 0$. This equation for v has the general solution $v = C_1 + C_2x$, so the given equation has the general solution

$$y = C_1e^{-2x} + C_2xe^{-2x} = C_1y_1(x) + C_2y_2(x),$$

where $y_2 = xe^{-2x}$ is a second solution of the DE, independent of y_1.

By Theorem 2 of Section 17.1, the general solution of the second-order, linear, nonhomogeneous equation (with $f(x) \neq 0$) is of the form

$$y = y_p(x) + y_h(x),$$

where $y_p(x)$ is any particular solution of the nonhomogeneous equation, and $y_h(x)$ is the general solution (as described above) of the corresponding homogeneous equation. In Section 17.6 we will discuss the solution of nonhomogeneous linear equations. First, however, in Section 17.5 we concentrate on some special classes of homogeneous, linear equations.

‖‖‖‖‖‖‖ Exercises 17.4 ‖‖‖‖‖‖‖

1. Show that $y = e^x$ is a solution of $y'' - 3y' + 2y = 0$, and find the general solution of this DE.

2. Show that $y = e^{-2x}$ is a solution of $y'' - y' - 6y = 0$, and find the general solution of this DE.

3. Show that $y = x$ is a solution of $x^2 y'' + 2xy' - 2y = 0$ on the interval $(0, \infty)$, and find the general solution on this interval.

4. Show that $y = x^2$ is a solution of $x^2 y'' - 3xy' + 4y = 0$ on the interval $(0, \infty)$, and find the general solution on this interval.

5. Show that $y = x$ is a solution of the differential equation $x^2 y'' - (2x + x^2)y' + (2 + x)y = 0$, and find the general solution of this equation.

6. Show that $y = x^{-1/2} \cos x$ is a solution of the Bessel equation with $\nu = 1/2$:

$$x^2 y'' + xy' + \left(x^2 - \frac{1}{4}\right)y = 0.$$

Find the general solution of this equation.

First-order systems

7. A system of n first-order, linear, differential equations in n unknown functions $y_1, \ y_2, \ \cdots, \ y_n$ is written

$$y_1' = a_{11}(x)y_1 + a_{12}(x)y_2 + \cdots + a_{1n}(x)y_n + f_1(x)$$
$$y_2' = a_{21}(x)y_1 + a_{22}(x)y_2 + \cdots + a_{2n}(x)y_n + f_2(x)$$
$$\vdots$$
$$y_n' = a_{n1}(x)y_1 + a_{n2}(x)y_2 + \cdots + a_{nn}(x)y_n + f_n(x).$$

Such a system is called an $n \times n$ **first-order linear system** and can be rewritten in vector-matrix form as $\mathbf{y}' = \mathcal{A}(x)\mathbf{y} + \mathbf{f}(x)$, where

$$\mathbf{y}(x) = \begin{pmatrix} y_1(x) \\ \vdots \\ y_n(x) \end{pmatrix}, \quad \mathbf{f}(x) = \begin{pmatrix} f_1(x) \\ \vdots \\ f_n(x) \end{pmatrix},$$

$$\mathcal{A}(x) = \begin{pmatrix} a_{11}(x) & \cdots & a_{1n}(x) \\ \vdots & \ddots & \vdots \\ a_{n1}(x) & \cdots & a_{nn}(x) \end{pmatrix}.$$

Show that the second-order, linear equation $y'' + a_1(x)y' + a_0(x)y = f(x)$ can be transformed into a 2×2 first-order system with $y_1 = y$ and $y_2 = y'$ having

$$\mathcal{A}(x) = \begin{pmatrix} 0 & 1 \\ -a_0(x) & -a_1(x) \end{pmatrix}, \quad \mathbf{f}(x) = \begin{pmatrix} 0 \\ f(x) \end{pmatrix}.$$

8. Generalize Exercise 7 to transform an nth-order linear equation

$$y^{(n)} + a_{n-1}(x)y^{(n-1)} + a_{n-2}(x)y^{(n-2)} + \cdots + a_0(x)y = f(x)$$

into an $n \times n$ first-order system.

9. If $\mathcal{A}$ is an $n \times n$ constant matrix, and if there exists a scalar λ and a nonzero constant vector $\mathbf{v}$ for which $\mathcal{A}\mathbf{v} = \lambda\mathbf{v}$, show that $\mathbf{y} = C_1 e^{\lambda x}\mathbf{v}$ is a solution of the homogeneous system $\mathbf{y}' = \mathcal{A}\mathbf{y}$.

10. Show that the determinant $\begin{vmatrix} 2 - \lambda & 1 \\ 2 & 3 - \lambda \end{vmatrix}$ is zero for two distinct values of λ. For each of these values find a nonzero vector $\mathbf{v}$ that satisfies the condition $\begin{pmatrix} 2 & 1 \\ 2 & 3 \end{pmatrix}\mathbf{v} = \lambda\mathbf{v}$. Hence solve the system

$$y_1' = 2y_1 + y_2, \qquad y_2' = 2y_1 + 3y_2.$$

17.5 Linear Differential Equations with Constant Coefficients

A differential equation of the form

$$a\,y'' + b\,y' + cy = 0, \qquad (\ast)$$

where a, b, and c are constants and $a \neq 0$, is called a **second-order, linear, homogeneous** differential equation with constant coefficients. The *second-order* refers to the presence of a second derivative; the terms *linear* and *homogeneous* refer to the fact that if $y_1(t)$ and $y_2(t)$ are two solutions of the equation, then so is $y(t) = Ay_1(t) + By_2(t)$ for any constants A and B:

If $ay_1''(t) + by_1'(t) + cy_1(t) = 0$ and $ay_2''(t) + by_2'(t) + cy_2(t) = 0$, and if $y(t) = Ay_1(t) + By_2(t)$, then $ay''(t) + by'(t) + cy(t) = 0$.

(See Section 17.1 for more details on this terminology.) Throughout this section we will assume that the independent variable in our functions is t rather than x, so the prime ($'$) refers to the derivative d/dt. This is because in most applications of such equations the independent variable is time.

Equations of type ($*$) arise in many applications of mathematics. In particular, they can model mechanical vibrations such as the motion of a mass suspended from an elastic spring or the current in certain electrical circuits. In most such applications the three constants a, b, and c are positive, although sometimes we may have $b = 0$.

Recipe for Solving $ay'' + by' + cy = 0$

In Section 3.4 we observed that the first-order, constant-coefficient equation $y' = ky$ has solution $y = Ce^{kt}$. Let us try to find a solution of equation ($*$) having the form $y = e^{rt}$. Substituting this expression into equation ($*$), we obtain

$$ar^2 e^{rt} + br e^{rt} + c e^{rt} = 0.$$

Since e^{rt} is never zero, $y = e^{rt}$ will be a solution of the differential equation ($*$) if and only if r satisfies the quadratic **auxiliary equation**

$$ar^2 + br + c = 0 \qquad (**)$$

which has roots given by the quadratic formula:

$$r = \frac{-b \pm \sqrt{b^2 - 4ac}}{2a} = -\frac{b}{2a} \pm \frac{\sqrt{D}}{2a},$$

where $D = b^2 - 4ac$ is called the **discriminant** of the auxiliary equation ($**$).

There are three cases to consider, depending on whether the discriminant D is positive, zero, or negative.

CASE I Suppose $D = b^2 - 4ac > 0$. Then the auxiliary equation has two different real roots, r_1 and r_2, given by

$$r_1 = \frac{-b - \sqrt{D}}{2a}, \qquad r_2 = \frac{-b + \sqrt{D}}{2a}.$$

(Sometimes these roots can be found easily by factoring the left side of the auxiliary equation.) In this case both $y = y_1(t) = e^{r_1 t}$ and $y = y_2(t) = e^{r_2 t}$ are solutions of the differential equation ($*$), and neither is a multiple of the other. As noted above, the function

$$y = A e^{r_1 t} + B e^{r_2 t}$$

is also a solution for any choice of the constants A and B. Since the differential equation is of second order and this solution involves two arbitrary constants, we suspect it is the **general solution**, that is, that every solution of the differential equation can be written in this form. Exercise 18 at the end of this section outlines a way to prove this.

CASE II Suppose $D = b^2 - 4ac = 0$. Then the auxiliary equation has two equal roots, $r_1 = r_2 = -b/(2a) = r$, say. Certainly $y = e^{rt}$ is a solution of ($*$). We can find the general solution by letting $y = e^{rt} u(t)$ and calculating:

$$y' = e^{rt}\big(u'(t) + ru(t)\big),$$
$$y'' = e^{rt}\big(u''(t) + 2ru'(t) + r^2 u(t)\big).$$

Substituting these expressions into ($*$), we obtain

$$e^{rt}\big(au''(t) + (2ar + b)u'(t) + (ar^2 + br + c)u(t)\big) = 0.$$

Since $e^{rt} \neq 0$, $2ar + b = 0$ and r satisfies (**), this equation reduces to $u''(t) = 0$, which has general solution $u(t) = A + Bt$ for arbitrary constants A and B. Thus the general solution of (*) in this case is

$$y = A\,e^{rt} + Bt\,e^{rt}.$$

CASE III Suppose $D = b^2 - 4ac < 0$. Then the auxiliary equation (**) has complex conjugate roots given by

$$r = \frac{-b \pm \sqrt{b^2 - 4ac}}{2a} = k \pm i\omega,$$

where $k = -b/(2a)$, $\omega = \sqrt{4ac - b^2}/(2a)$, and i is the imaginary unit ($i^2 = -1$; see Appendix I). As in Case I, the functions $y_1^*(t) = e^{(k+i\omega)t}$ and $y_2^*(t) = e^{(k-i\omega)t}$ are two independent solutions of (*), but they are not real-valued. However, since

$$e^{ix} = \cos x + i \sin x \qquad \text{and} \qquad e^{-ix} = \cos x - i \sin x$$

(as noted in the previous section and in Appendix II), we can find two real-valued functions that are solutions of (*) by suitably combining y_1^* and y_2^*:

$$y_1(t) = \frac{1}{2}y_1^*(t) + \frac{1}{2}y_2^*(t) = e^{kt} \cos(\omega t),$$

$$y_2(t) = \frac{1}{2i}y_1^*(t) - \frac{1}{2i}y_2^*(t) = e^{kt} \sin(\omega t).$$

Therefore, the general solution of (*) in this case is

$$y = A\,e^{kt}\,\cos(\omega t) + B\,e^{kt}\,\sin(\omega t).$$

The following examples illustrate the recipe for solving (*) in each of the three cases.

Example 1 Find the general solution of $y'' + y' - 2y = 0$.

Solution The auxiliary equation is $r^2 + r - 2 = 0$, or $(r + 2)(r - 1) = 0$. The auxiliary roots are $r_1 = -2$ and $r_2 = 1$, which are real and unequal. According to Case I, the general solution of the differential equation is
$$y = A\,e^{-2t} + Be^t.$$

Example 2 Find the general solution of $y'' + 6y' + 9y = 0$.

Solution The auxiliary equation is $r^2 + 6r + 9 = 0$, or $(r + 3)^2 = 0$, which has equal roots $r = -3$. According to Case II, the general solution of the differential equation is
$$y = A\,e^{-3t} + Bt\,e^{-3t}.$$

Example 3 Find the general solution of $y'' + 4y' + 13y = 0$.

Solution The auxiliary equation is $r^2 + 4r + 13 = 0$, which has solutions
$$r = \frac{-4 \pm \sqrt{16 - 52}}{2} = \frac{-4 \pm \sqrt{-36}}{2} = -2 \pm 3i.$$
Thus $k = -2$ and $\omega = 3$. According to Case III, the general solution of the given differential equation is
$$y = A\,e^{-2t}\cos(3t) + B\,e^{-2t}\sin(3t).$$

Initial-value problems for $ay'' + by' + cy = 0$ specify values for y and y' at an initial point. These values can be used to determine the values of the constants A and B in the general solution, so the initial-value problem has a unique solution.

Example 4   Solve the initial-value problem

$$
\begin{cases}
y'' + 2y' + 2y = 0 \\
y(0) = 2 \\
y'(0) = -3.
\end{cases}
$$

Solution   The auxiliary equation is $r^2 + 2r + 2 = 0$, which has roots

$$r = \frac{-2 \pm \sqrt{4 - 8}}{2} = -1 \pm i.$$

Thus Case III applies, $k = -1$ and $\omega = 1$. Thus, the differential equation has the general solution

$$y = A e^{-t} \cos t + B e^{-t} \sin t.$$

Also,

$$
\begin{aligned}
y' &= e^{-t}\bigl(-A \cos t - B \sin t - A \sin t + B \cos t\bigr) \\
&= (B - A) e^{-t} \cos t - (A + B) e^{-t} \sin t.
\end{aligned}
$$

Applying the initial conditions $y(0) = 2$ and $y'(0) = -3$, we obtain $A = 2$ and $B - A = -3$. Hence, $B = -1$ and the initial-value problem has the solution

$$y = 2 e^{-t} \cos t - e^{-t} \sin t.$$

Simple Harmonic Motion

Many natural phenomena exhibit periodic behaviour. The swinging of a clock pendulum, the vibrating of a guitar string or drum membrane, the altitude of a rider on a rotating ferris wheel, the motion of an object floating in wavy seas, and the voltage produced by an alternating current generator are but a few examples where quantities depend on time in a periodic way. Being periodic, the circular functions sine and cosine provide a useful model for such behaviour.

It often happens that a quantity displaced from an equilibrium value experiences a restoring force that tends to move it back in the direction of its equilibrium. Besides the obvious examples of elastic motions in physics, one can imagine such a model applying, say, to a biological population in equilibrium with its food supply or the price of a commodity in an elastic economy where increasing price causes decreasing demand and hence decreasing price. In the simplest models, the restoring force is proportional to the amount of displacement from equilibrium. Such a force causes the quantity to oscillate sinusoidally; we say that it executes *simple harmonic motion*.

As a specific example, suppose a mass m is suspended by an elastic spring so that it hangs unmoving in its equilibrium position. If it is displaced vertically by an amount y from this position, a force is exerted by the spring, directed to restore the mass to its equilibrium position. (See Figure 17.6.) This force is proportional to the displacement (Hooke's Law); its magnitude is $-ky$, where k is a positive constant called the **spring constant**. Assuming the spring is weightless, this force imparts to the mass m an acceleration d^2y/dt^2 that satisfies, by Newton's Second Law, $m(d^2y/dt^2) = -ky$ (mass × acceleration = force). Dividing this equation by m, we obtain the equation

$$\frac{d^2y}{dt^2} + \omega^2 y = 0, \qquad \text{where} \quad \omega^2 = \frac{k}{m}.$$

The second-order differential equation

$$\frac{d^2y}{dt^2} + \omega^2 y = 0$$

Figure 17.6

is called the **equation of simple harmonic motion**. Its auxiliary equation, $r^2 + \omega^2 = 0$, has complex roots $r = \pm i\omega$, so it has general solution

$$y = A \cos \omega t + B \sin \omega t,$$

where A and B are arbitrary constants.

For any values of the constants R and t_0, the function

$$y = R \cos\big(\omega(t - t_0)\big)$$

is also a general solution of the differential equation of simple harmonic motion. If we expand this formula using the addition formula for cosine, we get

$$y = R \cos \omega t_0 \cos \omega t + R \sin \omega t_0 \sin \omega t$$
$$= A \cos \omega t + B \sin \omega t,$$

where

$$A = R \cos(\omega t_0), \qquad\qquad B = R \sin(\omega t_0),$$
$$R^2 = A^2 + B^2, \qquad\qquad \tan(\omega t_0) = B/A.$$

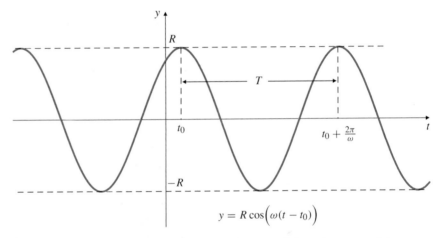

Figure 17.7 Simple harmonic motion

The constants A and B are related to the position y_0 and the velocity v_0 of the mass m at time $t = 0$:

$$y_0 = y(0) = A \cos 0 + B \sin 0 = A,$$
$$v_0 = y'(0) = -A\omega \sin 0 + B\omega \cos 0 = B\omega.$$

The constant $R = \sqrt{A^2 + B^2}$ is called the **amplitude** of the motion. Because $\cos x$ oscillates between -1 and 1, the displacement y varies between $-R$ and R. Note in Figure 17.7 that the graph of the displacement as a function of time is the curve $y = R \cos \omega t$ shifted t_0 units to the right. The number t_0 is called the **time-shift**. (The related quantity ωt_0 is called a **phase-shift**.) The **period** of this curve is $T = 2\pi/\omega$; it is the time interval between consecutive instants when the mass is at the same height moving in the same direction. The reciprocal $1/T$ of the period is called the **frequency** of the motion. It is usually measured in Hertz (Hz), that is, cycles per second. The quantity $\omega = 2\pi/T$ is called the **circular frequency**. It is measured in radians per second since 1 cycle = 1 revolution = 2π radians.

Example 5 Solve the initial-value problem

$$\begin{cases} y'' + 16y = 0 \\ y(0) = -6 \\ y'(0) = 32. \end{cases}$$

Find the amplitude, frequency, and period of the solution.

Solution   Here, $\omega^2 = 16$ so $\omega = 4$. The solution is of the form

$$y = A\cos(4t) + B\sin(4t).$$

Since $y(0) = -6$, we have $A = -6$. Also, $y'(t) = -4A\sin(4t) + 4B\cos(4t)$. Since $y'(0) = 32$, we have $4B = 32$, or $B = 8$. Thus, the solution is

$$y = -6\cos(4t) + 8\sin(4t).$$

The amplitude is $\sqrt{(-6)^2 + 8^2} = 10$, the frequency is $\omega/(2\pi) \approx 0.637$ Hz, and the period is $2\pi/\omega \approx 1.57$ s.

Example 6   (**Spring-mass problem**)   Suppose that a 100 g mass is suspended from a spring and that a force of 3×10^4 dynes (3×10^4 g-cm/s^2) is required to produce a displacement from equilibrium of 1/3 cm. At time $t = 0$ the mass is pulled down 2 cm below equilibrium and flicked upward with a velocity of 60 cm/s. Find its subsequent displacement at any time $t > 0$. Find the frequency, period, amplitude, and time-shift of the motion. Express the position of the mass at time t in terms of the amplitude and the time-shift.

Solution   The spring constant k is determined from Hooke's Law, $F = -ky$. Here $F = -3 \times 10^4$ g-cm/s^2 is the force of the spring on the mass displaced 1/3 cm:

$$-3 \times 10^4 = -\frac{1}{3}k,$$

so $k = 9 \times 10^4$ g/s^2. Hence, the circular frequency is $\omega = \sqrt{k/m} = 30$ rad/s, the frequency is $\omega/2\pi = 15/\pi \approx 4.77$ Hz, and the period is $2\pi/\omega \approx 0.209$ s.

Since the displacement at time $t = 0$ is $y_0 = -2$ and the velocity at that time is $v_0 = 60$, the subsequent displacement is $y = A\cos(30t) + B\sin(30t)$, where $A = y_0 = -2$ and $B = v_0/\omega = 60/30 = 2$. Thus

$$y = -2\cos(30t) + 2\sin(30t) \qquad (y \text{ in cm, } t \text{ in seconds}).$$

The amplitude of the motion is $R = \sqrt{(-2)^2 + 2^2} = 2\sqrt{2} \approx 2.83$ cm. The time-shift t_0 must satisfy

$$-2 = A = R\cos(\omega t_0) = 2\sqrt{2}\cos(30t_0),$$
$$2 = B = R\sin(\omega t_0) = 2\sqrt{2}\sin(30t_0),$$

so $\sin(30t_0) = 1/\sqrt{2} = -\cos(30t_0)$. Hence the phase-shift is $30t_0 = 3\pi/4$ radians, and the time-shift is $t_0 = \pi/40 \approx 0.0785$ s. The position of the mass at time $t > 0$ is also given by

$$y = 2\sqrt{2}\cos\left[30\left(t - \frac{\pi}{40}\right)\right].$$

Damped Harmonic Motion

If a and c are positive and $b = 0$, then equation

$$ay'' + by' + cy = 0$$

is the differential equation of simple harmonic motion and has oscillatory solutions of fixed amplitude as shown above. If $a > 0$, $b > 0$, and $c > 0$, then the roots of the auxiliary equation are either negative real numbers or, if $b^2 < 4ac$, complex numbers $k \pm i\omega$ with negative real parts $k = -b/(2a)$ (Case III). In this latter case the solutions still oscillate, but the amplitude diminishes exponentially as $t \to \infty$ because of the factor $e^{kt} = e^{-(b/2a)t}$. (See Exercise 17 below.) A system whose behaviour is modelled by such an equation is said to exhibit **damped harmonic motion**. If $b^2 = 4ac$ (Case II), the system is said to be **critically damped**, and if $b^2 > 4ac$ (Case I), it is **overdamped**. In these cases the behaviour is no longer oscillatory. (See Figure 17.8. Imagine a mass suspended by a spring in a jar of oil.)

Constant-Coefficient Equations of Higher Order

Because in most applications of equation ($*$) the dependent variable represents time, we will, as we did in Section 3.7, regard y as a function of t rather than x, so that the prime symbol ($'$) denotes the derivative d/dt. The basic result of Section 3.7 was that the function $y = e^{rt}$ was a solution of ($*$) provided that r satisfies the **auxiliary equation**

$$ar^2 + br + c = 0. \qquad (**)$$

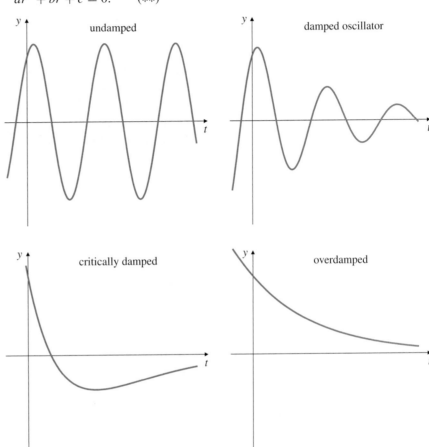

Figure 17.8

Undamped oscillator ($b = 0$)
Damped oscillator ($b > 0$, $b^2 < 4ac$)
Critically damped case ($b > 0$, $b^2 = 4ac$)
Overdamped case ($b > 0$, $b^2 > 4ac$)

The auxiliary equation is quadratic and can have either

(a) two distinct real roots, r_1 and r_2 (if $b^2 > 4ac$), in which case ($*$) has general solution $y = C_1 e^{r_1 t} + C_2 e^{r_2 t}$, or

(b) a single repeated real root r (if $b^2 = 4ac$), in which case ($*$) has general solution $y = (C_1 + C_2 t)e^{rt}$, or

(c) a pair of complex conjugate roots, $r = k \pm i\omega$ with k and ω real (if $b^2 < 4ac$), in which case ($*$) has general solution $y = e^{kt}\big(C_1 \cos(\omega t) + C_2 \sin(\omega t)\big)$.

The situation is analogous for higher-order linear, homogeneous DEs with constant coefficients. We describe the procedure without offering any proofs. If

$$P_n(r) = a_n r^n + a_{n-1} r^{n-1} + \cdots + a_2 r^2 + a_1 r + a_0$$

is a polynomial of degree n with constant coefficients a_j, $(0 \leq j \leq n)$, and $a_n \neq 0$, then the DE

$$P_n(D)y = 0, \qquad (\dagger)$$

where $D = d/dt$ can be solved by substituting $y = e^{rt}$ and obtaining the *auxiliary equation $P_n(r) = 0$.* This polynomial equation has n roots (see Appendix II) some of which may be equal and some or all of which can be complex. If the coefficients of the polynomial $P_n(r)$ are all real, then any complex roots must occur in complex conjugate pairs $k \pm i\omega$, where k and ω are real.

The general solution of (†) can be expressed as a *linear combination* of n independent particular solutions

$$y = C_1 y_1(t) + C_2 y_2(t) + \cdots + C_n y_n(t),$$

where the C_j are arbitrary constants. The independent solutions $y_1, y_2, \ldots, y_n$ are constructed as follows:

1. If r_1 is a k-fold real root of the auxiliary equation (i.e., if $(r - r_1)^k$ is a factor of $P_n(r)$), then

$$e^{r_1 t}, \quad te^{r_1 t}, \quad t^2 e^{r_1 t}, \quad \ldots, \quad t^{k-1} e^{r_1 t}$$

are k independent solutions of (†).

2. If $r = a + ib$ and $r = a - ib$ (where a and b are real) constitute a k-fold pair of complex conjugate roots of the auxiliary equation (i.e., if $[(r - a)^2 + b^2]^k$ is a factor of $P_n(r)$), then

$$e^{at} \cos bt, \quad te^{at} \cos bt, \quad \ldots, \quad t^{k-1} e^{at} \cos bt,$$
$$e^{at} \sin bt, \quad te^{at} \sin bt, \quad \ldots, \quad t^{k-1} e^{at} \sin bt$$

are $2k$ independent solutions of (†).

Example 7   Solve   (a) $y^{(4)} - 16y = 0$ and   (b) $y^{(5)} - 2y^{(4)} + y^{(3)} = 0$.

Solution   The auxiliary equation for (a) is $r^4 - 16 = 0$, which factors down to $(r - 2)(r + 2)(r^2 + 4) = 0$ and, hence, has roots $r = 2, -2, 2i$, and $-2i$. Thus the DE (a) has general solution

$$y = C_1 e^{2t} + C_2 e^{-2t} + C_3 \cos(2t) + C_4 \sin(2t)$$

for arbitrary constants C_1, C_2, C_3, and C_4.

The auxiliary equation for (b) is $r^5 - 2r^4 + r^3$, which factors to $r^3(r - 1)^2 = 0$, and so has roots $r = 0, 0, 0, 1, 1$. The general solution of the DE (b) is

$$y = C_1 + C_2 t + C_3 t^2 + C_4 e^t + C_5 t e^t,$$

where $C_1, \ldots, C_5$ are arbitrary constants.

Example 8   What are the order and the general solution of the constant-coefficient, linear, homogeneous DE whose auxiliary equation is

$$(r + 4)^3 (r^2 + 4r + 13)^2 = 0?$$

Solution   The auxiliary equation has degree 7, so the DE is of seventh order. Since $r^2 + 4r + 13 = (r + 2)^2 + 9$, which has roots $-2 \pm 3i$, the DE must have the general solution

$$y = C_1 e^{-4t} + C_2 t e^{-4t} + C_3 t^2 e^{-4t}$$
$$+ C_4 e^{-2t} \cos(3t) + C_5 e^{-2t} \sin(3t) + C_6 t e^{-2t} \cos(3t) + C_7 t e^{-2t} \sin(3t).$$

Euler (Equidimensional) Equations

A homogeneous, linear equation of the form

$$ax^2 \frac{d^2y}{dx^2} + bx \frac{dy}{dx} + cy = 0$$

is called an **Euler equation** or an **equidimensional equation**, the latter term being appropriate since all the terms in the equation have the same dimension (i.e., they are measured in the same units), provided that the constants a, b, and c all have the same dimension. The coefficients of an Euler equation are *not constant,* but there is a technique for solving these equations that is similar to that for solving equations with constant coefficients, so we include a brief discussion of these equations in this section. As in the case of constant coefficient equations, we assume that the constants a, b, and c are real numbers and that $a \neq 0$. Even so, the leading coefficient, ax^2, does vanish at $x = 0$ (which is called a **singular point** of the equation), and this can cause solutions to fail to be defined at $x = 0$. We will solve the equation in the interval $x > 0$; the same solution will also hold for $x < 0$ provided we replace x by $|x|$ in the solution.

Let us search for solutions in $x > 0$ given by powers of x; if

$$y = x^r, \qquad \frac{dy}{dx} = rx^{r-1}, \qquad \frac{d^2y}{dx^2} = r(r-1)x^{r-2},$$

then the Euler equation becomes

$$\big(ar(r-1) + br + c\big)x^r = 0.$$

This will be satisfied for all $x > 0$ provided that r satisfies the **auxiliary equation**

$$ar(r-1) + br + c = 0 \quad \text{or, equivalently} \quad ar^2 + (b-a)r + c = 0.$$

As for constant coefficient equations, there are three possibilities.

CASE I.   If $(b-a)^2 \geq 4ac$, then the auxiliary equation has two real roots:

$$r_1 = \frac{a - b + \sqrt{(b-a)^2 - 4ac}}{2a},$$
$$r_2 = \frac{a - b - \sqrt{(b-a)^2 - 4ac}}{2a}.$$

In this case, the Euler equation has the general solution

$$y = C_1 x^{r_1} + C_2 x^{r_2}, \qquad (x > 0).$$

The general solution is usually quoted in the form

$$y = C_1 |x|^{r_1} + C_2 |x|^{r_2},$$

which is valid in any interval not containing $x = 0$ and may even be valid on intervals containing the origin if, for example, r_1 and r_2 are nonnegative integers.

Example 9   Solve the initial-value problem

$$2x^2 y'' - xy' - 2y = 0, \qquad y(1) = 5, \qquad y'(1) = 0.$$

Solution   The auxiliary equation is $2r(r-1) - r - 2 = 0$, that is, $2r^2 - 3r - 2 = 0$, or $(r-2)(2r+1) = 0$, and has roots $r = 2$ and $r = -(1/2)$. Thus, the general solution of the differential equation (valid for $x > 0$) is

$$y = C_1 x^2 + C_2 x^{-1/2}.$$

The initial conditions imply that

$$5 = y(1) = C_1 + C_2 \quad \text{and} \quad 0 = y'(1) = 2C_1 - \frac{1}{2} C_2.$$

Therefore, $C_1 = 1$ and $C_2 = 4$, and the initial-value problem has solution

$$y = x^2 + \frac{4}{\sqrt{x}}, \qquad (x > 0).$$

CASE II.   If $(b-a)^2 = 4ac$, then the auxiliary equation has one double root, namely, the root $r = (a-b)/2a$. It is left to the reader to verify that in this case the transformation $y = x^r v(x)$ leads to the general solution

$$y = C_1 x^r + C_2 x^r \ln x, \qquad (x > 0),$$

or, more generally,

$$y = C_1 |x|^r + C_2 |x|^r \ln |x|, \qquad (x \neq 0).$$

CASE III.   If $(b-a)^2 < 4ac$, then the auxiliary equation has complex conjugate roots:

$$r = \alpha \pm i\beta, \qquad \text{where} \quad \alpha = \frac{a-b}{2a}, \quad \beta = \frac{\sqrt{4ac - (b-a)^2}}{2a}.$$

The corresponding powers x^r can be expressed in real form in a manner similar to that used for constant coefficient equations; we have

$$x^{\alpha \pm i\beta} = e^{(\alpha \pm i\beta) \ln x} = e^{\alpha \ln x} \left[\cos(\beta \ln x) \pm i \sin(\beta \ln x) \right]$$
$$= x^\alpha \cos(\beta \ln x) \pm i x^\alpha \sin(\beta \ln x).$$

Accordingly, the Euler equation has the general solution

$$y = C_1 |x|^\alpha \cos(\beta \ln |x|) + C_2 |x|^\alpha \sin(\beta \ln |x|).$$

Example 10   Solve the DE $x^2 y'' - 3xy' + 13y = 0$.

Solution   The DE has the auxiliary equation $r(r-1) - 3r + 13 = 0$, that is, $r^2 - 4r + 13 = 0$, which has roots $r = 2 \pm 3i$. The DE, therefore, has the general solution

$$y = C_1 x^2 \cos(3 \ln |x|) + C_2 x^2 \sin(3 \ln |x|).$$

Exercises 17.5

In Exercises 1–12, find the general solutions for the given equations.

1. $y'' + 7y' + 10y = 0$

2. $y'' - 2y' - 3y = 0$

3. $y'' + 2y' = 0$

4. $4y'' - 4y' - 3y = 0$

5. $y'' + 8y' + 16y = 0$

6. $y'' - 2y' + y = 0$

7. $y'' - 6y' + 10y = 0$

8. $9y'' + 6y' + y = 0$

9. $y'' + 2y' + 5y = 0$

10. $y'' - 4y' + 5y = 0$

11. $y'' + 2y' + 3y = 0$

12. $y'' + y' + y = 0$

In Exercises 13–15, solve the given initial-value problems.

13. $\begin{cases} 2y'' + 5y' - 3y = 0 \\ y(0) = 1 \\ y'(0) = 0. \end{cases}$

14. $\begin{cases} y'' + 10y' + 25y = 0 \\ y(1) = 0 \\ y'(1) = 2. \end{cases}$

15. $\begin{cases} y'' + 4y' + 5y = 0 \\ y(0) = 2 \\ y'(0) = 2. \end{cases}$

∗16. Show that if $\epsilon \neq 0$, the function $y_\epsilon(t) = \dfrac{e^{(1+\epsilon)t} - e^t}{\epsilon}$ satisfies the equation $y'' - (2 + \epsilon)y' + (1 + \epsilon)y = 0$. Caclulate $y(t) = \lim_{\epsilon \to 0} y_\epsilon(t)$ and verify that, as expected, it is a solution of $y'' - 2y' + y = 0$.

∗17. If $a > 0$, $b > 0$, and $c > 0$, prove that all solutions of the differential equation $ay'' + by' + cy = 0$ satisfy $\lim_{t \to \infty} y(t) = 0$.

∗18. Prove that the solution given in the discussion of Case I, namely, $y = A e^{r_1 t} + B e^{r_2 t}$, is the general solution for that case as follows: First, let $y = e^{r_1 t} u$ and show that u satisfies the equation

$$u'' - (r_2 - r_1)u' = 0.$$

Then let $v = u'$, so that v must satisfy $v' = (r_2 - r_1)v$. The general solution of this equation is $v = C e^{(r_2-r_1)t}$. Hence find u and y.

Find general solutions of the DEs in Exercises 19–22.

19. $y''' - 4y'' + 3y' = 0$

20. $y^{(4)} - 2y'' + y = 0$

21. $y^{(4)} + 2y'' + y = 0$

22. $y^{(4)} + 4y^{(3)} + 6y'' + 4y' + y = 0$

23. Show that $y = e^{2t}$ is a solution of

$$y''' - 2y' - 4y = 0$$

(where $'$ denotes d/dt), and find the general solution of this DE.

24. Write the general solution of the linear, constant-coefficient DE having auxiliary equation $(r^2 - r - 2)^2(r^2 - 4)^2 = 0$.

Find general solutions to the Euler equations in Exercises 25–30.

25. $x^2 y'' - xy' + y = 0$

26. $x^2 y'' - xy' - 3y = 0$

27. $x^2 y'' + xy' - y = 0$

28. $x^2 y'' - xy' + 5y = 0$

29. $x^2 y'' + xy' = 0$

30. $x^2 y'' + xy' + y = 0$

∗31. Solve the DE $x^3 y''' + xy' - y = 0$ in the interval $x > 0$.

Simple harmonic motion

In Exercises 32–33, solve the given initial-value problems. For each problem determine the circular frequency, the frequency, the period, and the amplitude of the solution.

32. $\begin{cases} y'' + 4y = 0 \\ y(0) = 2 \\ y'(0) = -5 \end{cases}$

33. $\begin{cases} y'' + 100y = 0 \\ y(0) = 0 \\ y'(0) = 3 \end{cases}$

34. Solve $\begin{cases} y'' + y = 0 \\ y(2) = 3 \\ y'(2) = -4 \end{cases}$

35. Solve $\begin{cases} y'' + \omega^2 y = 0 \\ y(a) = A \\ y'(a) = B \end{cases}$

∗36. Show that $y = \alpha \cos(\omega(t - c)) + \beta \sin(\omega(t - c))$ is a solution of the differential equation $y'' + \omega^2 y = 0$, and that it satisfies $y(c) = \alpha$ and $y'(c) = \beta\omega$. Express the solution in the form $y = A \cos(\omega t) + B \sin(\omega t)$ for certain values of the constants A and B depending on α, β, c, and ω.

Exercises 37–40 all refer to the differential equation of simple harmonic motion:

$$\frac{d^2 y}{dt^2} + \omega^2 y = 0, \qquad (\omega \neq 0). \qquad (†)$$

Together they show that $y = A \cos \omega t + B \sin \omega t$ is a *general solution* of this equation, that is, every solution is of this form for some choice of the constants A and B.

∗37. Show that $y = A \cos \omega t + B \sin \omega t$ is a solution of (†).

∗38. If $f(t)$ is any solution of (†), show that $\omega^2(f(t))^2 + (f'(t))^2$ is constant.

∗39. If $g(t)$ is a solution of (†) satisfying $g(0) = g'(0) = 0$, show that $g(t) = 0$ for all t.

∗40. Suppose that $f(t)$ is any solution of the differential equation (†). Show that $f(t) = A \cos \omega t + B \sin \omega t$, where $A = f(0)$ and $B\omega = f'(0)$.
(*Hint:* Let $g(t) = f(t) - A \cos \omega t - B \sin \omega t$.)

∗41. If $b^2 - 4ac < 0$, show that the substitution $y = e^{kt} u(t)$, where $k = -b/(2a)$, transforms $ay'' + by' + cy = 0$ into the equation $u'' + \omega^2 u = 0$, where $\omega^2 = (4ac - b^2)/(4a^2)$. Together with the result of Exercise 40, this confirms the recipe for Case III, in case you didn't feel comfortable with the complex number argument given in the text.

42. What mass should be suspended from the spring in Example 6 to provide a system whose natural frequency of oscillation is 10 Hz? Find the displacement of such a mass from its equilibrium position t s after it is pulled down 1 cm from equilibrium and flicked upward with a speed of 2 cm/s. What is the amplitude of this motion?

43. A mass of 400 g suspended from a certain elastic spring will oscillate with a frequency of 24 Hz. What would be the frequency if the 400 g mass were replaced with a 900 g mass? a 100 g mass?

∗44. Show that if t_0, A, and B are constants and $k = -b/(2a)$ and $\omega = \sqrt{4ac - b^2}/(2a)$, then

$$y = e^{kt}\left[A \cos\left(\omega(t - t_0)\right) + B \sin\left(\omega(t - t_0)\right)\right]$$

is an alternative to the general solution for $ay'' + by' + cy = 0$ for Case III ($b^2 - 4ac < 0$). This form of the general solution is useful for solving initial-value problems where $y(t_0)$ and $y'(t_0)$ are specified.

*45. Show that if t_0, A, and B are constants and $k = -b/(2a)$ and $\omega = \sqrt{b^2 - 4ac}/(2a)$, then

$$y = e^{kt}\left[A\cosh\Big(\omega(t - t_0)\Big) + B\sinh\Big(\omega(t - t_0)\Big)\right]$$

is an alternative to the general solution for $ay'' + by' + cy = 0$ for Case I ($b^2 - 4ac > 0$). This form of the general solution is useful for solving initial-value problems where $y(t_0)$ and $y'(t_0)$ are specified.

Use the forms of solution provided by Exercises 43 and 44 to solve the initial-value problems in Exercises 46–47.

46. $\begin{cases} y'' + 2y' + 5y = 0 \\ y(3) = 2 \\ y'(3) = 0 \end{cases}$ **47.** $\begin{cases} y'' + 4y' + 3y = 0 \\ y(3) = 1 \\ y'(3) = 0 \end{cases}$

48. By using the change of dependent variable $u(x) = c - k^2 y(x)$, solve the initial-value problem

$$\begin{cases} y''(x) = c - k^2 y(x) \\ y(0) = a \\ y'(0) = b \end{cases}$$

*49. A mass is attached to a spring mounted horizontally so the mass can slide along the top of a table. With a suitable choice of units, the position $x(t)$ of the mass at time t is governed by the differential equation

$$x'' = -x + F,$$

where the $-x$ term is due to the elasticity of the spring, and the F is due to the friction of the mass with the table. The frictional force should be constant in magnitude and directed opposite to the velocity of the mass when the mass is moving. When the mass is stopped, the friction should be constant and opposed to the spring force unless the spring force has the smaller magnitude, in which case the friction force should just cancel the spring force and the mass should remain at rest thereafter. For this problem, let the magnitude of the friction force be 1/5. Accordingly,

$$F = \begin{cases} -\dfrac{1}{5} & \text{if } x' > 0 \text{ or if } x' = 0 \text{ and } x < -\dfrac{1}{5} \\[2mm] \dfrac{1}{5} & \text{if } x' < 0 \text{ or if } x' = 0 \text{ and } x > \dfrac{1}{5} \\[2mm] x & \text{if } x' = 0 \text{ and } |x| \le \dfrac{1}{5}. \end{cases}$$

Find the position $x(t)$ of the mass at all times $t > 0$ if $x(0) = 1$ and $x'(0) = 0$.

17.6 Nonhomogeneous Linear Equations

We now consider the problem of solving the nonhomogeneous second-order differential equation

$$a_2(x)\frac{d^2 y}{dx^2} + a_1(x)\frac{dy}{dx} + a_0(x)y = f(x). \tag{$*$}$$

We assume that two independent solutions, $y_1(x)$ and $y_2(x)$, of the corresponding homogeneous equation

$$a_2(x)\frac{d^2 y}{dx^2} + a_1(x)\frac{dy}{dx} + a_0(x)y = 0$$

are known. The function $y_h(x) = C_1 y_1(x) + C_2 y_2(x)$, which is the general solution of the homogeneous equation, is called the **complementary function** for the nonhomogeneous equation. Theorem 2 of Section 17.1 suggests that the general solution of the nonhomogeneous equation is of the form

$$y = y_p(x) + y_h(x) = y_p(x) + C_1 y_1(x) + C_2 y_2(x),$$

where $y_p(x)$ is any **particular solution** of the nonhomogeneous equation. All we need to do is find *one solution* of the nonhomogeneous equation, and we can write the general solution.

There are two common methods for finding a particular solution y_p of the nonhomogeneous equation ($*$):

1. the method of undetermined coefficients and
2. the method of variation of parameters.

The first of these hardly warrants being called a *method*; it just involves making an educated guess about the form of the solution as a sum of terms with unknown coefficients and substituting this guess into the equation to determine the coefficients. This method works well for simple DEs, especially ones with constant coefficients. The nature of the *guess* depends on the nonhomogeneous term $f(x)$, but can also be affected by the solution of the corresponding homogeneous equation. A few examples will illustrate the ideas involved.

Example 1   Find the general solution of $y'' + y' - 2y = 4x$.

Solution   Because the nonhomogeneous term $f(x) = 4x$ is a first-degree polynomial, we "guess" that a particular solution can be found that is also such a polynomial. Thus we try

$$y = Ax + B, \qquad y' = A, \qquad y'' = 0.$$

Substituting these expressions into the given DE we obtain

$$0 + A - 2(Ax + B) = 4x \qquad \text{or}$$
$$-(2A + 4)x + (A - 2B) = 0.$$

This latter equation will be satisfied for all x provided $2A + 4 = 0$ and $A - 2B = 0$. Thus we require $A = -2$ and $B = -1$; a particular solution of the given DE is

$$y_p(x) = -2x - 1.$$

Since the corresponding homogeneous equation $y'' + y' - 2y = 0$ has auxiliary equation $r^2 + r - 2 = 0$ with roots $r = 1$ and $r = -2$, the given DE has the general solution

$$y = y_p(x) + C_1 e^x + C_2 e^{-2x} = -2x - 1 + C_1 e^x + C_2 e^{-2x}.$$

Example 2   Find general solutions of the equations (where $'$ denotes d/dt)
(a) $y'' + 4y = \sin t$,
(b) $y'' + 4y = \sin(2t)$,
(c) $y'' + 4y = \sin t + \sin(2t)$.

Solution

(a) Let us look for a particular solution of the form

$$y = A \sin t + B \cos t \qquad \text{so that}$$
$$y' = A \cos t - B \sin t$$
$$y'' = -A \sin t - B \cos t.$$

Substituting these expressions into the DE $y'' + 4y = \sin t$, we get

$$-A \sin t - B \cos t + 4A \sin t + 4B \cos t = \sin t,$$

which is satisfied for all x if $3A = 1$ and $3B = 0$. Thus $A = 1/3$ and $B = 0$. Since the homogeneous equation $y'' + 4y = 0$ has general solution $y = C_1 \cos(2t) + C_2 \sin(2t)$, the given nonhomogeneous equation has the general solution

$$y = \frac{1}{3} \sin t + C_1 \cos(2t) + C_2 \sin(2t).$$

(b) Motivated by our success in part (a), we might be tempted to try for a particular solution of the form $y = A\sin(2t) + B\cos(2t)$, but that won't work, because this function is a solution of the homogeneous equation, so we would get $y'' + 4y = 0$ for any choice of A and B. In this case it is useful to try

$$y = At\sin(2t) + Bt\cos(2t).$$

We have

$$\begin{aligned}
y' &= A\sin(2t) + 2At\cos(2t) + B\cos(2t) - 2Bt\sin(2t) \\
&= (A - 2Bt)\sin(2t) + (B + 2At)\cos(2t) \\
y'' &= -2B\sin(2t) + 2(A - 2Bt)\cos(2t) + 2A\cos(2t) \\
&\quad - 2(B + 2At)\sin(2t) \\
&= -4(B + At)\sin(2t) + 4(A - Bt)\cos(2t).
\end{aligned}$$

Substituting into $y'' + 4y = \sin(2t)$ leads to

$$-4(B + At)\sin(2t) + 4(A - Bt)\cos(2t) + 4At\sin(2t) + 4Bt\cos(2t)$$
$$= \sin(2t).$$

Observe that the terms involving $t\sin(2t)$ and $t\cos(2t)$ cancel out, and we are left with

$$-4B\sin(2t) + 4A\cos(2t) = \sin(2t),$$

which is satisfied for all x if $A = 0$ and $B = -1/4$. Hence, the general solution for part (b) is

$$y = -\frac{1}{4}t\cos(2t) + C_1\cos(2t) + C_2\sin(2t).$$

(c) Since the homogeneous equation is the same for (a), (b), and (c), and the non-homogeneous term in equation (c) is the sum of the nonhomogeneous terms in equations (a) and (b), the sum of particular solutions of (a) and (b) is a particular solution of (c). (This is because the equation is *linear*.) Thus, the general solution of equation (c) is

$$y = \frac{1}{3}\sin t - \frac{1}{4}t\cos(2t) + C_1\cos(2t) + C_2\sin(2t).$$

We summarize the appropriate forms to try for particular solutions of constant-coefficient equations as follows:

Trial solutions for constant-coefficient equations

Let $A_n(x)$, $B_n(x)$, and $P_n(x)$ denote the nth-degree polynomials

$$\begin{aligned}
A_n(x) &= a_0 + a_1x + a_2x^2 + \cdots + a_nx^n \\
B_n(x) &= b_0 + b_1x + b_2x^2 + \cdots + b_nx^n \\
P_n(x) &= p_0 + p_1x + p_2x^2 + \cdots + p_nx^n.
\end{aligned}$$

To find a particular solution $y_p(x)$ of the second-order linear, constant-coefficient, nonhomogeneous DE

$$a_2\frac{d^2y}{dx^2} + a_1\frac{dy}{dx} + a_0y = f(x),$$

use the following forms:

If $f(x) = P_n(x)$,   try $y_p = x^m A_n(x)$.

If $f(x) = P_n(x)e^{rx}$,   try $y_p = x^m A_n(x)e^{rx}$.

If $f(x) = P_n(x)e^{rx}\cos(kx)$, try $y_p = x^m e^{rx}[A_n(x)\cos(kx) + B_n(x)\sin(kx)]$.

If $f(x) = P_n(x)e^{rx}\sin(kx)$, try $y_p = x^m e^{rx}[A_n(x)\cos(kx) + B_n(x)\sin(kx)]$,

where m is the smallest of the integers 0, 1, and 2, that ensures that no term of y_p is a solution of the corresponding homogeneous equation

$$a_2\frac{d^2y}{dx^2} + a_1\frac{dy}{dx} + a_0 y = 0.$$

Resonance

For $\lambda > 0$, $\lambda \neq 1$, the solution $y_\lambda(t)$ of the initial-value problem

$$\begin{cases} y'' + y = \sin(\lambda t) \\ \quad y(0) = 0 \\ \quad y'(0) = 1 \end{cases}$$

can be determined by first looking for a particular solution of the DE having the form $y = A\sin(\lambda t)$, and then adding the complementary function $y = B\cos t + C\sin t$. The calculations give $A = 1/(1 - \lambda^2)$, $B = 0$, $C = (1 - \lambda - \lambda^2)/(1 - \lambda^2)$, so

$$y_\lambda(t) = \frac{\sin(\lambda t) + (1 - \lambda - \lambda^2)\sin t}{1 - \lambda^2}.$$

For $\lambda = 1$ the nonhomogeneous term in the DE is a solution of the homogeneous equation $y'' + y = 0$, so we must try for a particular solution of the form $y = At\cos t + Bt\sin t$. In this case, the solution of the initial-value problem is

$$y_1(t) = \frac{3\sin t - t\cos t}{2}.$$

(This solution can also be found by calculating $\lim_{\lambda \to 1} y_\lambda(t)$ using l'Hôpital's Rule.) Observe that this solution is unbounded; the amplitude of the oscillations becomes larger and larger as t increases. In contrast, the solutions $y_\lambda(t)$ for $\lambda \neq 1$ are bounded for all t, although they can become quite large for some values of t if λ is close to 1. The graphs of the solutions $y_{0.9}(t)$, $y_{0.95}(t)$, and $y_1(t)$ on the interval $-10 \leq t \leq 100$ are shown in Figure 17.9.

The phenomenon illustrated here is called **resonance**. Vibrating mechanical systems have natural frequencies at which they will vibrate. If you try to force them to vibrate at a different frequency, the amplitude of the vibrations will themselves vary sinusoidally over time, producing an effect known as **beats**. The amplitudes of the beats can grow quite large, and the period of the beats lengthens as the forcing frequency approaches the natural frequency of the system. If the system has no resistive damping (the one illustrated here has no damping), then forcing vibrations at the natural frequency will cause the system to vibrate at ever increasing amplitudes.

As a concrete example, if you push a child on a swing, the swing will rise highest if your pushes are timed to have the same frequency as the natural frequency of the swing. Resonance is used in the design of tuning circuits of radios; the circuit is tuned (ususally by a variable capacitor) so that its natural frequency of oscillation is the frequency of the station being tuned in. The circuit then responds much more strongly to the signal received from that station than to others on different frequencies.

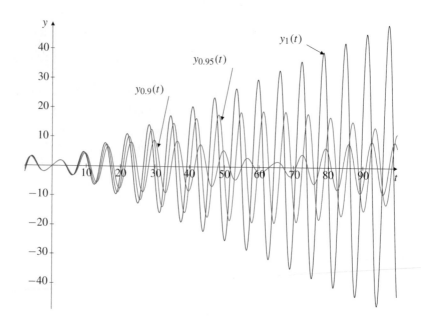

Figure 17.9 Resonance

Variation of Parameters

A more formal method for finding a particular solution $y_p(x)$ of the nonhomogeneous equation when we know two independent solutions, $y_1(x)$ and $y_2(x)$, of the homogeneous equation is to replace the constants in the complementary function by functions, that is, search for y_p in the form

$$y_p = u_1(x)y_1(x) + u_2(x)y_2(x).$$

Requiring y_p to satisfy the given nonhomogeneous DE provides one equation that must be satisfied by the two unknown functions u_1 and u_2. We are free to require them to satisfy a second equation also. To simplify the calculations below, we choose this second equation to be

$$u_1'(x)y_1(x) + u_2'(x)y_2(x) = 0.$$

Now we have

$$y_p' = u_1'y_1 + u_1y_1' + u_2'y_2 + u_2y_2' = u_1y_1' + u_2y_2'$$
$$y_p'' = u_1'y_1' + u_1y_1'' + u_2'y_2' + u_2y_2''.$$

Substituting these expressions into the given DE we obtain

$$a_2(u_1'y_1' + u_2'y_2') + u_1(a_2y_1'' + a_1y_1' + a_0y_1) + u_2(a_2y_2'' + a_1y_2' + a_0y_2)$$
$$= a_2(u_1'y_1' + u_2'y_2') = f(x),$$

because y_1 and y_2 satisfy the homogeneous equation. Therefore, u_1' and u_2' satisfy the pair of equations

$$u_1'(x)y_1(x) + u_2'(x)y_2(x) = 0,$$
$$u_1'(x)y_1'(x) + u_2'(x)y_2'(x) = \frac{f(x)}{a_2(x)}.$$

We can solve these two equations for the unknown functions u_1' and u_2' by Cramer's Rule (Theorem 5 of Section 10.6), or otherwise, and obtain

$$u_1' = -\frac{y_2(x)}{W(x)}\frac{f(x)}{a_2(x)}, \qquad u_2' = \frac{y_1(x)}{W(x)}\frac{f(x)}{a_2(x)},$$

where $W(x)$, called the **Wronskian** of y_1 and y_2, is the determinant

$$W(x) = \begin{vmatrix} y_1(x) & y_2(x) \\ y_1'(x) & y_2'(x) \end{vmatrix}.$$

Then u_1 and u_2 can be found by integration.

Example 3 Find the general solution of $y'' - 3y' + 2y = 4x$.

Solution First we solve the homogeneous equation $y'' - 3y' + 2y = 0$, which has auxiliary equation $r^2 - 3r + 2 = 0$ with roots $r = 1$ and $r = 2$. Therefore, two independent solutions of the homogeneous equation are $y_1 = e^x$ and $y_2 = e^{2x}$, and the complementary function is

$$y_h = C_1 e^x + C_2 e^{2x}.$$

A particular solution $y_p(x)$ of the nonhomogeneous equation can be found in the form

$$y_p = u_1(x)e^x + u_2(x)e^{2x},$$

where u_1 and u_2 satisfy

$$u_1' e^x + 2u_2' e^{2x} = 4x,$$
$$u_1' e^x + u_2' e^{2x} = 0.$$

We solve these linear equations for u_1' and u_2' and then integrate to obtain

$$u_1' = -4xe^{-x} \qquad u_2' = 4xe^{-2x}$$
$$u_1 = 4(x+1)e^{-x} \qquad u_2 = -(2x+1)e^{-2x}.$$

Hence, $y_p = 4x + 4 - (2x + 1) = 2x + 3$ is a particular solution of the nonhomogeneous equation, and the general solution is

$$y = 2x + 3 + C_1 e^x + C_2 e^{2x}.$$

Remark This method for solving the nonhomogeneous equation is called the **method of variation of parameters**. It is completely general and extends to higher-order equations in a reasonable way, but it is computationally somewhat difficult. We could have found y_p more easily had we "guessed" that it would be of the form $y_p = Ax + B$ and substituted this into the differential equation to get

$$-3A + 2(Ax + B) = 4x$$
$$\text{or} \quad 2Ax + (2B - 3A) = 4x.$$

The only way this latter equation can be satisfied for all x is to have $2A = 4$ and $2B - 3A = 0$, that is, $A = 2$ and $B = 3$.

Maple Calculations

Maple has a `dsolve` routine for solving (some) differential equations and initial-value problems. This routine takes as input a DE and, if desired, initial conditions for it. We illustrate for the equation $y'' + 2y' + 5y = 25t + 20$ (assuming that the independent variable is t):

```
>  DE := (D@@2)(y)(t)+2*D(y)(t)+5*y(t)=25*t+20;
```
$$DE := D^{(2)}(y)(t) + 2D(y)(t) + 5y(t) = 25t + 20$$

```
>  dsolve(DE, y(t));
```
$$y(t) = e^{(-t)}\sin(2t)\,\_C2 + e^{(-t)}\cos(2t)\,\_C1 + 2 + 5t$$

Note Maple's use of $\_C1$ and $\_C2$ for arbitrary constants. For an initial-value problem we supply the DE and its initial conditions to `dsolve` as a single list or set argument enclosed in square brackets or braces:

```
>  dsolve([DE, y(0)=3, D(y)(0)=-2], y(t));
```

$$y(t) = -3e^{(-t)}\sin(2t) + e^{(-t)}\cos(2t) + 2 + 5t$$

You might think that this output indicates that y has been defined as a function of t and you can find a decimal value for, say, $y(1)$ by giving the input `evalf(y(1))`. But this won't work. In fact, the output of the `dsolve` is just an equation with left side the symbol $y(t)$. We can, however, use this output to define y as a function of t as follows:

```
>  y := unapply(op(2,%),t);
```

$$y := t \rightarrow -3e^{(-t)}\sin(2t) + e^{(-t)}\cos(2t) + 2 + 5t$$

The `op(2,%)` in the `unapply` command refers to the second operand of the previous result (i.e., the right side of equation output from the `dsolve`). `unapply(f,t)` converts an expression `f` to a function of t. To confirm:

```
>  evalf(y(1));
```

$$5.843372646$$

Exercises 17.6

Find general solutions for the nonhomogeneous equations in Exercises 1–12 by the method of undetermined coefficients.

1. $y'' + y' - 2y = 1$

2. $y'' + y' - 2y = x$

3. $y'' + y' - 2y = e^{-x}$

4. $y'' + y' - 2y = e^x$

5. $y'' + 2y' + 5y = x^2$

6. $y'' + 4y = x^2$

7. $y'' - y' - 6y = e^{-2x}$

8. $y'' + 4y' + 4y = e^{-2x}$

9. $y'' + 2y' + 2y = e^x \sin x$

10. $y'' + 2y' + 2y = e^{-x} \sin x$

11. $y'' + y' = 4 + 2x + e^{-x}$

12. $y'' + 2y' + y = xe^{-x}$

13. Repeat Exercise 3 using the method of variation of parameters.

14. Repeat Exercise 4 using the method of variation of parameters.

15. Find a particular solution of the form $y = Ax^2$ for the Euler equation $x^2 y'' + xy' - y = x^2$, and hence obtain the general solution of this equation on the interval $(0, \infty)$.

16. For what values of r can the Euler equation $x^2 y'' + xy' - y = x^r$ be solved by the method of Exercise 15? Find a particular solution for each such r.

17. Try to guess the form of a particular solution for $x^2 y'' + xy' - y = x$, and hence obtain the general solution for this equation on the interval $(0, \infty)$.

18. Use variation of parameters to solve $x^2 y'' + xy' - y = x$.

19. Consider the nonhomogeneous, linear equation

$$x^2 y'' - (2x + x^2)y' + (2 + x)y = x^3.$$

Use the fact that $y_1(x) = x$ and $y_2(x) = xe^x$ are independent solutions of the corresponding homogeneous equation (see Exercise 5 of Section 17.4) to find the general solution of this nonhomogeneous equation.

20. Consider the nonhomogeneous, Bessel equation

$$x^2 y'' + xy' + \left(x^2 - \frac{1}{4}\right)y = x^{3/2}.$$

Use the fact that $y_1(x) = x^{-1/2}\cos x$ and $y_2(x) = x^{-1/2}\sin x$ are independent solutions of the corresponding homogeneous equation (see Exercise 6 of Section 17.4) to find the general solution of this nonhomogeneous equation.

17.7 Series Solutions of Differential Equations

In Section 17.5 we developed a recipe for solving second-order, linear, homogeneous differential equations with constant coefficients:

$$ay'' + by' + cy = 0$$

and Euler equations of the form

$$ax^2 y'' + bxy' + cy = 0.$$

Many of the second-order, linear, homogeneous differential equations that arise in applications do not have constant coefficients and are not of Euler type. If the coefficient functions of such an equation are sufficiently well-behaved, we can often find solutions in the form of power series (Taylor series). Such series solutions are frequently used to define new functions, whose properties are deduced partly from the fact that they solve particular differential equations. For example, Bessel functions of order v are defined to be certain series solutions of Bessel's differential equation

$$x^2 y'' + xy' + (x^2 - v^2)y = 0.$$

Series solutions for second-order homogeneous linear differential equations are most easily found near an **ordinary point** of the equation. This is a point $x = a$ such that the equation can be expressed in the form

$$y'' + p(x)y' + q(x)y = 0,$$

where the functions $p(x)$ and $q(x)$ are **analytic** at $x = a$. (Recall that a function f is analytic at $x = a$ if $f(x)$ can be expressed as the sum of its Taylor series in powers of $x - a$ in an interval of positive radius centred at $x = a$.) Thus we assume

$$p(x) = \sum_{n=0}^{\infty} p_n (x - a)^n,$$

$$q(x) = \sum_{n=0}^{\infty} q_n (x - a)^n,$$

with both series converging in some interval of the form $a - R < x < a + R$. Frequently $p(x)$ and $q(x)$ are polynomials so are analytic everywhere. A change of independent variable $\xi = x - a$ will put the point $x = a$ at the origin $\xi = 0$, so we can assume that $a = 0$.

The following example illustrates the technique of series solution around an ordinary point.

Example 1 Find two independent series solutions in powers of x for the Hermite equation

$$y'' - 2xy' + vy = 0.$$

For what values of v does the equation have a polynomial solution?

Solution We try for a power series solution of the form

$$y = \sum_{n=0}^{\infty} a_n x^n = a_0 + a_1 x + a_2 x^2 + a_3 x^3 + \cdots, \qquad \text{so that}$$

$$y' = \sum_{n=1}^{\infty} n a_n x^{n-1}$$

$$y'' = \sum_{n=2}^{\infty} n(n-1) a_n x^{n-2} = \sum_{n=0}^{\infty} (n+2)(n+1) a_{n+2} x^n.$$

(We have replaced n by $n + 2$ in order to get x^n in the sum for y''.) We substitute these expressions into the differential equation to get

$$\sum_{n=0}^{\infty} (n+2)(n+1) a_{n+2} x^n - 2 \sum_{n=1}^{\infty} n a_n x^n + v \sum_{n=0}^{\infty} a_n x^n = 0$$

$$\text{or} \quad 2a_2 + v a_0 + \sum_{n=1}^{\infty} \Big[(n+2)(n+1) a_{n+2} - (2n - v) a_n \Big] x^n = 0.$$

This identity holds for all x provided that the coefficient of every power of x vanishes; that is,

$$a_2 = -\frac{\nu a_0}{2}, \qquad a_{n+2} = \frac{(2n - \nu)a_n}{(n+2)(n+1)}, \qquad (n = 1, 2, \cdots).$$

The latter of these formulas is called a **recurrence relation**.

We can choose a_0 and a_1 to have any values; then the above conditions determine all the remaining coefficients a_n, $(n \geq 2)$. We can get one solution by choosing, for instance, $a_0 = 1$ and $a_1 = 0$. Then, by the recurrence relation,

$$a_3 = 0, \quad a_5 = 0, \quad a_7 = 0, \quad \cdots, \qquad \text{and}$$

$$a_2 = -\frac{\nu}{2}$$

$$a_4 = \frac{(4 - \nu)a_2}{4 \times 3} = -\frac{\nu(4 - \nu)}{2 \times 3 \times 4} = -\frac{\nu(4 - \nu)}{4!}$$

$$a_6 = \frac{(8 - \nu)a_4}{6 \times 5} = -\frac{\nu(4 - \nu)(8 - \nu)}{6!}$$

$$\cdots$$

The pattern is obvious here:

$$a_{2n} = -\frac{\nu(4 - \nu)(8 - \nu) \cdots (4n - 4 - \nu)}{(2n)!}, \qquad (n = 1, 2, \cdots).$$

One solution to the Hermite equation is

$$y_1 = 1 + \sum_{n=1}^{\infty} -\frac{\nu(4 - \nu)(8 - \nu) \cdots (4n - 4 - \nu)}{(2n)!} x^{2n}.$$

We observe that if $\nu = 4n$ for some non-negative integer n, then y_1 is an even polynomial of degree $2n$, because $a_{2n+2} = 0$ and all subsequent even coefficients therefore also vanish.

The second solution, y_2, can be found in the same way, by choosing $a_0 = 0$ and $a_1 = 1$. It is

$$y_2 = x + \sum_{n=1}^{\infty} \frac{(2 - \nu)(6 - \nu) \cdots (4n - 2 - \nu)}{(2n+1)!} x^{2n+1},$$

and it is an odd polynomial of degree $2n + 1$ if $\nu = 4n + 2$.

Both of these series solutions converge for all x. The ratio test can be applied directly to the recurrence relation. Since consecutive nonzero terms of each series are of the form $a_n x^n$ and $a_{n+2} x^{n+2}$, we calculate

$$\rho = \lim_{n \to \infty} \left| \frac{a_{n+2} x^{n+2}}{a_n x^n} \right| = |x|^2 \lim_{n \to \infty} \left| \frac{a_{n+2}}{a_n} \right| = |x|^2 \lim_{n \to \infty} \left| \frac{2n - \nu}{(n+2)(n+1)} \right| = 0$$

for every x, so the series converges by the ratio test. ∎

If $x = a$ is not an ordinary point of the equation

$$y'' + p(x)y' + q(x)y = 0,$$

then it is called a **singular point** of that equation. This means that at least one of the functions $p(x)$ and $q(x)$ is not analytic at $x = a$. If, however, $(x - a)p(x)$ and $(x - a)^2 q(x)$ are analytic at $x = a$, then the singular point is said to be a **regular singular point**. For example, the origin $x = 0$ is a regular singular point of Bessel's equation,

$$x^2 y'' + xy' + (x^2 - \nu^2)y = 0,$$

since $p(x) = 1/x$ and $q(x) = (x^2 - \nu^2)/x^2$ satisfy $xp(x) = 1$ and $x^2 q(x) = x^2 - \nu^2$, which are both polynomials and therefore analytic.

The solutions of differential equations are usually not analytic at singular points. However, it is still possible to find at least one series solution about such a point. The method involves searching for a series solution of the form x^μ times a power series, that is,

$$y = (x - a)^\mu \sum_{n=0}^{\infty} a_n(x - a)^n = \sum_{n=0}^{\infty} a_n(x - a)^{n+\mu}, \qquad \text{where } a_0 \neq 0.$$

Substitution into the differential equation produces a quadratic **indicial equation**, which determines one or two values of μ for which such solutions can be found, and a **recurrence relation** enabling the coefficients a_n to be calculated for $n \geq 1$. If the indicial roots are not equal and do not differ by an integer, two independent solutions can be calculated. If the indicial roots are equal or differ by an integer, one such solution can be calculated (corresponding to the larger indicial root), but finding a second independent solution (and so the general solution) requires techniques beyond the scope of this book. The reader is referred to standard texts on differential equations for more discussion and examples. We will content ourselves here with one final example.

Example 2   Find one solution, in powers of x, of Bessel's equation of order $\nu = 1$, namely,

$$x^2 y'' + xy' + (x^2 - 1)y = 0.$$

Solution   We try

$$y = \sum_{n=0}^{\infty} a_n x^{\mu+n}$$

$$y' = \sum_{n=0}^{\infty} (\mu + n) a_n x^{\mu+n-1}$$

$$y'' = \sum_{n=0}^{\infty} (\mu + n)(\mu + n - 1) a_n x^{\mu+n-2}.$$

Substituting these expressions into the Bessel equation, we get

$$\sum_{n=0}^{\infty} \Big[\big((\mu + n)(\mu + n - 1) + (\mu + n) - 1\big) a_n x^n + a_n x^{n+2} \Big] = 0$$

$$\sum_{n=0}^{\infty} \Big[(\mu + n)^2 - 1 \Big] a_n x^n + \sum_{n=2}^{\infty} a_{n-2} x^n = 0$$

$$(\mu^2 - 1)a_0 + \big((\mu + 1)^2 - 1\big)a_1 x + \sum_{n=2}^{\infty} \Big[\big((\mu + n)^2 - 1\big)a_n + a_{n-2} \Big] x^n = 0.$$

All of the terms must vanish. Since $a_0 \neq 0$ (we may take $a_0 = 1$), we obtain

$$\mu^2 - 1 = 0, \qquad\qquad \text{the indicial equation}$$

$$[(\mu + 1)^2 - 1]a_1 = 0,$$

$$a_n = -\frac{a_{n-2}}{(\mu + n)^2 - 1}, \quad (n \geq 2). \quad \text{the recurrence relation}$$

Evidently $\mu = \pm 1$; therefore $a_1 = 0$. If we take $\mu = 1$, then the recurrence relation is $a_n = -a_{n-2}/(n)(n+2)$. Thus,

$$a_3 = 0, \quad a_5 = 0, \quad a_7 = 0, \quad \cdots$$

$$a_2 = \frac{-1}{2 \times 4}, \quad a_4 = \frac{1}{2 \times 4 \times 4 \times 6}, \quad a_6 = \frac{-1}{2 \times 4 \times 4 \times 6 \times 6 \times 8}, \quad \cdots.$$

Again the pattern is obvious:

$$a_{2n} = \frac{(-1)^n}{2^{2n} n! (n+1)!},$$

and one solution of the Bessel equation of order 1 is

$$y = \sum_{n=0}^{\infty} \frac{(-1)^n}{2^{2n} n! (n+1)!} x^{2n+1}.$$

By the ratio test, this series converges for all x.

Remark   Observe that if we tried to calculate a second solution using $\mu = -1$ we would get the recurrence relation

$$a_n = -\frac{a_{n-2}}{n(n-2)},$$

and we would be unable to calculate a_2. This shows what can happen if the indicial roots differ by an integer.

Exercises 17.7

1. Find the general solution of $y'' = (x-1)^2 y$ in the form of a power series $y = \sum_{n=0}^{\infty} a_n (x-1)^n$.

2. Find the general solution of $y'' = xy$ in the form of a power series $y = \sum_{n=0}^{\infty} a_n x^n$ with a_0 and a_1 arbitrary.

3. Solve the initial-value problem

$$\begin{cases} y'' + xy' + 2y = 0 \\ y(0) = 1 \\ y'(0) = 2. \end{cases}$$

4. Find the solution of $y'' + xy' + y = 0$ that satisfies $y(0) = 1$ and $y'(0) = 0$.

5. Find the first three nonzero terms in a power series solution in powers of x for the initial-value problem $y'' + (\sin x)y = 0$, $y(0) = 1$, $y'(0) = 0$.

6. Find the solution, in powers of x, for the initial-value problem

$$(1-x^2)y'' - xy' + 9y = 0, \quad y(0) = 0, \quad y'(0) = 1.$$

7. Find two power series solutions in powers of x for $3xy'' + 2y' + y = 0$.

8. Find one power series solution for the Bessel equation of order $\nu = 0$, that is, the equation $xy'' + y' + xy = 0$.

Chapter Review

Key Ideas

• **What do the following phrases mean?**

◇ an ordinary DE   ◇ a partial DE

◇ the general solution of a DE

◇ a linear combination of solutions of a DE

◇ the order of a DE   ◇ a linear DE

◇ a separable DE   ◇ an exact DE

◇ an integrating factor   ◇ a constant coefficient DE

◇ an Euler equation   ◇ an auxiliary equation

• **Describe how to solve:**

◇ a separable DE   ◇ a first-order, linear DE

◇ a homogeneous, first-order DE

⋄ a constant coefficient DE   ⋄ an Euler equation

- **What conditions imply that an initial-value problem for a first-order DE has a unique solution near the initial point?**
- **Describe the following methods for solving first-order DEs numerically:**
  - ⋄ the Euler method   ⋄ the improved Euler method
  - ⋄ the fourth-order Runge–Kutta method
- **Describe the following methods for solving a nonhomogeneous, linear DE:**
  - ⋄ undetermined coefficients   ⋄ variation of parameters
- **What are an ordinary point and a regular singular point of a linear, second-order DE? Describe how series can be used to solve such an equation near such a point.**

Review Exercises

Find general solutions of the differential equations in Exercises 1–16.

1. $\dfrac{dy}{dx} = 2xy$

2. $\dfrac{dy}{dx} = e^{-y}\sin x$

3. $\dfrac{dy}{dx} = x + 2y$

4. $\dfrac{dy}{dx} = \dfrac{x^2 + y^2}{2xy}$

5. $\dfrac{dy}{dx} = \dfrac{x + y}{y - x}$

6. $\dfrac{dy}{dx} = -\dfrac{y + e^x}{x + e^y}$

7. $\dfrac{d^2y}{dt^2} = \left(\dfrac{dy}{dt}\right)^2$

8. $2\dfrac{d^2y}{dt^2} + 5\dfrac{dy}{dt} + 2y = 0$

9. $4y'' - 4y' + 5y = 0$

10. $2x^2y'' + y = 0$

11. $t^2\dfrac{d^2y}{dt^2} - t\dfrac{dy}{dt} + 5y = 0$

12. $\dfrac{d^3y}{dt^3} + 8\dfrac{d^2y}{dt^2} + 16\dfrac{dy}{dt} = 0$

13. $\dfrac{d^2y}{dx^2} - 5\dfrac{dy}{dx} + 6y = e^x + e^{3x}$

14. $\dfrac{d^2y}{dx^2} - 5\dfrac{dy}{dx} + 6y = xe^{2x}$

15. $\dfrac{d^2y}{dx^2} + 2\dfrac{dy}{dx} + y = x^2$ **16.** $x\dfrac{d^2y}{dx^2} - 2y = x^3$

Solve the initial-value problems in Exercises 17–26.

17. $\begin{cases} \dfrac{dy}{dx} = \dfrac{x^2}{y^2} \\ y(2) = 1 \end{cases}$

18. $\begin{cases} \dfrac{dy}{dx} = \dfrac{y^2}{x^2} \\ y(2) = 1 \end{cases}$

19. $\begin{cases} \dfrac{dy}{dx} = \dfrac{xy}{x^2 + y^2} \\ y(0) = 1 \end{cases}$

20. $\begin{cases} \dfrac{dy}{dx} + (\cos x)y = 2\cos x \\ y(\pi) = 1 \end{cases}$

21. $\begin{cases} y'' + 3y' + 2y = 0 \\ y(0) = 1 \\ y'(0) = 2 \end{cases}$

22. $\begin{cases} y'' + 2y' + (1 + \pi^2)y = 0 \\ y(1) = 0 \\ y'(1) = \pi \end{cases}$

23. $\begin{cases} y'' + 10y' + 25y = 0 \\ y(1) = e^{-5} \\ y'(1) = 0 \end{cases}$

24. $\begin{cases} x^2y'' - 3xy' + 4y = 0 \\ y(e) = e^2 \\ y'(e) = 0 \end{cases}$

25. $\begin{cases} \dfrac{d^2y}{dt^2} + 4y = 8e^{2t} \\ y(0) = 1 \\ y'(0) = -2 \end{cases}$

26. $\begin{cases} 2\dfrac{d^2y}{dx^2} + 5\dfrac{dy}{dx} - 3y = 6 + 7e^{x/2} \\ y(0) = 0 \\ y'(0) = 1 \end{cases}$

27. Find the equation of a curve that passes through the point $(2, 4)$ and has slope $3y/(x - 1)$ at any point (x, y) on it.

28. Find a family of curves that intersect every ellipse of the form $3x^2 + 4y^2 = C$ at right angles.

29. For what values of the constants A and B is the equation

$$[(x + A)e^x \sin y + \cos y]\,dx + x[e^x \cos y + B\sin y]\,dy = 0$$

exact? What is the general solution of the equation if A and B have these values?

30. Find a value of n for which x^n is an integrating factor for

$$(x^2 + 3y^2)\,dx + xy\,dy = 0$$

and solve the equation.

31. Show that $y = x$ is a solution of

$$x^2y'' - x(2 + x\cot x)y' + (2 + x\cot x)y = 0$$

and find the general solution of this equation.

32. Use the method of variation of parameters and the result of the previous exercise to find the general solution of the nonhomogeneous equation

$$x^2y'' - x(2 + x\cot x)y' + (2 + x\cot x)y = x^3\sin x.$$

33. Suppose that $f(x, y)$ and $\dfrac{\partial}{\partial y} f(x, y)$ are continuous on the whole xy-plane, and that $f(x, y)$ is bounded there, say $|f(x, y)| \le K$. Show that no solution of $y' = f(x, y)$ can have a vertical asymptote. Describe the region in the plane in which the solution to the initial-value problem

$$\begin{cases} y' = f(x, y) \\ y(x_0) = y_0 \end{cases}$$

must remain.

Appendix I

Complex Numbers

Old Macdonald had a farm,
Minus E-squared O

Mathematically simplified children's song

Many of the problems to which mathematics is applied involve the solution of equations. Over the centuries the number system had to be expanded many times to provide solutions for more and more kinds of equations. The natural numbers

$$\mathbb{N} = \{1, \, 2, \, 3, \, 4, \, \ldots\}$$

are inadequate for the solutions of equations of the form

$$x + n = m, \qquad (m, \, n \in \mathbb{N}).$$

Zero and negative numbers can be added to create the integers

$$\mathbb{Z} = \{\ldots, \, -3, \, -2, \, -1, \, 0, \, 1, \, 2, \, 3, \, \ldots\}$$

in which that equation has the solution $x = m - n$ even if $m < n$. (Historically, this extension of the number system came much later than some of those mentioned below.) Some equations of the form

$$nx = m, \qquad (m, \, n \in \mathbb{Z}, \quad n \neq 0),$$

cannot be solved in the integers. Another extension is made to include numbers of the form m/n, thus producing the set of rational numbers

$$\mathbb{Q} = \left\{ \frac{m}{n} \, : \, m, \, n \in \mathbb{Z}, \quad n \neq 0 \right\}.$$

Every linear equation

$$ax = b, \qquad (a, \, b \in \mathbb{Q}, \quad a \neq 0),$$

has a solution $x = b/a$ in $\mathbb{Q}$, but the quadratic equation

$$x^2 = 2$$

has no solution in $\mathbb{Q}$, as was shown in Section P.1. Another extension enriches the rational numbers to the real numbers $\mathbb{R}$ in which some equations like $x^2 = 2$ have solutions. However, other quadratic equations, for instance,

$$x^2 = -1$$

do not have solutions, even in the real numbers, so the extension process is not complete. In order to be able to solve any quadratic equation, we need to extend the real number system to a larger set, which we call the **complex number system**. In this appendix we will define complex numbers and develop some of their basic properties.

Definition of Complex Numbers

We begin by defining the symbol i, called **the imaginary unit,**[1] to have the property

$$i^2 = -1.$$

Thus, we could also call i the **square root of** -1 and denote it $\sqrt{-1}$. Of course, i is not a real number; no real number has a negative square.

DEFINITION 1

> A **complex number** is an expression of the form
>
> $$a + bi \qquad \text{or} \qquad a + ib,$$
>
> where a and b are *real numbers*, and i is the imaginary unit.

For example, $3+2i$, $\frac{7}{2} - \frac{2}{3}i$, $i\pi = 0+i\pi$, and $-3 = -3+0i$ are all complex numbers. The last of these examples shows that every real number can be regarded as a complex number. (We will normally use $a + bi$ unless b is a complicated expression, in which case we will write $a + ib$ instead. Either form is acceptable.)

It is often convenient to represent a complex number by a single letter; w and z are frequently used for this purpose. If a, b, x, and y are real numbers, and

$$w = a + bi \qquad \text{and} \qquad z = x + yi,$$

then we can refer to the complex numbers w and z. Note that $w = z$ if and only if $a = x$ and $b = y$. Of special importance are the complex numbers

$$0 = 0 + 0i, \qquad 1 = 1 + 0i, \qquad \text{and} \qquad i = 0 + 1i.$$

DEFINITION 2

> If $z = x + yi$ is a complex number (where x and y are real), we call x the **real part** of z and denote it $\text{Re}\,(z)$. We call y the **imaginary part** of z and denote it $\text{Im}\,(z)$:
>
> $$\text{Re}\,(z) = \text{Re}\,(x + yi) = x, \qquad \text{Im}\,(z) = \text{Im}\,(x + yi) = y.$$

Note that both the real and imaginary parts of a complex number are real numbers:

$$\text{Re}\,(3 - 5i) = 3 \qquad\qquad \text{Im}\,(3 - 5i) = -5$$
$$\text{Re}\,(2i) = \text{Re}\,(0 + 2i) = 0 \qquad\qquad \text{Im}\,(2i) = \text{Im}\,(0 + 2i) = 2$$
$$\text{Re}\,(-7) = \text{Re}\,(-7 + 0i) = -7 \qquad\qquad \text{Im}\,(-7) = \text{Im}\,(-7 + 0i) = 0.$$

Graphical Representation of Complex Numbers

Since complex numbers are constructed from pairs of real numbers (their real and imaginary parts), it is natural to represent complex numbers graphically as points in a Cartesian plane. We use the point with coordinates (a, b) to represent the complex number $w = a + ib$. In particular, the origin $(0, 0)$ represents the complex number 0, the point $(1, 0)$ represents the complex number $1 = 1 + 0i$, and the point $(0, 1)$ represents the point $i = 0 + 1i$. (See Figure I.1.)

---

[1] In some fields, for example, electrical engineering, the imaginary unit is denoted j instead of i. Like "negative," "surd," and "irrational," the term "imaginary" suggests the distrust that greeted the new kinds of numbers when they were first introduced.

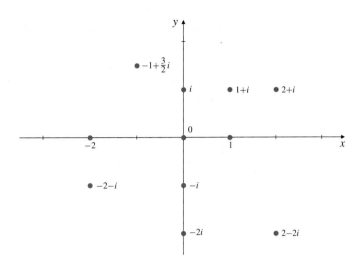

Figure I.1 An Argand diagram representing the complex plane

Such a representation of complex numbers as points in a plane is called an **Argand diagram**. Since each complex number is represented by a unique point in the plane, the set of all complex numbers is often referred to as the **complex plane**. The symbol $\mathbb{C}$ is used to represent the set of all complex numbers and, equivalently, the complex plane:

$$\mathbb{C} = \{x + yi \; : \; x, \; y, \; \in \mathbb{R}\}.$$

The points on the x-axis of the complex plane correspond to real numbers ($x = x + 0i$), so the x-axis is called the **real axis**. The points on the y-axis correspond to **pure imaginary** numbers ($yi = 0 + yi$), so the y-axis is called the **imaginary axis**.

It can be helpful to use the *polar coordinates* of a point in the complex plane.

DEFINITION 3

The distance from the origin to the point (a, b) corresponding to the complex number $w = a + bi$ is called the **modulus** of w and is denoted by $|w|$ or $|a + bi|$:

$$|w| = |a + bi| = \sqrt{a^2 + b^2}.$$

DEFINITION 4

If the line from the origin to (a, b) makes angle θ with the positive direction of the real axis (with positive angles measured counterclockwise), then we call θ an **argument** of the complex number $w = a + bi$ and denote it by $\arg(w)$ or $\arg(a + bi)$. (See Figure I.2.)

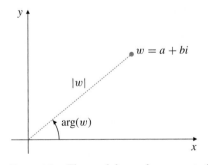

Figure I.2 The modulus and argument of a complex number

The modulus of a complex number is always real and non-negative. It is positive unless the complex number is 0. Modulus plays a similar role for complex numbers that absolute value does for real numbers. Indeed, sometimes modulus is called absolute value.

Arguments of complex numbers are not unique. If $w = a + bi \neq 0$, then any two possible values for $\arg(w)$ differ by an integer multiple of 2π. The symbol $\arg(w)$ actually represents not a single number, but a set of numbers. When we write $\arg(w) = \theta$, we are saying that the set $\arg(w)$ contains all numbers of the form $\theta + 2k\pi$, where k is an integer. Similarly, the statement $\arg(z) = \arg(w)$ says that two sets are identical.

If $w = a + bi$, where $a = \mathrm{Re}\,(w) \neq 0$, then

$$\tan \arg(w) = \tan \arg(a + bi) = \frac{b}{a}.$$

This means that $\tan\theta = b/a$ for every θ in the set $\arg(w)$.

It is sometimes convenient to restrict $\theta = \arg(w)$ to an interval of length 2π, say, the interval $0 \le \theta < 2\pi$, or $-\pi < \theta \le \pi$, so that nonzero complex numbers will have unique arguments. We will call the value of $\arg(w)$ in the interval $-\pi < \theta \le \pi$ the **principal argument** of w and denote it $\text{Arg}(w)$. Every complex number w except 0 has a unique principal argument $\text{Arg}(w)$.

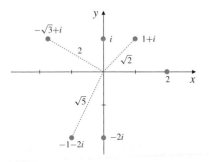

Figure I.3   Some complex numbers with their moduli

Example 1   (**Some moduli and principal arguments**) See Figure I.3.

$$|2| = 2 \qquad\qquad \text{Arg}(2) = 0$$
$$|1 + i| = \sqrt{2} \qquad\qquad \text{Arg}(1 + i) = \pi/4$$
$$|i| = 1 \qquad\qquad \text{Arg}(i) = \pi/2$$
$$|-2i| = 2 \qquad\qquad \text{Arg}(-2i) = -\pi/2$$
$$|-\sqrt{3} + i| = 2 \qquad\qquad \text{Arg}(-\sqrt{3} + i) = 5\pi/6$$
$$|-1 - 2i| = \sqrt{5} \qquad\qquad \text{Arg}(-1 - 2i) = -\pi + \tan^{-1}(2).$$

Remark   If $z = x + yi$ and $\text{Re}(z) = x > 0$, then $\text{Arg}(z) = \tan^{-1}(y/x)$. Many computer spreadsheets and mathematical software packages implement a two-variable arctan function denoted $\text{atan2}(x, y)$ which gives the polar angle of (x, y) in the interval $(-\pi, \pi]$. Thus

$$\text{Arg}(x + yi) = \text{atan2}(x, y).$$

> **⚠ BEWARE**
>
> Review the cautionary remark at the end of the discussion of the arctangent function in Section 3.5; different programs implement the two-variable arctangent using different notations and/or order of variables.

Given the modulus $r = |w|$ and any value of the argument $\theta = \arg(w)$ of a complex number $w = a + bi$, we have $a = r\cos\theta$ and $b = r\sin\theta$, so w can be expressed in terms of its modulus and argument as

$$w = r\cos\theta + i\,r\sin\theta.$$

The expression on the right side is called the **polar representation** of w.

DEFINITION 5

The **conjugate** or **complex conjugate** of a complex number $w = a + bi$ is another complex number, denoted $\overline{w}$, given by

$$\overline{w} = a - bi.$$

Example 2 $\overline{2 - 3i} = 2 + 3i, \qquad \overline{3} = 3, \qquad \overline{2i} = -2i.$

Observe that

$$\text{Re}(\overline{w}) = \text{Re}(w) \qquad\qquad |\overline{w}| = |w|$$
$$\text{Im}(\overline{w}) = -\text{Im}(w) \qquad \arg(\overline{w}) = -\arg(w).$$

In an Argand diagram the point $\overline{w}$ is the reflection of the point w in the real axis. (See Figure I.4.)

Note that w is real ($\text{Im}(w) = 0$) if and only if $\overline{w} = w$. Also, w is pure imaginary ($\text{Re}(w) = 0$) if and only if $\overline{w} = -w$. (Here, $-w = -a - bi$ if $w = a + bi$.)

Figure I.4   A complex number and its conjugate are mirror images of each other in the real axis

Complex Arithmetic

Like real numbers, complex numbers can be added, subtracted, multiplied, and divided. Two complex numbers are added or subtracted as though they are two-dimensional vectors whose components are their real and imaginary parts.

The sum and difference of complex numbers

If $w = a + bi$ and $z = x + yi$, where a, b, x, and y are real numbers, then

$$w + z = (a + x) + (b + y)i$$
$$w - z = (a - x) + (b - y)i.$$

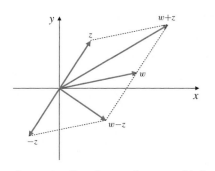

Figure I.5 Complex numbers are added and subtracted vectorially. Observe the parallelograms

In an Argand diagram the points $w + z$ and $w - z$ are the points whose position vectors are, respectively, the sum and difference of the position vectors of the points w and z. (See Figure I.5.) In particular, the complex number $a + bi$ is the sum of the real number $a = a + 0i$ and the pure imaginary number $bi = 0 + bi$.

Complex addition obeys the same rules as real addition: if w_1, w_2, and w_3 are three complex numbers, the following are easily verified:

$w_1 + w_2 = w_2 + w_1$	Addition is commutative.
$(w_1 + w_2) + w_3 = w_1 + (w_2 + w_3)$	Addition is associative.
$\lvert w_1 \pm w_2 \rvert \le \lvert w_1 \rvert + \lvert w_2 \rvert$	the triangle inequality

Note that $\lvert w_1 - w_2 \rvert$ is the distance between the two points w_1 and w_2 in the complex plane. Thus, the triangle inequality says that in the triangle with vertices w_1, $\mp w_2$ and 0, the length of one side is less than the sum of the other two.

It is also easily verified that the conjugate of a sum (or difference) is the sum (or difference) of the conjugates:

$$\overline{w + z} = \overline{w} + \overline{z}.$$

Example 3

(a) If $w = 2 + 3i$ and $z = 4 - 5i$, then

$$w + z = (2 + 4) + (3 - 5)i = 6 - 2i$$
$$w - z = (2 - 4) + (3 - (-5))i = -2 + 8i.$$

(b) $3i + (1 - 2i) - (2 + 3i) + 5 = 4 - 2i$.

Multiplication of the complex numbers $w = a + bi$ and $z = x + yi$ is carried out by formally multiplying the binomial expressions and replacing i^2 by -1:

$$wz = (a + bi)(x + yi) = ax + ayi + bxi + byi^2$$
$$= (ax - by) + (ay + bx)i.$$

The product of complex numbers

If $w = a + bi$ and $z = x + yi$, where a, b, x, and y are real numbers, then

$$wz = (ax - by) + (ay + bx)i.$$

Example 4

(a) $(2 + 3i)(1 - 2i) = 2 - 4i + 3i - 6i^2 = 8 - i$.

(b) $i(5 - 4i) = 5i - 4i^2 = 4 + 5i$.

(c) $(a + bi)(a - bi) = a^2 - abi + abi - b^2i^2 = a^2 + b^2$.

Part (c) of the example above shows that the square of the modulus of a complex number is the product of that number with its complex conjugate:

$$w\,\overline{w} = |w|^2.$$

Complex multiplication shares many properties with real multiplication. In particular, if w_1, w_2, and w_3 are complex numbers, then

$$w_1 w_2 = w_2 w_1 \qquad\qquad \text{Multiplication is commutative.}$$
$$(w_1 w_2) w_3 = w_1 (w_2 w_3) \qquad\qquad \text{Multiplication is associative.}$$
$$w_1 (w_2 + w_3) = w_1 w_2 + w_1 w_3 \qquad\qquad \text{Multiplication distributes over addition.}$$

The conjugate of a product is the product of the conjugates:

$$\overline{wz} = \overline{w}\,\overline{z}.$$

To see this, let $w = a + bi$ and $z = x + yi$. Then

$$\overline{wz} = \overline{(ax - by) + (ay + bx)i}$$
$$= (ax - by) - (ay + bx)i$$
$$= (a - bi)(x - yi) = \overline{w}\,\overline{z}.$$

It is particularly easy to determine the product of complex numbers expressed in polar form. If

$$w = r(\cos\theta + i\sin\theta) \qquad \text{and} \qquad z = s(\cos\phi + i\sin\phi),$$

where $r = |w|$, $\theta = \arg(w)$, $s = |z|$, and $\phi = \arg(z)$, then

$$wz = rs(\cos\theta + i\sin\theta)(\cos\phi + i\sin\phi)$$
$$= rs\big((\cos\theta\cos\phi - \sin\theta\sin\phi) + i(\sin\theta\cos\phi + \cos\theta\sin\phi)\big)$$
$$= rs\big(\cos(\theta + \phi) + i\sin(\theta + \phi)\big).$$

(See Figure I.6.) Since arguments are only determined up to integer multiples of 2π, we have proved that

The modulus and argument of a product

$$|wz| = |w||z| \qquad \text{and} \qquad \arg(wz) = \arg(w) + \arg(z).$$

The second of these equations says that the set $\arg(wz)$ consists of all numbers $\theta + \phi$, where θ belongs to the set $\arg(w)$ and ϕ to the set $\arg(z)$.

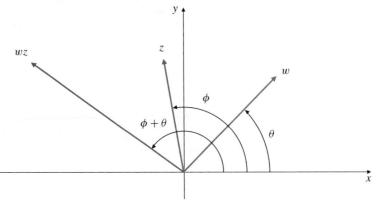

Figure I.6   The argument of a product is the sum of the arguments of the factors

More generally, if w_1, w_2, ..., w_n are complex numbers, then

$$|w_1 w_2 \cdots w_n| = |w_1||w_2| \cdots |w_n|$$
$$\arg(w_1 w_2 \cdots w_n) = \arg(w_1) + \arg(w_2) + \cdots + \arg(w_n).$$

Multiplication of a complex number by i has a particularly simple geometric interpretation in an Argand diagram. Since $|i| = 1$ and $\arg(i) = \pi/2$, multiplication of $w = a + bi$ by i leaves the modulus of w unchanged but increases its argument by $\pi/2$. (See Figure I.7.) Thus, multiplication by i rotates the position vector of w counterclockwise by 90° about the origin.

Let $z = \cos\theta + i\sin\theta$. Then $|z| = 1$ and $\arg(z) = \theta$. Since the modulus of a product is the product of the moduli of the factors and the argument of a product is the sum of the arguments of the factors, we have $|z^n| = |z|^n = 1$ and $\arg(z^n) = n\arg(z) = n\theta$. Thus,

$$z^n = \cos n\theta + i\sin n\theta,$$

and we have proved de Moivre's Theorem.

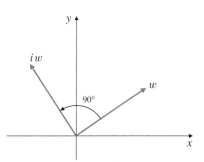

Figure I.7   Multiplication by i corresponds to counterclockwise rotation by 90°

THEOREM **1**

de Moivre's Theorem

$$\left(\cos\theta + i\sin\theta\right)^n = \cos n\theta + i\sin n\theta.$$

Remark   Much of the study of complex-valued functions of a complex variable is beyond the scope of this book. However, in Appendix II we will introduce a complex version of the exponential function having the following property: if $z = x + iy$ (where x and y are real), then

$$e^z = e^{x+iy} = e^x e^{iy} = e^x(\cos y + i\sin y).$$

Thus, the modulus of e^z is $e^{\text{Re}\,(z)}$, and Im (z) is a value of arg (e^z). In this context, de Moivre's Theorem just says

$$(e^{i\theta})^n = e^{in\theta}.$$

Example 5   Express $(1 + i)^5$ in the form $a + bi$.

Solution   Since $|(1 + i)^5| = |1 + i|^5 = \left(\sqrt{2}\right)^5 = 4\sqrt{2}$, and $\arg\left((1 + i)^5\right) = 5\arg(1 + i) = \dfrac{5\pi}{4}$, we have

$$(1 + i)^5 = 4\sqrt{2}\left(\cos\frac{5\pi}{4} + i\sin\frac{5\pi}{4}\right) = 4\sqrt{2}\left(-\frac{1}{\sqrt{2}} - \frac{1}{\sqrt{2}}i\right) = -4 - 4i.$$

de Moivre's Theorem can be used to generate trigonometric identities for multiples of an angle. For example, for $n = 2$ we have

$$\cos 2\theta + i\sin 2\theta = \left(\cos\theta + i\sin\theta\right)^2 = \cos^2\theta - \sin^2\theta + 2i\cos\theta\sin\theta.$$

Thus, $\cos 2\theta = \cos^2\theta - \sin^2\theta$, and $\sin 2\theta = 2\sin\theta\cos\theta$.

The **reciprocal** of the nonzero complex number $w = a + bi$ can be calculated by multiplying the numerator and denominator of the reciprocal expression by the conjugate of w:

$$w^{-1} = \frac{1}{w} = \frac{1}{a + bi} = \frac{a - bi}{(a + bi)(a - bi)} = \frac{a - bi}{a^2 + b^2} = \frac{\overline{w}}{|w|^2}.$$

Since $|\overline{w}| = |w|$, and $\arg(\overline{w}) = -\arg(w)$, we have

$$\left|\frac{1}{w}\right| = \frac{|\overline{w}|}{|w|^2} = \frac{1}{|w|} \qquad \text{and} \qquad \arg\left(\frac{1}{w}\right) = -\arg(w).$$

The **quotient** z/w of two complex numbers $z = x + yi$ and $w = a + bi$ is the product of z and $1/w$, so

$$\frac{z}{w} = \frac{z\overline{w}}{|w|^2} = \frac{(x + yi)(a - bi)}{a^2 + b^2} = \frac{xa + yb + i(ya - xb)}{a^2 + b^2}.$$

We have

> **The modulus and argument of a quotient**
>
> $$\left|\frac{z}{w}\right| = \frac{|z|}{|w|} \qquad \text{and} \qquad \arg\left(\frac{z}{w}\right) = \arg(z) - \arg(w).$$

The set $\arg(z/w)$ consists of all numbers $\theta - \phi$ where θ belongs to the set $\arg(z)$ and ϕ to the set $\arg(w)$.

Example 6 Simplify   (a) $\dfrac{2 + 3i}{4 - i}$ and   (b) $\dfrac{i}{1 + i\sqrt{3}}$.

Solution

(a) $\dfrac{2 + 3i}{4 - i} = \dfrac{(2 + 3i)(4 + i)}{(4 - i)(4 + i)} = \dfrac{8 - 3 + (2 + 12)i}{4^2 + 1^2} = \dfrac{5}{17} + \dfrac{14}{17}i.$

(b) $\dfrac{i}{1 + i\sqrt{3}} = \dfrac{i(1 - i\sqrt{3})}{(1 + i\sqrt{3})(1 - i\sqrt{3})} = \dfrac{\sqrt{3} + i}{1^2 + 3} = \dfrac{\sqrt{3}}{4} + \dfrac{1}{4}i.$

Alternatively, since $|1 + i\sqrt{3}| = 2$ and $\arg(1 + i\sqrt{3}) = \tan^{-1}\sqrt{3} = \dfrac{\pi}{3}$, the quotient in (b) has modulus $\dfrac{1}{2}$ and argument $\dfrac{\pi}{2} - \dfrac{\pi}{3} = \dfrac{\pi}{6}$. Thus

$$\frac{i}{1 + i\sqrt{3}} = \frac{1}{2}\left(\cos\frac{\pi}{6} + i\sin\frac{\pi}{6}\right) = \frac{\sqrt{3}}{4} + \frac{1}{4}i.$$

Roots of Complex Numbers

If a is a positive real number, there are two distinct real numbers whose square is a. These are usually denoted

$$\sqrt{a} \qquad \text{(the positive square root of } a\text{)} \quad \text{and}$$

$$-\sqrt{a} \qquad \text{(the negative square root of } a\text{)}.$$

Every nonzero complex number $z = x + yi$ (where $x^2 + y^2 > 0$) also has two square roots; if w_1 is a complex number such that $w_1^2 = z$, then $w_2 = -w_1$ also satisfies $w_2^2 = z$. Again, we would like to single out one of these roots and call it $\sqrt{z}$.

Let $r = |z|$, so that $r > 0$. Let $\theta = \text{Arg}(z)$. Thus $-\pi < \theta \le \pi$. Since

$$z = r(\cos\theta + i\sin\theta),$$

the complex number

$$w = \sqrt{r}\left(\cos\frac{\theta}{2} + i\sin\frac{\theta}{2}\right)$$

clearly satisfies $w^2 = z$. We call this w the **principal square root** of z and denote it $\sqrt{z}$. The two solutions of the equation $w^2 = z$ are, thus, $w = \sqrt{z}$ and $w = -\sqrt{z}$.

Observe that the real part of $\sqrt{z}$ is always non-negative since $\cos(\theta/2) \ge 0$ for $-\pi/2 < \theta \le \pi/2$. In this interval $\sin(\theta/2) = 0$ only if $\theta = 0$ in which case $\sqrt{z}$ is real and positive.

Example 7

(a) $\sqrt{4} = \sqrt{4(\cos 0 + i \sin 0)} = 2.$

(b) $\sqrt{i} = \sqrt{1\left(\cos\dfrac{\pi}{2} + i \sin\dfrac{\pi}{2}\right)} = \cos\dfrac{\pi}{4} + i \sin\dfrac{\pi}{4} = \dfrac{1}{\sqrt{2}} + \dfrac{1}{\sqrt{2}}\, i.$

(c) $\sqrt{-4i} = \sqrt{4\left[\cos\left(-\dfrac{\pi}{2}\right) + i \sin\left(-\dfrac{\pi}{2}\right)\right]} = 2\left[\cos\left(-\dfrac{\pi}{4}\right) + i \sin\left(-\dfrac{\pi}{4}\right)\right]$

$\qquad = \sqrt{2} - i\sqrt{2}.$

(d) $\sqrt{-\dfrac{1}{2} + i\,\dfrac{\sqrt{3}}{2}} = \sqrt{\cos\dfrac{2\pi}{3} + i \sin\dfrac{2\pi}{3}} = \cos\dfrac{\pi}{3} + i \sin\dfrac{\pi}{3} = \dfrac{1}{2} + \dfrac{\sqrt{3}}{2}\, i.$

Given a nonzero complex number z we can find n distinct complex numbers w that satisfy $w^n = z$. These n numbers are called nth roots of z. For example, if $z = 1 = \cos 0 + i \sin 0$, then each of the numbers

$$w_1 = 1$$
$$w_2 = \cos\frac{2\pi}{n} + i \sin\frac{2\pi}{n}$$
$$w_3 = \cos\frac{4\pi}{n} + i \sin\frac{4\pi}{n}$$
$$w_4 = \cos\frac{6\pi}{n} + i \sin\frac{6\pi}{n}$$
$$\vdots$$
$$w_n = \cos\frac{2(n-1)\pi}{n} + i \sin\frac{2(n-1)\pi}{n}$$

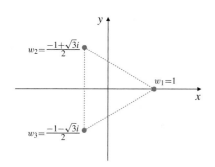

$w_2 = \dfrac{-1+\sqrt{3}i}{2}$

$w_1 = 1$

$w_3 = \dfrac{-1-\sqrt{3}i}{2}$

Figure I.8   The cube roots of unity

satisfies $w^n = 1$ so is an nth root of 1. (These numbers are usually called the nth roots of unity.) Figure I.8 shows the three cube roots of 1. Observe that they are at the three vertices of an equilateral triangle with centre at the origin and one vertex at 1. In general, the n nth roots of unity lie on a circle of radius 1 centred at the origin, at the vertices of a regular n-sided polygon with one vertex at 1.

If z is any nonzero complex number, and θ is the principal argument of z ($-\pi < \theta \le \pi$), then the number

$$w_1 = |z|^{1/n}\left(\cos\frac{\theta}{n} + i \sin\frac{\theta}{n}\right)$$

is called the **principal** nth root of z. All the nth roots of z are on the circle of radius $|z|^{1/n}$ centred at the origin and are at the vertices of a regular n-sided polygon with one vertex at w_1. (See Figure I.9.) The other nth roots are

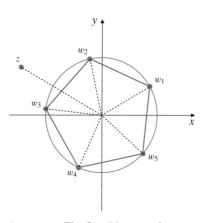

Figure I.9   The five 5th roots of z

$$w_2 = |z|^{1/n}\left(\cos\frac{\theta + 2\pi}{n} + i \sin\frac{\theta + 2\pi}{n}\right)$$
$$w_3 = |z|^{1/n}\left(\cos\frac{\theta + 4\pi}{n} + i \sin\frac{\theta + 4\pi}{n}\right)$$
$$\vdots$$
$$w_n = |z|^{1/n}\left(\cos\frac{\theta + 2(n-1)\pi}{n} + i \sin\frac{\theta + 2(n-1)\pi}{n}\right).$$

We can obtain all n of the nth roots of z by multiplying the principal nth root by the nth roots of unity.

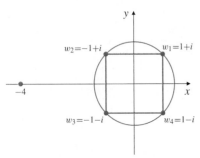

Example 8   Find the 4th roots of -4. Sketch them in an Argand diagram.

Solution   Since $|-4|^{1/4} = \sqrt{2}$ and $\arg(-4) = \pi$, the principal 4th root of -4 is

$$w_1 = \sqrt{2}\left(\cos\frac{\pi}{4} + i\sin\frac{\pi}{4}\right) = 1 + i.$$

The other three 4th roots are at the vertices of a square with centre at the origin and one vertex at $1 + i$. (See Figure I.10.) Thus the other roots are

$$w_2 = -1 + i, \qquad w_3 = -1 - i, \qquad w_4 = 1 - i.$$

Figure I.10   The four 4th roots of -4

Exercises: Appendix I	

In Exercises 1–4, find the real and imaginary parts (Re (z) and Im (z)) of the given complex numbers z, and sketch the position of each number in the complex plane (i.e., in an Argand diagram).

1. $z = -5 + 2i$

2. $z = 4 - i$

3. $z = -\pi i$

4. $z = -6$

In Exercises 5–15, find the modulus $r = |z|$ and the principal argument $\theta = \text{Arg}(z)$ of each given complex number z, and express z in terms of r and θ.

5. $z = -1 + i$

6. $z = -2$

7. $z = 3i$

8. $z = -5i$

9. $z = 1 + 2i$

10. $z = -2 + i$

11. $z = -3 - 4i$

12. $z = 3 - 4i$

13. $z = \sqrt{3} - i$

14. $z = -\sqrt{3} - 3i$

15. $z = 3\cos\frac{4\pi}{5} + 3i\sin\frac{4\pi}{5}$

16. If $\text{Arg}(z) = 3\pi/4$ and $\text{Arg}(w) = \pi/2$, find $\text{Arg}(zw)$.

17. If $\text{Arg}(z) = -5\pi/6$ and $\text{Arg}(w) = \pi/4$, find $\text{Arg}(z/w)$.

In Exercises 18–23, express in the form $z = x + yi$ the complex number z whose modulus and argument are given.

18. $|z| = 2$, $\arg(z) = \pi$

19. $|z| = 5$, $\arg(z) = \tan^{-1}\frac{3}{4}$

20. $|z| = 1$, $\arg(z) = \frac{3\pi}{4}$

21. $|z| = \pi$, $\arg(z) = \frac{\pi}{6}$

22. $|z| = 0$, $\arg(z) = 1$

23. $|z| = \frac{1}{2}$, $\arg(z) = -\frac{\pi}{3}$

In Exercises 24–27, find the complex conjugates of the given complex numbers.

24. $5 + 3i$

25. $-3 - 5i$

26. $4i$

27. $2 - i$

Describe geometrically (or make a sketch of) the set of points z in the complex plane satisfying the given equations or inequalities in Exercises 28–33.

28. $|z| = 2$

29. $|z| \leq 2$

30. $|z - 2i| \leq 3$

31. $|z - 3 + 4i| \leq 5$

32. $\arg z = \frac{\pi}{3}$

33. $\pi \leq \arg(z) \leq \frac{7\pi}{4}$

Simplify the expressions in Exercises 34–43.

34. $(2 + 5i) + (3 - i)$

35. $i - (3 - 2i) + (7 - 3i)$

36. $(4 + i)(4 - i)$

37. $(1 + i)(2 - 3i)$

38. $(a + bi)(\overline{2a - bi})$

39. $(2 + i)^3$

40. $\dfrac{2 - i}{2 + i}$

41. $\dfrac{1 + 3i}{2 - i}$

42. $\dfrac{1 + i}{i(2 + 3i)}$

43. $\dfrac{(1 + 2i)(2 - 3i)}{(2 - i)(3 + 2i)}$

44. Prove that $\overline{z + w} = \bar{z} + \bar{w}$.

45. Prove that $\overline{\left(\dfrac{z}{w}\right)} = \dfrac{\bar{z}}{\bar{w}}$.

46. Express each of the complex numbers $z = 3 + i\sqrt{3}$ and $w = -1 + i\sqrt{3}$ in polar form (i.e., in terms of its modulus and argument). Use these expressions to calculate zw and z/w.

47. Repeat Exercise 46 for $z = -1 + i$ and $w = 3i$.

48. Use de Moivre's Theorem to find a trigonometric identity for $\cos 3\theta$ in terms of $\cos\theta$ and one for $\sin 3\theta$ in terms of $\sin\theta$.

49. Describe the solutions, if any, of the equations (a) $\bar{z} = 2/z$ and (b) $\bar{z} = -2/z$.

50. For positive real numbers a and b it is always true that $\sqrt{ab} = \sqrt{a}\sqrt{b}$. Does a similar identity hold for $\sqrt{zw}$, where z and w are complex numbers? *Hint:* Consider $z = w = -1$.

51. Find the three cube roots of -1.

52. Find the three cube roots of $-8i$.

53. Find the three cube roots of $-1 + i$.

54. Find all the fourth roots of 4.

55. Find all complex solutions of the equation $z^4 + 1 - i\sqrt{3} = 0$.

56. Find all solutions of $z^5 + a^5 = 0$, where a is a positive real number.

∗57. Show that the sum of the n nth roots of unity is zero. *Hint:* Show that these roots are all powers of the principal root.

Appendix II

Complex Functions

The shortest path between two truths in the real domain passes through the complex domain

Jacques Hadamard   1865–1963
quoted in *The Mathematical Intelligencer, v 13, 1991*

Most of this book is concerned with developing the properties of **real functions**, that is, functions of one or more real variables, having values that are themselves real numbers or vectors with real components. The definition of *function* given in Section P.4 can be paraphrased to allow for complex-valued functions of a complex variable.

DEFINITION 1

> A **complex function** f is a rule that assigns a unique complex number $f(z)$ to each number z in some set of complex numbers (called the **domain** of the function).

Typically, we will use $z = x + yi$ to denote a general point in the domain of a complex function and $w = u + vi$ to denote the value of the function at z; if $w = f(z)$, then the real and imaginary parts of w ($u = \operatorname{Re}(w)$ and $v = \operatorname{Im}(w)$) are real-valued functions of z, and hence real-valued functions of the two real variables x and y:

$$u = u(x, y), \qquad v = v(x, y).$$

For example, the complex function $f(z) = z^2$, whose domain is the whole complex plane $\mathbb{C}$, assigns the value z^2 to the complex number z. If $w = z^2$ (where $w = u + vi$ and $z = x + yi$), then

$$u + vi = (x + yi)^2 = x^2 - y^2 + 2xyi,$$

so that

$$u = \operatorname{Re}(z^2) = x^2 - y^2 \qquad \text{and} \qquad v = \operatorname{Im}(z^2) = 2xy.$$

It is not convenient to draw the *graph* of a complex function. The graph of $w = f(z)$ would have to be drawn in a four-dimensional (real) space, since two dimensions (a z-plane) are required for the independent variable, and two more dimensions (a w-plane) are required for the dependent variable. Instead, we can graphically represent the behaviour of a complex function $w = f(z)$ by drawing the z-plane and the w-plane separately, and showing the image in the w-plane of certain, appropriately chosen sets of points in the z-plane. For example, Figure II.1 illustrates the fact that for the function $w = z^2$ the image of the quarter-disk $|z| \le a, 0 \le \arg(z) \le \frac{\pi}{2}$ is the half-disk $|w| \le a^2, 0 \le \arg(w) \le \pi$. To see why this is so, observe that if $z = r(\cos\theta + i\sin\theta)$, then $w = r^2(\cos 2\theta + i\sin 2\theta)$. Thus the function maps the circle $|z| = r$ onto the circle $|w| = r^2$ and the radial line $\arg(z) = \theta$ onto the radial line $\arg(w) = 2\theta$.

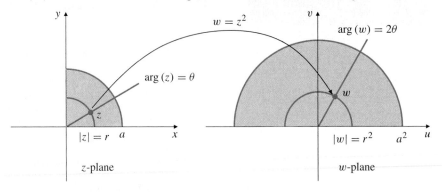

Figure II.1   The function $w = z^2$ maps a quarter-disk of radius a to a half-disk of radius a^2 by squaring the modulus and doubling the argument of each point z

Limits and Continuity

The concepts of limit and continuity carry over from real functions to complex functions in an obvious way providing we use $|z_1 - z_2|$ as the distance between the complex numbers z_1 and z_2. We say that

$$\lim_{z \to z_0} f(z) = \lambda$$

provided we can ensure that $|f(z) - \lambda|$ is as small as we wish by taking z sufficiently close to z_0. Formally,

DEFINITION 2

We say that $f(z)$ tends to the **limit** λ as z approaches z_0, and we write

$$\lim_{z \to z_0} f(z) = \lambda,$$

if for every positive real number ϵ there exists a positive real number δ (depending on ϵ), such that

$$0 < |z - z_0| < \delta \qquad \Longrightarrow \qquad |f(z) - \lambda| < \epsilon.$$

DEFINITION 3

The complex function $f(z)$ is **continuous** at $z = z_0$ if $\lim_{z \to z_0} f(z)$ exists and equals $f(z_0)$.

All the laws of limits and continuity apply as for real functions. Polynomials, that is, functions of the form

$$P(z) = a_0 + a_1 z + a_2 z^2 + \cdots + a_n z^n,$$

are continuous at every point of the complex plane. Rational functions, that is, functions of the form

$$R(z) = \frac{P(z)}{Q(z)},$$

where $P(z)$ and $Q(z)$ are polynomials, are continuous everywhere except at points where $Q(z) = 0$. Integer powers z^n are continuous except at the origin if $n < 0$. The situation for fractional powers is more complicated. For example, $\sqrt{z}$ (the principal square root) is continuous except at points $z = x < 0$. The function $f(z) = \overline{z}$ is continuous everywhere, because

$$|\overline{z} - \overline{z_0}| = |\overline{z - z_0}| = |z - z_0|.$$

The Complex Derivative

The definition of derivative is the same as for real functions:

DEFINITION **4**

> The complex function f is **differentiable** at z and has **derivative** $f'(z)$ there, provided
>
> $$\lim_{h \to 0} \frac{f(z+h) - f(z)}{h} = f'(z)$$
>
> exists.

Note, however, that in this definition h is a complex number. The limit must exist no matter how h approaches 0 in the complex plane. This fact has profound implications. The existence of a derivative in this sense forces the function f to be much better behaved than is necessary for a differentiable real function. For example, it can be shown that if $f'(z)$ exists for all z in an open region D in $\mathbb{C}$, then f has derivatives of *all* orders throughout D. Moreover, such a function is the sum of its Taylor series

$$f(z) = f(z_0) + f'(z_0)(z - z_0) + \frac{f''(z_0)}{2!}(z - z_0)^2 + \cdots$$

about any point z_0 in D; the series has positive radius of convergence R and converges in the disk $|z - z_0| < R$. For this reason, complex functions that are differentiable on open sets in $\mathbb{C}$ are usually called **analytic functions**. It is beyond the scope of this introductory appendix to prove these assertions. They are proved in courses and texts on complex analysis.

The usual differentiation rules apply:

$$\frac{d}{dz}\big(Af(z) + Bg(z)\big) = Af'(z) + Bg'(z)$$

$$\frac{d}{dz}\big(f(z)g(z)\big) = f'(z)g(z) + f(z)g'(z)$$

$$\frac{d}{dz}\left(\frac{f(z)}{g(z)}\right) = \frac{g(z)f'(z) - f(z)g'(z)}{\big(g(z)\big)^2}$$

$$\frac{d}{dz}f\big(g(z)\big) = f'\big(g(z)\big)g'(z).$$

As one would expect, the derivative of $f(z) = z^n$ is $f'(z) = nz^{n-1}$.

Example 1   Show that the function $f(z) = \overline{z}$ is not differentiable at any point.

Solution   We have

$$f'(z) = \lim_{h \to 0} \frac{\overline{z + h} - \overline{z}}{h}$$

$$= \lim_{h \to 0} \frac{\overline{z + h - z}}{h} = \lim_{h \to 0} \frac{\overline{h}}{h}.$$

But $\bar{h}/h = 1$ if h is real, and $\bar{h}/h = -1$ if h is pure imaginary. Since there are real and pure imaginary numbers arbitrarily close to 0, the limit above does not exist, so $f'(z)$ does not exist. ∎

The following theorem links the existence of the derivative of a complex function $f(z)$ with certain properties of its real and imaginary parts $u(x, y)$ and $v(x, y)$.

THEOREM ❶

The Cauchy–Riemann equations

If $f(z) = u(x, y) + iv(x, y)$ is differentiable at $z = x + yi$, then u and v satisfy the Cauchy-Riemann equations

$$\frac{\partial u}{\partial x} = \frac{\partial v}{\partial y}, \qquad \frac{\partial v}{\partial x} = -\frac{\partial u}{\partial y}.$$

Conversely, if u and v are sufficiently smooth (say, if they have continuous second partial derivatives near (x, y)), and if u and v satisfy the Cauchy–Riemann equations at (x, y), then f is differentiable at $z = x + yi$ and

$$f'(z) = \frac{\partial u}{\partial x} + i\frac{\partial v}{\partial x}.$$

PROOF First assume that f is differentiable at z. Letting $h = s + ti$, we have

$$f'(z) = \lim_{h \to 0} \frac{f(z + h) - f(z)}{h}$$

$$= \lim_{(s,t) \to (0,0)} \left[\frac{u(x + s, y + t) - u(x, y)}{s + it} + i\frac{v(x + s, y + t) - v(x, y)}{s + it}\right].$$

The limit must be independent of the path along which h approaches 0. Letting $t = 0$, so that $h = s$ approaches 0 along the real axis, we obtain

$$f'(z) = \lim_{s \to 0} \left[\frac{u(x + s, y) - u(x, y)}{s} + i\frac{v(x + s, y) - v(x, y)}{s}\right] = \frac{\partial u}{\partial x} + i\frac{\partial v}{\partial x}.$$

Similarly, letting $s = 0$, so that $h = ti$ approaches 0 along the imaginary axis, we obtain

$$f'(z) = \lim_{t \to 0} \left[\frac{u(x, y + t) - u(x, y)}{it} + i\frac{v(x, y + t) - v(x, y)}{it}\right]$$

$$= \lim_{t \to 0} \left[\frac{v(x, y + t) - v(x, y)}{t} - i\frac{u(x, y + t) - u(x, y)}{t}\right]$$

$$= \frac{\partial v}{\partial y} - i\frac{\partial u}{\partial y}.$$

Equating these two expressions for $f'(z)$ we see that

$$\frac{\partial u}{\partial x} = \frac{\partial v}{\partial y}, \qquad \frac{\partial v}{\partial x} = -\frac{\partial u}{\partial y}.$$

To prove the converse, we use the result of Exercise 19 of Section 12.6. Since u and v are assumed to have continuous second partial derivatives, we must have

$$u(x + s, y + t) - u(x, y) = s\frac{\partial u}{\partial x} + t\frac{\partial u}{\partial y} + O(s^2 + t^2)$$

$$v(x + s, y + t) - v(x, y) = s\frac{\partial v}{\partial x} + t\frac{\partial v}{\partial y} + O(s^2 + t^2),$$

where we have used the Big-O notation (see Definition 10 of Section 4.8); the expression $O(\lambda)$ denotes a term satisfying $|O(\lambda)| \leq K|\lambda|$ for some constant K. Thus, if u and v satisfy the Cauchy–Riemann equations, then

$$\frac{f(z+h) - f(z)}{h} = \frac{s\dfrac{\partial u}{\partial x} + t\dfrac{\partial u}{\partial y} + i\left(s\dfrac{\partial v}{\partial x} + t\dfrac{\partial v}{\partial y}\right) + O(s^2 + t^2)}{s + it}$$

$$= \frac{(s+it)\dfrac{\partial u}{\partial x} + i(s+it)\dfrac{\partial v}{\partial x}}{s + it} + O(\sqrt{s^2 + t^2})$$

$$= \frac{\partial u}{\partial x} + i\frac{\partial v}{\partial x} + O(\sqrt{s^2 + t^2}).$$

Thus we may let $h = s + ti$ approach 0 and obtain

$$f'(z) = \frac{\partial u}{\partial x} + i\frac{\partial v}{\partial x}.$$

It follows immediately from the Cauchy–Riemann equations that the real and imaginary parts of a differentiable complex function are real harmonic functions:

$$\frac{\partial^2 u}{\partial x^2} + \frac{\partial^2 u}{\partial y^2} = 0, \qquad \frac{\partial^2 v}{\partial x^2} + \frac{\partial^2 v}{\partial y^2} = 0.$$

(See Exercise 15 of Section 12.4.)

The Exponential Function

Consider the function

$$f(z) = e^x \cos y + ie^x \sin y,$$

where $z = x + yi$. The real and imaginary parts of $f(z)$,

$$u(x, y) = \text{Re}\,(f(z)) = e^x \cos y \qquad \text{and} \qquad v(x, y) = \text{Im}\,(f(z)) = e^x \sin y,$$

satisfy the Cauchy–Riemann equations

$$\frac{\partial u}{\partial x} = e^x \cos y = \frac{\partial v}{\partial y} \qquad \text{and} \qquad \frac{\partial v}{\partial x} = e^x \sin y = -\frac{\partial u}{\partial y}$$

everywhere in the z-plane. Therefore, $f(z)$ is differentiable (analytic) everywhere and satisfies

$$f'(z) = \frac{\partial u}{\partial x} + i\frac{\partial v}{\partial x} = e^x \cos y + ie^x \sin y = f(z).$$

Evidently $f(0) = 1$, and $f(z) = e^x$ if $z = x$ is a real number. It is therefore natural to denote the function $f(z)$ as the exponential function e^z.

> **The complex exponential function**
>
> $$e^z = e^x(\cos y + i \sin y) \qquad \text{for} \quad z = x + yi.$$

In particular, if $z = yi$ is pure imaginary, then

$$e^{yi} = \cos y + i \sin y,$$

a fact that can also be obtained by separating the real and imaginary parts of the Maclaurin series for e^{yi}:

$$e^{yi} = 1 + (yi) + \frac{(yi)^2}{2!} + \frac{(yi)^3}{3!} + \frac{(yi)^4}{4!} + \frac{(yi)^5}{5!} + \cdots$$

$$= \left(1 - \frac{y^2}{2!} + \frac{y^4}{4!} - \cdots\right) + i\left(y - \frac{y^3}{3!} + \frac{y^5}{5!} - \cdots\right)$$

$$= \cos y + i \sin y.$$

Observe that

$$|e^z| = \sqrt{e^{2x}\left(\cos^2 y + \sin^2 y\right)} = e^x,$$

$$\arg(e^z) = \arg(e^{yi}) = \arg(\cos y + i \sin y) = y,$$

$$\overline{e^z} = e^x \cos y - i e^x \sin y = e^x \cos(-y) + i e^x \sin(-y) = e^{\bar{z}}.$$

In summary:

Properties of the exponential function

If $z = x + yi$ then $\overline{e^z} = e^{\bar{z}}$. Also,

$$\operatorname{Re}(e^z) = e^x \cos y, \qquad\qquad |e^z| = e^x,$$

$$\operatorname{Im}(e^z) = e^x \sin y, \qquad\qquad \arg(e^z) = y.$$

Example 2   Sketch the image in the w-plane of the rectangle

$$R : a \le x \le b, \quad c \le y \le d$$

in the z-plane under the transformation $w = e^z$.

Solution   The vertical lines $x = a$ and $x = b$ get mapped to the concentric circles $|w| = e^a$ and $|w| = e^b$. The horizontal lines $y = c$ and $y = d$ get mapped to the radial lines $\arg(w) = c$ and $\arg(w) = d$. Thus, the rectangle R gets mapped to the polar region P shown in Figure II.2.

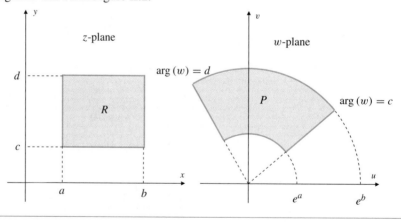

Figure II.2   Under the exponential function $w = e^z$ vertical lines get mapped to circles centred at the origin, and horizontal lines get mapped to half-lines radiating from the origin

Note that if $d - c \ge 2\pi$, then the image of R will be the entire annular region $e^a \le |w| \le e^b$, which may be covered more than once. The exponential function e^z is periodic with period $2\pi i$:

$$e^{z+2\pi i} = e^z \qquad \text{for all} \quad z,$$

and is therefore not one-to-one on the whole complex plane. However, $w = e^z$ is one-to-one from any horizontal strip of the form

$$-\infty < x < \infty, \qquad c < y \le c + 2\pi$$

onto the whole w-plane excluding the origin.

The Fundamental Theorem of Algebra

As observed at the beginning of Appendix I, extending the number system to include complex numbers allows larger classes of equations to have solutions. We conclude this appendix by verifying that polynomial equations always have solutions in the complex numbers.

A **complex polynomial** of degree n is a function of the form

$$P_n(z) = a_n z^n + a_{n-1} z^{n-1} + \cdots + a_2 z^2 + a_1 z + a_0,$$

where $a_0, a_1, \ldots, a_n$ are complex numbers and $a_n \neq 0$. The numbers a_i $(0 \leq i \leq n)$ are called the **coefficients** of the polynomial. If they are all real numbers, then $P_n(x)$ is called a real polynomial.

A complex number z_0 that satisfies the equation $P(z_0) = 0$ is called a **zero** or **root** of the polynomial. Every polynomial of degree 1 has a zero: if $a_1 \neq 0$, then $a_1 z + a_0$ has zero $z = -a_0/a_1$. This zero is real if a_1 and a_0 are both real.

Similarly, every complex polynomial of degree 2 has two zeros. If the polynomial is given by

$$P_2(z) = a_2 z^2 + a_1 z + a_0$$

(where $a_2 \neq 0$), then the zeros are given by the *quadratic formula*

$$z = z_1 = \frac{-a_1 - \sqrt{a_1^2 - 4a_2 a_0}}{2a_2} \quad \text{and} \quad z = z_2 = \frac{-a_1 + \sqrt{a_1^2 - 4a_2 a_0}}{2a_2}.$$

In this case $P_2(z)$ has two linear factors:

$$P_2(z) = a_2(z - z_1)(z - z_2).$$

Even if $a_1^2 - 4a_2 a_0 = 0$, so that $z_1 = z_2$, we still regard the polynomial as having two (equal) zeros, one corresponding to each factor. If the coefficients a_0, a_1, and a_2 are all real numbers, the zeros will be real provided $a_1^2 \geq 4a_2 a_0$. When real coefficients satisfy $a_1^2 < 4a_2 a_0$ then the zeros are complex, in fact, complex conjugates: $z_2 = \overline{z_1}$.

Example 3   Solve the equation $z^2 + 2iz - (1 + i) = 0$.

Solution   The zeros of this equation are

$$z = \frac{-2i \pm \sqrt{-4 + 4(1 + i)}}{2}$$

$$= -i \pm \sqrt{i}$$

$$= -i \pm \frac{1 + i}{\sqrt{2}} = \frac{1}{\sqrt{2}}\big(1 + (1 - \sqrt{2})i\big) \quad \text{or} \quad -\frac{1}{\sqrt{2}}\big(1 + (1 + \sqrt{2})i\big).$$

The Fundamental Theorem of Algebra asserts that every complex polynomial of positive degree has a complex zero.

THEOREM 2

The Fundamental Theorem of Algebra

If $P(z) = a_n z^n + a_{n-1} z^{n-1} + \cdots + a_1 z + a_0$ is a complex polynomial of degree $n \geq 1$, then there exists a complex number z_1 such that $P(z_1) = 0$.

PROOF   (We will only give an informal sketch of the proof.) We can assume that the coefficient of z^n in $P(z)$ is $a_n = 1$, since we can divide the equation $P(z) = 0$ by a_n without changing its solutions. We can also assume that $a_0 \neq 0$; if $a_0 = 0$, then $z = 0$ is certainly a zero of $P(z)$. Thus we deal with the polynomial

$$P(z) = z^n + Q(z),$$

where $Q(z)$ is a polynomial of degree less than n having a nonzero constant term. If R is sufficiently large, then $|Q(z)|$ will be less than R^n for all numbers z satisfying $|z| = R$. As z moves around the circle $|z| = R$ in the z-plane, $w = z^n$ moves around the circle $|w| = R^n$ in the w-plane (n times). Since the distance from z^n to $P(z)$ is equal to $|P(z) - z^n| = |Q(z)| < R^n$, it follows that the image of the circle $|z| = R$ under the transformation $w = P(z)$ is a curve that winds around the origin n times. (If you walk around a circle of radius r n times, with your dog on a leash of length less than r, and your dog returns to his starting point, then he must also go around the centre of the circle n times.) This situation is illustrated for the particular case

$$P(z) = z^3 + z^2 - iz + 1, \qquad |z| = 2$$

in Figure II.3. The image of $|z| = 2$ is the large curve in the w-plane that winds around the origin three times. As R decreases, the curve traced out by $w = P(z)$ for $|z| = R$ changes continuously. For R close to 0, it is a small curve staying close to the constant term a_0 of $P(z)$. For small enough R the curve will not enclose the origin. (In Figure II.3 the image of $|z| = 0.3$ is the small curve staying close to the point 1 in the w-plane.) Thus for some value of R, say $R = R_1$, the curve must pass through the origin. That is, there must be a complex number z_1, with $|z_1| = R_1$, such that $P(z_1) = 0$.

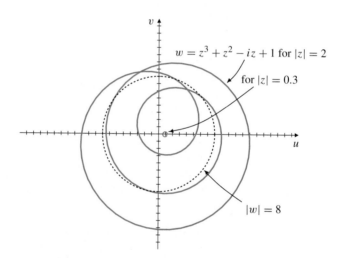

Figure II.3   The image of the circle $|z| = 2$ winds around the origin in the w-plane three times, but the image of $|z| = 0.3$ does not wind around the origin at all

*In the figure:* $w = z^3 + z^2 - iz + 1$ for $|z| = 2$; for $|z| = 0.3$; $|w| = 8$

Remark   The above proof suggests that there should be n such solutions of the equation $P(z) = 0$; the curve has to go from winding around the origin n times to winding around the origin 0 times as R decreases toward 0. We can establish this as follows. $P(z_1) = 0$ implies that $z - z_1$ is a factor of $P(z)$:

$$P(z) = (z - z_1)P_{n-1}(z),$$

where P_{n-1} is a polynomial of degree $n - 1$. If $n > 1$, then P_{n-1} must also have a zero, z_2, by the Fundamental Theorem. We can continue this argument inductively to obtain n zeros and factor $P(z)$ into a product of the constant a_n and n linear factors:

$$P(z) = a_n(z - z_1)(z - z_2) \cdots (z - z_n).$$

Of course, some of the zeros can be equal.

Remark   If P is a real polynomial, that is, one whose coefficients are all real numbers, then $\overline{P(z)} = P(\overline{z})$. Therefore, if z_1 is a nonreal zero of $P(z)$, then so is $z_2 = \overline{z_1}$:

$$P(z_2) = P(\overline{z_1}) = \overline{P(z_1)} = \overline{0} = 0.$$

Real polynomials can have complex zeros, but they must always occur in complex conjugate pairs. Every real polynomial of odd degree must have at least one real zero.

Example 4   Show that $z_1 = -i$ is a zero of the polynomial

$$P(z) = z^4 + 5z^3 + 7z^2 + 5z + 6,$$

and find all the other zeros of this polynomial.

Solution   First observe that $P(z_1) = P(-i) = 1 + 5i - 7 - 5i + 6 = 0$, so $z_1 = -i$ is indeed a zero. Since the coefficients of $P(z)$ are real, $z_2 = i$ must also be a zero. Thus $z + i$ and $z - i$ are factors of $P(z)$, and so is

$$(z + i)(z - i) = z^2 + 1.$$

Dividing $P(z)$ by $z^2 + 1$ we obtain

$$\frac{P(z)}{z^2 + 1} = z^2 + 5z + 6 = (z + 2)(z + 3).$$

Thus the four zeros of $P(z)$ are $z_1 = -i$, $z_2 = i$, $z_3 = -2$, and $z_4 = -3$.

Exercises: Appendix II

In Exercises 1–12, the z-plane region D consists of the complex numbers $z = x + yi$ that satisfy the given conditions. Describe (or sketch) the image R of D in the w-plane under the given function $w = f(z)$.

1. $0 \le x \le 1$, $0 \le y \le 2$; $w = \bar{z}$.

2. $x + y = 1$; $w = \bar{z}$.

3. $1 \le |z| \le 2$, $\frac{\pi}{2} \le \arg z \le \frac{3\pi}{4}$; $w = z^2$.

4. $0 \le |z| \le 2$, $0 \le \arg(z) \le \frac{\pi}{2}$; $w = z^3$.

5. $0 < |z| \le 2$, $0 \le \arg(z) \le \frac{\pi}{2}$; $w = \frac{1}{z}$.

6. $\frac{\pi}{4} \le \arg(z) \le \frac{\pi}{3}$; $w = -iz$.

7. $\arg(z) = -\frac{\pi}{3}$; $w = \sqrt{z}$.

8. $x = 1$; $w = z^2$.   **9.** $y = 1$; $w = z^2$.

10. $x = 1$; $w = \frac{1}{z}$.

11. $-\infty < x < \infty$, $\frac{\pi}{4} \le y \le \frac{\pi}{2}$; $w = e^z$.

12. $0 < x < \frac{\pi}{2}$, $0 < y < \infty$; $w = e^{iz}$.

In Exercises 13–16, verify that the real and imaginary parts of each function $f(z)$ satisfy the Cauchy–Riemann equations, and thus find $f'(z)$.

13. $f(z) = z^2$.   **14.** $f(z) = z^3$.

15. $f(z) = \frac{1}{z}$.   **16.** $f(z) = e^{z^2}$.

17. Use the fact that $e^{yi} = \cos y + i \sin y$ (for real y) to show that

$$\cos y = \frac{e^{yi} + e^{-yi}}{2} \quad \text{and} \quad \sin y = \frac{e^{yi} - e^{-yi}}{2i}.$$

Exercise 16 suggests that we define complex functions

$$\cos z = \frac{e^{zi} + e^{-zi}}{2} \quad \text{and} \quad \sin z = \frac{e^{zi} - e^{-zi}}{2i},$$

as well as extend the definitions of the hyperbolic functions to

$$\cosh z = \frac{e^z + e^{-z}}{2} \quad \text{and} \quad \sinh z = \frac{e^z - e^{-z}}{2}.$$

Exercises 18–26 develop properties of these functions and relationships between them.

18. Show that $\cos z$ and $\sin z$ are periodic with period 2π, and that $\cosh z$ and $\sinh z$ are periodic with period $2\pi i$.

19. Show that $(d/dz) \sin z = \cos z$ and $(d/dz) \cos z = -\sin z$. What are the derivatives of $\sinh z$ and $\cosh z$?

20. Verify the identities $\cos z = \cosh(iz)$ and $\sin z = -i \sinh(iz)$. What are the corresponding identities for $\cosh z$ and $\sinh(z)$ in terms of $\cos$ and $\sin$?

21. Find all complex zeros of $\cos z$ (i.e., all solutions of $\cos z = 0$).

22. Find all complex zeros of $\sin z$.

23. Find all complex zeros of $\cosh z$ and $\sinh z$.

24. Show that $\mathrm{Re}\,(\cosh z) = \cosh x \cos y$ and $\mathrm{Im}\,(\cosh z) = \sinh x \sin y$.

25. Find the real and imaginary parts of $\sinh z$.

26. Find the real and imaginary parts of $\cos z$ and $\sin z$.

Find the zeros of the polynomials in Exercises 27–32.

27. $P(z) = z^2 + 2iz$ **28.** $P(z) = z^2 - 2z + i$

29. $P(z) = z^2 + 2z + 5$ **30.** $P(z) = z^2 - 2iz - 1$

31. $P(z) = z^3 - 3iz^2 - 2z$ **32.** $P(z) = z^4 - 2z^2 + 4$

33. The polynomial $P(z) = z^4 + 1$ has two pairs of complex conjugate zeros. Find them, and hence express $P(z)$ as a product of two quadratic factors with real coefficients.

In Exercises 34–36, check that the given number z_1 is a zero of the given polynomial, and find all the zeros of the polynomial.

34. $P(z) = z^4 - 4z^3 + 12z^2 - 16z + 16$; $z_1 = 1 - \sqrt{3}\,i$.

35. $P(z) = z^5 + 3z^4 + 4z^3 + 4z^2 + 3z + 1$; $z_1 = i$.

36. $P(z) = z^5 - 2z^4 - 8z^3 + 8z^2 + 31z - 30$; $z_1 = -2 + i$.

37. Show that the image of the circle $|z| = 2$ under the mapping $w = z^4 + z^3 - 2iz - 3$ winds around the origin in the w-plane four times.

Appendix III

Continuous Functions

Geometry may sometimes appear to take the lead over analysis, but in fact precedes it only as a servant goes before his master to clear the path and light him on the way. The interval between the two is as wide as between empiricism and science, as between the understanding and the reason, or as between the finite and the infinite.

J. J. Sylvester   1814–1897
from *Philosophic Magazine, 1866*

The development of calculus depends in an essential way on the concept of the limit of a function and thereby on properties of the real number system. In Chapter 1 we presented these notions in an intuitive way and did not attempt to prove them except in Section 1.5, where the *formal* definition of limit was given and used to verify some elementary limits and prove some simple properties of limits.

Many of the results on limits and continuity of functions stated in Chapter 1 may seem quite obvious; most students and users of calculus are not bothered by applying them without proof. Nevertheless, mathematics is a highly logical and rigorous discipline, and any statement, however obvious, that cannot be proved by strictly logical arguments from acceptable assumptions must be considered suspect. In this appendix we build upon the formal definition of limit given in Section 1.5 and combine it with the notion of *completeness* of the real number system first encountered in Section P.1 to give formal proofs of the very important results about continuous functions stated in Theorems 8 and 9 of Section 1.4, the Max-Min Theorem and the Intermediate-Value Theorem. Most of our development of calculus in this book depends essentially on these two theorems.

The branch of mathematics that deals with proofs such as these is called mathematical analysis. This subject is usually not pursued by students in introductory calculus courses but is postponed to higher years and studied by students in majors or honours programs in mathematics. It is hoped that some of this material will be of value to honours-level calculus courses and individual students with a deeper interest in understanding calculus.

Limits of Functions

At the heart of mathematical analysis is the formal definition of limit, Definition 9 in Section 1.5, which we restate as follows:

The formal definition of limit

We say that $\lim_{x \to a} f(x) = L$ if for every positive number ϵ there exists a positive number δ, depending on ϵ (i.e., $\delta = \delta(\epsilon)$), such that

$$0 < |x - a| < \delta \quad \Longrightarrow \quad |f(x) - L| < \epsilon.$$

Section 1.5 was marked "optional" because understanding the material presented there was not essential for learning calculus. However, that material is an *essential* prerequisite for this appendix. It is highly recommended that you go back to Section 1.5 and read it carefully, paying special attention to Examples 2 and 4, and attempt at least Exercises 31–36. These exercises provide proofs for the standard laws of limits stated in Section 1.2.

Continuous Functions

Consider the following definitions of continuity, which are equivalent to those given in Section 1.4.

DEFINITION | **1**

Continuity of a function at a point

A function f, defined on an open interval containing the point a, is said to be continuous at the point a if

$$\lim_{x \to a} f(x) = f(a),$$

that is, if for every $\epsilon > 0$ there exists $\delta > 0$ such that if $|x - a| < \delta$, then $|f(x) - f(a)| < \epsilon$.

DEFINITION | **2**

Continuity of a function on an interval

A function f is continuous on an interval if it is continuous at every point of that interval. In the case of an endpoint of a closed interval, f need only be continuous on one side. Thus, f is continuous on the interval $[a, b]$ if

$$\lim_{t \to x} f(t) = f(x)$$

for each x satisfying $a < x < b$, and

$$\lim_{t \to a+} f(t) = f(a) \quad \text{and} \quad \lim_{t \to b-} f(t) = f(b).$$

These concepts are illustrated in Figure III.1.

Some important results about continuous functions are collected in Theorems 6 and 7 of Section 1.4, which we restate here:

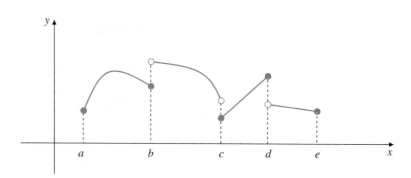

Figure III.1 f is continuous on the intervals $[a, b]$, (b, c), $[c, d]$, and $(d, e]$

THEOREM 1

Combining continuous functions

(a) If f and g are continuous at the point a, then so are $f + g$, $f - g$, fg, and, if $g(a) \neq 0$, f/g.

(b) If f is continuous at the point L and if $\lim_{x \to a} g(x) = L$, then we have

$$\lim_{x \to a} f(g(x)) = f(L) = f\left(\lim_{x \to a} g(x)\right).$$

In particular, if g is continuous at the point a (so that $L = g(a)$), then $\lim_{x \to a} f(g(x)) = f(g(a))$, that is, $f \circ g(x) = f(g(x))$ is continuous at $x = a$.

(c) The functions $f(x) = C$ (constant) and $g(x) = x$ are continuous on the whole real line.

(d) For any rational number r the function $f(x) = x^r$ is continuous at every real number where it is defined.

PROOF   Part (a) is just a restatement of various rules for combining limits; for example,

$$\lim_{x \to a} f(x)g(x) = (\lim_{x \to a} f(x))(\lim_{x \to a} g(x)) = f(a)g(a).$$

Part (b) can be proved as follows. Let $\epsilon > 0$ be given. Since f is continuous at L, there exists $k > 0$ such that $|f(g(x)) - f(L)| < \epsilon$ whenever $|g(x) - L| < k$. Since $\lim_{x \to a} g(x) = L$, there exists $\delta > 0$ such that if $0 < |x - a| < \delta$, then $|g(x) - L| < k$. Hence, if $0 < |x - a| < \delta$, then $|f(g(x)) - f(L)| < \epsilon$, and $\lim_{x \to a} f(g(x)) = f(L)$.

The proofs of (c) and (d) are left to the student in Exercises 3–9 at the end of this appendix.

Completeness and Sequential Limits

DEFINITION 3

A real number u is said to be an **upper bound** for a nonempty set S of real numbers if $x \leq u$ for every x in S.
The number u^* is called the **least upper bound** of S if u^* is an upper bound for S and $u^* \leq u$ for every upper bound u of S.
Similarly, ℓ is a **lower bound** for S if $\ell \leq x$ for every x in S. The number ℓ^* is the **greatest lower bound** of S if ℓ^* is a lower bound and $\ell \leq \ell^*$ for every lower bound ℓ of S.

Example 1   Set $S_1 = [2, 3]$ and $S_2 = (2, \infty)$. Any number $u \geq 3$ is an upper bound for S_1. S_2 has no upper bound; we say that it is not bounded above. The least upper bound of S_1 is 3. Any real number $\ell \leq 2$ is a lower bound for both S_1 and S_2. $\ell^* = 2$ is the greatest lower bound for each set. Note that the least upper bound and greatest lower bound of a set may or may not belong to that set.

We now recall the completeness axiom for the real number system, which we discussed briefly in Section P.1.

> **The completeness axiom for the real numbers**
>
> A nonempty set of real numbers that has an upper bound must have a least upper bound.
> Equivalently, a nonempty set of real numbers having a lower bound must have a greatest lower bound.

We stress that this is an *axiom* to be assumed without proof. It cannot be deduced from the more elementary algebraic and order properties of the real numbers. These other properties are shared by the rational numbers, a set that is not complete. The completeness axiom is essential for the proof of the most important results about continuous functions, in particular, for the Max-Min Theorem and the Intermediate-Value Theorem. Before attempting these proofs, however, we must develop a little more machinery.

In Section 9.1 we stated a version of the completeness axiom that pertains to *sequences* of real numbers; specifically, that an increasing sequence that is bounded above converges to a limit. We begin by verifying that this follows from the version stated above. (Both statements are, in fact, equivalent.) As noted in Section 9.1, the sequence

$$\{x_n\} = \{x_1, \ x_2, \ x_3, \ \ldots\}$$

is a function on the positive integers, that is, $x_n = x(n)$. We say that the sequence converges to the limit L, and we write $\lim x_n = L$, if the corresponding function $x(t)$ satisfies $\lim_{t \to \infty} x(t) = L$ as defined above. More formally,

DEFINITION 4

> **Limit of a sequence**
> We say that $\lim x_n = L$ if for every positive number ϵ there exists a positive number $N = N(\epsilon)$ such that $|x_n - L| < \epsilon$ holds whenever $n \geq N$.

THEOREM 2

If $\{x_n\}$ is an increasing sequence that is bounded above, that is,

$$x_{n+1} \geq x_n \qquad \text{and} \qquad x_n \leq K \qquad \text{for } n = 1, 2, 3, \ldots,$$

then $\lim x_n = L$ exists. (Equivalently, if $\{x_n\}$ is decreasing and bounded below, then $\lim x_n$ exists.)

PROOF Let $\{x_n\}$ be increasing and bounded above. The set S of real numbers x_n has an upper bound, K, and so has a least upper bound, say L. Thus $x_n \leq L$ for every n, and if $\epsilon > 0$, then there exists a positive integer N such that $x_N > L - \epsilon$. (Otherwise, $L - \epsilon$ would be an upper bound for S that is lower than the least upper bound.) If $n \geq N$, then we have $L - \epsilon < x_N \leq x_n \leq L$, so $|x_n - L| < \epsilon$. Thus $\lim x_n = L$. The proof for a decreasing sequence that is bounded below is similar.

THEOREM 3

If $a \leq x_n \leq b$ for each n, and if $\lim x_n = L$, then $a \leq L \leq b$.

PROOF Suppose that $L > b$. Let $\epsilon = L - b$. Since $\lim x_n = L$, there exists n such that $|x_n - L| < \epsilon$. Thus $x_n > L - \epsilon = L - (L - b) = b$, which is a contradiction, since we are given that $x_n \leq b$. Thus $L \leq b$. A similar argument shows that $L \geq a$.

THEOREM

If f is continuous on $[a, b]$, if $a \leq x_n \leq b$ for each n, and if $\lim x_n = L$, then $\lim f(x_n) = f(L)$.

The proof is similar to that of Theorem 1(b), and is left as Exercise 15 at the end of this appendix.

Continuous Functions on a Closed, Finite Interval

We are now in a position to prove the main results about continuous functions on closed, finite intervals.

THEOREM **5**

The Boundedness Theorem

If f is continuous on $[a, b]$, then f is bounded there; that is, there exists a constant K such that $|f(x)| \leq K$ if $a \leq x \leq b$.

PROOF   We show that f is bounded above; a similar proof shows that f is bounded below. For each positive integer n let S_n be the set of points x in $[a, b]$ such that $f(x) > n$:

$$S_n = \{x : a \leq x \leq b \quad \text{and} \quad f(x) > n\}.$$

We would like to show that S_n is empty for some n. It would then follow that $f(x) \leq n$ for all x in $[a, b]$; that is, n would be an upper bound for f on $[a, b]$.

Suppose, to the contrary, that S_n is nonempty for every n. We will show that this leads to a contradiction. Since S_n is bounded below (a is a lower bound), by completeness S_n has a greatest lower bound; call it x_n. (See Figure III.2.) Evidently $a \leq x_n$. Since $f(x) > n$ at some point of $[a, b]$ and f is continuous at that point, $f(x) > n$ on some interval contained in $[a, b]$. Hence $x_n < b$. It follows that $f(x_n) \geq n$. (If $f(x_n) < n$, then by continuity $f(x) < n$ for some distance to the right of x_n, and x_n could not be the greatest lower bound of S_n.)

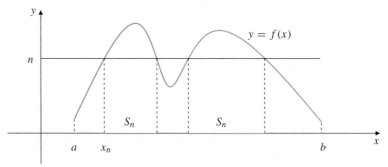

Figure III.2   The set S_n

For each n we have $S_{n+1} \subset S_n$. Therefore, $x_{n+1} \geq x_n$ and $\{x_n\}$ is an increasing sequence. Being bounded above (b is an upper bound) this sequence converges, by Theorem 2. Let $\lim x_n = L$. By Theorem 3, $a \leq L \leq b$. Since f is continuous at L, $\lim f(x_n) = f(L)$ exists by Theorem 4. But since $f(x_n) \geq n$, $\lim f(x_n)$ cannot exist. This contradiction completes the proof.

THEOREM

The Max-Min Theorem

If f is continuous on $[a, b]$, then there are points v and u in $[a, b]$ such that for any x in $[a, b]$ we have

$$f(v) \leq f(x) \leq f(u);$$

that is, f assumes maximum and minimum values on $[a, b]$.

PROOF   By Theorem 5 we know that the set $S = \{f(x) : a \leq x \leq b\}$ has an upper bound and, therefore, by the completeness axiom, a least upper bound. Call this least upper bound M. Suppose that there exists no point u in $[a, b]$ such that $f(u) = M$. Then by Theorem 1(a), $1/(M - f(x))$ is continuous on $[a, b]$. By Theorem 5, there exists a constant K such that $1/(M - f(x)) \leq K$ for all x in $[a, b]$. Thus $f(x) \leq M - 1/K$, which contradicts the fact that M is the *least* upper bound for the values of f. Hence, there must exist some point u in $[a, b]$ such that $f(u) = M$. Since M is an upper bound for the values of f on $[a, b]$, we have $f(x) \leq f(u) = M$ for all x in $[a, b]$.

The proof that there must exist a point v in $[a, b]$ such that $f(x) \geq f(v)$ for all x in $[a, b]$ is similar.

THEOREM 7

The Intermediate-Value Theorem

If f is continuous on $[a, b]$ and s is a real number lying between the numbers $f(a)$ and $f(b)$, then there exists a point c in $[a, b]$ such that $f(c) = s$.

PROOF   To be specific, we assume that $f(a) < s < f(b)$. (The proof for the case $f(a) > s > f(b)$ is similar.) Let $S = \{x : a \leq x \leq b$ and $f(x) \leq s\}$. S is nonempty (a belongs to S) and bounded above (b is an upper bound), so by completeness S has a least upper bound; call it c.

Suppose that $f(c) > s$. Then $c \neq a$ and, by continuity, $f(x) > s$ on some interval $(c - \delta, c]$ where $\delta > 0$. But this says $c - \delta$ is an upper bound for S lower than the least upper bound, which is impossible. Thus $f(c) \leq s$.

Suppose $f(c) < s$. Then $c \neq b$ and, by continuity, $f(x) < s$ on some interval of the form $[c, c + \delta)$ for some $\delta > 0$. But this says that $[c, c + \delta) \subset S$, which contradicts the fact that c is an upper bound for S. Hence we cannot have $f(c) < s$. Therefore, $f(c) = s$.

For more discussion of these theorems and some applications, see Section 1.4.

Exercises: Appendix III

1. Let $a < b < c$ and suppose that $f(x) \leq g(x)$ for $a \leq x \leq c$. If $\lim_{x \to b} f(x) = L$ and $\lim_{x \to b} g(x) = M$, prove that $L \leq M$. *Hint:* Assume that $L > M$ and deduce that $f(x) > g(x)$ for all x sufficiently near b. This contradicts the condition that $f(x) \leq g(x)$ for $a \leq x \leq b$.

2. If $f(x) \leq K$ on the intervals $[a, b)$ and $(b, c]$, and if $\lim_{x \to b} f(x) = L$, prove that $L \leq K$.

3. Use the formal definition of limit to prove that $\lim_{x \to 0+} x^r = 0$ for any positive, rational number r.

Prove the assertions in Exercises 4–9.

4. $f(x) = C$ (constant) and $g(x) = x$ are both continuous on the whole real line.

5. Every polynomial is continuous on the whole real line.

6. A rational function (quotient of polynomials) is continuous everywhere except where the denominator is 0.

7. If n is a positive integer and $a > 0$, then $f(x) = x^{1/n}$ is continuous at $x = a$.

8. If $r = m/n$ is a rational number, then $g(x) = x^r$ is continuous at every point $a > 0$.

9. If $r = m/n$, where m and n are integers and n is odd, show that $g(x) = x^r$ is continuous at every point $a < 0$. If $r \geq 0$, show that g is continuous at 0 also.

10. Prove that $f(x) = |x|$ is continuous on the real line.

Use the definitions from Chapter 3 for the functions in Exercises 11–14 to show that these functions are continuous on their respective domains.

11. $\sin x$

12. $\cos x$

13. $\ln x$

14. e^x

15. Prove Theorem 4.

16. Suppose that every function that is continuous and bounded on $[a, b]$ must assume a maximum value and a minimum value on that interval. Without using Theorem 5, prove that every function f that is continuous on $[a, b]$ must be bounded on that interval. *Hint:* Show that $g(t) = t/(1 + |t|)$ is continuous and increasing on the real line. Then consider $g\big(f(x)\big)$.

Appendix IV

The Riemann Integral

It seems to be expected of every pilgrim up the slopes of the mathematical Parnassus, that he will at some point or other of his journey sit down and invent a definite integral or two towards the increase of the common stock.

J. J. Sylvester   1814–1897

In Section 5.3 we defined the definite integral $\int_a^b f(x)\,dx$ of a function f that is continuous on the finite, closed interval $[a, b]$. The integral was defined as a kind of "limit" of Riemann sums formed by partitioning the interval $[a, b]$ into small subintervals. In this appendix we will reformulate the definition of the integral so that it can be used for functions that are not necessarily continuous; in the following discussion we assume only that f is **bounded** on $[a, b]$. Later we will prove Theorem 2 of Section 5.3, which asserts that any continuous function is integrable.

Recall that a **partition** P of $[a, b]$ is a finite, ordered set of points $P = \{x_0, x_1, x_2, \ldots, x_n\}$, where $a = x_0 < x_1 < x_2 < \cdots < x_{n-1} < x_n = b$. Such a partition subdivides $[a, b]$ into n subintervals $[x_0, x_1]$, $[x_1, x_2]$, $\ldots$, $[x_{n-1}, x_n]$, where $n = n(P)$ depends on the partition. The length of the jth subinterval $[x_{j-1}, x_j]$ is $\Delta x_j = x_j - x_{j-1}$.

Suppose that the function f is bounded on $[a, b]$. Given any partition P, the n sets $S_j = \{f(x) : x_{j-1} \le x \le x_j\}$ have least upper bounds M_j and greatest lower bounds m_j, $(1 \le j \le n)$, so that

$$m_j \le f(x) \le M_j \qquad \text{on} \qquad [x_{j-1}, x_j].$$

We define upper and lower Riemann sums for f corresponding to the partition P to be

$$U(f, P) = \sum_{j=1}^{n(P)} M_j \Delta x_j \qquad \text{and}$$

$$L(f, P) = \sum_{j=1}^{n(P)} m_j \Delta x_j.$$

(See Figure IV.1.) Note that if f is continuous on $[a, b]$, then m_j and M_j are, in fact, the minimum and maximum values of f over $[x_{j-1}, x_j]$ (by Theorem 6 of Appendix II); that is, $m_j = f(l_j)$ and $M_j = f(u_j)$, where $f(l_j) \le f(x) \le f(u_j)$ for $x_{j-1} \le x \le x_j$.

If P is any partition of $[a, b]$ and we create a new partition P^* by adding new subdivision points to those of P, thus subdividing the subintervals of P into smaller ones, then we call P^* a **refinement** of P.

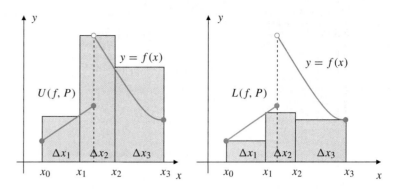

Figure IV.1 Upper and lower sums corresponding to the partition $P = \{x_0, x_1, x_2, x_3\}$

THEOREM **1**

If P^* is a refinement of P, then $L(f, P^*) \geq L(f, P)$ and $U(f, P^*) \leq U(f, P)$.

PROOF If S and T are sets of real numbers, and $S \subset T$, then any lower bound (or upper bound) of T is also a lower bound (or upper bound) of S. Hence, the greatest lower bound of S is at least as large as that of T, and the least upper bound of S is no greater than that of T.

Let P be a given partition of $[a, b]$ and form a new partition P' by adding one subdivision point to those of P, say, the point k dividing the jth subinterval $[x_{j-1}, x_j]$ of P into two subintervals $[x_{j-1}, k]$ and $[k, x_j]$. (See Figure IV.2.) Let m_j, m_j', and m_j'' be the greatest lower bounds of the sets of values of $f(x)$ on the intervals $[x_{j-1}, x_j]$, $[x_{j-1}, k]$, and $[k, x_j]$, respectively. Then $m_j \leq m_j'$ and $m_j \leq m_j''$. Thus $m_j(x_j - x_{j-1}) \leq m_j'(k - x_{j-1}) + m_j''(x_j - k)$, so $L(f, P) \leq L(f, P')$.

If P^* is a refinement of P, it can be obtained by adding one point at a time to those of P and thus $L(f, P) \leq L(f, P^*)$. We can prove that $U(f, P) \geq U(f, P^*)$ in a similar manner.

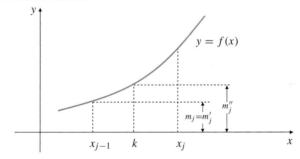

Figure IV.2 Adding one point to a partition

THEOREM **2**

If P and P' are any two partitions of $[a, b]$, then $L(f, P) \leq U(f, P')$.

PROOF Combine the subdivision points of P and P' to form a new partition P^*, which is a refinement of both P and P'. Then by Theorem 1,

$$L(f, P) \leq L(f, P^*) \leq U(f, P^*) \leq U(f, P').$$

No lower sum can exceed any upper sum.

Theorem 2 shows that the set of values of $L(f, P)$ for fixed f and various partitions P of $[a, b]$ is a bounded set; any upper sum is an upper bound for this set. By completeness, the set has a least upper bound, which we shall denote I_*. Thus, $L(f, P) \leq I_*$ for any partition P. Similarly, there exists a greatest lower bound I^* for the set of values of $U(f, P)$ corresponding to different partitions P. It follows that $I_* \leq I^*$. (See Exercise 4 at the end of this appendix.)

DEFINITION **1**

The Riemann integral

If f is bounded on $[a, b]$ and $I_* = I^*$, then we say that f is **Riemann integrable**, or simply **integrable** on $[a, b]$, and denote by

$$\int_a^b f(x)\, dx = I_* = I^*$$

the **(Riemann) integral** of f on $[a, b]$.

The following theorem provides a convenient test for determining whether a given bounded function is integrable:

THEOREM **3**

The bounded function f is integrable on $[a, b]$ if and only if for every positive number ϵ there exists a partition P of $[a, b]$ such that $U(f, P) - L(f, P) < \epsilon$.

PROOF   Suppose that for every $\epsilon > 0$ there exists a partition P of $[a, b]$ such that $U(f, P) - L(f, P) < \epsilon$, then

$$I^* \leq U(f, P) < L(f, P) + \epsilon \leq I_* + \epsilon.$$

Since $I^* < I_* + \epsilon$ must hold for every $\epsilon > 0$, it follows that $I^* \leq I_*$. Since we already know that $I^* \geq I_*$, we have $I^* = I_*$ and f is integrable on $[a, b]$.

Conversely, if $I^* = I_*$ and $\epsilon > 0$ are given, we can find a partition P' such that $L(f, P') > I_* - \epsilon/2$, and another partition P'' such that $U(f, P'') < I^* + \epsilon/2$. If P is a common refinement of P' and P'', then by Theorem 1 we have that $U(f, P) - L(f, P) \leq U(f, P'') - L(f, P') < (\epsilon/2) + (\epsilon/2) = \epsilon$, as required.

Example 1   Let $f(x) = \begin{cases} 0 & \text{if } 0 \leq x < 1 \text{ or } 1 < x \leq 2 \\ 1 & \text{if } x = 1. \end{cases}$

Show that f is integrable on $[0, 2]$ and find $\int_0^2 f(x)\, dx$.

Solution   Let $\epsilon > 0$ be given. Let $P = \{0,\ 1 - \epsilon/3,\ 1 + \epsilon/3,\ 2\}$. Then $L(f, P) = 0$ since $f(x) = 0$ at points of each of these subintervals into which P subdivides $[0, 2]$. (See Figure IV.3.) Since $f(1) = 1$, we have

$$U(f, P) = 0\left(1 - \frac{\epsilon}{3}\right) + 1\left(\frac{2\epsilon}{3}\right) + 0\left(2 - \left(1 + \frac{\epsilon}{3}\right)\right) = \frac{2\epsilon}{3}.$$

Hence, $U(f, P) - L(f, P) < \epsilon$ and f is integrable on $[0, 2]$. Since $L(f, P) = 0$ for every partition, $\int_0^2 f(x)\, dx = I_* = 0$.

Figure IV.3

Example 2   Let $f(x)$ be defined on $[0, 1]$ by

$$f(x) = \begin{cases} 1 & \text{if } x \text{ is rational} \\ 0 & \text{if } x \text{ is irrational.} \end{cases}$$

Show that f is not integrable on $[0, 1]$.

PROOF   Every subinterval of $[0, 1]$ having positive length contains both rational and irrational numbers. Hence, for any partition P of $[0, 1]$ we have $L(f, P) = 0$ and $U(f, P) = 1$. Thus $I_* = 0$ and $I^* = 1$, so f is not integrable on $[0, 1]$.

Uniform Continuity

When we assert that a function f is continuous on the interval $[a, b]$, we imply that for every x in that interval and every $\epsilon > 0$, we can find a positive number δ (depending on *both* x and ϵ) such that $|f(y) - f(x)| < \epsilon$ whenever $|y - x| < \delta$ and y lies in $[a, b]$. In fact, however, it is possible to find a number δ *depending only on* ϵ such that $|f(y) - f(x)| < \epsilon$ holds whenever x and y belong to $[a, b]$ and satisfy $|y - x| < \delta$. We describe this phenomenon by saying that f is **uniformly continuous** on the interval $[a, b]$.

THEOREM **4**

If f is continuous on the closed, finite interval $[a, b]$, then f is uniformly continuous on that interval.

PROOF   Let $\epsilon > 0$ be given. Define numbers x_n in $[a, b]$ and subsets S_n of $[a, b]$ as follows:

$$x_1 = a$$
$$S_1 = \left\{ x : x_1 < x \leq b \text{ and } |f(x) - f(x_1)| \geq \frac{\epsilon}{3} \right\}.$$

If S_1 is empty, stop; otherwise, let

$$x_2 = \text{ the greatest lower bound of } S_1$$
$$S_2 = \left\{ x : x_2 < x \leq b \text{ and } |f(x) - f(x_2)| \geq \frac{\epsilon}{3} \right\}.$$

If S_2 is empty, stop; otherwise, proceed to define x_3 and S_3 analogously. We proceed in this way as long as we can; if x_n and S_n have been defined and S_n is not empty, we define

$$x_{n+1} = \text{ the greatest lower bound of } S_n$$
$$S_{n+1} = \left\{ x : x_{n+1} < x \leq b \text{ and } |f(x) - f(x_{n+1})| \geq \frac{\epsilon}{3} \right\}.$$

At any stage where S_n is not empty, the continuity of f at x_n assures us that $x_{n+1} > x_n$ and $|f(x_{n+1}) - f(x_n)| = \epsilon/3$.

We must consider two possibilities for the above procedure: either S_n is empty for some n, or S_n is nonempty for every n.

Suppose S_n is nonempty for every n. Then we have constructed an infinite, increasing sequence $\{x_n\}$ in $[a, b]$ that, being bounded above (by b), must have a limit by completeness (Theorem 2 of Appendix II). Let $\lim x_n = x^*$. We have $a \leq x^* \leq b$. Since f is continuous at x^*, there exists $\delta > 0$ such that $|f(x) - f(x^*)| < \epsilon/8$ whenever $|x - x^*| < \delta$ and x lies in $[a, b]$. Since $\lim x_n = x^*$, there exists a positive integer N such that $|x_n - x^*| < \delta$ whenever $n \geq N$. For such n we have

$$\frac{\epsilon}{3} = |f(x_{n+1}) - f(x_n)| = |f(x_{n+1}) - f(x^*) + f(x^*) - f(x_n)|$$
$$\leq |f(x_{n+1}) - f(x^*)| + |f(x_n) - f(x^*)|$$
$$< \frac{\epsilon}{8} + \frac{\epsilon}{8} = \frac{\epsilon}{4},$$

which is clearly impossible. Thus S_n must, in fact, be empty for some n.

Suppose that S_N is empty. Thus, S_n is nonempty for $n < N$, and the procedure for defining x_n stops with x_N. Since S_{N-1} is not empty, $x_N < b$. In this case define $x_{N+1} = b$ and let

$$\delta = \min\{x_2 - x_1, \ x_3 - x_2, \ \ldots, \ x_{N+1} - x_N\}.$$

The minimum of a finite set of positive numbers is a positive number, so $\delta > 0$. If x lies in $[a, b]$, then x lies in one of the intervals $[x_1, x_2], [x_2, x_3], \ldots, [x_N, x_{N+1}]$. Suppose x lies in $[x_k, x_{k+1}]$. If y is in $[a, b]$ and $|y - x| < \delta$, then y lies in either the same subinterval as x or in an adjacent one; that is, y lies in $[x_j, x_{j+1}]$, where $j = k - 1, k$, or $k + 1$. Thus,

$$|f(y) - f(x)| = |f(y) - f(x_j) + f(x_j) - f(x_k) + f(x_k) - f(x)|$$
$$\leq |f(y) - f(x_j)| + |f(x_j) - f(x_k)| + |f(x_k) - f(x)|$$
$$< \frac{\epsilon}{3} + \frac{\epsilon}{3} + \frac{\epsilon}{3} = \epsilon,$$

which was to be proved.

We are now in a position to prove that a continuous function is integrable.

THEOREM 5

If f is continuous on $[a, b]$, then f is integrable on $[a, b]$.

PROOF   By Theorem 4, f is uniformly continuous on $[a, b]$. Let $\epsilon > 0$ be given. Let $\delta > 0$ be such that $|f(x) - f(y)| < \epsilon/(b-a)$ whenever $|x - y| < \delta$ and x and y belong to $[a, b]$. Choose a partition $P = \{x_0, x_1, \ldots, x_n\}$ of $[a, b]$ for which each subinterval $[x_{j-1}, x_j]$ has length $\Delta x_j < \delta$. Then the greatest lower bound, m_j and the least upper bound, M_j of the set of values of $f(x)$ on $[x_{j-1}, x_j]$ satisfy $M_j - m_j < \epsilon/(b - a)$. Accordingly,

$$U(f, P) - L(f, P) < \frac{\epsilon}{b - a} \sum_{j=1}^{n(P)} \Delta x_j = \frac{\epsilon}{b - a}(b - a) = \epsilon.$$

Thus f is integrable on $[a, b]$, as asserted.

Exercises: Appendix IV

1. Let $f(x) = \begin{cases} 1 & \text{if } 0 \leq x \leq 1 \\ 0 & \text{if } 1 < x \leq 2 \end{cases}$. Prove that f is integrable on $[0, 2]$ and find the value of $\int_0^2 f(x)\, dx$.

2. Let $f(x) = \begin{cases} 1 & \text{if } x = 1/n, \quad n = 1, 2, 3, \ldots \\ 0 & \text{for all other values of } x. \end{cases}$
    Show that f is integrable over $[0, 1]$ and find the value of the integral $\int_0^1 f(x)\, dx$.

* 3. Let $f(x) = 1/n$ if $x = m/n$, where m, n are integers having no common factors, and let $f(x) = 0$ if x is an irrational number. Thus, $f(1/2) = 1/2$, $f(1/3) = f(2/3) = 1/3$, $f(1/4) = f(3/4) = 1/4$, etc. Show that f is integrable on $[0, 1]$ and find $\int_0^1 f(x)\, dx$. *Hint:* Show that for any $\epsilon > 0$, only finitely many points of the graph of f over $[0, 1]$ lie above the line $y = \epsilon$.

4. Prove that I_* and I^* defined in the paragraph following Theorem 2 satisfy $I_* \leq I^*$ as claimed there.

5. Prove parts (c), (d), (e), (f), (g), and (h) of Theorem 3 in Section 5.4 for the Riemann integral.

6. Use the definition of uniform continuity given in the paragraph preceding Theorem 4 to prove that $f(x) = \sqrt{x}$ is uniformly continuous on $[0, 1]$. Do not use Theorem Theorem 4 itself.

7. Show directly from the definition of uniform continuity (without using Theorem 5 of Appendix II) that a function f uniformly continuous on a closed, finite interval is necessarily bounded there.

8. If f is bounded and integrable on $[a, b]$, prove that $F(x) = \int_a^x f(t)\, dt$ is uniformly continuous on $[a, b]$. (If f were continuous, we would have a stronger result; F would be differentiable on (a, b) and $F'(x) = f(x)$ (which is the Fundamental Theorem of Calculus).)

Appendix V
Doing Calculus with Maple

I think, therefore I am,

René Descartes 1596–1650
Discourse on Method

AI [Artificial Intelligences] think, therefore I am,

David Braue
APC Magazine, November 2003

Computer algebra systems like Maple and Mathematica are capable of doing most of the tedious calculations involved in doing calculus, especially the very intensive calculations required by many applied problems. (They cannot, of course, do the thinking for you; you must still fully understand what you are doing and what are the limitations of such programs.) Throughout this text we have inserted material illustrating how to use **Maple** to do common calculus-oriented calculations. These insertions range in length from single paragraphs and remarks to entire sections. To help you locate the Maple material appropriate for specific topics, we include below a list pointing to the text sections containing Maple examples and the pages on which they start.

Note, however, that this material assumes you are familiar with the basics of starting a Maple session, preferably with a graphical user interface which typically displays the prompt > when it is waiting for your input. In this book the input is shown in colour. It normally concludes with a semicolon (;) followed by pressing the <enter> key, which we omit from our examples. The output is typically printed by Maple centred in the window; we show it in black. For instance,

```
>   factor(x^2-x-2);
```

$$(x + 1)(x - 2)$$

Output can be supressed by using a colon (:) instead of a semicolon at the end of the input.

The author used Maple 9 for preparing the Maple examples in this edition. These examples are by no means complete or exhaustive. For a more complete treatment of Maple as a tool for doing calculus, the author highly recommends the excellent Maple lab manual *Calculus: The Maple Way*, written by his colleague, Professor Robert Israel of the University of British Columbia. Like this book, it is published by Pearson Canada under the Addison-Wesley logo.

List of Maple Examples and Discussion

Several of the topics in the above list are covered over several pages. Only the first page is listed.

Answers to Odd-Numbered Exercises

Chapter 9
Sequences, Series, and Power Series

Section 9.1   (page 478)

1. bounded, positive, increasing, convergent to 2
3. bounded, positive, convergent to 4
5. bounded below, positive, increasing, divergent to infinity
7. bounded below, positive, increasing, divergent to infinity
9. bounded, positive, decreasing, convergent to 0

11. divergent   13. divergent

15. ∞ 17. 0

19. 1   21. e^{-3}

23. 0   25. 1/2

27. 0   29. 0

31. $\lim_{n\to\infty} a_n = 5$
33. If $\{a_n\}$ is (ultimately) decreasing, then either it is bounded below, and therefore convergent, or else it is unbounded below and therefore divergent to negative infinity.

Section 9.2   (page 484)

1. $\dfrac{1}{2}$

3. $\dfrac{1}{(2+\pi)^8\big((2+\pi)^2 - 1\big)}$

5. $\dfrac{25}{4,416}$ 7. $\dfrac{8e^4}{e-2}$

9. diverges to ∞ 11. $\dfrac{3}{4}$

13. $\dfrac{1}{3}$ 15. div. to ∞

17. div. to ∞ 19. diverges

21. 14 m
25. If $\{a_n\}$ is ultimately negative, then the series $\sum a_n$ must either converge (if its partial sums are bounded below), or diverge to $-\infty$ (if its partial sums are not bounded below).

27. false, e.g. $\sum \dfrac{(-1)^n}{2^n}$ 29. true

31. true

Section 9.3   (page 494)

1. converges   3. diverges to ∞

5. converges   7. diverges to ∞

9. converges   11. diverges to ∞

13. diverges to ∞ 15. converges

17. converges   19. diverges to ∞

21. converges   23. converges

25. converges
27. $s_n + \dfrac{1}{3(n+1)^3} \le s \le s_n + \dfrac{1}{3n^3}$; $n = 6$

29. $s_n + \dfrac{2}{\sqrt{n+1}} \le s \le s_n + \dfrac{2}{\sqrt{n}}$; $n = 63$

31. $0 < s - s_n \le \dfrac{n+2}{2^n(n+1)!(2n+3)}$; $n = 4$

33. $0 < s - s_n \le \dfrac{2^n(4n^2 + 6n + 2)}{(2n)!(4n^2 + 6n)}$; $n = 4$

39. converges, $a_n^{1/n} \to (1/e) < 1$
41. no info from ratio test, but series diverges to infinity since all terms exceed 1.
43. (b) $s \le \dfrac{2}{k(1-k)}, k = \frac{1}{2},$

(c) $0 < s - s_n < \dfrac{(1+k)^{n+1}}{2^n k(1-k)}, k = \dfrac{n+2-\sqrt{n^2+8}}{2(n-1)}$
for $n \ge 2$
45. (a) 10, (b) 5, (c) 0.765

Section 9.4   (page 501)

1. conv. conditionally   3. conv. conditionally

5. diverges   7. conv. absolutely

9. conv. conditionally   11. diverges

13. 999   15. 13

17. converges absolutely if $-1 < x < 1$, conditionally if $x = -1$, diverges elsewhere
19. converges absolutely if $0 < x < 2$, conditionally if $x = 2$, diverges elsewhere
21. converges absolutely if $-2 < x < 2$, conditionally if $x = -2$, diverges elsewhere
23. converges absolutely if $-\frac{7}{2} < x < \frac{1}{2}$, conditionally if $x = -\frac{7}{2}$, diverges elsewhere
25. AST does not apply directly, but does if we remove all the 0 terms; series converges conditionally
27. (a) false, e.g. $a_n = \dfrac{(-1)^n}{n}$,

(b) false, e.g. $a_n = \dfrac{\sin(n\pi/2)}{n}$, (see Exercise 25),

(c) true
29. converges absolutely for $-1 < x < 1$, conditionally if $x = -1$, diverges elsewhere

Section 9.5   (page 511)

1. centre 0, radius 1, interval $(-1, 1)$
3. centre -2, radius 2, interval $[-4, 0)$
5. centre $\frac{3}{2}$, radius $\frac{1}{2}$, interval $(1, 2)$
7. centre 0, radius ∞, interval $(-\infty, \infty)$

9. $\dfrac{1}{(1-x)^3} = \displaystyle\sum_{n=0}^{\infty} \dfrac{(n+1)(n+2)}{2}\, x^n,\ (-1 < x < 1)$

11. $\dfrac{1}{(1-x)^2} = \displaystyle\sum_{n=0}^{\infty} (n+1)x^n,\quad (-1 < x < 1)$

13. $\dfrac{1}{(2-x)^2} = \displaystyle\sum_{n=0}^{\infty} \dfrac{n+1}{2^{n+2}}\, x^n,\quad (-2 < x < 2)$

15. $\ln(2-x) = \ln 2 - \displaystyle\sum_{n=1}^{\infty} \dfrac{x^n}{2^n n},\quad (-2 \le x < 2)$

17. $\dfrac{1}{x^2} = \displaystyle\sum_{n=0}^{\infty} \dfrac{n+1}{2^{n+2}}(x+2)^n,\quad (-4 < x < 0)$

19. $\dfrac{x^3}{1-2x^2} = \displaystyle\sum_{n=0}^{\infty} 2^n x^{2n+3},\quad \left(-\dfrac{1}{\sqrt{2}} < x < \dfrac{1}{\sqrt{2}}\right)$

21. $\left(-\tfrac{1}{4}, \tfrac{1}{4}\right);\quad \dfrac{1}{1+4x}$

23. $[-1, 1);\quad \tfrac{1}{3}$ if $x = 0$,

$\quad -\dfrac{1}{x^3}\ln(1-x) - \dfrac{1}{x^2} - \dfrac{1}{2x}$ otherwise

25. $(-1, 1);\quad \dfrac{2}{(1-x^2)^2}$ **27.** $3/4$

29. $\pi^2(\pi+1)/(\pi-1)^3$ **31.** $\ln(3/2)$

Section 9.6 (page 520)

1. $e^{3x+1} = \displaystyle\sum_{n=0}^{\infty} \dfrac{3^n e}{n!}\, x^n,\ (\text{all } x)$

3. $\sin\left(x - \dfrac{\pi}{4}\right)$
$= \dfrac{1}{\sqrt{2}} \displaystyle\sum_{n=0}^{\infty} (-1)^n \left[-\dfrac{x^{2n}}{(2n)!} + \dfrac{x^{2n+1}}{(2n+1)!}\right],\ (\text{all } x)$

5. $x^2 \sin\left(\dfrac{x}{3}\right) = \displaystyle\sum_{n=0}^{\infty} \dfrac{(-1)^n}{3^{2n+1}(2n+1)!}\, x^{2n+3},\ (\text{all } x)$

7. $\sin x \cos x = \displaystyle\sum_{n=0}^{\infty} \dfrac{(-1)^n 2^{2n}}{(2n+1)!}\, x^{2n+1},\ (\text{all } x)$

9. $\dfrac{1+x^3}{1+x^2} = 1 - x^2 + \displaystyle\sum_{n=2}^{\infty} (-1)^n\left(x^{2n-1} + x^{2n}\right),$
$\quad (-1 < x < 1)$

11. $\ln\dfrac{1+x}{1-x} = 2\displaystyle\sum_{n=1}^{\infty} \dfrac{x^{2n-1}}{2n-1},\quad (-1 < x < 1)$

13. $\cosh x - \cos x = 2\displaystyle\sum_{n=0}^{\infty} \dfrac{x^{4n+2}}{(4n+2)!},\ (\text{all } x)$

15. $e^{-2x} = e^2 \displaystyle\sum_{n=0}^{\infty} \dfrac{(-1)^n 2^n}{n!}(x+1)^n,\ (\text{all } x)$

17. $\cos x = \displaystyle\sum_{n=0}^{\infty} \dfrac{(-1)^{n+1}}{(2n)!}(x-\pi)^{2n},\ (\text{all } x)$

19. $\ln 4 + \displaystyle\sum_{n=1}^{\infty} \dfrac{(-1)^{n-1}}{4^n n}(x-2)^n,\quad (-2 < x \le 6)$

21. $\sin x - \cos x =$
$\quad \sqrt{2}\displaystyle\sum_{n=0}^{\infty} \dfrac{(-1)^n}{(2n+1)!}\left(x - \dfrac{\pi}{4}\right)^{2n+1},\ (\text{all } x)$

23. $\dfrac{1}{x^2} = \dfrac{1}{4}\displaystyle\sum_{n=0}^{\infty} \dfrac{n+1}{2^n}(x+2)^n,\quad (-4 < x < 0)$

25. $(x-1) + \displaystyle\sum_{n=2}^{\infty} \dfrac{(-1)^n}{n(n-1)}(x-1)^n,\quad (0 \le x \le 2)$

27. $1 + \dfrac{x^2}{2} + \dfrac{5x^4}{24}$ **29.** $x + \dfrac{x^2}{2} - \dfrac{x^3}{6}$

31. $1 + \dfrac{x}{2} - \dfrac{x^2}{8}$ **33.** e^{x^2} (all x)

35. $\dfrac{e^x - e^{-x}}{2x} = \dfrac{\sinh x}{x}$ if $x \ne 0$, 1 if $x = 0$

37. (a) $1 + x + x^2$, (b) $3 + 3(x-1) + (x-1)^2$

Section 9.7 (page 524)

1. $\dfrac{1}{720}(0.2)^7$ **3.** 1.22140

5. 3.32011 **7.** 0.99619

9. -0.10533 **11.** 0.42262

13. 1.54306

15. $I(x) = \displaystyle\sum_{n=0}^{\infty} \dfrac{(-1)^n}{(2n+1)(2n+1)!}\, x^{2n+1},\ (\text{all } x)$

17. $K(x) = \displaystyle\sum_{n=0}^{\infty} \dfrac{(-1)^n}{(n+1)^2}\, x^{n+1},\quad (-1 \le x \le 1)$

19. $M(x) = \displaystyle\sum_{n=0}^{\infty} \dfrac{(-1)^n}{(2n+1)(4n+1)}\, x^{4n+1},$
$\quad (-1 \le x \le 1)$

21. 0.946 **23.** 2

25. $-3/25$ **27.** 0

Section 9.8 (page 528)

1. $\sqrt{1+x}$
$=$
$1 + \displaystyle\sum_{n=1}^{\infty} \dfrac{(-1)^{n-1}\, 1 \times 3 \times 5 \times \cdots \times (2n-3)}{2^n n!}\, x^n$
$|x| < 1$

3. $\sqrt{4+x}$
$= 2 + \dfrac{x}{4} + 2\displaystyle\sum_{n=2}^{\infty} (-1)^{n-1}\dfrac{1 \times 3 \times 5 \times \cdots \times (2n-3)}{2^{3n} n!}\, x^n,$
$\quad (-4 < x \le 4)$

5. $\displaystyle\sum_{n=0}^{\infty}(n+1)x^n,\quad |x| < 1$

Section 9.9 (page 534)

1. $2\pi/3$ **3.** π

5. $2\displaystyle\sum_{n=1}^{\infty} (-1)^{n-1}(\sin(nt))/n$

7. $\dfrac{1}{4} - \displaystyle\sum_{n=1}^{\infty} \left(\dfrac{2\cos((2n-1)\pi t)}{(2n-1)^2\pi^2} + \dfrac{(-1)^n \sin(n\pi t)}{n\pi}\right)$

9. 1

11. $2\displaystyle\sum_{n=1}^{\infty} \dfrac{(-1)^n}{n\pi}\sin(n\pi t)$

13. $\pi^2/8$

Review Exercises (page 534)

1. conv. to 0 **3.** div. to ∞

5. $\lim_{n\to\infty} a_n = \sqrt{2}$ **7.** $4\sqrt{2}/(\sqrt{2}-1)$

9. 2

11. converges

13. converges

15. converges

17. conv. abs.

19. conv. cond.

21. conv. abs. for x in $(-1, 5)$, cond. for $x = -1$, div. elsewhere

23. 1.202

25. $\sum_{n=0}^{\infty} x^n/3^{n+1}$, $|x| < 3$

27. $1 + \sum_{n=1}^{\infty} (-1)^{n-1} x^{2n}/(ne^n)$, $-\sqrt{e} < x \leq \sqrt{e}$

29. $x + \sum_{n=1}^{\infty} (-1)^n 2^{2n-1} x^{2n+1}/(2n)!$, all x

31. $(1/2) + \sum_{n=1}^{\infty} \dfrac{(-1)^n 1 \times 4 \times 7 \times \cdots \times (3n - 2)x^n}{2 \times 24^n n!}$, $-8 < x \leq 8$

33. $\sum_{n=0}^{\infty} (-1)^n (x - \pi)^n/\pi^{n+1}$, $0 < x < 2\pi$

35. $1 + 2x + 3x^2 + \frac{10}{3}x^3$

37. $1 - \frac{1}{2}x^2 + \frac{5}{24}x^4$

39. $\begin{cases} \cos \sqrt{x} & \text{if } x \geq 0 \\ \cosh \sqrt{|x|} & \text{if } x < 0 \end{cases}$

41. $\pi^2/(\pi - 1)^2$

43. $\ln(e/(e - 1))$

45. 1/14

47. 3,   0.49386

49. $\sum_{n=1}^{\infty} \dfrac{2}{n} \sin(nt)$

Challenging Problems   (page 535)

5. (c) 1.645

7. (a) ∞, (c) e^{-x^2}, (d) $f(x) = e^{x^2} \int_0^x e^{-t^2}\, dt$

Chapter 10
Vectors and Coordinate Geometry in 3-Space

Section 10.1   (page 542)

1. 3 units

3. $\sqrt{6}$ units

5. $|z|$ units; $\sqrt{y^2 + z^2}$ units

7. $\cos^{-1}(-4/9) \approx 116.39°$

9. $\sqrt{3}/2$ sq. units

11. $\sqrt{n - 1}$ units

13. the half-space containing the origin and bounded by the plane passing through $(0, -1, 0)$ perpendicular to the y-axis.

15. the vertical plane (parallel to the z-axis) passing through $(1, 0, 0)$ and $(0, 1, 0)$.

17. the sphere of radius 2 centred at $(1, -2, 3)$.

19. the solid circular cylinder of radius 2 with axis along the x-axis.

21. the parabolic cylinder generated by translating the parabola $z = y^2$ in the yz-plane in the direction of the x-axis .

23. the plane through the points $(6, 0, 0)$, $(0, 3, 0)$ and $(0, 0, 2)$.

25. the straight line through $(1, 0, 0)$ and $(1, 1, 1)$.

27. the circle in which the sphere of radius 2 centred at the origin intersects the sphere of radius 2 with centre $(2, 0, 0)$.

29. the ellipse in which the plane $z = x$ intersects the circular cylinder of radius 1 and axis along the z-axis.

31. the part of the solid circular cylinder of radius 1 and axis along the z-axis lying above or on the plane $z = y$.

33. bdry $(0, 0)$ and $x^2 + y^2 = 1$; interior= S; S open

35. bdry of S is S; interior empty; S is closed

37. bdry — the spheres $x^2 + y^2 + z^2 = 1$ and $x^2 + y^2 + z^2 = 4$; interior — points between these spheres; S is closed

39. bdry of S is S, namely the line $x = y = z$; interior is empty; S closed

Section 10.2   (page 551)

1. (a) $3\mathbf{i} - 2\mathbf{j}$, (b) $-3\mathbf{i} + 2\mathbf{j}$, (c) $2\mathbf{i} - 5\mathbf{j}$, (d) $-2\mathbf{i} + 4\mathbf{j}$, (e) $-\mathbf{i} - 2\mathbf{j}$, (f) $4\mathbf{i} + \mathbf{j}$, (g) $-7\mathbf{i} + 20\mathbf{j}$, (h) $2\mathbf{i} - (5/3)\mathbf{j}$

3. a) $6\mathbf{i} - 10\mathbf{k}$, $8\mathbf{j}$, $-3\mathbf{i} + 20\mathbf{j} + 5\mathbf{k}$
   b) $5\sqrt{2}$, $5\sqrt{2}$
   c) $\frac{3}{5\sqrt{2}}\mathbf{i} \pm \frac{4}{5\sqrt{2}}\mathbf{j} - \frac{1}{\sqrt{2}}\mathbf{k}$ d) 18
   e) $\cos^{-1}(9/25) \approx 68.9°$ f) $18/5\sqrt{2}$
   g) $(27/25)\mathbf{i} + (36/25)\mathbf{j} - (9/5)\mathbf{k}$

9. from southwest at $50\sqrt{2}$ km/h.

11. head at angle θ to the east of AC, where
$$\theta = \sin^{-1} \frac{3}{2\sqrt{1 + 4k^2}}.$$
The trip not possible if $k < \frac{1}{4}\sqrt{5}$. If $k > \frac{1}{4}\sqrt{5}$ there is a second possible heading, $\pi - \theta$, but the trip will take longer.

13. $t = 2$

15. $\cos^{-1}(2/\sqrt{6}) \approx 35.26°$, $90°$

17. $(\mathbf{i} + \mathbf{j} + \mathbf{k})/\sqrt{3}$

19. $\lambda = 1/2$, midpoint, $\lambda = 2/3$, 2/3 of way from P_1 to P_2, $\lambda = -1$, P_1 is midway between this point and P_2.

21. plane through point with position vector $(b/|\mathbf{a}|^2)\mathbf{a}$ perpendicular to $\mathbf{a}$.

23. $\mathbf{x} = 2\mathbf{i} - 3\mathbf{j} - 4\mathbf{k}$

25. $(|\mathbf{u}|\mathbf{v} + |\mathbf{v}|\mathbf{u})/\||\mathbf{u}|\mathbf{v} + |\mathbf{v}|\mathbf{u}\|$

31. $\mathbf{u} = (\mathbf{w} \bullet \mathbf{a}/|\mathbf{a}|^2)\mathbf{a}$, $\mathbf{v} = \mathbf{w} - \mathbf{u}$

33. $\mathbf{x} = (\mathbf{a} + K\hat{\mathbf{u}})/(2r)$, $\mathbf{y} = (\mathbf{a} - K\hat{\mathbf{u}})/(2s)$, where $K = \sqrt{|\mathbf{a}|^2 - 4rst}$ and $\hat{\mathbf{u}}$ is any unit vector

35. about 12.373 m   **37.** about 19 m

Section 10.3   (page 559)

1. $5\mathbf{i} + 13\mathbf{j} + 7\mathbf{k}$

3. $\sqrt{6}$ sq. units

5. $\pm\frac{1}{3}(2\mathbf{i} - 2\mathbf{j} + \mathbf{k})$

15. 4/3 cubic units

17. $k = -6$

19. $\lambda = \dfrac{\mathbf{x} \bullet (\mathbf{v} \times \mathbf{w})}{\mathbf{u} \bullet (\mathbf{v} \times \mathbf{w})}$, $\mu = \dfrac{\mathbf{x} \bullet (\mathbf{w} \times \mathbf{u})}{\mathbf{u} \bullet (\mathbf{v} \times \mathbf{w})}$, $\nu = \dfrac{\mathbf{x} \bullet (\mathbf{u} \times \mathbf{v})}{\mathbf{u} \bullet (\mathbf{v} \times \mathbf{w})}$

21. $\mathbf{u} \times (\mathbf{v} \times \mathbf{w}) = -2\mathbf{i} + 7\mathbf{j} - 4\mathbf{k}$, $(\mathbf{u} \times \mathbf{v}) \times \mathbf{w} = \mathbf{i} + 9\mathbf{j} + 9\mathbf{k}$; the first is in the plane of $\mathbf{v}$ and $\mathbf{w}$, the second is in the plane of $\mathbf{u}$ and $\mathbf{v}$.

Section 10.4   (page 567)

1. a) $x^2 + y^2 + z^2 = z^2$;   b) $x + y + z = x + y + z$; c) $x^2 + y^2 + z^2 = -1$

3. $x - y + 2z = 0$

5. $7x + 5y - z = 12$

7. $x - 5y - 3z = -7$

9. $x + 6y - 5z = 17$

11. $(\mathbf{r}_1 - \mathbf{r}_2) \bullet [(\mathbf{r}_1 - \mathbf{r}_3) \times (\mathbf{r}_1 - \mathbf{r}_4)] = 0$

13. planes passing through the line $x = 0$, $y + z = 1$ (except the plane $y + z = 1$ itself)

15. $\mathbf{r} = (1 + 2t)\mathbf{i} + (2 - 3t)\mathbf{j} + (3 - 4t)\mathbf{k}$,
$(-\infty < t < \infty)$
$x = 1 + 2t, \ y = 2 - 3t, z = 3 - 4t, \quad (-\infty < t < \infty)$
$\dfrac{x-1}{2} = \dfrac{y-2}{-3} = \dfrac{z-3}{-4}$

17. $\mathbf{r} = t(7\mathbf{i} - 6\mathbf{j} - 5\mathbf{k}); \quad x = 7t, \quad y = -6t,$
$z = -5t; \quad x/7 = -y/6 = -z/5$

19. $\mathbf{r} = \mathbf{i} + 2\mathbf{j} - \mathbf{k} + t(\mathbf{i} + \mathbf{j} + \mathbf{k});$
$x = 1 + t, \quad y = 2 + t, \quad z = -1 + t;$
$x - 1 = y - 2 = z + 1$

21. $\frac{x-4}{-5} = \frac{y}{3}, \quad z = 7$

25. $\mathbf{r}_i \neq \mathbf{r}_j, (i, j = 1, \cdots, 4, \ i \neq j),$
$\mathbf{v} = (\mathbf{r}_1 - \mathbf{r}_2) \times (\mathbf{r}_3 - \mathbf{r}_4) \neq 0, (\mathbf{r}_1 - \mathbf{r}_3) \bullet \mathbf{v} = 0.$

27. $7\sqrt{2}/10$ units   **29.** $18/\sqrt{69}$ units

31. all lines parallel to the xy-plane and passing through (x_0, y_0, z_0).

33. (x, y, z) satisfies the quadratic if either
$A_1 x + B_1 y + C_1 z = D_1$ or $A_2 x + B_2 y + C_2 z = D_2.$

Section 10.5   *(page 570)*

1. ellipsoid centred at the origin with semiaxes 6, 3 and 2 along the x-, y- and z-axes respectively.

3. sphere with centre $(1, -2, 3)$ and radius $1/\sqrt{2}$.

5. elliptic paraboloid with vertex at the origin, axis along the z-axis, and cross-section $x^2 + 2y^2 = 1$ in the plane $z = 1$.

7. hyperboloid of two sheets with vertices $(\pm 2, 0, 0)$ and circular cross-sections in planes $x = c, (c^2 > 4)$.

9. hyperbolic paraboloid — same as $z = x^2 - y^2$ but rotated $45°$ about the z-axis (counterclockwise as seen from above).

11. hyperbolic cylinder parallel to the y-axis, intersecting the xz-plane in the hyperbola $(x^2/4) - z^2 = 1$.

13. parabolic cylinder parallel to the y-axis.

15. circular cone with vertex $(2, 3, 1)$, vertical axis, and semi-vertical angle $45°$.

17. circle in the plane $x + y + z = 1$ having centre $(1/3, 1/3, 1/3)$ and radius $\sqrt{11/3}$.

19. a parabola in the plane $z = 1 + x$ having vertex at $(-1/2, 0, 1/2)$ and axis along the line $z = 1 + x$, $y = 0$.

21. $\dfrac{y}{b} - \dfrac{z}{c} = \lambda\left(1 - \dfrac{x}{a}\right), \quad \dfrac{y}{b} + \dfrac{z}{c} = \dfrac{1}{\lambda}\left(1 + \dfrac{x}{a}\right);$
$\dfrac{y}{b} - \dfrac{z}{c} = \mu\left(1 + \dfrac{x}{a}\right), \quad \dfrac{y}{b} + \dfrac{z}{c} = \dfrac{1}{\mu}\left(1 - \dfrac{x}{a}\right)$

23. $\mathbf{a} = \mathbf{i} \pm \mathbf{k}$ (or any multiple)

Section 10.6   *(page 579)*

1. $\begin{pmatrix} 6 & 7 \\ 5 & -3 \\ 1 & 1 \end{pmatrix}$ **3.** $\begin{pmatrix} aw + by & ax + bz \\ cw + dy & cx + dz \end{pmatrix}$

5. $\mathscr{A}\mathscr{A}^T = \begin{pmatrix} 4 & 3 & 2 & 1 \\ 3 & 3 & 2 & 1 \\ 2 & 2 & 2 & 1 \\ 1 & 1 & 1 & 1 \end{pmatrix}$ $\mathscr{A}^2 = \begin{pmatrix} 1 & 2 & 3 & 4 \\ 0 & 1 & 2 & 3 \\ 0 & 0 & 1 & 2 \\ 0 & 0 & 0 & 1 \end{pmatrix}$

7. 36   **15.** $\begin{pmatrix} 1 & -1 & 0 \\ 0 & 1 & -1 \\ 0 & 0 & 1 \end{pmatrix}$

17. $x = 1, \ y = 2, \ z = 3$

19. $x_1 = 1, \ x_2 = 2, \ x_3 = -1, \ x_4 = -2$

21. neg. def.   **23.** pos. def.

25. indefinite

Section 10.7   *(page 588)*

1. 2 units

5. `sp:=(U,V)->DotProduct(`
`U,Normalize(V,2),conjugate=false)`

7. `ang := (u,v) -> evalf(`
`(180/Pi)*VectorAngle(U,V))`

9. `VolT:=(U,V,W)->(1/6)*abs(`
`DotProduct(U,(V &x W),conjugate=false))`

11. $(u, v, x, y, z) = (1, 0, -1, 3, 2)$

13. -935

15. $\begin{bmatrix} 9 & -36 & 30 \\ -36 & 192 & -180 \\ 30 & -180 & 180 \end{bmatrix}$

Review Exercises   *(page 589)*

1. plane parallel to y-axis through $(3, 0, 0)$ and $(0, 0, 1)$

3. all points on or above the plane through the origin with normal $\mathbf{i} + \mathbf{j} + \mathbf{k}$

5. circular paraboloid with vertex at $(0, 1, 0)$ and axis along the y-axis, opening in the direction of increasing y

7. hyperbolic paraboloid

9. points inside the ellipsoid with vertices at $(\pm 2, 0, 0)$, $(0, \pm 2, 0,$ and $(0, 0, \pm 1)$

11. cone with axis along the x-axis, vertex at the origin, and elliptical cross-sections perpendicular to its axis

13. oblique circular cone (elliptic cone). Cross-sections in horizontal planes $z = k$ are circles of radius 1 with centres at $(k, 0, k)$

15. horizontal line through $(0, 0, 3)$ and $(2, -1, 3)$

17. circle of radius 1 centred at $(1, 1, 1)$ in plane normal to $\mathbf{i} + \mathbf{j} + \mathbf{k}$

19. $2x - y + 3z = 0$ **21.** $2x + 5y + 3z = 2$

23. $7x + 4y - 8z = 6$

25. $\mathbf{r} = (2 + 3t)\mathbf{i} + (1 + t)\mathbf{j} - (1 + 2t)\mathbf{k}$

27. $x = 3t, \ y = -2t, \ z = 4t$

29. $(\mathbf{r}_2 - \mathbf{r}_1) \times (\mathbf{r}_3 - \mathbf{r}_1) = 0$

31. $(3/2)\sqrt{34}$ sq. units

33. $\mathscr{A}^{-1} = \begin{pmatrix} 1 & 0 & 0 & 0 \\ -2 & 1 & 0 & 0 \\ 1 & -2 & 1 & 0 \\ 0 & 1 & -2 & 1 \end{pmatrix}$

35. *pos.def.*

Challenging Problems   (page 589)

5. condition: $\mathbf{a} \bullet \mathbf{b} = 0$,

$\mathbf{x} = \dfrac{\mathbf{b} \times \mathbf{a}}{|\mathbf{a}|^2} + t\mathbf{a}$ (for any scalar t)

Chapter 11
Vector Functions and Curves

Section 11.1   (page 597)

1. $\mathbf{v} = \mathbf{j}$, $v = 1$, $\mathbf{a} = \mathbf{0}$, path is the line $x = 1$, $z = 0$

3. $\mathbf{v} = 2t\mathbf{j} + \mathbf{k}$, $v = \sqrt{4t^2 + 1}$, $\mathbf{a} = 2\mathbf{j}$, path is the parabola $y = z^2$, in the yz-plane

5. $\mathbf{v} = 2t\mathbf{i} - 2t\mathbf{j}$, $v = 2\sqrt{2}t$, $\mathbf{a} = 2\mathbf{i} - 2\mathbf{j}$, path is the straight half-line $x + y = 0$, $z = 1$, $(x \geq 0)$

7. $\mathbf{v} = -a\sin t\mathbf{i} + a\cos t\mathbf{j} + c\mathbf{k}$, $v = \sqrt{a^2 + c^2}$, $\mathbf{a} = -a\cos t\mathbf{i} - a\sin t\mathbf{j}$, path is a circular helix

9. $\mathbf{v} = -3\sin t\,\mathbf{i} - 4\sin t\,\mathbf{j} + 5\cos t\,\mathbf{k}$, $v = 5$, $\mathbf{a} = -\mathbf{r}$, path is the circle of intersection of the plane $4x = 3y$ with the sphere $x^2 + y^2 + z^2 = 25$

11. $\mathbf{a} = \mathbf{v} = \mathbf{r}$, $v = \sqrt{a^2 + b^2 + c^2}\, e^t$, path is the straight line $\dfrac{x}{a} = \dfrac{y}{b} = \dfrac{z}{c}$

13. $\mathbf{v} = -(e^{-t}\cos e^t + \sin e^t)\mathbf{i}$
$\quad + (-e^{-t}\sin e^t + \cos e^t)\mathbf{j} - e^t\mathbf{k}$
$v = \sqrt{1 + e^{-2t} + e^{2t}}$
$\mathbf{a} = [(e^{-t} - e^t)\cos e^t + \sin e^t]\mathbf{i}$
$\quad + [(e^{-t} - e^t)\sin e^t - \cos e^t]\mathbf{j} - e^t\mathbf{k}$
The path is a spiral lying on the surface $z = -1/\sqrt{x^2 + y^2}$

15. $\mathbf{a} = -3\pi^2\mathbf{i} - 4\pi^2\mathbf{j}$ **17.** $\sqrt{3/2}(-\mathbf{i} + \mathbf{j} - 2\mathbf{k})$

19. $\mathbf{v} = 2\mathbf{i} + 4\mathbf{j} + 4\mathbf{k}$, $\mathbf{a} = -\frac{8}{9}(2\mathbf{i} + \mathbf{j} - 2\mathbf{k})$

29. $\dfrac{d}{dt}\big(\mathbf{u} \times (\mathbf{v} \times \mathbf{w})\big) = \dfrac{d\mathbf{u}}{dt} \times (\mathbf{v} \times \mathbf{w})$
$\quad + \mathbf{u} \times \left(\dfrac{d\mathbf{v}}{dt} \times \mathbf{w}\right) + \mathbf{u} \times \left(\mathbf{v} \times \dfrac{d\mathbf{w}}{dt}\right)$

31. $\mathbf{u}''' \bullet (\mathbf{u} \times \mathbf{u}')$

33. $\mathbf{r} = \mathbf{r}_0 e^{2t}$, $\mathbf{a} = 4\mathbf{r}_0 e^{2t}$; the path is a straight line through the origin in the direction of $\mathbf{r}_0$

35. $\mathbf{r} = \mathbf{r}_0 + \dfrac{1 - e^{-ct}}{c}\mathbf{v}_0 - \dfrac{g}{c^2}(ct + e^{-ct} - 1)\mathbf{k}$

Section 11.2   (page 604)

1. $\dfrac{e-1}{e}$, $\dfrac{e^2 - 1}{e^2}$

3. $\mathbf{r} = \cos t\mathbf{i} + \sin t\mathbf{j} + \mathbf{k}$; the curve is a circle of radius 1 in the plane $z = 1$

5. $4.76°$ west of south; $\dfrac{\pi^2 R}{72}$ towards the ground, where R is the radius of the earth

7. (a) tangential only, $90°$ counterclockwise from $\mathbf{v}$.
(b) tangential only, $90°$ clockwise from $\mathbf{v}$.
(c) normal only

9. 16.0 hours, $52.7°$

Section 11.3   (page 611)

1. $x = \sqrt{a^2 - t^2}$, $y = t$, $0 \leq t \leq a$

3. $x = a\sin\theta$, $y = -a\cos\theta$, $\frac{\pi}{2} \leq \theta \leq \pi$

5. $\mathbf{r} = -2t\mathbf{i} + t\mathbf{j} + 4t^2\mathbf{k}$

7. $\mathbf{r} = 3\cos t\mathbf{i} + 3\sin t\mathbf{j} + 3(\cos t + \sin t)\mathbf{k}$

9. $\mathbf{r} = (1 + 2\cos t)\mathbf{i} - 2(1 - \sin t)\mathbf{j}$
$\quad + (9 + 4\cos t - 8\sin t)\mathbf{k}$

11. Choice (b) leads to $\mathbf{r} = \dfrac{t^2 - 1}{2}\mathbf{i} + t\mathbf{j} + \dfrac{t^2 + 1}{2}\mathbf{k}$, which represents the whole parabola. Choices (a) and (c) lead to separate parametrizations for the halves $y \geq 0$ and $y \leq 0$ of the parabola. For (a) these are $\mathbf{r} = t\mathbf{i} \pm \sqrt{1 + 2t}\mathbf{j} + (1 + t)\mathbf{k}$, $(t \geq -1/2)$

13. $(17\sqrt{17} - 16\sqrt{2})/27$ units

15. $\displaystyle\int_1^T \dfrac{\sqrt{4a^2t^4 + b^2t^2 + c^2}}{t}\,dt$ units;
$a(T^2 - 1) + c\ln T$ units

17. $\pi\sqrt{2 + 4\pi^2} + \ln(\sqrt{2}\pi + \sqrt{1 + 2\pi^2})$ units

19. $\sqrt{2e^{4\pi} + 1} - \sqrt{3} + \frac{1}{2}\ln\dfrac{e^{4\pi} + 1 - \sqrt{2e^{4\pi} + 1}}{e^{4\pi}}$
$\quad - \frac{1}{2}\ln(2 - \sqrt{3})$ units

21. straight line segments from $(0, 0)$ to $(1, 1)$, then to $(0, 2)$

23. $\mathbf{r} = \dfrac{1}{\sqrt{A^2 + B^2 + C^2}}(As\mathbf{i} + Bs\mathbf{j} + Cs\mathbf{k})$

25. $\mathbf{r} = a\left(1 - \dfrac{s}{K}\right)^{3/2}\mathbf{i} + a\left(\dfrac{s}{K}\right)^{3/2}\mathbf{j} + b\left(1 - \dfrac{2s}{K}\right)\mathbf{k}$,
$0 \leq s \leq K$, $K = (\sqrt{9a^2 + 16b^2})/2$

Section 11.4   (page 619)

1. $\hat{\mathbf{T}} = \dfrac{1}{\sqrt{1 + 16t^2 + 81t^4}}(\mathbf{i} - 4t\mathbf{j} + 9t^2\mathbf{k})$

3. $\hat{\mathbf{T}} = \dfrac{1}{\sqrt{1 + \sin^2 t}}(\cos 2t\mathbf{i} + \sin 2t\mathbf{j} - \sin t\mathbf{k})$

Section 11.5   (page 625)

1. $1/2$, $27/2$ **3.** $27/(4\sqrt{2})$

5. $\hat{\mathbf{T}} = (\mathbf{i} + 2\mathbf{j})/\sqrt{5}$, $\hat{\mathbf{N}} = (-2\mathbf{i} + \mathbf{j})/\sqrt{5}$, $\hat{\mathbf{B}} = \mathbf{k}$

7. $\hat{\mathbf{T}} = \dfrac{1}{\sqrt{1 + t^2 + t^4}}(\mathbf{i} + t\mathbf{j} + t^2\mathbf{k})$,
$\hat{\mathbf{B}} = \dfrac{1}{\sqrt{t^4 + 4t^2 + 1}}(t^2\mathbf{i} - 2t\mathbf{j} + \mathbf{k})$,
$\hat{\mathbf{N}} = \dfrac{-(t + 2t^3)\mathbf{i} + (1 - t^4)\mathbf{j} + (t^3 + 2t)\mathbf{k}}{\sqrt{t^4 + 4t^2 + 1}\sqrt{1 + t^2 + t^4}}$,
$\kappa = \dfrac{\sqrt{t^4 + 4t^2 + 1}}{(t^4 + t^2 + 1)^{3/2}}$, $\tau = \dfrac{2}{t^4 + 4t^2 + 1}$

9. $\kappa(t) = 1/\sqrt{2}$, $\tau(t) = 0$, curve is a circle in the plane $y + z = 4$, having centre $(2, 1, 3)$ and radius $\sqrt{2}$

11. (a) $\hat{\mathbf{T}} = \mathbf{i}$, $\hat{\mathbf{N}} = \dfrac{2\mathbf{j} - \mathbf{k}}{\sqrt{5}}$,
$\hat{\mathbf{B}} = \dfrac{\mathbf{j} + 2\mathbf{k}}{\sqrt{5}}$, $\kappa = \sqrt{5}$, $\tau = 0$

(b) $\hat{\mathbf{T}} = \sqrt{\frac{2}{3}}(\mathbf{j} - \frac{1}{\sqrt{2}}\mathbf{k})$, $\hat{\mathbf{B}} = \frac{1}{\sqrt{13}}(-\mathbf{i} + 2\mathbf{j} + 2\sqrt{2}\mathbf{k})$,
$\hat{\mathbf{N}} = -\frac{1}{\sqrt{39}}(6\mathbf{i} + \mathbf{j} + \sqrt{2}\mathbf{k})$, $\kappa = \frac{2\sqrt{39}}{9}$, $\tau = -\frac{6\sqrt{2}}{13}$

13. max a/b^2, min b/a^2

15. $\kappa = \dfrac{e^x}{(1+e^{2x})^{3/2}}$, $\mathbf{r} = (x - 1 - e^{2x})\mathbf{i} + (2e^x + e^{-x})\mathbf{j}$

17. $\dfrac{3}{2\sqrt{2}ar}$

21. $\mathbf{r} = -4x^3\mathbf{i} + (3x^2 + \frac{1}{2})\mathbf{j}$

23. $f(x) = \frac{1}{8}(15x - 10x^3 + 3x^5)$

Section 11.6   (page 634)

3. velocity: $1/\sqrt{2}$, $1/\sqrt{2}$; acceleration: $-e^{-\theta}/2$, $e^{-\theta}/2$.

5. $|a_r| = \dfrac{v_0^2}{5}\left(\dfrac{2}{r^2} + \dfrac{1}{r^3}\right)$

7. 42,777 km, the equatorial plane

9. $\dfrac{T}{4\sqrt{2}}$ **13.** $3/4$

15. $(1/2) - (\epsilon/\pi)$

19. $r = A \sec\omega(\theta - \theta_0)$, $\omega^2 = 1 - (k/h^2)$ if $k < h^2$,
$r = 1/(A + B\theta)$ if $k = h^2$,
$r = A e^{\omega\theta} + B e^{-\omega\theta}$, $\omega^2 = (k/h^2) - 1$,   if $k > h^2$;
there are no bounded orbits that do not approach the origin except in the case $k = h^2$ if $B = 0$ when there are circular orbits. (Now aren't you glad gravitation is an inverse square rather than an inverse cube attraction?)

21. centre $\left(\dfrac{\ell\epsilon}{\epsilon^2 - 1}, 0\right)$;

asymptotes in directions $\theta = \pm \cos^{-1}\left(-\dfrac{1}{\epsilon}\right)$;

semi-transverse axis $a = \dfrac{\ell}{\epsilon^2 - 1}$;

semi-conjugate axis $b = \dfrac{\ell}{\sqrt{\epsilon^2 - 1}}$;

semi-focal separation $c = \dfrac{\ell\epsilon}{\epsilon^2 - 1}$.

Review Exercises   (page 636)

3. $\mathbf{v} = 2(\mathbf{i} + 2\mathbf{j} + 2\mathbf{k})$, $\mathbf{a} = (8/3)(-2\mathbf{i} - \mathbf{j} + 2\mathbf{k})$

5. $\kappa = \tau = \sqrt{2}/(e^t + e^{-t})^2$

9. $4a(1 - \cos(T/2))$ units

11. $\mathbf{r}_C(t) = a(t - \sin t)\mathbf{i} + a(1 - \cos t)\mathbf{j}$

13. $\hat{\boldsymbol{\rho}} = \sin\phi\cos\theta\mathbf{i} + \sin\phi\sin\theta\mathbf{j} + \cos\phi\mathbf{k}$ right-handed
$\hat{\boldsymbol{\phi}} = \cos\phi\cos\theta\mathbf{i} + \cos\phi\sin\theta\mathbf{j} - \sin\phi\mathbf{k}$
$\hat{\boldsymbol{\theta}} = -\sin\theta\mathbf{i} + \cos\theta\mathbf{j}$

Challenging Problems   (page 636)

1. (a) $\boldsymbol{\Omega} = \Omega\dfrac{\mathbf{j} + \mathbf{k}}{\sqrt{2}}$, $\Omega \approx 7.272 \times 10^{-5}$.
(b) $\mathbf{a}_C = -\sqrt{2}\Omega v\mathbf{i}$.
(c) about 15.5 cm west of P.

3.
(c) $\mathbf{v}(t) = (\mathbf{v}_0 - (\mathbf{v}_0 \bullet \mathbf{k})\mathbf{k})\cos(\omega t) + (\mathbf{v}_0 \times \mathbf{k})\sin(\omega t) + (\mathbf{v}_0 \bullet \mathbf{k})\mathbf{k}$.
(d) Straight line if $\mathbf{v}_0$ is parallel to $\mathbf{k}$, circle if $\mathbf{v}_0$ is perpendicular to $\mathbf{k}$.

5. (a) $y = (48 + 24x^2 - x^4)/64$

7. (a) Yes, time $\pi a/(v\sqrt{2})$, (b) $\phi = \dfrac{\pi}{2} - \dfrac{vt}{a\sqrt{2}}$,

$\theta = \ln\left[\sec\left(\dfrac{vt}{a\sqrt{2}}\right) + \tan\left(\dfrac{vt}{a\sqrt{2}}\right)\right]$.
(c) infinitely often

Chapter 12

Partial Differentiation

Section 12.1   (page 645)

1. all (x, y) with $x \neq y$ **3.** all (x, y) except $(0, 0)$

5. all (x, y) satisfying $4x^2 + 9y^2 \geq 36$

7. all (x, y) with $xy > -1$

9. all (x, y, z) except $(0, 0, 0)$

11. $z = f(x, y) = x$ **13.** $z = f(x, y) = y^2$

15. $f(x, y) = \sqrt{x^2 + y^2}$ **17.** $f(x, y) = |x| + |y|$

19. $f(x, y) = x - y = C$ **21.** $f(x, y) = xy = C$

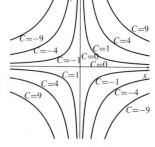

23. $f(x, y) = \dfrac{x - y}{x + y} = C$ **25.** $f(x, y) = xe^{-y} = C$

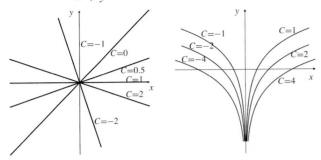

27. At B, because the contours are closer together there

29. a plane containing the y-axis, sloping uphill in the x direction

31. a right-circular cone with base in the xy-plane and vertex at height 5 on the z-axis

33. No, different curves of the family must not intersect in the region.

35. (a) $\sqrt{x^2 + y^2}$, (b) $(x^2 + y^2)^{1/4}$,
(c) $x^2 + y^2$, (d) $e^{\sqrt{x^2 + y^2}}$

37. spheres centred at the origin

39. circular cylinders with axis along the z-axis

41. regular octahedra with vertices on the coordinate axes

Section 12.2 (page 650)

1. 2

3. does not exist

5. -1

7. 0

9. does not exist

11. 0

13. $f(0, 0) = 1$

15. all (x, y) such that $x \neq \pm y$; yes; yes $f(x, x) = \dfrac{1}{2x}$ makes f continuous at (x, x) for $x \neq 0$; no, f has no continuous extension to the line $x + y = 0$.

17. no, yes   **19.** $a = c = 0$, $b \neq 0$

23. a surface having no tears in it, meeting vertical lines through points of the region exactly once

Section 12.3 (page 656)

1. $f_1(x, y) = f_1(3, 2) = 1$, $f_2(x, y) = f_2(3, 2) = -1$

3. $f_1 = 3x^2 y^4 z^5$, $f_2 = 4x^3 y^3 z^5$, $f_3 = 5x^3 y^4 z^4$
All three vanish at $(0, -1, -1)$.

5. $\dfrac{\partial z}{\partial x} = \dfrac{-y}{x^2 + y^2}$, $\dfrac{\partial z}{\partial y} = \dfrac{x}{x^2 + y^2}$,
At $(-1, 1)$: $\dfrac{\partial z}{\partial x} = -\dfrac{1}{2}$, $\dfrac{\partial z}{\partial y} = -\dfrac{1}{2}$

7. $f_1 = \sqrt{y} \cos(x\sqrt{y})$, $f_2 = \dfrac{x \cos(x\sqrt{y})}{2\sqrt{y}}$,
At $(\pi/3, 4)$: $f_1 = -1$, $f_2 = -\pi/24$

9. $\dfrac{\partial w}{\partial x} = y \ln z \, x^{(y \ln z - 1)}$, $\dfrac{\partial w}{\partial y} = \ln x \, \ln z \, x^{y \ln z}$,
$\dfrac{\partial w}{\partial z} = \dfrac{y \ln x}{z} x^{y \ln z}$
At $(e, 2, e)$: $\dfrac{\partial w}{\partial x} = \dfrac{\partial w}{\partial z} = 2e$, $\dfrac{\partial w}{\partial y} = e^2$.

11. $f_1(0, 0 = 2$, $f_2(0, 0) = -1/3$

13. $z = -4x - 2y - 3$; $\dfrac{x+2}{-4} = \dfrac{y-1}{-2} = \dfrac{z-3}{-1}$

15. $z = \dfrac{1}{\sqrt{2}} \left(1 - \dfrac{x - \pi}{4} + \dfrac{\pi}{16}(y - 4) \right)$;
$\dfrac{x - \pi}{-1/4\sqrt{2}} = \dfrac{y - 4}{\pi/16\sqrt{2}} = \dfrac{z - 1/\sqrt{2}}{-1}$

17. $z = \dfrac{2}{5} + \dfrac{3x}{25} - \dfrac{4y}{25}$; $\dfrac{x-1}{3} = \dfrac{y-2}{-4} = \dfrac{z-1/5}{-25}$

19. $z = \ln 5 + \dfrac{2}{5}(x - 1) - \dfrac{4}{5}(y + 2)$;
$\dfrac{x - 1}{2/5} = \dfrac{y + 2}{-4/5} = \dfrac{z - \ln 5}{-1}$

21. $z = \dfrac{x + y}{2} - \dfrac{\pi}{4}$; $2(x - 1) = 2(y + 1) = -z - \dfrac{\pi}{4}$

23. $(0, 0)$, $(1, 1)$, $(-1, -1)$

33. $w = f(a, b, c) + f_1(a, b, c)(x - a) + f_2(a, b, c)(y - b) + f_3(a, b, c)(z - c)$

35. $\sqrt{7}/4$ units

37. $f_1(0, 0) = 1$, $f_2(0, 0)$ does not exist.

39. f is continuous at $(0, 0)$; f_1 and f_2 are not.

Section 12.4 (page 662)

1. $\dfrac{\partial^2 z}{\partial x^2} = 2(1 + y^2)$, $\dfrac{\partial^2 z}{\partial x \partial y} = 4xy$, $\dfrac{\partial^2 z}{\partial y^2} = 2x^2$

3. $\dfrac{\partial^2 w}{\partial x^2} = 6xy^3 z^3$, $\dfrac{\partial^2 w}{\partial y^2} = 6x^3 yz^3$,
$\dfrac{\partial^2 w}{\partial z^2} = 6x^3 y^3 z$, $\dfrac{\partial^2 w}{\partial x \partial y} = 9x^2 y^2 z^3$,
$\dfrac{\partial^2 w}{\partial x \partial z} = 9x^2 y^3 z^2$, $\dfrac{\partial^2 w}{\partial y \partial z} = 9x^3 y^2 z^2$

5. $\dfrac{\partial^2 z}{\partial x^2} = -y e^x$, $\dfrac{\partial^2 z}{\partial x \partial y} = e^y - e^x$, $\dfrac{\partial^2 z}{\partial y^2} = x e^y$

7. 27, 10, $x^2 e^{xy}\big(xz \sin xz - (3 + xy) \cos xz\big)$

19. $u(x, y, z, t) = t^{-3/2} e^{-(x^2 + y^2 + z^2)/4t}$

Section 12.5 (page 671)

1. $\dfrac{\partial w}{\partial t} = f_1 g_2 + f_2 h_2 + f_3 k_2$

3. $\dfrac{\partial z}{\partial u} = g_1 h_1 + g_2 f' h_1$

5. $\dfrac{dw}{dz} = f_1 g_1 h' + f_1 g_2 + f_2 h' + f_3$,
$\dfrac{\partial w}{\partial z}\Big|_x = f_2 h' + f_3$,
$\dfrac{\partial w}{\partial z}\Big|_{x,y} = f_3$

7. $\dfrac{\partial z}{\partial x} = \dfrac{-5y}{13x^2 - 2xy + 2y^2}$

9. $2 f_1(2x, 3y)$ **11.** $2x \, f_2(y^2, x^2)$

13. $dT/dt = e^{-t}\big(f'(t) - f(t)\big)$; $dT/dt = 0$ if $f(t) = e^t$: in this case the decrease in T with time (at fixed depth) is exactly balanced by the increase in T with depth.

15. $4 f_{11} + 12 f_{12} + 9 f_{22}$, $6 f_{11} + 5 f_{12} - 6 f_{22}$,
$9 f_{11} - 12 f_{12} + 4 f_{22}$

17. $f_1 \cos s - f_2 \sin s + f_{11} t \cos s \sin s + f_{12} t (\cos^2 s - \sin^2 s) - f_{22} t \sin s \cos s$

19. $f_2 + 2y^2 f_{12} + xy f_{22} - 4xy f_{31} - 2x^2 f_{32}$;
all derivatives at $(y^2, xy, -x^2)$

27. $\sum_{i,j=1}^{n} x_i x_j f_{ij}(x_1, \cdots, x_n) = k(k-1) f(x_1, \cdots, x_n)$

31. $u(x, y) = f(x + ct)$

Section 12.6   (page 679)

1. 6.9   **3.** 0.0814

5. 2.967   **7.** (a) 3%, (b) 2%, (c) 1%

9. 8.88 ft^2

11. 169 m, 24 m, most sensitive to angle at B

13. $\begin{pmatrix} \cos\theta & -r\sin\theta \\ \sin\theta & r\cos\theta \end{pmatrix}$

15. $\begin{pmatrix} 2x & z & y \\ -\ln z & 2y & -x/z \end{pmatrix}$,   (5.99, 3.98)

Section 12.7   (page 688)

1. $4\mathbf{i} + 2\mathbf{j}$; $z = 4x + 2y - 3$; $2x + y = 3$

3. $(3\mathbf{i} - 4\mathbf{j})/25$; $3x - 4y - 25z + 10 = 0$;
$3x - 4y + 5 = 0$

5. $(2\mathbf{i} - 4\mathbf{j})/5$; $2x - 4y - 5z = 10 - 5\ln 5$; $x - 2y = 5$

7. $x + y - 3z = -3$ **9.** $\sqrt{3}y + z = \sqrt{3} + \pi/3$

11. $\dfrac{4}{\sqrt{5}}$ **13.** $1 - 2\sqrt{3}$

17. in directions making angles $-30°$ or $-150°$ with positive x-axis; no; $-\mathbf{j}$.

19. $7\mathbf{i} - \mathbf{j}$

21. a)

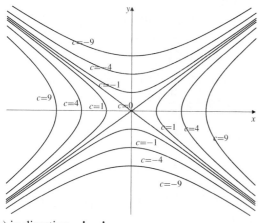

b) in direction $-\mathbf{i} - \mathbf{j}$
c) $4\sqrt{2}k$ deg/unit time
d) $12k/\sqrt{5}$ deg/unit time
e) $x^2 y = -4$

23. $3x^2 - 2y^2 = 10$ **25.** $-4/3$

27. $\mathbf{i} - 2\mathbf{j} + \mathbf{k}$

33. $D_{\mathbf{v}}(D_{\mathbf{v}}f) = v_1^2 f_{11} + v_2^2 f_{22} + v_3^2 f_{33} + 2v_1 v_2 f_{12} + 2v_1 v_3 f_{13}$

$+ 2v_2 v_3 f_{23}$.
This is the second derivative of f as measured by an observer moving with velocity $\mathbf{v}$.

35. $\dfrac{\partial^2 T}{\partial t^2} + 2D_{\mathbf{v}(t)}\left(\dfrac{\partial T}{\partial t}\right) + D_{\mathbf{a}(t)}T + D_{\mathbf{v}(t)}(D_{\mathbf{v}(t)}T)$

Section 12.8   (page 698)

1. $-\dfrac{x^4 + 3xy^2}{y^3 + 4x^3 y}$, $y \neq 0$, $y^2 \neq -4x^3$

3. $\dfrac{3xy^4 + xz}{xy - 2y^2 z}$, $y \neq 0$, $x \neq 2yz$

5. $\dfrac{x - 2t^2 w}{2xy^2 - w}$, $w \neq 2xy^2$ **7.** $-\dfrac{\partial G/\partial x}{\partial G/\partial u}$, $\dfrac{\partial G}{\partial u} \neq 0$

9. $-\dfrac{v^2 H_2 + w H_3}{u^2 H_1 + t H_3}$, $u^2 H_1 + t H_3 \neq 0$,
all derivatives at $(u^2 w, v^2 t, wt)$

11. $\dfrac{2w - 4y}{4x - w}$, $4x \neq w$ **13.** $\dfrac{1}{6}, \dfrac{1}{2}, \dfrac{1}{6}, -\dfrac{1}{2}, -\dfrac{1}{6}$

15. r; all points except the origin

17. $-3/2$

19. $-\dfrac{\partial(F, G, H)}{\partial(y, z, w)} \Big/ \dfrac{\partial(F, G, H)}{\partial(x, z, w)}$

21. 15; $-\dfrac{\partial(F, G, H)}{\partial(x_2, x_3, x_5)} \Big/ \dfrac{\partial(F, G, H)}{\partial(x_1, x_3, x_5)}$

23. $2(u + v)$, -2, 0

Section 12.9   (page 704)

1. $\sum_{n=0}^{\infty} (-1)^n \dfrac{x^n y^{2n}}{2n + 1}$

3. $\sum_{n=0}^{\infty} (-1)^n \dfrac{x^{2n+1}(y + 1)^{2n+1}}{2n + 1}$

5. $\sum_{n=0}^{\infty} \sum_{k=0}^{n} \dfrac{1}{k!(n - k)!} x^{2k} y^{2n-2k}$

7. $\dfrac{1}{2} - \dfrac{1}{4}(x - 2) + \dfrac{1}{2}(y - 1) + \dfrac{1}{8}(x - 2)^2$
$- \dfrac{1}{2}(x - 2)(y - 1) + \dfrac{1}{2}(y - 1)^2 - \dfrac{1}{16}(x - 2)^3$
$+ \dfrac{3}{8}(x - 2)^2(y - 1) - \dfrac{3}{4}(x - 2)(y - 1)^2 + \dfrac{1}{2}(y - 1)^3$

9. $x + y^2 - \dfrac{x^3}{3}$

11. $1 - (y - 1) + (y - 1)^2 - \dfrac{1}{2}(x - \dfrac{\pi}{2})^2$

13. $-x - x^2 - (5/6)x^3$

15. $-\dfrac{x}{3} - \dfrac{2y}{3} - \dfrac{2x^2}{27} - \dfrac{8xy}{27} - \dfrac{8y^2}{27}$

17. $\dfrac{[(2n)!]^3}{(n!)^2}$

Review Exercises   (page 704)

1.

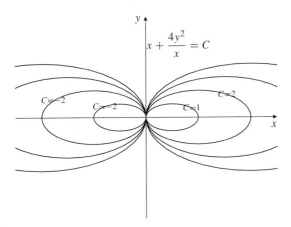

$$x + \frac{4y^2}{x} = C$$

$C = -2$, $C = -2$, $C = 1$, $C = 2$

3.

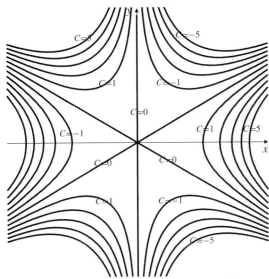

$C = 3$, $C = -5$, $C = 1$, $C = -1$, $C = 0$, $C = -1$, $C = 1$, $C = 5$, $C = 0$, $C = 0$, $C = -1$, $C = -1$, $C = -5$

5. cont. except on lines $x = \pm y$; can be extended to $x = y$ except at the origin; if $f(0,0) = 0$ then $f_1(0,0) = f_2(0,0) = 1$

7. (a) $ax + by + 4cz = 16$,
(b) the circle $z = 1$, $x^2 + y^2 = 12$, (c) $\pm(2, 2, \sqrt{2})$

9. 7,500 m², 7.213%

11. (a) $-1/\sqrt{2}$, (b) dir. of $\pm(\mathbf{i} + 3\mathbf{j} - 4\mathbf{k})$, (c) dir. of $-7\mathbf{i} + 5\mathbf{j} + 2\mathbf{k}$

15. (a) $\partial u/\partial x = -5$, $\partial u/\partial y = 1$, (b) -1.13

Chapter 13
Applications of Partial Derivatives

Section 13.1   (page 714)

1. $(2, -1)$, loc. (abs) min.

3. $(0,0)$, saddle pt; $(1,1)$, loc. min.

5. $(-4, 2)$, loc. max.

7. $(0, n\pi)$, $n = 0, \pm1, \pm2, \cdots$, all saddle points

9. $(0, a)$, $(a > 0)$, loc min; $(0, a)$, $(a < 0)$, loc max; $(0,0)$ saddle point; $(\pm1, 1/\sqrt{2})$, loc. (abs) max; $(\pm1, -1/\sqrt{2})$, loc. (abs) min.

11. $(3^{-1/3}, 0)$, saddle pt.

13. max at (x, x), min at $(x, -x)$, $x \neq 0$

15. $(-1, -1)$, $(1, -1)$, $(-1, 1)$, saddle pts; $(-3, -3)$, loc. min.

17. $(1, 1, \frac{1}{2})$, saddle pt.

19. $(0,0)$, saddle pt; $(\frac{1}{\sqrt{2}}, \frac{1}{\sqrt{2}})$, $(-\frac{1}{\sqrt{2}}, -\frac{1}{\sqrt{2}})$, loc. (abs) max; $(\frac{1}{\sqrt{2}}, -\frac{1}{\sqrt{2}})$, $(-\frac{1}{\sqrt{2}}, \frac{1}{\sqrt{2}})$, loc. (abs) min.

21. max $e^{-3/2}/2\sqrt{2}$, min $-e^{-3/2}/2\sqrt{2}$; f is continuous everywhere, and $f(x, y, z) \to 0$ as $x^2 + y^2 + z^2 \to \infty$.

23. $L^3/108$ cu. units   **25.** $8abc/(3\sqrt{3})$ cu. units

27. CPs are $(\sqrt{\ln 3}, -\sqrt{\ln 3})$ and $(-\sqrt{\ln 3}, \sqrt{\ln 3})$.

29. f does not have a local minimum at $(0, 0)$; the second derivative test is inconclusive ($B^2 = AC$).

Section 13.2   (page 720)

1. max $5/4$, min -2

3. max $(\sqrt{2} - 1)/2$, min $-(\sqrt{2} + 1)/2$.

5. max $2/3\sqrt{3}$, min 0 **7.** max 1, min -1

9. max $1/\sqrt{e}$, min $-1/\sqrt{e}$ **11.** max $4/9$, min $-4/9$

13. no limit; yes, max $f = e^{-1}$ (at all points of the curve $xy = 1$)

15. \$625,000,   \$733,333

17. max $37/2$ at $(7/4, 5)$

19. 6667 kg deluxe, 6667 kg standard

Section 13.3   (page 728)

1. $84, 375$ **3.** 1 unit

5. max 4 units, min 2 units

7. $a = \pm\sqrt{3}$, $b = \pm2\sqrt{3}$, $c = \pm\sqrt{3}$

9. max 8, min -8 **11.** max 2, min -2

13. max 7, min -1 **15.** $\dfrac{2\sqrt{6}}{3}$ units

17. $\pm\sqrt{\dfrac{3n(n+1)}{2(2n+1)}}$ **19.** $\frac{1}{6} \times \frac{1}{3} \times \frac{2}{3}$

21. width $= \left(\dfrac{2V}{15}\right)^{1/3}$, depth $= 3 \times$ width, height $= \dfrac{5}{2} \times$ width

23. max 1, min $-\frac{1}{2}$

27. method will not fail if $\nabla f = \mathbf{0}$ at extreme point; but we will have $\lambda = 0$.

Section 13.4   (page 734)

1. at $(\bar{x}, \bar{y})$ where $\bar{x} = \left(\sum_{i=1}^{n} x_i\right)/n$, $\bar{y} = \left(\sum_{i=1}^{n} y_i\right)/n$

3. $a = \left(\sum_{i=1}^{n} y_i e^{x_i}\right) / \left(\sum_{i=1}^{n} e^{2x_i}\right)$

5. If $A = \sum x_i^2$, $B = \sum x_i y_i$, $C = \sum x_i$, $D = \sum y_i^2$,
$E = \sum y_i$, $F = \sum x_i z_i$, $G = \sum y_i z_i$,
and $H = \sum z_i$, then

$$\Delta = \begin{vmatrix} A & B & C \\ B & D & E \\ C & E & n \end{vmatrix}, \qquad a = \frac{1}{\Delta}\begin{vmatrix} F & B & C \\ G & D & E \\ H & E & n \end{vmatrix},$$

$$b = \frac{1}{\Delta}\begin{vmatrix} A & F & C \\ B & G & E \\ C & H & n \end{vmatrix}, \qquad c = \frac{1}{\Delta}\begin{vmatrix} A & B & F \\ B & D & G \\ C & E & H \end{vmatrix}.$$

7. Use linear regression to fit $\eta = a + bx$ to the data $(x_i, \ln y_i)$. Then $p = e^a$, $q = b$. These are not the same values as would be obtained by minimizing the expression $\sum(y_i - pe^{qx_i})^2$.

9. Use linear regression to fit $\eta = a + b\xi$ to the data $\left(x_i, \dfrac{y_i}{x_i}\right)$. Then $p = a, q = b$. Not the same as minimizing $\sum(y_i - px_i - qx_i^2)^2$.

11. Use linear regression to fit $\eta = a + b\xi$ to the data $\left(e^{-2x_i}, \dfrac{y_i}{e^{x_i}}\right)$. Then $p = a, q = b$. Not the same as minimizing $\sum(y_i - pe^{x_i} - qe^{-x_i})^2$. Other answers possible.

13. If $A = \sum x_i^4$, $B = \sum x_i^3$, $C = \sum x_i^2$, $D = \sum x_i$, $H = \sum x_i^2 y_i$, $I = \sum x_i y_i$, and $J = \sum y_i$, then

$$\Delta = \begin{vmatrix} A & B & C \\ B & C & D \\ C & D & n \end{vmatrix}, \qquad a = \frac{1}{\Delta}\begin{vmatrix} H & B & C \\ I & C & D \\ J & D & n \end{vmatrix},$$

$$b = \frac{1}{\Delta}\begin{vmatrix} A & H & C \\ B & I & D \\ C & J & n \end{vmatrix}, \qquad c = \frac{1}{\Delta}\begin{vmatrix} A & B & H \\ B & C & I \\ C & D & J \end{vmatrix}.$$

15. $a = 5/6$, $I = 1/252$

17. $a = 15/16$, $b = -1/16$, $I = 1/448$

19. $a = \frac{20}{\pi^5}(\pi^2 - 16)$, $b = \frac{12}{\pi^4}(20 - \pi^2)$

21. $a_k = \frac{2}{\pi}\int_0^\pi f(x)\cos kx\,dx$, $(k = 0, 1, 2, \cdots)$

23. $\pi - \frac{4}{\pi}\sum_{k=0}^{\infty}\frac{\cos((2k+1)x)}{(2k+1)^2}$; $-x$

Section 13.5   (page 743)

1. $\dfrac{(-1)^n n!}{(x+1)^{n+1}}$

3. $2\sqrt{\pi}(\sqrt{y} - \sqrt{x})$

5. $\dfrac{2x}{(1+x^2)^2}$; $\dfrac{(6x^2-2)}{(1+x^2)^3}$

7. $\dfrac{\pi}{2x}$, assume $x > 0$; $\dfrac{\pi}{4x^3}$; $\dfrac{3\pi}{16x^5}$

9. $n!$

11. $f(x) = \int_0^x e^{-t^2/2}\,dt$

13. $y = x^2$

15. $x^2 + y^2 = 1$

17. $y = x - \frac{1}{4}$

19. no

21. no; a line of singular points

23. $x^2 + y^2 + z^2 = 1$

25. $y = x - \epsilon\sin(\pi x) + \frac{\pi\epsilon^2}{2}\sin(2\pi x) + \cdots$

27. $y = \frac{1}{2} - \frac{2}{5}\epsilon x - \frac{16}{125}\epsilon^2 x^2 + \cdots$

29. $x \approx 1 - \dfrac{1}{100e} - \dfrac{1}{30000e^2}$, $y \approx 1 - \dfrac{1}{30000e^2}$

Section 13.6   (page 746)

1. $(0.797105, 2.219107)$

3. $(\pm 0.2500305, \pm 3.9995115)$, $(\pm 1.9920783, \pm 0.5019883)$

5. $(0.3727730, 0.3641994)$, $(-1.4141606, -0.9877577)$

7. $x = x_0 - \frac{\Delta_1}{\Delta}$, $y = y_0 - \frac{\Delta_2}{\Delta}$, $z = z_0 - \frac{\Delta_3}{\Delta}$,
where $\Delta = \dfrac{\partial(f, g, h)}{\partial(x, y, z)}\Big|_{(x_0, y_0, z_0)}$

and Δ_i is Δ with the ith column replaced with $\begin{matrix} f \\ g \\ h \end{matrix}$

9. 18 iterations near $(0, 0)$, 4 iterations near $(1, 1)$; the two curves are tangent at $(0,0)$, but not at $(1,1)$.

Section 13.7   (page 751)

1. $(\pm.45304, .81204, \pm.36789)$, $(\pm.96897, .17751, \pm.17200)$

3. local and absolute max 0.81042 at $(-0.33853, -0.52062)$; local and absolute min -0.66572 at $(0.13319, 0.53682)$

5. -4.5937

Review Exercises   (page 752)

1. $(0, 0)$ saddle pt., $(1, -1)$ loc. min.

3. $(2/3, 4/3)$ loc. min; $(2, -4)$ and $(-1, 2)$ saddle points

5. yes, 2, on the sphere $x^2 + y^2 + z^2 = 1$

7. max $1/(4e)$, min $-1/(4e)$

9. (a) $L^2/48$ cm^2, (b) $L^2/16$ cm^2

11. 4π sq. units   **13.** 16π cu. units

15. 1,688 widgets, \$2.00 each

17. $y \approx -2x - \epsilon xe^{-2x} + \epsilon^2 x^2 e^{-4x}$

Challenging Problems   (page 753)

3. $\frac{1}{2}\ln(1 + x^2)\tan^{-1}x$

Chapter 14
Multiple Integration

Section 14.1   (page 759)

1. 15   **3.** 21

5. 15   **7.** 96

9. 80   **11.** 36.6258

13. 20   **15.** 0

17. 5π **19.** $\frac{\pi a^3}{3}$

21. $\frac{1}{6}$

Section 14.2   (page 766)

1. $5/24$ **3.** 4

5. $\dfrac{ab(a^2+b^2)}{3}$ **7.** π

9. $\dfrac{3}{56}$ **11.** $\frac{33}{8}\ln 2 - \frac{45}{16}$

13. $\dfrac{e-2}{2}$

15. $\dfrac{1}{2}\left(1 - \dfrac{1}{e}\right)$; region is a triangle with vertices $(0, 0)$, $(1, 0)$ and $(1, 1)$

17. $\frac{\pi}{4\lambda}$; region is a triangle with vertices $(0, 0)$, $(0, 1)$ and $(1, 1)$

19. 1/4 cu. units   **21.** 1/3 cu. units

23. ln 2 cu. units   **25.** $\frac{\pi}{2\sqrt{2}}$ cu. units

27. $\frac{16a^3}{3}$ cu. units

Section 14.3   (page 771)

1. converges to 1   **3.** converges to $\pi/2$

5. diverges to ∞ **7.** converges to 4

9. converges to $1 - \frac{1}{e}$ **11.** diverges to ∞

13. converges to $2 \ln 2$ **15.** $k > a - 1$

17. $k < -1 - a$

19. $k > -\frac{1+a}{1+b}$ (provided $b > -1$)

21. $\frac{1}{2}, -\frac{1}{2}$ (different answers are possible because the *double integral* does not exist.)

23. $\frac{a^2}{3}$ **25.** $\frac{4\sqrt{2}a}{3\pi}$

27. yes, $1/(2\pi)$

Section 14.4   (page 780)

1. $\pi a^4/2$ **3.** $2\pi a$

5. $\pi a^4/4$ **7.** $a^3/3$

9. $\pi(e^{a^2} - 1)/4$ **11.** $\frac{(\sqrt{3}+1)a^3}{6}$

13. $\frac{1}{3}$ **15.** $\frac{2a}{3}$

17. $k < 1$; $\frac{\pi}{1-k}$ **19.** $\frac{a^4}{16}$

21. $\frac{2\pi}{3}$ cu. units   **23.** $\frac{4\pi(2\sqrt{2}-1)a^3}{3}$ cu. units

25. $16[1 - (1/\sqrt{2})]a^3$ cu. units

27. $1 - \frac{4\sqrt{2}}{3\pi}$ units   **29.** $\frac{4}{3}\pi abc$ cu. units

31. $2a \sinh a$ **33.** $\frac{3 \ln 2}{2}$ sq. units

35. $\frac{1}{4}(e - e^{-1})$

Section 14.5   (page 787)

1. $8abc$ **3.** 16π

5. $2/3$ **7.** $1/15$

9. $2/(3\pi)$ **11.** $\frac{3}{16} \ln 2$

13. $\pi\sqrt{\frac{\pi}{6}}$ **15.** $1/8$

17. $\int_0^1 dx \int_0^1 dy \int_0^{1-y} f(x, y, z)\, dz$

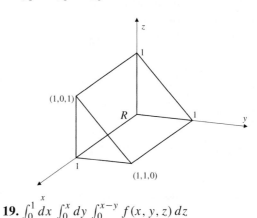

19. $\int_0^1 dx \int_0^x dy \int_0^{x-y} f(x, y, z)\, dz$

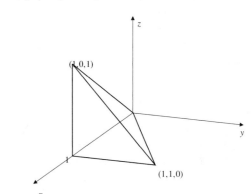

27. $(e - 1)/3$

29. $\overline{f} = \frac{1}{\text{vol}(R)} \iiint_R f\, dV$; 1

Section 14.6   (page 794)

1. Cartesian: $(-\sqrt{3}, 3, 2)$; cylindrical: $[2\sqrt{3}, 2\pi/3, 2]$

3. Cartesian: $(\sqrt{3}, 1, -2)$; spherical: $[2\sqrt{2}, 3\pi/4, \pi/6]$

5. the half-plane $x = 0$, $y > 0$

7. the xy-plane

9. the circular cylinder of radius 4 with axis along the z-axis

11. the xy-plane

13. sphere of radius 1 with centre $(0, 0, 1)$

15. $\frac{2}{3}\pi a^3 \left(1 - \frac{1}{\sqrt{2}}\right)$ cu. units

17. 24π cu. units   **19.** $\left(2\pi - \frac{32}{9}\right)a^3$ cu. units.

21. $\frac{abc}{3} \tan^{-1}\frac{a}{b}$ cu. units   **23.** $\frac{\pi ab}{2}$ cu. units

25. $\frac{8\pi a^5}{15}$ **27.** $\frac{2\pi a^5}{5}\left(1 - \frac{c}{\sqrt{c^2+1}}\right)$

29. $\frac{7\pi}{12}$ **31.** $\frac{ha^3}{12}$; $\frac{\pi a^2 h^2}{48}$

Section 14.7   (page 803)

1. 3π sq. units   **3.** $2\pi a^2$ sq. units

5. $24\pi/\sqrt{3}$ sq. units   **7.** $(5\sqrt{5} - 1)/12$ sq. units

9. 4 sq. units   **11.** 5.123

13. $4\pi A\left[a - \sqrt{B}\tan^{-1}\left(\frac{a}{\sqrt{B}}\right)\right]$ units

15. $2\pi km\delta(h + \sqrt{a^2 + (b-h)^2} - \sqrt{a^2 + b^2})$

17. $2\pi km\delta(h + \sqrt{a^2 + (b-h)^2} - \sqrt{a^2 + b^2})$

19. $\left(\frac{1}{3}, \frac{1}{3}, \frac{1}{2}\right)$ 21. $\left(\frac{3a}{8}, \frac{3a}{8}, \frac{3a}{8}\right)$

23. The model still involves angular acceleration to spin the ball—it doesn't just fall. Part of the gravitational energy goes to producing this spin even in the limiting case.

25. $I = \pi\delta a^2 h\left(\frac{h^2}{3} + \frac{a^2}{4}\right), \quad \bar{D} = \left(\frac{h^2}{3} + \frac{a^2}{4}\right)^{1/2}$

27. $I = \frac{\pi\delta a^2 h}{3}\left(\frac{2h^2 + 3a^2}{20}\right), \quad \bar{D} = \left(\frac{2h^2 + 3a^2}{20}\right)^{1/2}$

29. $I = \frac{5a^5\delta}{12}, \quad \bar{D} = \sqrt{\frac{5}{12}}\,a$

31. $I = \frac{8}{3}\delta abc(a^2 + b^2), \quad \bar{D} = \sqrt{\frac{a^2 + b^2}{3}}$

33. $m = \frac{4\pi}{3}\delta(a^2 - b^2)^{3/2}, \quad I = \frac{1}{5}m(2a^2 + 3b^2)$

35. $\frac{5a^2 g\sin\alpha}{7a^2 + 3b^2}$

39. The moment of inertia about the line
$\mathbf{r}(t) = At\mathbf{i} + Bt\mathbf{j} + Ct\mathbf{k}$ is

$$\frac{1}{A^2 + B^2 + C^2}\big((B^2 + C^2)P_{xx} + (A^2 + C^2)P_{yy}$$
$$+ (A^2 + B^2)P_{zz} - 2ABP_{xy} - 2ACP_{xz} - 2BCP_{yz}\big).$$

Review Exercises   (page 804)

1. $3/10$ 3. $\ln 2$

5. $k = 1/\sqrt{3}$

7. $\displaystyle\int_0^1 dx \int_x^1 dy \int_y^1 f(x, y, z)\,dz$

9. $(1 - e^{-a^2})/(2a)$ 11. $\dfrac{8\pi}{15}(18\sqrt{6} - 41)a^5$

13. vol $= 7/12, \bar{z} = 11/28$ 15. $17/24$

17. $\dfrac{1}{6}\displaystyle\int_0^{\pi/2}\left[(1 + 16\cos^2\theta)^{3/2} - 1\right]d\theta \approx 7.904$ sq. units

Challenging Problems   (page 805)

1. $\pi abc\left(\dfrac{2}{3} - \dfrac{8}{9\sqrt{3}}\right)$ cu. units

3. (b) (i) $\sum_{n=1}^{\infty}(-1)^{n-1}1/n^2$, (ii) $\sum_{n=1}^{\infty}1/n^3$, (iii) $\sum_{n=1}^{\infty}(-1)^{n-1}1/n^3$

5. $4 - \tan^{-1}(\sqrt{2}) + \dfrac{32}{3}\tan^{-1}\left(\dfrac{5}{\sqrt{2}}\right) - \dfrac{4}{3}(7\pi + 2\sqrt{2})$
≈ 18.9348 cu. units

7. $a^3/210$ cu. units

Chapter 15

Vector Fields

Section 15.1   (page 811)

1. field lines: $y = x + C$

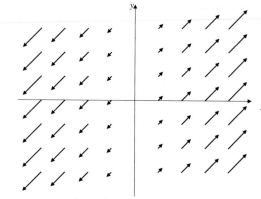

3. field lines: $y^2 = x^2 + C$

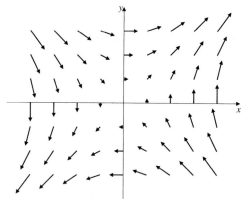

5. field lines: $y = -\frac{1}{2}e^{-2x} + C$

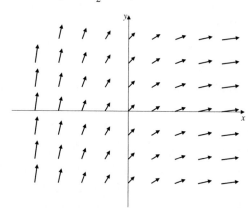

7. field lines: $y = Cx$

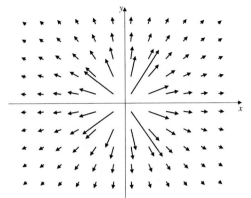

9. streamlines are lines parallel to $\mathbf{i} - \mathbf{j} - \mathbf{k}$

11. streamlines: $x^2 + y^2 = a^2$, $x = a\sin(z - b)$ (spirals)

13. $y = C_1 x$, $2x = z^2 + C_2$

15. $y = Ce^{1/x}$ **17.** $r = \theta + C$

19. $r = C\theta^2$

Section 15.2 (page 819)

1. conservative; $\dfrac{x^2}{2} - y^2 + \dfrac{3z^2}{2}$

3. not conservative

5. conservative; $x^2 y + y^2 z - z^2 x$

7. $-2\dfrac{\mathbf{r} - \mathbf{r}_0}{|\mathbf{r} - \mathbf{r}_0|^4}$

9. $(x^2 + y^2)/z$; equipotential surfaces are paraboloids $z = C(x^2 + y^2)$; field lines are ellipses $x^2 + y^2 + 2z^2 = A$, $y = Bx$ in vertical planes through the origin

11. $\mathbf{v} = \dfrac{m(x\mathbf{i} + y\mathbf{j} + (z - \ell)\mathbf{k})}{[x^2 + y^2 + (z - \ell)^2]^{3/2}} + \dfrac{m(x\mathbf{i} + y\mathbf{j} + (z + \ell)\mathbf{k})}{[x^2 + y^2 + (z + \ell)^2]^{3/2}}$,

$\mathbf{v} = \mathbf{0}$ only at the origin; $\mathbf{v}(x, y, 0) = \dfrac{2m(x\mathbf{i} + y\mathbf{j})}{(x^2 + y^2 + \ell^2)^{3/2}}$;

speed maximum on the circle $x^2 + y^2 = \ell^2/2$, $z = 0$

15. $\phi = -\dfrac{\mu y}{r^2}$, $\mathbf{F} = \dfrac{\mu(2xy\mathbf{i} + (y^2 - x^2)\mathbf{j})}{r^4}$, $(r^2 = x^2 + y^2)$

21. $\phi = \frac{1}{2}r^2 \sin 2\theta$

Section 15.3 (page 824)

1. $\frac{a^2}{2}\left(\sqrt{2} + \ln(1 + \sqrt{2})\right)$ **3.** 8 gm

5. $\frac{\delta}{6}\left((2e^{4\pi} + 1)^{3/2} - 3^{3/2}\right)$

7. $3\sqrt{14}$

9. $m = 2\sqrt{2}\pi^2$, $(0, -1/\pi, 4\pi/3)$

11. $(e^6 + 3e^4 - 3e^2 - 1)/(3e^3)$

13. $\left(\sqrt{2} + \ln(\sqrt{2} + 1)\right)a^2/2$

15. $\pi/\sqrt{2}$

17. $4\sqrt{b^2 + c^2}\, E\left(\sqrt{\dfrac{b^2 - a^2}{b^2 + c^2}}\right)$;

$\sqrt{b^2 + c^2}\, E\left(\sqrt{\dfrac{b^2 - a^2}{b^2 + c^2}}, T\right)$

Section 15.4 (page 831)

1. $-1/4$ **3.** $1/2$

5. 0 **7.** $19/2$

9. $e^{1 + (\pi/4)}$

11. $A = 2$, $B = 3$; $4\ln 2 - \frac{1}{2}$

13. $-13/2$ **15.** a) πa^2, b) $-\pi a^2$

17. a) $\dfrac{\pi a^2}{2}$, b) $-\dfrac{\pi a^2}{2}$ **19.** a) $ab/2$, b) $-ab/2$

23. The plane with origin removed is not simply connected.

Section 15.5 (page 842)

1. $dS = ds\,dz = \sqrt{(g(\theta))^2 + (g'(\theta))^2}\,d\theta\,dz$

3. $\dfrac{\pi ab\sqrt{A^2 + B^2 + C^2}}{|C|}$ sq. units $(C \neq 0)$

5. (a) $dS = |\nabla F/F_2|\,dx\,dz$,   (b) $dS = |\nabla F/F_1|\,dy\,dz$

7. $\frac{\pi}{8}$ **9.** $16a^2$ sq. units

13. 2π **15.** $1/96$

17. $\pi(3e + e^3 - 4)/3$

19. $2\pi a^2 + \dfrac{2\pi ac^2}{\sqrt{a^2 - c^2}}\ln\left(\dfrac{a + \sqrt{a^2 - c^2}}{c}\right)$ sq. units

21. $2\pi\sqrt{A^2 + B^2 + C^2}/|D|$

23. one-third of the way from the base to the vertex on the axis

25. $2\pi k\sigma ma\left(\dfrac{1}{\sqrt{a^2 + (b - h)^2}} - \dfrac{1}{\sqrt{a^2 + b^2}}\right)$

27. $I = \frac{8}{3}\pi\sigma a^4$; $\bar{D} = \sqrt{\frac{2}{3}}\,a$

29. $\frac{3}{5}g\sin\alpha$

Section 15.6 (page 848)

1. 6 **3.** $3abc$

5. $\pi(3a^2 - 4ab + b^2)/2$ **7.** 4π

9. $2\sqrt{2}\pi$ **11.** $4\pi/3$

13. $4\pi m$ **15.** a) $2\pi a^2$,   b) 8

Review Exercises (page 848)

1. $(3e/2) - (3/(2e))$ **3.** $8\sqrt{2}/15$

5. 1

7. (a) $6\pi mgb$, (b) $6\pi R\sqrt{a^2 + b^2}$

9. (b) e^2 **11.** $(x\mathbf{i} - y\mathbf{j})/\sqrt{x^2 + y^2}$

Challenging Problems (page 849)

1. centroid $(0, 0, 2/\pi)$; upper half of the surface of the torus obtained by rotating the circle $(x - 2)^2 + z^2 = 1$, $y = 0$, about the z-axis

Chapter 16
Vector Calculus

Section 16.1 (page 858)

1. $\operatorname{div}\mathbf{F} = 2$, $\operatorname{curl}\mathbf{F} = 0$
3. $\operatorname{div}\mathbf{F} = 0$, $\operatorname{curl}\mathbf{F} = -\mathbf{i} - \mathbf{j} - \mathbf{k}$
5. $\operatorname{div}\mathbf{F} = 1$, $\operatorname{curl}\mathbf{F} = -\mathbf{j}$
7. $\operatorname{div}\mathbf{F} = f'(x) + g'(y) + h'(z)$, $\operatorname{curl}\mathbf{F} = 0$
9. $\operatorname{div}\mathbf{F} = \cos\theta\left(1 + \dfrac{1}{r}\cos\theta\right)$;

$$\operatorname{curl}\mathbf{F} = -\sin\theta\left(1 + \dfrac{1}{r}\cos\theta\right)\mathbf{k}$$

11. $\operatorname{div}\mathbf{F} = 0$; $\operatorname{curl}\mathbf{F} = (1/r)\mathbf{k}$

Section 16.2 (page 864)

7. $\operatorname{div}\mathbf{F}$ can have any value, $\operatorname{curl}\mathbf{F}$ must be normal to $\mathbf{F}$
9. $f(r) = Cr^{-3}$
15. If $\mathbf{F} = \nabla\phi$ and $\mathbf{G} = \nabla\psi$ then $\nabla \times (\phi\nabla\psi) = \mathbf{F} \times \mathbf{G}$.
17. $\mathbf{G} = ye^{2z}\mathbf{i} + xye^{2z}\mathbf{k}$ is one possible vector potential.

Section 16.3 (page 868)

1. $\pi a^2 - 4a^3$ 3. 9
5. $\dfrac{3\pi ab}{8}$ sq. units
7. 0

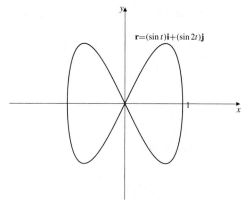

$\mathbf{r} = (\sin t)\mathbf{i} + (\sin 2t)\mathbf{j}$

Section 16.4 (page 873)

1. $4\pi a^3$ 3. $(4/3)\pi a^3$
5. 360π 7. $81/4$
11. $\frac{2}{3}\pi a^2 b + \frac{3}{10}\pi a^4 b + \pi a^2$
13. (a) $12\sqrt{3}\pi a^3$, (b) $-4\sqrt{3}\pi a^3$, (c) $16\sqrt{3}\pi a^3$
15. $(6 + 2\bar{x} + 4\bar{y} - 2\bar{z})V$ 17. $9\pi a^2$

Section 16.5 (page 878)

1. $1/2$ 3. $-3\pi a^2$
7. 9π
9. $\alpha = -\frac{1}{2}$, $\beta = -3$, $I = -\frac{3}{8}\pi$
11. yes, $\phi\nabla\psi$

Section 16.6 (page 885)

(none)

Section 16.7 (page 896)

1. $\nabla f = \theta z\hat{\mathbf{r}} + z\hat{\boldsymbol{\theta}} + r\theta\mathbf{k}$ 3. $\operatorname{div}\mathbf{F} = 2$, $\operatorname{curl}\mathbf{F} = 0$
5. $\operatorname{div}\mathbf{F} = \dfrac{2\sin\phi}{\rho}$, $\operatorname{curl}\mathbf{F} = -\dfrac{\cos\phi}{\rho}\hat{\boldsymbol{\theta}}$
7. $\operatorname{div}\mathbf{F} = 0$, $\operatorname{curl}\mathbf{F} = \cot\phi\,\hat{\boldsymbol{\rho}} - 2\hat{\boldsymbol{\phi}}$
9. scale factors: $h_u = \left|\dfrac{\partial\mathbf{r}}{\partial u}\right|$, $h_v = \left|\dfrac{\partial\mathbf{r}}{\partial v}\right|$

local basis: $\hat{\mathbf{u}} = \dfrac{1}{h_u}\dfrac{\partial\mathbf{r}}{\partial u}$, $\hat{\mathbf{v}} = \dfrac{1}{h_v}\dfrac{\partial\mathbf{r}}{\partial v}$
area element: $dA = h_u h_v\, du\, dv$

11. $\nabla f(r, \theta) = \dfrac{\partial f}{\partial r}\hat{\mathbf{r}} + \dfrac{1}{r}\dfrac{\partial r}{\partial\theta}\hat{\boldsymbol{\theta}}$

$$\nabla \bullet \mathbf{F}(r, \theta) = \dfrac{\partial F_r}{\partial r} + \dfrac{1}{r}F_r + \dfrac{1}{r}\dfrac{\partial F_\theta}{\partial\theta}$$

$$\nabla \times \mathbf{F}(r, \theta) = \left(\dfrac{\partial F_\theta}{\partial r} + \dfrac{1}{r}F_\theta - \dfrac{1}{r}\dfrac{\partial F_r}{\partial\theta}\right)\mathbf{k}$$

13. u-surfaces: vertical elliptic cylinders with focal axes at $x = \pm a$, $y = 0$
v-surfaces: vertical hyperbolic cylinders with focal axes at $x = \pm a$, $y = 0$
z-surfaces: horizontal planes
u-curves: horizontal hyperbolas with foci $x = \pm a$, $y = 0$
v-curves: horizontal ellipses with foci $x = \pm a$, $y = 0$
z-curves: vertical straight lines

15. $\nabla f = \dfrac{\partial^2 f}{\partial\rho^2} + \dfrac{2}{\rho}\dfrac{\partial f}{\partial\rho} + \dfrac{1}{\rho^2}\dfrac{\partial^2 f}{\partial\phi^2}$

$$+ \dfrac{\cot\phi}{\rho^2}\dfrac{\partial f}{\partial\phi} + \dfrac{1}{\rho^2\sin^2\phi}\dfrac{\partial^2 f}{\partial\theta^2}$$

Review Exercises (page 896)

1. 128π 3. -6
5. $3/4$ 7. $\lambda = -3$, no
11. the ellipsoid $x^2 + 4y^2 + z^2 = 4$ with outward normal

Challenging Problems (page 897)

1. $\operatorname{div}\mathbf{v} = 3C$

Chapter 17
Ordinary Differential Equations

Section 17.1 (page 902)

1. 1, linear, homogeneous   3. 1, nonlinear
5. 2, linear, homogeneous
7. 3, linear, nonhomogeneous
9. 4, linear, homogeneous
11. (a) and (b) are solutions, (c) is not
13. $y_2 = \sin(kx)$, $y = -3(\cos(kx) + (3/k)\sin(kx))$

15. $y = \sqrt{2}(\cos x + 2 \sin x)$
17. $y = x + \sin x + (\pi - 1) \cos x$

Section 17.2   (page 913)

1. $y^2 = Cx$
3. $x^3 - y^3 = C$

5. $Y = Ce^{t^2/2}$
7. $y = \dfrac{Ce^{2x} - 1}{Ce^{2x} + 1}$

9. $y = -\ln\left(Ce^{-2t} - \frac{1}{2}\right)$
11. $2\tan^{-1}(y/x) = \ln(x^2 + y^2) + C$
13. $y = x\tan(\ln|x| + C)$ **15.** $y = x\tan^{-1}(\ln|Cx|)$
17. $y = x^3 + Cx^2$ **19.** $y = \frac{3}{2} + Ce^{-2x}$
21. $y = x - 1 + Ce^{-x}$ **23.** $y = (1 + e^{1-10t})/10$
25. $y = (x + 2)e^{1/x}$ **27.** $y = \sqrt{4 + x^2}$
29. $y = \dfrac{2x}{1 + x}, \ (x > 0)$ **31.** $y^3 + 3y - 3x^2 = 24$
35. $2xy + x^2y^2 = C$ **37.** $xe^{xy} = C$
39. $\ln|x| - \frac{y}{x^2} = C$ **41.** b

43. If $a = b$ the given solution is indeterminate $0/0$; in this case the solution is $x = a^2kt/(1 + akt)$.

45. $v = \sqrt{\dfrac{mg}{k}} \ \dfrac{e^{2\sqrt{kg/mt}} - 1}{e^{2\sqrt{kg/mt}} + 1}, \quad v \to \sqrt{\dfrac{mg}{k}}$

47. the hyperbolas $x^2 - y^2 = C$

49. $\dfrac{1}{M}\left(\dfrac{\partial N}{\partial x} - \dfrac{\partial M}{\partial y}\right)$ must depend only on y

51. $\dfrac{1}{M}\left(\dfrac{\partial N}{\partial x} - \dfrac{\partial M}{\partial y}\right)$ must depend only on y.
$x - y^2e^y = Cy^2$

53. $\dfrac{1}{\mu}\dfrac{d\mu}{dx} = \dfrac{\dfrac{\partial N}{\partial x} - \dfrac{\partial M}{\partial y}}{xM - yN}$ must depend only on xy;
$\dfrac{\sin x}{y} - \dfrac{y}{x} = C$

Section 17.3   (page 921)

1. (a) 1.97664, (b) 2.187485, (c) 2.306595
3. (a) 2.436502, (b) 2.436559, (c) 2.436563
5. (a) 1.097897, (b) 1.098401
7. (a) 0.89441, (b) 0.87996, (c) 0.872831
9. (a) 0.865766, (b) 0.865769, (c) 0.865769
11. (a) 0.898914, (b) 0.903122, (c) 0.904174
13. $y = 2/(3 - 2x)$
17. (b) $u = 1/(1 - x)$, $v = \tan(x + \frac{\pi}{4})$. $y(x)$ is defined at least on $[0, \pi/4)$ and satisfies $1/(1-x) \le y(x) \le \tan(x+\frac{\pi}{4})$ there.

Section 17.4   (page 924)

1. $y = C_1e^x + C_2e^{2x}$ **3.** $y = C_1x + \dfrac{C_2}{x^2}$

5. $y = C_1x + C_2xe^x$

Section 17.5   (page 934)

1. $y = Ae^{-5t} + Be^{-2t}$ **3.** $y = A + Be^{-2t}$
5. $y = (A + Bt)e^{-4t}$
7. $y = (A\cos t + B\sin t)e^{3t}$
9. $y = (A\cos 2t + B\sin 2t)e^{-t}$
11. $y = (A\cos\sqrt{2}t + B\sin\sqrt{2}t)e^{-t}$
13. $y = \frac{6}{7}e^{t/2} + \frac{1}{7}e^{-3t}$
15. $y = e^{-2t}(2\cos t + 6\sin t)$
19. $y = C_1 + C_2e^t + C_3e^{3t}$
21. $y = C_1\cos t + C_2\sin t + C_3t\cos t + C_4t\sin t$
23. $y = C_1e^{2t} + C_2e^{-t}\cos t + C_3e^{-t}\sin t$
25. $y = Ax + Bx\ln x$ **27.** $y = Ax + \dfrac{B}{x}$
29. $y = A + B\ln x$
31. $y = C_1x + C_2x\ln x + C_3x(\ln x)^2$
33. $y = \frac{3}{10}\sin(10t)$, circ freq 10, freq $\frac{10}{2\pi}$, per $\frac{2\pi}{10}$, amp $\frac{3}{10}$
35. $y = A\cos(\omega(t - a)) + \dfrac{B}{\omega}\sin(\omega(t - a))$
43. 16 Hz for 900gm; 48 Hz for 100 gm
47. $y = e^{6-2t}[\cosh(t - 3) + 2\sinh(t - 3)]$
49. $x(t) = \begin{cases} \frac{4}{5}\cos t + \frac{1}{5} & \text{if } 0 \le t \le \pi \\ \frac{2}{5}\cos t - \frac{1}{5} & \text{if } \pi < t \le 2\pi \\ \frac{1}{5} & \text{if } t > 2\pi \end{cases}$

Section 17.6   (page 941)

1. $y = -\dfrac{1}{2} + C_1e^x + C_2e^{-2x}$

3. $y = -\dfrac{1}{2}e^{-x} + C_1e^x + C_2e^{-2x}$

5. $y = -\dfrac{2}{125} - \dfrac{4x}{25} + \dfrac{x^2}{5} + C_1e^{-x}\cos(2x) + C_2e^{-x}\sin(2x)$

7. $y = -\dfrac{1}{5}xe^{-2x} + C_1e^{-2x} + C_2e^{3x}$

9. $y = \dfrac{1}{8}e^x(\sin x - \cos x) + e^{-x}(C_1\cos x + C_2\sin x)$

11. $y = 2x + x^2 - xe^{-x} + C_1 + C_2e^{-x}$

15. $y_p = \dfrac{x^2}{3}, \quad y = \dfrac{x^2}{3} + C_1x + \dfrac{C_2}{x}$

17. $y = \dfrac{1}{2}x\ln x + C_1x + \dfrac{C_2}{x}$

19. $y = -x^2 + C_1x + C_2xe^x$

Section 17.7   (page 945)

1. $y = a_0\left(1 + \displaystyle\sum_{k=1}^{\infty} \dfrac{(x - 1)^{4k}}{4(k!)(3)(7)\cdots(4k - 1)}\right)$
$+ a_1\left(x - 1 + \displaystyle\sum_{k=1}^{\infty} \dfrac{(x - 1)^{4k+1}}{4(k!)(5)(9)\cdots(4k + 1)}\right)$

3. $y = \displaystyle\sum_{n=0}^{\infty}(-1)^n\left[\dfrac{2^n n!}{(2n)!}x^{2n} + \dfrac{1}{2^{n-1}n!}x^{2n+1}\right]$

5. $y = 1 - \frac{1}{6}x^3 + \frac{1}{120}x^5 + \cdots$

7. $y_1 = 1 + \displaystyle\sum_{k=1}^{\infty} \frac{(-1)^k x^k}{(k!)(2)(5)(8)\cdots(3k-1)},$

$y_2 = x^{1/3}\left(1 + \displaystyle\sum_{k=0}^{\infty} \frac{(-1)^k x^k}{(k!)(4)(7)\cdots(3k+1)}\right)$

Review Exercises (page 945)

1. $y = Ce^{x^2}$　　　　　　**3.** $y = Ce^{2x} - \dfrac{x}{2} - \dfrac{1}{4}$

5. $x^2 + 2xy - y^2 = C$　　**7.** $y = C_1 - \ln|t + C_2|$

9. $y = e^{x/2}(C_2 \cos x + C_2 \sin x)$

11. $y = C_1 t \cos(2\ln|t|) + C_2 t \sin(2\ln|t|)$

13. $y = \frac{1}{2}e^x + xe^{3x} + C_1 e^{2x} + C_2 e^{3x}$

15. $y = x^2 - 4x + 6 + C_1 e^{-x} + C_2 x e^{-x}$

17. $y = (x^3 - 7)^{1/3}$　　　**19.** $y = e^{x^2/2y^2}$

21. $y = 4e^{-t} - 3e^{-2t}$　　**23.** $y = (5t - 4)e^{-5t}$

25. $y = e^{2t} - 2\sin(2t)$　　**27.** $y = 4(x-1)^3$

29. $A = 1,\ B = -1,\quad x(e^x \sin y + \cos y) = C$

31. $y = C_1 x + C_2 x \cos x$

Appendix I   Complex Numbers
(page A-10)

1. $\Re(z) = -5,\ \Im(z) = 2$　　**3.** $\Re(z) = 0,\ \Im(z) = -\pi$

5. $|z| = \sqrt{2},\ \theta = 3\pi/4$　　**7.** $|z| = 3,\ \theta = \pi/2$

9. $|z| = \sqrt{5},\ \theta = \tan^{-1} 2$

11. $|z| = 5,\ \theta = -\pi + \tan^{-1}(4/3)$

13. $|z| = 2,\ \theta = -\pi/6$　　**15.** $|z| = 3,\ \theta = 4\pi/5$

17. $11\pi/12$　　　　　　　**19.** $4 + 3i$

21. $\frac{\pi\sqrt{3}}{2} + \frac{\pi}{2}i$　　　　　**23.** $\frac{1}{4} - \frac{\sqrt{3}}{4}i$

25. $-3 + 5i$　　　　　　　**27.** $2 + i$

29. closed disk, radius 2, centre 0

31. closed disk, radius 5, centre $3 - 4i$

33. closed plane sector lying under $y = 0$ and to the left of $y = -x$

35. 4　　　　　　　　　　**37.** $5 - i$

39. $2 + 11i$　　　　　　　**41.** $-\frac{1}{5} + \frac{7}{5}i$

43. 1

47. $zw = -3 - 3i,\ \dfrac{z}{w} = \dfrac{1+i}{3}$

49. (a) circle $|z| = \sqrt{2}$, (b) no solutions

51. $-1,\ \frac{1}{2} \pm \frac{\sqrt{3}}{2}i$

53. $2^{1/6}(\cos\theta + i\sin\theta)$ where $\theta = \pi/4,\ 11\pi/12,\ 19\pi/12$

55. $\pm 2^{1/4}\left(\frac{\sqrt{3}}{2} + \frac{1}{2}i\right),\ \pm 2^{1/4}\left(\frac{1}{2} - \frac{\sqrt{3}}{2}i\right)$

Appendix II   Complex Functions
(page A-19)

1. $0 \le \Re(w) \le 1,\ -2 \le \Im(w) \le 0$

3. $1 \le |w| \le 4,\ \pi \le \arg w \le \dfrac{3\pi}{2}$

5. $\dfrac{1}{2} \le |w| < \infty,\ -\dfrac{\pi}{2} \le \arg w \le 0$

7. $\arg(w) = 5\pi/6$　　　　**9.** parabola $v^2 = 4u + 4$

11. $u \ge 0,\ v \ge u$　　　　**13.** $f'(z) = 2z$

15. $f'(z) = -1/z^2$

19. $\dfrac{d}{dz}\sinh z = \cosh z,\ \dfrac{d}{dz}\cosh z = \sinh z$

21. $z = \dfrac{\pi}{2} + k\pi,\ (k \in \mathbb{Z})$

23. zeros of $\cosh z$: $z = i\left(\dfrac{\pi}{2} + k\pi\right)\ (k \in \mathbb{Z})$
zeros of $\sinh z$: $z = k\pi i\ (k \in \mathbb{Z})$

25. $\Re(\sinh z) = \sinh x \cos y,\ \Im(\sinh z) = \cosh x \sin y$

27. $z = 0,\ -2i$　　　　　　**29.** $z = -1 \pm 2i$

31. $z = 0,\ i,\ 2i$

33. $z = \dfrac{1 \pm i}{\sqrt{2}},\ z = \dfrac{-1 \pm i}{\sqrt{2}}$
$z^4 + 1 = (z^2 + \sqrt{2}z + 1)(z^2 - \sqrt{2}z + 1)$

35. $z = -1,\ -1,\ -1,\ i,\ -i$

Index

INTEGRALS INVOLVING $\sqrt{x^2 \pm a^2}$ $(a > 0)$

(If $\sqrt{x^2 - a^2}$, assume $x > a > 0$.)

$$\int \sqrt{x^2 \pm a^2}\, dx = \frac{x}{2}\sqrt{x^2 \pm a^2} \pm \frac{a^2}{2}\ln|x + \sqrt{x^2 \pm a^2}| + C$$

$$\int \frac{dx}{\sqrt{x^2 \pm a^2}} = \ln|x + \sqrt{x^2 \pm a^2}| + C$$

$$\int \frac{\sqrt{x^2 + a^2}}{x}\, dx = \sqrt{x^2 + a^2} - a\ln\left|\frac{a + \sqrt{x^2 + a^2}}{x}\right| + C$$

$$\int \frac{\sqrt{x^2 - a^2}}{x}\, dx = \sqrt{x^2 - a^2} - a\tan^{-1}\frac{\sqrt{x^2 - a^2}}{a} + C$$

$$\int x^2\sqrt{x^2 \pm a^2}\, dx = \frac{x}{8}(2x^2 \pm a^2)\sqrt{x^2 \pm a^2} - \frac{a^4}{8}\ln|x + \sqrt{x^2 \pm a^2}| + C$$

$$\int \frac{x^2}{\sqrt{x^2 \pm a^2}}\, dx = \frac{x}{2}\sqrt{x^2 \pm a^2} \mp \frac{a^2}{2}\ln|x + \sqrt{x^2 \pm a^2}| + C$$

$$\int \frac{\sqrt{x^2 \pm a^2}}{x^2}\, dx = -\frac{\sqrt{x^2 \pm a^2}}{x} + \ln|x + \sqrt{x^2 \pm a^2}| + C$$

$$\int \frac{dx}{x^2\sqrt{x^2 \pm a^2}} = \mp\frac{\sqrt{x^2 \pm a^2}}{a^2 x} + C$$

$$\int \frac{dx}{(x^2 \pm a^2)^{3/2}} = \frac{\pm x}{a^2\sqrt{x^2 \pm a^2}} + C$$

$$\int (x^2 \pm a^2)^{3/2}\, dx = \frac{x}{8}(2x^2 \pm 5a^2)\sqrt{x^2 \pm a^2} + \frac{3a^4}{8}\ln|x + \sqrt{x^2 \pm a^2}| + C$$

INTEGRALS INVOLVING $\sqrt{a^2 - x^2}$ $(a > 0, |x| < a)$

$$\int \sqrt{a^2 - x^2}\, dx = \frac{x}{2}\sqrt{a^2 - x^2} + \frac{a^2}{2}\sin^{-1}\frac{x}{a} + C$$

$$\int \frac{\sqrt{a^2 - x^2}}{x}\, dx = \sqrt{a^2 - x^2} - a\ln\left|\frac{a + \sqrt{a^2 - x^2}}{x}\right| + C$$

$$\int \frac{x^2}{\sqrt{a^2 - x^2}}\, dx = -\frac{x}{2}\sqrt{a^2 - x^2} + \frac{a^2}{2}\sin^{-1}\frac{x}{a} + C$$

$$\int x^2\sqrt{a^2 - x^2}\, dx = \frac{x}{8}(2x^2 - a^2)\sqrt{a^2 - x^2} + \frac{a^4}{8}\sin^{-1}\frac{x}{a} + C$$

$$\int \frac{dx}{x^2\sqrt{a^2 - x^2}} = -\frac{\sqrt{a^2 - x^2}}{a^2 x} + C$$

$$\int \frac{\sqrt{a^2 - x^2}}{x^2}\, dx = -\frac{\sqrt{a^2 - x^2}}{x} - \sin^{-1}\frac{x}{a} + C$$

$$\int \frac{dx}{x\sqrt{a^2 - x^2}} = -\frac{1}{a}\ln\left|\frac{a + \sqrt{a^2 - x^2}}{x}\right| + C$$

$$\int \frac{dx}{(a^2 - x^2)^{3/2}} = \frac{x}{a^2\sqrt{a^2 - x^2}} + C$$

$$\int (a^2 - x^2)^{3/2}\, dx = \frac{x}{8}(5a^2 - 2x^2)\sqrt{a^2 - x^2} + \frac{3a^4}{8}\sin^{-1}\frac{x}{a} + C$$

INTEGRALS OF INVERSE TRIGONOMETRIC FUNCTIONS

$$\int \sin^{-1}x\, dx = x\sin^{-1}x + \sqrt{1 - x^2} + C$$

$$\int \tan^{-1}x\, dx = x\tan^{-1}x - \frac{1}{2}\ln(1 + x^2) + C$$

$$\int \sec^{-1}x\, dx = x\sec^{-1}x - \ln|x + \sqrt{x^2 - 1}| + C \quad (x > 1)$$

$$\int x\sin^{-1}x\, dx = \frac{1}{4}(2x^2 - 1)\sin^{-1}x + \frac{x}{4}\sqrt{1 - x^2} + C$$

$$\int x\tan^{-1}x\, dx = \frac{1}{2}(x^2 + 1)\tan^{-1}x - \frac{x}{2} + C$$

$$\int x\sec^{-1}x\, dx = \frac{x^2}{2}\sec^{-1}x - \frac{1}{2}\sqrt{x^2 - 1} + C \quad (x > 1)$$

$$\int x^n\sin^{-1}x\, dx = \frac{x^{n+1}}{n+1}\sin^{-1}x - \frac{1}{n+1}\int \frac{x^{n+1}}{\sqrt{1 - x^2}}\, dx + C \text{ if } n \neq -1$$

$$\int x^n\tan^{-1}x\, dx = \frac{x^{n+1}}{n+1}\tan^{-1}x - \frac{1}{n+1}\int \frac{x^{n+1}}{1 + x^2}\, dx + C \text{ if } n \neq -1$$

$$\int x^n\sec^{-1}x\, dx = \frac{x^{n+1}}{n+1}\sec^{-1}x - \frac{1}{n+1}\int \frac{x^n}{\sqrt{x^2 - 1}}\, dx + C$$
$$(n \neq -1, \ x > 1)$$

EXPONENTIAL AND LOGARITHMIC INTEGRALS

$$\int xe^x\, dx = (x - 1)e^x + C$$

$$\int x^n e^x\, dx = x^n e^x - n\int x^{n-1}e^x\, dx$$

$$\int \ln x\, dx = x\ln x - x + C$$

$$\int x^n \ln x\, dx = \frac{x^{n+1}}{n+1}\ln x - \frac{x^{n+1}}{(n+1)^2} + C, \quad (n \neq -1)$$

$$\int x^n(\ln x)^m\, dx = \frac{x^{n+1}}{n+1}(\ln x)^m - \frac{m}{n+1}\int x^n(\ln x)^{m-1}\, dx \quad (n \neq -1)$$

$$\int e^{ax}\sin bx\, dx = \frac{e^{ax}}{a^2 + b^2}(a\sin bx - b\cos bx) + C$$

$$\int e^{ax}\cos bx\, dx = \frac{e^{ax}}{a^2 + b^2}(a\cos bx + b\sin bx) + C$$

INTEGRALS OF HYPERBOLIC FUNCTIONS

$$\int \sinh x\, dx = \cosh x + C$$

$$\int \cosh x\, dx = \sinh x + C$$

$$\int \tanh x\, dx = \ln(\cosh x) + C$$

$$\int \coth x\, dx = \ln|\sinh x| + C$$

$$\int \operatorname{sech} x\, dx = 2\tan^{-1}(e^x) + C$$

$$\int \operatorname{csch} x\, dx = \ln\left|\tanh\frac{x}{2}\right| + C$$

$$\int \sinh^2 x\, dx = \frac{1}{4}\sinh 2x - \frac{x}{2} + C$$

$$\int \cosh^2 x\, dx = \frac{1}{4}\sinh 2x + \frac{x}{2} + C$$

$$\int \tanh^2 x\, dx = x - \tanh x + C$$

$$\int \coth^2 x\, dx = x - \coth x + C$$

$$\int \operatorname{sech}^2 x\, dx = \tanh x + C$$

$$\int \operatorname{csch}^2 x\, dx = -\coth x + C$$

$$\int \operatorname{sech} x\tanh x\, dx = -\operatorname{sech} x + C$$

$$\int \operatorname{csch} x\coth x\, dx = -\operatorname{csch} x + C$$

INTEGRATION RULES

$$\int (Af(x) + Bg(x))\, dx = A \int f(x)\, dx + B \int g(x)\, dx$$

$$\int f'(g(x))\, g'(x)\, dx = f(g(x)) + C$$

$$\int U(x)\, dV(x) = U(x)\, V(x) - \int V(x)\, dU(x)$$

$$\int_a^b f'(x)\, dx = f(b) - f(a)$$

$$\frac{d}{dx} \int_a^x f(t)\, dt = f(x)$$

ELEMENTARY INTEGRALS

$$\int x^r\, dx = \frac{1}{r+1} x^{r+1} + C \text{ if } r \neq -1$$

$$\int \frac{dx}{x} = \ln|x| + C$$

$$\int e^x\, dx = e^x + C$$

$$\int a^x\, dx = \frac{a^x}{\ln a} + C$$

$$\int \sin x\, dx = -\cos x + C$$

$$\int \cos x\, dx = \sin x + C$$

$$\int \sec^2 x\, dx = \tan x + C$$

$$\int \csc^2 x\, dx = -\cot x + C$$

$$\int \sec x \tan x\, dx = \sec x + C$$

$$\int \csc x \cot x\, dx = -\csc x + C$$

$$\int \tan x\, dx = \ln|\sec x| + C$$

$$\int \cot x\, dx = \ln|\sin x| + C$$

$$\int \sec x\, dx = \ln|\sec x + \tan x| + C$$

$$\int \csc x\, dx = \ln|\csc x - \cot x| + C$$

$$\int \frac{dx}{\sqrt{a^2 - x^2}} = \sin^{-1}\frac{x}{a} + C \quad (a > 0,\ |x| < a)$$

$$\int \frac{dx}{a^2 + x^2} = \frac{1}{a}\tan^{-1}\frac{x}{a} + C \quad (a > 0)$$

$$\int \frac{dx}{a^2 - x^2} = \frac{1}{2a}\ln\left|\frac{x+a}{x-a}\right| + C \quad (a > 0)$$

$$\int \frac{dx}{x\sqrt{x^2 - a^2}} = \frac{1}{a}\sec^{-1}\left|\frac{x}{a}\right| + C \quad (a > 0,\ |x| > a)$$

TRIGONOMETRIC INTEGRALS

$$\int \sin^2 x\, dx = \frac{x}{2} - \frac{1}{4}\sin 2x + C$$

$$\int \cos^2 x\, dx = \frac{x}{2} + \frac{1}{4}\sin 2x + C$$

$$\int \tan^2 x\, dx = \tan x - x + C$$

$$\int \cot^2 x\, dx = -\cot x - x + C$$

$$\int \sec^3 x\, dx = \frac{1}{2}\sec x \tan x + \frac{1}{2}\ln|\sec x + \tan x| + C$$

$$\int \csc^3 x\, dx = -\frac{1}{2}\csc x \cot x + \frac{1}{2}\ln|\csc x - \cot x| + C$$

$$\int \sin ax\, \sin bx\, dx = \frac{\sin(a-b)x}{2(a-b)} - \frac{\sin(a+b)x}{2(a+b)} + C \text{ if } a^2 \neq b^2$$

$$\int \cos ax\, \cos bx\, dx = \frac{\sin(a-b)x}{2(a-b)} + \frac{\sin(a+b)x}{2(a+b)} + C \text{ if } a^2 \neq b^2$$

$$\int \sin ax\, \cos bx\, dx = -\frac{\cos(a-b)x}{2(a-b)} - \frac{\cos(a+b)x}{2(a+b)} + C \text{ if } a^2 \neq b^2$$

$$\int \sin^n x\, dx = -\frac{1}{n}\sin^{n-1} x \cos x + \frac{n-1}{n}\int \sin^{n-2} x\, dx$$

$$\int \cos^n x\, dx = \frac{1}{n}\cos^{n-1} x \sin x + \frac{n-1}{n}\int \cos^{n-2} x\, dx$$

$$\int \tan^n x\, dx = \frac{1}{n-1}\tan^{n-1} x - \int \tan^{n-2} x\, dx \text{ if } n \neq 1$$

$$\int \cot^n x\, dx = \frac{-1}{n-1}\cot^{n-1} x - \int \cot^{n-2} x\, dx \text{ if } n \neq 1$$

$$\int \sec^n x\, dx = \frac{1}{n-1}\sec^{n-2} x \tan x + \frac{n-2}{n-1}\int \sec^{n-2} x\, dx \text{ if } n \neq 1$$

$$\int \csc^n x\, dx = \frac{-1}{n-1}\csc^{n-2} x \cot x + \frac{n-2}{n-1}\int \csc^{n-2} x\, dx \text{ if } n \neq 1$$

$$\int \sin^n x \cos^m x\, dx = -\frac{\sin^{n-1} x \cos^{m+1} x}{n+m} + \frac{n-1}{n+m}\int \sin^{n-2} x \cos^m x\, dx \text{ if } n \neq -m$$

$$\int \sin^n x \cos^m x\, dx = \frac{\sin^{n+1} x \cos^{m-1} x}{n+m} + \frac{m-1}{n+m}\int \sin^n x \cos^{m-2} x\, dx \text{ if } m \neq -n$$

$$\int x \sin x\, dx = \sin x - x \cos x + C$$

$$\int x \cos x\, dx = \cos x + x \sin x + C$$

$$\int x^n \sin x\, dx = -x^n \cos x + n \int x^{n-1} \cos x\, dx$$

$$\int x^n \cos x\, dx = x^n \sin x - n \int x^{n-1} \sin x\, dx$$